Nonlinear Ocean Waves
and the Inverse Scattering Transform

This is Volume 97 in the
INTERNATIONAL GEOPHYSICS SERIES
A series of monographs and textbooks
Edited by RENATA DMOWSKA, DENNIS HARTMANN and H.THOMAS ROSSBY
A complete list of books in this series appears at the end of this volume.

Nonlinear Ocean Waves and the Inverse Scattering Transform 1 ed.

Alfred R. Osborne

ELSEVIER

AMSTERDAM • BOSTON • HEIDELBERG • LONDON
NEW YORK • OXFORD • PARIS • SAN DIEGO
SAN FRANCISCO • SINGAPORE • SYDNEY • TOKYO

Academic Press is an imprint of Elsevier

Academic Press is an imprint of Elsevier
30 Corporate Drive, Suite 400, Burlington, MA 01803, USA
525 B Street, Suite 1900, San Diego, CA 92101-4495, USA
84 Theobald's Road, London WC1X 8RR, UK

First edition 2010

Library of Congress Cataloging-in-Publication Data
A catalog record for this book is available from the Library of Congress

British Library Cataloguing in Publication Data
A catalogue record for this book is available from the British Library

ISBN: 978-0-12-528629-9

For information on all Academic Press publications
visit our website at books.elsevier.com

Printed and bound by CPI Group (UK) Ltd, Croydon, CR0 4YY
Transferred to Digital Printing 2013

**Working together to grow
libraries in developing countries**

www.elsevier.com | www.bookaid.org | www.sabre.org

ELSEVIER   BOOK AID
            International   Sabre Foundation

Cover Caption: Life emerged from the world's oceans. Much of our modern scientific
knowledge has been stimulated by this fact and by the eternal impact that ocean waves
have on human existence. The cover shows a relatively new concept, a numerical
simulation of a large nonlinear "rogue" wave, which is shown emerging from the sea of
modern knowledge for the dynamics of ocean waves. This new knowledge, called the
inverse scattering transform, has been used to numerically simulate the monster wave on
the cover. Amazingly, this knowledge describes a kind of *nonlinear Fourier analysis* and
the wave is a kind of *nonlinear Fourier component* in the inverse scattering transform.
This book gives a brief overview of some aspects of this theory and its application to the
field of physical oceanography as tools for enhanced physical understanding, data
analysis and assimilation, and hyperfast modeling of ocean waves.

Talia iactanti stridens Aquilone procella velum adversa ferit, fluctusque ad sidera tollit. Franguntur remi, tum prora avertit et undis dat latus, insequitur cumulo praeruptus aquae mons. Hi summo in fluctu pendent; his unda dehiscens terram inter fluctus aperit, furit aestus harenis.

Aeneis—Vergili—19 BC (Original Latin)

...una stridente raffica d'Aquilon coglie d'un tratto la vela in mezzo e, alzando I flutti al cielo, schianta di colpo I remi, volge il legno offrendo il fianco ai flutti, e tosto un monte d'acqua sovrasta, immenso, smisurato. Sulla cresta dell'onde questi pendono; a quelli, spalancandosi fra I flutti, l'onda discopre il fondo ove l'arena al vortice mulina.

Eneide—Virgilio—19 AC (Italian Translation)

...a squall came howling from the north-east, catching the sail full on, raising the waves to the sky, breaking the oars in a single blow, wrenching the boat around to offer its flank to the waves as a mountain of water rose above them, immense and immeasurable. Some of the ships rocked on the crests of the waves; the other ships watched in the troughs as the sea parted, exposing the sands on the bottom as they whirled in the furious winds.

Aeneid—Virgil—19 BC (English Translation, Francesco Osborne)

... one can only comment again on the remarkable ingenuity of the various investigators involved in these recent developments. The results have given a tremendous boost to the study of nonlinear waves and nonlinear phenomena in general. Doubtless much more of value will be discovered, and the different approaches have added enormously to the arsenal of "mathematical methods." Not least is the lesson that exact solutions are still around and one should not always turn too quickly to a search for the ε.

Whitham, 1973

The scientist does not study nature because it is useful; he studies it because he delights in it and he delights in it because it is beautiful. If nature were not beautiful it would not be worth knowing and if nature were not worth knowing, life would not be worth living.

Henri Poincaré

Table of Contents

Preface

The field of physical oceanography owes a great debt to the work of Joseph Fourier (1822). The Fourier transform, for nearly 200 years, has provided one of the most important *mathematical tools* for understanding the dynamics of *linear wave trains* that are described by *linear partial differential equations* with well-defined dispersion relations. One of the important results of Fourier analysis is the *principle of linear superposition* in which any function can be viewed as a sum of sinusoidal waves with different amplitudes, phases, and frequencies. In modern times the application of the Fourier transform to the analysis of *measured wave trains* has evolved into well-known and standard techniques for the analysis *of space and time series*. The Fourier method has provided the experimentalist with a marvelous tool for analyzing data not only in terms of the Fourier modes themselves but also as a technique for computing power spectra, transfer functions, bi- and tri-spectra, digital filtering, multi-channel analysis, the wavelet transform, and many other aspects of data analysis and interpretation. Another aspect of the Fourier transform is its ubiquitous use as a tool for the *numerical modeling* of both linear and nonlinear wave equations.

The major aim of the present work is to take a significant step toward applications of the *nonlinear Fourier analysis* of measured space and time series and for the *nonlinear numerical modeling of wave trains*. The approach is based upon a generalization of linear Fourier analysis referred to as the *inverse scattering transform* (IST) and its generalizations. In particular, I emphasize the role of the *Gel'fand-Levitan-Marchenko* (GLM) integral equation (for *infinite-line boundary conditions*) and the *Riemann theta function* (for *periodic boundary conditions*). Just as linear Fourier analysis provides sine wave basis functions onto which data may be projected, so does IST provide *nonlinear basis functions* for a similar purpose. Examples of these basis functions include the ordinary sine wave, the Stokes wave, solitons, shock waves, etc.

This book essentially uses the inverse scattering transform (IST) to study nonlinear properties of ocean waves. This nonlinear Fourier approach is based upon Riemann theta functions, a kind of multi-dimensional Fourier series. Applications are given for surface and internal soliton dynamics, rogue waves, acoustic waves and vortex dynamics. Specific arguments discussed in the book are:

(1) Applications of the physics of nonlinear waves and their coherent structures, as solutions of the IST, are discussed for many problems of interest in physical oceanography. The IST spectral decomposition is a nonlinear superposition law of waves

and various types of coherent structures or basis functions such as Stokes waves, solitons, unstable "rogue" modes, shock waves (fronts) and vortices.

(2) Development of hyperfast algorithms for numerically integrating nonlinear wave equations. These numerical algorithms are perfectly parallelizable and are roughly $1000\,N$ times faster than conventional fast Fourier transform (FFT) solutions, where N is the number of processors or cores in the system. For a computer system with 1000 cores the new algorithm is about one million times faster than traditional FFT numerical implementations on a single core.

(3) The algorithms do not blow up or degrade numerically as FFT solutions to nonlinear Hamiltonian systems often do for large values of time. This is because the numerical solutions of the IST are evaluated explicitly and exactly at each value of time.

(4) Development of time series analysis algorithms for analyzing field or laboratory data. The spectral decomposition is in terms of the nonlinear basis set of coherent structures mentioned above. Nonlinear filtering is an important feature of the method. Many examples are given for the analysis of nonlinear, oceanic wave data.

From a mathematical point of view, IST solves particular "integrable" *nonlinear partial differential wave equations* such as the Korteweg-deVries (KdV), the nonlinear Schroedinger (NLS), and the Kadomtsev-Petviashvili (KP) equations. Because of the mathematical complexity of these theories of nonlinear wave propagation, one cannot expect to bridge all the physical possibilities for the analysis of nonlinear wave data or modeling in a single monograph. Nevertheless, it is hoped that the present work will provide important source material for a fresh, new, and exciting area of numerical and experimental research.

The search for integrability in nonlinear wave equations using IST has been the major theoretical focus of the field of "soliton physics." A list of important results found in this field over the past 50 years includes: (1) discovery of the soliton by Zabusky and Kruskal (1965), (2) discovery of the IST solution of the KdV equation for infinite-line boundary conditions by Gardner et al. (1967), (3) discovery of the Zakharov equation and the NLS equation for deep-water wave trains (Zakharov, 1968), (4) integration of the NLS equation by Zakharov and Shabat (1972), (5) integration of the KdV equation for periodic boundary conditions by Dubrovin and Novikov (1975a,b), Dubrovin et al. (1976), (6) integration of the periodic NLS equation by Kotljarov and Its (1976), and (7) integration of the KP equation for periodic boundary conditions by Krichever (1988). The results for *periodic* boundary conditions are fundamental for this book, for they form the fundamental core of knowledge from which nonlinear time series analysis and modeling techniques have been developed.

This is an unusual book, on the one hand because of its broad nonlinear mathematical and physical perspective and, on the other hand, because of its review and presentation of new and novel nonlinear methods. The main focus relates to applications in a wide variety of physical situations including surface water waves, internal waves, plasma physics, equatorial Rossby waves, nonlinear optics, etc. Applications of the IST require results from many fields including pure and applied mathematics, theoretical physics, numerical analysis, experimental measurements, and the (space and) time series analysis of nonlinear wave data. Hence, many different fields are involved and several dozen scientific journals have reported significant results. It goes without saying that the evolution of the developments described herein have been substantially delayed over the past 25 years (a) due to the mathematical richness of the theoretical formalisms, (b) due to the complex interplay (or not) among the various fields, (c) due to subsequent lengthy efforts to address the myriad new problems relating to the application of the methods to the specific fields, and (d) due to the large number of unique challenges in the development of numerical algorithms. It is likely that important developments will continue to occur in the near future as the complexities of nonlinear Fourier analysis are further clarified. The ultimate challenges are (1) the further theoretical development of new mathematical and physical situations in which the IST applies, (2) the continued development of new and innovative nonlinear data analysis procedures, (3) improved understanding of physical processes in terms of IST variables, and (4) rapid evolution of numerical algorithms for the hyperfast simulation of wave fields.

Potential users of the material in this book include those who are interested in improving their knowledge of nonlinear wave motion, those who are interested in applying the methods to the nonlinear time series analysis of data with nonlinear filtering, and those who are interested in numerical modeling of nonlinear wave motion, including phase resolving, spectral and stochastic models. An important perspective is that one is able to analyze data *at the same order* as the numerical simulation of a particular nonlinear partial differential equation (PDE), that is, one chooses a particular PDE model and then does data analysis and hyperfast numerical modeling directly from the spectral structure of the PDE.

The following individuals provided stimulating comments and conversation over the years: Simonetta Abenda, Mark J. Ablowitz, Nail Akhmediev, Julius Bendat, Marco Boiti, Alan Bishop, John Boyd, Mario Bruschi, Terry Burch, Annalisa Calini, Francesco Calogero, Roberto Camassa, Gigi Cavaleri, Robert Conte, Bob Dean, Benard Deconinck, Toni Degasperis, Phillip Drazin, Marie Farge, David Farmer, Hermann Flaschka, Thanasis Fokas, Allan Fordy, Gregory Forest, Chris Garrett, Annalisa Griffa, Roger Grimshaw, Jeff Hanson, Joe Hammack, Diane Henderson, Darryl Holm, Dave Kaup, Yuji Kodama, Martin Kruskal, Bill Kuperman, Kevin Lamb, Peter Lax, Decio Levi, Michael Longuet-Higgins, Jim Lynch, Anne Karin Magnusson, V. B. Matveev, Ken Melville, David W. McLaughlin, Kenneth D. McLaughlin, Richard McLaughlin, Chang Mei, Sonja Nikolić, Tony Maxworthy, Jim McWilliams, John Miles, Walter Munk, Steve Murray, Alan Newell, Lev Ostrovsky, Paul Palo, Joe Pedlosky, Germana Peggion, Flora Pempinelli, Howell Peregrine, Stefano Pierini, Robert Pinkel, Andrei Pushkarev, Orlando Ragnisco, Donald Resio, Paola Malanotte Rizzoli, Allan Robinson, Pierre Sabatier, Phillip Saffman, Paolo Santini, Connie Schober, Alwyn Scott, Alberto Scotti, Harvey Segur, Jane Smith, Carl Trygve Stansberg,

Michael Stiassnie, Harry Swinney, Bob Taylor, Gene Tracy, Val Swail, Alex Warn-Varnas, Bruce West, Dick Yue, Henry Yuen, Norm Zabusky, Jerzy Zagrodzinski, Vladimir Zakharov.

I would also like to sincerely thank John Fedor and Sara Pratt of Elsevier for their fine efforts in the editing of this book. Karthikeyan Murthy expertly handled the typesetting.

This work has been supported over the past 20 years by the Office of Naval Research (Tom Curtin, Manny Fiadeiro, Scott Harper, Frank Herr, Ellen Livingston, Steve Murray, Terry Paluszkiewicz, Steve Ramberg, Michael Shlesinger, Jeffrey Simmen, Tom Swean, Linwood Vincent) and more recently by the Naval Facilities Engineering Service Center (Bob Taylor and Paul Palo) and the Army Corp of Engineers of the United States of America (Donald Resio, Jeff Hanson).

Alfred R. Osborne
Torino, Italy
Arlington, Virginia, USA

Part One

Introduction: Nonlinear Waves

Conventional physical oceanography emphasizes *measurements* and *modeling* efforts that can go hand in hand to extend and enhance our understanding of physical processes in the ocean. The physics comes in at the level of the order of approximation of the nonlinear partial differential equations (PDEs) that are chosen as candidates to describe the processes in a particular data set. The role of *ordinary linear Fourier analysis* is fundamental in all studies, not only for data analysis but also for modeling. Of course one must include external effects such as the wind, bathymetry, dissipation, stratification, shape of the coastline, etc. Here we are primarily concerned with surface and internal waves and acoustic wave propagation in the ocean.

The *inverse scattering transform* (IST) described herein provides additional possibilities for research that may be useful to the investigator: (1) The physical structure of a PDE can often be described by a *nonlinear spectral theory* (inverse scattering transform, IST) which emphasizes the role of *coherent structures* such as positive and negative solitons, shocks, kinks, table-top solitons, vortices, fronts, unstable modes, etc. Nonlinear spectral theory and nonlinear modes contrast to linear Fourier analysis that uses sine waves. (2) The spectral structure of the nonlinear PDE provides numerical tools to *nonlinearly analyze time series data*. (3) The IST allows one to develop *hyperfast numerical models*. (4) In all of these contexts the concept of *nonlinear filtering* is important, that is, at any moment in the analysis one may focus upon certain nonlinear Fourier components (coherent structures, say) and extract them from the spectrum to see how they behave in the absence of the others. Thus, we get the detailed physics of coherent structures, nonlinear time series analysis tools, hyperfast modeling and nonlinear filtering, all associated with our choice of a particular nonlinear PDE for the situation at hand. The method can also be extended to the *assimilation of data* in real time. This book gives an overview of these additional possibilities for research using IST and how to apply them primarily in the areas of surface, internal waves, acoustic waves and vortex dynamics.

Doi: 10.1016/S0074-6142(10)97039-3

It is important to distinguish the present approach from other approaches that give alternative decompositions to linear Fourier analysis (empirical eigenfunction analysis, wavelet transforms, etc.). In the present work we are dealing with nonlinear modes that are *solutions to nonlinear PDEs*. Nonlinear interactions among these nonlinear modes are a natural part of the formulation. Thus, the IST provides the most natural set of modes for a particular kind of nonlinear wave motion. Other approaches are certainly useful for many different reasons, but they do not in general solve nonlinear PDEs and hence do not contain the spectral decomposition of the nonlinear physics. Of course the IST reduces to the linear Fourier transform in the small-amplitude, sinusoidal linear limit: sine wave modes solve linear PDEs.

How complex are the nonlinear wave equations that can be described by the methods given herein? An increasing battery of numerical and theoretical methods is ensuring that the order of approximation and number of applicable equations will continue to increase apparently without bound. Thus, the applicability of the method apparently has endless possibilities for present and future research in many and other areas of ocean dynamics such as geophysical fluid dynamics and turbulence, both of which are described herein. The ideas presented here will insure a place for this research in a wide variety of other fields such as nonlinear optics, plasma physics, solid state physics, etc.

This book offers several pathways to follow for those interested in particular areas of research. A *first reading* of the book might include all or parts of the following chapters: 1, 2, 5, 8, 9, 24–34. If you are interested in an overview of some of the essential ideas of the *inverse scattering method* see Chapters 2, 3, 9–16. *Numerical methods* are confined primarily to Chapters 3, 9, 17–23.

The preliminary version of this book contained about 1500 pages, far too large for a single volume. The decision was made to truncate the book to its present size and to place the remaining material into a later volume. As a consequence the infinite number of classes of nonlinear, integrable wave equations are addressed by the generic IST method herein, primarily with periodic/quasi-periodic boundary conditions. Nonintegrable equations, including variable bathymetry, wind forcing, variable shaped coastline, dissipation etc. will be addressed in a sequel to this volume. However, the methods of this volume, based upon Riemann theta functions, are also applicable to nonintegrable model equations as well.

1 Brief History and Overview of Nonlinear Water Waves

1.1 Linear and Nonlinear Fourier Analysis

Man has long been intrigued by the study of water waves, one of the most ubiquitous of all known natural phenomena. Who has not been fascinated by the rolling and churning of the surf on a beach or the often-imposing presence of large waves at sea? How many countless times have ship captains logged the treacherous encounters with high waves in the deep ocean or later reported (if they were lucky) the damage to their ships? Man's often strained friendship with the world's oceans, and its waves and natural resources, has endured at least since the beginning of recorded history and perhaps even to the invention of ocean going vessels thousands of years ago. But it is only in the last 200 years that the study of water waves has been placed on a firm foundation, not only from the point of view of the physics and mathematics, but also from the perspective of experimental science and engineering.

While water waves are one of the most common of all natural phenomena, they possess an extremely rich mathematical structure. Water waves belong to one of the most difficult areas of fluid dynamics (Batchelor, 1967; Lighthill, 1986) and wave mechanics (Whitham, 1974; Stoker, 1957; LeBlond and Mysak, 1978; Lighthill, 1978; Mei, 1983; Drazin and Johnson, 1989; Johnson, 1997); Craik, 2005, namely the study of nonlinear, dispersive waves in two-space and one-time dimensions. The governing equations of motion are coupled nonlinear partial differential equations in two fields: the *surface elevation*, $\eta(x, t)$, and the *velocity potential*, $\phi(x, t)$. Analytically, these equations are difficult to solve because of the *nonlinear boundary conditions* that are imposed on an *unknown free surface*. This set of equations is known as the *Euler equations*, which are based upon several physical assumptions: (1) the waves are irrotational, (2) the motion is inviscid, (3) the fluid is incompressible, (4) surface tension effects are negligible, and (5) the pressure over the free surface is a constant. While one may question a number of these assumptions, it is safe to say that they allow us to study a wide variety of wave phenomena to an excellent order of approximation.

Generally speaking, the Euler equations of motion (Chapter 2) which govern the behavior of water waves are *highly nonlinear* and *nonintegrable*. The term "nonlinear" implies that the larger the waves are, the more their shapes deviate from simple sinusoidal behavior. The term "integrable" means that the equations of

Doi: 10.1016/S0074-6142(10)97001-0

motion can be *exactly solved* for particular boundary conditions. It is often fashionable in modern times to discuss higher-order "nonintegrability" in terms of such exotic phenomena as bifurcations, singular perturbation theory, and chaos.

Clearly, the special case of *linear wave motion*, for a well-defined *dispersion relation*, can be solved exactly by the method of the *Fourier transform* (Chapter 2). The Fourier method allows one to project the free surface elevation (and other dynamical properties such as the velocity potential) onto *linear modes* that are simple *sinusoidal waves. Linear superposition* of the sine waves gives the exact solution for the wave dynamics for all space and time. Modern research developments have led to the development of the *discrete Fourier transform* and its much celebrated and accelerated algorithm, the fast Fourier transform (FFT). These developments of course emphasize the importance of *periodic boundary conditions* in the analysis of time series data and in numerical modeling situations, because the discrete Fourier transform is a periodic function.

A large number of scientific fields have embraced the Fourier approach. These include the study of laboratory water waves, oceanic surface and internal waves, light waves in fiber optics, acoustic waves, mechanical vibrations, etc. Both scientists and engineers in such diverse fields as optics, ocean engineering, communications engineering, spectroscopy, image analysis, remotely sensed satellite data acquisition, plasma physics, etc., have all benefited from the use of Fourier methods. Tens of thousands of scientific papers have contributed to the various fields and a number of books have provided a clear pathway through the difficulties and pitfalls of linear (space and) time series analysis, not only from the point of view of data analysis procedures, but also from the point of view of numerical algorithms. Clearly, linear Fourier analysis is one of the most important tools ever developed for the scientific and engineering study of wave-like phenomena.

The power of the Fourier method for determining the exact solution of linear wave equations is often cast in terms of the *Cauchy problem* for one-space and one-time dimensions: Given the wave profile as a function of space, x, at some initial value of time, $t = 0$, determine the solution of the surface wave dynamics for all values of x for all future (and past) times, t, that is, given the *initial surface elevation* $\eta(x, 0)$ compute $\eta(x, t)$ for all t. In two-space dimensions (x, y, t), this perspective has the obvious generalization. Of course, the major goal of the field of nonlinear wave mechanics is to fully describe the surface elevation, $\eta(x, y, t)$, and the velocity potential, $\phi(x, y, z, t)$, for all space and time.

Within this theoretical context, an important aspect of the Fourier transform is the extension of the approach to the *analysis of experimental data*. Typically, (1) the wave amplitude is *measured as a function of the spatial variable*, x, at some fixed time, $t = 0$ (this approach is often discussed in terms of *remote sensing methods*) or (2) the amplitude is *measured as a function of time*, t, at some fixed spatial location, $x = 0$ (for which one obtains *a time series*). Clearly, one may also consider an *array* of fixed locations at which the wave amplitude is measured as a function of time. From a mathematical point of view, the first of these approaches is naturally associated with the *Cauchy problem* (one measures *space series* and Fourier analysis is defined over the *spatial variable* in terms of *wavenumber*) while the second method is associated with a *boundary value problem* (one measures *time series* and Fourier analysis is defined over the *time variable* and the associated *frequency*). Extension of the Fourier method to other aspects of the data analysis problem, such as the filtering of data and the analysis of random data, are also well known and are used often by researchers whose goal is to better understand wave-like phenomena.

For those familiar with the analysis of measured space or time series the most often used numerical tool is the FFT, a *discrete algorithm* that obeys *periodic boundary conditions*. The Fourier transform for *infinite-line* or *infinite-space* boundary conditions has also been an important mathematical development; it solves the famous "rock-in-a-pond" problem. For most data analysis purposes, the *discrete, periodic* Fourier transform is most often preferred.

As simple as the picture is for linear, dispersive wave motion, the extension of the Fourier approach to *nonlinear wave dynamics* has followed a long and difficult road. Analytical approaches for solving nonlinear wave equations have been slow to evolve and it is only in the last 50 years that general methods have become available. This theoretical work was a natural evolution that began, at least in modern terms, with the work of Fermi et al. (1955) who discovered a marvelous temporal recurrence property for a chain of nonlinearly connected oscillators. A few years later, Zabusky and Kruskal (1965) discovered the *soliton* in numerical solutions of the Korteweg-deVries (KdV) equation (small-but-finite amplitude, long waves in shallow water). Then the exact solution of the Cauchy problem for the KdV equation was found for infinite-line boundary conditions (Gardner, Green, Kruskal, and Miura, GGKM, 1967) using a new mathematical method now known as the *inverse scattering transform* (IST). This work was only the beginning of many new approaches for integrating nonlinear wave equations and for discovering their physical properties (Leibovich and Seebass, 1974; Lonngren and Scott, 1978; Lamb, 1980; Ablowitz and Segur, 1981; Eilenberger, 1981; Calogero and Degasperis, 1982; Newell, 1983; Matsuno, 1984; Novikov et al., 1984; Tracy, 1984; Faddeev and Takhtajan, 1987; Drazin and Johnson, 1989; Fordy, 1990; Infeld and Rowlands, 1990; Makhankov, 1990; Ablowitz and Clarkson, 1991; Dickey, 1991; Gaponov-Grekhov and Rabinovich, 1992; Newell and Moloney, 1992; Belokolos et al., 1994; Ablowitz and Fokas, 1997; Johnson, 1997; Remoissenet, 1999; Polishchuk, 2003; Ablowitz et al., 2004; Hirota, 2004).

From *data analysis and numerical modeling points of view*, the IST plays a role in the study of nonlinear wave dynamics similar to the linear, periodic Fourier transform *provided that the IST exists for periodic boundary conditions* for a physically suitable *nonlinear wave equation*. One motivation for periodic boundary conditions for nonlinear equations rests with the fact that most applications of linear Fourier analysis are based upon the FFT, a periodic algorithm. The *periodic* formulation for the IST was discovered for the KdV equation in the mid-1970s (see Belokolos et al., 1994 and cited references) and subsequently applied to a number of other physically important wave equations. In this chapter, the *Riemann theta function* plays the central theoretical and experimental roles.

Of course, one can see that the nonlinear Fourier analysis of time series data must contain a number of pit falls. Understanding *how to project the right data onto the right basis functions* becomes a major part of the data analysis regimen. To this end, one must be sure to understand the underlying physical formulation of the governing wave equations for a particular experimental situation. But, given the recent developments of numerical algorithms and data analysis procedures, one can certainly be tempted to use them to improve our understanding of the nonlinear dynamics of water waves. The main goals of this chapter are to (1) provide a body of knowledge that will improve our ability to analyze space and time series of measurements of nonlinear laboratory and oceanic wave trains and how to (2) develop hyperfast nonlinear numerical wave models. In this way we hope to enhance our understanding of nonlinear water wave dynamics.

1.2   The Nineteenth Century

It is safe to say that the systematic study of water waves was one of the first fluid-mechanical problems to be approached using the modern formulation of the Navier-Stokes type of equations. I recount a number of early investigations that employed the analytical technique together with experimental methods to better understand water wave dynamics.

1.2.1   *Developments During the First Half of the Nineteenth Century*

One of the important early problems related to the so-called "pebble-in-a-pond" problem: one launches a pebble into a pond and then observes the waves that emanate from the disturbance. This problem was formulated by the French Academy of Sciences in 1806: A prize was offered for the solution of the wave pattern evolving from a point source in one spatial dimension. Amazingly, both Cauchy and Poisson solved this problem independently (and shared the prize) using the Fourier transform.

With the success of Cauchy and Poisson, the linearization of water wave dynamics became an important area of research. Both Airy (1845) and Stokes (1847) provided summaries of the theory of linear and nonlinear waves and tides.

One of the most important contributions of the first half of the nineteenth century was the work of John Scott Russell (1838) who published a comprehensive study of laboratory wave measurements for the British Association for the Advancement of Science. His work, titled *Report on Waves*, is without doubt one of the greatest early contributions to water wave mechanics. Not the least of his accomplishments was his ability to accurately measure wave motion in a period before the development of modern sensors and electronic equipment. One of his major results was the discovery of the "great wave of translation" or *solitary wave*, as it is known today. It would be 120 years before the important discovery of the *soliton*, a mathematical-physical abstraction of Russell's work (Zabusky and Kruskal, 1965). Russell's personal comments about his discovery of the phenomenon (Russell, 1838, p. 319) are of historical interest. The scene is a canal, still existing today, near Edinburgh, Scotland:

> *I was observing the motion of a boat which was rapidly drawn along a narrow channel by a pair of horses, when the boat suddenly stopped—not so the mass of water in the channel which it had put in motion; it accumulated round the prow of the vessel in a state of violent agitation, then suddenly leaving it behind, rolled forward with great velocity, assuming the form of a large solitary elevation, a rounded, smooth and well-defined heap of water, which continued its course along the channel apparently without change of form or diminution of speed. I followed it on horseback, and overtook it still rolling on at the rate of some eight or nine miles an hour, preserving its figure some thirty feet long and a foot to a foot and a half in height. Its height gradually diminished and after a chase of one or two miles I lost it in the windings of the channel. Such, in the month of August 1834, was my first chance interview with that singular and beautiful phenomenon.*

The boats on these canals were often referred to as "fly boats." These were long (21 m), narrow boats (1.5 m) that were horse-drawn. An interesting recounting of their operation was discussed by Forester (1953) in the novel *Hornblower and the Atropos*. Hornblower, on the way to London to take command of his new ship the *Atropos*, was onboard a fly boat, in the first class cabin, with his wife and son, speeding down a canal:

> *Hornblower noticed that the boatmen had the trick of lifting the bows, by a sudden acceleration, onto the crest of the bow raised by her passage, and retaining them there. This reduced the turbulence in the canal to a minimum; it was only when he looked aft that he could see, far back, the reeds at the banks bowing and straightening again long after they had gone by. It was this trick that made the fantastic speed possible. The cantering horses maintained their nine miles an hour, being changed every half hour.*

It seems that the canal companies had learned to "lift" the fly boats (with an energetic application of a whip to the horses) up on top of the "bow wave" or solitary wave created when the boat was set in motion. In this way, their ordinary procedure was to "surf" on the solitary waves. Of course, trains were invented only a few years later and the definition of "fantastic speed" was raised.

Russell later conducted laboratory experiments to better understand the solitary waves and described them thusly (Emmerson, 1977):

> *I made a little reservoir of water at the end of the trough, and filled this with a little heap of water, raised above the surface of the fluid in the trough. The reservoir was fitted with a movable side or partition; on removing which, the water within the reservoir was released. It will be supposed by some that on the removal of the partition the little heap of water settled itself down*

in some way in the end of the trough beneath it, and that this end of the trough became fuller than the other, thereby producing an inclination of the water's surface, which gradually subsided till the whole got level again. No such thing. The little released heap of water acquired life, and commenced a performance of its own, presenting one of the most beautiful phenomena that I ever saw. The heap of water took a beautiful shape of its own; and instead of stopping, ran along the whole length of the channel to the other end, leaving the channel as quiet and as much at rest as it had been before. If the end of the channel had just been so low that it could have jumped over, it would have leaped out, disappeared from the trough, and left the whole canal at rest just as it was before.

This is the most beautiful and extraordinary phenomenon; the first day I saw it was the happiest day of my life. Nobody had ever had the good fortune to see it before, or, at all events, to know what it meant. It is now known as the solitary wave of translation.

The book by Emmerson (1977) gives a complete overview of the life of John Scott Russell and his contributions to science, engineering, and naval architecture. It is worth mentioning that Russell's study of solitary waves consisted also in the design of the shapes of ship hulls. In fact, he provided some of the first analytical designs of hulls ever devised, largely based on the interactions of the hull with solitary waves. A lovely account of this entire story, including Russell's interplay with others in the field such as Airy, is given in the book by Darrigol (2005) (see also Bullough (1988); Zabusky, 2005).

Russell's *Report on Waves* see also Russell, 1885 was credited with having motivated Stokes (1847) work and the subsequent publication of his treatise *Theory of Oscillatory Waves*. In this important work, Stokes summarized the known results for linear wave theory and then introduced his now famous expansion (the so-called Stokes wave), which today is viewed as one of the cornerstones of modern methods for the study of weakly nonlinear wave theory and to the method of multiple scales (Whitham, 1974). A modern perspective on the physics of solitary waves and solitons is given by Miles (1977, 1979, 1980, 1981, 1983). The physics of highly nonlinear waves is treated by Longuet-Higgins (1961, 1962, 1964, 1974), Longuet-Higgins and Fenton (1974).

1.2.2   The Latter Half of the Nineteenth Century

Russell's discovery of the solitary wave subsequently led to successful theoretical formulations of nonlinear waves. Work by Stokes (1847), Boussinesq (1872), and Korteweg and deVries (1895) provided the appropriate perspective. Essentially, the (lowest order) solitary wave has the following analytical form for a single, positive pulse:

$$\eta(x, t) = \eta_0 \text{sech}^2[(x - ct)/L], \tag{1.1}$$

where the phase speed, c, and pulse width, L, are given by

$$c = c_0(1 + \eta_0/2h), \tag{1.2}$$

$$L = \sqrt{4h^3/3\eta_0}. \tag{1.3}$$

Here, h is the water depth, g is the acceleration of gravity, and $c_0 = \sqrt{gh}$ is the linear phase speed, that is, the velocity of an infinitesimal linear sine wave. Note that the phase speed, c, of the solitary wave (1.2) is proportional to its amplitude, η_0; larger solitary waves travel faster than their smaller counterparts.

Korteweg and deVries (1895) found the above formula as an exact solution to the following nonlinear wave equation:

$$\eta_t + c_0\eta_x + \alpha\eta\eta_x + \beta\eta_{xxx} = 0, \tag{1.4}$$

which they discovered and which now bears their name, the KdV equation. Here, $\alpha = 3c_0/2h$ and $\beta = c_0h^2/6$. The free surface elevation, $\eta(x, t)$, is a function of space x and time t. Equation (1.4) describes the weakly nonlinear evolution of long, unidirectional surface waves in shallow water. The KdV equation is the first of the so-called "soliton" equations and is integrable by the IST (Gardner et al., 1967). Nonlinear Fourier analysis and numerical modeling for the KdV and other equations, and how to implement the approach in the analysis of data, are central topics of this book.

To get a preliminary idea about how nonlinear Fourier methods have arisen, consider the following traveling-wave periodic solution to the KdV equation (Korteweg and deVries, 1895):

$$\eta(x, t) = \frac{4k^2}{\lambda} \sum_{n=1}^{\infty} \frac{n(-1)^n q^n}{1 - q^{2n}} \cos[nk_0(x - Ct) + \phi_0]$$
$$= 2\eta_0 cn^2\{(K(m)/\pi)[k_0 x - \omega_0 t + \phi_0]; m\}, \tag{1.5}$$

where $\lambda = \alpha/6\beta = 3/2h^3$. The modulus, m, of the Jacobian elliptic function, cn, the nonlinear phase speed, C, and the nome, q, depend explicitly on the amplitude, η_0 (see Chapter 8). The dispersion relation is $\omega_0 = Ck_0$. Because of the presence of the elliptic function, cn, the above expression has come to be known as a *cnoidal wave*. Note that the series in Equation (1.5), suitably truncated to N terms, is the shallow-water, *Nth-order Stokes wave* (Whitham, 1974). In the limit as the modulus $m \to 0$, the cnoidal wave reduces to a sine wave; when $m \to 1$, the cnoidal wave approaches a solitary wave or soliton (1.1). Intermediate values of the modulus correspond to the Stokes wave with various levels of nonlinearity. An example of several cnoidal waves (with differing moduli and wavenumbers) is shown in Figure 1.1.

As will be discussed in detail herein the cnoidal wave is the *nonlinear basis function* for the periodic IST for the KdV equation (Chapter 10). The cnoidal wave is the basis function onto which measured, unidirectional shallow-water time series may be projected (Chapters 10, 20–23, 28, 30, and 31).

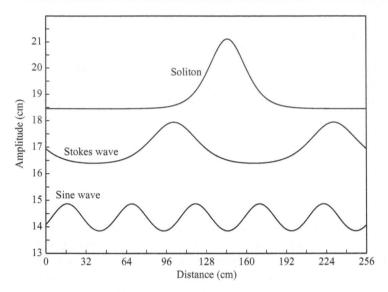

Figure 1.1 Examples of cnoidal waves.

Other contributions important for the study of water waves, but little known to many researchers in the field, include the seminal works by Poincaré, Riemann, Weierstrauss, Frobenius, Baker, Lie, and Akhiezer, just to name a few (Baker, 1897). Many of the important results in various areas of the field of pure mathematics were developed by these and others in the last half of the nineteenth century. Seminal breakthroughs in algebraic geometry, group theory, and Riemann theta functions have led to important applications in the modern formulations of water waves. These works have led to the discovery of the Riemann theta functions as a descriptor of the nonlinear spectral theory for water wave dynamics in both shallow and deep water. The theta function is the primary tool for the time series analysis of nonlinear wave trains and for numerical modeling as discussed in this monograph.

1.3 The Twentieth Century

The observations of solitary waves by John Scott Russell and the subsequent theoretic description by Stokes, Boussinesq, and Korteweg and deVries constituted the extent of physical understanding of solitary waves at the beginning of the twentieth century.

For nearly 70 years after the work of Korteweg and deVries, the solitary wave was considered to be a relatively unimportant curiosity in the field of nonlinear wave theory (Miura, 1974), although one application to shallow-water ocean waves remains a remarkable exception (Munk, 1949). Nevertheless, from a mathematical point of view, it was generally thought that the collision of two solitary waves would result in a strong nonlinear interaction and would ultimately end in their destruction (Scott et al., 1973). That this was not true left many surprises for future workers in the field (Zabusky and Kruskal, 1965).

It is fair to say that the study of nonlinear waves, for the first half of the twentieth century, was not viewed as an important area of research by physicists or mathematicians. Fields such as quantum mechanics and nuclear physics took the attention of many researchers. Practical applications of water waves were enhanced by activities during the Second World War and a subsequent upsurge in activity came with the invention of the electronic computer and the use of linear Fourier analysis to spectrally analyze measured wave trains for the first time (Kinsman, 1965). However, the study of the solitary wave was still an important and unfinished area of research.

One of the most important contributions came in one of the last papers of Enrico Fermi (Fermi et al., 1955). This work is now referred to as the *Fermi-Pasta-Ulam problem* and the phenomenon that these investigators discovered is known as *FPU recurrence*. The research was motivated by the suggestion of Debye (1914) that, in an anharmonic lattice, the finite value of the thermal conductivity arises in consequence of nonlinear effects. Thus, just at the dawn of the computer age, Fermi, Pasta, and Ulam decided to conduct a numerical experiment to study the nonlinear behavior of the anharmonic lattice. They were guided by the (incorrect) assumption that, since the lattice elements were connected nonlinearly, any smooth initial condition for the lattice member positions, over large enough times, might evolve toward a final *ergodic state* consisting of an *equipartition of energy* among the Fourier modes of the system. They considered a line of equal mass points connected with one another by nonlinear springs with the force law $F(\Delta x) = K[\Delta x + \rho(\Delta x)^2]$, where K is the linear spring constant and ρ multiplies the nonlinear part of the force law. The equations of motion are given by (x_i is the excursion of the point mass m from it equilibrium value)

$$x_{i,tt} = \frac{K}{m}\{(x_{i+1} + x_{i-1} - 2y_i) + \rho[(x_{i+1} - x_i)^2 - (x_i - x_{i-1})^2]\}$$

for $i = 1, 2, \ldots, N-1$ with the boundary conditions $x_0 = x_N = 0$. They chose $N = 64$ and used a sinusoidal initial condition, $x_i(0) = \sin(i\pi/N)$, where $x_{i,t}(0) = 0$ (the subscript i refers to the lattice point and t refers to the temporal derivative). The workers had anticipated that equipartitioning of the modes implied that the Fourier spectrum of the initial sine wave (a Dirac delta function) would tend toward white noise as $t \to \infty$. However, in consequence of their numerical study, FPU found that there was no tendency for the system to thermalize, that is, no equipartition occurred during the dynamical evolution. Instead, the system tended to share its initial energy with only a few linear Fourier modes and to eventually (almost) return to the sinusoidal initial condition (e.g., *FPU recurrence*).

Zabusky and Kruskal (1965) revisited the FPU problem and found that the lattice equations used by FPU (provided that one restricts the dynamics to unidirectional motion) reduce, at leading order, to the KdV equation! They then conducted numerical experiments on this equation and discovered solitary wave-like solutions that interacted *elastically* with each other and they coined the word *soliton* to describe them. In their work, they found that two solitons interact with one another and experience a constant phase shift (a displacement of their relative positions) after the collision dynamics are complete, but the fundamental soliton properties (height and speed) remained the same after the interaction, independent of the collision process.

The next important discovery was made by GGKM (1967) who discovered the IST solution of the KdV equation for infinite-line boundary conditions $(|\eta(x, t)| \to 0$ as $|x| \to \infty)$. The Cauchy problem evolves as shown in Figure 1.2. An initial, localized waveform evolves into well-separated, *rank-ordered solitons* and a *trailing radiation tail*. Of course, it was clear that this scenario resembles the nuclear fission process, in that a nucleus fissions into its constituent particles and radiation.

Within 5 years of the discovery of the IST by GGKM, the nonlinear Schrödinger equation (NLS) was solved for infinite-line boundary conditions by Zakharov and Shabat (1972). Shortly thereafter the work of Ablowitz, Kaup, Newell, and Segur (AKNS) (1974) extended IST to an infinite number of integrable wave equations. Since that time, there has been an ever-expanding effort to discover integrable wave equations for other mathematical and physical contexts including higher dimensions. Overviews of nonlinear science, including the field of solitons, are given in Scott (2003, 2005).

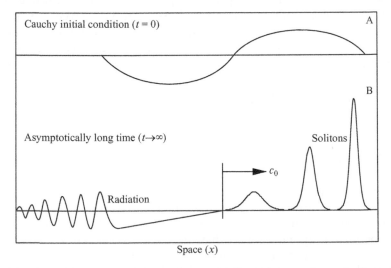

Figure 1.2 An arbitrary waveform at time $t = 0$ (here shown schematically to be a simple, truncated oscillatory wave) (A) evolves into a sequence of rank-ordered solitons plus a radiation tail as $t \to \infty$ (B).

1.4 Physically Relevant Nonlinear Wave Equations

There are a number of physically important nonlinear wave equations that play an important role in the work described in this book. I now briefly discuss some of these: the Korteweg-deVries (KdV), the Kadomtsev-Petviashvili (KP), and the nonlinear Schrödinger (NLS) equations. I emphasize the role of *Riemann theta functions* in the solutions to these three equations for the important case with periodic boundary conditions.

1.4.1 The Korteweg-deVries Equation

The KdV equation (1.4) describes the motion of small-but-finite amplitude shallow-water waves that propagate in the positive x direction. Rather general solutions to KdV, for periodic boundary conditions, can be written in terms of Riemann theta functions, $\Theta_N(x, t)$, where N refers to the number of *modes, degrees of freedom,* or *cnoidal waves* in the spectrum:

$$\eta(x,t) = \frac{2}{\lambda}\frac{\partial^2}{\partial x^2}\ln\Theta_N(x,t), \quad \lambda = \alpha/6\beta, \tag{1.6}$$

$$\Theta_N(x,t) = \sum_{M_1=-\infty}^{\infty}\sum_{M_2=-\infty}^{\infty}\cdots\sum_{M_N=-\infty}^{\infty}\exp\left[i\sum_{n=1}^{N}m_nX_n(x,t)+\frac{1}{2}\sum_{m=1}^{N}\sum_{n=1}^{N}m_mm_nB_{mn}\right]. \tag{1.7}$$

Here, the phases are given by $X_n = k_nx - \omega_nt + \phi_n$, the k_n are wavenumbers, the ω_n are frequencies, and the ϕ_n are phases (see Chapters 5, 10–12, and 14–16 for additional discussion of these parameters). Note that the Riemann theta function consists of N nested summations over a complex exponential. The imaginary part of the argument of the exponential behaves like $\sim k_nx - \omega_nt + \phi_n$, just as with the ordinary linear Fourier transform. The second term in the argument of the exponential is a double sum over the *interaction* or *period matrix*, B_{mn}, which is $N \times N$. Because B_{mn} is a *Riemann* matrix it is *symmetric* and *negative definite*. These properties guarantee mathematical convergence of Equation (1.7).

To better understand what the Riemann theta function means physically, the solution to the KdV equation (1.4) can be written in the following way (Osborne, 1995a,b):

$$\eta(x,t) = \frac{2}{\lambda}\frac{\partial^2}{\partial x^2}\ln\Theta_N(x,t)$$

$$= \underbrace{\eta_{cn}(x,t)}_{\substack{\text{Linear superposition}\\\text{of cnoidal waves}}} + \underbrace{\eta_{int}(x,t)}_{\substack{\text{Nonlinear interactions}\\\text{among the cnoidal waves}}}, \tag{1.8}$$

where

$$\eta_{cn}(x,t) = 2\sum_{n=1}^{N} \eta_n cn^2\{(K(m_n)/\pi)[k_n x - \omega_n t + \phi_n]; m_n\}.$$

Thus, the solution to the KdV equation can be constructed as the linear superposition of N cnoidal waves (Equation (1.8)) plus mutual interactions among the cnoidal waves (see Chapters 5 and 10–12 for additional details). One should note, however, that the nonlinear interactions are not necessarily small. Indeed in the large amplitude, soliton limit they are quite large.

A simple example of the spectral decomposition of a wave train can be seen in Figure 1.3. There are five cnoidal waves in the spectrum. Note that the wave labeled m_1 is a soliton, while those labeled m_2 and m_3 are Stokes waves; the waves labeled m_4 and m_5 are sine waves. By summing the cnoidal waves and adding the nonlinear interactions, one obtains an exact solution to the KdV equation (bottom curve in Figure 1.3). The main influence of the "nonlinear interactions" is to introduce *phase shifting* into the cnoidal wave positions. While it is tempting to think of the interactions as being perturbative in nature, this is an incorrect perspective due to the fact that for very nonlinear waves (generally when there are many solitons in the spectrum), the interaction contribution can be as large as the summed cnoidal waves themselves. One should

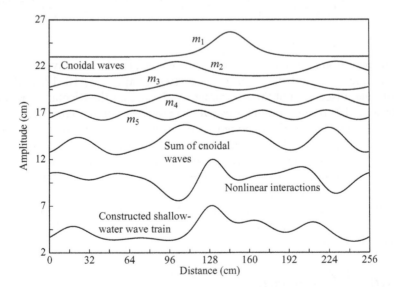

Figure 1.3 The cnoidal wave components in the spectrum of a simple example for the KdV equation are shown, together with the sum of the cnoidal waves, nonlinear interactions, and synthesized five-component wave train. The linear superposition of the cnoidal waves plus interactions yields the synthesized wave train at the bottom of the panel.

think of Figure 1.3 as a prototypical example of the *nonlinear spectral decomposition* of a shallow-water wave train in one-space and one-time $(1 + 1)$ dimensions.

1.4.2 The Kadomtsev-Petviashvili Equation

The KP equation is a generalization of the KdV equation to $2 + 1$ dimensions and describes the motion of shallow-water waves when *directional spreading* is important. One assumes that the y motion (transverse to the dominant direction, x) is small and one finds (Chapters 2, 11, and 32)

$$\frac{\partial}{\partial x}\left[\eta_t + c_0\eta_x + \alpha\eta\eta_x + \left(\beta - \frac{c_0 T}{18\rho g}\right)\eta_{xxx}\right] + \frac{c_0}{2}\eta_{yy} = 0, \qquad (1.9)$$

where $\eta(x, y, t)$ is the surface elevation, T is the surface tension, g is the acceleration of gravity, h is the depth, and ρ is the water density; the parameters c_0, α, and β are the same as those for the KdV equation.

When the surface tension dominates, for water depths less than about a centimeter ($h < \sqrt{T/3\rho g}$), Equation (1.9) is referred to as KPI. When the surface tension is negligible, for depths much larger than a centimeter, Equation (1.9) is called KPII. Note that KPII reduces to the KdV equation when the y coordinate motions are negligible, that is, when there is no directional spreading in the wave train and the motion is essentially unidirectional.

Rather general solutions to the KPII equation, for periodic boundary conditions, can be written in terms of the Riemann theta function, $\Theta_N(x, y, t)$, where again N refers to the number of modes, degrees of freedom, or cnoidal waves in the spectrum:

$$\lambda\eta(x, y, t) = 2\frac{\partial^2}{\partial x^2}\ln\Theta_N(x, y, t), \qquad (1.10)$$

where the theta function has now been generalized to two spatial dimensions:

$$\Theta_N(x, y, t) = \sum_{m_1=-\infty}^{\infty}\sum_{m_2=-\infty}^{\infty}\cdots\sum_{m_N=-\infty}^{\infty}\exp\left[i\sum_{n=1}^{N}m_n X_n(x, y, t) + \frac{1}{2}\sum_{m=1}^{N}\sum_{n=1}^{N}m_m m_n B_{mn}\right].$$
$$(1.11)$$

The phases are given by $X_n = k_n x + l_n y - \omega_n t + \phi_n$, where k_n and l_n are wavenumbers in the x and y directions, respectively; the ω_n are frequencies; and the ϕ_n are phases. Note that the Riemann theta function resembles that previously discussed for the KdV equation with the addition of the term $l_n y$ in the argument. In this case, the imaginary part of the argument of the exponential behaves like $\sim k_n x + l_n y - \omega_n t + \phi_n$. Once again the second term in the argument of the exponential is a double sum over the *interaction* or *period matrix*, B_{mn}, which, even for this higher dimensional wave equation, is still $N \times N$. The Riemann matrix B_{mn} is symmetric and negative definite, the latter property being necessary to ensure convergence of the series (1.11). The solution to the KP equation can then be written in the following way:

$$\eta(x, y, t) = \frac{2}{\lambda} \frac{\partial^2}{\partial x^2} \ln \Theta_N(x, y, t)$$

$$= \underbrace{\eta_{cn}(x, y, t)}_{\substack{\text{Linear superposition} \\ \text{of cnoidal waves}}} + \underbrace{\eta_{int}(x, y, t)}_{\substack{\text{Nonlinear interactions} \\ \text{among the cnoidal waves}}} , \qquad (1.12)$$

where

$$\eta_{cn}(x, y, t) = 2 \sum_{n=1}^{N} \eta_n cn^2 \{ (K(m_n)/\pi)[k_n x + l_n y - \omega_n t + \phi_n]; m_n \}.$$

Thus, the solution to the KPII equation can be constructed as the linear superposition of N cnoidal waves, each with its own direction in the x-y plane, plus mutual interactions among the cnoidal waves (see Chapter 11 for discussion of the periodic KPII equation and Chapter 32 for a hyperfast numerical simulation). Elsewhere in this book, I often refer to the KPII equation as just the KP equation for short. This is because ocean surface waves are, to leading order, described by the KPII equation and I do not further consider the KPI equation outside of this chapter.

An example of a solution of the KPII equation is shown in Figures 1.4 and 1.5. The Riemann spectrum is chosen to have four cnoidal waves, each of which has its own individual amplitude, phase, and direction as shown in Figure 1.4. In Figure 1.5, I give the spectral construction of the solution to the KPII equation using these four cnoidal waves. First, in Figure 1.5A, I show the sum of the cnoidal waves in the upper panel. Beneath this figure is shown the nonlinear interaction contribution (Figure 1.5B). Finally, I give the sum of the cnoidal waves plus the interactions in Figure 1.5C. The result in Figure 1.5C is the actual solution of the KPII equation at time $t = 0$. This waveform is physically the solution of the shallow-water wave problem and is a result applicable to shallow-water coastal zones. Chapter 32 gives a full explanation of a hyperfast numerical model for the KPII equation.

A solution to KPI is given in Figure 1.6. This amazing solution is found by literally "pasting together" two of the "cnoidal waves" (Riemann theta function modes) in this fully three-dimensional case. Note that both the surface elevation and its contours are shown in the figure. From the contours, it is easy to interpret this solution as a "tripole." This particular case, KPI, corresponds to water depths less than about a centimeter. Up to the present time,

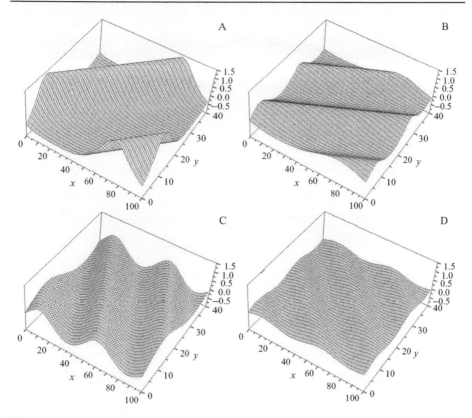

Figure 1.4   Four cnoidal waves in the example solution of the KP equation. The wave moduli are: (A) $m = 0.98$, (B) $m = 0.88$, (C) $m = 0.70$, and (D) $m = 0.37$. The directions of the cnoidal waves, however, are not collinear in this fully three-dimensional case.

I know of no experiments that have verified the presence of the tripole solution in very shallow-water waves.

1.4.3 The Nonlinear Schrödinger Equation

The nonlinear Schrödinger equation describes the dynamics of waves in infinitely deep water in $1 + 1$ dimensions. In dimensional form, it is given by

$$i\left(\frac{\partial\psi}{\partial t} + C_g\frac{\partial\psi}{\partial x}\right) + \mu\frac{\partial^2\psi}{\partial x^2} + v|\psi|^2\psi = 0, \tag{1.13}$$

where $C_g = \omega_0/2k_0$, $\mu = -\omega_0/8k_0^2$, and $v = -\omega_0 k_0^2/2$ (Yuen and Lake, 1982). The associated linear, deep-water dispersion relation is given by $\omega_0^2 = gk_0$. Here, $\psi(x, t)$ is the complex envelope function of a narrow-banded wave train whose amplitude $\eta(x, t)$ is given by

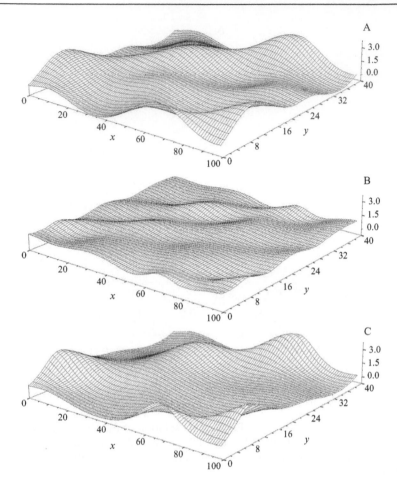

Figure 1.5 Example solution to the KP equation based upon the cnoidal waves in Figure 1.4. (A) The linear superposition of these cnoidal waves. (B) The nonlinear interactions. (C) The solution to KP is the sum of (A) and (B).

$$\eta(x, t) = \psi(x, t)e^{ik_0 x - i\omega_0 t} + \text{c.c.,} \tag{1.14}$$

where c.c. means complex conjugate. Thus, we see that the surface elevation is written as the complex modulation $\psi(x, t)$ of a carrier wave $e^{ik_0 x - i\omega_0 t}$. Here, k_0 and ω_0 are the wavenumber and frequency of the carrier wave, respectively, and C_g is the linear group speed. The NLS equation has an exact IST solution on the infinite line (Zakharov and Shabat, 1972) and a typical evolution for the Cauchy problem is shown in Figure 1.7. Here, the solution to the NLS equation is represented in terms of its modulus, $A(x, t) = |\psi(x, t)|$, and its phase, $\phi(x, t)$:

$$\psi(x, t) = A(x, t)e^{i\phi(x, t)}. \tag{1.15}$$

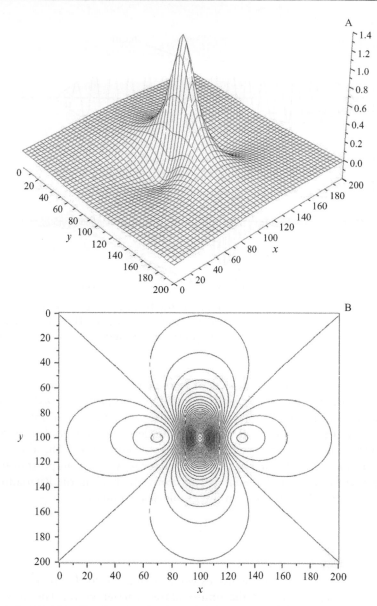

Figure 1.6   (A) Tripole solution of KPI and (B) contours of the solution.

$A(x, t)$ is the *real envelope function* that one observes by eye as shown in Figure 1.7A and B. In the figure the envelope graphed is $A(x, 0)$ (Figure 1.7A). We see that an *initially localized wave train* at time $t = 0$ evolves into a sequence of *envelope solitons* plus a *background radiation field* as $t \to \infty$.

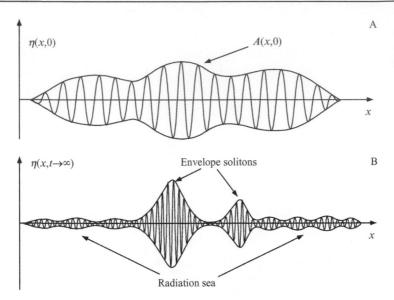

Figure 1.7 (A) The evolution of an initial narrow-banded wave train for which the carrier has fast oscillations with respect to the envelope. (B) The long-time evolution of the initial wave train into envelope solitons and background radiation.

One of the more important aspects of water wave dynamics governed by the NLS equation is the fact that it experiences the *Benjamin-Feir instability* (BF) (Benjamin and Feir, 1967). Thus, an initial sine wave modulated by *very small variations* in the envelope will eventually undergo exponential growth and deviate very much from the sinusoidal shape (Chapters 12, 18, and 29). To study the influence of the BF instability on water waves, it is important to consider the case for *periodic boundary conditions*. Once again the Riemann theta functions are useful for solving an integrable wave equation:

$$\psi(x,t) = a_0 \frac{\Theta_N(x,t|\phi^-)}{\Theta_N(x,t|\phi^+)} e^{-(1/2)i\omega_0 k_0^2 a_0^2 t}. \tag{1.16}$$

Here, $a_0 e^{-i\omega_0 k_0^2 a_0^2 t/2}$ provides the Stokes wave *correction to the frequency* (often called the *frequency shift*) of the unmodulated carrier wave. This is seen by using Equation (1.16) in Equation (1.14) to find the explicit form of the free surface elevation:

$$\eta(x,t) = a_0 \frac{\Theta_N(x,t|\phi^-)}{\Theta_N(x,t|\phi^+)} e^{ik_0 x - i\omega_0' t} + \text{c.c.}, \quad \omega_0' = \omega_0 \left(1 + \frac{1}{2} k_0^2 a_0^2\right).$$

The modulation in Equation (1.16) is constructed from the Riemann theta function:

$$\Theta_N(x,t|\phi^\pm) = \sum_{m_1=-\infty}^{\infty}\sum_{m_2=-\infty}^{\infty}\cdots\sum_{m_N=-\infty}^{\infty} \exp\left\{i\sum_{n=1}^{N}m_n X_n^\pm(x,t) + \pi i\sum_{m=1}^{N}\sum_{n=1}^{N}m_m m_n B_{mn}\right\}$$

$$(1.17)$$

for $X_n^\pm(x,t) = K_n x - \Omega_n t + \phi_n^\pm$. In contrast to the results for the KdV equation, this latter expression is more general, in that the interaction matrix is a *complex* quantity. Furthermore, the dispersion relation can also give imaginary frequency:

$$\Omega = \pm\frac{\omega_0}{8k_0^2}K(K^2 - 8k_0^4 a^2)^{1/2}. \tag{1.18}$$

Note that the frequency is imaginary for $K < 2\sqrt{2}k_0^2 a$, that is, for long wave modulations. An imaginary frequency ensures at least one solution that *exponentially grows in time*; this is the mechanism of the BF instability. Thus, a small modulation, no matter how small, will explode exponentially in time. To illustrate this point, I have conducted a simple simulation in which the modulation is taken to be a small-amplitude sine wave with amplitude 10^{-5}. The results are shown in Figure 1.8A. The flat plane for early times ($t \to -\infty$) is just the modulation envelope $A(x, t)$ as it was originally defined. However, after a while exponential growth dominates and a sharp peak in the modulation envelope forms; this soon disappears and the unmodulated state returns. A more complex evolution is shown in Figure 1.8B where multiple peaks (unstable modes) form in this rather complicated solution of the NLS equation. In spite of the quite unusual nature of these solutions, it is important to realize that the IST provides exact analytic, periodic solutions in terms of Riemann theta functions.

In what way is the numerical simulation in Figure 1.8A related to the exact periodic solutions described by Equation (1.16)? Chapters 12, 18, and 24 discuss that this numerical situation is described exactly by the *homoclinic solution* to NLS (Akhmediev et al., 1987):

$$\psi(x,t) = A\left[\frac{\cos[\sqrt{2}\lambda A(x - C_g t)]\mathrm{sech}[2\lambda^2 A^2\mu t] + i\sqrt{2}\tanh[2\lambda^2 A^2\mu t]}{\sqrt{2} - \cos[\sqrt{2}\lambda A(x - C_g t)]\mathrm{sech}[2\lambda^2 A^2\mu t]}\right]e^{2i\lambda^2 A^2\mu t},$$

$$(1.19)$$

where $\lambda = \sqrt{2}k_0^2$ and $\mu = -\omega_0/8k_0^2$. It is worth noting that this formula has a very interesting physical interpretation. Together with Equation (1.14), we see that the nonlinear dynamics in this case consist of a slowly modulated carrier wave as $t \to -\infty$. As time increases toward $t \sim 0$, the wave amplitude rises up to about 2.4 times the carrier wave amplitude. This "rogue" wave slowly disappears beneath the background carrier once again as $t \to \infty$. Thus, a relatively benign sea state, once in its lifetime, according to Equation (1.19), rises up to its full glory at $t = 0$ and is then subsides once again into the background waves.

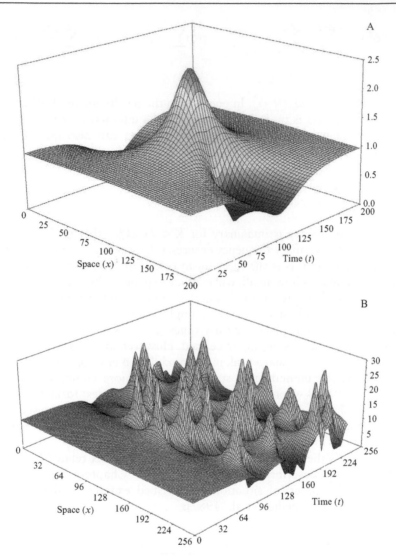

Figure 1.8   (A) Graph of the modulus of the space/time evolution of the simplest "rogue wave" solution to the sNLS equation given by Equation (1.19). (B) Graph of the modulus of the space/time evolution of a multimodal initial modulation that leads to the generation of many "rogue waves" in a solution of the sNLS equation given by Equation (1.16). (See color plate).

Such an amazing solution, easily derived from Equation (1.16), deserves careful attention in this monograph. It and an infinite class of other "rogue" wave solutions are studied both theoretically and experimentally herein. Of course, Equation (1.16) is a *nonlinear Fourier component* in the IST formulation of the NLS equation. This perspective provides the connection to time series analysis for deep-water wave trains. Figure 1.8B provides a multimodal solution to the Schrödinger equation in which many "rogue" waves are seen to appear from a more complex small-amplitude modulation at $t = 0$. Indeed, solutions of this type might be referred to as a "rogue sea."

1.4.4 Numerical Examples of Nonlinear Wave Dynamics

There are countless examples of nonlinear wave dynamics governed by integrable wave equations. There are two examples that are favorites of mine and I would like to briefly discuss them in this introductory chapter. The first is the dynamics of *equatorial Rossby waves* (Boyd, 1983). Nonlinear Rossby waves are governed by the equatorial channel that restricts their motion to lie along the equator, propagating from East to West. While the East-West dynamics is governed by the KdV equation, the North-South shape of the waves is given by an eigenfunction. Consequently, a soliton is found to have the form shown in Figure 1.9, where two recirculating, vortical regions are found, one above and the other below the equator. One thus has a double-vortex solution as shown graphically from the contours as given in Figure 1.9B. Recent interest in the dynamics of equatorial Rossby waves has arisen thanks to their importance in climate dynamics, particularly with regard to the *spatial-temporal evolution of El Niño*.

Another example of nonlinear, integrable dynamics is that shown in Figure 1.10. Here, I address the space-time evolution of a *random highly nonlinear,*

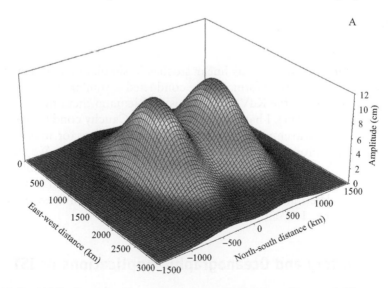

A

Figure 1.9 (A) Surface elevation of an equatorial Rossby soliton and (B) contours of the Rossby soliton. Note that the single soliton dynamics are equivalent to a double vortex that sweeps (transports) passive tracers from the East to the West along the equator. (See color plate).

Continued

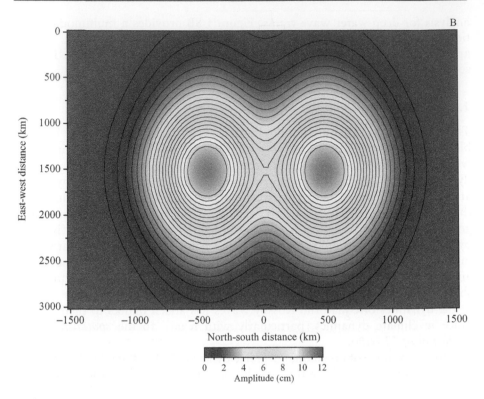

Figure 1.9—cont'd

shallow-water wave train. Just as linear stochastic simulations are commonly made using the Fourier transform, I have conducted a similar stochastic simulation using the IST for the KdV equation using Riemann theta functions. The results are quite surprising. I have defined the initial Cauchy condition as a random function with wavenumber spectrum k^{-2} (appropriate for internal wave dynamics) with uniformly distributed random Fourier phases. We therefore have a fully *stochastic* nonlinear system that evolves into a number of *solitons* and background radiation. This is an important instance when a stochastic system behaves deterministically, that is, the motion is dominated by soliton dynamics (Osborne, 1995a,b).

1.5 Laboratory and Oceanographic Applications of IST

I now give a brief discussion of the IST analysis of time series of experimentally measured data to familiarize the reader with some of the aspects of the work presented herein. I consider three data sets: (1) laboratory measurements in the wave tank facility at the Hydraulic Section of the Department of Civil

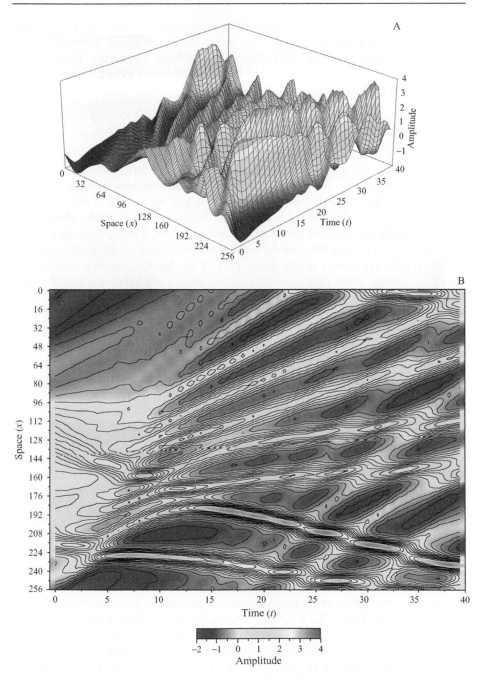

Figure 1.10 Space-time evolution of a random initial condition for the KdV equation. (See color plate).

Engineering in Florence, (2) surface wave measurements in the Adriatic Sea on a fixed offshore platform in 16.5 m water depth, and (3) internal wave measurements made in the Andaman Sea, offshore Thailand. These three examples serve as a brief introduction to the application of the IST as a time series analysis tool.

1.5.1   Laboratory Investigations

The wave tank at the University of Florence is 1 m × 1 m × 50 m and is computer-driven via a control and feedback loop of a hydraulically actuated paddle. In the present simple case, the paddle motion was programmed to generate a simple sine wave of amplitude 2 cm and period 4 s in 40 cm water depth (Chapter 31). Figure 1.11 shows the measured wave train about 4 m from the paddle (see bottom curve) (Osborne and Petti, 1998). This time series has been projected onto the cnoidal wave basis functions of the KdV equation and the results are shown in the upper part of the figure. The first 12 cnoidal waves are shown. Note that the odd number modes are relatively small while the even modes are relatively large (numbering from top to bottom). This occurs because we have taken *two periods* of the measured wave train that is not perfectly periodic, but only quasiperiodic. In fact, a perfectly periodic wave train would result in the odd modes all having zero amplitude. In Figure 1.11, the first mode is a low-amplitude solitary wave while the other odd modes are small-amplitude sine waves. The even modes, however, are more interesting as they are larger and more nonlinear. The second mode is in fact a large Stokes wave with height 4 cm. The forth mode is a smaller amplitude Stokes wave with 2.3 cm height. By summing the cnoidal waves, we get the signal shown in the middle of the figure. This linear sum of the nonlinear modes does not recover the measured wave train very well. Only by including the nonlinear interactions do we exactly recover the measured time series. The "nonlinear interactions" might better be labeled "interaction phase shifts" because that is exactly what they do, that is, they globally shift the phases of the cnoidal waves in exactly the right way to account for the quadratic nonlinearity in the leading order nonlinear water wave dynamics of the KdV equation.

1.5.2   Surface Waves in the Adriatic Sea

The measurement program in the Adriatic Sea resulted in the time series shown at the bottom of Figure 1.12 (Osborne and Burch, 1980) (Chapter 25). This wave train has been low-pass filtered in the frequency interval (0-0.2 Hz). The resultant time series has been projected onto cnoidal wave modes and the results are shown in the upper part of the figure. Fifty cnoidal waves are graphed from low frequency (upper) to high frequency (lower). Note the band of rather large cnoidal waves near the center of the figure. The linear superposition of the cnoidal waves gives the signal labeled "sum of cnoidal

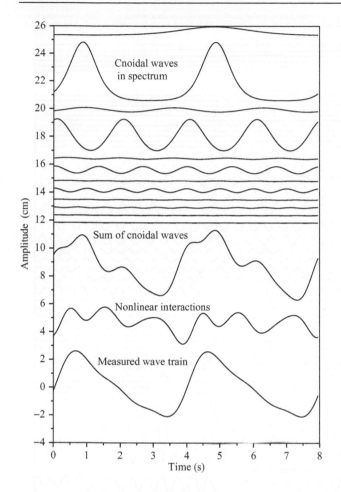

Figure 1.11 Results of inverse scattering transform analysis of wave data obtained in the wave tank in Florence. Shown in vertical order are the 12 cnoidal waves in the spectrum, the sum of the cnoidal waves, the nonlinear interactions, and the measured wave train.

waves." The nonlinear interactions are also shown. Summation of the cnoidal wave contribution plus the nonlinear interactions recovers the low-pass-filtered wave train at the bottom of the figure. An important aspect of these results is that the nonlinear interactions are out of phase with the summed cnoidal waves by about 180°. This surprising fact, together with other aspects of the nonlinear dynamics of shallow-water waves are discussed in Chapters 29 and 31.

1.6   Hyperfast Numerical Modeling

I now give an example of a hyperfast numerical simulation using the Davey-Stewartson equations (see Chapter 34 for a discussion). These are coupled equations which describe the nonlinear interactions between the free surface

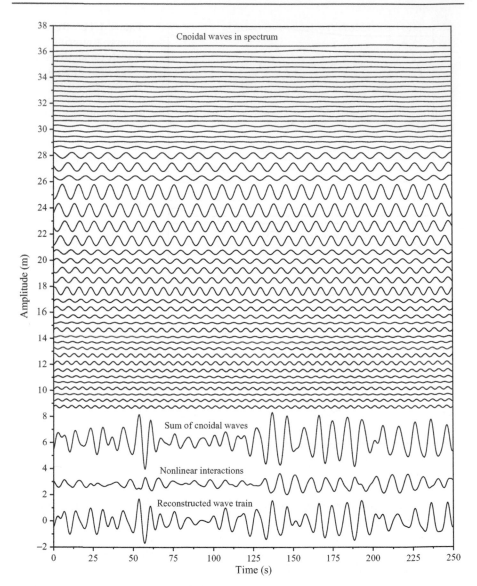

Figure 1.12 Nonlinear Fourier decomposition of an Adriatic Sea time series. The first 50 cnoidal waves in the spectrum are shown (corresponding to nonlinear low-pass filtering of the measured wave data from 0 to 0.2 Hz). Also shown are the wave trains corresponding to the sum of the cnoidal waves, the nonlinear interactions, and the reconstructed low-pass-filtered input time series.

elevation in 2 + 1 dimensions, $\eta(x, y, t)$, and the velocity potential, $\phi(x, y, t)$. The physical form for the *surface elevation* is given by the modulated carrier wave:

$$\eta(x, y, t) \simeq \frac{i\omega_0}{\sqrt{gk_0^3}} \Psi(x, y, t) e^{ikx_0 - i\omega_0 t} + \text{c.c.} \tag{1.20}$$

The physical form for the *velocity potential* has the form:

$$\phi(x,y,t) \simeq \sqrt{\frac{g}{k_0^3}}\left(\Phi(x,y,t) + \frac{\cosh k_0(z+h)}{\cosh k_0 h}\Psi(x,y,t)e^{ik_0x-i\omega_0 t} + \text{c.c.}\right).$$

(1.21)

We see that the velocity potential has two contributions at this order: (1) the slowly varying potential field, $\Phi(x, y, t)$, associated with the *radiation stress term* in the Stokes wave expansion and (2) an oscillatory part, $\Psi(x, y, t)$, which is related to the surface wave oscillations. The coupled wave equations for these two fields are given by (Chapter 2)

$$i\Psi_\tau + \lambda\Psi_{XX} + \mu\Psi_{YY} + \chi|\Psi|^2\Psi = \chi_0\Psi\Phi_X,$$
$$\alpha\Phi_{XX} + \Phi_{YY} = -\beta(|\Psi|^2)_X.$$

(1.22)

Using the methods of Chapter 34, we arrive at the beautiful simulation of a rogue wave as shown in Figure 1.13, a 44 m wave with a slope of about 1/10. This result corresponds very nearly to a single, nonlinear Fourier component of the Davey-Stewartson equations.

I now discuss a hyperfast numerical model of the nonlinear Schrödinger equation with periodic boundary conditions in one-space x and one-time t dimensions for a *random wave train*. I show a simple evolution of a JONSWAP sea state using the NLS equation for infinite water depth. In this case, I have taken the significant wave height to be $H_s = 3$ m and the spectral enhancement factor "gamma" to be $\gamma = 3$. Shown in Figure 1.14 is the *envelope* of the wave trains as a function of space and time. It is for this reason that the vertical coordinate is positive definite, because the envelope is by definition above zero. The envelope has been normalized by the standard deviation of the wave train at zero time. Large peaks in this graph are the envelopes of individual wave packets. A large packet must therefore have a largest wave that coincides with the peak of the envelope; large envelopes mean large waves. There are typically three to five waves in each packet in these simulations. The initial condition is therefore the envelope of the initial JONSWAP wave train. The initial wave train was formed by the usual linear Fourier method with random phases. This implies that the initial wave train is a Gaussian field. In Figure 1.14, we can see emerging, indeed "exploding," from the initial condition at $t = 0$ a number of extreme packets that might be interpreted as "rogue" in character. The tremendous and rapid growth of the wave trains is easily interpreted in terms of the so-called "unstable modes" in the nonlinear Schrödinger equation (see Chapters 12, 18, 24, and 29).

To better see the growth of the extreme waves due to the cubic nonlinearity in the nonlinear Schrödinger equation, I give an alternative graph of the results in Figure 1.15. Shown are space series of the envelope function for the initial condition ($t = 0$) and for the wave train at the end of the simulation for

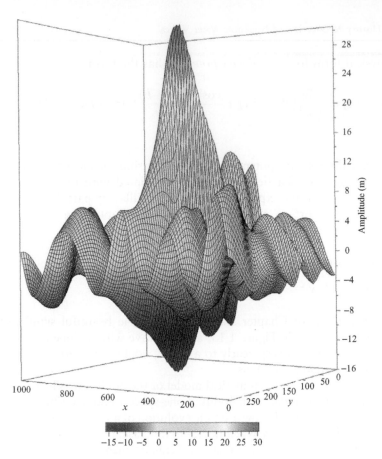

Figure 1.13 Rogue wave simulation using the Davey-Stewartson equations (Chapter 34). (See color plate).

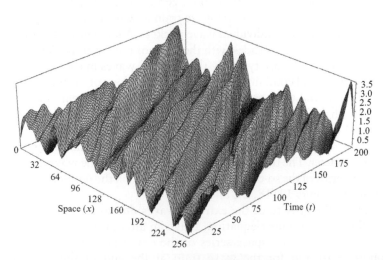

Figure 1.14 Simulation of the NLS equation for a JONSWAP power spectrum with $H_s = 3$ m and $\gamma = 3$. Extreme waves are shown as they emerge in red. (See color plate).

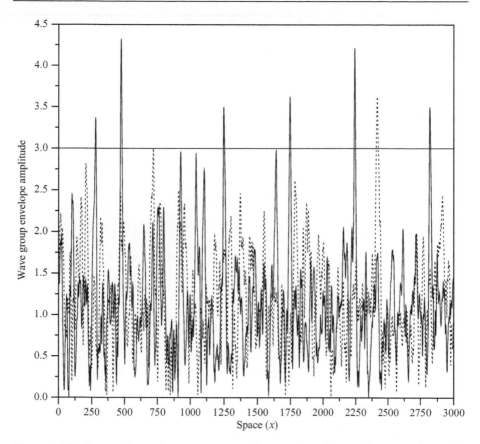

Figure 1.15 Comparison of wave group envelope amplitudes as a function of time for JONSWAP spectrum initial conditions for a linear wave simulation (dotted line) and for a nonlinear simulation based on the cubic nonlinear Schrödinger equation (solid line).

large time. Again the amplitude of the envelope has been normalized by the number of standard deviations for the initial conditions. Note that the linear Gaussian initial conditions exceed three standard deviations only once, while after the nonlinear simulation there are six wave packets that exceed three standard deviations. Chapter 29 shows the results of wave tank experiments at Marintek, SINTEF, Trondheim, where similar kinds of rogue wave evolutions are observed experimentally and are analyzed using the IST.

Additional examples of nonlinear wave motion could almost be added without limit to this introductory chapter. Perhaps, it is best to begin the journey to see a number of others.

2 Nonlinear Water Wave Equations

2.1 Introduction

The first successful wave theories, discovered at the beginning of the nineteenth century by Cauchy and Laplace, were linear and dispersive and solvable by the ordinary, linear Fourier transform (Whitham, 1974). Higher-order theories such as the KdV (Korteweg-deVries, 1895) and nonlinear Schrödinger (NLS) (Zakharov, 1968) equations arise from nonlinear *singular perturbations* of these leading order linear theories using the Euler equations as the natural (nonlinear) starting point. Many of the simpler derived nonlinear partial differential equations have been found to be *integrable* and are solvable by a relatively new method of mathematical physics known as the *inverse scattering transform* (IST) (Leibovich and Seebass, 1974; Lonngren and Scott, 1978; Lamb, 1980; Ablowitz and Segur, 1981; Eilenberger, 1981; Calogero and Degasperis, 1982; Matsuno, 1984; Novikov et al., 1984; Tracy, 1984; Newell, 1985; Faddeev and Takhtajan, 1987; Drazin and Johnson, 1989; Fordy, 1990; Infeld and Rowlands, 1990; Makhankov, 1990; Ablowitz and Clarkson, 1991; Dickey, 1991; Gaponov-Grekhov and Rabinovich, 1992; Newell and Moloney, 1992; Belokolos et al., 1994; Ablowitz and Fokas, 1997; Johnson, 1997; Remoissenet, 1999; Polishchuk, 2003; Ablowitz et al., 2004; Hirota, 2004).

IST is a natural *nonlinear generalization of the linear Fourier transform.* The solutions of these nonlinear wave equations typically include solitons, and the equations and methods of solution are often referred to as "soliton theories." These theories are natural generalizations of linear wave theory to nonlinear wave motion, that is, by allowing a suitable nonlinear parameter to become small, the linear dispersive wave theories are naturally recovered. The soliton theories have many kinds of *coherent structures*, that is, they include solitons, negative solitons ("holes"), shocks, vortices, unstable "rogue" modes, etc. These structures are typically *nonlinear Fourier components* in the IST theory.

Because of the ubiquitous nature of the IST in the study of nonlinear wave equations, it is natural to think of studying various kinds of natural wave motions using this formulation. Understanding the physics, providing nonlinear Fourier data analysis tools, nonlinear data assimilation, and nonlinear modeling of waves are among the possible applications. This chapter summarizes a number of the equations that are important in the fields of surface and internal waves. The list given here is far from complete and has been truncated to keep the book finite in size.

Doi: 10.1016/S0074-6142(10)97002-2

2.2   Linear Equations

The simplest linear equations are found by eliminating the nonlinear terms from the Euler equations (Whitham, 1974). The simplest *linear wave equation* has the form:

$$\eta_{tt} - c^2 \eta_{xx} = 0, \tag{2.1}$$

where c is the linear phase speed. While all information about nonlinearity is lost, what remains is analytically tractable by elementary means.

The range of validity for linear equations can be extended by including dispersion, perhaps to some truncated order or to infinite order. For water waves, the dispersion relation has the familiar form:

$$\omega^2 = gk \tanh kh, \quad k = \sqrt{k_x^2 + k_y^2}. \tag{2.2}$$

where g is the acceleration of gravity and h is the water depth. The resultant dispersive wave solutions are given by the linear Fourier transform. A list of linear wave equations and a determination of their solutions as a Cauchy initial value problem using linear Fourier analysis can be found for example in Whitham (1974) and Ablowitz and Segur (1981).

Some of the simpler linear wave equations are now briefly discussed. The linearized Korteweg-deVries equation is given by

$$\eta_t + c_0 \eta_x + \beta \eta_{xxx} = 0, \quad \eta(x, 0) \to \eta(x, t)$$

and its directionally spread counterpart is the linearized Kadomtsev-Petviashvili (KP):

$$(\eta_t + c_0 \eta_x + \beta \eta_{xxx})_x + \frac{c_0}{2} \eta_{yy} = 0, \quad \eta(x, y, 0) \to \eta(x, y, t).$$

Here, the constant coefficients are given by: $c_0 = \sqrt{gh}$ is the linear phase speed and $\beta = c_0 h^2/6$ is the dispersive coefficient. The leading order motion in the above two equations has the form $\eta_t + c_0 \eta_x \simeq 0$ and therefore the equations describe only rightward-moving wave trains. To the right of the above two equations, $\eta(x, 0) \to \eta(x, t)$ and $\eta(x, y, 0) \to \eta(x, y, t)$ imply that we are addressing the Cauchy initial value problem, that is, given the waveform over the spatial variables at $t = 0$, we determine the wave motion for all future space and time. Note that the linearized KdV equation arises from a simple expansion of Equation (2.2) about $k \simeq 0$ (shallow water) in one spatial dimension, while the linearized KP equation results from a similar expansion in two dimensions.

The linearized Boussinesq equation is given by

$$\eta_{tt} - c_0^2 \eta_{xx} - 2c_0 \beta \eta_{xxxx} = 0.$$

While this equation formally has both left- and right-moving solutions, the dynamics are consistent with the order of the KdV equation and include only rightward-moving wave trains by physical assumption (Miles, 1981).

Of course, one can think of the general theories for linear water waves as describing the dynamics of surface waves with the dispersion relation:

$$\omega = \pm\sqrt{gk \tanh kh}.$$

Thus, there are right- and left-moving components in the wave field with phase speeds:

$$c = \frac{\omega}{k} = \pm\sqrt{\frac{g}{k} \tanh kh}.$$

The ordinary linear Fourier transform describes the evolution of wave trains described by this dispersion relation. Whitham (1974) gives an excellent discussion. Of course, the Fourier transform, given the dispersion relation, provides a general theory for solving linear wave equations (Ablowitz and Segur, 1981 give a fine overview).

2.3 The Euler Equations

It is convenient to assume that the wave motion is described by the Euler equations (Whitham, 1974). One then derives from them a variety of nonlinear, *approximate* wave equations using the method of multiple scales. Many of the equations found are integrable by IST. The application of this body of theoretical methods based upon IST (and related approaches) to ocean waves is the main topic of this book. A number of nonintegrable equations will be treated in a future sequel to this volume.

Shallow-water approximations of the Euler equations focus around the Korteweg-deVries and Kadomtsev-Petviashvili equations. In deep water, the NLS equation and other envelope-type equations are common. We give a brief summary of some of the more important of these equations, with emphasis on integrable equations.

We assume that the fluid is homogeneous, incompressible, and inviscid. Surface tension is neglected. The fluid is subject to a constant, vertical gravitational force g and lies on a constant horizontal bottom located at $z = -h$. The free surface wave motion oscillates around $z \sim 0$ and is designated $z = \eta(x, y, t)$. A velocity potential $\phi(x, y, t)$ can be introduced since the velocity field is assumed to be curl-free. The governing equations of fluid motion are those due to Euler (Lamb, 1932; Whitham, 1974):

$$\nabla^2\phi = 0, \; -h < z < \eta(x, y, t) \quad \text{(Laplace equation)}, \tag{2.3a}$$

$$w = \frac{\partial \phi}{\partial z} = 0, \ z = -h \quad \text{(bottom boundary condition)}, \tag{2.3b}$$

where on the free surface $z = \eta(x, y, t)$

$$\eta_t + \phi_x \eta_x + \phi_y \eta_y = \phi_z \quad \text{(kinematic boundary condition)}, \tag{2.4}$$

$$\phi_t + \frac{1}{2}|\nabla \phi|^2 + g\eta = 0 \quad \text{(dynamical boundary condition)}, \tag{2.5}$$

and the particle velocity components due to the wave motion are given by

$$\mathbf{u}(x, y, z) = \nabla \phi = [u, v, w]. \tag{2.6}$$

The Euler equations are the starting place for a wide variety of investigations of nonlinear wave motion. The major mathematical difficulty with this formulation lies in the free surface boundary conditions, which are the source of nonlinearity in the problem. Indeed, the solution to the Laplace equation for the velocity potential occurs with an *unknown* boundary (the free surface) as seen in Equation (2.3a). Note that the Euler equations can be viewed as a Cauchy problem in the nonlinear, coupled fields $\eta(x, y, t)$ and $\phi(x, y, z, t)$. Given the spatial behavior of these fields at time zero $[\eta(x, y, 0), \phi(x, y, z, 0)]$, Euler's equations allow one to determine the behavior over all future times $[\eta(x, y, t), \phi(x, y, z, t)]$, provided we specify the boundary conditions for η and ϕ in terms of x, y: normally, either *infinite-plane* $(|\nabla \phi| \to 0, \eta \to 0$ as $(x^2 + y^2) \to \infty)$ or *periodic boundary conditions* (η and ϕ are assumed spatially periodic on selected spatial intervals $0 \le x \le L_x, 0 \le y \le L_y$). The Euler equations are often programmed as a *higher-order method* (Dommermuth and Yue, 1987; West et al., 1987; Choi, 1995). The importance of periodic boundary conditions is emphasized by Bryant (1973).

2.4   Wave Motion in 2 + 1 Dimensions

Starting with the Euler equations, there are many approximations of wave motion in $2 + 1$ dimensions. Some of these are briefly discussed in this section.

2.4.1   The Zakharov Equation

In an important paper, Zakharov (1968) derived from the Euler equations a wave equation for broad-banded waves of moderate amplitude. In the spectral representation of the sea surface given by the Zakharov equation, the free surface elevation $\eta(\mathbf{x}, t)$ is related to a spectral function $b(\mathbf{k}, t)$ by the expression

$$\eta(\mathbf{x}, t) = \frac{1}{2\pi} \int_{-\infty}^{\infty} \left(\frac{|\mathbf{k}|}{2\omega(\mathbf{k})} \right)^{1/2} \left\{ b(\mathbf{k}, t) e^{i(\mathbf{k} \cdot \mathbf{x} - \omega t)} + b^*(\mathbf{k}, t) e^{-i(\mathbf{k} \cdot \mathbf{x} - \omega t)} \right\} d\mathbf{k}. \tag{2.7a}$$

If the motion is separated into slow and fast oscillations, and upon introducing a small parameter ε to characterize the nonlinearity, one may write

$$b(\mathbf{k}, t) = [\varepsilon A(\mathbf{k}, \tau) + \varepsilon^2 A'(\mathbf{k}, t')]e^{-i\omega(\mathbf{k})t}. \tag{2.7b}$$

The equation governing the motion of $A(\mathbf{k}, t)$ is the *Zakharov equation*:

$$i\frac{\partial A(\mathbf{k}, t)}{\partial t} = \int_{-\infty}^{\infty} \int_{-\infty}^{\infty} \int_{-\infty}^{\infty} T(\mathbf{k}, \mathbf{k}_1, \mathbf{k}_2, \mathbf{k}_3)\delta(\mathbf{k} + \mathbf{k}_1 - \mathbf{k}_2 - \mathbf{k}_3) \times \exp\{i[\omega(\mathbf{k})$$
$$+ \omega(\mathbf{k}_1) - \omega(\mathbf{k}_2) - \omega(\mathbf{k}_3)]t\}A^*(\mathbf{k}_1)A(\mathbf{k}_2)A(\mathbf{k}_3)d\mathbf{k}_1 d\mathbf{k}_2 d\mathbf{k}_3. \tag{2.8}$$

Here, $T(\mathbf{k}, \mathbf{k}_1, \mathbf{k}_2, \mathbf{k}_3)$ is a real coupling coefficient first derived by Zakharov (1968; see also Crawford et al., 1981); δ is the Dirac delta function that selects the resonant wave vectors. The Zakharov equation describes the slightly detuned resonant interaction of all wave components satisfying

$$\mathbf{k}_0 + \mathbf{k}_1 = \mathbf{k}_2 + \mathbf{k}_3, \quad \omega(\mathbf{k}_0) + \omega(\mathbf{k}_1) = \omega(\mathbf{k}_2) + \omega(\mathbf{k}_3), \tag{2.9}$$

which are expressions first found by Phillips (1960).

2.4.2 The Davey-Stewartson Equations

The study of the so-called *envelope equations* for nonlinear waves is based upon the NLS equation that has been derived in many areas of physics over the last several decades. Indeed, the essential physics can be found in the work of Ginzburg and Landau (1950) and Ginzburg (1956) in their study of a macroscopic theory of superconductivity. Subsequently, Ginzburg and Pitaevskii (1958) studied the theory of superfluidity. In a fundamental study of nonlinear water waves, the 2 + 1 NLS equation was first found by Zakharov (1968).

We now turn to an NLS-type system that is given by a kind of *nonlocal* NLS equation that is coupled to a mean term in the velocity potential. Physically, the NLS equation (at the carrier frequency for short waves) is coupled to the slowly varying contribution of the velocity potential (at zero frequency for long waves). The system of equations was first derived by Benney and Roskes (1969) and then rederived by Davey and Stewartson (1974; see also Zakharov and Rubenchik, 1972; Ablowitz and Segur, 1981).

In 2 + 1 dimensions, the Euler equations can be reduced to the *Benney-Roskes-Davey-Stewartson* (DS) equations in terms of the fields $\Phi(x, y, t)$ and $\Psi(x, y, t)$, where $\Phi(x, y, t)$ is the normalized slowly varying part of the velocity potential and $\Psi(x, y, t)$ is the normalized complex envelope of a narrow-banded wave train (with wavenumber k_0 and frequency ω_0) which may be modulated in both the x and y directions. As a result, one finds the DS equation that describes a directional, narrow-banded, sea state:

$$i\Psi_\tau + \lambda\Psi_{XX} + \mu\Psi_{YY} + \chi|\Psi|^2\Psi = \chi_0\Phi_X\Psi,$$

$$\alpha\Phi_{XX} + \Phi_{YY} = -\beta(|\Psi|^2)_X. \tag{2.10}$$

where

$$\sigma = \tanh(\kappa h), \quad \kappa = \sqrt{k_0^2 + l^2}, \tag{2.11}$$

$$\omega^2 = gh\sigma \geq 0, \tag{2.12}$$

$$\omega_0^2 = g\kappa \tag{2.13}$$

$$\lambda = \frac{\kappa^2\left(\dfrac{\partial^2\omega}{\partial\kappa^2}\right)}{2\omega_0}, \tag{2.14}$$

$$\mu = \frac{\kappa^2\left(\dfrac{\partial^2\omega}{\partial l^2}\right)}{2\omega_0} = \frac{\kappa C_g}{2\omega_0} \geq 0, \tag{2.15}$$

$$\chi = -\left(\frac{\omega_0}{4\omega}\right)\left\{\frac{\left(1-\sigma^2\right)\left(9-\sigma^2\right)}{\sigma^2} + 8\sigma^2 - 2\left(1-\sigma^2\right)^2\right\}, \tag{2.16}$$

$$\chi_0 = 1 + \frac{\kappa C_g}{2\omega}(1-\sigma^2) \geq 0, \tag{2.17}$$

$$\alpha = \frac{gh - C_g^2}{gh}, \tag{2.18}$$

$$\beta = \left(\frac{\omega}{\omega_0 k_0 h}\right)\left(\frac{\kappa C_g}{\omega}(1-\sigma^2) + 2\right) \geq 0, \tag{2.19}$$

$$\nu = \chi - \frac{\chi_1\beta}{\alpha}. \tag{2.20}$$

The following dimensionless variables have been used:

$$X = \varepsilon k_0(x - C_g t), \quad Y = \varepsilon k_0 y, \quad \tau = \varepsilon^2(gk_0)^{1/2}t,$$

$$\Psi = k_0^2(gk_0)^{-1/2}\psi, \quad \Phi = k_0^2(gk_0)^{-1/2}\phi, \tag{2.21}$$

where $\phi(x, y, z, t)$ is the dimensional velocity potential and $\Psi(x, y, t)$ is the dimensional envelope function.

The physical form for the *surface elevation* is given by the modulated carrier wave:

$$\eta(x,y,t) \approx i\omega \sqrt{\frac{1}{gk_0^3}} \Psi(x,y,t) e^{ik_0 x - i\omega_0 t} + \text{c.c.} \tag{2.22}$$

The physical form for the *velocity potential* has the form:

$$\phi(x,y,t) \approx \sqrt{\frac{g}{k_0^3}} \left(\Phi(x,y,t) + \frac{\cosh k_0(z+h)}{\cosh k_0 h} \Psi(x,y,t) e^{ik_0 x - i\omega_0 t} + \text{c.c.} \right) \tag{2.23}$$

The DS equations in the *infinite depth limit* become the NLS equation in $2+1$ dimensions (in this case the mean flow vanishes):

$$i(\psi_t + C_g \psi_x) + \mu \psi_{xx} + \rho \psi_{yy} + \nu |\psi|^2 \psi = 0. \tag{2.24}$$

The associated linear, deep-water dispersion relation is given by $\omega_0^2 = gk_0$, $C_g = \omega_0/2k_0$, $\mu = -\omega_0/8k_0^2$, $\rho = \omega_0/4k_0^2$, and $\nu = -\omega_0 k_0^2/2$. This equation is *not* integrable by IST. In the limit that the motion becomes unidirectional one obtains the usual one-dimensional nonlinear Schrödinger equation, (2.31) below, which is an integrable equation.

2.4.3 The Davey-Stewartson Equations in Shallow Water

In shallow water ($kh \to 0$), the DS equations have the following simple form:

$$\begin{aligned}
i\Psi_\tau - \sigma\Psi_{XX} + \Psi_{YY} - \sigma|\Psi|^2\Psi &= \Psi\Phi_X, \\
\sigma\Phi_{XX} + \Phi_{YY} &= -2(|\Psi|^2)_X,
\end{aligned} \tag{2.25}$$

where $\sigma = 1/3$. The scattering transform for these equations has been studied by Boiti et al. (1988, 1989, 1990, 1991) and Fokas and Santini (1989, 1990). The invention of the term *dromion* was an important step in the mathematical and physical understanding of these equations.

2.4.4 The Kadomtsev-Petviashvili Equation

The Kadomtsev-Petviashvili (1970) equation is a $2+1$ generalization of the KdV equation (in scaled form):

$$(u_t + 6uu_x + \sigma_0 u_{xxx})_x + u_{yy} = 0, \quad \sigma_0 = \pm 1. \tag{2.26}$$

Physically, this equation describes the propagation of shallow-water wave trains whose dominant direction lies along the x-axis, but whose energy is spread somewhat about this dominant direction (see Chapters 11 and 32).

2.4.5 The KP-Gardner Equation

The so-called KP-Gardner equation consists of the KP equation plus the cubic Gardner term $\delta\eta^2\eta_x$:

$$\eta_t + c_0\eta_x + \alpha\eta\eta_x + \beta\eta_{xxx} + \gamma\partial_x^{-1}\eta_{yy} = \delta\eta^2\eta_x. \tag{2.27}$$

This equation is integrable by IST (see Chapter 34).

2.4.6 The 2 + 1 Gardner Equation

The 2 + 1 Gardner equation adds to the KP equation the cubic Gardner term $\delta\eta^2\eta_x$ plus a nonlinear spreading term $\rho\eta_x\partial_x^{-1}\eta_y$:

$$\eta_t + c_0\eta_x + \alpha\eta\eta_x + \beta\eta_{xxx} + \gamma\partial_x^{-1}\eta_{yy} = \delta\eta^2\eta_x + \rho\eta_x\partial_x^{-1}\eta_y. \tag{2.28}$$

This equation is integrable by IST (see discussion in Chapter 34). This is a relatively new shallow-water wave equation that is characterized by enhanced nonlinearity and improved, nonlinear directional spreading (Konopelchenko and Dubrovsky, 1984; Konopelchenko, 1991). See Chapter 33 for the physical modeling of (2.28).

2.4.7 The 2 + 1 Boussinesq Equation

The 2 + 1 Boussinesq equation (1871, 1872, 1877) is found by adding the spreading term η_{yy} to the 1 + 1 Boussinesq equation:

$$\eta_{tt} - c_0^2(\eta_{xx} + \eta_{yy}) - \alpha'(\eta^2)_{xx} - 2c_0\beta\eta_{xxxx} = 0. \tag{2.29}$$

This equation is integrable by the inverse scattering method in 1+1 dimensions (McKean, 1981) (see also Chen, 1998; Clarkson and Mansfield, 1994). Here $\alpha' = c_0\alpha$.

2.5 Wave Motion in 1 + 1 Dimensions

In 1 + 1 dimensions, we discuss the Zakharov equation and its lower-order envelope equations, which include the deep- and shallow-water NLS equations.

2.5.1 The Zakharov Equation

The Zakharov equation in 1 + 1 dimensions is derived by assuming that all wave vectors lie along a single direction. The quartet interaction still works but there is no wave spreading. We thus set $\mathbf{k} = [k, 0]$ in the Zakharov equation, which assumes wave propagation along the x direction:

$$i\frac{\partial A(\mathbf{k}, t)}{\partial t} = \int_{-\infty}^{\infty} \int_{-\infty}^{\infty} \int_{-\infty}^{\infty} T(\mathbf{k}, \mathbf{k}_1, \mathbf{k}_2, \mathbf{k}_3)\delta(\mathbf{k} + \mathbf{k}_1 - \mathbf{k}_2 - \mathbf{k}_3) \times \exp\{i[\omega(\mathbf{k})$$
$$+ \omega(\mathbf{k}_1) - \omega(\mathbf{k}_2) - \omega(\mathbf{k}_3)]t\}A^*(\mathbf{k}_1)A(\mathbf{k}_2)A(\mathbf{k}_3)d\mathbf{k}_1 d\mathbf{k}_2 d\mathbf{k}_3.$$

$$(2.30)$$

Here, $T(\mathbf{k}, \mathbf{k}_1, \mathbf{k}_2, \mathbf{k}_3)$ is a real, unidirectional interaction coefficient, obtained from the usual one by setting $\mathbf{k} = [k, 0]$.

2.5.2 The Nonlinear Schrödinger Equation for Arbitrary Water Depth

The Zakharov equation is of course at higher order than the NLS equation. In fact, the NLS equation was first derived by Zakharov (1968) from his equation. This is easily done in two or three dimensions by expanding the frequencies $\omega(\mathbf{k}_i)$ about a carrier wave vector to second order, and then replacing the interaction coefficient $T(\mathbf{k}, \mathbf{k}_1, \mathbf{k}_2, \mathbf{k}_3)$ by its value when all four arguments are evaluated at the carrier wave vector $(k_0, 0)$. The integral equation that results is the Fourier-transformed equation for the complex envelope function $\psi(x, y, t)$ that satisfies the NLS equation. The resultant simplification captured by NLS results by retaining only the leading order terms in nonlinearity and dispersion; furthermore, NLS describes only wave trains with narrow-banded spectra. This is a drastic simplification over the spectrally broad-banded Zakharov equation.

We summarize the NLS equation valid for all water depths (see, e.g., Zakharov, 1968; Hasimoto and Ono, 1972; Yuen and Lake, 1982):

$$i(\psi_t + C_g\psi_x) + \mu\psi_{xx} + \nu|\psi|^2\psi = 0. \tag{2.31}$$

The constant, real coefficients are given by

$$C_g = \frac{\partial\omega_0}{\partial k_0} = \frac{c_0}{2}\left[1 + \frac{(1 - \sigma^2)k_0 h}{\sigma}\right] \quad \text{(Group velocity)}, \tag{2.32}$$

$$\mu = \frac{1}{2}\frac{\partial^2\omega_0}{\partial k_0^2} = -\frac{g}{8k_0\sigma\omega_0}\{[\sigma - k_0 h(1 - \sigma^2)]^2 + 4k_0^2 h^2\sigma^2(1 - \sigma^2)\}, \tag{2.33}$$

$$\nu = -\frac{k_0^4}{2\omega_0}\left(\frac{c}{2\sigma}\right)^2\left\{\frac{1}{C_g^2 - gh}[4c^2 + 4(1 - \sigma^2)cC_g + gh(1 - \sigma^2)^2]\right.$$
$$\left. + \frac{(9 - 10\sigma^2 + 9\sigma^4)}{2\sigma^2}\right\}, \tag{2.34}$$

with the dispersion relation:

$$\omega_0^2 = gk_0\sigma, \quad \sigma = \tanh(k_0 h) \tag{2.35}$$

and the linear phase speed:

$$c_0 = \frac{\omega_0}{k_0} = \left(\frac{g\sigma}{k_0}\right)^{1/2}. \tag{2.36}$$

NLS has the complex envelope solution:

$$\psi(x,t) = A(x,t)e^{-i\omega't+i\phi(x,t)}, \tag{2.37}$$

where $A(x,t)$ is the real envelope and $\phi(x,t)$ is the real phase. The associated *Stokes field* approximation to the *free surface elevation* is given to second order by

$$\eta(x,t) = -\frac{\gamma A^2}{4k_0\sigma} + A\left[1 + \frac{C_g A\phi_x}{\omega_0}\right]\cos\theta + \frac{C_g A_x}{\omega_0}\sin\theta + \frac{\delta A^2}{4k_0\sigma}\cos 2\theta + \cdots, \tag{2.38}$$

where $A = A(x,t)$ is the real *modulation envelope* and

$$\theta = \theta(x,t) = k_0 x - (\omega_0 + \omega')t + \phi(x,t) \tag{2.39}$$

is the *total phase*. Here

$$\theta_0(x,t) = k_0 x - \omega_0 t \tag{2.40}$$

is the *carrier phase* and $\phi(x,t)$ is the *modulation phase*. The real constants γ and δ are given by

$$\gamma = \frac{2\omega_0 k_0 C_g + (1 - \sigma^2)ghk_0^2}{gh - C_g^2}, \tag{2.41}$$

$$\delta = \frac{(3 - \sigma^2)k_0^2}{\sigma^2}, \tag{2.42}$$

and the *nonlinear, amplitude-dependent frequency correction* is

$$\omega' = \nu\frac{g^2\bar{A}^2}{4\omega_0^2}. \tag{2.43}$$

The term $-\gamma A^2/4k_0\sigma$ in the free surface elevation (2.38) corresponds to slow, long wave variations referred to as *radiation stress* (Longuet-Higgins and Stewart, 1960). Radiation stress depresses the mean sea level beneath a packet of surface waves.

2.5.3   The Deep-Water Nonlinear Schrödinger Equation

The $1 + 1$ NLS equation in deep water has been studied extensively by Yuen and Lake (1982):

$$i(\psi_t + C_g\psi_x) + \mu\psi_{xx} + \nu|\psi|^2\psi = 0. \tag{2.44}$$

The associated linear, deep-water dispersion relation is given by $\omega_0^2 = gk_0$, $C_g = \omega_0/2k_0$, $\mu = -\omega_0/8k_0^2$, and $\nu = -\omega_0 k_0^2/2$. This equation is integrable by IST by the method of (Zakharov and Shabat, 1972; Ablowitz et al., 1974; see also Ablowitz and Segur, 1981).

2.5.4   The KdV Equation

A singular perturbation expansion of the Euler equations about $k \sim 0$ in $1 + 1$ dimensions yields the (space-like) KdV equation (Ablowitz and Segur, 1981):

$$\eta_t + c_0\eta_x + \alpha\eta\eta_x + \beta\eta_{xxx} = 0. \tag{2.45}$$

$\eta(x, t)$ is the wave amplitude as a function of space x and time t, $\alpha = 3c_0/2h$ and $\beta = c_0h^2/6$. Note that Equation (2.45) has the linear dispersion relation $\omega = c_0k - \beta k^3$. KdV solves the Cauchy problem: Given the wave train at $t = 0$, $\eta(x, 0)$, Equation (2.42) determines the motion for all time thereafter, $\eta(x, t)$.

2.5.5   The KdV Equation Plus Higher-Order Terms

If one expands the Euler equations in $1 + 1$ dimensions to one order of approximation higher than the KdV equation, one obtains what has been called the second Whitham equation (W2) (Whitham, 1974):

$$\eta_t + c_0\eta_x + \alpha\eta\eta_x + \beta\eta_{xxx} = \lambda_1\eta_{xxxxx} + \lambda_2\eta\eta_{xxx} + \lambda_3\eta_x\eta_{xx} + \lambda_4\eta^2\eta_x, \tag{2.46}$$

where c_0, α, and β are the same as for the KdV equation. The other constants have the physical values:

$$\lambda_1 = \frac{c_0h^4}{36}, \quad \lambda_2 = \frac{5c_0h}{12}, \quad \lambda_3 = \frac{5c_0h}{6}, \quad \lambda_4 = \frac{15c_0}{8h^2}. \tag{2.47}$$

This equation is *not* integrable by IST. Perturbation solutions by the method of *Lie-Kodama transforms* (Kodama, 1985a,b; Fokas and Liu, 1996) are quite useful for applications (Osborne, 1997).

If we rescale W2 ($u = (\alpha/6\beta)\eta, t = \beta t'$ and then drop the primes), we have a normalized form for the equation:

$$u_t + 6uu_x + u_{xxx} + \varepsilon(u_{5x} + \alpha_2 uu_{xxx} + \alpha_3 u_x u_{xx} + \alpha_4 u^2 u_x) = 0, \tag{2.48}$$

for which $\alpha_1 = 1$, $\alpha_2 = 100/19$, $\alpha_3 = 230/19$, and $\alpha_4 = -60/19$.

An integrable equation at the same order, part of the so-called *KdV hierarchy*, is that found by Lax (1968):

$$u_t + 6uu_x + u_{xxx} + u_{5x} + 10uu_{xxx} + 20u_x u_{xx} + 30u^2 u_x = 0. \tag{2.49}$$

The *Camassa-Holm (CH) equation* (1993, 1994) is given by

$$u_t + 6uu_x + u_{xxx} - \varepsilon(u_{xxt} + 2uu_{xxx} + 4u_x u_{xx}) = 0, \tag{2.50}$$

which is one of the more exciting soliton equations found in recent years; the CH equation is integrable by IST (Constantin and McKean, 1999; Constantin, 2001).

The *modified KdV equation* has the form:

$$u_t + 6u^2 u_x + u_{xxx} = 0 \tag{2.51}$$

and is integrable (Ablowitz and Segur, 1981).

The *Gardner equation* has the form:

$$\eta_t + c_0 \eta_x + \alpha \eta \eta_x + \beta \eta_{xxx} = \lambda_4 \eta^2 \eta_x. \tag{2.52}$$

The Gardner equation is integrable due in part to its intimate relationship with the mKdV equation.

The *Boussinesq equation* is given by

$$\eta_{tt} - c_0^2 \eta_{xx} - \alpha'(\eta^2)_{xx} - 2c_0 \beta \eta_{xxxx} = 0, \tag{2.53}$$

where c_0 and β are the same as for the KdV equation and $\alpha' = 2\alpha c_0$. This equation is also integrable by IST (McKean, 1981).

The *Ostrovsky equation* has the form:

$$(\eta_t + c_0 \eta_x + \alpha \eta \eta_x + \beta \eta_{xxx})_x = \frac{f_0^2}{2c_0} \eta \tag{2.54}$$

or

$$\eta_t + c_0 \eta_x + \alpha \eta \eta_x + \beta \eta_{xxx} = \frac{f_0^2}{2c_0} \partial_x^{-1} \eta. \tag{2.55}$$

This equation describes internal wave propagation in the presences of the Earth's rotation (Ostrovsky, 1978). The linear dispersion relation is

$$\omega = c_0 k - \beta k^3 + \frac{f_0^2}{2c_0} \frac{1}{k}. \tag{2.56}$$

The Ostrovsky equation is not integrable and has the amazing property that it creates long wave components that are much longer than the solitons of the KdV equation (Grimshaw, 1985; Grimshaw et al., 1998; Boyd, 2005).

2.6 Perspective in Terms of the Inverse Scattering Transform

In Tables 2.1 and 2.2, we summarize several nonlinear wave equations that are integrable by IST and whether they are stable in the Benjamin-Feir sense. Table 2.1 indicates the situation in $1 + 1$ dimensions.

The Zakharov equation, which describes the nonlinear evolution of the deep-water *envelope* function, is not integrable by IST. The Hasimoto and Ono form of the $1 + 1$ NLS equation, which describes envelope dynamics in arbitrary depth, is stable in shallow water, is Benjamin-Feir unstable in deep water, and is integrable by IST for all depths.

In $2 + 1$ dimensions (see Table 2.2) the KP equation, which describes a directionally spread sea surface elevation in shallow water, is stable to perturbations

Table 2.1 $1 + 1$ Dimensional Equations

Equation	Stable?	Integrable by IST?
KdV	Yes	Yes
mKdV	No, Benjamin-Feir	Yes
Gardner	No, Benjamin-Feir	Yes
Boussinesq	Yes	Yes
Camassa-Holm	Yes	Yes
Second Whitham equation	No, Benjamin-Feir	No
Second Lax equation	Yes	Yes
Deep-water NLS	No, Benjamin-Feir	Yes
Shallow-water NLS	Yes	Yes
Variable depth NLS	Yes in shallow water No, BF in deep water	Yes

Table 2.2 $2 + 1$ Dimensional Equations

Equation	Stable?	Integrable by IST?
KP	Yes	Yes
KP-Gardner	No, Benjamin-Feir	Yes
$2 + 1$ Gardner	No, Benjamin-Feir	Yes
$2 + 1$ Boussinesq	Yes	Yes
Davey-Stewartson, shallow water	Yes	Yes
Davey-Stewartson, deep water	No, Benjamin-Feir	No
$2 + 1$ NLS	No, Benjamin-Feir	No

and is integrable by IST. The Zakharov equation, and its higher-order extensions (Shermer and Stiassnie, 1991), has both type I (Benjamin-Feir) and type II (transversal) instabilities (Shrira et al., 1996) and is not integrable by known methods (Crawford et al., 1981). The Davey-Stewartson equations, which describe the nonlinear interactions between the surface elevation and the slowly varying part of the velocity potential to leading order, is integrable in the shallow-water limit, but not in intermediate or deep water.

This completes our discussion of several partial differential wave equations that describe surface wave motion in $1 + 1$ and $2 + 1$ dimensions.

2.7 Characterizing Nonlinearity

The purpose of this section is to discuss a way to characterize nonlinearity in ocean waves over arbitrary depths. The results serve to provide perspective for both theoretical and experimental studies.

We propose to characterize nonlinearity by the constant coefficient in front of the cos 2θ term in the Stokes field in Equation (2.38):

$$I = \frac{A^2 \delta}{4\sigma k_0} = \frac{A^2(3 - \sigma^2)k_0}{4\sigma^3}, \tag{2.57}$$

where $\sigma = \tanh(k_0 h)$; $\omega_0^2 = gk_0\sigma$. In the *shallow-water limit*, $k_0 h \ll 1$, $\sigma \cong k_0 h$ and we find

$$I = \frac{3A^2}{4k_0^2 h^3} = AU. \tag{2.58}$$

Thus in shallow water I is the wave amplitude A times the Ursell number $U = 3A/4k_0^2 h^3$. In the *deep-water limit* $\sigma = 1$ and we find

$$I = \frac{1}{2}k_0 A^2 = AU_d. \tag{2.59}$$

In deep water I is the amplitude A times a deep-water Ursell number U_d. It is worth noting that $U_d = k_0 A/2$ (=the sea slope) is the nonlinear parameter that the NLS spectral (inverse scattering) problem sees in the deep-water limit of the theory, whereas in shallow water it is the usual Ursell number that scales the scattering transform.

We are thus motivated to introduce a *generalized Ursell number* U_g, based upon Equation (2.57), valid for all water depths. Hence $I = AU_g$, where

$$U_g = \frac{A\delta}{4\sigma k_0} = \frac{A(3 - \sigma^2)k_0}{4\sigma^3} \tag{2.60}$$

In *shallow water* U_g is the usual *Ursell number* (Ursell, 1953):

$$\lim_{h \to 0} U_g \approx \frac{3A}{4k_0^2 h^3} = U. \tag{2.61}$$

In *deep water* U_g is the deep-water Ursell number or *sea slope*:

$$\lim_{h \to \infty} U_g \approx \frac{1}{2} k_0 A = U_d. \tag{2.62}$$

For experimental or theoretical purposes, we can plot the depth-dependent parameter U_g as a function of the wavelength-to-depth ratio (2.60). Figure 2.1 gives several plots of the generalized Ursell number for several values of the amplitude-to-depth ratio.

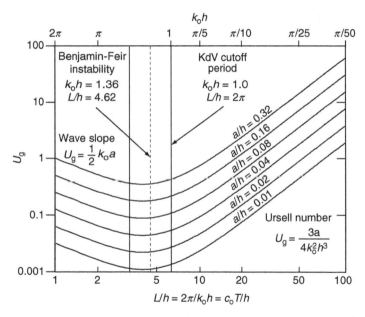

Figure 2.1 Ursell number diagram illustrating how nonlinear a wave train is as a function of depth.

3 The Infinite-Line Inverse Scattering Transform

3.1   Introduction

The focus of this chapter is on the inverse scattering transform (IST) solution of the infinite-line Korteweg-deVries (KdV) equation and on the numerical analysis of this problem. The goal is to provide insight about the classical problem of nonlinear, shallow-water wave motion and its IST solution for infinite-line boundary conditions where the concept of the *soliton* has its roots. This chapter, therefore, does not discuss the problem of periodic boundary conditions that is of most interest to physical oceanographers. Read this chapter only if you need an introduction to *soliton theory* and if you want to get perspective about the *spirit* of the numerical methods given herein. Then read later chapters on the study of nonlinear wave motion with periodic boundary conditions. This chapter is a truncated version of Osborne (1991a,b); see also Provenzale and Osborne (1991). The work in this chapter is easily applied to tsunami dynamics, that is, when an earthquake creates an initial waveform whose long-time evolution can be studied (Hammack, 1973).

The nonlinear Fourier analysis of wave motion governed approximately by the KdV equation on the infinite line is the central point of discussion. I assume that the wave amplitude is recorded in the form of a discrete space or time series that is determined either by experimental measurement or by computer simulation of the physical system of interest. I develop numerical data analysis procedures based upon the scattering transform solution to the KdV equation as given by Gardner et al. (1967). I am motivated by the observation that historically the Fourier transform has been ubiquitously used to spectrally analyze linear wave data; methods are developed for employing the scattering transform as a tool to similarly analyze nonlinear wave data. Specifically, I develop numerical methods to evaluate the direct scattering transform (DST) of a space or time series. Our approach thus provides a basis for analyzing and interpreting nonlinear wave behavior in the wavenumber or frequency domain. The DST spectrum separates naturally into soliton and radiation components and may be simply interpreted in terms of the large time asymptotic state of the infinite-line KdV equation.

The study of the physics of wave motion has historically benefited from analytical, numerical, and experimental exploitation of the linear *Fourier transform* (FT) (see, e.g., Ablowitz and Segur, 1981; Bendat and Piersol, 1986 and references cited therein). One reason for the wide applicability of this method is that many *linear partial differential equations* (i.e., linear wave equations, LWEs) may be solved exactly using Fourier methods. Thus while most physical systems are nonlinear, the fact that a closely related linear system may be exactly solvable by the Fourier transform often provides key insight. An important intermediate step in these calculations is the appearance of the Fourier wavenumber or frequency spectrum. While the wave motion itself may be a rather complicated function of space and time, the time evolution of the Fourier spectrum for linear wave motion is quite simple: *the Fourier amplitudes are constants, while the phases vary* sinusoidally as $e^{-i\omega t}$. Because of this simple behavior, the Fourier spectrum is often viewed as more fundamental than the wave motion itself; the Fourier components constitute the *phase space* of the system. Furthermore, given the Fourier spectrum and the dispersion relation (easily found from the LWE), the wave motion is known for all time and is represented as a *linear superposition of the sinusoidal normal modes*.

For experimental or numerical studies, observation of the time evolution of the Fourier spectrum has long provided a useful means for probing system behavior. The study of nonlinear systems (which may also include the presence of dissipation, external forcing, etc.) has benefited from use of Fourier methods; the motion of the Fourier spectrum is often still "simple" in some sense and this then provides useful information about the physics of higher-order effects. In numerical studies and in the analysis of data the use of the FT has been aided by advances in numerical methods, most prominent of which is the discovery of the fast Fourier transform (FFT, see, e.g., Press et al., 1992 and cited references).

In the last 20 years, there has been considerable progress in the understanding of certain nonlinear wave equations (NLWEs). Beginning with the work of Gardner et al. (1967), who found the exact solution to the Korteweg-deVries equation on the infinite interval for a suitably localized initial wave, a major revolution has occurred in mathematical physics. Solutions to entire classes of NLWEs on the infinite interval have been found (Gardner et al., 1967; Ablowitz et al., 1974; Lamb, 1980; Novikov et al., 1984; Ablowitz and Segur, 1981; Calogero and Degasperis, 1982; Dodd et al., 1982; Newell, 1985; these include in particular the KdV, nonlinear Schrödinger, sine-Gordon, and modified KdV equations) and the methods have been christened the IST. It can be shown that IST is a nonlinear generalization of the Fourier transform (Ablowitz et al., 1974; Ablowitz and Segur, 1981). Several wave equations are now known to have solutions not only on the infinite interval, but also on the periodic domain (these include the KdV: Dubrovin and Novikov, 1975a,b; Its and Matveev, 1975; Lax, 1975; Dubrovin et al., 1976; McKean and Trubowitz, 1976; nonlinear Schrödinger: Kotljarov and Its, 1976; Ma and Ablowitz, 1981; Tracy, 1984; Tracy and Chen, 1988; and sine-Gordon equations: Forest and McLaughlin, 1982). Some of these periodic solutions have been shown to be nonlinear generalizations of Fourier series (Flaschka, 1974; Flaschka and McLaughlin, 1976; Osborne and Bergamasco, 1985, 1986). Certain discrete wave equations on the infinite line (continuous in time, discrete in space) also have exact solutions given by the IST (Ablowitz and Ladik, 1975, 1976a,b, 1977; see Ablowitz and Segur, 1981 for a review). One of the key features of the IST is that it approaches the Fourier transform in the small-amplitude, linear limit; in this way, the IST solution covers not only the nonlinear problem, but also the associated linearized problem.

Given the availability of these new mathematical methods, in analogy with historical use of the Fourier transform, we have suggested that the IST be applied to the study of various problems in nonlinear wave physics. The feature that I have exploited is that the DST is a wavenumber representation of a nonlinear signal (assumed governed by some NLWE) just as the direct Fourier transform (DFT) is a wavenumber representation of a linear signal. One advantage of the DST is that it provides a spectral representation of the wave motion at one (singular-perturbative) order of approximation higher than the associated linear problem and, hence, one order of approximation higher than the linear Fourier transform. Thus, one is "closer" (in the wavenumber domain) to the actual nonlinear physics of a particular system.

Application of the algorithm developed herein has led to the resolution of the Zabusky and Kruskal problem (Osborne and Bergamasco, 1986), which addresses how infinite-line solitons are related to solitons on the periodic domain. Other investigators have studied periodic spectral problems for the Toda lattice (Ferguson et al., 1982), and the sine-Gordon and nonlinear Schrödinger equations (Bishop and Lomdahl, 1986; Bishop et al., 1986a,b). These important papers have considered nonlinear problems with a small number of excited degrees of freedom (i.e., nonlinear Fourier modes); this contrasts with the present monograph that emphasizes the study of oceanic systems that may range up to several thousand degrees of freedom.

The objective of the present chapter is to give the formal derivation and an error analysis of a numerical algorithm for the DST for the KdV equation. One goal of this chapter is to document theoretical, numerical, and interpretive approaches necessary for the practical implementation of this kind of nonlinear Fourier analysis. A large portion of this chapter is devoted to the development of a foundation, language, and philosophy for the potential user of this approach. An algorithm for nonlinear Fourier analysis is developed together with consistent procedures for application to the analysis of data. In this context, I emphasize the need for generalization of the methods and at the same time the need to embrace existing numerical approaches and procedures for the Fourier transform. To this end, I develop a DST algorithm that has the same two-point recursion structure as the discrete Fourier transform and which reduces identically to the DFT in the small-amplitude, linear limit.

This chapter has been written not only to give the development of an algorithm for the DST, but also to provide the reader with the tools for rapid implementation of the procedure and for understanding the physical implications of its use. I do not assume that the reader is an expert in scattering transform theory. Three principal uses of the algorithms are suggested:

(1) *Generation of a wavenumber (or frequency) domain spectrum.* Given a computer-generated signal or a measured space or time series of some localized or periodic wave motion, compute the DST, that is, generate the wavenumber (or frequency) space representation of the input signal.

(2) *Filtering.* The inverse scattering transform is the inverse of the DST operation, that is, the IST reconstructs the signal given the DST spectrum. To filter a signal of certain wavenumber or frequency components, one first eliminates from the spectrum the unwanted components and then the nonlinear, filtered signal is reconstructed with the IST.

(3) *Time evolution.* The wave motion may be evolved forward or backward in time t simply by selecting the value of t desired and then executing the IST algorithm. The output of the IST operation is the signal advanced to the specified time. In the filtering operation of (2) above, the time t is normally taken to be zero.

It is worthwhile at this point to list several reasons why I feel that nonlinear signal processing methods may be useful in the study of nonlinear wave motion:

(1) *Frequency domain as well as time domain analysis.* A common way to investigate the behavior of a particular NLWE is to directly integrate the equation by numerical methods (Zabusky, 1981). Such an approach has proven successful in the study of nonlinear waves because it allows determination of the behavior of the system in the space-time domain where the physical motion is easily and directly observable. Our motivation for developing nonlinear signal processing techniques is that the DST can be used to observe the associated behavior of the system in the wavenumber or frequency domain. Observation of the much simpler motion of the DST spectrum (whose amplitudes are constant in time if the system is integrable, or slowly varying in time if it is "nearly" integrable) can offer additional insight into the behavior of the system.

(2) *Testing how well a NLWE describes a particular system.* To find out if some physical system is described by a certain NLWE, the DST of the measured signal at two spatially (or temporally) separated points can be compared. If the corresponding DST spectral amplitudes are equal to within their estimated precision, the NLWE describes the system to a good order of approximation.

(3) *The investigation of nonintegrable NLWEs.* The numerical study of nonintegrable systems is an important area of research. Very often, the equation may be nearly integrable and the DST can serve as a useful tool for understanding small deviations from integrability (due, say, to external forcing or dissipation). In this case, the output of the numerical simulation of the NLWE is the input signal to the DST algorithm. The motion of the spectral components provides information about nonintegrable (possibly perturbative) behavior.

(4) *Means for quantitatively determining higher-order nonlinear effects in an experimental context.* Often, measurements are made of a physical system that has

approximate behavior governed by some integrable NLWE, but higher-order effects may also be present. The motion of the DST spectrum can provide clues about these physical effects.

(5) *The investigation of chaotic behavior.* Some systems exhibit chaotic behavior when, say, driven by some external force and damped by (say) viscosity in the fluid. By observing the motion of the DST spectrum, the onset of chaotic or turbulent behavior may be investigated. The possibility for the exchange of energy among coherent states (solitons) and the spectrum of the background radiation is a natural problem to investigate by the nonlinear Fourier approach.

(6) *Nonlinear filtering of signals to determine the time domain behavior of certain spectral components.* By selectively removing certain DST components (discrete or continuous) and returning to the space-time domain, the behavior of the filtered system may be observed for $t \neq 0$. This may be important in understanding how certain components change in time and how they undergo mutual interactions. This application also allows for the possibility of isolating the (hopefully small) number of nonlinear degrees of freedom, which actually dominate the motion.

(7) *Searching for physical effects by filtering noise from a measured signal.* Very often measured space or time series contain unwanted noise which obscures the physical behavior under investigation. Nonlinear filtering of high-frequency noise, for example, is straightforward using the IST.

This chapter discusses algorithms for computing the direct and inverse scattering transforms for the whole-line KdV equation. The motivation for the development of the algorithms is to provide tools for wavenumber or frequency domain analysis of nonlinear experimental data and of computer-generated nonlinear wave motion; both of these problems are assumed to be governed approximately by, but to evolve at an order somewhat greater than, the Korteweg-deVries equation. The DST and IST for the periodic KdV equation are developed in Chapters 10, 17, and 32.

The rest of this chapter is organized as follows. In Section 3.2, we review the Fourier transform solution to the linearized KdV equation. These results provide a basis for discussing the scattering transform solution to the KdV equation on the infinite interval (Section 3.3). The relationship between the linear Fourier transform and the scattering transform is discussed in Section 3.4, where particular emphasis is placed on those results necessary for development of the numerical algorithm for the DST, and for physically interpreting nonlinear wave motion governed by the KdV equation. Section 3.5 reviews some of the important assumptions leading to discrete Fourier methods. Section 3.6 discusses how I modify and elaborate on these to develop a set of assumptions that the DST algorithm must satisfy.

Actual details of the numerical algorithm for the direct scattering transform for infinite-line boundary conditions are given in Osborne (1991). The assumptions discussed above provide guidelines used throughout this book for developing discrete algorithms for the inverse scattering transform with periodic boundary conditions.

3.2 The Fourier Transform Solution to the Linearized KdV Equation

One can describe the approximate motion of infinitesimal amplitude, long dispersive waves in shallow water by the linearized KdV equation (set $\alpha = 0$ in Equation (2.45)):

$$\eta_t + c_0\eta_x + \beta\eta_{xxx} = 0. \tag{3.1}$$

I write the equation in dimensional form where $\eta(x, t)$ is the amplitude of the free surface, $c_0 = \sqrt{gh}$ is the linear phase speed, g is the acceleration of gravity, h is the water depth, and $\beta = c_0h^2/6$ is the constant coefficient of the dispersive term. Subscripts refer to partial derivatives with respect to space x or time t. Equation (3.1) is written in laboratory coordinates and has the dispersion relation:

$$\omega = c_0k - \beta k^3. \tag{3.2}$$

The KdV equation (Equation (2.45)) and its linearized form (Equation (3.1)) appear in many other physical contexts (see Chapter 4 of Ablowitz and Segur, 1981). The constant coefficients change their form depending upon the physics of the problem; to apply the spectral analysis techniques in this chapter, one uses constants suitable for a particular application or one rescales the KdV equation to give only numerical constants.

One can use the Fourier transform to find the solution to Equation (3.1) on the infinite interval $(-\infty < x < \infty)$ for the Cauchy initial value problem, that is, for $\eta(x, t = 0)$ given, one seeks $\eta(x, t)$ for all time t. This may be done by first forming the DFT of the initial wave $\eta(x, 0)$:

$$F(k) = \int_{-\infty}^{\infty} \eta(x, 0)e^{-ikx}dx. \tag{3.3}$$

The Fourier spectrum changes in time by the simple relation:

$$F(k, t) = F(k)e^{-i\omega t}. \tag{3.4}$$

The time evolution of the initial wave is then described by the inverse Fourier transform (IFT):

$$\eta(x, t) = \frac{1}{2\pi}\int_{-\infty}^{\infty} F(k, t)e^{ikx}dk. \tag{3.5}$$

Validity of the Fourier method requires that the usual Dirichlet conditions be satisfied and that

$$\int_{-\infty}^{\infty} |\eta(x,0)|\mathrm{d}x < \infty. \tag{3.6}$$

For our own purposes, the essential features of the Fourier transform are (1) the DFT (Equation (3.3)) generates a wavenumber domain representation of the initial wave which is called the Fourier spectrum, (2) The Fourier spectrum has simple time evolution given by Equation (3.4), and (3) The IFT (Equation (3.5)) evolves the initial wave in space and time. Analogs to these three steps are found in the structure of the scattering transform solution to the KdV equation as discussed in the following section.

3.3   The Scattering Transform Solution to the KdV Equation

The KdV equation describes the motion of small-but-finite amplitude, long waves in shallow water:

$$\eta_t + c_0\eta_x + \alpha\eta\eta_x + \beta\eta_{xxx} = 0, \quad -\infty < x < \infty. \tag{3.7}$$

Here, $\alpha = 3c_0/2h$; the other variables are defined with respect to Equation (3.1).

The direct scattering problem for a localized nonlinear signal that evolves according to Equation (3.7) is the Schrödinger eigenvalue problem (Gardner et al., 1967):

$$\psi_{xx} + [\lambda\eta(x,0) + \kappa^2]\psi(x) = 0, \quad -\infty < x < \infty. \tag{3.8}$$

The constant parameter λ, a measure of nonlinearity to dispersion, is given by $\lambda = \alpha/6\beta$. The solutions to Equation (3.8), with infinite-line boundary conditions, correspond to both real and imaginary wavenumber κ. When the wavenumber takes on some real value $\kappa = k/2$ (division by the arbitrary factor of two insures that the definition of wavenumber is compatible with that in the Fourier transform (3.3), see Section 3.4), then the eigensolutions have the following asymptotic boundary conditions:

$$\lim_{x\to-\infty} \psi(x) = a(k)\mathrm{e}^{-ikx/2}, \tag{3.9}$$

$$\lim_{x\to\infty} \psi(x) = \mathrm{e}^{-ikx/2} + b(k)\mathrm{e}^{ikx/2}. \tag{3.10}$$

The coefficient $b(k)$ in Equation (3.10) is referred to as the *DST continuous spectrum*.

When the wavenumber is imaginary $\kappa = iK_n$ the eigenfunction solutions to Equation (3.8), $\psi_n(x, K_n)$, are bounded only for a finite set of discrete eigenvalues K_n, where $1 \leq n \leq N$. Each eigenvalue corresponds to one of the N solitons in the discrete part of the DST spectrum. Each soliton is uniquely related to its amplitude by $\eta_n = 2K_n^2/\lambda$. The remaining part of the discrete spectrum is determined by taking the following normalization for the discrete eigenfunctions:

$$\int_{-\infty}^{\infty} \psi_n^2(x, K_n)dx = 1. \tag{3.11}$$

Associated phase coefficients are then found from

$$C_n = \lim_{x \to \infty} e^{K_n x} \psi_n(x, K_n). \tag{3.12}$$

The collection of information $\{K_n, C_n, N\}$ is the DST *soliton (discrete) spectrum*. The complete *DST spectrum* is given by the following set of information:

$$\text{DST} = \{K_n, C_n, N; b(k)\}. \tag{3.13}$$

Thus, Equation (3.13) is the DST of a nonlinear wave evolving by Equation (3.7), just as Equation (3.3) is the DFT of a linear wave evolving by Equation (3.1). The connection between the DST and the Fourier transform is discussed in Section 3.4.

The time evolution of the DST spectrum is simple:

$$\begin{aligned} K_n(t) &= K_n, \\ C_n(t) &= C_n e^{\Omega_n t}, \\ N(t) &= N, \\ b(k, t) &= b(k)e^{-i\omega t}. \end{aligned} \tag{3.14}$$

Note the similarity of these equations with the time evolution of the Fourier spectrum (Equation (3.4)). The linearized dispersion relation for the KdV equation (set $\alpha = 0$ in Equation (3.7) to get Equation (3.1)) can be written in terms of the wavenumber κ of the Schrödinger eigenvalue problem: $\Omega = \kappa(c_0 - 4\beta\kappa^2)$. Then the dispersion relations for the discrete ($\kappa = iK_n$) and continuous spectra ($\kappa = k/2$) follow:

$$\Omega_n = c_0 K_n + 4\beta K_n^3, \tag{3.15a}$$

$$\omega = c_0 k - \beta k^3. \tag{3.15b}$$

These relations are used in Equation (3.14) to evolve the spectrum in time.

The IST (in analogy with the IFT for linear wave motion) evolves the wave in space and time. I now outline the mathematical structure of IST. One solves the Gelfan'd-Levitan-Marchenko (GLM) integral equation:

$$K(x, y) + B(x + y) + \int_x^\infty K(x, z)B(z + y)\mathrm{d}z = 0, \quad y > x, \tag{3.16}$$

whose solution $K(x, y)$ is used to find the wave amplitude for all x and t:

$$\eta(x, t) = \frac{2}{\lambda}\frac{\mathrm{d}K(x, x)}{\mathrm{d}x}. \tag{3.17}$$

The kernel of the GLM equation is given by

$$B(r, t) = \sum_{n=1}^N C_n^2(t)\mathrm{e}^{-K_n r} + \frac{1}{4\pi}\int_{-\infty}^\infty b(k, t)\mathrm{e}^{\mathrm{i}kr/2}\mathrm{d}k. \tag{3.18}$$

I have for convenience suppressed the time dependence in $B(r, t)$ and $K(x, y, t)$ in Equations (3.16) and (3.17). This is possible because the solution to the GLM equation $K(x, y, t)$ (Equation (3.16)) may be considered to be a function of y only; x and t simply play the role of parameters in the formulation.

The phase coefficients C_n in Equation (3.12) depend upon the normalization (Equation (3.11)) for the eigenfunction solutions $\psi_n(x, K_n)$ to Equation (3.8). This implies that knowledge of these functions is necessary before the C_n may be found. An alternative expression for the C_n which is independent of the $\psi_n(x, K_n)$ is given by (Kay and Moscs, 1955; Ablowitz and Segur, 1981):

$$C_n^2 = -\mathrm{i}r_n, \tag{3.19}$$

where the r_n are the residues of the reflection coefficient $b(k)$ at the poles $\kappa = \mathrm{i}K_n$:

$$r_n = \frac{1}{2\pi\mathrm{i}}\oint b(k)\mathrm{d}k = \mathrm{i}\lim_{K \to K_n}(K - K_n)b(\mathrm{i}K). \tag{3.20}$$

The right-hand side of this expression obtains because the poles of $b(\kappa)$ are simple. Then the phase coefficients are given by

$$C_n^2 = \lim_{K \to K_n}(K - K_n)b(\mathrm{i}K). \tag{3.21}$$

Thus, the latter expression makes computation of the C_n independent of the $\psi_n(x, K_n)$ and their integrability condition (3.11), a point exploited in the numerical methods.

I also note that to compute the number of solitons a convenient formula is (Zakharov et al., 1980)

$$N = \lim_{k \to \infty} [\arg(a(k)) - \arg(a(-k))]. \tag{3.22}$$

This turns out to be a good way to compute N because it too is independent of the eigenfunctions. The implementation of Equation (3.22) is discussed in Provenzale and Osborne (1991).

For the scattering transform to be valid, the following integral condition must hold:

$$\int_{-\infty}^{\infty} (1 + |x|) |\eta(x, 0)| dx < \infty. \tag{3.23}$$

Thus, a localized wave field that vanishes rapidly as $|x| \to \infty$ satisfies Equation (3.23). In numerical applications, the field $\eta(x, 0)$ is contained in some finite array (i.e., a space series); $\eta(x, 0)$ is assumed to be identically zero outside the confines of this array, hence Equation (3.23) is always satisfied.

In summary, the features of the scattering transform solution to the KdV equation which are of interest are (1) the DST (3.8)–(3.12) generates a spectral representation (3.13) of the input wave $\eta(x, 0)$, (2) evolution of the spectrum in time is simple (Equations (3.14) and (3.15)), and (3) the IST (3.16)–(3.18) evolves the wave in time. These three steps are analogous to those for the linear Fourier transform discussed in the last section.

3.4 The Relationship Between the Fourier Transform and the Scattering Transform

Here, I briefly discuss the fact that scattering transform theory approaches Fourier transform theory as the wave amplitude becomes small. The inference is that for sufficiently small waves, the effect of nonlinearity becomes insignificant and linear Fourier theory is recovered. I emphasize those aspects of the theory that are important to the development of the DST and IST algorithms and to a physical interpretation of the nonlinear wave motion.

The Fourier (small-amplitude) limit occurs in the absence of solitons, that is, when there are no discrete eigenvalues. A necessary and sufficient condition for no solitons is that $\eta(x, 0)$ be negative definite. I first write the Schrödinger eigenvalue problem as an integral equation (Morse and Feshbach, 1953):

$$\psi(x) = e^{-ikx/2} - \frac{\lambda}{ik} \int_{-\infty}^{\infty} e^{ik|x-x'|/2} \eta(x', 0)\psi(x') dx'. \tag{3.24}$$

This expression implicitly contains the boundary conditions (Equations (3.9) and (3.10)) for the continuous spectrum. To see this, I write Equation (3.24) as

$$\psi(x) = e^{-ikx/2} - \frac{\lambda}{ik} \int_{-\infty}^{x} e^{ik(x-x')/2} \eta(x',0)\psi(x')dx'$$

$$- \frac{\lambda}{ik} \int_{x}^{\infty} e^{ik(x'-x)/2} \eta(x',0)\psi(x')dx'.$$

(3.25)

If I take the limit as $x \to \pm\infty$, I recover Equations (3.3) and (3.4) with explicit expressions for $a(k)$ and $b(k)$ in terms of integrals over the initial wave $\eta(x,0)$ and eigenfunction $\psi(x)$:

$$a(k) = 1 - \frac{\lambda}{ik} \int_{-\infty}^{\infty} e^{ikx'/2} \eta(x',0)\psi(x')dx',$$

(3.26)

$$b(k) = -\frac{\lambda}{ik} \int_{-\infty}^{\infty} e^{-ikx'/2} \eta(x',0)\psi(x')dx'.$$

(3.27)

Thus after solving Equation (3.2) for $\psi(x)$, one can formally obtain $a(k)$ and the continuous DST spectrum $b(k)$ by Equations (3.26) and (3.27). As previously pointed out, however, I shall avoid computation of the eigenfunction $\psi(x)$ (they are not part of the DST spectrum and hence need not be computed) in the numerical methods (Osborne (1989)); simpler means for computing $a(k)$ and $b(k)$ are used instead.

Equations (3.26) and (3.27) are nevertheless useful in establishing the behavior of the DST in the small-amplitude limit; the integrals are assumed to be small, so that a Neumann series expansion is possible and I take as a first approximation $\psi^{(0)}(x) \cong \exp\{-ikx/2\}$, and this, when inserted back into Equation (3.24), gives

$$\psi^{(1)}(x) \cong \exp\{-ikx/2\} - \frac{\lambda}{ik} \int_{-\infty}^{\infty} e^{ik|x-x'|/2} \eta(x',0)\psi^{(0)}(x')dx'$$

(3.28)

a result often referred to as the Born approximation. To this order approximate expressions for $a(k)$ and $b(k)$ are

$$a^{(1)}(k) \cong 1 - \frac{\lambda}{ik} \int_{-\infty}^{\infty} \eta(x,0)dx,$$

(3.29)

$$-ikb^{(1)}(k) \cong \lambda \int_{-\infty}^{\infty} \eta(x,0)e^{-ikx}dx.$$

(3.30)

Thus, $a^{(1)}(k)$ is related to the area under the initial waveform, while $-ikb^{(1)}(k)$ is just the Fourier transform of $\lambda\eta(x, 0)$. This establishes the connection between the continuous part of the scattering transform and the Fourier transform. Formally, I write for sufficiently small η

$$-ikb(k) \cong \lambda \int_{-\infty}^{\infty} \eta(x, 0)e^{-ikx}dx. \tag{3.31}$$

Note that our use of $\kappa = k/2$ in the Schrödinger eigenvalue problem (3.2) has resulted in a DFT limit (Equation (3.31)) consistent with the notation used in our definition of the DFT (Equation (2.3)); this motivates division by an additional factor of 2 in the wavenumber.

It is important to establish under what specific conditions I can expect the DST to give results equivalent to the Fourier transform. A crude estimate can be made by letting $\eta(x, 0)$ be a square wave centered on the origin, with amplitude η_0 and half-width L, and then Equations (3.29) and (3.30) may be integrated exactly. Given that these integrals must be small for the small-amplitude (Born) approximation to be valid, I have

$$\lambda\eta_0/k^2 \ll 1. \tag{3.32}$$

Thus for sufficiently large wavenumber or small amplitude, the DST approaches the Fourier transform. To consider what happens at long wavelength (small k), I note that for some fixed η_0, the Fourier transform limit (Equation (3.32)) must fail as $k \to 0$. Details of this effect and its physical consequences on shallow-water wave motion are discussed in Osborne (1983) and cited references. The small-amplitude limit condition (3.32) is necessary for demonstrating that the DST numerical algorithm has the Fourier limit for sufficiently small amplitudes (Osborne (1989)).

To obtain additional physical insight, I put the condition (3.32) into dimensionless form. In the Schrödinger problem (3.8), I set $\eta(x, 0) = \eta_{max}u(x)$ and $x = Lr$, where η_{max} is the maximum value of $\eta(x, 0)$ and L is a characteristic length of the initial wave. This gives the dimensionless Schrödinger problem:

$$\psi_{rr} + [\lambda_U u(r) + \chi^2]\psi = 0, \tag{3.33}$$

where $\lambda_U = \lambda\eta_{max}L^2$ is the Ursell number (Miles, 1980 and cited references) and $\chi = kL$ is a dimensionless wavenumber. Then condition (3.32) is equivalent to

$$\lambda_U = 3\eta_{max}L^2/2h^3 \ll \chi^2. \tag{3.34}$$

Thus, the Born approximation of quantum mechanics corresponds to small Ursell number in nonlinear, shallow-water wave dynamics. Physically, a small Ursell number occurs when nonlinear effects are small; conversely, large

Ursell number corresponds to the case when nonlinear interactions among spectral components are enhanced due to large waves. Note also that Equation (3.32) may be thought of as a *spectral Ursell number*, $\lambda_k = \lambda \eta_0 / k^2$, that is, an Ursell number associated with each radiation component in the spectrum. Computation of λ_k provides an estimate of how strongly a spectral component interacts with its neighbors: $\lambda_k \ll 1$ for small interactions, $\lambda_k \sim 1$ for moderate interactions, and $\lambda_k \gg 1$ for strong interactions. The linear Fourier transform limit of the DST (3.31) occurs for small spectral Ursell number, $\lambda_k \ll 1$.

I now consider the small-amplitude limit of the *inverse scattering transform*. Since no solitons are present, Equation (3.12) can be written:

$$B(2x, t) = \frac{1}{4\pi} \int_{-\infty}^{\infty} b(k) e^{ikx - i\omega t} dk. \tag{3.35}$$

Segur (1972) has discussed how the GLM equation may be expanded in a Neumann series for small-amplitude waves; the first term in this expansion results from ignoring the integral term of the GLM equation (3.16). With Equation (3.35) this leads to (for small η)

$$\lambda \eta(x, t) \cong \frac{1}{2\pi} \int_{-\infty}^{\infty} [-ikb(k)] e^{ikx - i\omega t} dk. \tag{3.36}$$

Thus for sufficiently small amplitude, the wave motion is recovered at $t = 0$ and is evolved in time thereafter by the IFT. This is clearly the inverse of Equation (3.31) (compare to Equations (3.3) and (3.5)) and the connection of IST with linear Fourier theory is evident. Note that $F(k) \cong -ikb(k)/\lambda$ in the small-amplitude limit. The linear Fourier transform limit (3.36) of the IST (3.10)–(3.12) is seen to occur generally when the spectral Ursell number is small, $\lambda_k \ll 1$.

3.5 Review of Assumptions Implicit in the Discrete, Finite Fourier Transform

I now review some of the important considerations that historically lead to discrete algorithms for the Fourier transform. Two of the most important signal processing problems relate to the use of *Fourier series* (for signals periodic on an interval $(0, L)$) and the *Fourier transform* (for nonperiodic signals on the infinite interval). The theoretical properties of Fourier series and Fourier transforms differ somewhat, but, fortunately for most practical problems in data analysis, their digital computational procedures are normally the same (Bendat and Piersol, 1986). This results from the fact that only a finite range Fourier series or transform can be computed from digital signals and normally this finite range is taken to be the period of an associated Fourier series. The above observation ultimately motivated detailed study of *periodic nonlinear Fourier analysis*, the main topic of this book.

For linear wave motion, the problem consists of the numerical integration of the Fourier transform (3.3). One is commonly confronted with two kinds of distinct problems: (1) numerical evaluation of the integral is required for some *analytic function* $\eta(x, 0)$ and (2) evaluation of the integral is required given some *discrete function* whose intermediate values are often unknown. While there is some overlap between the two, it is fair to say that the first case tends to rely on rigorous numerical integration of Equation (3.3) by some high-order integrator such as fifth- or sixth-order Adams-Bashforth or Runge-Kutta. The latter case (which occurs most often in the analysis of data) tends to rely on discrete methods based upon the finite Fourier transform. Use of the finite transform is standard in the analysis of time series because (1) the method is computationally efficient (FFT algorithm) and (2) it is based upon a theory that is exact for discrete signals (Cooley, 1961; Cooley and Tukey, 1965; Cooley et al., 1969; Singleton, 1969). To see how the finite Fourier transform arises, I restrict the limits of Equation (3.3) to a finite spatial interval $(0, L)$ on which there lies a localized initial wave $\eta(x, 0)$ that is taken to be zero outside this interval:

$$F(k) = \int_0^L \eta(x, 0) e^{-ikx} dx. \tag{3.37}$$

I further assume that $\eta(x, 0)$ is sampled at M equally spaced values of x a distance ($\Delta x = L/M$) apart so that $x_m = m\Delta x$ and

$$\eta_m = \eta(m\Delta x, 0), \quad m = 0, 1, 2, \ldots, M - 1. \tag{3.38}$$

For arbitrary wavenumber k, the discrete version (rectangular approximation) of Equation (3.37) is

$$F(k) = \Delta x \sum_{m=0}^{M-1} \eta_m e^{-ikx_m}. \tag{3.39}$$

One then selects discrete wavenumbers for evaluation of $F(k)$:

$$k_j = j\Delta k = \frac{2\pi j}{L} = \frac{2\pi j}{M\Delta x}, \quad j = 0, 1, 2, \ldots, M - 1. \tag{3.40}$$

The Fourier components are then found from

$$F(k_j) = \Delta x \sum_{m=0}^{M-1} \eta_m e^{-ik_j x_m} = \frac{L}{M} \sum_{m=0}^{M-1} \eta_m e^{-2\pi i(jm/M)}, \tag{3.41}$$

where $\Delta k = 2\pi/\Delta x$. The right-hand side of Equation (3.41) is often referred to as the definition of the discrete, finite Fourier transform whose mathematical properties have been studied in detail (Cooley, 1961; Cooley and Tukey, 1965; Cooley et al., 1969; Singleton, 1969). The components of $F(k_j)$ are unique only out to the Nyquist frequency that occurs for $j = M/2$.

The easiest way to evaluate Equation (3.39) or Equation (3.41) numerically is through the simple recursion relation:

$$F_m(k) = F_{m-1}(k) + \eta_m z^{-mk} \Delta x, \qquad (3.42)$$

where $F_{-1}(k) = 0$ and $z = \exp(i\Delta x)$. It is clear that such an algorithm requires M^2 operations (ranging over both space x and wavenumber k) where an operation is defined as "multiplication of an exponential times an amplitude, followed by a summation." FFT techniques allow the $F(k_j)$ to be computed by an algorithm equivalent to Equations (3.41) and (3.42) but which requires only $M \log M$ operations. Herein I discuss an algorithm for the DST that is analogous to Equation (3.42), that is, two-point recursive. Development of this DST algorithm, however, is not as simple or obvious as the step from Equation (3.39) or Equation (3.41) to Equation (3.42) for the Fourier transform. A "fast" DST algorithm must await future developments. Nevertheless, it is important to note that I am seeking a numerical algorithm that has discrete Fourier structure.

Some points are worth noting about the assumptions implicit in the above procedures. First, in the context of numerical analysis (as opposed to data analysis), note that while the rectangular approximation was used for evaluation of the integral (3.37), corrections to Equation (3.37) for trapezoidal and higher-order approximations can be made if desired (for a discussion and references see Ng, 1974). The usual procedure in signal processing applications, however, is to employ the FFT algorithm (based upon Equation (3.41) which is equivalent to the two-point recursive formula (3.42)) for the spectral analysis of data and of other (say computer-generated) digital signals (Cooley, 1961; Cooley and Tukey, 1965; Cooley et al., 1969; Singleton, 1969; Bendat and Piersol, 1971). No corrections are normally considered at higher order, primarily because of ignorance about the behavior of the signal at intermediate locations between the discrete points. In the analysis of data, one normally measures several thousand or ten thousand points in a time series with a small discretization interval (in either time or space) and this implies that higher-order corrections are likely not very significant.

An important consideration about the infinite-interval Fourier transform $F(k)$ relates to how the wavenumbers are selected for numerical computation. Based upon known results for the periodic Fourier transform, a convenient wavenumber resolution is (see Equation (3.40))

$$\Delta k = \frac{2\pi}{L}, \qquad (3.43)$$

where L is the period. The upper frequency cutoff is given by the Nyquist wavenumber:

$$k_N = \frac{\pi}{\Delta x}.$$ (3.44)

Theoretically, one assumes knowledge of the wave amplitude over the entire (infinite) real axis. This implies that knowledge of all wavenumbers is necessary to reconstruct the wave. In numerical computation of the *infinite-interval Cauchy problem*, one normally measures or generates a discrete signal which is appreciably different from zero only on some interval $(0, L)$ and which is *assumed* to be zero everywhere outside this interval. The implication is that the signal has infinite length and according to Equation (3.43), I have $\Delta k \rightarrow 0$. Thus, the wavenumber k is essentially continuous, a known result of Fourier transform theory. What this means in a practical sense is that while Equation (3.43) forms some basis for resolving the wavenumbers, one can deviate from this and use a smaller wavenumber resolution in the spectrum if desired. The Nyquist cutoff given by Equation (3.44) must also be selected with care. One must use a sufficiently small Δx to insure that the interval $(0, k_N)$ contains most of the spectral energy. Otherwise, aliasing of spectral components may occur just as in the linear problem. Furthermore, one must keep in mind that the finite Fourier transform (3.41) is a periodic algorithm; the true infinite-line Cauchy problem is approached only as $L \rightarrow \infty$.

3.6   Assumptions Leading to a Discrete Algorithm for the Direct Scattering Transform

In the previous section, it was seen that the assumptions leading to the discrete, finite Fourier transform include (1) truncation of a localized discrete signal to some finite interval $(0, L)$ and (2) rectangular approximation of the Fourier integral. These result in Equation (3.39) and, with an appropriate selection for the wavenumbers, one finds Equation (3.41), which is the finite Fourier transform. To develop a numerical algorithm for the DST, I shall consider Equation (3.39) as fundamental (thus deferring selection of the wavenumbers) and proceed to find an algorithm that approaches Equation (3.39) when the wave amplitude becomes sufficiently small. Thus, Equation (3.39) is an important guiding criterion in what follows. In this way, I develop nonlinear spectral methods that are compatible with known discrete Fourier methods.

To pursue an algorithm for the DST, I consider, as before, a discrete signal with amplitudes at coordinate positions $x_m = m\Delta x$, where Δx is a constant spatial interval between points and $0 \leq m \leq M - 1$ (one could also generalize the results to a finite element analysis in which Δx is not constant, but I do not pursue this here since Δx is normally constant for measured data and for computer-generated space or time series). To truncate the signal to some interval $(0, L) = (0, M\Delta x)$, I assume that all wave amplitudes are zero to the left of and including x_0 and to the right of and including x_M. Recall that the rectangular approximation (3.39) for the Fourier integral (3.37) implies $\eta(x, 0)\exp(-ikx)$ is a (complex) constant in each interval Δx. Since this product of amplitude and exponential do not appear explicitly in the Schrödinger eigenvalue problem, it does not appear straightforward to use the same form of rectangular approximation here.

However, I have been able to develop an alternate discretization that, for practical purposes, works rather well (see Figure 3.1). A continuous wave amplitude function $\eta(x, 0)\exp(-ikx)$ is discretized at intervals $x_m = m\Delta x$ ($\eta(x_m, 0)$, Figure 3.1B). The discrete signal is then replaced by a piecewise constant function as shown in Figure 3.1C. Each constant partition has width Δx centered on coordinate x_m; the constant amplitude η_m in each interval is

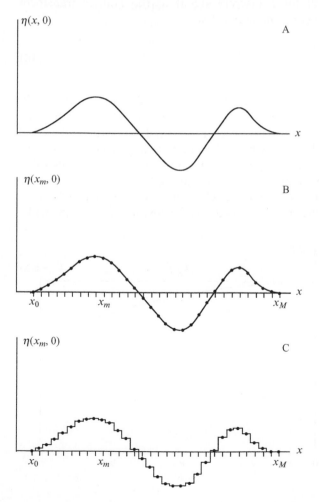

Figure 3.1 An initial wave amplitude function $\eta(x, 0)$ that varies continuously as a function of x is shown in (A) and is discretized at intervals Δx in (B). In (C), one associates a piecewise constant function with the discrete array (B). This latter function is then used in the development of the numerical algorithm for the direct spectral transform (DST).

assumed to be the same as the amplitudes of the previous discretization procedure for the Fourier transform. In selecting this form for the wave amplitude function, I am also motivated by the fact that the Schrödinger eigenvalue problem has an exact solution for functions of this type. Since I consider a signal of this kind to be almost as good as the original rectangular approximation to Equation (3.37), I would like to compare the Fourier transform of the two types of wave forms. For the piecewise constant signal, one easily finds for the Fourier integral

$$F_p(k) = \Delta x \left[\frac{\sin(k\Delta x/2)}{k\Delta x/2}\right] \sum_{m=0}^{M-1} \eta_m e^{-ikx_m} \qquad (3.45)$$

a result, which is amazingly close to Equation (3.39), differing only by the factor in square brackets. This factor acts essentially as a filter which relates the Fourier spectrum $F(k)$(3.39) for a discrete signal to the Fourier transform $F_p(k)$ (3.45) for a piecewise constant signal. I set

$$S(k\Delta x) = \left[\frac{\sin(k\Delta x/2)}{k\Delta x/2}\right]. \qquad (3.46)$$

A graph of this function is shown in Figure 3.2. For $k\Delta x$ sufficiently small, $S(k\Delta x) \cong 1$ and I have $F_p(k) \cong F(k)$. Thus, the Fourier transform for the discrete and piecewise constant signals is essentially equal for sufficiently small wavenumber. The filter $S(k\Delta x)$ slowly and monotonically decreases toward its first zero at $k_0 = 2\pi/\Delta x$. The Nyquist wavenumber occurs at half this value $k_N = k_0/2 = \pi/\Delta x$ where $S(k\Delta x)$ has decreased to 0.64.

It is important at this point to understand what influence the filter (3.46) may have on the physics of the DST spectrum (as derived below in Sections 3.8–3.11).

Figure 3.2 Shape of the filter that relates the direct Fourier transform (DFT) of a discrete function to the DFT of a piecewise constant function.

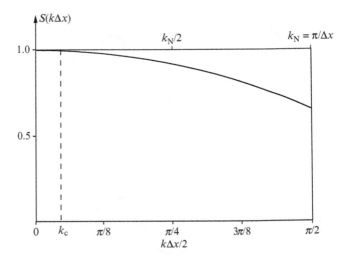

In problems of water wave motion, a wavenumber cutoff beyond which KdV evolution does not apply is given approximately by $k_c = 1/h$. This is because the long wave assumption is no longer valid for larger wavenumbers much greater than k_c. Normally $k_c/k_N \sim 0.01-0.1$ (Osborne, 1983) and for $k \gg k_c$ the DST spectrum approaches the Fourier transform. Thus in the wavenumber range in which KdV physics is important $(0, k_c)$, one can normally arrange for $S(k\Delta x)$ to be near one (by making Δx sufficiently small so that $S(k\Delta x) \cong 1$) and one need not concern himself with the presence of the filter. One can always remove the filter from the spectrum by dividing the continuous DST spectrum by Equation (3.46) if a direct comparison with the FFT is desired. This latter procedure is recommended in practical implementations of the algorithm. Another important point is that since the filter differs from 1 only at high wavenumbers (if the Nyquist wavenumber is chosen large enough) where the physics is essentially linear, it is unlikely that nonlinear effects can strongly influence the actual shape of the filter for the DST spectrum.

Another additional concern is to establish what effect the selection of a piecewise constant waveform has on the algorithm for the IST, that is, for the selection of a numerical method for the solution of the GLM equation (3.10). The relationship between the solutions of GLM, $K(x, x)$, and the solution of KdV, $\eta(x, 0)$, can be seen by rewriting Equation (3.11) as

$$K(x, x) = -\frac{\lambda}{2} \int_x^\infty \eta(x, 0) \mathrm{d}x. \tag{3.47}$$

I write the discrete (piecewise constant) form for Equation (3.47):

$$K_{m+1/2} = -\frac{\lambda}{2} \Delta x \sum_{j=m}^{M} \eta_{j+1}. \tag{3.48}$$

This implies that to be consistent with our selection of a piecewise constant wave, the solution of the GLM equation must be trapezoidal as shown in Figure 3.3. $K(x, x)$ is a piecewise linear function evaluated at the positions $x_{m+1/2}$, that is, at the half-integers $m + 1/2 = 1/2, 3/2, 5/2, \ldots, M + 1/2$. The initial (piecewise constant) wave may be recovered by the obvious discrete formula (see Figure 3.3B):

$$\eta_m = \frac{2}{\lambda} \frac{(K_{m+1/2} - K_{m-1/2})}{\Delta x}. \tag{3.49}$$

Thus, our discretization procedure for the DST (developed by analogy with Fourier analysis) has led to the need for an IST procedure, which is trapezoidal. A trapezoidal algorithm for solving the GLM equation has previously been given by Hald (1979).

The above considerations are important for the development of a discrete algorithm for the solution of the Schrödinger eigenvalue problem on the infinite

Figure 3.3 Discretization
of a wave train (A) and the
associated discretization of
the Gelfan'd-Levitan-
Marchenko equation (B).

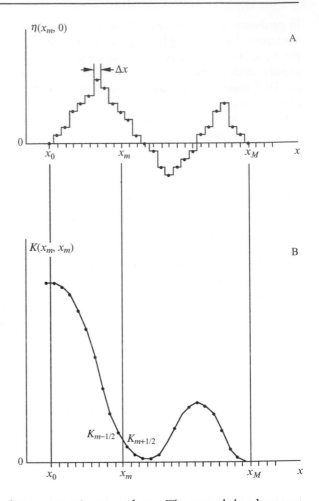

line, referred to here as the direct scattering transform. The actual development
of the algorithm itself is straightforward and is discussed in detail in Osborne
(1991) (see also the companion paper Provenzale and Osborne (1991)). It will
not go unnoticed to the reader of these papers that some attempt was made
to formulate the algorithm in analogy with the Schrödinger equation for periodic boundary conditions. Due to the fact that the formulation was based on
infinite-line boundary conditions, the results are somewhat limited. It is for this
reason that the ensuing research has been directed toward the problem with
periodic boundary conditions, something that is no surprise to oceanographers.
This Chapter serves as an introduction to most of the assumptions necessary
for the development of discrete algorithms. Extension to the periodic problem
is the main topic in the remainder of this book. Chapter 17 discusses the periodic
problem for the Schrödinger eigenvalue problem. Thus actual data analysis
applications should be addressed with the results of this latter chapter. Numerical results for the IST for the nonlinear Schrödinger equation on the infinite line
have been addressed by Boffetta and Osborne, 1992.

4 The Infinite-Line Hirota Method

4.1 Introduction

The inverse scattering transform (IST) is a marvelous mathematical method, but it does not necessarily help us to determine if an equation of immediate interest is integrable. The "direct" method of Hirota instead provides a straightforward approach that allows one to determine if an equation has soliton solutions. It is thought that one-, two-, and three-soliton solutions are necessary to conclude that the equation is integrable (see, e.g., Hietarinta, 2002; Hirota, 2004, also for a long list of references). The direct method can be programmed for symbolic computation with REDUCE, Mathematica, and Maple so that much of the algebra can be accomplished with symbolic computation Baldwin, et al, 2004.

4.2 The Hirota Method

The *Hirota direct method* was developed in the 1970s to study soliton solutions and integrability in nonlinear wave equations. The approach provided an alternative theoretical tool for attacking soliton equations. Not many investigators appreciated the fundamental role played by the method at first, but it soon became a kind of entertaining game to send a postcard to Hirota in Japan with an equation written on it. In a couple of weeks, he would usually respond with his observations about the possible integrability of the equation. The Hirota method is often applied to new equations whose integrability is uncertain, before applying the IST via Lax pairs, the inverse problem, etc. Indeed, the Hirota method provides tools for actually deriving the Lax pair directly, providing a direct link to IST. The method has been documented in a recent book (Hirota, 2004), but you will also enjoy reading the works of Hietarinta (2002). The references by Ablowitz and Segur (1981) and Drazin and Johnson (1989) are also a good place to start.

4.3 The Korteweg-deVries Equation

The Korteweg-deVries (KdV) equation has the simple scaled form:

$$u_t + 6uu_x + u_{xxx} = 0. \tag{4.1}$$

Doi: 10.1016/S0074-6142(10)97004-6

Make the "dependent variable (Hirota)" transformation

$$u(x,t) = 2\partial_{xx} \ln F(x,t) \tag{4.2}$$

and find

$$F_{xt}F - F_x F_t + F_{xxxx}F - 4F_{xxx}F_x + 3F_{xx}^2 = 0. \tag{4.3}$$

This is a *homogeneous, bilinear form*. It is homogeneous because each term is itself a bilinear form. While one can think that the latter equation is more complex than the KdV equation, there is some utility in examining the solutions in light of the *Hirota method*. Note that it is important that the correct dependent variable transformation be selected to obtain a homogeneous, bilinear form. See Chapter 33 for further discussion and references.

It is useful to introduce the following *Hirota operator*:

$$D_x^m D_t^n G \cdot F = \left(\frac{\partial}{\partial x} - \frac{\partial}{\partial x'}\right)^m \left(\frac{\partial}{\partial t} - \frac{\partial}{\partial t'}\right)^n G(x,t)F(x',t')\Big|_{\substack{x'=x \\ t'=t}}. \tag{4.4}$$

Using Equation (4.4) in Equation (4.3), we have the shorthand expression for the bilinear form:

$$(D_x D_t + D_x^4)F \cdot F = 0. \tag{4.5}$$

The following Hirota operator identities hold:

$$\begin{aligned}
D_x D_t F \cdot F &= 2(FF_{xt} - F_x F_t), \\
D_x^2 F \cdot F &= 2(FF_{xx} - F_x^2), \\
D_x^2 F \cdot F &= 2(FF_{xxxx} - 4F_x F_{xxx} + 3F_{xx}^2).
\end{aligned} \tag{4.6}$$

We now consider the dimensional KdV equation:

$$\eta_t + c_0 \eta_x + \alpha \eta \eta_x + \beta \eta_{xxx} = 0. \tag{4.7}$$

Make the dimensional dependent variable transformation

$$\eta(x,t) = \frac{2}{\lambda} \partial_{xx} \ln \theta(x,t), \quad \lambda = \frac{\alpha}{6\beta}. \tag{4.8}$$

This is useful for identifying the physical terms in the bilinear form. We first make the substitution $\eta = w_x$ and integration once in x to find

$$w_t + c_0 w_x + \frac{\alpha}{2} w_x^2 + \beta w_{xxx} + c = 0. \tag{4.9}$$

In the search for soliton solutions, where infinite-line boundary conditions hold, it is natural to take $c = 0$. However, in later chapters we will be concerned with periodic boundary conditions and therefore c will be kept finite. Now, make the final substitution $w = (2/\lambda)\partial_x \ln F$ and obtain

$$FF_{xt} - F_xF_t + c_0(FF_{xx} - F_x^2) + \beta(3F_{xx}^2 - 4F_xF_{xxx} + FF_{xxxx}) = 0, \qquad (4.10)$$

which, using the Hirota bilinear operator, has the form:

$$(D_xD_t + c_0D_x^2 + \beta D_x^4)F\cdot F = 0 \qquad (4.11)$$

Miraculous cancellation of terms renders the bilinear form without the nonlinear KdV term with coefficient α. Such is the miracle of integrability.

Nota Bene. The bilinear form does not have the nonlinear coefficient, α, but only the linear phase speed, c_0, and the dispersion coefficient, β. Therefore, the bilinear form does not have a term equivalent to the nonlinear term in the KdV equation. Inspection of the bilinear reform reveals the linear dispersion relation: $\omega = c_0k^2 - \beta k^3$. Thus, the bilinear form has in some sense "cancelled" nonlinearity. Indeed, the nonlinear coefficient, α, enters in the formulation only through the transformation $(\alpha/12\beta)\eta(x,t) = \partial_{xx} \ln \theta(x,t)$; α rescales the physical amplitude of the waves to include nonlinearity.

It is convenient to derive several additional properties of the operator notation used in Equation (4.4). These follow in a natural way from this definition:

$$D_x^m G\cdot F = (-1)^m D_x^m F\cdot G,$$

$$D_x^m F\cdot F = 0 \quad \text{for } m \text{ odd},$$

$$D_x^m F\cdot 1 = \partial_x^m F, \qquad (4.12)$$

$$D_x^m D_t^n e^{k_1x - \omega_1t}\cdot e^{k_2x - \omega_2t} = (k_1 - k_2)^m(\omega_2 - \omega_1)^n e^{(k_1+k_2)x - (\omega_1+\omega_2)t}.$$

We are now ready to proceed with the Hirota method by assuming the following form for the function $F(x, t)$:

$$F(x,t) = 1 + \varepsilon f^{(1)}(x,t) + \varepsilon^2 f^{(2)}(x,t) + \cdots, \qquad (4.13)$$

where

$$f^{(1)} = \sum_{n=1}^{N} e^{X_n}, \qquad X_n = k_nx - \omega_nt + \phi_n. \qquad (4.14)$$

It should be noted that Equation (4.13) is a sum of exponentials, not unlike terms in a Fourier series, but the exponents are taken to be real; this restricts the approach to soliton solutions, not oscillatory ones. The beauty of the Hirota procedure is that the series (4.13) truncates as we see below. Here, we recognize the wave number, frequency, and phase: k_n, ω_n, and ϕ_n. Use Equations (4.13) and (4.14) in Equation (4.11) and get

$$(D_x D_t + D_x^4)(1 + \varepsilon f^{(1)}(x,t) + \varepsilon^2 f^{(2)}(x,t) + \ldots) \cdot (1 + \varepsilon f^{(1)}(x,t) + \varepsilon^2 f^{(2)}(x,t) + \ldots) = 0.$$
$$(4.15)$$

Apply the properties (4.12) and equate like powers of ε to zero to obtain:

$$
\begin{aligned}
&O(\varepsilon^0 = 1):\ 0 = 0, \\
&O(\varepsilon^1):\ 2(\partial_x \partial_t + \partial_x^4)f^{(1)} = 0, \\
&O(\varepsilon^2):\ 2(\partial_x \partial_t + \partial_x^4)f^{(2)} = -(D_x D_t + D_x^4)f^{(1)} \cdot f^{(1)}, \\
&O(\varepsilon^3):\ 2(\partial_x \partial_t + \partial_x^4)f^{(2)} = -2(D_x D_t + D_x^4)f^{(1)} \cdot f^{(2)}.
\end{aligned}
\qquad (4.16)
$$

The $O(\varepsilon)$ equation above is homogeneous and has the solution $f^{(1)} = e^{X_1}$, where $\omega_1 = -k_1^3$ for $N = 1$. The equation for $f^{(2)}$ is given by the $O(\varepsilon^2)$ equation of (4.16). Use of the last equation of (4.16) reduces the $O(\varepsilon^2)$ equation to

$$(\partial_x \partial_t + \partial_x^4)f^{(2)} = 0,$$

so that $f^{(2)} = 0$ and the expansion (4.13) terminates.

Therefore, the $N = 1$ solution is given by

$$F_1 = 1 + e^{X_1}, \quad \omega_1 = -k_1^3,$$

and by Equation (4.2) we have a solution of KdV:

$$u(x,t) = \frac{1}{2}k_1^2 \operatorname{sech}^2 \frac{1}{2}(k_1 x - \omega_1 t + \phi_1),$$

which is a single soliton.

For the case $N = 2$, we set

$$f^{(2)} = e^{X_1} + e^{X_2}, \quad X_n = k_n x + k_n^3 t + \phi_n.$$

Then the $O(\varepsilon^2)$ equation of (4.16) becomes

$$2(\partial_x \partial_t + \partial_x^4)f^{(2)} = -2[(k_1 - k_2)(\omega_2 - \omega_1) + (k_1 - k_2)^4]e^{X_1 + X_2}.$$

This, then, has the simple solution:

$$f^{(2)} = e^{X_1 + X_2 + B_{12}}, \quad e^{B_{mn}} = \left(\frac{k_m - k_n}{k_m + k_n}\right)^2,$$

where it is clear that $k_m \neq k_n$. If we use $f^{(1)}$ and $f^{(2)}$ in the $O(\varepsilon^3)$ equation of (4.16), we see that the right-hand side vanishes so that $f^{(3)} = 0$. Hence for $N = 2$, we have

$$F_2 = 1 + e^{X_1} + e^{X_2} + e^{X_1+X_2+B_{12}}.$$

The two-soliton solution then is computed from Equation (4.2).

Continuing to the case for $N = 3$, we find

$$F_2 = 1 + e^{X_1} + e^{X_2} + e^{X_3} + e^{X_1+X_2+B_{12}} + e^{X_1+X_3+B_{13}} + e^{X_2+X_3+B_{23}}$$

$$+ e^{X_1+X_2+X_3+B_{12}+B_{13}+B_{23}}.$$

Finally, the N-soliton solution for the KdV equation is given by

$$u_t + 6uu_x + u_{xxx} = 0, \quad u(x,t) = 2\partial_{xx} \ln F_N(x,t),$$

$$F_N = \sum_{\boldsymbol{\mu}=0,1} \exp\left\{ \sum_{n=1}^{N} \mu_n X_n + \sum_{1 \leq m < n}^{N} \mu_m \mu_n B_{mn} \right\}, \quad B_{mn} = \ln\left(\frac{k_m - k_n}{k_m + k_n}\right)^2,$$

$$X_n = k_n x - \omega_n t + \phi_n.$$

$$(4.17)$$

Here $\boldsymbol{\mu} = 0, 1$ refers to each of the μ_n, $n = 1, 2 \ldots N$.

The periodic Hirota method is discussed in Chapter 6 and is based upon a Fourier series, not with real exponentials as in the soliton expansion above, but with complex exponentials as Riemann theta functions.

4.4   The Hirota Method for Solving the KP Equation

A two-dimensional extension of the KdV equation is the KP equation:

$$(u_t + 6uu_x + u_{xxx})_x + \alpha u_{yy} = 0, \quad \alpha = \pm 1. \tag{4.18}$$

The Hirota N-soliton solution is given by (Ablowitz and Segur, 1981; Hirota, 2004)

$$f(x,y) = \sum_{\boldsymbol{\nu}=0,1} \exp\left(\sum_{i=1}^{N} \nu_i \eta_i + \sum_{i \geq 1}^{N} \sum_{j > i}^{N} \nu_i \nu_j A_{ij} \right) \tag{4.19}$$

for

$$u(x,y) = 2(\log f)_{xx}, \tag{4.20}$$

where the phases are given by

$$\eta_i = k_i(x + p_i y - C_i t + \phi_i) \tag{4.21}$$

for the phase speeds:

$$C_i = k_i^2 + \alpha p_i^2 \tag{4.22}$$

and phase shifts:

$$e^{A_{ij}} = \left(\frac{3(k_i - k_j)^2 - \alpha(p_i - p_j)^2}{3(k_i + k_j)^2 - \alpha(p_i - p_j)^2} \right). \tag{4.23}$$

4.5   The Nonlinear Schrödinger Equation

Given the nonlinear Schrödinger (NLS) equation, with complex solution $\psi(x, t)$

$$i\psi_t + \mu\psi_{xx} + v|\psi|^2\psi = 0 \tag{4.24}$$

seek a solution as the ratio of two functions:

$$\psi(x, t) = \frac{G(x, t)}{F(x, t)}. \tag{4.25}$$

As will be seen below, we can treat $F(x, t)$ as a real function and $G(x, t)$ as a complex function. Insert Equation (4.25) into Equation (4.24) to get

$$\frac{G}{F^3}(2\mu F_x^2 + v|G|^2)$$
$$+ \frac{i(FG_t - GF_t) - 2\mu F_x G_x - \mu GF_{xx} + \mu FG_{xx}}{F^2} = 0. \tag{4.26}$$

Now add $2\mu GF_{xx}/F^2$ to the second term and subtract $2\mu GF_{xx}/F^2$ from the first term to get

$$\frac{G}{F^3}(2\mu F_x^2 + v|G|^2) - \frac{2\mu GF_{xx}}{F^2}$$
$$+ \frac{i(FG_t - GF_t) - 2\mu F_x G_x - \mu GF_{xx} + \mu FG_{xx}}{F^2} + \frac{2\mu GF_{xx}}{F^2} = 0.$$

At this stage, we can "separate the variables" and set

$$i(G_tF - GF_t) + \mu(G_{xx}F - 2G_xF_x + GF_{xx}) = 0$$
$$2\mu(FF_{xx} - F_x^2) - v|G|^2 = 0$$

(4.27)

Equations (4.27) can be put in the standard Hirota operator form:

$$(iD_t + \mu D_x^2)G{\cdot}F = 0,$$
$$\mu D_x^2 F{\cdot}F - v|G|^2 = 0,$$

(4.28)

where the operators are

$$D_t G{\cdot}F = G_t F - GF_t,$$

$$D_x G{\cdot}F = G_x F - GF_x,$$

$$D_x^2 G{\cdot}F = G_{xx}F - 2G_xF_x + GF_{xx},$$

$$D_x^2 F{\cdot}F = 2(FF_{xx} - F_x^2).$$

From the second of Equation (4.28), get

$$|\psi|^2 = \frac{GG^*}{F^2} = \frac{2\mu}{v}\partial_{xx}\ln F.$$

(4.29)

At this point, the N-soliton solution arises as before by suitable exponential expansions (set $\mu = v = 1$):

$$F = \sum_{\mu=0,1} D_1(\mu_1, \mu_2)\exp\left\{\sum_{1\leq i<j}\mu_i\mu_j A_{ij} + \sum_{i=1}^{2N}\mu_i(k_i x - \omega_i t + \phi_i)\right\},$$

$$G = \sum_{\mu=0,1} D_2(\mu_1, \mu_2)\exp\left\{\sum_{1\leq i<j}\mu_i\mu_j A_{ij} + \sum_{i=1}^{2N}\mu_i(k_i x - \omega_i t + \phi_i)\right\},$$

where

$$k_{i+N} = k_i^*, \quad \omega_{i+N} = \omega_i^*, \quad \phi_{i+N} = \phi_i^*, \quad \omega_i = -ik_i^2, \quad i = 1, 2, \ldots, N,$$

$$A_{ij} = \ln\left[\frac{1}{2}(k_i + k_j)^{-2}\right] \quad \text{for} \quad i = 1, 2, \ldots, N \quad \text{and}$$

$$j = N + 1, N + 2, \ldots, 2N,$$

$$A_{ij} = \ln\left[\frac{1}{2}(k_i - k_j)^{-2}\right] \quad \text{for} \quad i = N+1, N+2, \ldots, 2N \quad \text{and}$$

$$j = N+1, N+2, \ldots, 2N,$$

and

$$D_1(\mu_1, \mu_2) = \begin{cases} 1, & \text{when } \sum_{i=1}^{N} \mu_i = \sum_{i=1}^{N} \mu_{i+N}, \\ 0, & \text{otherwise,} \end{cases}$$

$$D_2(\mu_1, \mu_2) = \begin{cases} 1, & \text{when } 1 + \sum_{i=1}^{N} \mu_{i+N} = \sum_{i=1}^{N} \mu_i, \\ 0, & \text{otherwise,} \end{cases}$$

for which the N-soliton solution of the NLS equation is given by

$$\psi(x,t) = \frac{G(x,t)}{F(x,t)}.$$

4.6 The Modified KdV Equation

The modified KdV (mKdV) equation has the form:

$$u_t + 6u^2 u_x + u_{xxx} = 0.$$

A first step is to let

$$u = \frac{G}{F},$$

for which one finds decoupled bilinear forms:

$$(D_t + D_x^3)G{\cdot}F = 0, \quad D_x^2 F{\cdot}F = 2G^2.$$

With a little work, one is able to find the additional dependent variable transformation:

$$u = i\left(\ln\frac{f^*}{f}\right)_x,$$

which has the bilinear equations:

$$(D_t + D_x^3)f^*{\cdot}f = 0, \quad D_x^2 f^*{\cdot}f = 0.$$

Finally, the exponential expansion gives the N-soliton solution:

$$f_N = \sum_{\mu=0,1} \exp\left\{ \sum_{1 \leq i < j} \mu_i \mu_j A_{ij} + \sum_{i=1}^{2N} \mu_i(k_i x - \omega_i t + \phi_i + i\pi/2) \right\},$$

where

$$A_{ij} = \ln\left[\left(\frac{k_i - k_j}{k_i + k_j} \right)^2 \right], \quad \omega_i = -k_i^3$$

These simple examples only briefly introduce the method. Hirota's book (2004) is the next step for the interested reader.

Part Two

Periodic Boundary Conditions

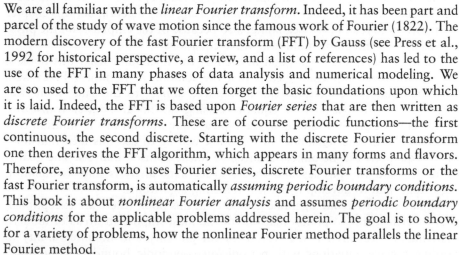

We are all familiar with the *linear Fourier transform*. Indeed, it has been part and parcel of the study of wave motion since the famous work of Fourier (1822). The modern discovery of the fast Fourier transform (FFT) by Gauss (see Press et al., 1992 for historical perspective, a review, and a list of references) has led to the use of the FFT in many phases of data analysis and numerical modeling. We are so used to the FFT that we often forget the basic foundations upon which it is laid. Indeed, the FFT is based upon *Fourier series* that are then written as *discrete Fourier transforms*. These are of course periodic functions—the first continuous, the second discrete. Starting with the discrete Fourier transform one then derives the FFT algorithm, which appears in many forms and flavors. Therefore, anyone who uses Fourier series, discrete Fourier transforms or the fast Fourier transform, is automatically *assuming periodic boundary conditions*. This book is about *nonlinear Fourier analysis* and assumes *periodic boundary conditions* for the applicable problems addressed herein. The goal is to show, for a variety of problems, how the nonlinear Fourier method parallels the linear Fourier method.

This book is about applications of the inverse scattering transform (IST; Fourier analysis for *nonlinear* problems) to surface and internal waves, ocean acoustics, and vortex dynamics. However, in its original context, the IST was and is primarily theoretical research of nonlinear wave equations with infinite-line or infinite-plane boundary conditions (Ablowitz and Segur, 1981; Ablowitz and Clarkson, 1991; Ablowitz et al., 2004). As seen in the paragraph above, however, oceanographic applications most often require periodic boundary conditions. Over 95% of the literature in the field of the IST is for infinite-line boundary conditions and the remaining different area of the literature explores periodic and/or quasi-periodic boundary conditions (Belokolos et al., 1994). Periodic boundary conditions require the application of *algebraic geometry* to solve the basic mathematical structure of nonlinear wave equations. The often-felt opinion that this area of mathematics is very esoteric means that the field is often avoided even by many mathematicians. Thus, it is no surprise that the *interest* of most researchers in physical oceanography is soon damped when introduced to nonlinear Fourier analysis.

This book in part forms a bridge from the esoteric field of algebraic geometry (for the study of the solutions of nonlinear partial differential equations (PDEs) with periodic/quasi-periodic boundary conditions) to the field of oceanography. It will come as no surprise that my goal in writing this book has been to spare the field of oceanography some of the pain that comes in learning significant results from this field. You are quite welcome to read the literature on your own in this regard and ample references are given.

To begin Section 2 on "Periodic Boundary Conditions" I discuss rather superficially the mathematics, and try instead to physically motivate the formulations in the context of data analysis, simulation, and modeling (Chapter 5). This chapter should definitely be read on a first reading of the book. Chapter 6 tells you how to compute the spectrum (Riemann spectrum) for a nonlinear wave equation *without algebraic geometry*, using only algebra. So, this chapter is important to read in the beginning, also because of the emphasis on the physical aspects of integrable nonlinear wave equations.

These two chapters are a good first effort to understand what this book is about. The most important idea is that the solution of nonlinear PDEs with periodic boundary conditions can be reduced *almost* to the same numerical form as that for linear wave equations using the linear Fourier transform. In Chapter 5 I give flowcharts of the typical computation of the solution to a nonlinear PDE. The remainder of the book will be to use this idea that nonlinear equations can be almost reduced to linear Fourier analysis. The goal is to understand the physics and to construct data analysis and assimilation algorithms and to develop hyperfast modeling algorithms.

While I tend to avoid algebraic geometry in this book (there are exceptions), it is nevertheless my opinion that one of the most important areas of modern mathematical research is the search for integrability and nonintegrability in nonlinear wave equations with periodic/quasi-periodic boundary conditions. Why? Because the results of these researches provide ways to do nonlinear Fourier analysis which are immediately applicable to the various fields of wave physics, including physical oceanography. The products are mainly two: (1) rules to construct the Riemann matrix and phases (the *nonlinear Fourier spectrum*) to solve a particular nonlinear wave equation and (2) a so-called *nonlinear Hirota transformation*. Once you have these two things you can understand the physics, analyze and assimilate data, and build hyperfast numerical models of nonlinear wave equations!

5 Periodic Boundary Conditions: Physics, Data Analysis, Data Assimilation, and Modeling

5.1 Introduction

The theoretical formulation known as the *inverse scattering transform* (IST) was found first for the Korteweg-deVries equation by Kruskal and coworkers (Gardner et al., 1967) many decades ago. The method, based primarily *on infinite-line boundary conditions*, has since been applied to very large classes of nonlinear wave equations, often referred to as being "integrable" or "solvable" by IST. In spite of the fundamental role that infinite-line/plane IST has played in the theoretical development of nonlinear water wave theories, we do not directly use these theories here. However, some knowledge of them is a prerequisite for this book. They are covered in great detail in many monographs and constitute some of the most important work in mathematical physics of the twentieth century (Leibovich and Seebass, 1974; Newell, 1974; Lonngren and Scott, 1978; Lamb, 1980; Ablowitz and Segur, 1981; Eilenberger, 1981; Calogero and Degasperis, 1982; Matsuno, 1984; Novikov et al., 1984; Tracy, 1984; Faddeev and Takhtajan, 1987; Drazin and Johnson, 1989; Fordy, 1990; Infeld and Rowlands, 1990; Makhankov, 1990; Ablowitz and Clarkson, 1991; Dickey, 1991; Gaponov-Grekhov and Rabinovich, 1992; Newell and Moloney, 1992; Belokolos et al., 1994; Ablowitz and Fokas, 1997; Johnson, 1997; Remoissenet, 1999; Polishchuk, 2003; Ablowitz et al., 2004; Hirota, 2004).

This monograph instead focuses on IST for *periodic boundary conditions*. The main reasons for the selection of these boundary conditions are clear to a physical oceanographer, but are I think less clear to the general theoretical community. Periodic boundary conditions allow the oceanographer or fluid dynamicist to develop data analysis procedures and to model ocean waves, both surface and internal, in a quite transparent physical manner consistent with the generation and dissipation mechanisms. Furthermore, periodic boundary conditions are consistent with the assumptions of stationarity and ergodicity often assumed in the analysis of data. Modeling nonlinear waves with periodic boundary conditions is one of the most common activities. The Fourier series

Doi: 10.1016/S0074-6142(10)97005-8

and Fourier transform of time series differ in their theoretical properties but not, for most practical purposes, in their digital computational details. This is because only a finite length Fourier series or transform can actually be computed with discretized data and the finite range can always be considered as the period of an associated Fourier series. For the fast computation of periodic nonlinear wave motion, we require the Riemann theta function and its reduction to an ordinary Fourier series that in turn can be computed with the fast Fourier transform (FFT) (Chapters 9 and 22). This contrasts to the computation of the Gelfand-Levitan-Marchenko integral equation for the infinite-line problem, which requires much longer execution times (Ablowitz and Ladik, 1975, 1976a,b, 1977; Hald, 1979).

I now briefly discuss linear wave motion. The *linear Fourier transform* is an invaluable aid for data analysis and modeling. Indeed, the *fast Fourier transform* algorithm is essentially a *discrete, periodic Fourier series* and is one of the most useful numerical tools for the study of wave-like phenomena. The *linear Fourier series* is given, in two-space and one-time dimensions (2 + 1), by the formula

$$\eta(x, y, t) = \sum_{m=-\infty}^{\infty} \sum_{n=-\infty}^{\infty} \eta_{mn}(t) e^{ik_m x + il_n y}. \tag{5.1}$$

Here, $\eta(x, y, t)$ is the surface elevation and the $\eta_{mn}(t)$ are the *time-varying Fourier amplitudes* in wavenumber space $(k_m, l_n) = 2\pi(m/L_x, n/L_y)$, where (L_x, L_y) define the domain of the periodic box. Equation (5.1) solves *linear partial differential wave equations* with *well-defined dispersion*. In this case, the time-varying coefficients have the elementary form: $\eta_{mn}(t) = \eta_{mn}(0)e^{-i\omega_{mn}t + \phi_n}$.

Figure 5.1 gives a flowchart of a typical linear model for wave motion defined by an arbitrary dispersion relation, $\omega_{mn} = \omega_{mn}(k_m, l_n)$. I have designed the flowchart in a particular way so that it can be easily extrapolated to the *nonlinear problem* shown in Figure 5.2. The linear problem shown in Figure 5.1 has three steps:

(1) Read in the Fourier spectrum and dispersion relation. Also defined is the length of time of the simulation, T, and the number of time points desired, N_t.
(2) Compute the time evolution of the linear Fourier coefficients $\eta_{mn}(t) = \eta_{mn}(0)e^{-i\omega_{mn}t + \phi_n}$ for all $-N_x \leq m \leq N_x$, $-N_y \leq n \leq N_y$ (this is labeled "preprocessor" in the figure).
(3) Use the two-dimensional FFT to compute the surface elevation $\eta(x, y, t)$ from the time-varying Fourier coefficients $\eta_{mn}(t)$ (this is labeled "solution of linear PDE" in the figure). The "preprocessor" computes the time-varying Fourier coefficients and the "solution of the linear wave equation" computes the surface elevation by the FFT, iterated over all values of time.

In Section 5.6, I show how nonlinear integrable equations can be numerically modeled by a simple modification of the flowchart in Figure 5.1.

Nonlinearity is an important additional ingredient in the study of ocean waves. The *main focus of this chapter* is the development of numerical

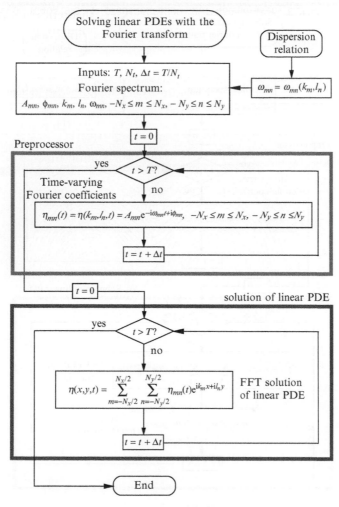

Figure 5.1 Flowchart for numerically solving linear partial differential wave equations.

procedures that allow us to describe the *nonlinear physics*, to analyze *nonlinear data*, and to *build models of nonlinear wave dynamics* using the *periodic IST*, a theory which extends infinite-line (or plane) IST to *periodic boundary conditions*. The goal of this chapter is to provide the researcher with numerical tools to address nonlinear wave problems with the same facility as one has addressed linear problems in the past.

What advantages does the integrable, nonlinear periodic approach have over linear methods? There are several, but some of the most important are:

(1) *Physics.* Development of the *nonlinear wave physics* in terms of *nonlinear basis functions* and their nonlinear interactions with one other. Examples of nonlinear basis functions are *Stokes waves, solitons, shock waves, nonlinear (rogue) modes,*

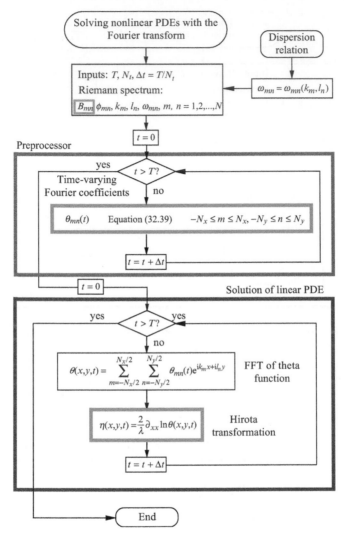

Figure 5.2 Numerically solving nonlinear, integrable partial differential wave equations.

vortices, *holes, tabletop solitons, kinks,* etc. These are essentially *nonlinear Fourier components* in a theory that solves *integrable, nonlinear wave equations.* The important mathematical tool in this endeavor is the *Riemann theta function,* a kind of *multidimensional Fourier series.* The nonlinear basis functions are generally characterized with theta functions (e.g., think of a soliton or vortex) and one often refers to them as *coherent structures.* Many examples are given herein.

(2) *Data analysis and assimilation.* Development of *data analysis* and *data assimilation* *algorithms* for *nonlinear wave dynamics.* These approaches allow for the computation of the so-called Riemann spectrum of nonlinear wave trains, a kind of spectrum appropriate for nonlinear wave motions that reduces to the ordinary linear Fourier

transform in the small-amplitude limit. An important development has been the construction of *nonlinear filtering* methods in which one is able to establish the role that the various nonlinear modes have in a particular data set. Are there solitons in a measured wave train? Can they be classified in terms of physical modes of the system? Can they be extracted (filtered out of) a time series? These are the kinds of questions addressed and answered herein.

(3) *Numerical modeling.* Development of *new nonlinear numerical models of ocean waves.* Models of this type are often referred to as "hyperfast" in this chapter, primarily because they are very fast relative to the so-called *first-generation* or brute force algorithms (see Chapter 20) for theta functions and because the new methods are typically two or more orders of magnitude faster than the well-known FFT approaches such as Tappert (1977) and Fornberg and Whitham (1978). The nonlinear numerical models based upon the IST describe the dynamics of nonlinear (coherent) modes as they evolve and interact with one another. One of the advantages of models of this type is that they do not experience numerical "blow-ups" often experienced by FFT algorithms when used to solve nonlinear wave equations for large times. Because time is just a parameter in the periodic IST, one can extrapolate to any desired value of t without the intermediate integration time steps.

The fundamental theory for the numerical methods given here is the *periodic inverse scattering transform* (for a first look at the literature see, e.g., Baker, 1897; Belokolos et al., 1994). The Riemann theta function, $\theta(x, y, t)$, forms the central theme of these integrable theories and hence a major effort has been to develop numerical methods for computing these functions, but with the physics, data analysis, and ocean wave modeling defining the basic needs. The relationships of the theta functions to the physical fields (e.g., the surface elevation and velocity potential) are provided by so-called Hirota transformations, some of which are described in this chapter and in Chapters 4 and 6 (Hirota, 2004).

Of course, life would be simpler if one could represent a particular nonlinear physical field described by a single theta function, $\theta(x, t)$, that is, by a simple *space series*, $\theta(x, 0)$, or *time series*, $\theta(0, t)$. While this does not turn out to be the case, one often finds physical fields that can be described by simple algebraic manipulations of theta functions (dependent variable transformations of Hirota type), or those described by *differential polynomials* of $\theta(x, t)$ known as Lie-Kodama transformations (Kodama, 1985a,b).

5.2 Riemann Theta Functions as Ordinary Fourier Analysis

Before proceeding it is worthwhile giving the form of the *Riemann theta function*:

$$\theta(x, y, t) = \sum_{m \in \mathbb{Z}} q_m e^{i m \cdot X(x, y, t)}, \quad q_m = e^{-(1/2) m \cdot B m},$$

$$X(x, y, t) = kx + ly - \omega t + \phi$$

(5.2)

or the alternative equivalent form

$$\theta(x, y, t) = \sum_{m \in \mathbb{Z}} Q_m(t) e^{im \cdot kx}, \quad Q_m(t) = e^{-(1/2)m \cdot Bm - im \cdot \omega t + im \cdot \phi}. \tag{5.3}$$

Here, the matrix **B** is known as the *Riemann* or *period matrix*, where **k** is a wavenumber vector in the x-direction, **l** is a wavenumber vector in the y-direction, the vector **ω** constitutes the *frequencies* and the vector **φ** constitutes the *phases*. Herein I refer to **B**, **ω**, **φ** as the *Riemann spectrum*. This is in analogy to the linear Fourier amplitude and phases. The Riemann spectrum (together with a specific Hirota dependent variable transformation) is *particular* for solving a given nonlinear wave equation. The vector **m** is N-dimensional and covers the set $\mathbb{Z}$, that is, over the negative, zero, and positive integers. Given the vector notation for the Riemann theta function as written above, we see a striking resemblance to ordinary linear Fourier analysis. Chapters 7–9 give many of the mathematical and physical properties of theta functions.

Given a particular nonlinear wave equation with periodic boundary conditions, perhaps also known to be integrable by some of the many theoretical methods, one is faced with the beautiful mathematical problem of determining the most general Riemann spectrum that solves the nonlinear wave equation. This is the essential problem of the mathematical physics and is discussed in a large body of the literature (see list of monographs above). Brief overviews of the literature are given for the KdV equation (Chapters 10, 14–17, 19, and 20), the KP equation (Chapters 11 and 32), the nonlinear Schrödinger equation (Chapters 12, 18, and 24), plus many others.

One of the most important results for physical applications of the method is the *Novikov conjecture* that states that the theta function (essentially its Riemann spectrum) must be associated with an underlying Riemann surface to solve a particular nonlinear wave equation. I revisit this topic in Chapters 14–16 and 32 and in Section 5.6 on hyperfast numerical modeling.

It is shown here (Chapters 8 and 9) that the theta function may be written as an ordinary linear, periodic Fourier series:

$$\theta(x, y, t) = \sum_{m=-\infty}^{\infty} \sum_{n=-\infty}^{\infty} \theta_{mn}(t) e^{ik_m x + il_n y}, \tag{5.4}$$

where $k_m = 2\pi m / L_x$ and $l_n = 2\pi n / L_y$ are the x and y wavenumbers. Periodic boundary conditions imply: $\theta(x, y, t) = \theta(x + L_x, y + L_y, t)$. The periodic domain is defined on the rectangle (L_x, L_y) in the (x, y) domain. The time-dependent Fourier coefficients $\theta_{mn}(t)$ are determined in terms of the *Riemann spectrum* **B**, **φ** and the frequency **ω** in the theta function (see Chapters 9 and 32 for a fuller discussion). One should not think that because of the simple form of Equation (5.4) that the latter function is equivalent to the linear Fourier series (Equation (5.1)). The properties of Equations (5.2) and (5.4) are quite special with respect to Equation (5.1). Indeed, the Fourier coefficients $\theta_{mn}(t)$ are *not* simple sinusoids in time, but instead are themselves Fourier series with *incommensurable* frequencies. Equations (5.2) and (5.4) have *many* wonderful properties: Not only can they be added, subtracted, and multiplied but they can also be *divided* (Mumford, 1983, 1984, 1991). This latter property, among many others, makes the theta function one of the most important and useful functions in all of pure and applied mathematics and mathematical physics. With the work accomplished over the past few decades, these functions have become important for solving nonlinear wave equations and for applications in the analysis of data and modeling.

The ordinary linear, periodic Fourier series form of the Riemann theta function (Equation (5.4)) is important for applications to the field of physical oceanography. This is because once one numerically reduces the theta function to an ordinary Fourier series, the algorithmic computation of the solutions to nonlinear wave equations can be accomplished with the two-dimensional FFT. Reduction of the theta function to the ordinary Fourier series is central to the numerical methods given herein and is discussed in some detail in Chapters 9 and 32.

5.3 The Use of Generalized Fourier Series to Solve Nonlinear Wave Equations

A brief discussion of several nonlinear wave equations is now given in terms of theta functions.

5.3.1 Near-Shore, Shallow-Water Regions

The *Korteweg and deVries (KdV) equation* describes the unidirectional propagation of small-but-finite-amplitude surface and internal waves in shallow water:

$$\eta_t + c_0\eta_x + \alpha\eta\eta_x + \beta\eta_{xxx} = 0 \tag{5.5}$$

Here, $\eta(x,t)$ is the wave amplitude as a function of space and time. The coefficients for *shallow-water surface waves* are given by $c_0 = \sqrt{gh}$, $\alpha = 3c_0/2h$, and $\beta = c_0h^2/6$; h is the water depth and c_0 is the linear phase speed. Coefficients for the propagation of *internal waves* are computed from the density stratification and are discussed in Chapter 25. The KdV equation has the generalized Fourier solution:

$$\eta(x,t) = \frac{2}{\lambda}\frac{\partial^2}{\partial x^2}\ln\theta(x,t|\mathbf{B},\boldsymbol{\phi}) \tag{5.6}$$

for $\lambda = \alpha/6\beta$. The theta function, $\theta(x,t|\mathbf{B},\boldsymbol{\phi})$, is discussed in detail in Chapters 7–9. This is a first example of a *Hirota transformation*: the second derivative of the log of a theta function gives the solution to the KdV equation. The *inverse problem* associated with Equation (5.6) is discussed in Chapters 10–12, that is, one determines the period matrix, wavenumbers, frequencies, and phases appropriate to solving the *Cauchy problem* for KdV: Given the initial condition, $\eta(x, t = 0)$, compute for all time the solution, $\eta(x, t)$.

It is important to recognize that Equation (5.6), in the limit of small-amplitude waves, gives the usual ordinary Fourier series solution to the *linearized* KdV equation: $\eta_t + c_0\eta_x + \beta\eta_{xxx} = 0$. Thus, the generalized Fourier expression (Equation (5.6)), in the small-amplitude limit, is just ordinary Fourier analysis used for everyday data analysis problems. However, the main advantage of Equation (5.6) is that when the waves are *not* small in amplitude, one is able to fully generalize to a nonlinear basis set of *cnoidal waves* and to include nonlinear interactions among the cnoidal waves.

The *Kadomtsev-Petviashvili (KP) equation* is given by

$$\eta_t + c_0\eta_x + \alpha\eta\eta_x + \beta\eta_{xxx} + \gamma\partial_x^{-1}\eta_{yy} = 0 \tag{5.7}$$

The first three constant coefficients are the same as those for the KdV equation while the last is given by $\gamma = c_0/2$. The notation in the last term of the KP equation, ∂_x^{-1}, is the antipartial derivative in x, for indeed the equation may also be written: $(\eta_t + c_0\eta_x + \alpha\eta\eta_x + \beta\eta_{xxx})_x + \gamma\eta_{yy} = 0$. More specifically,

$$(\partial_x^{-1}f)(x) = \int_{-\infty}^{x} f(x')\mathrm{d}x'$$

for $-\infty < x < \infty$ (the infinite plane problem) and

$$(\partial_x^{-1}f)(x) = \int_{0}^{x} f(x')\mathrm{d}x'$$

for $0 \leq x \leq L$ (periodic boundary conditions).

In Equation (5.6), $\eta(x, y, t)$ is the wave amplitude as a function of the two spatial variables, x, y and time, t. The KP equation (Equation (5.7)) is a natural two-space dimensional extension of the KdV equation (Equation (5.5)). The periodic KP solutions thus include *directional spreading* in the wave field:

$$\eta(x, y, t) = 2\frac{\partial^2}{\partial x^2} \ln\theta(x, y, t | \mathbf{B}, \boldsymbol{\phi}) \tag{5.8}$$

Here, the generalized Fourier series has the same form as above, but the phase has the *two-dimensional* expression:

$$\mathbf{X}(x, y, t) = \mathbf{k}x + \mathbf{l}y - \boldsymbol{\omega}t + \boldsymbol{\phi}. \tag{5.9}$$

Here, the spatial term $\mathbf{k}x$ (corresponding to the dominant wave direction) has been joined by the lateral spatial term $\mathbf{l}y$ (perpendicular to the dominant direction), which allows wave spreading to be taken into account. Note that

Equation (5.8) solves the Cauchy problem for the KP equation: Given the spatial variation of the surface $\eta(x, y, 0)$ at time $t = 0$, compute the solution to KP for all space and time, $\eta(x, y, t)$.

5.3.2 Shallow- and Deep-Water Nonlinear Wave Dynamics for Narrow-Banded Wave Trains

The study of nonlinear, narrow-banded wave trains has a long and rich history. Basically wave motion of this type has different behaviors depending upon the value of $k_0 h_0$, where k_0 is the dominant wavenumber of the wave train and h_0 is the water depth. The wave dynamics are *stable* for the case $k_0 h_0 < 1.36$. Conversely, the wave trains can become *unstable* for $k_0 h_0 > 1.36$, an effect known as the *modulational or Benjamin-Feir instability*. One physical application relates to *rogue wave dynamics* in ocean waves (see Chapters 12, 18, 24, and 29). The physics of this problem to leading order is governed by the *nonlinear Schrödinger (NLS) equation*, which is the *simplest* of the possible *nonlinear deep-water wave equations*. The NLS equation in $1 + 1$ dimensions (one-space and one-time dimensions (x, t)) is given by

$$i(\psi_t + C_g \psi_x) + \mu \psi_{xx} + v|\psi|^2 \psi = 0 \tag{5.10}$$

The coefficients for deep-water waves are given by $C_g = \omega_0/2k_0$; $\mu = -\omega_0/8k_0^2$; $v = -\omega_0 k_0^2/2$; and a_0, k_0, and ω_0 are, respectively, the amplitude, wavenumber, and frequency of the *carrier wave*. Modification of these coefficients to include dependence on the water depth is discussed in Chapter 2.

An important class of *spectral solutions* of Equation (5.10) can be written in terms of generalized Fourier series:

$$\psi(x, t) = a_0 \frac{\theta(x, t|\mathbf{B}, \mathbf{\Phi}^-)}{\theta(x, t|\mathbf{B}, \mathbf{\Phi}^+)} e^{i\omega_0 k_0^2 a_0^2 t/2} \tag{5.11}$$

Two different sets of phases (determined by the periodic IST), $\mathbf{\Phi}^\pm$, are required to specify a solution. To understand this solution suppose that the "modulation function" is unity, that is, assume that there is no physical modulation in the problem, so that $\theta(x, t|\mathbf{B}, \mathbf{\Phi}^-)/\theta(x, t|\mathbf{B}, \mathbf{\Phi}^+) = 1$. Then the complex modulation becomes $\psi(x, t) = a_0 \exp\{i\omega_0 k_0^2 a_0^2 t/2\}$; this latter expression may then be interpreted as the *complex carrier wave* in the absence of modulation. The Expression (5.11) contains the ratio of two theta functions. As mentioned above, the properties of theta functions are so special as to allow for their division of one by the other.

The *free surface elevation* has the following expression:

$$\eta(x, t) = \text{Re}[\psi(x, t)e^{ik_0 x - i\omega_0 t}]. \tag{5.12}$$

Using the simple unmodulated carrier wave (insert $\psi(x,t) = a_0 \exp\{i\omega_0 k_0^2 a_0^2 t/2\}$ into Equation (5.11)), it is straightforward to show that the free surface elevation (Equation (5.12)) is just the *leading order Stokes wave* (see details in Chapter 2). Thus, the theory of the nonlinear Schrödinger equation includes the description of the Stokes wave. Of course, Equations (5.11) and (5.12) may be viewed as an approach for the nonlinear Fourier analysis of deep-water wave data, an area where there are still many challenges for the development of numerical methods and data analysis procedures. Formally speaking, Equation (5.10) solves the Cauchy problem for the NLS equation: Given $\psi(x, t = 0)$, compute the solution for all space and time, $\psi(x, t)$.

In 2 + 1 dimensions (two-space and one-time dimensions (x, y, t)), NLS is given by

$$i(\psi_t + C_g \psi_x) - \mu \psi_{xx} + \rho \psi_{yy} - \nu |\psi|^2 \psi = 0 \tag{5.13}$$

for $\rho = \omega_0/4k_0^2$. While Equation (5.13) is not strictly integrable by the IST, there is good reason to believe that there exist large classes of approximate solutions of the form

$$\psi(x, y, t) \cong a_0 \frac{\theta(x, y, t | \mathbf{B}, \mathbf{\Phi}^-)}{\theta(x, y, t | \mathbf{B}, \mathbf{\Phi}^+)} e^{i\omega_0 k_0^2 a_0^2 t/2} \tag{5.14}$$

These solutions may be thought of as modulations in many directions with respect to the dominant wave direction, so that *directional spreading* is taken into account. Once again there are interesting challenges to develop this technology for data analysis purposes.

The surface elevation of the sea surface is governed by the expression:

$$\eta(x, y, t) = \text{Re}[\psi(x, y, t) \exp\{ik_0 x - i\omega_0 t\}].$$

5.4   Dynamical Applications of Theta Functions

As mentioned in Chapters 7 and 8 in most of the *dynamical (modeling) applications* discussed in this chapter, the dimensions, X_n, are *not* the spatial dimensions (x, y, z) of traditional applications of the N-dimensional Fourier transform. Instead, the generalized dimensions have the *physical form*:

$$X_n(x, t) = k_n x - \omega_n t + \phi_n, \quad n = 1, 2, \ldots, N.$$

This form of course arises from the algebro-geometric solution of nonlinear, integrable wave equations. The *dimension vector* is given by

$$\mathbf{X} = \mathbf{X}(x, t) = \mathbf{k}x - \boldsymbol{\omega}t + \boldsymbol{\phi}. \tag{5.15}$$

Here, the *wavenumber vector,* **k**, the *frequency vector,* **ω**, and the *phase vector,* **φ** are given by explicit formulas in inverse scattering theory (depending upon the particular wave equation being studied in terms of algebraic-geometric loop integrals; see Chapters 10, 12, 14, and 19). One has

$$
\begin{aligned}
\mathbf{X}(x, t) &= [k_1 x - \omega_1 t + \phi_1, k_2 x - \omega_2 t + \phi_2, \ldots, k_N x - \omega_N t + \phi_N], \\
\mathbf{k} &= [k_1, k_2, \ldots, k_N], \\
\boldsymbol{\omega} &= [\omega_1, \omega_2, \ldots, \omega_N], \\
\boldsymbol{\phi} &= [\phi_1, \phi_2, \ldots, \phi_N].
\end{aligned}
\tag{5.16}
$$

The identification of the generalized dimensions with the *dynamical variables* x, t is a crucial step in the application of Equations (5.2) and (5.3) to the solution of *particular nonlinear wave equations* and to the *space/time series analysis of data.* Numerical determination of the parameters B_{ij}, k_i, ω_i, and ϕ_i is a major accomplishment for the application of Equation (5.2) in the numerical integration of nonlinear PDEs, in the analysis of data and in modeling. For much of this chapter, these parameters will be treated as known constants.

The *Riemann matrix,* **B**, has the form

$$
\mathbf{B} = \begin{bmatrix}
B_{11} & B_{12} & B_{13} & \ldots & B_{1N} \\
B_{12} & B_{22} & B_{23} & \ldots & B_{2N} \\
B_{13} & B_{23} & B_{33} & \ldots & B_{3N} \\
\vdots & \vdots & \vdots & \ddots & \vdots \\
B_{1N} & B_{2N} & B_{3N} & \ldots & B_{NN}
\end{bmatrix}.
\tag{5.17}
$$

The $N \times N$ (complex) matrix, **B**, is *symmetric* and its real part must be *positive definite* to ensure mathematical convergence of the generalized Fourier series (Equations (5.2) and (5.3)) (in this context, note the presence of the minus sign before the double sum in the argument of the exponential in Equations (5.2) and (5.3)).

A major difference between ordinary Fourier analysis and generalized Fourier analysis is that one has a *vector of Fourier amplitudes,* $C = \{C_n, n = 1, 2, \ldots, N\}$, for the first and a *matrix of Fourier amplitudes,* $\mathbf{B} = \{B_{ij} = B_{ji}, i, j = 1, 2, \ldots, N\}$, for the second.

Of course, in the case of the generalized Fourier transform, the rules for computing the matrix of amplitudes and the various vectors for wavenumber, frequency, and phases are different from ordinary Fourier analysis, for indeed the generalized Fourier formulation is applied to solve problems in *nonlinear wave dynamics.* Note that the generalized Fourier series (Equation (5.2)) is a *nested sum* over exponentials (one summation for each degree of freedom) rather than a single summation for ordinary Fourier analysis (Equation (5.1)) for each spatial dimension.

Physically the *Riemann matrix* generalizes linear Fourier analysis in two ways:

- The *diagonal terms* (or more generally submatrices centered on the diagonal) are *not* associated with a simple sinusoidal basis, but are instead associated with more complex functions such as Stokes waves, cnoidal waves, solitons, unstable modes, etc. Thus, the diagonal terms (or submatrices along the diagonal) are associated with a set of *nonlinear basis functions*.
- The *off-diagonal terms* account for *nonlinear interactions* among the basis functions associated with the diagonal elements.

Thus, the periodic IST, with its *matrix of Fourier amplitudes*, allows a simple sinusoidal basis to be generalized to a basis of nonlinear functions, while providing for nonlinear interactions among these basis functions.

5.5 Data Analysis and Data Assimilation

For many data analysis purposes, it is more convenient to record data as time series rather than as space series. For the analysis of *time series data* in shallow water, it is more appropriate to consider the *time-like* KdV equation (tKdV) (Karpman, 1975; Ablowitz and Segur, 1981):

$$\eta_x + c_0' \eta_t + \alpha' \eta \eta_x + \beta' \eta_{xxx} = 0, \quad \eta(x,t) = \eta(x, t+T), \tag{5.18}$$

where $\eta(0,t)$ is assumed given and $c_0' = 1/c_0$, $\alpha' = -\alpha/c_0^2$, and $\beta' = -\beta/c_0$; Equation (5.18) has the linearized dispersion relation $k = \omega/c_0 + (\beta/c_0^4)\omega^3$. The inverse scatting transform of Equation (5.18) is easily obtained from Equation (5.5) by a simple change of variables (Osborne, 1993a,b,c,d,e): $x \to t$, $t \to x$, $k \to \omega$, and $\omega \to k$. Equation (5.18) solves the *boundary value problem*: Given the solution of the equation at some specific spatial location, $x = 0$, namely $\eta(0,t)$, Equation (5.18) determines the solutions at all other spatial points $x \neq 0$: $\eta(x,t)$. In an experimental context, we call $\eta(0,t)$ a *time series*. Therefore, when we analyze data we will always use the periodic IST associated with a "time-like" equation such as Equation (5.18) rather than the "space-like" equation (Equation (5.5)) (unless of course we have to analyze a space series in which case we analyze with a "space-like" equation). Each of the nonlinear wave equations analyzed herein has a "time-like" form and these are treated in detail when data analysis applications are discussed.

Data assimilation problems are in general very difficult even for linear dynamical motions, but for nonlinear dynamics they can be very challenging. It is for this reason that I have added Chapter 23 on nonlinear adiabatic annealing, which allows us to determine the Riemann spectrum of data from a time series, or to *iterate* on incoming data to re-estimate or improve the constantly changing Riemann spectrum during a field exercise. This is an important area of research because it allows one to treat problems in real time. An example would be to take radar data in real time aboard a ship in high seas, to process the data to determine the possibility of rogue waves, and to provide to the captain and crew estimates of possible rogue wave encounters.

5.6 Hyperfast Modeling of Nonlinear Waves

The previous generation of brute force algorithms is *exponential* in character, that is, the computer time increases exponentially with the number of nonlinear Fourier modes (number of degrees of freedom) being computed. The brute force algorithms suffice for up to 10 or 20 degrees of freedom, but even with supercomputers the problems for larger degrees of freedom became intractable. Indeed, it was easy to encounter problems for which universal lifetimes would be necessary using these older algorithms. This is because the computer time $\sim 10^N$. Thus if we are able, for a particular problem, to compute 10 degrees of freedom in 1 h of computer time, then 11 degrees of freedom will take 10 h, 12 degrees of freedom 100 h, etc.

This impossible situation led to the development of newer algorithms that now make large numbers of degrees of freedom possible, because computations are *polynomial* in computer time, $a + bN + cN^2 + \dots$. There are two approaches given in this book that are polynomial in computer time:

(1) The first class of "hyperfast" algorithms uses Equation (5.4) to compute the theta functions. This provides huge improvements in computer time, but there remains the "preprocessor" function which can be very slow. Two other steps improve computer time substantially. First one computes the modular transformation of the Riemann matrix (Chapter 8), which "compactifies" it, especially in the soliton limit. Secondly, to reduce the number of operations in the "preprocessor" part of algorithm, it is important to sum over an *n*-ellipsoid in the integer "summation" space of the theta functions. The resultant algorithm is then a major improvement over previous methods that I have used. Chapters 8, 9, 22, and 32 give some perspective on this class of algorithms.

(2) The second class of algorithms arose from an attempt to develop a "fast theta function transform" in analogy with the "fast Fourier transform." This approach requires discretization of the theta function and development of some of its properties (Chapter 21). Implementation of the algorithm, however, requires a merging of discrete properties with the ordinary FFT (for the accounting) and the material of Chapters 9 and 22 (to reduce the number of computations over the integer lattice).

Both of these algorithms have proven practical for working on many problems not possible in previous generations. Recent progress, together with several dozen false attempts made over the last two decades, has added to the knowledge base of practical numerical applications of theta functions.

Figure 5.2 is a flowchart of some of the essential features of the first algorithm for modeling nonlinear equations given above. It is worthwhile to compare Figure 5.2 to Figure 5.1, for the nonlinear problem differs from the linear problem only by the contents of the preprocessor box in Figure 5.2. First, the input spectrum is now the Riemann matrix instead of the two-dimensional Fourier transform. Second, the rule for computing the time-varying Fourier coefficients is different, that is, it is given by Equations (9.0c), (9.2), or (32.39) instead of the elementary rule for linear Fourier analysis $\eta_{mn}(t) = \eta_{mn}(0)e^{-i\omega_{mn}t+\phi_n}$. Note that we are dealing with the theta function at this point. Third one has to make a Hirota dependent variable transformation to compute the surface elevation from the theta function. While it is easy to give a simple flowchart describing the nonlinear numerical modeling of integrable wave equations, the programming is not so simple as can be seen in the rest of this book. But the goal is to leap from Figure 5.1 to Figure 5.2 in as simple a way as possible.

The advantage of using theta functions is that they work in a very special "lattice space" that makes computations fast. The disadvantage is that their study has necessarily been described in the language of algebraic geometry for over the last century and a half, and this is an area of mathematics where not too many individuals enjoy going to work. My personal view is that applications of theta functions are just beginning and that many frustrations lie ahead for those who would like to move into this territory. However, there will be many satisfactions for the few who do venture forward.

6 The Periodic Hirota Method

6.1 Introduction

The present chapter is fundamental for the numerical work in this book. The Hirota method for testing integrability of nonlinear wave equations is important for deriving the so-called infinite-line, N-soliton solutions (Chapter 4). However, for the purposes of most of the work in the field of physical oceanography we require *periodic boundary conditions*. This requirement is in complete analogy with the linear Fourier series that is of course a periodic algorithm. Indeed, use of the fast Fourier transform (FFT) means that one is assuming periodic boundary conditions, be it for data analysis, for data assimilation or for modeling purposes.

This chapter discusses some of the aspects of using the Hirota method for periodic boundary conditions. A sequel to this discussion is Chapter 16 (on the Nakamura-Boyd method) that begins with the results given in this chapter and shows how to use the periodic Hirota method to determine the *Riemann spectrum* (the nonlinear Fourier spectrum) of the solution of a particular nonlinear wave equation, one of the most important aspects of nonlinear Fourier methods. This chapter forms a foundation for the results given in Chapter 16 and for other aspects of the periodic problem.

6.2 The Hirota Method

The Hirota method essentially transforms a nonlinear wave equation into an N-linear form. If this N-linear form has one, two, and three soliton solutions, then one can be fairly sure that the equation is integrable. A preliminary analysis using the Hirota method often provides us with the insight necessary to begin a full analysis of the problem using the inverse scattering transform.

The dependent variable transformation of Hirota, for nonlinear, integrable wave equations, results in an almost miraculous cancellation of opposing forces such as nonlinearity and dispersion. This procedure therefore also provides a physical interpretation for soliton solutions. I give some examples below to illustrate the periodic Hirota method.

Doi: 10.1016/S0074-6142(10)97006-X

6.3 The Burgers Equation

The Burgers equation (Whitham, 1974) is given by

$$u_t + uu_x = \mu u_{xx}. \tag{6.1}$$

This is a one-dimensional "Navier-Stokes" type of equation with nonlinearity and dissipation. This is normally solved with the *Cole-Hopf transformation*:

$$u = -2\mu \partial_x \ln\theta(x, t). \tag{6.2}$$

To carry this out first let

$$u = \partial_x w(x, t).$$

After applying this expression and integrating once we have

$$w_t + \frac{1}{2}w_x^2 - \mu w_{xx} + c = 0, \tag{6.3}$$

where c is an integration constant. Now apply

$$w = -2\mu \ln\theta(x, t)$$

and find

$$w_t = -2\mu \frac{\theta_t}{\theta},$$

$$\frac{1}{2}w_x^2 = 2\mu^2 \left(\frac{\theta_x}{\theta}\right)^2,$$

$$-\mu w_{xx} = -2\mu^2 \left(\frac{\theta_x}{\theta}\right)^2 + 2\mu^2 \frac{\theta_{xx}}{\theta},$$

so that

$$\underbrace{-2\mu \frac{\theta_t}{\theta}}_{w_t} + \underbrace{2\mu^2 \left(\frac{\theta_x}{\theta}\right)^{\!\!2}}_{\frac{1}{2}w_x^2} \underbrace{- 2\mu^2 \left(\frac{\theta_x}{\theta}\right)^{\!\!2} + 2\mu^2 \frac{\theta_{xx}}{\theta}}_{-\mu w_{xx}} + c = 0.$$

We see that the nonlinear term $w_x^2/2 = 2\mu^2(\theta_x/\theta)^2$ partially balances with the dissipation term $-\mu w_{xx} = -2\mu^2[(\theta_x/\theta)^2 + (\theta_{xx}/\theta)]$! This suggests that *the physics of the Burgers equation has coherent structures that result from the balance of nonlinearity and dissipation*. These are of course the *shock-wave solutions*. The final reduced equation, valid for periodic boundary conditions, is given by

$$\theta_t - \mu\theta_{xx} - \frac{c}{2\mu}\theta = 0.$$

Of course when we take infinite-line boundary conditions, $c = 0$, and we have the *heat equation*:

$$\theta_t - \mu\theta_{xx} = 0 \tag{6.4}$$

The Cole-Hopf transformation has *linearized* the Burgers equation by balancing nonlinearity and dissipation.

An interesting approach is to derive directly the bilinear form for the Burgers equation. This is done by inserting Equation (6.2) into Equation (6.1) to get

$$\theta_x\theta_t - \theta\theta_{xt} - \mu(\theta_x\theta_{xx} - \theta\theta_{xxx}) - \rho\theta^2 = 0. \tag{6.5}$$

Using the Hirota operator notation, one finds

$$(D_xD_t - \mu D_{xxx})\theta\cdot\theta = \rho\theta\cdot\theta, \tag{6.6}$$

where

$$D_xD_t\theta\cdot\theta = \theta_x\theta_t - \theta\theta_{xt}, \quad D_{xxx}\theta\cdot\theta = \theta_x\theta_{xx} - \theta\theta_{xxx}.$$

A trial solution follows the usual Hirota search for solitons (set $\rho = 0$):

$$\theta(x,t) = 1 + e^{kx-\omega t}. \tag{6.7}$$

Inserting this into Equation (6.5) gives the dispersion relation:

$$\omega = -\mu k^2.$$

The *soliton solution* to the Burgers equation is just a *shock-wave* solution:

$$u(x,t) = -\mu k \frac{e^{kx-\omega t}}{1 + e^{kx-\omega t}}. \tag{6.8}$$

Two-shock waves take the form:

$$\theta(x,t) = 1 + e^{k_1x-\omega_1 t} + e^{k_2x-\omega_2 t},$$

which when substituted into Equation (6.2) gives the *two-shock solution* of Equation (6.1). See Wang et al. (2004) for more details. N-shock solutions follow in a natural way. The periodic solutions with Riemann theta functions can be found directly from Equation (6.5). The notion that coherent structures occur naturally in dissipative systems has been reviewed in Balmforth, 1995.

6.4 The Korteweg-de Vries Equation

The *dimensional KdV equation* has the form:

$$\eta_t + c_0\eta_x + \alpha\eta\eta_x + \beta\eta_{xxx} = 0.$$

Here, $\eta(x, t)$ is the amplitude of the free surface, $c_0 = \sqrt{gh}$ is the linear phase speed, g is the acceleration of gravity, h is the water depth, $\alpha = 3c_0/2h$ is the coefficient of the nonlinear term, and $\beta = c_0 h^2/6$ is the coefficient of the dispersive term. We then seek to make the transformation:

$$\eta(x, t) = \frac{2}{\lambda}\partial_{xx}\ln\theta(x, t), \quad \lambda = \frac{\alpha}{6\beta}. \tag{6.9}$$

We first make the substitution $\eta = w_x$ and integrate once in x to find

$$w_t + c_0 w_x + \frac{\alpha}{2}w_x^2 + \beta w_{xxx} + \rho = 0. \tag{6.10}$$

In the search for soliton solutions, where infinite-line boundary conditions hold, it is natural to take $\rho = 0$. However, we are concerned with periodic boundary conditions in the present chapter and therefore ρ must be kept finite. Now, make the final substitution $w = (2/\lambda)\partial_x \ln\theta$ and obtain for the terms in Equation (6.10):

$$\lambda\frac{\theta^2}{2}w_t = \theta\theta_{xt} - \theta_x\theta_t, \tag{6.11}$$

$$\lambda\frac{\theta^2}{2}c_0 w_x = c_0(\theta\theta_{xx} - \theta_x^2), \tag{6.12}$$

$$\lambda\frac{\theta^2}{2}\frac{\alpha}{2}w_x^2 = \frac{6\beta(\theta_x^2 - \theta\theta_{xx})^2}{\theta^2} = -12\beta\frac{\theta_x^2\theta_{xx}}{\theta} + 6\beta\frac{\theta_x^4}{\theta^2} + 6\beta\theta_{xx}^2, \tag{6.13}$$

$$\lambda\frac{\theta^2}{2}\beta w_{xxx} = \beta\left(\frac{12\theta_x^2\theta_{xx}}{\theta} - \frac{6\theta_x^4}{\theta^2} - 3\theta_{xx}^2 - 4\theta_x\theta_{xxx} + \theta\theta_{xxxx}\right). \tag{6.14}$$

Inserting Equations (6.11)–(6.14) into Equation (6.10) gives

$$\underbrace{\theta\theta_{xt} - \theta_x\theta_t}_{w_t} + \underbrace{c_0(\theta\theta_{xx} - \theta_x^2)}_{c_0 w_x}$$

$$+ \underbrace{\beta\left(-\frac{12\theta_x^2\theta_{xx}}{\theta} + \frac{6\theta_x^4}{\theta^2} + 6\theta_{xx}^2\right)}_{\alpha w_x^2/2} + \underbrace{\beta\left(\frac{12\theta_x^2\theta_{xx}}{\theta} - \frac{6\theta_x^4}{\theta^2} - 3\theta_{xx}^2 - 4\theta_x\theta_{xxx} + \theta\theta_{xxx}\right)}_{\beta w_{xxx}} + \frac{\alpha c}{12\beta}\theta^2 = 0.$$

Simplifying the last equation, we have the following bilinear form:

$$\theta\theta_{xt} - \theta_x\theta_t + c_0(\theta\theta_{xx} - \theta_x^2) + \beta(3\theta_{xx}^2 - 4\theta_x\theta_{xxx} + \theta\theta_{xxxx}) + \frac{\alpha c}{12\beta}\theta^2 = 0$$

$$(6.15)$$

Notice that in the determination of Equation (6.15), the higher-order terms in Equations (6.13) and (6.14) cancel and we are left with a *bilinear form*. Thus, much of the nonlinearity is canceled by dispersion upon substitution of the Hirota transformation in the KdV equation. This is important, because without this cancellation we would be faced with solving a quadrilinear form instead of the bilinear form (Equation (6.15)). This amazing simplification is characteristic of integrable wave equations. In this way, the balance of nonlinearity and dispersion provides a physical description of solitons. Using the Hirota bilinear operator notation, Equation (6.15) takes the form:

$$(D_x D_t + c_0 D_x^2 + \beta D_x^4 + \rho)\theta\cdot\theta = 0 \tag{6.16}$$

where

$$\rho = \frac{1}{2}\lambda c = \frac{\alpha c}{12\beta} \tag{6.17}$$

and where we have used

$$D_x D_t \theta\cdot\theta = 2(\theta\theta_{xt} - \theta_x\theta_t),$$
$$D_x^2 \theta\cdot\theta = 2(\theta\theta_{xx} - \theta_x^2), \tag{6.18}$$
$$D_x^4 \theta\cdot\theta = 2(\theta\theta_{xxxx} - 4\theta_x\theta_{xxx} + 3\theta_{xx}^2).$$

Note that the shorthand symbol $\theta\cdot\theta$ just means θ^2.

In the *soliton* case, $\rho = 0$ and we have only the bilinearized equation:

$$(D_x D_t + c_0 D_x^2 + \beta D_x^4)\theta\cdot\theta = 0 \tag{6.19}$$

Nota Bene. It is worthwhile asking how does the bilinear form change if the *Hirota derivatives* (Equation (6.18)) are replaced with normal partial derivatives? Hence we form:

$$(\partial_x \partial_t + c_0 \partial_x^2 + \beta \partial_x^4 + \rho)\theta \cdot \theta = 0$$

Here, we have

$$\partial_x \partial_t \theta \cdot \theta = 2(\theta \theta_{xt} + \theta_x \theta_t),$$

$$\partial_x^2 \theta \cdot \theta = 2(\theta \theta_{xx} + \theta_x^2), \tag{6.20}$$

$$\partial_x^2 \theta \cdot \theta = 2(\theta \theta_{xxxx} + 4\theta_x \theta_{xxx} + 3\theta_{xx}^2).$$

Thus, the Hirota derivatives have a minus sign before the terms for which the second argument has an odd number of derivatives, compare Equation (6.18) with Equation (6.20). This means that the bilinear form of the KdV equation with Hirota derivatives has much richer solutions than the associated bilinear form with ordinary partial derivatives.

Nota Bene. Inspection of the bilinear form reveals the linear dispersion relation: $k\omega - c_0 k^2 + \beta k^4 = 0$ or $\omega = c_0 k^2 - \beta k^3$. The bilinear form does not depend explicitly on the nonlinear coefficient, α, but only the linear phase speed, c_0, the dispersion coefficient, β, and the normalized integration constant, ρ. Therefore, the theta functions do not have the nonlinear coefficient in their bilinear form representation. Indeed, the nonlinear coefficient, α, enters in the formulation only through the transformation $(\alpha/12\beta)\eta(x,t) = \partial_{xx} \ln\theta(x,t)$; α rescales the physical amplitude of the waves to include nonlinearity. Thus, wave motion associated with large α "scales up" the theta function to give smaller diagonal elements (larger nonlinear degrees of freedom); this is the physical basis for the requisite large nonlinearity and the soliton limit of the theory.

6.5   The KP Equation

The KP equation is given by

$$\eta_t + c_0 \eta_x + \alpha \eta \eta_x + \beta \eta_{xxx} + \gamma \partial^{-1} \eta_{yy} = 0, \tag{6.21}$$

or

$$(\eta_t + c_0 \eta_x + \alpha \eta \eta_x + \beta \eta_{xxx})_x + \gamma \eta_{yy} = 0,$$

or

$$\eta_{xt} + c_0 \eta_{xx} + \alpha(\eta \eta_x)_x + \beta \eta_{xxxx} + \gamma u_{yy} = 0.$$

The goal is to make the following transformation of the KP equation:

$$\eta(x,t) = \frac{2}{\lambda} \partial_{xx} \ln\theta(x,t), \quad \lambda = \frac{\alpha}{6\beta}. \tag{6.22}$$

To this end, we first make the substitution $u = w_x$ to get

$$w_{xxt} + c_0 w_{xxx} + \alpha(w_x w_{xx})_x + \beta w_{xxxxx} + \gamma w_{xyy} = 0,$$

and integrate once in x to find the final result:

$$w_{xt} + c_0 w_{xx} + \alpha(w_x w_{xx}) + \beta w_{xxxx} + \gamma w_{yy} + c_1 = 0.$$

Now, make the further transformation $w = v_x$ to find

$$v_{xxt} + c_0 v_{xxx} + \alpha(v_{xx} v_{xxx}) + \beta v_{xxxxx} + \gamma v_{xyy} + c_1 = 0,$$

or

$$v_{xxt} + c_0 v_{xxx} + \frac{1}{2}\alpha(v_{xx}^2)_x + \beta v_{xxxxx} + \gamma v_{xyy} + c_1 = 0.$$

Integrate this to get

$$v_{xt} + c_0 v_{xx} + \frac{1}{2}\alpha v_{xx}^2 + \beta v_{xxxx} + \gamma v_{yy} + c_1 x + c_2 = 0.$$

Finally, make the transformation

$$v = \frac{12\beta}{\alpha}\ln\theta,$$

and we have for the individual terms

$$\frac{\alpha}{12\beta}\theta^2 v_{xt} = \theta\theta_{xt} - \theta_t\theta_x,$$

$$\frac{\alpha}{12\beta}\theta^2 c_0 v_{xx} = c_0(\theta\theta_{xx} - \theta_x^2),$$

$$\frac{\alpha}{12\beta}\theta^2 \frac{1}{2}\alpha v_{xx}^2 = \beta\left(\frac{6\theta_x^4}{\theta^2} - \frac{12\theta_x^2\theta_{xx}}{\theta} + 6\theta_{xx}^2\right),$$

$$\frac{\alpha}{12\beta}\theta^2 \beta v_{xxxx} = \beta\left(-\frac{6\theta_x^4}{\theta^2} + \frac{12\theta_x^2\theta_{xx}}{\theta} - 3\theta_{xx}^2 - 4\theta_x\theta_{xxx} + \theta\theta_{xxxx}\right),$$

$$\frac{\alpha}{12\beta}\theta^2 \gamma v_{yy} = \gamma(\theta\theta_{yy} - \theta_y^2).$$

Summing all these contributions (the slashes are for the terms which identically cancel) gives the bilinear form for the KP equation:

$$(\theta\theta_{xt} - \theta_x\theta_t) + c_0(\theta\theta_{xx} - \theta_x^2) + \beta(\theta\theta_{xxxx} - 4\theta_x\theta_{xxx} + 3\theta_{xx}^2) + \gamma(\theta\theta_{yy} - \theta_y^2) + \frac{\rho}{2} = 0$$
(6.23)

Here, we have set $c_1 = 0$ to insure periodic boundary conditions and let $c_2 = \rho$. The similarity to this bilinear form and that for the KdV equation is obvious, that is, we have essentially added the term $\gamma(\theta\theta_{yy} - \theta_y^2)$ to the KdV bilinear form.

In Hirota operator notation,

$$(D_x D_t + c_0 D_x^2 + \beta D_x^4 + \gamma D_y^2 + \rho)\theta\cdot\theta = 0$$
(6.24)

Here, the Hirota derivatives are given by

$$\begin{aligned}
D_x D_t \theta\cdot\theta &= 2(\theta\theta_{xt} - \theta_x\theta_t), \\
D_x^2 \theta\cdot\theta &= 2(\theta\theta_{xx} - \theta_x^2), \\
D_x^4 \theta\cdot\theta &= 2(\theta\theta_{xxxx} - 4\theta_x\theta_{xxx} + 3\theta_{xx}^2), \\
D_y^2 \theta\cdot\theta &= 2(\theta\theta_y - \theta_y^2),
\end{aligned}$$
(6.25)

and we have by inspection the linear dispersion relation from Equation (6.24): $k\omega - c_0 k^2 + \beta k^4 - \gamma l^2 = 0$ or $\omega = c_0 k^2 - \beta k^4 + \gamma l^2/k$, where k is the x wave number and l is the y wave number.

To use the KP equation for the analysis of data, for data assimilation, and for hyperfast modeling, one must make a detailed analysis of Equation (6.24) using Chapters 14–16. However, I would like to show you how, literally on the back of an envelope, to get into the hyperfast modeling business. Essentially, I want to show you how to get estimates of the Riemann spectrum that will allow you to get started; then, you can later do the higher-order details. First recall the Hirota infinite-line soliton solution of KP discussed in Chapter 4. I now give the formulation of Section 4.5 for KP (Equation (6.21)) in physical units:

$$f(x, y) = \sum_{\nu=0,1} \exp\left(\sum_{i=1}^{N} \nu_i X_i + \sum_{i\geq 1}^{N}\sum_{j>i}^{N} \nu_i \nu_j A_{ij}\right)$$
(6.26)

for

$$u(x, y) = \frac{12\beta}{\alpha}(\log f)_{xx},$$
(6.27)

where the phases are given by

$$X_i = k_i x + l_i y - \omega_i t + \phi_i \tag{6.28}$$

for the phase speeds

$$\omega_i = c_0 k_i + \beta k_i^3 + \frac{c_0}{2} \frac{l_i^2}{k_i} \tag{6.29}$$

and phase shifts

$$e^{A_{ij}} = \left(\frac{(k_i - k_j)^2 - \left(\frac{l_i k_j - l_j k_i}{k_i k_j h}\right)^2}{(k_i + k_j)^2 - \left(\frac{l_i k_j - l_j k_i}{k_i k_j h}\right)^2} \right). \tag{6.30}$$

This is the soliton solution of the KP equation in physical units and it is valid for the infinite-line problem. To get to the periodic problem in a quick and easy step, let

$$k_i \rightarrow i\kappa_i, \quad l_i \rightarrow i\lambda_i, \quad \omega_i \rightarrow i\omega_i. \tag{6.31}$$

This means in the formulation above we now get oscillatory solutions, for indeed

$$e^{X_i} = e^{k_i x + l_i y - \omega_i t + \phi_i} \longrightarrow e^{i k_i x + i l_i y - i\omega_i t + i\phi_i}. \tag{6.32}$$

Then, after a careful analysis (Belokolos et al., 1994), we see that Equation (6.26) becomes a theta function, $f \rightarrow \theta$:

$$\theta(x, y, t) = \sum_{m_1 = -M}^{M} \sum_{m_2 = -M}^{M} \cdots \sum_{m_N = -M}^{M} \exp \left\{ -\frac{1}{2} \sum_{m=1}^{N} \sum_{n=1}^{N} m_m m_n B_{mn} \right\}$$

$$\exp \left\{ i \sum_{n=1}^{N} m_n \kappa_n x + i \sum_{n=1}^{N} m_n \lambda_n y - i \sum_{n=1}^{N} m_n \omega_n t + i \sum_{n=1}^{N} m_n \phi_n \right\},$$

with frequency given by (use Equation (6.31) in Equation (6.29)):

$$\omega_n \simeq c_0 k - \beta \kappa_n^3 + \frac{c_0}{2} \frac{\lambda_n^2}{\kappa_n} + \cdots \tag{6.33}$$

and period matrix given by (use Equation (6.31) in Equation (6.30)):

$$B_{mn} \cong \ln \left[\frac{(\kappa_m - \kappa_n)^2 + \left(\frac{\kappa_n \lambda_m - \kappa_m \lambda_n}{\kappa_m \kappa_n h}\right)^2}{(\kappa_m + \kappa_n)^2 + \left(\frac{\kappa_n \lambda_m - \kappa_m \lambda_n}{\kappa_m \kappa_n h}\right)^2} \right] + \cdots, \quad m \neq n \tag{6.34}$$

Here is the miracle: Use Equations (6.33) and (6.34) in the theta function for the KP equation and you have a hyperfast model for KP (Chapter 32)! Of course, I have added "+···" to Equations (6.32) and (6.34) to indicate the higher-order terms not considered in this simple calculation; the higher terms are discussed in detail in Chapter 32. But the fact is that you can now do hyperfast modeling by using Equations (6.33) and (6.34) together with the fast numerical algorithms for the KP equation given in this book (Chapters 9, 21–23, and 32).

Here is how you get onboard for new nonlinear wave equations:

(1) Decide which integrable equation you want to model.
(2) Find the N-soliton solution for the equation in this book or in the literature.
(3) Convert the N-soliton solution to dimensional units, since the previous formulations will probably be in normalized form.
(4) Set $k_i \to i\kappa_i$, $l_i \to i\lambda_i$, and $\omega_i \to i\omega_i$ to get the leading order (linear) dispersion relation and Riemann matrix.
(5) Build the model using fast numerical methods given herein (see Figure 32.2 for a flowchart of the program for the KP equation or (for example) Chapter 33 for the 2 + 1 Gardner equation).

This approach should give you fast positive feedback on the development of hyperfast numerical models. Then of course you can go to Chapters 14–16 to add in the amplitude-dependent terms in Equations (6.33) and (6.34).

6.6 The Nonlinear Schrödinger Equation

Given the nonlinear Schrödinger equation, with complex solution $\psi(x, t)$

$$i\psi_t + \mu\psi_{xx} + v|\psi|^2\psi = 0 \tag{6.35}$$

seek a solution as the ratio of two functions:

$$\psi(x,t) = \frac{G(x,t)}{F(x,t)}. \tag{6.36}$$

We then get

$$\psi_t = -\frac{GF_t}{F^2} + \frac{G_t}{F} = \frac{G_tF - GF_t}{F^2},$$

$$\psi_{xx} = \frac{2GF_x^2}{F^3} - \frac{2F_xG_x}{F^2} - \frac{GF_{xx}}{F^2} + \frac{FG_{xx}}{F^2} = \frac{2GF_x^2}{F^3} + \frac{FG_{xx} - GF_{xx} - 2F_xG_x}{F^2},$$

$$|\psi|^2\psi = \frac{|G|^2}{F^3}.$$

Insert these into the nonlinear Schrödinger equation and get

$$\underbrace{\frac{i(G_t F - G F_t)}{F^2}}_{i\psi_t} + \underbrace{\mu\left[\frac{2GF_x^2}{F^3} + \frac{FG_{xx} - GF_{xx} - 2F_x G_x}{F^2}\right]}_{\psi_{xx}} + \underbrace{v\frac{|G|^2}{F^3}}_{|\psi|^2\psi} = 0.$$

As will be seen below, we can treat $F(x, t)$ as a real function and $G(x, t)$ as a complex function. This result then becomes

$$\frac{G}{F^3}\left(2\mu F_x^2 + v|G|^2\right) + \frac{i(FG_t - GF_t) - 2\mu F_x G_x - \mu GF_{xx} + \mu FG_{xx}}{F^2} = 0.$$

(6.37)

At this stage, we could naively multiply Equation (6.28) by F^3 to get a trilinear form. An alternative route would be to add $2\mu GF_{xx}/F^2$ to the second term and subtract $2\mu GF_{xx}/F^2$ from the first term to get

$$\frac{G}{F^3}\left(2\mu F_x^2 + v|G|^2\right) - \frac{2\mu GF_{xx}}{F^2} + \frac{i(FG_t - GF_t) - 2\mu F_x G_x - \mu GF_{xx} + \mu FG_{xx}}{F^2} + \frac{2\mu GF_{xx}}{F^2} = 0,$$

which gives

$$\frac{G}{F^3}\left(2\mu(F_x^2 - FF_{xx}) + v|G|^2\right) + \frac{i(FG_t - GF_t) - 2\mu F_x G_x + \mu GF_{xx} + \mu FG_{xx}}{F^2} = 0.$$

At this stage, we can "separate the variables" by setting

$$i(G_t F - G F_t) + \mu(G_{xx} F - 2G_x F_x + GF_{xx}) = \lambda FG \qquad (6.38a)$$

$$2\mu(FF_{xx} - F_x^2) - v|G|^2 = \lambda F^2 \qquad (6.38b)$$

for λ a constant. The nonlinear term in the NLS equation $|\psi|^2$ has been isolated in Equation (6.38b). Thus, the major source of nonlinearity has been separated out of NLS and the result is the bilinear form that is Equation (6.38a). Notice that the separation of variables reduces the trilinear form mentioned above to a bilinear form!

Equations (6.38a) and (6.38b) can be put in Hirota operator form:

$$(iD_t + \mu D_x^2)G{\cdot}F = \lambda G{\cdot}F \qquad (6.39a)$$

$$\mu D_x^2 F{\cdot}F - v|G|^2 = \lambda F^2 \qquad (6.39b)$$

The Hirota operators are

$$D_t G \cdot F = G_t F - G F_t,$$

$$D_x G \cdot F = G_x F - G F_x,$$

$$D_x^2 G \cdot F = G_{xx} F - 2 G_x F_x + G F_{xx},$$

$$D_x^2 F \cdot F = 2(F F_{xx} - F_x^2).$$

From Equation (6.39b), we can solve for $\psi \psi^* = |\psi|^2$

$$\psi \psi^* = |\psi|^2 = \frac{G G^*}{F^2} = \frac{|G|^2}{F^2} = \frac{2\mu}{\nu} \left(\frac{F F_{xx} - F_x^2}{F^2} \right) - \frac{\lambda}{\nu}. \tag{6.40}$$

Then

$$|\psi|^2 = \frac{G G^*}{F^2} = \frac{2\mu}{\nu} \partial_{xx} \ln F - \frac{\lambda}{\nu} \tag{6.41}$$

This latter expression is interpreted in the following way. The modulus of the solution of the NLS equation is related to the function $F(x, t)$. The implication of this observation is that the function $G(x, t)$ has the phase of $\psi(x, t)$. In quantum mechanical terms, the function $F(x, t)$ can be viewed as providing the probability $|\psi|^2$ for the probability amplitude $\psi(x, t)$.

Notice that the nonlinear separation of variables could have been motivated from the start by seeking the linearized dispersion relation in Equation (6.39a).

At this point, the emphasis switches to Equation (6.39a) which can be used to solve for the function G given the function F:

$$i(G_t F - G F_t) + \mu(G_{xx} F - 2 G_x F_x + G F_{xx}) = \lambda F G \tag{6.42}$$

to give the solution of NLS (Equation (6.35)).

Note that we could set

$$\lambda \to 2\mu\lambda$$

and get

$$|\psi|^2 = \frac{G G^*}{F^2} = \frac{2\mu}{\nu} (\partial_{xx} \ln F - \lambda) \tag{6.43}$$

and

$$i(F G_t - G F_t) + \mu(G F_{xx} - 2 F_x G_x + F G_{xx}) = 2\mu\lambda F G \tag{6.44}$$

Details of the actual theta function solutions are discussed in Chapters 12, 18, and 24, but this is the starting point.

6.7 The KdV-Burgers Equation

The equation is given by

$$u_t + 6uu_x + u_{xxx} = \mu u_{xx}. \tag{6.45}$$

The term on the right-hand side includes dissipation in a way similar to that in the Navier-Stokes equations. Make the transformation

$$u = w_x$$

and get, after integrating once,

$$w_t + 3w_x^2 + w_{xxx} + c = \mu w_{xx},$$

where c is an integration constant. Make the further transformation

$$w(x,t) = 2\partial_x \ln\theta(x,t) = 2\frac{\theta_x}{\theta}$$

and, after multiplying by θ^2, get

$$\theta\theta_{xt} - \theta_x\theta_t + 3\theta_{xx}^2 - 4\theta_x\theta_{xxx} + \theta\theta_{xxxx} - c\theta^2 + \mu\left[3\theta_x\theta_{xx} - \theta\theta_{xxx} - 2\frac{\theta_x^3}{\theta}\right] = 0. \tag{6.46}$$

It will be noticed that this is the *KdV-Hirota equation* with a perturbation for dissipation with coefficient, μ.

In the present case, all but the last term in the equation above is a quadratic form. The dissipation term of course means that the KdV-Burgers equation is not integrable. Multiplying by an additional factor of theta gives

$$\theta^2\theta_{xt} - \theta\theta_x\theta_t + 3\theta\theta_{xx}^2 - 4\theta\theta_x\theta_{xxx} + \theta^2\theta_{xxxx} - c\theta^3 + \mu[3\theta\theta_x\theta_{xx} - \theta^2\theta_{xxx} - 2\theta_x^3] = 0, \tag{6.47}$$

which is a cubic form. To my knowledge, Equation (6.47) has not been studied in detail for the physics of the KdV-Burgers equation. This problem is left to the reader to work as a nontrivial exercise.

6.8 The Modified KdV Equation

The modified KdV (mKdV) equation is given by (Ablowitz and Segur, 1981)

$$u_t + 6u^2 u_x + u_{xxx} = 0. \tag{6.48}$$

The appropriate dependent variable transformation is given by

$$u = i\left(\ln\frac{g^*}{g} \right)_x, \tag{6.49}$$

for which one finds the bilinear forms:

$$(D_t + D_x^3)g^* \cdot g = \rho g^* \cdot g, \tag{6.50}$$
$$D_x^2 g^* \cdot g = \rho g^* \cdot g,$$

where ρ is the integration constant. The motivation for this decoupling is to preserve the linear dispersion relation in the first of these equations.

Note that if $u(x, t)$ is a solution of Equation (6.48), then $-u(x, t)$ is also a solution; this is consistent with the known positive and negative soliton solutions of the equation. Furthermore, Equation (6.48) is known to reduce to the NLS equation if one averages over the fast oscillations (Lamb, 1980; Bishop and Lomdahl, 1986; Bishop et al., 1986a,b) of Equation (6.48). Thus, Equation (6.48) also has unstable mode solutions that are similar to those of the NLS equation. While there are few direct applications of the mKdV equation, it is also true that mKdV is a fundamental building block on the way to higher-order behavior in shallow-water waves which leads to the Gardner equation, the KP equation, the KP-Gardner, and the $2 + 1$ Gardner equations (Chapter 33).

6.9 The Boussinesq Equation

The Boussinesq equation in normalized form is given by (Nakamura, 1980)

$$u_{tt} - u_{xx} - 3(u^2)_{xx} - u_{xxxx} = 0. \tag{6.51}$$

If we make the dependent variable transformation

$$u(x, t) = 2\partial_{xx} \ln\theta(x, t), \tag{6.52}$$

then we get the bilinear form:

$$(D_t^2 - D_x^2 - D_x^4 + \rho)\theta \cdot \theta = 0, \tag{6.53}$$

which of course is seen to preserve the form of the linear dispersion relation for the Boussinesq equation.

The dimensional form of the Boussinesq equation is given by

$$u_{tt} - c_0 u_{xx} - \frac{3c_0^2}{2h}(u^2)_{xx} - \frac{c_0^2 h^2}{3} u_{xxxx} = 0.$$

This equation can be written in terms of the KdV coefficients:

$$u_{tt} - c_0 u_{xx} - c_0 \alpha (u^2)_{xx} - 2c_0 \beta u_{xxxx} = 0,$$

which has the bilinear form:

$$(D_t^2 - c_0 D_x^2 - 2c_0 \beta D_x^4 + \rho)\theta \cdot \theta = 0.$$

Formally speaking, the Boussinesq equation describes shallow-water waves that propagate both to the right and to the left. However, as Miles (1981) has pointed out, Boussinesq assumed rightward moving waves in his derivation, so that the equation is *physically limited* to rightward moving waves, while *mathematically* it can describe waves moving in both directions. This is a point to be remembered in applications, where Boussinesq is then physically equivalent to the KdV equation.

6.10 The 2 + 1 Boussinesq Equation

The Boussinesq equation in normalized form is given by

$$u_{tt} - u_{xx} - u_{yy} - 3(u^2)_{xx} - u_{xxxx} = 0. \tag{6.54}$$

If we make the dependent variable transformation

$$u(x, y, t) = 2\partial_{xx} \ln\theta(x, y, t), \tag{6.55}$$

then we get the bilinear form:

$$(D_t^2 - D_x^2 - D_y^2 - D_x^4 + \rho)\theta \cdot \theta = 0, \tag{6.56}$$

which of course is seen to preserve the form of the linear dispersion relation for the 2 + 1 Boussinesq equation.

6.11 The 2 + 1 Gardner Equation

The 2 + 1 Gardner equation contains a number of other integrable equations: The *KdV equation*:

$$\eta_t + c_0 \eta_x + \alpha \eta \eta_x + \beta \eta_{xxx} = 0.$$

The *modified KdV*:

$$\eta_t + c_0\eta_x + \beta\eta_{xxx} = \delta\eta^2\eta_x.$$

The *Gardner equation*:

$$\eta_t + c_0\eta_x + \alpha\eta\eta_x + \beta\eta_{xxx} = \delta\eta^2\eta_x.$$

The *KP equation*:

$$\eta_t + c_0\eta_x + \alpha\eta\eta_x + \beta\eta_{xxx} + \gamma\partial_x^{-1}\eta_{yy} = 0.$$

The *KP-Gardner equation*:

$$\eta_t + c_0\eta_x + \alpha\eta\eta_x + \beta\eta_{xxx} + \gamma\partial_x^{-1}\eta_{yy} = \delta\eta^2\eta_x.$$

It is indeed a modern miracle of the inverse scattering transform and the Hirota method that all of the above five equations are integrable. Additionally, the 2 + 1 Gardner equation is also integrable.

The *2 + 1 Gardner equation*:

$$\eta_t + c_0\eta_x + \alpha\eta\eta_x + \beta\eta_{xxx} + \gamma\partial_x^{-1}\eta_{yy} = \delta\eta^2\eta_x + \rho\eta_x\partial_x^{-1}\eta_y,$$

where

$$c_0 = \sqrt{gh}, \quad \alpha = \frac{3c_0}{2h}, \quad \beta = \frac{c_0h^2}{6}, \quad \gamma = \frac{c_0}{2}, \quad \delta = \frac{15c_0}{8h^2}, \quad \rho = \sqrt{\frac{15}{8}}\left(\frac{c_0}{h}\right),$$

where g is the gravitational acceleration and h is the water depth.

The bilinear forms for the 2 + 1 Gardner equation are found by rewriting the equation as a system (Zhang et al., 2008):

$$u_t - u_{xxx} - 6\beta uu_x = \frac{3}{2}\alpha^2 u^2 u_x - 3v_y + 3\alpha u_x v = 0,$$

$$v_x = u_y,$$

for which the Hirota type transformations are made:

$$u = \frac{2}{\alpha}\left[\ln\left(\frac{g}{f}\right)\right]_x,$$

$$v = \frac{2}{\alpha}\left[\ln\left(\frac{g}{f}\right)\right]_y.$$

The bilinear forms are given by

$$(D_y + D_x^2 - 2(\beta/\alpha)D_y + \rho)g{\cdot}f = 0,$$

$$(D_t - D_x^3 + 3D_xD_y - 6(\beta/\alpha)D_y)g{\cdot}f = 0.$$

Details of a numerical model for the 2 + 1 Gardner equation are given in Chapter 33.

Part Three

Multidimensional Fourier Analysis

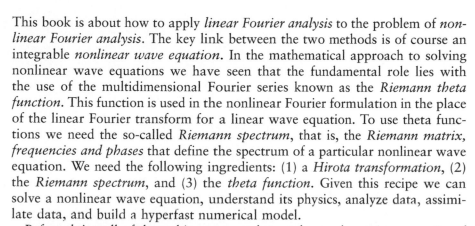

This book is about how to apply *linear Fourier analysis* to the problem of *nonlinear Fourier analysis*. The key link between the two methods is of course an integrable *nonlinear wave equation*. In the mathematical approach to solving nonlinear wave equations we have seen that the fundamental role lies with the use of the multidimensional Fourier series known as the *Riemann theta function*. This function is used in the nonlinear Fourier formulation in the place of the linear Fourier transform for a linear wave equation. To use theta functions we need the so-called *Riemann spectrum*, that is, the *Riemann matrix, frequencies and phases* that define the spectrum of a particular nonlinear wave equation. We need the following ingredients: (1) a *Hirota transformation*, (2) the *Riemann spectrum*, and (3) the *theta function*. Given this recipe we can solve a nonlinear wave equation, understand its physics, analyze data, assimilate data, and build a hyperfast numerical model.

Before doing all of these things we need to understand certain properties of theta functions. To this end I give an overview of multidimensional Fourier series in Chapter 7. Then I look at the properties of Riemann theta functions in Chapter 8. I show how the Riemann theta function can be reduced to an *ordinary Fourier transform with time-varying coefficients* (Chapter 9). This result is one of the most important for applications: *The esoteric theta function of algebraic geometry can be reduced to the ordinary Fourier transform with time-varying coefficients* for which classical analysis holds and for which we have 50 years of data analysis and modeling experience! This idea will provide a *crucial tool for implementation of theta functions in computer codes* and provides a *major step for computing theta functions in finite (polynomial rather than exponential) time*.

7 Multidimensional Fourier Series

7.1 Introduction

Multidimensional Fourier series have been a fundamental concept in pure and applied mathematics and various branches of the physical sciences for over a century (Baker, 1897; Bellman, 1961). In this chapter, I discuss multidimensional Fourier series and attempt to familiarize the reader with a number of their fascinating properties. The focus of this chapter is to prepare the reader for the introduction of *Riemann theta functions* in Chapter 8. *Theta functions are reductions of multidimensional Fourier series* and are the primary *tool for the solution of integrable, nonlinear wave equations*, the main subject of this book.

7.2 Linear Fourier Series

Ordinary Fourier series have the following simple form:

$$F(x) = \sum_{n=-\infty}^{\infty} F_n e^{ik_n x}, \tag{7.1}$$

where the *inverse Fourier transform* is given by

$$F_n \equiv F(k_n) = \frac{1}{L} \int_0^L F(x) e^{-ik_n x} \mathrm{d}x, \tag{7.2}$$

where $k_n = 2\pi n/L$, L the period of the wave train. I treat *Fourier series* (Equation (7.1)) here, rather than infinite-line *Fourier* integrals, because most data analysis and modeling applications assume $F(x)$ to be a *periodic (or at most quasiperiodic) function of space* $(F(x) = F(x + L), 0 \leq x \leq L)$ and/or *time* $(0 \leq t \leq T)$.

The *Fourier transform* of the function $F(x)$ (Equation (7.1)) consists of the *vector of real Fourier amplitudes*, $\mathbf{A} = \{A_n, n = 1, 2, \ldots, N\}$, and *vector of Fourier phases*, $\mathbf{\phi} = \{\phi_n, n = 1, 2, \ldots, N\}$, which are related to the complex coefficients, F_n, in Equation (7.1) by

$$F_n = A_n e^{i\phi_n}.$$

For *dynamical applications,* for example in the solution of *linear partial differential wave equations,* one also requires the vector of Fourier wavenumbers, $\mathbf{k} = \{k_n, n = 1, 2, \ldots, N\}$, and frequencies, $\boldsymbol{\omega} = \{\omega_n, n = 1, 2, \ldots, N\}$, for which the Fourier transform can be written:

$$F(x, t) = \sum_{n=-\infty}^{\infty} A_n e^{i(k_n x - \omega_n t + \phi_n)}. \tag{7.3}$$

The wavenumbers and frequencies are related by the *linear dispersion relation,* $\omega_n = \omega_n(k_n)$. One often assumes that the ϕ_n are uniformly distributed random numbers.

Generally speaking, one selects either wavenumbers, k_n, or frequencies, ω_n, to be *commensurable,* that is, to be *equally spaced* in the wavenumber or frequency domains, thus leading to periodic boundary conditions. One is thus led naturally to the *discrete Fourier transform* (DFT) (Chapter 21). We are familiar with the equally spaced frequencies often assumed when we Fourier analyze a time series with the DFT algorithm (Bendat and Piersol, 1986). Likewise, equally spaced (commensurable) wavenumbers are often assumed when we analyze space series, typically obtained from remotely sensed data.

Application of Fourier series has been made practical by the development of the *fast Fourier transform* (FFT) (Cooley and Tukey, 1965). Periodicity is of course a prime requirement of the FFT, often used for space/time series analysis purposes (for a deeper discussion of the FFT, see, e.g., Bendat and Piersol, 1986). An additional requirement made for the FFT is that $F(x)$ must be a *discrete function,* that is, $F(x_m), 0 \leq m \leq M - 1$, for m an integer. Discrete theta functions are discussed in detail in Chapter 21.

A list of modern uses of the linear Fourier transform is quite long and includes (1) solving linear partial differential (wave) equations with a well-defined dispersion relation, (2) the analysis of space and time series data (Fourier analysis, power spectral analysis, cross- and bispectral analysis, etc.), and (3) use in the fast numerical simulation of higher-order nonlinear wave equations.

Example 7.1: Constructing Stokes Waves

A *Stokes wave* occurs when we constrain Equation (7.3) by the relation $k_n x - \omega_n t + \phi_n = n(k_0 x - \omega_0 t + \phi_0)$, where, say, $k_0 = 2\pi/L_0, \omega_0 = 2\pi/T_0$, and $\phi_0 = $ constant (L_0 is the wavelength, T_0 is the period, and ϕ_0 is an arbitrary phase), that is, the Stokes wave arises when the components are *phase locked.* We are able to describe *only a single Stokes wave* using the Fourier series (Equation (7.3)). This simple result illustrates why ordinary linear Fourier analysis is unable to simultaneously describe sine

Continued

Example 7.1: Constructing Stokes Waves—Cont'd

waves and Stokes waves, or indeed, multiple Stokes waves in the same wave train.

The focus of this chapter is to remove this limit on the use of Fourier analysis, that is, multidimensional Fourier analysis is used to address the nonlinear dynamics of waves. This enables us to describe wave trains that simultaneously have sine waves, Stokes waves, solitons, unstable modes, shock waves, vortices, etc., all of which may be interacting nonlinearly with each other.

7.3   Multidimensional or *N*-Dimensional Fourier Series

A natural generalization of the ordinary Fourier series discussed above is the *multidimensional (N-dimensional) Fourier series* given by

$$\Phi(X_1, X_2, \ldots, X_N) = \sum_{m_1=-\infty}^{\infty} \sum_{m_2=-\infty}^{\infty} \cdots \sum_{m_N=-\infty}^{\infty} C_{m_1, m_2, \ldots, m_N} \exp\left\{ i \sum_{n=1}^{N} m_n X_n \right\}, \quad (7.4)$$

where the associated *inverse multidimensional Fourier transform* has the expression:

$$C_{m_1, m_2, \ldots, m_N} = \frac{1}{(2\pi)^N} \int_0^{2\pi} \int_0^{2\pi} \cdots \int_0^{2\pi} \Phi(X_1, X_2, \ldots, X_N) \exp\left\{ -i \sum_{n=1}^{N} m_n X_n \right\} dX_1 dX_2 \cdots dX_N.$$

$$(7.5)$$

Equations (7.4) and (7.5) are thus an *N-dimensional Fourier transform pair*. The integers $[m_1, m_2, \ldots, m_N]$ are *summation indices* associated with each of the nested sums of Equation (7.4). The $X_1, X_2, \ldots, X_N$ are the N *dimensions* on which the rather general and arbitrary multidimensional field $\Phi(X_1, X_2, \ldots, X_N)$ is defined. This fundamental result was discovered in the nineteenth century as an exercise in complex analysis and was at that time of considerable interest in pure mathematics (see, e.g., Baker, 1897). It is quite common to apply multidimensional analysis in one, two, or three dimensions in the geophysical sciences. During the past few decades, multidimensional Fourier series have been related to the solutions of the so-called nonlinear integrable (soliton) equations (as discussed in Chapters 2–6). For this purpose, a simple reduction of the series (Equation (7.4)) referred to as Riemann theta functions has been used (see Chapter 8).

We now divide multidimensional Fourier analysis into two separate categories: (1) ordinary or *conventional multidimensional Fourier analysis* (already used in many physical and geophysical applications) and (2) *dynamical multi-dimensional Fourier analysis* (the main scope and focus of this chapter).

7.4 Conventional Multidimensional Fourier Series

Below are examples that illustrate the conventional way that multidimensional Fourier series are applied to physical and geophysical applications. The examples given are in one, two, and three dimensions. Furthermore, the dimensions are related to the spatial coordinates x, y, and z. It is important to note that we do *not* generally constrain the phase to be locked in any of these examples. Phase locking will be crucial in the *dynamical applications* given in later sections.

Example 7.2: One-Dimensional Ordinary Linear Fourier Analysis

If we restrict Equation (7.4) to *one spatial dimension* and set $X_1 = k_x x$, then we get

$$\Phi(x) = \sum_{m=-\infty}^{\infty} C_m e^{ik_m x},$$

where $k_m = 2\pi m/L$. This is just the one-dimensional Fourier transform (Equation (7.1)) with inverse

$$C_m = \frac{1}{2\pi}\int_0^{2\pi} \Phi(X_1)e^{-imX_1}dX_1 = \frac{1}{L}\int_0^L \Phi(x)e^{-ik_m x}dx.$$

Example 7.3: Two-Dimensional Ordinary Linear Fourier Analysis

Now, let us suppose that we choose the dimensions, $X_1, X_2, \ldots, X_N$, to *coincide with two spatial dimensions*, so that $X_1 = k_x x$ and $X_2 = k_y y$. In this case one has

$$\Phi(x,y) = \sum_{m_1=-\infty}^{\infty}\sum_{m_2=-\infty}^{\infty} C_{m_1,m_2} e^{i(m_1 X_1 + m_2 X_2)}$$
$$= \sum_{m=-\infty}^{\infty}\sum_{n=-\infty}^{\infty} C_{m,n} e^{i(m\Delta k_x x + n\Delta k_y y)}.$$

Continued

Example 7.3: Two-Dimensional Ordinary Linear Fourier Analysis—Cont'd

Let $k_m = m\Delta k_x = 2\pi m/L_x$ and $k_n = n\Delta k_y = 2\pi n/L_y$ (L_x and L_y are the spatial periods in the x and y directions, $0 \le x \le L_x, 0 \le y \le L_y$) and we get the two-dimensional Fourier transform:

$$\Phi(x,y) = \sum_{m=-\infty}^{\infty} \sum_{n=-\infty}^{\infty} C(k_m, k_n) e^{i(k_m x + k_n y)}. \tag{7.6}$$

Thus, it is clear that the two-dimensional wave field, $\Phi(x,y)$, has the two-dimensional wavenumber Fourier transform, $C(k_m, k_n)$.

The *inverse two-dimensional Fourier transform* is

$$C(k_m, k_n) = \frac{1}{L_x L_y} \int_0^{L_y} \int_0^{L_x} \Phi(x,y) e^{-i(k_m x + k_n y)} \, dx \, dy.$$

Example 7.4: Three-Dimensional Ordinary Linear Fourier Analysis

Now, let us suppose that we choose the dimensions, $X_1, X_2, \ldots, X_N$, to *coincide with three spatial dimensions*, so that $X_1 = k_x x, X_2 = k_y y$, and $X_3 = k_z z$. In this case one has

$$\Phi(x,y) = \sum_{m_1=-\infty}^{\infty} \sum_{m_2=-\infty}^{\infty} \sum_{m_3=-\infty}^{\infty} C_{m_1, m_2, m_3} e^{i(m_1 X_1 + m_2 X_2 + m_3 X_3)}$$

$$= \sum_{l=-\infty}^{\infty} \sum_{m=-\infty}^{\infty} \sum_{n=-\infty}^{\infty} C_{l,m,n} e^{i(l\Delta k_x x + m\Delta k_y y + n\Delta k_y y)}.$$

As before, $k_l = l\Delta k_x = 2\pi l/L_x, k_m = m\Delta k_y = 2\pi m/L_y$, and $k_n = n\Delta k_z = 2\pi n/L_z$ (L_x, L_y, and L_z are the spatial periods in the x, y, and z directions, $0 \le x \le L_x, 0 \le y \le L_y, 0 \le z \le L_z$) and get the three-dimensional Fourier transform:

$$\Phi(\mathbf{x}) = \sum_{l=-\infty}^{\infty} \sum_{m=-\infty}^{\infty} \sum_{n=-\infty}^{\infty} C(k_l, k_m, k_n) e^{i(k_l x + k_m y + k_n y)} = \sum_{\mathbf{m}=-\infty}^{\infty} C(\mathbf{k}) e^{i\mathbf{k_m} \cdot \mathbf{x}}. \tag{7.7}$$

Thus, it is clear that the three-dimensional wave field, $\Phi(\mathbf{x})$, has the three-dimensional wavenumber Fourier transform, $C(\mathbf{k})$.

Continued

Example 7.4: Three-Dimensional Ordinary Linear Fourier Analysis—Cont'd

The inverse three-dimensional transform is in vector notation:

$$C(\mathbf{k}) = \frac{1}{L^3} \int_0^{L_y} \Phi(\mathbf{x}) e^{-i\mathbf{k_m \cdot x}} d\mathbf{x}^3.$$

7.5 Dynamical Multidimensional Fourier Series

We have just rediscovered ordinary linear Fourier analysis in one, two, and three dimensions. In this way, the Fourier approach is typically applied to a wide range of physical problems.

The above route to ordinary linear Fourier analysis is the one normally traveled, that is, such that the dimensions, X_n, of the N-dimensional Fourier series (Equation (7.4)) are identified with the one, two, or three spatial dimensions in which data are most often recorded ($X_1 = k_x x$, $X_2 = k_y y$, and $X_3 = k_z z$). A wide range of applications is commonly pursued in the analysis of space series (remote sensing applications, crystallography, etc.) and time series (single or multiple probes at particular spatial locations to form an antenna).

The route to nonlinear Fourier analysis as followed in this chapter does *not* associate the dimensions, X_n, of the N-dimensional Fourier series (Equation (7.4)) with the ordinary spatial dimensions, but instead associates them with the *dynamical dimensions*:

$$X_n = k_n x - \omega_n t + \phi_n, \quad n = 1, 2, \ldots, N.$$

Generally, there are N of the dynamical dimensions (one for each nonlinear mode in the system), that is, to be concrete I assume that there are N degrees of freedom, where N is a relatively large number, that is, $N \sim 10, 100$ or 1000. In this case, with the inclusion of the time, t, we require a well-behaved dispersion relation, $\omega_n = \omega_n(k_n)$. The important new ingredient here is the inclusion of the phase, ϕ_n, in the definition of X_n. This allows *multiple Stokes waves and their nonlinear interactions* to be described by a single multidimensional Fourier series. To see how N-dimensional analysis works in this dynamical case, consider the following two examples in one and two dimensions.

Example 7.5: One-Dimensional Dynamical Fourier Analysis

If we restrict Equation (7.4) to *one dynamical dimension*, $X_1 = k_1 x - \omega_1 t + \phi_1$ ($k_1 = 2\pi/L$, $\Delta k = 2\pi/L$), where $0 \le x \le L$), then we get a linear Fourier series. This is seen in the following steps:

$$\Phi(x, t) = \sum_{m=-\infty}^{\infty} A_m e^{im(k_1 x - \omega_1 t + \phi_1)} = \sum_{m=-\infty}^{\infty} A_m e^{i(k_m x - \omega_m t + \phi_m)},$$

Continued

Example 7.5: One-Dimensional Dynamical Fourier Analysis—Cont'd

where $k_m = 2\pi m/L$, $\omega_m = m\omega_1(k_1)$, and $\phi_m = m\phi_1$. This simple example *appears* to be similar to the case of *one spatial dimension* where we include the time, t, and the phase, ϕ_m. Note, however, that all linear Fourier components in this example are *phase locked*, that is, $\phi_m = n\phi_1$, thus allowing for the existence of a *bound mode* for a nonlinear wave train. Due to the phase locking, the above result is a kind of *Stokes wave* (here "Stokes wave" just means that the higher harmonics are phase locked with the primary harmonic; it will be shown how *this* Stokes wave is related to *the* Stokes wave solution of nonlinear partial differential equations in later chapters). This is an *essential difference* between nonlinear dynamical Fourier analysis and ordinary linear Fourier analysis, that is, the idea that phase locking can be accounted for in a systematic way, an idea that becomes clearer in the example which follows.

The inverse transform is given by

$$A_m = \frac{1}{2\pi} \int_0^{2\pi} \Phi(X_1) e^{-imX_1} dX_1 = \frac{1}{L} \int_0^L \Phi(x) e^{-i(k_m x - \omega_m t + \phi_m)} dx.$$

Example 7.6: Two-Dimensional Dynamical Fourier Analysis

Now, let us suppose that we choose the dimensions, $X_1, X_2, \ldots, X_N$, to *coincide with the two dynamical dimensions*, so that $X_1 = k_1 x - \omega_1 t + \phi_1$ and $X_2 = k_2 x - \omega_2 t + \phi_2$. In this case one has

$$\Phi(x, t) = \sum_{m_1=-\infty}^{\infty} \sum_{m_2=-\infty}^{\infty} C_{m_1, m_2} \exp\{i[(m_1 k_1 + m_2 k_2)x$$
$$- (m_1 \omega_1 + m_2 \omega_2)t + (m_1 \phi_1 + m_2 \phi_2)]\}.$$

In this example, there are two modes, one labeled "1" and the other "2." Each mode has its own phase-locked components, $m_1\phi_1$, $m_2\phi_2$, and the modes may be viewed as being made up of these bound components, that is, each of the two modes is a Stokes wave. Each Stokes mode is free to move at its own phase velocity, $c_1 = \omega_1/k_1$, $c_2 = \omega_2/k_2$, and in this sense the modes are "free." Thus, each Stokes mode consists of a collection of phase-locked (bound) components. The two Stokes modes are free to independently propagate relative to each other while simultaneously nonlinearly interacting with each other. Stated another way the two Stokes (bound) modes (degrees of freedom) occur because of nonlinearity (leading to the phase locking) and any interaction among these modes is also nonlinear (effects contained in the coefficients, C_{m_1, m_2}). The inverse transform in two dimensions is given by

Continued

Example 7.6: Two-Dimensional Dynamical Fourier Analysis—Cont'd

$$C_{m_1, m_2} = \frac{1}{(2\pi)^2} \int_0^{2\pi} \int_0^{2\pi} \Phi(x, t) \exp\{-i[(m_1 k_1 + m_2 k_2)x$$
$$- (m_1 \omega_1 + m_2 \omega_2)t + (m_1 \phi_1 + m_2 \phi_2)]\} d(k_1 x) d(k_2 x).$$

This latter relation is not very useful for the study of nonlinear wave dynamics for determining the coefficients of the multidimensional Fourier series because we are generally unable to measure a wave amplitude, $\Phi(x, t)$, in terms of the dimensions $k_n x - \omega_n t + \phi_n$. Indeed, a major portion of this book is dedicated to the more complex, but highly rigorous inverse scattering transform and other related methods, which are necessary for inverting the multidimensional Fourier transform for a particular integrable, nonlinear wave dynamics.

Applications of dynamical multidimensional Fourier series are considered in later chapters. One example is that for shallow-water wave dynamics (described by the KdV or KP equations) where each Stokes wave has phase-locked components (bound modes) which interact nonlinearly with each other (see Chapters 8–11, 32, and 33). One can generalize this approach to unstable mode ("rogue wave") dynamics governed by the Benjamin-Feir instability (Chapters 12, 18, and 24), acoustic wave propagation (Chapter 26), internal wave propagation (Chapter 25), and vortex dynamics (Chapter 27).

7.6 Alternative Notations for Multidimensional Fourier Series

The N-dimensional Fourier transform may also be written in short-hand *vector notation*:

$$\Phi(\mathbf{X}) = \sum_{\mathbf{m}=-\infty}^{\infty} C_{\mathbf{m}} e^{i \mathbf{m} \cdot \mathbf{X}}, \tag{7.8}$$

where the associated inverse N-dimensional Fourier transform has the expression:

$$C_{\mathbf{m}} = \frac{1}{(2\pi)^N} \int_0^{2\pi} e^{i \mathbf{m} \cdot \mathbf{X}} d\mathbf{X}. \tag{7.9}$$

The integer vector is $\mathbf{m} = [m_1, m_2, \ldots, m_N]$ and the dimension vector is $\mathbf{X} = [X_1, X_2, \ldots, X_N]$.

7.6.1 Baker's Notation

The *generalized Fourier series* for an arbitrary function of N variables is given, for an *N-degree-of-freedom system*, in Baker's notation (Baker, 1897) by:

$$\phi(u_1, u_2, \ldots, u_N) = \sum_{m_1=-\infty}^{\infty} \sum_{m_2=-\infty}^{\infty} \cdots \sum_{m_N=-\infty}^{\infty} A_{m_1, m_2, \ldots, m_N} \exp\left\{ 2\pi i \sum_{n=1}^{N} m_n u_n \right\}.$$

$$(7.10)$$

The *inverse generalized Fourier transform* determines the *generalized Fourier coefficients*, $A_{m_1, m_2, \ldots, m_N}$, which are given by

$$A_{m_1, m_2, \ldots, m_N} = \int_0^1 \int_0^1 \cdots \int_0^1 \phi(u_1, u_2, \ldots, u_N) \exp\left\{ -2\pi i \sum_{n=1}^{N} m_n u_n \right\} du_1 du_2 \cdots du_N.$$

$$(7.11)$$

In *vector notation*,

$$\phi(\mathbf{u}) = \sum_{\mathbf{m}=-\infty}^{\infty} A_{\mathbf{m}} e^{2\pi i \mathbf{m} \cdot \mathbf{u}}$$

$$(7.12)$$

$$A_{\mathbf{m}} = \int_0^1 \phi(\mathbf{u}) e^{-2\pi i \mathbf{m} \cdot \mathbf{u}} d\mathbf{u}$$

$$(7.13)$$

7.6.2 Inverse Scattering Transform Notation

The *generalized Fourier series* for an arbitrary function of N variables is given, for the *N-degree-of-freedom system*, by ($X_n = 2\pi u_n$, *IST notation*, see Chapter 8)

$$\phi(X_1, X_2, \ldots, X_N) = \sum_{m_1=-\infty}^{\infty} \sum_{m_2=-\infty}^{\infty} \cdots \sum_{m_N=-\infty}^{\infty} A_{m_1, m_2, \ldots, m_N} \exp\left\{ i \sum_{n=1}^{N} m_n X_n \right\}$$

$$(7.14)$$

The *inverse generalized Fourier series* determines the *generalized Fourier coefficients*, $A_{m_1, m_2, \ldots, m_N}$, which are given by

$$A_{m_1, m_2, \ldots, m_N} = \frac{1}{(2\pi)^N} \int_0^{2\pi} \int_0^{2\pi} \cdots \int_0^{2\pi} \phi(X_1, X_2, \ldots, X_N) \exp\left\{ -i \sum_{n=1}^{N} m_n X_n \right\} dX_1 dX_2 \cdots dX_N$$

$$(7.15)$$

It will be convenient, in analogy with ordinary Fourier analysis, to call the $A_{m_1, m_2, ..., m_N}$ the *phase space* of the system.

In *vector notation*,

$$\phi(\mathbf{X}) = \sum_{\mathbf{m}=-\infty}^{\infty} A_{\mathbf{m}} e^{i\mathbf{m} \cdot \mathbf{X}} \tag{7.16}$$

$$A_{\mathbf{m}} = \frac{1}{(2\pi)^N} \int_0^{2\pi} \phi(\mathbf{X}) e^{-i\mathbf{m} \cdot \mathbf{X}} d\mathbf{X} \tag{7.17}$$

We now have the usual (IST) variables for the argument:

$$X_n = k_n x - \phi_n, \quad 1 \le n \le N,$$

where

$$dX_n = k_n dx = \frac{2\pi}{L} n dx$$

and

$$k_n = n\Delta k \quad \text{for } \Delta k = \frac{2\pi}{L}.$$

where L is the *spatial period* of the wave train and $k_n = [1, 2, \ldots, N]\Delta k$ (*standard Fourier notation*):

$$\phi(x) = \sum_{m_1=-\infty}^{\infty} \sum_{m_2=-\infty}^{\infty} \cdots \sum_{m_N=-\infty}^{\infty} A_{m_1, m_2, ..., m_N} \exp\left\{ i\frac{2\pi}{L} \sum_{n=1}^{N} m_n(nx) + i\sum_{n=1}^{N} m_n\phi_n \right\} \tag{7.18}$$

$$A_{m_1, m_2, ..., m_N} = \frac{1}{L^N} \int_0^L \int_0^L \cdots \int_0^L \phi(x, 2x, 3x, \ldots, Nx)$$

$$\exp\left\{ -i\frac{2\pi}{L} \sum_{n=1}^{N} m_n(nx) - i\sum_{n=1}^{N} m_n\phi_n \right\} dx(2dx)(3dx) \cdots (Ndx) \tag{7.19}$$

so that

$$\phi(x) = \sum_{m_1=-\infty}^{\infty} \sum_{m_2=-\infty}^{\infty} \cdots \sum_{m_N=-\infty}^{\infty} A_{m_1, m_2, ..., m_N} \exp\left\{ i\sum_{n=1}^{N} m_n\phi_n \right\} \exp\left\{ i\frac{2\pi}{L} \sum_{n=1}^{N} m_n(nx) \right\} \tag{7.20}$$

$$A_{m_1, m_2, \ldots, m_N} \exp\left\{ i \sum_{n=1}^{N} m_n \phi_n \right\} = \frac{1}{L^N} \int_0^L \int_0^L \cdots \int_0^L \phi(x, 2x, 3x, \ldots, Nx)$$

$$\times \exp\left\{ -i \frac{2\pi}{L} \sum_{n=1}^{N} m_n(nx) \right\} dx(2dx)(3dx)\cdots(Ndx) \tag{7.21}$$

7.6.3 Relationship to Riemann Theta Functions

It is rather easy to write the *generalized Fourier coefficients* for the special case of the *Riemann theta functions* (see detailed discussion in Chapter 8):

$$A_{m_1, m_2, \ldots, m_N} = \exp\left\{ \frac{1}{2} \sum_{j=1}^{N} \sum_{k=1}^{N} B_{jk} m_j m_k \right\}$$

$$= \frac{1}{(2\pi)^N} \int_0^{2\pi} \int_0^{2\pi} \cdots \int_0^{2\pi} \phi(X_1, X_2, \ldots, X_N) \tag{7.22}$$

$$\exp\left\{ -i \sum_{n=1}^{N} m_n X_n \right\} dX_1 dX_2 \cdots dX_N.$$

This defines the generalized Fourier coefficients in terms of the so-called *Riemann matrix*, B_{jk}:

$$A_{m_1, m_2, \ldots, m_N} = \exp\left\{ \frac{1}{2} \sum_{j=1}^{N} \sum_{k=1}^{N} B_{jk} m_j m_k \right\}. \tag{7.23}$$

This expression is a Gaussian in lattice space. Now write the above results in *vector notation*:

$$\phi(\mathbf{X}) = \sum_{\mathbf{m}=-\infty}^{\infty} A_{\mathbf{m}} e^{2\pi i \mathbf{m} \cdot \mathbf{X}}, \tag{7.24}$$

$$A_{\mathbf{m}} = \int_0^1 \phi(\mathbf{X}) e^{-2\pi i \mathbf{m} \cdot \mathbf{X}} d^N \mathbf{X}. \tag{7.25}$$

Exact Fourier series for theta functions:

$$\theta(\mathbf{X}) = \sum_{\mathbf{m}=-\infty}^{\infty} e^{\frac{1}{2} \mathbf{m} \cdot \mathbf{Bm}} e^{2\pi i \mathbf{m} \cdot \mathbf{X}} \tag{7.26}$$

7.7 Simple Examples of Dynamical Multidimensional Fourier Series

Now we can write the *one-degree-of-freedom case* directly from the IST form for the generalized Fourier transform (Equations (7.14) and (7.15)):

$$\phi(X_1) = \sum_{m_1=-\infty}^{\infty} A_{m_1} e^{im_1 X_1}.$$

Here, we know that

$$X_1 = k_1 x + \phi_1, \quad k_1 = \frac{2\pi}{L}$$

and get

$$\phi(x) = \sum_{n=-\infty}^{\infty} A_n \exp\left\{i\left(\frac{2\pi}{L} nx + n\phi\right)\right\}$$

The inverse problem is just

$$A_n = \frac{1}{2\pi} \int_0^{2\pi} \phi(X) e^{-inX} dX$$

for

$$X = kx + \phi, \quad k = \frac{2\pi}{L}.$$

This gives

$$A_n = \frac{1}{L} \int_{-\phi/k}^{L-(\phi/k)} \phi(x) e^{-in(kx+\phi)} dx$$

or

$$A_n = \frac{1}{L} \int_0^L \phi(x) e^{-in(kx+\phi)} dx$$

since $\phi(x)$ is periodic $\phi(x + L) = \phi(x)$.

Now we can write the *two-degree-of-freedom* *case* directly from the IST form for the generalized Fourier transform (Equations (7.14) and (7.15)):

$$\phi(X_1, X_2) = \sum_{m_1=-\infty}^{\infty} \sum_{m_2=-\infty}^{\infty} A_{m_1, m_2} \exp\left\{ i \sum_{n=1}^{2} m_n X_n \right\}.$$

Here, we know that

$$X_n = k_n x + \phi_n, \quad k_n = \frac{2\pi n}{L}, \quad n = 1, 2$$

and get

$$\phi(X_1, X_2) = \sum_{m_1=-\infty}^{\infty} \sum_{m_2=-\infty}^{\infty} A_{m_1, m_2} \exp\left\{ i \left[m_1 \left(\frac{2\pi}{L} x + \phi_1 \right) + m_2 \left(\frac{4\pi}{L} x + \phi_2 \right) \right] \right\}$$

Now, let us suppose the degrees of freedom are uncoupled, so that

$$A_{m_1, m_2} = A_{m_1} A_{m_2}.$$

Then we have

$$\phi(X_1, X_2) = \sum_{m_1=-\infty}^{\infty} A_{m_1} \exp\left\{ i m_1 \left(\frac{2\pi}{L} x + \phi_1 \right) \right\} \sum_{m_2=-\infty}^{\infty} A_{m_2} \exp\left\{ i m_2 \left(\frac{4\pi}{L} x + \phi_2 \right) \right\},$$

which can be written as phase-locked series:

$$\phi(X_1) = \sum_{m_1=-\infty}^{\infty} A_{m_1} \exp\left\{ i m_1 \left(\frac{2\pi}{L} x + \phi_1 \right) \right\},$$

$$\phi(X_2) = \sum_{m_2=-\infty}^{\infty} A_{m_2} \exp\left\{ i m_2 \left(\frac{4\pi}{L} x + \phi_2 \right) \right\},$$

and finally

$$\phi(X_1, X_2) = \phi(X_1)\phi(X_2),$$

that is, the product of two series.

The inverse problem is just

$$A_{m_1,m_2} = \frac{1}{(2\pi)^2} \int_0^{2\pi} \int_0^{2\pi} \phi(X_1, X_2) \exp\left\{ -i \sum_{n=1}^2 m_n X_n \right\} dX_1 dX_2$$

or

$$A_{m_1,m_2} = \frac{1}{L^2} \int_0^L \int_0^L \phi(x, 2x) \exp\left\{ -i \left[m_1 \left(\frac{2\pi}{L} x + \phi_1 \right) + m_2 \left(\frac{2\pi}{L} (2x) + \phi_2 \right) \right] \right\} dx d(2x)$$

If the system is uncoupled, then

$$\phi(x, 2x) = \phi(x)\phi(2x)$$

and

$$A_{m_1,m_2} = \frac{1}{(2\pi)^2} \int_0^{2\pi} \phi(X_1) e^{-im_1 X_1} dX_1 \int_0^{2\pi} \phi(X_2) e^{-im_2 X_2} dX_2$$

use

$$X_n = k_n x + \phi_n, \quad k_n = \frac{2\pi n}{L} = n\Delta kx, \quad \Delta k = \frac{2\pi}{L},$$

so that

$$A_{m_1,m_2} = \frac{2\pi}{L} \frac{4\pi}{L} \frac{1}{(2\pi)^2} e^{-i(m_1\phi_1 + m_2\phi_2)} \int_0^{2\pi} \phi(k_1 x) e^{-im_1 k_1 x} dx \int_0^{2\pi} \phi(k_2 x) e^{-im_2 k_2 x} dx,$$

which can also be written

$$A_{m_1,m_2} = \frac{1}{L^2} \int_0^L \phi(x) \exp\left\{ -im_1 \left(\frac{2\pi}{L} x + \phi_1 \right) \right\} dx \int_0^L \phi(2x)$$

$$\exp\left\{ -im_2 \left(\frac{2\pi}{L} (2x) + \phi_2 \right) \right\} d(2x).$$

Finally, we have

$$A_{m_1,m_2} = A_{m_1} A_{m_2} \tag{7.27}$$

where

$$A_{m_1} = \frac{1}{L} \int_0^L \phi(x) \exp\left\{ -im_1 \left(\frac{2\pi}{L} x + \phi_1 \right) \right\} dx,$$

$$A_{m_2} = \frac{1}{L} \int_0^L \phi(2x) \exp\left\{ -im_2 \left(\frac{2\pi}{L}(2x) + \phi_2 \right) \right\} d(2x).$$

Let us now consider the case for *three degrees of freedom*, $N = 3$. We have

$$\mathbf{m} = [m_1, m_2, m_3].$$

Thus, we could entertain the possibility of getting the generalized Fourier coefficients, $A_{m_1, m_2, \ldots, m_N}$, for some space series. We have for three degrees of freedom

$$\phi(X_1, X_2, X_3) = \phi(x)$$
$$= \sum_{m_1=-\infty}^{\infty} \sum_{m_2=-\infty}^{\infty} \sum_{m_3=-\infty}^{\infty} A_{m_1, m_2, m_3} \exp\left\{ \frac{2\pi i x}{L} \sum_{n=1}^{N} n m_n + \phi_n \right\}$$

$$(7.28)$$

and

$$A_{m_1, m_2, m_3} = \int_0^L \int_0^L \int_0^L \phi(X_1, X_2, X_3) \exp\left\{ -2\pi i \sum_{n=1}^{N} n m_n \right\} dX_1 dX_2 dX_3 \quad (7.29)$$

If we uncouple the degrees of freedom, we have

$$A_{m_1, m_2, m_3} = A_{m_1} A_{m_2} A_{m_3}.$$

Nota Bene. While we have uncoupled the simple cases for two and three degrees of freedom we must in reality face the eventual truth: Coupling among the degrees of freedom is generic behavior for nonlinear systems. We can uncouple multidimensional Fourier series only for the uninteresting problem that occurs for linear systems.

7.8 General Rules for Dealing with Dynamical Multidimensional Fourier Series

Here are some helpful rules for using multidimensional Fourier series in a physical context:

- If we set $A_{m_1, m_2, m_3, \ldots, m_N} = A_{m_1} A_{m_2} A_{m_3} \cdots A_{m_N}$, then the MFT separates into N independent *modes* or *degrees of freedom*. They can be viewed as kinds of *Stokes waves* or as *bound modes* but they do not interact with one another.

- If we set $A_{m_1,m_2,m_3,...,m_N} \neq 0$ for $m_1 = m_2 = m_3 = \cdots = m_N = n$ and $A_{m_1,m_2,m_3,...,m_N} = 0$ otherwise, then the GFT becomes an *ordinary Fourier series*. In this case, each mode becomes an ordinary Fourier amplitude, A_n.
- When $A_{m_1,m_2,m_3,...,m_N} \neq A_{m_1} A_{m_2} A_{m_3} \cdots A_{m_N}$, the *modes* are *nonlinearly coupled* and *undergo nonlinear interactions* with each other. Each of the Stokes modes may be viewed as *free* because they move relative to one another in the wave field while interacting nonlinearly with one another.
- When we have the particular coefficients $A_{m_1,m_2,...,m_N} = \exp\left(\frac{1}{2}\sum_{j=1}^{N}\sum_{k=1}^{N} B_{jk} m_j m_k\right)$, the MFT reduces to a *theta function* (Chapter 8) and we can solve *integrable wave equations* such as KdV, KP, and NLS.
- Generally speaking, the *GFT is much more general* than the *Riemann theta function* and hence provides for additional possibilities for the study of nonlinear problems. One can think of the coefficients $A_{m_1,m_2,...,m_N}$ as the *phase space* for the *dynamical system* under study. This phase space lies on a *hypercube* of dimension N.

7.9   Reductions of Multidimensional Fourier Series

It is worth considering possible extensions of theta functions to "higher order." Suppose we address the N-dimensional Fourier series (Equation (7.4)) as a likely starting point. To this end, consider a particular form for the coefficients $C_{m_1,m_2,...,m_N}$ as the exponential of a Taylor series of a function of the vector **m**. First, let us consider how such a function $U(q_1, q_2, \ldots, q_N)$ could be expanded in some generalized variable q_j, $j = 1, N$, so that

$$C_{q_1,q_2,...,q_N} = e^{U(q_1,q_2,...,q_N)}.$$

We write the Taylor series of the "potential" in the exponential as

$$
\begin{aligned}
U(q_1, q_2, \ldots, q_N) = {} & f_0 + \sum_{k=1}^{N} q_k \frac{\partial U}{\partial q_k}\bigg|_0 + \frac{1}{2}\sum_{j=1}^{N}\sum_{k=1}^{N} q_j q_k \frac{\partial^2 U}{\partial q_j \partial q_k}\bigg|_0 \\
& + \frac{i}{3!}\sum_{i=1}^{N}\sum_{j=1}^{N}\sum_{k=1}^{N} q_i q_j q_k \frac{\partial^3 U}{\partial q_i \partial q_j \partial q_k}\bigg|_0 \\
& + \frac{1}{4!}\sum_{i=1}^{N}\sum_{j=1}^{N}\sum_{k=1}^{N}\sum_{l=1}^{N} q_i q_j q_k q_l \frac{\partial^4 U}{\partial q_i \partial q_j \partial q_k \partial q_l}\bigg|_0 + \cdots.
\end{aligned}
\tag{7.30}
$$

We are all familiar with normal mode expansions in the theory of *harmonic oscillations*. We therefore recognize the first term as just a constant (which we take to be zero), the second term is an ordinary displacement in the mean equilibrium position of the oscillations, the third term is the potential of a linear oscillation, the forth term is the first correction to the oscillation potential which includes nonlinear effects (cubic correction), and the fifth term is another correction for nonlinearity (quartic correction).

We can replace the derivative terms by simplifying the notation:

$$U(q_1, q_2, \ldots, q_N) = \sum_{i=1}^{N} q_i V_i + \frac{1}{2} \sum_{i=1}^{N} \sum_{j=1}^{N} q_i q_j D_{ij}$$

$$+ \frac{i}{3!} \sum_{i=1}^{N} \sum_{j=1}^{N} \sum_{k=1}^{N} q_i q_j q_k S_{ijk} + \frac{1}{3!} \sum_{i=1}^{N} \sum_{j=1}^{N} \sum_{k=1}^{N} \sum_{l=1}^{N} q_i q_j q_k q_l R_{ijkl} + \cdots,$$

(7.31)

where

$$V_i \equiv \frac{\partial U}{\partial q_i} \Big|_0, \qquad\qquad D_{ij} \equiv \frac{\partial^2 U}{\partial q_i \partial q_j} \Big|_0 \equiv -B_{ij},$$

$$S_{ijk} \equiv \frac{\partial^3 U}{\partial q_i \partial q_j \partial q_k} \Big|_0 \equiv -T_{ijk}, \quad R_{ijkl} \equiv \frac{\partial^4 U}{\partial q_i \partial q_j \partial q_k \partial q_l} \Big|_0 \equiv -K_{ijkl}.$$

(7.32)

Here, V_i is a vector, B_{ij} is a (negative definite) matrix, T_{ijk} is a (negative definite) tensor of third rank, and K_{ijkl} is a (negative definite) tensor of forth rank. Note that the matrix B_{ij} is symmetric ($B_{ij} = B_{ji}$), since the order of differentiation is immaterial if U has continuous second derivatives. The tensor T_{ijk} is also symmetric (in all three indices, $T_{ijk} = T_{jik} = T_{kij} = \cdots$) again because the order of differentiation is immaterial. Clearly, the tensor K_{ijkl} is also symmetric for the same reason. For most applications, it is natural to consider a "potential" about the equilibrium position (a minimum), so that V_i can be taken to be zero.

Since we are dealing with problems on the lattice, and since we generally deal with periodic and quasiperiodic boundary conditions, we can replace the components q_j, $j - 1$, N with the integer lattice vector $\mathbf{m} = [m_1, m_2, \ldots, m_N]$. This gives the expression:

$$U(m_1, m_2, \ldots, m_N) = -\frac{1}{2} \sum_{i=1}^{N} \sum_{j=1}^{N} m_i m_j B_{ij}$$

$$- \frac{1}{3!} \sum_{i=1}^{N} \sum_{j=1}^{N} \sum_{k=1}^{N} m_i m_j m_k S_{ijk}$$

(7.33)

$$- \frac{1}{3!} \sum_{i=1}^{N} \sum_{j=1}^{N} \sum_{k=1}^{N} \sum_{l=1}^{N} m_i m_j m_k m_l K_{ijkl} + \cdots.$$

Let us now write the exponential of this expression:

$$q_{m_1, m_2, \ldots, m_N} = \exp\left\{ -\frac{1}{2} \sum_{i=1}^{N} \sum_{j=1}^{N} m_i m_j B_{ij} - \frac{1}{3!} \sum_{i=1}^{N} \sum_{j=1}^{N} \sum_{k=1}^{N} m_i m_j m_k S_{ijk} \right.$$

$$\left. -\frac{1}{4!} \sum_{i=1}^{N} \sum_{j=1}^{N} \sum_{k=1}^{N} \sum_{l=1}^{N} m_i m_j m_k m_l K_{ijkl} + \cdots \right\}.$$

We now can construct the following Fourier series (see Equation (7.4)):

$$\Theta(X_1, X_2, \ldots, X_N) = \sum_{m_1=-\infty}^{\infty} \sum_{m_2=-\infty}^{\infty} \cdots \sum_{m_N=-\infty}^{\infty} q_{m_1, m_2, \ldots, m_N} \exp\left\{ i \sum_{n=1}^{N} m_n X_n \right\},$$

for which we have a kind of *generalized theta function*:

$$\Theta(X_1, X_2, \ldots, X_N)$$
$$= \sum_{m_1=-\infty}^{\infty} \sum_{m_2=-\infty}^{\infty} \cdots \sum_{m_N=-\infty}^{\infty} \exp\left\{ -\frac{1}{2} \sum_{i=1}^{N} \sum_{j=1}^{N} m_i m_j B_{ij} - \frac{1}{3!} \sum_{i=1}^{N} \sum_{j=1}^{N} \sum_{k=1}^{N} m_i m_j m_k S_{ijk} \right.$$
$$\left. -\frac{1}{4!} \sum_{i=1}^{N} \sum_{j=1}^{N} \sum_{k=1}^{N} \sum_{l=1}^{N} m_i m_j m_k m_l K_{ijkl} \right\} \exp\left\{ i \sum_{n=1}^{N} m_n X_n \right\} \qquad (7.34)$$

It is natural to refer to the term

$$\exp\left\{ -\frac{1}{2} \sum_{i=1}^{N} \sum_{j=1}^{N} m_i m_j B_{ij} \right\}$$

as a "Gaussian on the lattice" characterized by the *Riemann matrix* B_{ij}. Likewise

$$\exp\left\{ -\frac{1}{3!} \sum_{i=1}^{N} \sum_{j=1}^{N} \sum_{k=1}^{N} m_i m_j m_k S_{ijk} \right\}$$

suggests that S_{ijk} be called a *skewness tensor*, whereas the expression

$$\exp\left\{ -\frac{1}{4!} \sum_{i=1}^{N} \sum_{j=1}^{N} \sum_{k=1}^{N} \sum_{l=1}^{N} m_i m_j m_k m_l K_{ijkl} \right\}$$

suggests that K_{ijkl} is a "kurtosis tensor." Equation (7.34) reduces to the ordinary theta function when the skewness tensor $S_{ijk} = 0$ and the kurtosis tensor $K_{ijkl} = 0$. Otherwise we have a generalization of the theta function, which may be a legitimate candidate for working with higher-order nonlinearities in wave dynamics.

In the *one-dimensional extended theta function* Equation (7.34) has the form

$$\Theta(X) = \sum_{m=-\infty}^{\infty} \exp\left\{ -\frac{1}{2} m^2 B - \frac{1}{6} m^3 S - \frac{1}{24} m^4 K \right\} \exp\{imX\} \qquad (7.35)$$

The coefficient is after all the exponential of a polynomial in the integer m. This expression can be written:

$$\Theta(X|q,p,r) = \sum_{m=-\infty}^{\infty} q^{m^2} p^{m^3} r^{m^4} e^{imX},$$

where $q = e^{-(1/2)B}$, $p = e^{-(1/6)S}$, and $r = e^{-(1/24)K}$ with $0 \leq q < 1, 0 \leq p < 1$, and $0 \leq r < 1$, respectively.

If the tensors S_{ijk} and K_{ijkl} are small, $S_{ijk} \ll 1$ and $K_{ijk} \ll 1$, we can write

$$\Theta(X_1, X_2, \ldots, X_N) = \sum_{m_1=-\infty}^{\infty} \sum_{m_2=-\infty}^{\infty} \cdots \sum_{m_N=-\infty}^{\infty} \left(1 - \frac{1}{3!} \sum_{i=1}^{N} \sum_{j=1}^{N} \sum_{k=1}^{N} m_i m_j m_k S_{ijk} \right.$$

$$\left. - \frac{1}{4!} \sum_{i=1}^{N} \sum_{j=1}^{N} \sum_{k=1}^{N} \sum_{l=1}^{N} m_i m_j m_k m_l K_{ijkl} \right) \exp\left\{ -\frac{1}{2} \sum_{i=1}^{N} \sum_{j=1}^{N} m_i m_j B_{ij} \right\} \exp\left\{ i \sum_{n=1}^{N} m_n X_n \right\}.$$

$$(7.36)$$

This may be interpreted as a theta function (essentially a Gaussian on the lattice defined by the vector **m**) that has been extended to include skewness (breaking of left-right asymmetry) and kurtosis (symmetric variations in the tails of the Gaussian). Future study may determine whether extended theta functions can play a significant role in pure and applied mathematics, in theoretical physics and in data analysis applications.

7.10   Theta Functions Solve a Diffusion Equation

Theta functions

$$\Theta(X_1, X_2, \ldots, X_N | \mathbf{B}) = \sum_{m_1=-\infty}^{\infty} \sum_{m_2=-\infty}^{\infty} \cdots \sum_{m_N=-\infty}^{\infty}$$

$$\exp\left\{ -\frac{1}{2} \sum_{i=1}^{N} \sum_{j=1}^{N} m_i m_j B_{ij} \right\} \exp\left\{ i \sum_{i=1}^{N} m_i X_i \right\}$$

have the property that one can take derivatives with respect to both the dimensions, X_m, and the period matrix elements, B_{mn}. For example,

$$\frac{\partial \theta}{\partial X_n} = i \sum_{m_1=-\infty}^{\infty} \sum_{m_2=-\infty}^{\infty} \cdots \sum_{m_N=-\infty}^{\infty} m_n \exp\left\{ -\frac{1}{2} \sum_{i=1}^{N} \sum_{j=1}^{N} m_i m_j B_{ij} \right\} \exp\left\{ i \sum_{i=1}^{N} m_i X_i \right\},$$

$$\frac{\partial^2 \theta}{\partial X_m X_n} = -\sum_{m_1=-\infty}^{\infty} \sum_{m_2=-\infty}^{\infty} \cdots \sum_{m_N=-\infty}^{\infty} m_m m_n \exp\left\{-\frac{1}{2}\sum_{i=1}^{N}\sum_{j=1}^{N} m_i m_j B_{ij}\right\} \exp\left\{i\sum_{i=1}^{N} m_i X_i\right\}.$$

Likewise

$$\frac{\partial \theta}{\partial B_{mn}} = -\frac{1}{2}\sum_{m_1=-\infty}^{\infty} \sum_{m_2=-\infty}^{\infty} \cdots \sum_{m_N=-\infty}^{\infty} m_m m_n \exp\left\{-\frac{1}{2}\sum_{i=1}^{N}\sum_{j=1}^{N} m_i m_j B_{ij}\right\} \exp\left\{i\sum_{i=1}^{N} m_i X_i\right\}.$$

And therefore

$$\frac{\partial^2 \theta}{\partial X_m X_n} = 2\frac{\partial \theta}{\partial B_{mn}} \tag{7.37}$$

which is a kind of diffusion equation. Using the explicit form for the dimensions, $X_m = k_m x - \phi_m$, we have

$$\frac{\partial^2 \theta}{\partial x^2} = 2k_m k_n \frac{\partial \theta}{\partial B_{mn}} \tag{7.38}$$

We will return to these diffusion equations at a later stage in the development. It is also clear that

$$\Theta(X_1, X_2, \ldots, X_N, T_1, T_2, \ldots, T_N | \mathbf{B})$$
$$= \sum_{m_1=-\infty}^{\infty} \sum_{m_2=-\infty}^{\infty} \cdots \sum_{m_N=-\infty}^{\infty} \exp\left\{-\frac{1}{2}\sum_{i=1}^{N}\sum_{j=1}^{N} m_i m_j B_{ij}\right\} \exp\left\{i\sum_{i=1}^{N} m_i X_i - i\sum_{i=1}^{N} m_i T_i\right\},$$

where $X_m = k_m x + \phi_m$ and $T_m = \omega_m x$. Then

$$\frac{\partial \theta}{\partial X_n} = i\sum_{m_1=-\infty}^{\infty} \sum_{m_2=-\infty}^{\infty} \cdots \sum_{m_N=-\infty}^{\infty} m_n$$
$$\exp\left\{-\frac{1}{2}\sum_{i=1}^{N}\sum_{j=1}^{N} m_i m_j B_{ij}\right\} \exp\left\{i\sum_{i=1}^{N} m_i X_i - i\sum_{i=1}^{N} m_i T_i\right\},$$

$$\frac{\partial \theta}{\partial T_n} = -i\sum_{m_1=-\infty}^{\infty} \sum_{m_2=-\infty}^{\infty} \cdots \sum_{m_N=-\infty}^{\infty} m_n$$
$$\exp\left\{-\frac{1}{2}\sum_{i=1}^{N}\sum_{j=1}^{N} m_i m_j B_{ij}\right\} \exp\left\{i\sum_{i=1}^{N} m_i X_i - i\sum_{i=1}^{N} m_i T_i\right\}.$$

So that

$$\frac{\partial \theta}{\partial X_n} + \frac{\partial \theta}{\partial T_n} = 0$$

and finally

$$\frac{\partial \theta}{\partial t} + c_n \frac{\partial \theta}{\partial x} = 0 \quad \text{for} \quad c_n = \frac{\omega_n}{k_n}. \tag{7.39}$$

Note that the wavenumber, k_n, and frequency, ω_n, are constant parameters in the solution of this simple case.

7.11 Multidimensional Fourier Series Solve Linear Wave Equations

Multidimensional Fourier series can be used to solve linear wave equations, written in the form

$$\Theta(X_1, X_2, \ldots, X_N) = \sum_{m_1=-\infty}^{\infty} \sum_{m_2=-\infty}^{\infty} \cdots \sum_{m_N=-\infty}^{\infty} q_{m_1, m_2, \ldots, m_N} \exp\left\{ i \sum_{n=1}^{N} m_n X_n \right\}. \tag{7.40}$$

Then it is natural to solve *linear wave equations* of the form

$$\eta_t + \alpha_1 \eta_x + \alpha_2 \eta_{xx} + \alpha_3 \eta_{xxx} + \alpha_4 \eta_{xxxx} + \cdots = 0, \tag{7.41}$$

where $\alpha_n, n = 1, 2, \cdots$ are constants and the dimensions have the dynamical form:

$$X_n = k_n x - \omega_n t + \phi_n.$$

For linear equations of this type, one normally substitutes an ordinary linear Fourier series of the form

$$\eta(x, t) = \sum_{n=-\infty}^{\infty} \eta_n e^{ik_n x - i\omega_n t},$$

which for Equation (7.41) one gets the dispersion relation:

$$\omega = \alpha_1 k + i\alpha_2 k^2 - \alpha_3 k^3 - i\alpha_4 k^4 + \cdots \tag{7.42}$$

Thus, a Fourier series solves Equation (7.41) with this dispersion relation for all Cauchy problems, that is, given $\eta(x, 0)$ we can compute $\eta(x, t)$ for all t.

Now, let us insert the multidimensional Fourier series into the above wave equation and upon eliminating the summations over the integer indices, we have

$$-i\sum_{n=1}^{N} m_n\omega_n + i\alpha_1 \sum_{n=1}^{N} m_n k_n - \alpha_2 \left(\sum_{n=1}^{N} m_n k_n\right)^2 - i\alpha_3 \left(\sum_{n=1}^{N} m_n k_n\right)^3 + \alpha_4 \left(\sum_{n=1}^{N} m_n k_n\right)^4 = 0.$$

Making the square, cubic, and quartic terms explicit, we have

$$i\sum_{n=1}^{N} m_n\omega_n = i\alpha_1 \sum_{n=1}^{N} m_n k_n - \alpha_2 \sum_{i=1}^{N}\sum_{j=1}^{N} m_i m_j B_{ij} - i\alpha_3 \sum_{i=1}^{N}\sum_{j=1}^{N}\sum_{k=1}^{N} m_i m_j m_k S_{ijk}$$
$$+ \alpha_4 \sum_{i=1}^{N}\sum_{j=1}^{N}\sum_{k=1}^{N}\sum_{l=1}^{N} m_i m_j m_k m_l K_{ijkl}$$

$$(7.43)$$

where a *matrix* B_{ij}, *skewness tensor* S_{ijk}, and *kurtosis tensor* K_{ijkl} are given by

$$B_{ij} = k_i k_j, \quad S_{ijk} = k_i k_j k_k, \quad K_{ijkl} = k_i k_j k_k k_l \tag{7.44}$$

for

$$\mathbf{k} = [k_1, k_2, \ldots, k_N] = [1, 2, \ldots, N]\Delta k = 2\pi[1, 2, \ldots, N]/L.$$

Finally, we have the solution

$$\Theta(x,t) = \sum_{m_1=-\infty}^{\infty}\sum_{m_2=-\infty}^{\infty}\cdots\sum_{m_N=-\infty}^{\infty} q_{m_1, m_2, \ldots, m_N}$$
$$\exp\left\{i\sum_{n=1}^{N} m_n k_n x - i\sum_{n=1}^{N} m_n \omega_n t + i\sum_{n=1}^{N} m_n \phi_n\right\}.$$

Using the dispersion relation (Equation (7.42)) given above, we have

$$\Theta(x,t) = \sum_{m_1=-\infty}^{\infty}\sum_{m_2=-\infty}^{\infty}\cdots\sum_{m_N=-\infty}^{\infty} q_{m_1, m_2, \ldots, m_N}$$
$$\exp\left\{i\sum_{n=1}^{N} m_n k_n x - i\alpha_1 \sum_{n=1}^{N} m_n k_n t + \alpha_2 \sum_{i=1}^{N}\sum_{j=1}^{N} m_i m_j B_{ij} t\right.$$
$$\left. + i\alpha_3 \sum_{i=1}^{N}\sum_{j=1}^{N}\sum_{k=1}^{N} m_i m_j m_k S_{ijk} t - i\alpha_4 \sum_{i=1}^{N}\sum_{j=1}^{N}\sum_{k=1}^{N}\sum_{l=1}^{N} m_i m_j m_k m_k K_{ijkl} t\right\}$$

$$(7.45)$$

The solution of a simple linear equation with differential polynomial terms leads to a solution in terms of multidimensional Fourier series with *matrix* ($B_{ij}t$), *skewness* ($S_{ijk}t$), and *kurtosis tensors* ($K_{ijkl}t$). Here, the coefficients, $q_{m_1, m_2, ..., m_N}$, are determined from the Cauchy problem (see Bellman, 1961 on heat equation).

It is natural to assume that dissipation occurs in the even derivative terms and dispersion in the odd derivative terms. Perhaps the simplest example, illustrating *dissipation*, is to take $\alpha_3 = \alpha_4 = 0$ and $\alpha_2 = \nu$ (real), so that the resulting solution is

$$\eta(x,t) = \sum_{m_1=-\infty}^{\infty} \sum_{m_2=-\infty}^{\infty} \cdots \sum_{m_N=-\infty}^{\infty} q_{m_1, m_2, ..., m_N}$$

$$\exp\left\{-\nu \sum_{i=1}^{N} \sum_{j=1}^{N} m_i m_j B_{ij} t\right\} \exp\left\{i \sum_{n=1}^{N} m_n k_n x - i\alpha_1 \sum_{n=1}^{N} m_n k_n t\right\}.$$

Therefore, the Riemann matrix plays the role of dissipation because the term $-\nu B_{ij}$ is assumed to be real and negative definite. In this problem, it is clear that the coefficients are necessary to solve the Cauchy problem.

Generally speaking, we must have expressions of the form (by inserting an ordinary Fourier series into the above wave equation)

$$k_m^2 = \sum_{ij} B_{ij} m_i m_j,$$

$$k_m^3 = \sum_{ijk} S_{ijk} m_i m_j m_k,$$

$$k_m^4 = \sum_{ijkl} K_{ijk} m_i m_j m_k m_l.$$

These expressions are found by recognizing when

$$\sum_{j=1}^{N} m_j k_j = \frac{2\pi}{L} \sum_{j=1}^{N} j m_j = \frac{2\pi}{L} I_m$$

for $I_m = n$, that is, is an integer multiple of the wavenumber. It is not hard to show that the above multidimensional Fourier series, summed appropriately, gives back the ordinary Fourier series as a result. We are left with the question: Why use a multidimensional Fourier series Equation (7.1) to solve a linear equation? The answer is that one should not. Only for *nonlinear* wave equations are multidimensional Fourier series useful.

7.12   Details for Two Degrees of Freedom

Here is the two-dimensional Fourier transform

$$\phi(X_1, X_2) = \sum_{m_1=-\infty}^{\infty} \sum_{m_2=-\infty}^{\infty} A_{m_1, m_2} \exp\left\{ i \sum_{n=1}^{2} m_n X_n \right\} \tag{7.46}$$

$$A_{m_1, m_2} = \frac{1}{(2\pi)^2} \int_0^{2\pi} \int_0^{2\pi} \phi(X_1, X_2) e^{-i[m_1 X_1 + m_2 X_2]} dX_1 dX_2 \tag{7.47}$$

Let us now consider the simple case (assumed to be rapidly convergent) where one sums from -1 to 1 over the summation indices:

$$\phi(X_1, X_2) = \sum_{m_1=-1}^{1} \sum_{m_2=-1}^{1} A_{m_1, m_2} \exp\left\{ i \sum_{n=1}^{2} m_n X_n \right\}. \tag{7.48}$$

This summation has specific terms, denoted below, where I relate theta function parameters (second column) to generalized Fourier parameters (third column). The summation index pair (m_1, m_2) is given in the first column:

$$\begin{array}{lll}
(-1, -1) & qpre^{-iX_1-iX_2} & A_{-1,-1}e^{-iX_1-iX_2} \\
(-1, 0) & qe^{-iX_1} & A_{-1,0}e^{-iX_1} \\
(-1, 1) & qpr^{-1}e^{-iX_1+iX_2} & A_{-1,1}e^{-iX_1+iX_2} \\
(0, -1) & pe^{-iX_2} & A_{0,-1}e^{-iX_2} \\
(0, 0) & 1 & A_{0,0} \; (0, +1)\, pe^{iX_2}\, A_{0,1}e^{iX_2} \\
(1, -1) & qpr^{-1}e^{iX_1-iX_2} & A_{1,-1}e^{iX_1-iX_2} \\
(1, 0) & qe^{iX_1} & A_{1,0}e^{iX_1} \\
(1, 1) & qpre^{iX_1+iX_2} & A_{1,1}e^{iX_1+iX_2}
\end{array} \tag{7.49}$$

So that we get

$$\phi(X_1, X_2) = A_{0,0} + 2A_{1,0} \cos X_1 + 2A_{0,1} \cos X_2 + 2A_{1,1} \cos(X_1 + X_2)$$
$$+ 2A_{1,-1} \cos(X_1 - X_2) \tag{7.50}$$

and where

$$A_{0,0} = 1, \; A_{1,0} = q, \; A_{0,1} = p, \; A_{1,-1} = qpr^{-1}, \; A_{1,1} = qpr,$$

so that

$$\phi(X_1, X_2) = 1 \quad + 2q\cos X_1 + 2p\cos X_2 + 2qpr^{-1}\cos(X_1 - X_2) + 2qpr\cos(X_1 + X_2)$$

$$A_{0,0} \qquad A_{1,0} \qquad\quad A_{0,1} \qquad\qquad A_{1,-1} \qquad\qquad\qquad A_{1,1}$$

$$A_{-1,0} \qquad A_{0,-1} \qquad\quad A_{-1,1} \qquad\qquad\qquad A_{-1,-1}$$

$$(7.51)$$

where the parameters p, q, and r are related to the period matrix of the theta functions by

$$q = e^{-(1/2)B_{11}}, \quad p = e^{-(1/2)B_{22}}, \quad r = e^{-B_{12}}.$$

Notice that

$$A_{m_1, m_2} = A_{-m_1, -m_2} \qquad\qquad\qquad\qquad\qquad\qquad\qquad (7.52)$$

Now insert

$$\phi(X_1, X_2) = 1 + 2q\cos X_1 + 2p\cos X_2 + 2qpr^{-1}\cos(X_1 - X_2)$$
$$+ 2qpr\cos(X_1 + X_2)$$

into the inverse problem

$$A_{m_1, m_2} = \frac{1}{(2\pi)^2} \int_0^{2\pi} \int_0^{2\pi} \phi(X_1, X_2) e^{-i[m_1 X_1 + m_2 X_2]} dX_1 dX_2$$

and get

$$A_{m_1, m_2} = \frac{1}{(2\pi)^2} \int_0^{2\pi} \int_0^{2\pi} [1 + 2q\cos X_1 + 2p\cos X_2] e^{-i[m_1 X_1 + m_2 X_2]} dX_1 dX_2$$

$$+ \frac{1}{(2\pi)^2} \int_0^{2\pi} \int_0^{2\pi} [2qpr^{-1}\cos(X_1 - X_2)$$

$$+ 2qpr\cos(X_1 + X_2)] e^{-i[m_1 X_1 + m_2 X_2]} dX_1 dX_2 \qquad\qquad (7.53)$$

$$A_{m_1,m_2} = \frac{1}{(2\pi)^2}\int_0^{2\pi}\int_0^{2\pi}[1 + 2q\cos X_1 + 2p\cos X_2]e^{-i[m_1 X_1 + m_2 X_2]}dX_1 dX_2$$

$$+ \frac{1}{(2\pi)^2}\int_0^{2\pi}\int_0^{2\pi}[2qpr^{-1}\cos(X_1 - X_2)$$

$$+ 2qpr\cos(X_1 + X_2)]e^{-i[m_1 X_1 + m_2 X_2]}dX_1 dX_2,$$

$$A_{m_1,m_2} = \frac{1}{(2\pi)^2}\int_0^{2\pi}\int_0^{2\pi}e^{-i[m_1 X_1 + m_2 X_2]}dX_1 dX_2$$

$$+ \frac{2q}{(2\pi)^2}\int_0^{2\pi}\int_0^{2\pi}\cos X_1 e^{-i[m_1 X_1 + m_2 X_2]}dX_1 dX_2$$

$$+ \frac{2p}{(2\pi)^2}\int_0^{2\pi}\int_0^{2\pi}\cos X_2 e^{-i[m_1 X_1 + m_2 X_2]}dX_1 dX_2 \qquad (7.54)$$

$$+ \frac{2qpr^{-1}}{(2\pi)^2}\int_0^{2\pi}\int_0^{2\pi}\cos(X_1 - X_2) e^{-i[m_1 X_1 + m_2 X_2]}dX_1 dX_2$$

$$+ \frac{2qpr}{(2\pi)^2}\int_0^{2\pi}\int_0^{2\pi}\cos(X_1 + X_2) e^{-i[m_1 X_1 + m_2 X_2]}dX_1 dX_2$$

Now introduce the following useful notation:

Relating Trigonometric Identities and MFA Coefficients

Define the following function:

$$B_{m_1,m_2/n_1,n_2} = \frac{1}{(2\pi)^2}\int_0^{2\pi}\int_0^{2\pi}\cos(n_1 X_1 + n_2 X_2) e^{-i[m_1 X_1 + m_2 X_2]}dX_1 dX_2,$$

$$(7.55)$$

for which

$$B_{m_1,m_2/n_1,n_2} = \begin{cases} 1, & \text{for } m_1 = n_1 \text{ and } m_2 = n_2, \\ 0, & \text{for } m_1 \neq n_1 \text{ and } m_2 \neq n_2, \end{cases}$$

and get

$$A_{m_1,m_2} = B_{m_1,m_2/0,0} + 2qB_{m_1,m_2/1,0} + 2pB_{m_1,m_2/0,1}$$
$$+ 2pqr^{-1}B_{m_1,m_2/1,-1} + 2qprB_{m_1,m_2/1,1}. \qquad (7.56)$$

Therefore,

$$A_{0,0} = B_{0,0/0,0} = 1,$$

$$A_{1,0} = 2qB_{1,0/1,0} = 2q,$$

$$A_{0,1} = 2pB_{0,1/0,1} = 2p,$$

$$A_{1,-1} = 2qpr^{-1}B_{1,-1/1,-1} = 2qpr^{-1},$$

$$A_{1,-1} = 2qpr^{-1}B_{1,-1/1,-1} = 2qpr^{-1},$$

On the basis of this result, we can easily compute analytic expressions for all of the coefficients.

7.13 Converting Multidimensional Fourier Series to Ordinary Fourier Series

The *multidimensional Fourier series* is given by

$$\Phi(X_1, X_2, \ldots, X_N) = \sum_{m_1=-\infty}^{\infty} \sum_{m_2=-\infty}^{\infty} \cdots \sum_{m_N=-\infty}^{\infty} A_{m_1, m_2, \ldots, m_N} \exp\left\{ i \sum_{j=1}^{N} m_j X_j \right\} \tag{7.57}$$

where it is assumed

$$X_j = k_j x - \omega_j t + \phi_j$$

can be written in the following form:

$$\Phi(x, t) = \sum_{j=-\infty}^{\infty} C_j e^{i(K_j x - \Omega_j t + \phi_j)}, \quad C_j = A_{m_1^j m_2^j \cdots m_N^j}, \tag{7.58}$$

where j is an ordering parameter associated with each vector $\mathbf{m}_j = [m_1^j, m_2^j, \ldots, m_N^j]$ in the summation. One can think of Equation (7.58) as the reduction of Equation (7.57) from a nested sum to a single sum. We have the following parameters:

$$K_j = \mathbf{m}_j \cdot \mathbf{k} = \sum_{n=1}^{N} m_n^j k_n,$$

$$\Omega_j = \mathbf{m}_j \cdot \boldsymbol{\omega} = \sum_{n=1}^{N} m_n^j \omega_n, \qquad\qquad (7.59)$$

$$\Phi_j = \mathbf{m}_j \cdot \boldsymbol{\phi} = \sum_{n=1}^{N} m_n^j \phi_n.$$

For a number of reasons (primarily for numerical and data analysis purposes) *the wavenumbers are often assumed to be commensurable* (and, therefore, to consider a simple example, the KdV equation and its IST are perfectly periodic in the spatial variable, $0 \leq x \leq L$, L the period of the wave train):

$$\mathbf{k} = [1, 2, \ldots, N]\Delta k,$$

where as usual $k_n = n\Delta k$ and $\Delta k = 2\pi/L$ is the wavenumber interval in the spectrum. Of course, due to the complex nature of the nonlinear interactions (see Equation 7.59) the frequencies are *never* commensurable; in the linear limit (for, say, the KdV equation) $\omega_j = c_0 k_j - \beta k_j^3$ and clearly commensurable frequencies cannot exist even there. In Equations (7.59), the multidimensional Fourier formulation given by Equation (7.58) resembles linear Fourier analysis. Chapter 9 gives additional details on this perspective.

How then do the wavenumbers K_j, behave for the class of θ-functions given by Equation (7.58)? Clearly their formulation is different than those for the linear Fourier transform, that is,

$$K_j = \mathbf{m}_j \cdot \mathbf{k} = [m_1^j, m_2^j, \ldots, m_N^j] \cdot [1, 2, \ldots, N]\Delta k = \Delta k \sum_{n=1}^{N} n m_n^j. \qquad (7.60)$$

Hence, the multidimensional Fourier series wavenumbers, K_j, are *integer multiples* I_j of Δk:

$$K_j = I_j \Delta k = \frac{2\pi}{L} I_j, \quad I_j = \sum_{n=1}^{N} n m_n^j \qquad\qquad (7.61)$$

and are therefore *commensurable, often duplicated, and not ordered with the integers*. According to Equation (7.61), the wavenumbers in Equation (7.57) fall on the ordinary Fourier wavenumbers $k_n = 2\pi n \Delta k$. In fact, an *infinite number of terms* in Equation (7.57) fall on each k_n. See Figure 7.1 for an example.

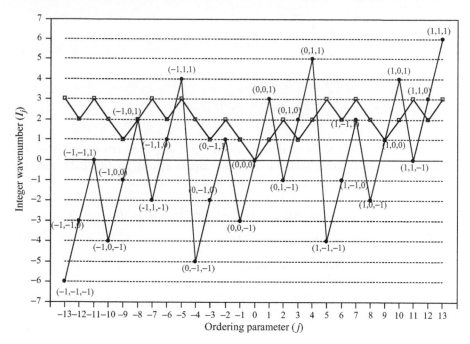

Figure 7.1   Theta function wavenumbers and partial sum for $M = 1$ and $N = 3$.

Since all the wavenumbers in the *multidimensional Fourier series* fall on the k_j and since the wavenumbers are duplicated, we can write from Equation (7.58) (for the special case $t = 0$):

$$\Phi(x, 0) = \sum_{j=-\infty}^{\infty} C_j e^{i(K_j x + \Phi_j)} = \sum_{n=-\infty}^{\infty} c_n e^{i k_n x} \qquad (7.62)$$

where

$$c_n = \sum_{\substack{\text{sum } j \text{ over subset} \\ \text{of } j \text{ for which } I_j = n}} C_j e^{i\Phi_j}, \ 1 \leq n \leq \infty \qquad (7.63)$$

The above theta series (Equation (7.62)) (for $t = 0$) is nothing more than an ordinary Fourier transform! The Fourier coefficients, c_n, in Equation (7.62) are given by the *series* (Equation (7.63)).

Now, let us investigate the ordinary Fourier representation for the theta function. We begin with the right-hand side of Equation (7.62):

$$\Phi(x, 0) = \sum_{n=-\infty}^{\infty} c_n e^{ik_n x}$$

$$= a_0 + 2\sum_{n=1}^{\infty} d_n \cos(k_n x + \phi_n) \tag{7.64}$$

$$= a_0 + 2\sum_{n=1}^{\infty} a_n \cos(k_n x) + b_n \sin(k_n x),$$

where we have used the facts that $k_{-n} = k_n$ and $c_{-n} = c_n^*$ (to ensure that the multidimensional Fourier series is real for applications to the KdV and KP equations), so that $d_n = d_{-n}, \phi_{-n} = -\phi_n$, and $c_0 = d_0$ and

$$c_n = d_n e^{i\phi_n} = a_n - ib_n \text{ and } c_{-n} = c_n^* = d_n e^{-i\phi_n} = a_n + ib_n,$$

$$d_n = \sqrt{a_n^2 + b_n^2} \text{ and } \tan\phi_n = -b_n/a_n, \tag{7.65}$$

$$a_n = d_n \cos\phi_n \text{ and } b_n = -d_n \sin\phi_n.$$

The coefficients are given in terms of inverse scattering variables by

$$c_n = \sum_{\substack{\text{sum } j \text{ over subset} \\ \text{of } j \text{ for which } I_j = n}} C_j e^{i\Phi_j}, \ 1 \le n \le \infty, \tag{7.66}$$

so that

$$c_n = \sum_{\substack{\text{sum } j \text{ over subset} \\ \text{of } j \text{ for which } I_j = n}} C_j \cos(\Phi_j) + iC_j \sin(\Phi_j) = a_n - ib_n, \tag{7.67}$$

and

$$a_n = \sum_{\substack{\text{sum } j \text{ over subset} \\ \text{of } j \text{ for which } I_j = n}} C_j \cos(\Phi_j), \ 1 \le n \le \infty, \tag{7.68}$$

$$b_n = -\sum_{\substack{\text{sum } j \text{ over subset} \\ \text{of } j \text{ for which } I_j = n}} C_j \sin(\Phi_j), \ 1 \le n \le \infty, \tag{7.69}$$

where C_j and Θ_j are given in Equations (7.58) and (7.59), respectively. We also have

$$d_n = \sqrt{a_n^2 + b_n^2} \text{ and } \tan\phi_n = -b_n/a_n,$$

so that

$$
d_n = \sqrt{\left(\sum_{\substack{\text{sum } j \text{ over subset} \\ \text{of } j \text{ for which } I_j = n}} C_j \cos(\Phi_j) \right)^2 + \left(\sum_{\substack{\text{sum } j \text{ over subset} \\ \text{of } j \text{ for which } I_j = n}} C_j \sin(\Phi_j) \right)^2},
$$

$$(7.70)$$

$$
\tan \phi_n = \frac{\displaystyle\sum_{\substack{\text{sum } j \text{ over subset} \\ \text{of } j \text{ for which } I_j = n}} C_j \sin(\Phi_j)}{\displaystyle\sum_{\substack{\text{sum } j \text{ over subset} \\ \text{of } j \text{ for which } I_j = n}} C_j \cos(\Phi_j)}.
$$

$$(7.71)$$

We have therefore reduced the multidimensional Fourier series to an ordinary Fourier transform with coefficients given by c_n, a_n, b_n, d_n, and phase ϕ_n as given above in terms of inverse scattering transform variables.

8 Riemann Theta Functions

8.1 Introduction

An important ingredient in the field of *nonlinear Fourier analysis* is that of the *generalized Fourier series* (Chapter 7) and *its simplified reduction* known as *Riemann theta functions* (this chapter). In Chapter 2, I have discussed how the nonlinear dynamics of water waves are governed, to leading order in nonlinearity, by particular wave equations that are *integrable by the inverse scattering transform*. Examples are the KdV, KP, and NLS equations. Details of the IST solution of these equations are given in Chapters 10–12. Fundamental in the IST formulation, for periodic or quasiperiodic boundary conditions, is the role of *Riemann theta functions*, the subject of this chapter (Baker, 1897, 1907; Whittaker and Watson, 1902; Coble, 1929; Ford, 1929; Bateman and Erdelyi, 1955; Bellman, 1961; Abramowitz and Stegun, 1964; Siegel, 1969a, b,c; Igusa, 1972; Fay, 1973; Mumford, 1983, 1984, 1991, 2004; Zagrodzinski, J. A., 1983; Griffiths, 1989; Griffiths and Harris, 1994). I discuss a number of useful properties of these functions. These are the fundamental series first introduced in the later half of the nineteenth century by Riemann and are related to the so-called *Jacobian inverse problem*. At that time, this now-classical problem was an exercise in pure mathematics and influenced much of the modern-day understanding of many branches of mathematics and physics (Baker, 1897; Baker, 1907; Belokolos et al., 1994). A major focus in this chapter is the use of theta functions as a tool of the *nonlinear space/time series analysis of data and for nonlinear modeling*.

8.2 Riemann Theta Functions

It is often convenient to write the multidimensional Fourier series in the form

$$\theta(X_1, X_2, \ldots, X_N) = \sum_{m_1=-\infty}^{\infty} \sum_{m_2=-\infty}^{\infty} \cdots \sum_{m_N=-\infty}^{\infty} C_{m_1, m_2, \ldots, m_N} \exp\left\{ i \sum_{n=1}^{N} m_n X_n \right\} \quad (8.1)$$

and then the *inverse problem* (the inverse generalized Fourier transform) is given by

$$C_{m_1, m_2, \ldots, m_N} = \exp\left\{-\frac{1}{2}\sum_{m=1}^{N}\sum_{n=1}^{N} m_m m_n B_{mn}\right\}$$

$$= \frac{1}{(2\pi)^N} \int_0^{2\pi} \int_0^{2\pi} \cdots \int_0^{2\pi} \theta(X_1, X_2, \ldots, X_N) \qquad (8.2)$$

$$\times \exp\left\{-i\sum_{n=1}^{N} m_n X_n\right\} dX_1 dX_2 \cdots dX_N,$$

where we have introduced the latter relation as a simple reduction of multidimensional Fourier series to the *Riemann theta function* (Baker, 1897) that is then given by

$$\theta(X_1, X_2, \ldots, X_N) = \sum_{m_1=-\infty}^{\infty} \sum_{m_2=-\infty}^{\infty} \cdots \sum_{m_N=-\infty}^{\infty}$$

$$\times \exp\left\{-\frac{1}{2}\sum_{m=1}^{N}\sum_{n=1}^{N} m_m m_n B_{mn}\right\} \exp\left\{i\sum_{n=1}^{N} m_n X_n\right\}. \qquad (8.3)$$

Thus, formally speaking, the period matrix, B_{ij}, can be computed from the theta function, $\theta(X_1, X_2, \ldots, X_N)$, by Equation (8.2). The X_k are the *generalized dimensions* $(1 \leq k \leq N)$, so that the Riemann theta function (Equation (8.3)) may be thought of as a kind of *generalized Fourier analysis* in N dimensions, with its generality somewhat reduced from the N-dimensional Fourier series (Equation (8.1)). This is because the coefficients $C_{m_1, m_2, \ldots, m_N}$ in Equation (8.1) correspond to a *tensor of rank N* while the B_{jk} are components of a *matrix* in Equation (8.3). Thus, generalized Fourier analysis in terms of Riemann theta functions (Equation (8.3)) is less general than N-dimensional Fourier analysis, the latter of which is perhaps the most general of all Fourier techniques. Despite their limitations, Riemann theta functions are among the most important functions in pure and applied mathematics and form the basis for most of the generalized (nonlinear) Fourier numerical analysis procedures presented in this chapter. The B_{jk} are the elements of the *complex Riemann matrix* which is $N \times N$, symmetric and has positive definite real part to insure convergence of the series.

The theta function (Equation (8.3)) can also be written in the following vector form:

$$\theta(\mathbf{X}) = \sum_{m=-\infty}^{\infty} e^{-(1/2)\mathbf{m} \cdot \mathbf{Bm}} e^{i\mathbf{m} \cdot \mathbf{X}}, \qquad (8.4)$$

where the generalized dimensions are $\mathbf{X} = [X_1, X_2, \ldots, X_N]$ and the vector of integers is $\mathbf{m} = [m_1, m_2, \ldots, m_N]$. The Riemann theta function is unfamiliar to many investigators in the physical sciences and the purpose of this chapter is to provide a user's perspective of this function as a numerical tool primarily for the space/time series analysis of discrete data. Those interested in the mathematical details of the theta function as an exercise in algebraic geometry are referred to the literature (Novikov et al., 1984; Tracy, 1984; Belokolos et al., 1994; Deconinck, 1998; Polishchuk, 2003; Gesztesy et al., 2003).

8.3 Simple Properties of Theta Functions

The applications of theta functions as tools for solving integrable, nonlinear wave equations for the analysis of data and for hyperfast modeling are the major topics of this chapter. However, the formidable expression (Equation (8.3)) for these generalized Fourier series requires that we make some effort to understand a number of their properties. The focus is on properties that will help us to better understand how they may be used numerically. The results are based upon identities derived from the theta functions themselves.

8.3.1 Symmetry of the Riemann Matrix

The Riemann theta function requires a symmetric Riemann matrix, $\mathbf{B}$. The reasons are related to a fundamental problem in pure and applied mathematics (that of the Jacobian inverse problem; Baker, 1897), but we can easily see why symmetry of the Riemann matrix is natural. Suppose that $\mathbf{B}$ is *not* symmetric; it can then be written as the sum of *symmetric* $(\mathbf{s} = (\mathbf{B} + \mathbf{B}^{\mathrm{T}})/2)$ and *antisymmetric* parts $(\mathbf{a} = (\mathbf{B} - \mathbf{B}^{\mathrm{T}})/2)$. Consider a simple three-dimensional case, $N = 3$ (genus 3 in the terminology of algebraic geometry):

$$\mathbf{s} = \frac{1}{2}(\mathbf{B} + \mathbf{B}^{\mathrm{T}}) = \begin{bmatrix} B_{11} & \frac{1}{2}(B_{12} + B_{21}) & \frac{1}{2}(B_{13} + B_{31}) \\ \frac{1}{2}(B_{12} + B_{21}) & B_{22} & \frac{1}{2}(B_{32} + B_{23}) \\ \frac{1}{2}(B_{13} + B_{31}) & \frac{1}{2}(B_{32} + B_{23}) & B_{33} \end{bmatrix},$$

so that

$$\mathbf{s} = \begin{bmatrix} s_{11} & s_{12} & s_{13} \\ s_{12} & s_{22} & s_{23} \\ s_{13} & s_{23} & s_{33} \end{bmatrix}, \quad s_{ij} = \frac{1}{2}(B_{ij} + B_{ji}).$$

Furthermore,

$$\mathbf{a} = \frac{1}{2}(\mathbf{B} - \mathbf{B}^{\mathrm{T}}) = \begin{bmatrix} 0 & \frac{1}{2}(B_{12} - B_{21}) & \frac{1}{2}(B_{13} - B_{31}) \\ \frac{1}{2}(B_{21} - B_{12}) & 0 & \frac{1}{2}(B_{32} - B_{23}) \\ \frac{1}{2}(B_{31} - B_{13}) & \frac{1}{2}(B_{23} - B_{32}) & 0 \end{bmatrix},$$

so that

$$\mathbf{a} = \begin{bmatrix} 0 & a_{12} & a_{13} \\ -a_{12} & 0 & a_{23} \\ -a_{13} & -a_{23} & 0 \end{bmatrix}, \quad a_{ij} = \frac{1}{2}(B_{ij} - B_{ji}).$$

Now carry out the double sum in the Riemann theta function:

$$\sum_{j=1}^{N}\sum_{k=1}^{N} m_j m_k B_{jk} = \sum_{j=1}^{N}\sum_{k=1}^{N} m_j m_k s_{jk} + \sum_{j=1}^{N}\sum_{k=1}^{N} m_j m_k a_{jk}.$$

The sum over the symmetric part is clearly finite. However, for the antisymmetric part, the diagonal terms are zero and the off-diagonal terms have opposite signs; the double sum over the antisymmetric part thus cancels out, so that

$$\sum_{j=1}^{N}\sum_{k=1}^{N} m_j m_k B_{jk} = \sum_{j=1}^{N}\sum_{k=1}^{N} m_j m_k s_{jk}.$$

Therefore, even if the matrix **B** has an antisymmetric part, only the symmetric part appears in the final computation of the theta function. This is not a proof that the Riemann matrix must be symmetric, but only that any antisymmetric part makes no contribution to the calculation of the theta function.

8.3.2   One-Dimensional Theta Functions: Connection to Classical Elliptic Functions

Let us now explore the theta function (Equation (8.3)) in order to see why it is referred to as a "generalized Fourier series." First notice that the number of dimensions N of the $N \times N$ Riemann matrix is exactly the number of nested sums before the exponential. What happens when $N = 1$? In this case, the Riemann matrix is just a simple scalar $B_{11} = b$ (for b a complex constant) and the theta function can be written:

$$\theta(x) = \sum_{n=-\infty}^{\infty} \exp\left\{ inkx - \frac{1}{2}bn^2 \right\}, \tag{8.5a}$$

where for physical reasons it is natural to set $X = kx$, where $k = 2\pi/L$, for L the period of the theta function. This is the classical function often referred to as the Jacobian θ-function, θ_3, which is normally rewritten (Abramowitz and Stegun, 1964):

$$\theta_3(x) \equiv \theta_3(x, q) = 1 + 2\sum_{n=1}^{\infty} q^{n^2} \cos(nkx) \tag{8.5b}$$

where $q = \exp(-b/2)$ is referred to as the *nome*. Thus, a single-degree-of-free-dom theta function has the simple Fourier series representation.

As discussed elsewhere (see, e.g., Abramowitz and Stegun, 1964), it is convenient to set $b = 2\pi K'/K$, so that

$$q = e^{-\pi K'/K}, \tag{8.6}$$

where K and K' are elliptic integrals:

$$K(m) = K = \int_0^1 [(1-t^2)(1-mt^2)]^{-1/2} dt = \int_0^{\pi/2} \frac{d\theta}{[1-m\sin^2\theta]^{1/2}}, \tag{8.7}$$

$$K'(m) = K(m_1) = \int_0^1 [(1-t^2)(1-m_1 t^2)]^{-1/2} dt = \int_0^{\pi/2} \frac{d\theta}{[1-m_1\sin^2\theta]^{1/2}}, \tag{8.8}$$

with $m + m_1 = 1$ and $K(m) = K'(m_1) = K'(1-m)$.

The *single-degree-of-freedom* θ-*function* (Equation (8.4)) may be written in terms of the Jacobi θ_3-function (Equation (8.5)) or in terms of another Jacobian theta function θ_4, since $\theta_3(kx + \pi, q) = \theta_4(kx, q)$. This can be seen by setting

$$b = -b_0 + 2\pi i,$$

where the number b_0 is real and positive. The factor of 2π will be justified in the following way. Note that the imaginary part of the exponent in Equation (8.4) is just

$$\exp\{i(nX + \pi n^2)\} = \exp\{in(X + \pi n)\} = \exp\{in(X + \pi)\},$$

so the resultant wave is phase shifted by π (one half the period of 2π). Set $q = \exp(-b_0/2)$ (the *nome*) and use $e^{-i\pi n^2} = (-1)^n$ to find

$$\theta_4(x, q) = 1 + 2\sum_{n=1}^{\infty} (-1)^n q^{n^2} \cos(nkx), \tag{8.9}$$

where we have identified the series with the Jacobi theta function, θ_4, and set $X = kx$, $k = 2\pi/L$.

8.3.3   Multiple, Noninteracting Degrees of Freedom

It is instructive to see what happens to the theta function in the absence of interactions, that is, when we *artificially* set the off-diagonal terms to zero in the period matrix in Equation (8.3):

$$\theta_N(X_1, X_2, \ldots, X_N) = \sum_{m_1=-\infty}^{\infty} \exp\left\{im_1X_1 - \frac{1}{2}m_1^2B_{11}\right\}$$

$$\times \sum_{m_2=-\infty}^{\infty} \exp\left\{im_2X_2 - \frac{1}{2}m_2^2B_{22}\right\} \cdots \qquad (8.10)$$

$$\times \sum_{m_N=-\infty}^{\infty} \exp\left\{im_NX_N - \frac{1}{2}m_N^2B_{NN}\right\}.$$

Hence

$$\theta_N(X_1, X_2, \ldots, X_N) = \theta_1(X_1)\theta_2(X_2)\cdots\theta_N(X_N) = \prod_{n=1}^{N}\theta_n(X_n), \qquad (8.11)$$

where

$$\theta_n(X_n) = \sum_{m=-\infty}^{\infty} \exp\left\{imX_n - \frac{1}{2}m^2B_{nn}\right\}. \qquad (8.12)$$

Therefore, in the absence of interactions, the N-dimensional theta function reduces to the product of N one-dimensional theta functions. I have abused the notation here some, since these latter functions, θ_n, are just the function θ_3 in Jacobi's original notation.

Notice that if we set $X_n = k_nx - \omega_nt + \phi_n$, then Equation (8.12) becomes a kind of Stokes wave, that is, the components in the linear Fourier spectrum are *phase locked*. Notice that I stated above that I am *artificially* setting the off-diagonal elements of the period matrix to zero. Formally speaking, as will be seen below this is not a legal operation; I have done it here only to illustrate the contributions of the various terms in the theta function. In reality, a particular limiting procedure is required which *linearizes* the theta function and results in the linear Fourier transform as seen below.

8.3.4 A Theta Function Identity

It is possible to rewrite the complete θ-function theory, for arbitrary N, in a general form that allows for easy physical understanding (Osborne, 1995a,b). To do this, the vector form of the theta functions is given by

$$\theta(\mathbf{X}) = \sum_{\mathbf{m}} e^{i\mathbf{m}\cdot\mathbf{X}-(1/2)\mathbf{m}\cdot\mathbf{Bm}},$$

where as in $\mathbf{m} = [m_1, m_2, \ldots, m_N]$ and $\mathbf{X} = [X_1, X_2, \ldots, X_N]$. The summation over all negative, zero and positive integers is implied. Separating the period matrix $\mathbf{B}$ into diagonal ($\mathbf{D}$) and off-diagonal ($\mathbf{O}$) parts

$$\mathbf{B} = \mathbf{D} + \mathbf{O},$$

one can write

$$\theta(\mathbf{X}) = \sum_{\mathbf{m}} e^{i\mathbf{m}\cdot\mathbf{X}-(1/2)\mathbf{m}\cdot\mathbf{D}\mathbf{m}} + \left\{ \sum_{\mathbf{m}} e^{i\mathbf{m}\cdot\mathbf{X}-(1/2)\mathbf{m}\cdot\mathbf{B}\mathbf{m}} - \sum_{\mathbf{m}} e^{i\mathbf{m}\cdot\mathbf{X}-(1/2)\mathbf{m}\cdot\mathbf{D}\mathbf{m}} \right\},$$

$$\theta(\mathbf{X}) = \sum_{\mathbf{m}} e^{i\mathbf{m}\cdot\mathbf{X}-(1/2)\mathbf{m}\cdot\mathbf{D}\mathbf{m}} + \left\{ \sum_{\mathbf{m}} \left[e^{-(1/2)\mathbf{m}\cdot\mathbf{O}\mathbf{m}} - 1 \right] e^{i\mathbf{m}\cdot\mathbf{X}-(1/2)\mathbf{m}\cdot\mathbf{D}\mathbf{m}} \right\},$$

$$\theta(\mathbf{X}) = \sum_{\mathbf{m}} e^{i\mathbf{m}\cdot\mathbf{X}-(1/2)\mathbf{m}\cdot\mathbf{D}\mathbf{m}} \left\{ 1 + \frac{\sum_{\mathbf{m}} \left[e^{-(1/2)\mathbf{m}\cdot\mathbf{O}\mathbf{m}} - 1 \right] e^{i\mathbf{m}\cdot\mathbf{X}-(1/2)\mathbf{m}\cdot\mathbf{D}\mathbf{m}}}{\sum_{\mathbf{m}} e^{i\mathbf{m}\cdot\mathbf{X}-(1/2)\mathbf{m}\cdot\mathbf{D}\mathbf{m}}} \right\}.$$

$$(8.13)$$

Equation (8.13) can also be written in the following notation:

$$\theta(\mathbf{X}) = F(\mathbf{X}) \left\{ 1 + \frac{F(\mathbf{X}, G)}{F(\mathbf{X})} \right\}, \qquad (8.14)$$

where

$$F(\mathbf{X}, G) = \sum_{\mathbf{m}} G(\mathbf{O}) e^{i\mathbf{m}\cdot\mathbf{X}-(1/2)\mathbf{m}\cdot\mathbf{D}\mathbf{m}}, \quad G(\mathbf{O}) = e^{-(1/2)\mathbf{m}\cdot\mathbf{O}\mathbf{m}} - 1, \qquad (8.15)$$

$$F(\mathbf{X}) \equiv F(\mathbf{X}, 1) = \prod_{n=1}^{N} \theta_n(X_n). \qquad (8.16)$$

Here, the θ_n in Equation (8.16) are the ordinary Jacobian theta functions θ_3 (Equation (8.12)), one for each of the N degrees of freedom. Alternatively,

$$\theta(\mathbf{X}) = \theta_{\text{int}}(\mathbf{X}|\mathbf{O}) \prod_{n=1}^{N} \theta_n(X_n) \theta_{\text{int}}(\mathbf{X}) = \theta_1(X_1)\theta_2(X_2)\cdots\theta_N(X_N) \qquad (8.17)$$

where

$$\theta_n(X_n) = \sum_{m=-\infty}^{\infty} \exp\left\{ imX_n - \frac{1}{2}B_{nn}m^2 \right\} \qquad (8.18)$$

$$\theta_{\text{int}}(\mathbf{X}) = 1 + \frac{\sum_{\mathbf{m}} \left[e^{-(1/2)\mathbf{m}\cdot\mathbf{O}\mathbf{m}} - 1 \right] e^{i\mathbf{m}\cdot\mathbf{X}-(1/2)\mathbf{m}\cdot\mathbf{D}\mathbf{m}}}{\prod_{n=1}^{N} \theta_n(X_n)} \qquad (8.19)$$

Thus, we have another (identical) expression (Equation (8.17)) for the Riemann theta function (Equation (8.3)) which consists of the product of N single-degree-of-freedom theta functions, $\theta_n(X_n)$ (which depend only on the diagonal elements, $\mathbf{D}$, of the Riemann matrix) times an "interaction" term, $\theta_{\text{int}}(\mathbf{X})$, which depends also on the off-diagonal elements.

The identity (Equations (8.17)–(8.19)) provides us with physical perspective about Riemann theta functions, but to my knowledge provides no improvement in computational speed over the expression (Equation (8.3)).

8.3.5   Relationship of Generalized Fourier Series to Ordinary Fourier Series

Let us now explore the theta function in order to see why it is referred to as a "generalized Fourier series." To this end, assume that the elements of the diagonal terms are large with respect to the off-diagonal terms. This means that the interaction term is simple, $\theta_{int} \approx 1$. The theta function then has the form

$$\theta(\mathbf{X}) = \prod_{n=1}^{N} \theta_n(X_n) = \theta_1(X_1)\theta_2(X_2)\cdots\theta_N(X_N). \tag{8.20}$$

Thus the theta function is a product of simple one-degree-of-freedom theta functions. Since the diagonal terms are large we have

$$\theta_n \cong 1 + 2q_n \cos X_n,$$

because in this case the coefficients are small:

$$q_n = e^{-(1/2)B_{nn}} \ll 1.$$

With the above assumptions, to leading order we then have

$$\theta(\mathbf{X}) \cong (1 + C_1 \cos X_1)(1 + C_2 \cos X_2)\cdots(1 + C_N \cos X_N),$$

$$\theta(\mathbf{X}) \cong 1 + C_1 \cos X_1 + C_2 \cos X_2 + \cdots + C_N \cos X_N = 1 + \sum_{n=1}^{N} C_n \cos X_n.$$

This is just an ordinary Fourier series, where

$$C_n = 2q_n = 2e^{-(1/2)B_{nn}}.$$

Thus, the Riemann theta function approaches an ordinary linear Fourier series when the off-diagonal terms become small with respect to the diagonal terms, $B_{nn} \gg B_{mn}$ and when the diagonal terms themselves are sufficiently large so that $q_n \ll 1$. The appropriate *physical perspective*, as will be seen below, is that when the amplitudes

$$e^{-(1/2)\mathbf{m}\cdot\mathbf{Bm}} \approx e^{-(1/2)\mathbf{m}\cdot\mathbf{Dm}} \ll 1$$

(D is the diagonal part of the matrix **B**) become small, the theta function reduces to an ordinary Fourier series. Generally speaking, the theta function describes N wave modes (the one-degree-of-freedom theta functions), including their mutual pairwise interactions.

8.3.6 Alternative Form for Theta Functions in Terms of Cosines

If we write the trigonometric form for the complex exponential in the theta function, we have

$$
\theta_N(X_1, X_2, \ldots, X_N) = \sum_{m_1=-\infty}^{\infty} \sum_{m_2=-\infty}^{\infty} \cdots \sum_{m_N=-\infty}^{\infty} \exp\left[-\frac{1}{2}\sum_{j=1}^{N}\sum_{k=1}^{N} m_j B_{jk} m_k\right] \cos\left[\sum_{k=1}^{N} m_k X_k\right]
$$

$$
+ i \sum_{m_1=-\infty}^{\infty} \sum_{m_2=-\infty}^{\infty} \cdots \sum_{m_N=-\infty}^{\infty} \exp\left[-\frac{1}{2}\sum_{j=1}^{N}\sum_{k=1}^{N} m_j B_{jk} m_k\right] \sin\left[\sum_{k=1}^{N} m_k X_k\right].
$$

The summation with the sine function is odd and therefore zero, so we obtain

$$
\theta_N(X_1, X_2, \ldots, X_N)
$$
$$
= \sum_{m_1=-\infty}^{\infty} \sum_{m_2=-\infty}^{\infty} \cdots \sum_{m_N=-\infty}^{\infty} \exp\left[-\frac{1}{2}\sum_{j=1}^{N}\sum_{k=1}^{N} m_j B_{jk} m_k\right] \cos\left[\sum_{k=1}^{N} m_k X_k\right] \quad (8.21)
$$

For computer programming purposes, this latter formula can be replaced by another that is twice as efficient. To this end, the rule for summing the theta functions can be modified (1) to include all integer vectors up to $(0, 0, \ldots, 0)$, provided (2) that the exponentials get replaced by cosines:

$$
\theta_N(X_1, X_2, \ldots, X_N)
$$
$$
= 1 + 2 \sum_{m_1=-\infty}^{0} \sum_{m_2=-\infty}^{0} \cdots \sum_{m_N=-\infty}^{-1} \exp\left[-\frac{1}{2}\sum_{j=1}^{N}\sum_{k=1}^{N} m_j B_{jk} m_k\right] \cos\left[\sum_{k=1}^{N} m_k X_k\right] \quad (8.22)
$$

This expression reduces computational effort in two ways: (1) complex exponentials need not be computed, only cosines; (2) half as many terms need to be calculated because the summation is stopped at $\mathbf{M} = (0, 0, \ldots, -1)$. Hence, a factor of 4 in total computational effort is obtained.

A shorthand notation for Equation (8.22) is given by

$$
\theta_N(\mathbf{X}) = 1 + 2 \sum_{m_1=-\infty}^{0} \sum_{m_2=-\infty}^{0} \cdots \sum_{m_N=-\infty}^{-1} q_{\mathbf{m}} \cos[\mathbf{m} \cdot \mathbf{X}],
$$

$$
q_{\mathbf{m}} = \exp\left[-\frac{1}{2}\mathbf{m} \cdot \mathbf{Bm}\right]. \quad (8.23)
$$

This latter expression can also be written in the following form:

$$\theta_N(\mathbf{X}) = 1 + 2 \sum_{\mathbf{m}=-\infty}^{0^-} q_{\mathbf{m}} \cos[\mathbf{m}\cdot\mathbf{X}], \tag{8.24}$$

where $0^- = [0, 0, \ldots, -1]$ (the "-1" rather than "0" is to exclude the term "1" in the theta sum) and

$$\mathbf{X} = [X_1, X_2, \ldots, X_N],$$

where

$$X_n = [k_1 x - \omega_1 t + \phi_1, k_2 x - \omega_2 t + \phi_2, \ldots, k_N x - \omega_N t + \phi_N].$$

As per usual

$$\mathbf{m} = [m_1, m_2, \ldots, m_N],$$

$$\mathbf{k} = [k_1, k_2, \ldots, k_N],$$

$$\boldsymbol{\omega} = [\omega_1, \omega_2, \ldots, \omega_N],$$

$$\boldsymbol{\phi} = [\phi_1, \phi_2, \ldots, \phi_N].$$

Finally, in the interest of completeness, note that we can also write the results in terms of theta functions in space, x, and time, t. The scalar form of Equation (8.24) is

$$\theta_N(x, t)$$

$$= 1 + 2 \sum_{m_1=-\infty}^{0} \sum_{m_2=-\infty}^{0} \cdots \sum_{m_N=-\infty}^{-1} \exp\left\{ -\frac{1}{2} \sum_{i=1}^{N} \sum_{j=1}^{N} m_i m_j B_{ij} \right\} \tag{8.25}$$

$$\times \cos\left[\left(\sum_{i=1}^{N} m_i k_i \right) x - \left(\sum_{i=1}^{N} m_i \omega_i \right) t + \sum_{i=1}^{N} m_i \phi_i \right]$$

This is the *dynamical form* of the theta function that will be used for many applications in this book. Generally speaking, the *Riemann matrix is complex* and the *real part* must be *positive definite* to insure convergence of the theta function series.

8.3.7 Partial Sums of Theta Functions

Partial theta summations are written in the following form:

$$\theta_{MN}(X_1, X_2, \ldots, X_N) = \sum_{m_1=-M}^{M} \sum_{m_2=-M}^{M} \cdots \sum_{m_N=-M}^{M} \exp\left\{ i\sum_{k=1}^{N} m_k X_k - \frac{1}{2}\sum_{j=1}^{N}\sum_{k=1}^{N} m_j B_{jk} m_k \right\}.$$

(8.26)

The θ-function is subscripted with M (which defines a partial sum as an approximation for the theta function) and N (the number of degrees of freedom in the theta function). One could generalize Equation (8.26) to have different limits for each of the nested sums; this idea is given a practical basis in the numerical examples discussed in the sequel.

Now consider the specific case for the partial sum of two degrees of freedom, $N = 2$:

$$\theta_{M2}(X_1, X_2) = \sum_{m_1=-M}^{M} \sum_{m_2=-M}^{M} q^{m_1^2} p^{m_2^2} r^{m_1 m_2} e^{im_1 X_1 + im_2 X_2},$$

(8.27)

or

$$\theta_{M2}(X_1, X_2) = \sum_{m_1=-M}^{M} \sum_{m_2=-M}^{M} q^{m_1^2} p^{m_2^2} r^{m_1 m_2} \cos(m_1 X_1 + m_2 X_2),$$

(8.28)

or

$$\theta_{M2}(X_1, X_2) = 1 + 2 \sum_{m_1=-M}^{0} \sum_{m_2=-M}^{-1} q^{m_1^2} p^{m_2^2} r^{m_1 m_2} \cos(m_1 X_1 + m_2 X_2),$$

(8.29)

where

$$q = e^{-(1/2)B_{11}}, \quad p = e^{-(1/2)B_{22}}, \quad r = e^{-B_{12}}.$$

(8.30)

When $M = 1$, the following partial sum arises:

$$\theta_{12} = 1 + 2q\cos X_1 + 2p\cos X_2 + 2qpr^{-1}\cos(X_1 - X_2)$$
$$+ 2qpr\cos(X_1 + X_2).$$

(8.31)

Table 8.1 illustrates this summation.

Table 8.1 Partial theta function summation for $M = 1$, $N = 2$

l	(m_1, m_2)	Theta summation elements	I_l
1	$(-1, -1)$	$2qpr \cos(X_1 + X_2)$	-3
2	$(-1, 0)$	$2q \cos(X_1)$	-1
3	$(-1, 1)$	$2qpr^{-1} \cos(X_1 - X_2)$	1
4	$(0, -1)$	$2p \cos(X_2)$	-2
5	$(0, 0)$	1	0

For $M = 2$, the partial sum at next order is given by

$$
\begin{aligned}
\theta_{22} = \theta_{12} &+ 2q^4 \cos 2X_1 + 2p^4 \cos 2X_2 \\
&+ 2q^4 p^4 r^4 \cos 2(X_1 + X_2) + 2q^4 pr^2 \cos(2X_1 + X_2) + 2q^4 pr^{-2} \cos(2X_1 - X_2) \\
&+ 2q^4 p^4 r^{-4} \cos 2(X_1 - X_2) + 2qp^4 r^2 \cos(X_1 + 2X_2) + 2qp^4 r^{-2} \cos(X_1 - 2X_2).
\end{aligned}
$$
(8.32)

Table 8.2 shows how this summation is made.

For $M = 3$:

$$
\begin{aligned}
\theta_{32} = \theta_{12} &+ \theta_{22} + 2q^9 \cos 3X_1 + 2p^9 \cos 3X_2 \\
&+ 2q^9 p^9 r^9 \cos(3X_1 + 3X_2) + 2q^9 p^4 r^6 \cos(3X_1 + 2X_2) + 2q^9 pr^3 \cos(3X_1 + X_2) \\
&+ 2q^9 pr^{-3} \cos(3X_1 - X_2) + 2q^9 p^9 r^{-9} \cos(3X_1 - 3X_2) + 2q^4 p^9 r^6 \cos(2X_1 + 3X_2) \\
&+ 2q^4 p^9 r^{-6} \cos(2X_1 - 3X_2) + 2qp^9 r^3 \cos(X_1 + 3X_2) + 2qp^9 r^{-3} \cos(X_1 - 3X_2).
\end{aligned}
$$
(8.33)

Table 8.2 Partial theta function summation for $M = 2$, $N = 2$

l	(m_1, m_2)	Theta summation elements	I_l
1	$(-2, -2)$	$2q^4 p^4 r^4 \cos(2X_1 + 2X_2)$	-6
2	$(-2, -1)$	$2q^4 pr^2 \cos(2X_1 + X_2)$	-4
3	$(-2, 0)$	$2q^4 \cos(2X_1)$	-2
4	$(-2, 1)$	$2q^4 pr^{-2} \cos(2X_1 - X_2)$	0
5	$(-2, 2)$	$2q^4 p^4 r^{-4} \cos(2X_1 - 2X_2)$	2
6	$(-1, -2)$	$2qp^4 r^2 \cos(X_1 + 2X_2)$	-5
7	$(-1, -1)$	$2qpr \cos(X_1 + X_2)$	-3
8	$(-1, 0)$	$2q \cos(X_1)$	-1
9	$(-1, 1)$	$2qpr^{-1} \cos(X_1 - X_2)$	1
10	$(-1, 2)$	$2qp^4 r^{-2} \cos(X_1 - 2X_2)$	3
11	$(0, -2)$	$2p^4 \cos(2X_2)$	-4
12	$(0, -1)$	$2p \cos(X_2)$	-2
13	$(0, 0)$	1	0

The latter expression in complete form (using Equations (8.31) and (8.32)) is

$$
\begin{aligned}
\theta_{32} = {} & 1 + 2q\cos X_1 + 2p\cos X_2 + 2q^4\cos 2X_1 + 2p^4\cos 2X_2 + 2q^9\cos 3X_1 + 2p^9\cos 3X_2 \\
& + 2qpr^{-1}\cos(X_1 - X_2) + 2qpr\cos(X_1 + X_2) \\
& + 2q^4p^4r^4\cos 2(X_1 + X_2) + 2q^4pr^2\cos(2X_1 + X_2) + 2q^4pr^{-2}\cos(2X_1 - X_2) \\
& + 2q^4p^4r^{-4}\cos 2(X_1 - X_2) + 2qp^4r^2\cos(X_1 + 2X_2) + 2qp^4r^{-2}\cos(X_1 - 2X_2) \\
& + 2q^9p^9r^9\cos(3X_1 + 3X_2) + 2q^9pr^3\cos(3X_1 + X_2) + 2q^9pr^{-3}\cos(3X_1 - X_2) \\
& + 2q^9p^4r^6\cos(3X_1 + 2X_2) + 2q^9p^9r^{-9}\cos(3X_1 - 3X_2) + 2q^4p^9r^6\cos(2X_1 + 3X_2) \\
& + 2q^4p^9r^{-6}\cos(2X_1 - 3X_2) + 2qp^9r^3\cos(X_1 + 3X_2) + 2qp^9r^{-3}\cos(X_1 - 3X_2).
\end{aligned}
$$
$$(8.34)$$

Hence by increasing M, one finds higher-order partial-sum approximations of the theta functions, including their mutual nonlinear interaction terms. Note that by going to higher order, one also finds all possible combinations of the sums and differences of the arguments $(X_n = k_n x + \phi_n)$ in the IST spectrum. This result should come as no surprise to those who have experience in computing bi- and trispectra of nonlinear wave trains; in the periodic IST, however, one is able to determine *all* N-spectra of the wave train, provided of course that one has the patience to compute the appropriate partial sums.

In the above expressions, it is clear that by setting $r = 1$, the interactions are excluded and the θ-functions reduce identically to the factored forms:

$$
\begin{aligned}
\theta_{12} = {} & (1 + 2q\cos X_1)(1 + 2p\cos X_2), \\
\theta_{22} = {} & (1 + 2q\cos X_1 + 2q^4\cos 2X_1)(1 + 2p\cos X_2 + 2p^4\cos 2X_2), \\
\theta_{32} = {} & (1 + 2q\cos X_1 + 2q^4\cos 2X_1 + 2q^9\cos 3X_1) \\
& \times (1 + 2p\cos X_2 + 2p^4\cos 2X_2 + 2p^9\cos 3X_2).
\end{aligned}
$$
$$(8.35)$$

In these latter expressions, there are no nonlinear interactions between the two components m_1, m_2. Nonlinear interactions are taken into account only when $B_{12} \neq 0$ such that $r \neq 1$.

Another partial sum which is useful for the study of three degrees of freedom, $N = 3$, is given by

$$
\theta_{M3} = \sum_{m_1 = -M}^{M} \sum_{m_2 = -M}^{M} \sum_{m_3 = -M}^{M} C(m_1, m_2, m_3)\exp\{im_1 X_1 + im_2 X_2 + im_3 X_3\}, \quad (8.36)
$$

$$
\theta_{M3} = 1 + 2\sum_{m_1 = -M}^{0} \sum_{m_2 = -M}^{0} \sum_{m_3 = -M}^{-1} C(m_1, m_2, m_3)\cos(m_1 X_1 + m_2 X_2 + m_3 X_3). \quad (8.37)
$$

The coefficients have the following form:

$$C(m_1, m_2, m_3) = q_1^{m_1^2} q_2^{m_2^2} q_3^{m_3^2} r_{12}^{m_1 m_2} r_{13}^{m_1 m_3} r_{23}^{m_2 m_3}, \tag{8.38}$$

where

$$q_1 = e^{-(1/2)B_{11}}, \quad q_2 = e^{-(1/2)B_{22}}, \quad q_3 = e^{-(1/2)B_{33}},$$
$$r_{12} = e^{-B_{12}}, \quad r_{13} = e^{-B_{13}}, \quad r_{23} = e^{-B_{23}}. \tag{8.39}$$

For the specific case $M = 1$:

$$
\begin{aligned}
\theta_{13} = {} & 1 + 2q_1 \cos X_1 + 2q_2 \cos X_2 + 2q_3 \cos X_3 \\
& + 2q_1 q_2 r_{12} \cos(X_1 + X_2) + 2q_1 q_2 r_{12}^{-1} \cos(X_1 - X_2) \\
& + 2q_1 q_3 r_{13} \cos(X_1 + X_3) + 2q_1 q_3 r_{13}^{-1} \cos(X_1 - X_3) \\
& + 2q_2 q_3 r_{23} \cos(X_2 + X_3) + 2q_2 q_3 r_{23}^{-1} \cos(X_2 - X_3) \\
& + 2q_1 q_2 q_3 r_{12} r_{13} r_{23} \cos(X_1 + X_2 + X_3) + 2q_1 q_2 q_3 r_{12} r_{13}^{-1} r_{23}^{-1} \cos(X_1 + X_2 - X_3) \\
& + 2q_1 q_2 q_3 r_{12}^{-1} r_{13} r_{23}^{-1} \cos(X_1 - X_2 + X_3) + 2q_1 q_2 q_3 r_{12}^{-1} r_{13}^{-1} r_{23} \cos(X_1 - X_2 - X_3).
\end{aligned}
\tag{8.40}
$$

In the absence of interactions, one has $r_{12} = 1$, $r_{13} = 1$, and $r_{23} = 1$ so that

$$\theta_{13} = (1 + 2q_1 \cos X_1)(1 + 2q_2 \cos X_2)(1 + 2q_3 \cos X_3). \tag{8.41}$$

8.3.8 Examples of Simple Partial Theta Sums

Let us now simplify the notation of the theta partial sums and retain *commensurable wavenumbers*. The theta functions are therefore periodic in space, $\theta(x, t) = \theta(x + L, t)$. This is the most important case for the analysis of data and for modeling purposes. Consider the vector form:

$$\theta_{MN}(x|\mathbf{B}, \boldsymbol{\phi}) = 1 + 2 \sum_{\mathbf{m}=-\mathbf{m}}^{0^-} q_{\mathbf{m}} \cos[\mathbf{m} \cdot \mathbf{k} x + \mathbf{m} \cdot \boldsymbol{\phi}],$$

where

$$q_{\mathbf{m}} = e^{-(1/2)\mathbf{m} \cdot \mathbf{B} \mathbf{m}}.$$

Of course, the Riemann matrix, $\mathbf{B}$, and the phases, $\boldsymbol{\phi}$, are assumed known and given. Here 0^- means sum up to $\mathbf{m} = \{0, 0, \ldots, -1\}$ as before. We may also set for convenience

$$C_{\mathbf{m}} = C_{\mathbf{m}}(x) = q_{\mathbf{m}} \cos[\mathbf{m} \cdot \mathbf{k} x + \mathbf{m} \cdot \boldsymbol{\phi}],$$

In this notation we can look at a simple example:

$$\theta_{12}(x) = 1 + 2C_{-1,-1} + 2C_{-1,0} + 2C_{-1,1} + 2C_{0,-1},$$

where

$$C_{m_1,m_2} = q_{m_1,m_2}\cos[(m_1,m_2)\cdot(1,2)\Delta kx + (m_1,m_2)\cdot(\phi_1,\phi_2)]$$

or

$$C_{m_1,m_2} = q_{m_1,m_2}\cos[(m_1 + 2m_2)\Delta kx + m_1\phi_1 + m_2\phi_2],$$

so that

$$C_{-1,-1} = q_{-1,-1}\cos[3\Delta kx + \phi_1 + 2\phi_2],$$

$$C_{-1,0} = q_{-1,0}\cos[\Delta kx + \phi_1],$$

$$C_{-1,1} = q_{-1,1}\cos[\Delta kx - \phi_1 + \phi_2],$$

$$C_{0,-1} = q_{0,-1}\cos[2(\Delta kx + \phi_2)],$$

where

$$q_{m_1,m_2} = \exp\left\{-\frac{1}{2}\sum_{i=1}^{2}\sum_{j=1}^{2}m_i m_j B_{ij}\right\} = \exp\left\{-\frac{1}{2}(m_1^2 B_{22} + m_2^2 B_{22} + m_1 m_2 B_{12})\right\}.$$

Thus, we have

$$q_{-1,-1} = \exp\left\{-\frac{1}{2}(B_{11} + 2B_{12} + B_{22})\right\},$$

$$q_{-1,0} = \exp\left\{-\frac{1}{2}B_{11}\right\},$$

$$q_{-1,1} = \exp\left\{-\frac{1}{2}(B_{11} - 2B_{12} + B_{22})\right\},$$

$$q_{0,-1} = \exp\left\{-\frac{1}{2}B_{22}\right\}.$$

Finally,

$$\theta_{12}(x|\mathbf{B}) = 1 + 2\exp\left\{-\frac{1}{2}B_{11}\right\}\cos[\Delta kx + \phi_1]$$

$$+ 2\exp\left\{-\frac{1}{2}B_{22}\right\}\cos[2(\Delta kx + \phi_2)]$$

$$+ 2\exp\left\{-\frac{1}{2}(B_{11} - 2B_{12} + B_{22})\right\}\cos[\Delta kx - \phi_1 + \phi_2]$$

$$+ 2\exp\left\{-\frac{1}{2}(B_{11} + 2B_{12} + B_{22})\right\}\cos[3\Delta kx + \phi_1 + 2\phi_2].$$

In this example I have used the relation:

$$(k_1, k_2) = (1, 2)\Delta k, \quad \Delta k = 2\pi/L$$

These results consist of two Stokes wave modes at wavenumbers 1 and 2 plus interaction terms. The wavenumbers of these terms are $m_1 k_1 + m_2 k_2 \Rightarrow -(k_1 + k_2), -k_1, -(k_1 - k_2), -k_2$. Thus, we have the first Stokes wave at k_1, the second Stokes wave at k_2 and two interaction terms, the first at the summed wavenumbers $k_1 + k_2$ and the second at the differenced wavenumbers $k_1 - k_2$.

Table 8.3 gives the actual terms in the partial theta summation for this latter case. Figure 8.1 graphs the integer wavenumber I_l (for a full discussion and definition see Section 8.4 below) and the number of waves that are interacting for each value of the ordering parameter. Note that Table 8.3 and Figure 8.1 consider the partial theta summation using (8.28) rather than (8.29); the results are however equivalent.

Table 8.3 Theta function partial sum for $M = 1$, $N = 2$

l	(m_1, m_2)	Theta summation elements	I_l	N-wise interacts
1	$(-1, -1)$	$qpre^{-iX_1 - iX_2}$	-3	2
2	$(-1, 0)$	qe^{-iX_1}	-1	1
3	$(-1, 1)$	$qpr^{-1}e^{-iX_1 + iX_2}$	1	2
4	$(0, -1)$	pe^{-iX_2}	-2	1
5	$(0, 0)$	1	0	0
6	$(0, 1)$	pe^{iX_2}	2	1
7	$(1, -1)$	$qpr^{-1}e^{iX_1 - iX_2}$	-1	2
8	$(1, 0)$	qe^{iX_1}	1	1
9	$(1, 1)$	$qpre^{iX_1 + iX_2}$	3	2

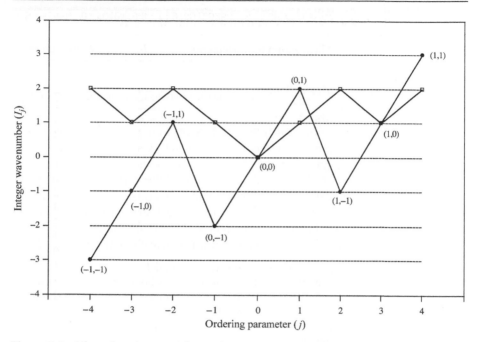

Figure 8.1 Theta function partial sum for $M = 1$, $N = 2$. Here $j = -4, -3, \ldots 3, 4$ and $l = 1, 2 \ldots 9$.

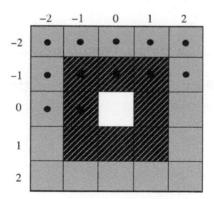

Figure 8.2 Summation of theta function for $M = 1$, $N = 2$ and for $M = 2$, $N = 2$.

Let us now consider the partial summation of the thetas over $-2, 2$ (Figure 8.3). The explicit terms are shown in Table 8.4. Figure 8.2 summarizes the asymmetric terms in the partial theta summations as a block diagram for the cases $M = 1$, $N = 2$ and $M = 2$, $N = 2$.

Figure 8.3 graphs the values of the integer wavenumber and the number of interactions for each value of the ordering parameter. Again the symmetric form of the partial summation (8.28) is used rather than the asymmetric form (8.29). Figure 8.4 provides a more complex example for which $M = 1$, $N = 3$.

Table 8.4 Theta function partial sum for $M = 2$, $N = 2$

l	(m_1, m_2)	Theta summation elements	I_l	N-wise interacts
1	$(-2, -2)$	$q^4 p^4 r^4 e^{-2iX_1 - 2iX_2}$	-6	2
2	$(-2, -1)$	$q^4 p r^2 e^{-2iX_1 - iX_2}$	-4	2
3	$(-2, 0)$	$q^4 e^{-2iX_1}$	-2	1
4	$(-2, 1)$	$q^4 p r^{-2} e^{-2iX_1 + iX_2}$	0	2
5	$(-2, 2)$	$q^4 p^4 r^{-4} e^{-2iX_1 + 2iX_2}$	2	2
6	$(-1, -2)$	$q p^4 r^2 e^{-iX_1 - 2iX_2}$	-5	2
7	$(-1, -1)$	$q p r e^{-iX_1 - iX_2}$	-3	2
8	$(-1, 0)$	$q e^{-iX_1}$	-1	1
9	$(-1, 1)$	$q p r^{-1} e^{-iX_1 + iX_2}$	1	2
10	$(-1, 2)$	$q p^4 r^{-2} e^{-iX_1 + 2iX_2}$	3	2
11	$(0, -2)$	$p^4 e^{-2iX_2}$	-4	1
12	$(0, -1)$	$p e^{-iX_2}$	-2	1
13	$(0, 0)$	1	0	0
14	$(0, 1)$	$p e^{iX_2}$	2	1
15	$(0, 2)$	$p^4 e^{2iX_2}$	4	1
16	$(1, -2)$	$q p^4 r^{-2} e^{iX_1 - 2iX_2}$	-3	2
17	$(1, -1)$	$q p r^{-1} e^{iX_1 - iX_2}$	-1	2
18	$(1, 0)$	$q e^{iX_1}$	1	1
19	$(1, 1)$	$q p r e^{iX_1 + iX_2}$	3	2
20	$(1, 2)$	$q p^4 r^2 e^{iX_1 + 2iX_2}$	5	2
21	$(2, -2)$	$q^4 p^4 r^{-4} e^{2iX_1 - 2iX_2}$	-2	2
22	$(2, -1)$	$q^4 p r^{-2} e^{2iX_1 - iX_2}$	0	2
23	$(2, 0)$	$q^4 e^{2iX_1}$	2	1
24	$(2, 1)$	$q^4 p r^2 e^{2iX_1 + iX_2}$	4	2
25	$(2, 2)$	$q^4 p^4 r^4 e^{2iX_1 + 2iX_2}$	6	2

8.4   Statistical Properties of Theta Function Parameters

The θ-function

$$\theta_N(x, t) = \sum_{m_1 = -\infty}^{\infty} \sum_{m_2 = -\infty}^{\infty} \cdots \sum_{m_N = -\infty}^{\infty} \exp \left\{ i \sum_{j=1}^{N} m_j X_j - \frac{1}{2} \sum_{i=1}^{N} \sum_{j=1}^{N} m_i m_j B_{ij} \right\}, \quad (8.42a)$$

where

$$X_j = k_j x - \omega_j t + \phi_j$$

can be written

$$\theta_N(x, t) = \sum_{l=1}^{\infty} q_l e^{i(K_l x - \Omega_l t + \Phi_l)},$$

$$\qquad\qquad\qquad\qquad\qquad\qquad\qquad\qquad (8.42b)$$

$$q_l = e^{-(1/2) \mathbf{m} \cdot \mathbf{B} \mathbf{m}} = \exp \left\{ -\frac{1}{2} \sum_{i=1}^{N} \sum_{j=1}^{N} m_i^l m_j^l B_{ij} \right\},$$

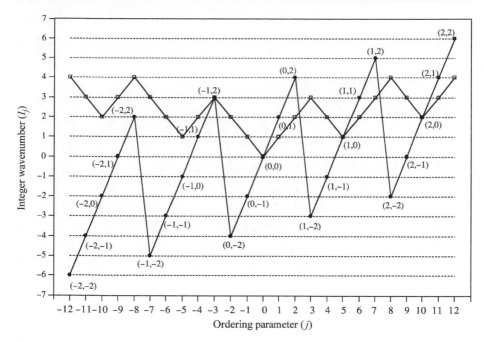

Figure 8.3 Theta function partial sum for $M = 2$, $N = 2$. Here $j = -12, -11, \ldots 11$, 12 and $l = 1, 2 \ldots 25$.

where l is an ordering parameter associated with each vector $\mathbf{m} \equiv \mathbf{m}^l = [m^l_1, m^l_2, \ldots, m^l_N]$ in the θ-sum. One can think of Equation (8.42b) as the reduction of Equation (8.42a) from a nested sum to a single sum. We have the following theta function parameters that can be written in the form:

$$K_l = \mathbf{m}_l \cdot \mathbf{k} = \sum_{n=1}^{N} m^l_n k_n,$$

$$\Omega_l = \mathbf{m}_l \cdot \boldsymbol{\omega} = \sum_{n=1}^{N} m^l_n \omega_n, \qquad (8.43)$$

$$\Phi_l = \mathbf{m}_l \cdot \boldsymbol{\phi} = \sum_{n=1}^{N} m^l_n \phi_n.$$

For a number of reasons (primarily for numerical and data analysis purposes), *the wavenumbers are often assumed to be commensurable* (and, therefore, e.g., the KdV equation and its IST are perfectly periodic in the spatial variable, $0 \leq x \leq L$, L the period of the wave train):

$$\mathbf{k} = [1, 2, \ldots, N]\Delta k,$$

where as usual $k_n = n\Delta k$ and $\Delta k = 2\pi/L$ is the wavenumber interval in the spectrum. Of course, due to the complex nature of the nonlinear

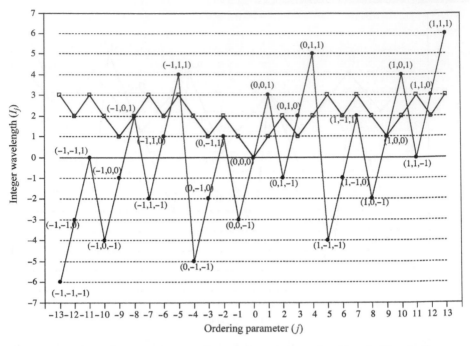

Figure 8.4 Theta function wavenumbers and partial sum for $M = 1$, $N = 3$. Here $j = -13, -12, \ldots 12, 13$ and $l = 1, 2 \ldots 27$.

interactions, the frequencies are *never* commensurable; in the linear limit $\omega_j = c_0 k_j - \beta_j k_j^3$ (for the KdV equation, see Chapter 10) and clearly commensurable frequencies cannot exist even there. In the variables (Equation (8.43)), the inverse scattering transform in the θ-function formulation given by Equation (8.42b) resembles linear Fourier analysis.

How then do the wavenumbers for periodic KdV, K_l, behave for the class of θ-functions given by Equation (8.42b)? Clearly, their formulation is different than those for the linear Fourier transform, that is,

$$K_l = \mathbf{m}_l \cdot \mathbf{k} = [m_1^l, m_2^l, \ldots, m_N^l] \cdot [1, 2, \ldots, N] \Delta k = \Delta k \sum_{j=1}^{N} j m_j^l. \qquad (8.44)$$

Hence the θ-function wavenumbers, K_l, for the KdV equation are *integer multiples* I_l of Δk:

$$K_l = I_l \Delta k = \frac{2\pi}{L} I_l, \quad I_l = \sum_{j=1}^{N} j m_j^l \qquad (8.45)$$

and are therefore *commensurable, often duplicated, and not ordered with the integers*. According to Equation (8.44), the wavenumbers in Equation (8.45) fall on the ordinary Fourier wavenumbers $k_j = 2\pi j \Delta k$. In fact, an *infinite number of terms* in Equation (8.45) fall on each k_j. An example for $M = 3$, $N = 5$ is given in Figure 8.5 where the integer wavenumbers of (8.45) are graphed as a function of the ordering parameter. This leads to the idea of a histogram of occurrence for each k_j. See Figure 8.6 for an example, where again $M = 3$, $N = 5$.

Since all the wavenumbers in the θ-sum fall on the k_j and since the wavenumbers are duplicated, we can write from Equation (8.46) (for the special case $t = 0$) (Osborne, 1995a,b)

$$\theta(x, 0) = \sum_{l=1}^{\infty} q_l e^{i(K_l x + \Phi_l)} = \sum_{n=-\infty}^{\infty} \theta_n e^{ik_n x} \tag{8.46}$$

where

$$\theta_n = \sum_{\substack{\text{sum } l \text{ over subset} \\ \text{of } l \text{ for which } I_l = n}} q_l e^{i\Phi_l}, \quad 1 \le n \le \infty \tag{8.47}$$

The above theta series (Equation (8.46)) (for $t = 0$) is nothing more than an ordinary Fourier transform! The Fourier coefficients, θ_n, in Equation (8.46) are given by the *temporal Fourier series* (Equation (8.47)).

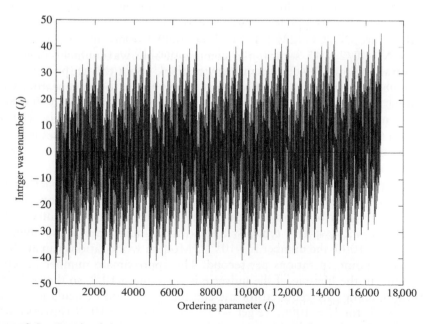

Figure 8.5 Graph of the integer function (Equation (8.45)) for a theta function summation for $M = 3$, $N = 5$.

Let us now discuss how the summation in Equation (8.47) is carried out. Figure 8.5 shows a graph of the integer function (Equation (8.46)) as a function of the ordering parameter, l. Note that the function is fractal and is highly oscillating. As a result, there are many values of the ordering parameter for each value of the Fourier parameter, n. In the figure, the horizontal line at $n = 20$ crosses the integer wavenumber at a large number of places corresponding to different sets of the summation integers m_j. The summation in Equation (8.47) includes all of the terms for which the horizontal line at $n = 20$ crosses I_l. Full details are given in Chapter 9.

Partial θ-sums over the individual indices M_n may be taken over arbitrary limits $(-M, M)$, rather than $(-\infty, \infty)$. Then the number of *terms* in the theta summation is $(2M + 1)^N$ and the number of commensurable *wavenumbers* in the spectrum is given by $2J + 1$. The integer I_l is bounded, $-J \leq I_l \leq J$, where

$$J = M \sum_{n=1}^{N} n = \frac{1}{2} MN(N + 1). \tag{8.48}$$

Figure 8.5 presents a graph of I_l as given by Equations (8.44) and (8.45), that is, a graph of the integer commensurable wavenumbers $I_l = K_l/\Delta k$ as a function of the ordering parameter $l, 1 \leq l \leq N_{\max}$ (the parameter j is also an ordering parameter given in the interval $-J_{\max} < j < J_{\max}$, where $J_{\max} = (N_{\max} - 1)/2$). For this particular case, the summation in the θ-function is made over $-3 \leq M \leq 3$ for a system with $N = 5$ degrees of freedom, so that there are $N_{\max} = (2M + 1)^N = 7^5 = 16,807$ terms in the θ-function; $J = MN(N + 1)/2 = 45$ so that the commensurable wavenumber range has the interval $-45 \leq I_l \leq 45$. A useful way to think about this graph is that the wavenumbers in the theta function have many values, often repeating with increasing ordering parameter. It is easy to see that by summing the theta function amplitudes, q_l, for a particular I_l gives the ordinary linear Fourier transform for that particular I_l (Equation (8.45))! Chapter 9 discusses in detail this "reduction" of the theta function, an important aspect of the numerical methods for computing theta functions.

The properties of the integer function I_l have a substantial impact on the numerical computation of θ-function solutions of the KdV equation. This is because the number of complex exponentials which must be computed in the θ-function is $\sim(2M + 1)^N$, typically an enormous number. For example, for $M = 10$, $N = 1000$, the number is $\sim 10^{1320}$. Modern workstations run at about 10^9 floating point operations per second. The approximate number of terms computable in the lifetime of the universe is therefore $\sim 10^{26}$, so that $\sim 10^{1294}$ Universal lifetimes are required for the complete θ-function calculation! Even if M is only 1, for $N = 1000$, we get $\sim 10^{477}$ operations or 10^{451} Universal lifetimes. These estimates alone suggest rather emphatically that θ-functions are

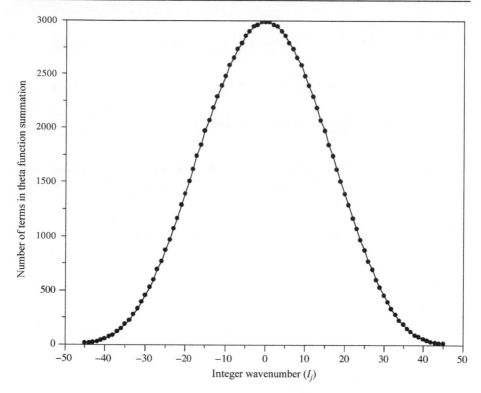

Figure 8.6 Histogram of number of terms in one-dimensional partial theta summation as a function of integer wavenumber for $M = 3$, $N = 5$.

not very useful for physical applications. It is fair to say that the enormous amounts of computer time required, based upon these simple estimates, have discouraged large N applications of θ-functions in soliton systems for many years.

The wavenumbers that appear in a summation of terms in a theta partial sum are given by the binomial distribution. For a sufficiently large number of trials this approaches a Gaussian distribution. A case is shown in Figure 8.6 for $M = 3$, $N = 5$ where the total number of terms in the theta function is given by $(2M + 1)^N = 16,807$. Each point in this histogram gives the number of wavenumbers associated with wavenumber.

8.5 Theta Functions as Ordinary Fourier Series

Now, let us investigate the ordinary Fourier representation for the theta function. We begin with the right-hand side of Equation (8.50):

$$\theta_N(x, 0) = \sum_{n=-\infty}^{\infty} \theta_n e^{ik_n x}$$

$$= a_0 + 2 \sum_{n=1}^{\infty} d_n \cos(k_n x + \phi_n) \qquad (8.49)$$

$$= a_0 + 2 \sum_{n=1}^{\infty} a_n \cos(k_n x) + b_n \sin(k_n x),$$

where we have used the facts that $k_{-n} = k_n$ and $\theta_{-n} = \theta_n^*$ (to ensure that $\theta_N(x, 0)$ is real for applications to the KdV and KP equations) so that $d_n = d_{-n}$, $\phi_{-n} = -\phi_n$ and

$$\theta_n = d_n e^{i\phi_n} = a_n - ib_n \quad \text{and} \quad \theta_{-n} = \theta_n^* = d_n e^{-i\phi_n} = a_n + ib_n,$$

$$d_n = \sqrt{a_n^2 + b_n^2} \quad \text{and} \quad \tan\phi_n = -b_n/a_n,$$

$$a_n = d_n \cos\phi_n \quad \text{and} \quad b_n = -d_n \sin\phi_n.$$

The coefficients are given in terms of inverse scattering variables by

$$\theta_n = \sum_{\substack{\text{sum } l \text{ over subset} \\ \text{of } l \text{ for which } I_l = n}} q_l e^{i\Phi_l}, \quad 1 \le n \le \infty, \qquad (8.50)$$

so that

$$\theta_n = \sum_{\substack{\text{sum } l \text{ over subset} \\ \text{of } l \text{ for which } I_l = n}} q_l \cos(\Phi_l) + iq_l \sin(\Phi_l) = a_n - ib_n,$$

and

$$a_n = \sum_{\substack{\text{sum } l \text{ over subset} \\ \text{of } l \text{ for which } I_l = n}} q_l \cos(\Phi_l), \quad 1 \le n \le \infty,$$

$$b_n = -\sum_{\substack{\text{sum } l \text{ over subset} \\ \text{of } l \text{ for which } I_l = n}} q_l \sin(\Phi_l), \quad 1 \le n \le \infty,$$

where q_l and Φ_l are given in Equations (8.42b) and (8.43), respectively.

We also have

$$d_n = \sqrt{a_n^2 + b_n^2} \quad \text{and} \quad \tan\phi_n = -b_n/a_n,$$

so that

$$d_n = \sqrt{\left(\sum_{\substack{\text{sum } l \text{ over subset} \\ \text{of } l \text{ for which } I_l=n}} q_l \cos(\Phi_l)\right)^2 + \left(\sum_{\substack{\text{sum } l \text{ over subset} \\ \text{of } l \text{ for which } I_l=n}} q_l \sin(\Phi_l)\right)^2},$$

$$\tan\phi_n = \frac{\displaystyle\sum_{\substack{\text{sum } l \text{ over subset} \\ \text{of } l \text{ for which } I_l=n}} q_l \sin(\Phi_l)}{\displaystyle\sum_{\substack{\text{sum } l \text{ over subset} \\ \text{of } l \text{ for which } I_l=n}} q_l \cos(\Phi_l)}.$$

We have therefore reduced the θ-series to an ordinary Fourier transform with coefficients given by θ_n, a_n, b_n, d_n, and phase ϕ_n as given above in terms of inverse scattering transform variables.

Let us discuss some examples of the computation of Equation (8.50) for several simple cases. First rewrite Equation (8.50) as

$$\theta_n = \sum_{\substack{\text{sum } l \text{ over subset} \\ \text{of } l \text{ for which } I_l=n}} q_l e^{i\Phi_l} = \sum_{\substack{\text{sum } l \text{ over subset} \\ \text{of } l \text{ for which } I_l=n}} \exp\left\{i\sum_{i=1}^{N} m_i^l \phi_i - \frac{1}{2}\sum_{i=1}^{N}\sum_{j=1}^{N} m_i^l m_j^l B_{ij}\right\},$$

$$\theta_n = \sum_{\substack{\text{sum } l \text{ over subset} \\ \text{of } l \text{ for which } I_l=n}} q_l \cos(\Phi_l) + iq_l \sin(\Phi_l)$$

$$= \sum_{\substack{\text{sum } l \text{ over subset} \\ \text{of } l \text{ for which } I_l=n}} \exp\left\{-\frac{1}{2}\sum_{i=1}^{N}\sum_{j=1}^{N} m_i^l m_j^l B_{ij}\right\}\left[\cos\left(\sum_{i=1}^{N} m_i^l \phi_i\right) + i\sin\left(\sum_{i=1}^{N} m_i^l \phi_i\right)\right].$$

Written in this fashion, we see that the phase and the period matrix are separated. For applications to the KdV and KP equations, the period matrix is real and hence the modulus of the argument of the above sum is related uniquely to the (real) period matrix, while the phase comes from the phase of the argument. The theta function, in the form of an ordinary Fourier series with commensurable wavenumbers, has the form

$$\theta(x, 0) = \sum_{n=-\infty}^{\infty} \theta_n e^{i k_n x} = a_0 + 2 \sum_{n=1}^{\infty} c_n \cos(k_n x + \phi_n)$$

Now what happens when there is time dependence in the theta function? We have

$$\theta_N(x, t) = \sum_{l=1}^{\infty} q_l e^{i(K_l x - \Omega_l t)} = \sum_{n=1}^{\infty} \theta_n(t) e^{i k_n x} \qquad (8.51)$$

for which

$$\theta_n(t) = \sum_{\substack{\text{sum } l \text{ over subset} \\ \text{of } l \text{ for which } I_l = n}} q_l e^{-i(\Omega_l t - \Phi_l)} \qquad (8.52)$$

Alternatively, we could write Equation (8.51) as

$$\theta_N(x, t) = \sum_{n=1}^{\infty} \theta'_n(t) e^{i(k_n x - \omega_n t + \phi_n)},$$

with the appropriate definition for the parameters $\theta'_n = \theta_n \exp(i \omega_n t - i \phi_n)$.

Now here are the other forms of the Fourier series that have time-varying coefficients:

$$\theta_N(x, t) = \sum_{n=-\infty}^{\infty} \theta_n(t) e^{i k_n x} = a_0 + 2 \sum_{n=1}^{\infty} d_n(t) \cos(k_n x + \phi_n)$$

$$= a_0 + 2 \sum_{n=1}^{\infty} a_n(t) \cos(k_n x) + b_n(t) \sin(k_n x),$$

where

$$\theta_n(t) = d_n(t) e^{i \phi_n} = a_n(t) - i b_n(t),$$

$$d_n(t) = \sqrt{a_n^2(t) + b_n^2(t)} \quad \text{and} \quad \tan(\phi_n) = -b_n(t) / a_n(t),$$

$$a_n(t) = d_n(t) \cos(\phi_n) \quad \text{and} \quad b_n(t) = -d_n(t) \sin(\phi_n).$$

The coefficients are

$$\theta_n(t) = \sum_{\substack{\text{sum } l \text{ over subset} \\ \text{of } l \text{ for which } I_l = n}} q_l e^{-i(\Omega_l t - \Phi_l)}, \quad 1 \leq n \leq \infty,$$

$$a_n(t) = \sum_{\substack{\text{sum } l \text{ over subset} \\ \text{of } l \text{ for which } I_l = n}} q_l \cos(\Omega_l t - \Phi_l), \quad 1 \leq n \leq \infty,$$

$$b_n(t) = \sum_{\substack{\text{sum } l \text{ over subset} \\ \text{of } l \text{ for which } I_l = n}} q_l \sin(\Omega_l t - \Phi_l), \quad 1 \leq n \leq \infty.$$

The θ-functions are different, in that the time-varying coefficients *do not vary sinusoidally in time, but in general have much more complex behavior.* Indeed, the time-varying coefficients are themselves Fourier series over time with incommensurable frequencies. Chapter 9 discusses these issues in greater detail.

8.6 Perturbation Expansion of Theta Functions in Terms of an Interaction Parameter

Recall the theta function in vector form:

$$\theta(\mathbf{X}) = \sum_{\mathbf{m} = -\infty}^{\infty} e^{i\mathbf{m}\cdot\mathbf{X} - (1/2)\mathbf{m}\cdot\mathbf{Bm}},$$

where $\mathbf{X} = [X_1, X_2, \ldots, X_N]$. Let us look at the theta function for only two degrees of freedom and simultaneously make the notation explicit:

$$\theta(X_1, X_2) = \sum_{m_1 = -\infty}^{\infty} \sum_{m_2 = -\infty}^{\infty} \exp\left\{ i(m_1 X_1 + m_2 X_2) - \frac{1}{2} m_1^2 B_{11} - \frac{1}{2} m_2^2 B_{22} - m_1 m_2 B_{12} \right\}.$$

Now set

$$q = e^{-(1/2)B_{11}}, \quad p = e^{-(1/2)B_{22}}, \quad r = e^{-B_{12}}, \tag{8.53}$$

so that

$$\theta(X_1, X_2) = \sum_{m_1 = -\infty}^{\infty} \sum_{m_2 = -\infty}^{\infty} q^{m_1^2} p^{m_2^2} r^{m_1 m_2} e^{i(m_1 X_1 + m_2 X_2)}.$$

Replace r by

$$r = 1 - \varepsilon.$$

Then use the series expansion

$$(1 - \varepsilon)^m = 1 + \sum_{n=1}^{\infty} (-1)^n \binom{m}{n} \varepsilon^n$$

$$= 1 - m\varepsilon + \frac{1}{2!} m(m-1)\varepsilon^2 - \frac{1}{3!} m(m-1)(m-2)\varepsilon^3 + \cdots.$$

Here, we think of ε as being small. This is motivated by the fact that the theta function, for $r = 1$, has no interaction terms:

$$\theta(X_1, X_2) = \sum_{m_1=-\infty}^{\infty} q^{m_1^2} e^{im_1 X_1} \sum_{m_2=-\infty}^{\infty} p^{m_2^2} e^{im_2 X_2} = \theta_1(X_1)\theta_2(X_2).$$

Thus when there are no interactions, the theta function is reduced to the product of two elementary theta functions (called θ_3 by Jacobi).

Now, let us replace $r = 1 - \varepsilon$ by the above series expansion in the theta function:

$$\theta(X_1, X_2) = \sum_{m_1=-\infty}^{\infty} \sum_{m_2=-\infty}^{\infty} q^{m_1^2} p^{m_2^2} (1-\varepsilon)^{m_1 m_2} e^{i(m_1 X_1 + m_2 X_2)}.$$

This gives

$$\theta(X_1, X_2) = \sum_{m_1=-\infty}^{\infty} \sum_{m_2=-\infty}^{\infty} q^{m_1^2} p^{m_2^2} e^{i(m_1 X_1 + m_2 X_2)}$$

$$- \varepsilon \sum_{m_1=-\infty}^{\infty} \sum_{m_2=-\infty}^{\infty} m_1 m_2 q^{m_1^2} p^{m_2^2} e^{i(m_1 X_1 + m_2 X_2)}$$

$$+ \frac{\varepsilon^2}{2} \sum_{m_1=-\infty}^{\infty} \sum_{m_2=-\infty}^{\infty} m_1 m_2 (m_1 m_2 - 1) q^{m_1^2} p^{m_2^2} e^{i(m_1 X_1 + m_2 X_2)}$$

$$- \frac{\varepsilon^3}{6} \sum_{m_1=-\infty}^{\infty} \sum_{m_2=-\infty}^{\infty} m_1 m_2 (m_1 m_2 - 1)(m_1 m_2 - 2) q^{m_1^2} p^{m_2^2} e^{i(m_1 X_1 + m_2 X_2)} + \cdots.$$

This leads to considerable simplification, since

$$m_1 m_2 (m_1 m_2 - 1) = m_1^2 m_2^2 - m_1 m_2,$$

$$m_1 m_2 (m_1 m_2 - 1)(m_1 m_2 - 2) = m_1^3 m_2^3 - 3 m_1^2 m_2^2 + 2 m_1 m_2.$$

Hence, the theta function expansion becomes

$$\theta(X_1, X_2) = \sum_{m_1=-\infty}^{\infty} q^{m_1^2} e^{im_1 X_1} \sum_{m_2=-\infty}^{\infty} p^{m_2^2} e^{im_2 X_2}$$

$$-\varepsilon \sum_{m_1=-\infty}^{\infty} m_1 q^{m_1^2} e^{im_1 X_1} \sum_{m_2=-\infty}^{\infty} m_2 p^{m_2^2} e^{im_2 X_2}$$

$$+\frac{\varepsilon^2}{2} \sum_{m_1=-\infty}^{\infty} \sum_{m_2=-\infty}^{\infty} (m_1^2 m_2^2 - m_1 m_2) q^{m_1^2} p^{m_2^2} e^{i(m_1 X_1 + m_2 X_2)}$$

$$-\frac{\varepsilon^3}{6} \sum_{m_1=-\infty}^{\infty} \sum_{m_2=-\infty}^{\infty} (m_1^3 m_2^3 - 3m_1^2 m_2^2 + 2m_1 m_2) q^{m_1^2} p^{m_2^2} e^{i(m_1 X_1 + m_2 X_2)} + \cdots.$$

Note that

$$\partial_{X_1} \theta_1(X_1) \partial_{X_2} \theta_2(X_2) = -\sum_{m_1=-\infty}^{\infty} \sum_{m_2=-\infty}^{\infty} m_1 m_2 q^{m_1^2} p^{m_2^2} e^{i(m_1 X_1 + m_2 X_2)}$$

$$= -\sum_{m_1=-\infty}^{\infty} m_1 q^{m_1^2} e^{im_1 X_1} \sum_{m_2=-\infty}^{\infty} m_2 p^{m_2^2} e^{im_2 X_2}.$$

Other simplifying relations also follow. Introduce the new notation

$$\partial_{X_1} \theta_1(X_1) \equiv \theta_1', \quad \partial_{X_2} \theta_2(X_2) \equiv \theta_2'$$

and get the following result for the above theta function expansion:

$$\theta(X_1, X_2) = \theta_1 \theta_2 + \varepsilon \theta_1' \theta_2' + \frac{1}{2} \varepsilon^2 [\theta_1'' \theta_2'' - \theta_1' \theta_2'] + \frac{1}{6} \varepsilon^3 [\theta_1''' \theta_2''' - 3\theta_1'' \theta_2'' + 2\theta_1' \theta_2'] + \cdots$$

$$(8.54)$$

This compact result indicates that the full two-dimensional theta function, for small interactions, can be written explicitly as a series in the two *one-dimensional* theta functions and its derivatives.

8.7 N-Mode Interactions

The term "N-mode interactions" refers to the number of nonzero elements in the **m** vector in the theta function summation. Consider the integer vector **m** = $\{m_1, m_2, \ldots, m_N\}$ of the forms $[\cdot, 0, 0\ldots0]$, $[0, \cdot, 0\ldots0]$, $[0, 0, \cdot\ldots0],\ldots,$ $[0, 0, 0\ldots\cdot]$, that is, all the elements in the dimension vector **m** of the lattice are zero except for one. In the same way, two-mode interactions occur when two elements of the integer vector **m** are nonzero: $[\cdot, \cdot, 0, 0\ldots0]$, $[0, \cdot, \cdot, 0\ldots0]$,

$[0, \cdot, 0, \cdot \ldots 0], \ldots, [0, 0, 0 \ldots \cdot, \cdot]$. In this way, one can decompose the theta function solution into a sum of one-mode, two-mode, three-mode, etc., interactions. The single modes correspond to the sum-of-cnoidal-waves contribution discussed in a previous section. The sum of the two-mode interactions corresponds to all possible combinations of paired cnoidal wave interactions. We can write this symbolically in the following way:

$$
\theta = 1 + 2 \sum_{\substack{[\cdot, 0 \ldots 0] \\ [0, \cdot \ldots 0] \\ \vdots \\ [0, 0 \ldots \cdot]}} + 2 \sum_{\substack{[\cdot, \cdot \ldots 0] \\ [\cdot, 0, \cdot \ldots 0] \\ \vdots \\ [0, 0 \ldots \cdot, \cdot]}} + 2 \sum_{\substack{[\cdot, \cdot, \cdot \ldots 0] \\ [\cdot, 0, \cdot, \cdot \ldots 0] \\ \vdots \\ [0, 0 \ldots \cdot, \cdot, \cdot 1]}} + 2 + \cdots.
$$

The theta function can thus be written as a sum over wavenumber singles, pairs, triples, etc. This representation is shown in Figures 8.1, 8.3, and 8.4 (open squares denote points of spectrum) where the number of modal interactions is graphed as a function of the ordering parameter, l.

8.8   Poisson Summation for Theta Functions

Gaussian series (also known as *Poisson summation*) are also useful for computing theta functions. I first discuss the one-dimensional case and then address N degrees of freedom.

8.8.1   Gaussian Series for One-Degree-of-Freedom Theta Functions

The simple Poisson summation formula is given by Bellman (1961) and Boyd (1984a, b, c). The classical theta function

$$
\theta_4(x) = 1 + 2 \sum_{n=1}^{\infty} (-1)^n q^{n^2} \cos(2nkx) \tag{8.55}
$$

can be rewritten using *Poisson summation* to give a Gaussian series for the simple theta function, θ_4:

$$
\theta_4(x) = s^{1/2} \sum_{n=-\infty}^{\infty} \exp\left\{ -s\left[kx/2 - \frac{\pi}{2}(2n+1) \right]^2 / \pi \right\}. \tag{8.56}
$$

Note this latter formula consists of a sequence of *Gaussians* on the periodic interval and hence for the KdV or KP equations when one takes the second derivative of the logarithm of the theta function, one maps an infinite number of Gaussians (over an infinite number of spatially periodic intervals) to a soliton on $(0, L)$.

One can also write the formulas for the Jacobian θ_3 function:

$$\theta_3(x) = 1 + 2\sum_{n=1}^{\infty} q^{n^2} \cos(2nkx),$$

for which the Poisson summation is given by

$$\theta_3 = s^{1/2} \sum_{n=-\infty}^{\infty} \exp\{-s[kx/2 - \pi n]^2/\pi\}$$

or

$$\theta_3 = s^{1/2} \sum_{n=-\infty}^{\infty} \exp\left\{-\frac{1}{2B}[kx - 2\pi n]^2\right\},$$

where

$$q = e^{-(1/2)B} = e^{-\pi/s},$$

so that

$$B^{-1} = \frac{s}{2\pi} = \frac{1}{2\pi}\frac{K}{K'}, \quad B = \frac{2\pi}{s} = 2\pi\frac{K'}{K}.$$

This is exactly what we expect for a single degree of freedom. Bellman points out the importance of using the Gaussian series when the parameter B is small, that is, in the soliton limit for the KdV or KP equations. Indeed when one is near the soliton limit, a summation over $n = -1, 0, 1$ can be a quite accurate approximation to the soliton.

Another important relation for the single-degree-of-freedom theta function is

$$\theta_4 = 2s^{1/2} \exp\{-s(kx)^2/\pi\} \sum_{n=0}^{\infty} q'^{(n+1/2)^2} \cosh[(2n+1)skx]. \tag{8.57}$$

This can be easily derived from the Poisson summation formula above. In these formulas the following relations hold (Abramowitz and Stegun, 1964):

$$q = e^{-\pi/s} \text{ (the nome)},$$

$$q' = e^{-\pi s} \text{ (the complementary nome)}.$$

Hence

$$q' = e^{\pi^2/\ln q},$$

$$s = -\pi/\ln q.$$

Furthermore,

$$s = K/K',$$

where K and K' are the usual elliptic integrals.

As $q \to 1$ $(q' \to 0)$, the Gaussian series gives a single soliton solution of the KdV equation. In the limit as $q \to 0$ $(q' \to 1)$, the Gaussian series gives a sine wave.

A main goal at this point is to convert the usual N-degree-of-freedom theta summation into a Gaussian series. Then, in the presence of many solitons, one expects that the convergence of the Gaussian series will be much more rapid than the usual N-degree-of-freedom theta series.

Nota Bene. Return to the case of the theta function converging to a Gaussian series. Recall that because the second derivative of the log of a Gaussian is a constant, nothing very important occurs in terms of nonlinear wave propagation. To this end we return to the Gaussian series:

$$\theta_4(x) = s^{1/2} \sum_{n=-\infty}^{\infty} \exp\left\{-\frac{s}{4\pi}[kx - (2n+1)\pi]^2\right\}.$$

Now if we want to better understand a single soliton, note that $q \to 1$ $(q' \to 0)$, which implies that the above series is rapidly convergent. How does this occur? Consider the summation over only 0, 1. Then we get

$$\theta_4 = s^{1/2}\exp\left\{-\frac{s}{4\pi}[kx - \pi]^2\right\} \qquad (n = 0)$$

$$+ s^{1/2}\exp\left\{-\frac{s}{4\pi}[kx - 3\pi]^2\right\} \qquad (n = 1).$$

So what we seem to have is our fundamental Gaussian inside the periodic interval plus two other Gaussians outside the interval (immediately to the right and the left) whose tails contribute to the non-Gaussian behavior inside the periodic interval. It is this difference between Gaussian and non-Gaussian behavior which gives rise to a single soliton solution. To better understand this point take a look at the argument of the Gaussians in the Poisson series above:

$$kx - (2n+1)\pi = \frac{2\pi}{L}[x - x_n],$$

where $k = 2\pi/L$ and

$$x_n = (2n + 1)\frac{L}{2}.$$

Furthermore, the argument of the exponential is given by

$$\frac{s}{4\pi}[kx - (2n + 1)\pi]^2 = \frac{1}{2}\left[\frac{x - x_n}{\sigma}\right]^2, \quad \sigma = \sqrt{\frac{2\pi}{sk^2}}.$$

Hence x_n lies at the center of each Gaussian in the Poisson summation formula for the Jacobi θ_4 notation. This emphasizes that the Poisson summation is an array of equally spaced Gaussians on the real axis.

It is worthwhile looking at the above results from the point of view of successive approximations to the infinite Poisson series. Summing over only the $n = 0$ term gives the Gaussian at the center of the interval $(0, L)$, for which $x_0 = L/2$; taking the second derivative of the logarithm of this gives a constant for the solution to KdV, clearly a poor approximation! The next approximation is for $n = -1, 0, 1$ for which $x_{-1} = -L/2$, $x_0 = L/2$, and $x_1 = 3L/2$. The three Gaussians here provide a fine approximation for the soliton limit; it is the effect of the overlapping tails of the Gaussians from adjacent periods that generates the solitons quite nicely. These results can be seen in Figure 8.8, where the spatial period of interest for solutions of KdV (see Chapter 10 for further discussion) is $(0, L) = (0, 256 \text{ cm})$ and the total interval graphed is $(-L, 2L) = (-256, 512)$; the three Gaussians can be seen to lie at the center of each of these three spatial periods.

Now note from Figure 8.7 that the solitons occur in the intervals between the Gaussians, where the tails overlap. Near the peaks of the Gaussians, the second derivative of the logarithms gives a constant, while in the intermediate regions where the tails dominate, the solitons occur! On the basis of these results, it behoves one to think about using Poisson summation to approximate soliton behavior on the periodic interval; clearly for moduli close to 1, we can use three Gaussians in the series to good approximation. Note that in Figure 8.8, even though there are two soliton peaks, there is in fact only one soliton on the periodic interval of interest $(0, L)$. Thus, the three Gaussians on three adjacent periodic intervals work together to make one soliton on the fundamental interval $(0, L)$. In reality, it takes an infinite number of Gaussians over an infinite number of adjacent periods to make one soliton in each of these periods or, more importantly, to make one soliton on the fundamental period $(0, L)$. The bottom line in all this is that it is nice to know that the approximation $n = -1, 0, 1$ works so well numerically near the soliton limit (Figure 8.8).

Figure 8.7 Poisson series of Gaussians for a soliton with period of 256 cm.

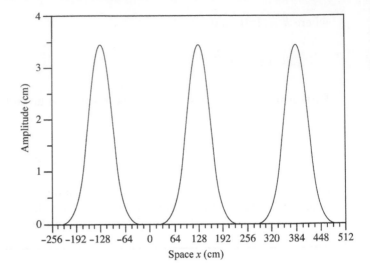

Figure 8.8 Single soliton solution on interval (0, 256 cm) are actually two solitons on (−256, 512 cm).

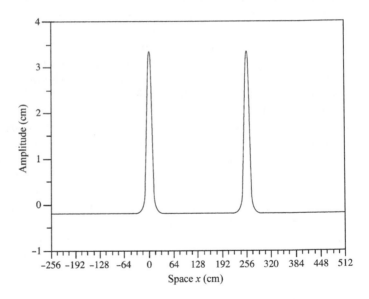

8.8.2 The Infinite-Line Limit

The infinite-line limit occurs for only two Gaussians; the overlapping tails of the two Gaussians give the soliton after taking the second derivative of the log of the sum of the Gaussians. These results are shown in Figures 8.9 and 8.10 where two Gaussians are shown symmetrically placed about the origin. This perspective should be compared to *Hirota's method* for the infinite line (see Section 8.9).

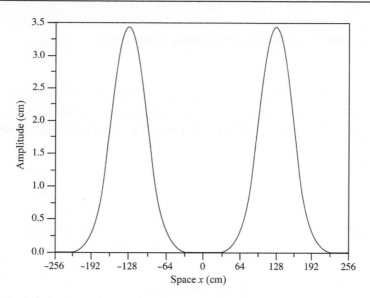

Figure 8.9 Infinite-line soliton arises from two Gaussians.

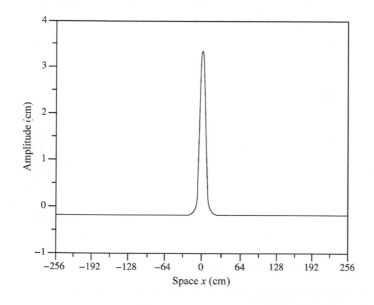

Figure 8.10 An infinite-line soliton arises from the two Gaussians in Figure 8.9.

8.8.3 Fourier and Gaussian Series for N-Dimensional Theta Functions

The θ-function in two dimensions is given by

$$\theta(X_1, X_2) = \sum_{m_1=-\infty}^{\infty} \sum_{m_2=-\infty}^{\infty} \exp\left\{ -\frac{1}{2}[B_{11}m_1^2 + 2B_{12}m_1m_2 + B_{22}m_2^2] + i(m_1X_1 + m_2X_2) \right\}.$$

$$(8.58)$$

Let us now look at the corresponding Poisson summation formula for two degrees of freedom (Bellman, 1961; Boyd, 1984a, b, c)

$$\theta(X_1, X_2) = \sum_{m_1=-\infty}^{\infty} \sum_{m_2=-\infty}^{\infty}$$

$$\exp\left\{-\frac{1}{2}[F_{11}(X_1 - \pi m_1)^2 + 2F_{12}(X_1 - \pi m_1)(X_2 - \pi m_2) + F_{22}(X_2 - \pi m_2)^2]\right\}$$

$$(8.59)$$

for

$$\mathbf{F} = \mathbf{B}^{-1}. \tag{8.60}$$

8.8.4 Gaussian Series for Theta Functions

It is straightforward to show that the N-dimensional Poisson summation is

$$\theta(\mathbf{X}) = \sum_{\mathbf{m}} \exp\left\{-\frac{1}{2}\mathbf{\chi}^{\mathsf{T}}\mathbf{F}\mathbf{\chi}\right\}, \tag{8.61}$$

where

$$\begin{aligned}
&\mathbf{\chi} = \mathbf{X} - \pi\mathbf{m}, \\
&\mathbf{X} = [X_1, X_2, \dots, X_N], \quad \mathbf{m} = [m_1, m_2, \dots, m_N], \\
&\chi_i = X_i - \pi m_i, \quad X_i = k_i x - \omega_i t + \phi_i, \\
&\mathbf{F} = \mathbf{B}^{-1}.
\end{aligned} \tag{8.62}$$

We see that we now have an N-dimensional Gaussian series for the theta function. The summation (8.61) is over the positive, zero and negative integers.

8.8.5 One-Degree-of-Freedom Gaussian Series

From the above, it is clear that for a single degree of freedom we have

$$\theta(X_1) = \sum_{m=-\infty}^{\infty} \exp\left\{-\frac{1}{2}F_{11}(X_1 - \pi m)^2\right\}. \tag{8.63}$$

From previous considerations, near the soliton limit, we have for $m = 0, 1$

$$\theta(X_1) = e^{-(1/2)F_{11}X_1^2} + e^{-(1/2)F_{11}(X_1-\pi)^2},$$

where $X_1 = k_1 x - \phi_1$. Generally speaking,

$$\theta(x) = \sum_{m=-\infty}^{\infty} \exp\left\{-\frac{1}{2}F_{11}(k_1 x - \pi m - \phi_1)^2\right\},$$

where

$$F_{11} = 1/B_{11}.$$

8.8.6 Many-Degree-of-Freedom Gaussian Series

In Osborne (1995a,b), I showed that the solutions to KdV can be represented as a linear superposition of cnoidal waves plus nonlinear interactions among the cnoidal waves. Can this idea, based upon the Fourier series for theta functions, be extended to theta functions represented by *Gaussian series*? The answer is yes and the proof follows easily. We have

$$\theta(\mathbf{X}) = \sum_{\mathbf{m}} \exp\left[-\frac{1}{2}\boldsymbol{\chi}^{T}\mathbf{F}\boldsymbol{\chi}\right],$$

where the summation is over the integers as before and

$$\mathbf{F} = \mathbf{B}^{-1}.$$

Suppose that there are no interactions, so that the off-diagonal terms in the period matrix, $\mathbf{B}$, are zero, that is,

$$\mathbf{B} = \mathbf{D} + \mathbf{O} = \mathbf{D},$$

so that

$$\mathbf{F} = \mathbf{B}^{-1} = \mathbf{D}^{-1} = \begin{bmatrix} \dfrac{1}{D_{11}} & 0 & 0 & 0 \\[2mm] 0 & \dfrac{1}{D_{22}} & 0 & 0 \\[2mm] \vdots & \vdots & \ddots & \vdots \\[2mm] 0 & 0 & 0 & \dfrac{1}{D_{NN}} \end{bmatrix}.$$

Therefore, we can write

$$\theta(x,t) = \sum_m \exp\left\{-\frac{1}{2}\mathbf{x} \cdot \mathbf{F}\mathbf{x}\right\} = \sum_m \exp\left\{-\frac{1}{2}\mathbf{x} \cdot \mathbf{B}^{-1}\mathbf{x}\right\},$$

$$\theta = \sum_m \exp\left\{-\frac{1}{2}\mathbf{x} \cdot \mathbf{D}^{-1}\mathbf{x}\right\} + \left[\sum_m \exp\left\{-\frac{1}{2}\mathbf{x} \cdot \mathbf{B}^{-1}\mathbf{x}\right\} - \sum_m \exp\left\{-\frac{1}{2}\mathbf{x} \cdot \mathbf{D}^{-1}\mathbf{x}\right\}\right],$$

$$\theta = \sum_m \exp\left\{-\frac{1}{2}\mathbf{x} \cdot \mathbf{D}^{-1}\mathbf{x}\right\}\left[1 + \frac{\sum_m \exp\left\{-\frac{1}{2}\mathbf{x} \cdot \mathbf{B}^{-1}\mathbf{x}\right\} - \sum_m \exp\left\{-\frac{1}{2}\mathbf{x} \cdot \mathbf{D}^{-1}\mathbf{x}\right\}}{\sum_m \exp\left\{-\frac{1}{2}\mathbf{x} \cdot \mathbf{D}^{-1}\mathbf{x}\right\}}\right].$$

Hence

$$\ln\theta = \ln\sum_m \exp\left\{-\frac{1}{2}\mathbf{x} \cdot \mathbf{D}^{-1}\mathbf{x}\right\}$$

$$+ \ln\left[1 + \frac{\sum_m \exp\left\{-\frac{1}{2}\mathbf{x} \cdot \mathbf{B}^{-1}\mathbf{x}\right\} - \sum_m \exp\left\{-\frac{1}{2}\mathbf{x} \cdot \mathbf{D}^{-1}\mathbf{x}\right\}}{\sum_m \exp\left\{-\frac{1}{2}\mathbf{x} \cdot \mathbf{D}^{-1}\mathbf{x}\right\}}\right].$$

Take the second derivative of the logarithm of this latter expression to find the solution of the KdV equation:

$$u(x,t) = \lambda\eta(x,t) = 2\partial_{xx}\ln\theta$$

$$= 2\partial_{xx}\ln\sum_m \exp\left\{-\frac{1}{2}\mathbf{x} \cdot \mathbf{D}^{-1}\mathbf{x}\right\} +$$

$$2\frac{\partial^2}{\partial x^2}\ln\left[1 + \frac{\sum_m \exp\left\{-\frac{1}{2}\mathbf{x} \cdot \mathbf{B}^{-1}\mathbf{x}\right\} - \sum_m \exp\left\{-\frac{1}{2}\mathbf{x} \cdot \mathbf{D}^{-1}\mathbf{x}\right\}}{\sum_m \exp\left\{-\frac{1}{2}\mathbf{x} \cdot \mathbf{D}^{-1}\mathbf{x}\right\}}\right].$$

This reduces to

$$u(x,t) = \lambda\eta(x,t) = u_{\text{sol}}(x,t) + u_{\text{int}}(x,t),$$

where

$$u_{sol}(x, t) = 2\partial_{xx} \ln \sum_m \exp\left\{-\frac{1}{2}\chi \cdot \mathbf{D}^{-1}\chi\right\},$$

$$u_{int}(x, t) = 2\partial_{xx} \ln \left[1 + \frac{\sum_m \exp\left\{-\frac{1}{2}\chi \cdot \mathbf{B}^{-1}\chi\right\} - \sum_m \exp\left\{-\frac{1}{2}\chi \cdot \mathbf{D}^{-1}\chi\right\}}{\sum_m \exp\left\{-\frac{1}{2}\chi \cdot \mathbf{D}^{-1}\chi\right\}}\right].$$

$$(8.64)$$

The first term above is just the linear superposition of the soliton components. The interaction term is required to include the soliton phase shifts. This latter expression provides the general formula for soliton phase shifts on the periodic interval.

8.8.7 Comments on Numerical Analysis

I have implemented the N-dimensional Gaussian series given above numerically and have found it to be wanting. The biggest problem occurs because we are using Gaussians to simulate *periodic* functions, that is, we are using a localized function to represent an oscillatory function. While the Gaussian series converge rapidly in the soliton limit they converge slowly in the linear limit. The main problem with Gaussian series is that I have not been able to find a way to compute them rapidly. This contrasts with the linear Fourier series representation for the theta function which in Chapter 9 I show can lead to rapid execution times.

8.8.8 Modular Transformations for Computing Theta Function Parameters

One can transform the *oscillatory basis* to the *soliton basis* by a modular transform using the following matrix:

$$\mathbf{A} = \begin{bmatrix} 1 & 0 & 0 & 0 & 0 \\ -1 & 1 & 0 & 0 & 0 \\ 0 & -1 & 1 & 0 & 0 \\ 0 & 0 & -1 & 1 & 0 \\ 0 & 0 & 0 & -1 & 1 \end{bmatrix}. \tag{8.65}$$

Thus, the matrix $\mathbf{A}$ has 1 on the diagonal terms and -1 on the lower off-diagonal terms, with zeroes everywhere else. See Siegel (1969c) for a rigorous derivation and an extended definition of the modular transformation.

The *special modular transformation* of the θ-variables is written as follows:

$$\hat{\mathbf{X}} = \mathbf{A}\mathbf{X}$$

for

$$\mathbf{X} = [X_1, X_2, \ldots, X_N],$$
$$X_i = k_i x - \omega_i t + \phi_i.$$

Therefore,

$$\hat{\mathbf{k}} = \mathbf{A}\mathbf{k},$$
$$\hat{\omega} = \mathbf{A}\omega, \tag{8.66}$$
$$\hat{\phi} = \mathbf{A}\phi.$$

The period matrix has the transformed expression:

$$\hat{\mathbf{B}} = \mathbf{A}\mathbf{B}\mathbf{A}^{\mathrm{T}}. \tag{8.67}$$

The modular transformation considered here is that referred to as $\mathbf{A}_2^{-1}$ by Boyd (1984a, b, c).

For the case $N = 2$,

$$\hat{\mathbf{k}} = \mathbf{A}\mathbf{k} = \Delta k \begin{bmatrix} 1 & 0 \\ -1 & 1 \end{bmatrix} \begin{bmatrix} 1 \\ 2 \end{bmatrix} = \Delta k \begin{bmatrix} 1 \\ 1 \end{bmatrix}$$

as required for the *soliton basis*, that is, all cnoidal wave components in the spectrum have period $L = 2\pi/\Delta k$. Here, I am assuming commensurable wavenumbers as in Osborne (1995a,b):

$$\mathbf{k} = \Delta k[1, 2, \ldots, N].$$

In the two dimensional case we have:

$$\hat{\mathbf{X}} = \mathbf{A}\mathbf{X} = \begin{bmatrix} 1 & 0 \\ -1 & 1 \end{bmatrix} \begin{bmatrix} X_1 \\ X_2 \end{bmatrix} = \begin{bmatrix} X_1 \\ X_2 - X_1 \end{bmatrix},$$

for which

$$X_1 = k_1 x - \omega_1 t + \phi_1,$$
$$X_2 = k_2 x - \omega_2 t + \phi_2.$$

Further,

$$X_2 - X_1 = (k_2 - k_1)x - (\omega_2 - \omega_1)t + \phi_2 - \phi_1.$$

Therefore, the modular transformed wavenumbers, frequencies, and phases are given by

$$\hat{k}_1 = k_1,$$

$$\hat{k}_2 = k_2 - k_1,$$

$$\hat{\omega}_1 = \omega_1,$$

$$\hat{\omega}_2 = \omega_2 - \omega_1,$$

$$\hat{\phi}_1 = \phi_1,$$

$$\hat{\phi}_2 = \phi_2 - \phi_1.$$

Now consider the *phase speed* of the transformed components. We have

$$\omega_i = c_i k_i,$$

$$X_i = k_i(x - c_i t) + \phi_i.$$

Hence

$$\hat{\omega}_2 = \omega_2 - \omega_1 = k_2 c_2 - k_1 c_1.$$

The transformed phase speeds are

$$\hat{c}_1 = c_1,$$

$$\hat{c}_2 = \frac{\hat{\omega}_2}{\hat{k}_2} = \frac{k_2 c_2 - k_1 c_1}{k_2 - k_1}.$$

For the period matrix, we find

$$\hat{\mathbf{B}} = \mathbf{A}\mathbf{B}\mathbf{A}^{\mathrm{T}},$$

$$\hat{\mathbf{B}} = \begin{bmatrix} 1 & 0 \\ -1 & 1 \end{bmatrix} \begin{bmatrix} B_{11} & B_{12} \\ B_{12} & B_{22} \end{bmatrix} \begin{bmatrix} 1 & -1 \\ 0 & 1 \end{bmatrix}, \tag{8.68}$$

$$\hat{\mathbf{B}} = \begin{bmatrix} B_{11} & B_{12} - B_{11} \\ B_{12} - B_{11} & B_{22} - 2B_{12} + B_{11} \end{bmatrix}$$

The modular transformation for *Gaussian series* has the simple form:

$$\hat{R} = \left[A^T\right]^{-1} R A^{-1},$$

$$\hat{R} = \begin{bmatrix} 1 & 1 \\ 0 & 1 \end{bmatrix} \begin{bmatrix} R_{11} & R_{12} \\ R_{12} & R_{22} \end{bmatrix} \begin{bmatrix} 1 & 0 \\ 1 & 1 \end{bmatrix}, \tag{8.69}$$

$$\hat{R} = \begin{bmatrix} R_{11} + 2R_{12} + R_{22} & R_{12} + R_{22} \\ R_{12} + R_{22} & R_{22} \end{bmatrix}$$

8.9   Solitons on the Infinite Line and on the Periodic Interval

Consider the KdV equation in normalized form: $u_t + 6uu_x + u_{xxx} = 0$. The soliton solution (at $t = 0$) is given by

$$u(x, 0) = 2K^2 \text{sech}^2(Kx).$$

We know from the method of Hirota (2004) and from the N-soliton solution of the KdV equation that

$$u(x, 0) = 2\partial_{xx} F(x).$$

This means that an appropriate choice for the function $F(x)$ is

$$F(x) = \cosh(Kx).$$

An intermediate step is

$$\partial_x \ln \theta(x) = \tanh(Kx).$$

Now, suppose we take a "truncated" version of the single soliton solution on some periodic interval, $-L/2 \le x \le L/2$. Since the soliton decays exponentially to the left and right of the peak, we can represent this curve to "graphical accuracy" by retaining enough of the "tails," so that, while the function does not go exactly to zero, it is very small in the interval $(0, L)$. Furthermore, on this interval let us remove the mean to conserve water mass and then we have

$$u(x,0) = 2K^2\text{sech}^2(kx) - 2K/\pi, \quad -L/2 \le x \le L/2, \quad \int_{-L/2}^{L/2} u(x,0)dx = 0 \quad (8.70)$$

To arrive at this result, it has been assumed that

$$\int_{-\infty}^{\infty} u(x,0)dx = 2K^2 \int_{-\infty}^{\infty} \text{sech}^2(Kx)dx = 4K \cong 2K^2 \int_{-L/2}^{L/2} \text{sech}^2(Kx)dx.$$

Dividing this result by 2π gives the mean, which I have removed from the soliton above. The soliton on the interval $-L/2 \le x \le L/2$ with its mean removed is described by the approximate theta function:

$$\theta(x) \cong e^{-Kx^2/2\pi} \cosh(Kx)$$

This latter expression is just the sum of two Gaussians. It is easy to show that

$$u(x,t) = 2\partial_{xx} \ln \theta(x) = 2K^2\text{sech}^2(kx) - 2K/\pi$$

An intermediate step is

$$\partial_x \ln \theta(x) = K \tanh(Kx) - Kx/\pi.$$

Viewed in this way, the above expression is just the sum of two Gaussians, as anticipated.

8.10 N-Dimensional Theta Functions as a Sum of One-Degree-of-Freedom Thetas

Here is a useful identity relating different forms of theta functions:

$$\theta(x,0) = \sum_{m=-\infty}^{\infty} \exp\left\{ -\frac{1}{2}\mathbf{m}\cdot\mathbf{Bm} + i\mathbf{m}\cdot\mathbf{k}x + i\mathbf{m}\cdot\boldsymbol{\phi} \right\}$$

$$= 1 + 2\sum_{m=-\infty}^{0^-} q_m \cos[\mathbf{m}\cdot\mathbf{k}x + \mathbf{m}\cdot\boldsymbol{\phi}] = \sum_{m'=-\infty}^{0^-} \theta_m(x,0) - P \qquad (8.71)$$

The *first form* is the usual definition of the theta function as a Fourier series, a linear superposition of complex exponentials. The *second form* is a cosine series, that is, a linear superposition of trigonometric functions. The *third form* is in terms of the *single-degree-of-freedom theta function*, a linear superposition of one-dimensional theta functions. The one-dimensional theta functions have the expression

$$\theta_{\mathbf{m}}(x) = 1 + 2\sum_{n=1}^{\infty} q_{\mathbf{m}}^{n^2} \cos[n(\mathbf{m}\cdot\mathbf{X})] = 1 + 2\sum_{n=1}^{\infty} q_{\mathbf{m}}^{n^2} \cos[n(\mathbf{m}\cdot\mathbf{k}x + \mathbf{m}\cdot\boldsymbol{\phi})] \quad (8.72)$$

In Equation (8.71), the *last summation is special*: It is over the integer vectors in the range between minus infinity and $\mathbf{m} = \{0, 0, \ldots, -1\}$ for which the sum is over integer vectors *not divisible by any positive integer except one*. The prime on the integer vector $\mathbf{m}$ ($\mathbf{m}'$) signifies this summation over the restricted integer set: All integer vectors divisible by the positive integers, $n \neq 1$, are excluded from the summation. This means that vectors of the form, say, of $\mathbf{m} = \{1, 2, 3\}$ are included in the summation set, but not $\mathbf{m} = 2\{1, 2, 3\}=\{2, 4, 6\}$ or $\mathbf{m} = 5\{1, 2, 3\}=\{5, 10, 15\}$, that is, these are harmonics and should not be included in the summation as they are already included in the one-degree-of-freedom theta summations (Equation (8.72)) themselves.

Another observation about Equations (8.71) and (8.72) is in order. The one-degree-of-freedom theta functions are kinds of *Stokes waves* since the harmonics are all phase locked with each other. Note further that the last summation in Equation (8.71) says that N-dimensional theta functions can be viewed as an infinite summation of one-degree-of-freedom theta functions, that is, an infinite *summation of free waves* $\theta_{\mathbf{m}}(x|q)$. Each $\theta_{\mathbf{m}}(x|q)$, however, is a Stokes wave that may be viewed as an *infinite sum of bound waves* due to phase locking. Thus, free (one-degree-of-freedom theta functions) and bound modes (phase-locked components of one-degree-of-freedom theta functions) are automatically included in the Fourier analysis of N-dimensional theta functions.

Furthermore in Equation (8.71) P, a constant, is the number of *hidden dimensions* in the sum. Formally P is *infinity*, but practically speaking for numerical calculations it is given by

$$P = \frac{1}{2}[(2M + 1)^N - 1] - 1 \qquad (8.73)$$

where M is the limit in the summations (numerically finite, see Equation (8.76)) and N is the number of degrees of freedom (number of nested summations in the first form of Equation (8.71)).

Example for a finite summation (partial sum): If $M = 1$, $N = 2$, we get $P = 3$.

In the theta functions, I am assuming that the Riemann matrix, $\mathbf{B}$, and the phases, $\boldsymbol{\phi}$, are given. Here 0^- means sum up to $\mathbf{m} = \{0, 0, \ldots, -1\}$ and then stop. The term $\mathbf{m} = \{0, 0, \ldots, 0\}$ has already been included in the thetas, that is, this is just the "1" in the second and third forms of Equation (8.71).

Equation (8.72) is a single-degree-of-freedom theta function with *nome*:

$$q_\mathrm{m} = e^{-(1/2)\mathbf{m}\cdot\mathbf{Bm}} \tag{8.74}$$

and *argument*:

$$\mathbf{m}\cdot\mathbf{X} = \mathbf{m}\cdot\mathbf{k}x + \mathbf{m}\cdot\boldsymbol{\phi}. \tag{8.75}$$

Note that Equations (8.74) and (8.75) are *scalar inputs* to the single-degree-of-freedom theta (Equation (8.72)). Equation (8.72) is just the theta function referred to by Jacobi as θ_3 after suitable changes of definition for the nome and argument.

8.11 N-Dimensional Partial Theta Sums over One-Degree-of-Freedom Theta Functions

Let us consider the following examples of *partial theta sums* as determined from Equation (8.71) by summing over *particular terms* in the series (see Equation (8.76)). The first partial sum in Equation (8.76) is already used in my book. The second type of partial sum in Equation (8.76) is new, that is, it applies the notion that a linear superposition of single-degree-of-freedom theta functions can be used to express the more complex N-dimension theta function:

$$
\begin{aligned}
\theta_{MN}(x,0) &= 1 + 2 \sum_{\mathbf{m}=-\mathbf{M}}^{0^-} q_\mathrm{m} \cos[\mathbf{m}\cdot\mathbf{k}x + \mathbf{m}\cdot\boldsymbol{\phi}] \\
&= \sum_{\mathbf{m}'=-\mathbf{M}}^{0^-} \theta_\mathrm{m}(x) - P = 1 + \sum_{\mathbf{m}'=-\mathbf{M}}^{0^-} [\theta_\mathrm{m}(x) - 1]
\end{aligned}
\tag{8.76}
$$

To be concrete, I use here the lower limit $\mathbf{m} = -\mathbf{M} = -[M, M, \ldots, M]$ for which all of the elements of the vector $\mathbf{m}$ are the same integer constant, M.

A computer program to compute N-dimensional theta functions using one-dimensional theta functions seems a laudable goal, and the rest of this section is designed to aid the reader in this regard. Let us see how the partial summations in Equation (8.76) work. First address the *first sum* on the right of Equation (8.76):

$$\theta_{MN}(x|\mathbf{B}, \boldsymbol{\phi}) = 1 + 2 \sum_{\mathbf{m}=-\mathbf{M}}^{0^-} q_\mathrm{m} \cos[\mathbf{m}\cdot\mathbf{k}x + \mathbf{m}\cdot\boldsymbol{\phi}]. \tag{8.77}$$

We may also set for convenience

$$C_\mathrm{m} = C_\mathrm{m}(x) = q_\mathrm{m} \cos[\mathbf{m}\cdot\mathbf{k}x + \mathbf{m}\cdot\boldsymbol{\phi}]. \tag{8.78}$$

Now, let us consider some examples. Let us choose $\mathbf{M} = \{1, 1, \ldots, 1\}$ and $N = 2$, so that we are summing over Table 8.5 for which we find

$$\theta_{1,2}(x) = 1 + 2C_{-1,-1} + 2C_{-1,0} + 2C_{-1,1} + 2C_{0,-1}, \tag{8.79}$$

where

$$C_{m_1, m_2} = q_{m_1, m_2} \cos[(m_1, m_2) \cdot (1, 2)\Delta kx + (m_1, m_2) \cdot (\phi_1, \phi_2)] \tag{8.80}$$

or

$$C_{m_1, m_2} = q_{m_1, m_2} \cos[(m_1 + 2m_2)\Delta kx + m_1\phi_1 + m_2\phi_2], \tag{8.81}$$

so that

$$\begin{aligned}
C_{-1,-1} &= q_{-1,-1} \cos[3\Delta kx + \phi_1 + 2\phi_2], \\
C_{-1,0} &= q_{-1,0} \cos[\Delta kx + \phi_1], \\
C_{-1,1} &= q_{-1,1} \cos[\Delta kx - \phi_1 + \phi_2], \\
C_{0,-1} &= q_{0,-1} \cos[2(\Delta kx + \phi_2)].
\end{aligned} \tag{8.82}$$

Also,

$$\begin{aligned}
q_{m_1, m_2} &= \exp\left\{ -\frac{1}{2} \sum_{i=1}^{2} \sum_{j=1}^{2} m_i m_j B_{ij} \right\} \\
&= \exp\left\{ -\frac{1}{2} (m_1^2 B_{22} + m_2^2 B_{22} + m_1 m_2 B_{12}) \right\},
\end{aligned}$$

so that

$$\begin{aligned}
q_{-1,-1} &= \exp\left\{ -\frac{1}{2} (B_{11} + 2B_{12} + B_{22}) \right\}, \\
q_{-1,0} &= \exp\left\{ -\frac{1}{2} B_{11} \right\}, \\
q_{-1,1} &= \exp\left\{ -\frac{1}{2} (B_{11} - 2B_{12} + B_{22}) \right\}, \\
q_{0,-1} &= \exp\left\{ -\frac{1}{2} B_{22} \right\}.
\end{aligned} \tag{8.83}$$

Finally,

$$\theta_{12}(x|\mathbf{B}) = 1 + 2\exp\left\{-\frac{1}{2}B_{11}\right\}\cos[\Delta kx + \phi_1]$$

$$+ 2\exp\left\{-\frac{1}{2}B_{22}\right\}\cos[2(\Delta kx + \phi_2)]$$

$$+ 2\exp\left\{-\frac{1}{2}(B_{11} - 2B_{12} + B_{22})\right\}\cos[\Delta kx - \phi_1 + \phi_2] \qquad (8.84)$$

$$+ 2\exp\left\{-\frac{1}{2}(B_{11} + 2B_{12} + B_{22})\right\}\cos[3\Delta kx + \phi_1 + 2\phi_2]$$

Now, let us go to the second summation in (8.76) over one-degree-of-freedom thetas:

$$\theta_{MN}(x|\mathbf{B}) = \sum_{\mathbf{m}=-\mathbf{m}}^{0^-} \theta_{\mathbf{m}}(x) - P. \qquad (8.85)$$

Here P, given by (8.73), is the number of hidden dimensions 4 minus 1, or 3. We have

$$\theta_{12}(x|q) = \theta_{-1,-1}\left(3\Delta kx + \phi_1 + 2\phi_2 \bigg| \exp\left\{-\frac{1}{2}(B_{11} + 2B_{12} + B_{22})\right\}\right)$$

$$+ \theta_{-1,0}\left(\Delta kx + \phi_1 \bigg| \exp\left\{-\frac{1}{2}B_{11}\right\}\right)$$

$$+ \theta_{-1,1}\left(\Delta kx - \phi_1 \bigg| \exp\left\{-\frac{1}{2}(B_{11} - 2B_{12} + B_{22})\right\}\right) \qquad (8.86)$$

$$+ \theta_{0,-1}\left(2\Delta kx + 2\phi_2 \bigg| \exp\left\{-\frac{1}{2}B_{22}\right\}\right) - 3$$

For a really nonlinear case, that is for two solitons, we clearly need more terms, in the theta function.

Now, let us convert this last expression to a form with two degree of freedom plus nonlinear interactions. First reverse the sign of $\mathbf{m}$; this is legal as the expressions (8.76) and (8.77) are invariant to such a sign reversal:

Table 8.5 Theta function partial sum for $M = 1$, $N = 2$

l	(m_1, m_2)	Theta summation elements	I_l
1	$(-1, -1)$	$\theta_{-1,-1}(3\Delta kx + \phi_1 + 2\phi_2 \mid \exp\{-\frac{1}{2}(B_{11} + 2B_{12} + B_{22})\})$	-3
2	$(-1, 0)$	$\theta_{-1,0}(\Delta kx + \phi_1 \mid \exp\{-\frac{1}{2}B_{11}\})$	-1
3	$(-1, 1)$	$\theta_{-1,1}(\Delta kx - \phi_1 \mid \exp\{-\frac{1}{2}(B_{11} - 2B_{12} + B_{22})\})$	1
4	$(0, -1)$	$\theta_{0,-1}(2\Delta kx + 2\phi_2 \mid \exp\{-\frac{1}{2}B_{22}\})$	-2
P		-3	

$$\theta_{MN}(x|q) = \theta_{1,1}\left(3\Delta kx + \phi_1 + 2\phi_2 \middle| \exp\left\{-\frac{1}{2}(B_{11} + 2B_{12} + B_{22})\right\}\right)$$

$$+ \theta_{1,0}\left(\Delta kx + \phi_1 \middle| \exp\left\{-\frac{1}{2}B_{11}\right\}\right)$$

$$+ \theta_{1,-1}\left(\Delta kx - \phi_1 \middle| \exp\left\{-\frac{1}{2}(B_{11} - 2B_{12} + B_{22})\right\}\right)$$

$$+ \theta_{0,1}\left(2\Delta kx + 2\phi_2 \middle| \exp\left\{-\frac{1}{2}B_{22}\right\}\right) - 3$$

Now separate into single degree of freedom plus interactions:

$$\theta_{MN}(x|q) = \theta_{1,0}\left(\Delta kx + \phi_1 \middle| \exp\left\{-\frac{1}{2}B_{11}\right\}\right)$$

$$+ \left[\theta_{0,1}\left(2\Delta kx + 2\phi_2 \middle| \exp\left\{-\frac{1}{2}B_{22}\right\}\right) - 1\right]$$

$$+ \theta_{1,1}\left(3\Delta kx + \phi_1 + 2\phi_2 \middle| \exp\left\{-\frac{1}{2}(B_{11} + 2B_{12} + B_{22})\right\}\right)$$

$$+ \theta_{1,-1}\left(\Delta kx - \phi_1 \middle| \exp\left\{-\frac{1}{2}(B_{11} - 2B_{12} + B_{22})\right\}\right) - 2,$$

$$\theta_{MN}(x|q) = \left[\theta_{1,0}\left(\Delta kx + \phi_1 \,\middle|\, \exp\left\{-\frac{1}{2}B_{11}\right\}\right) + \theta_{0,1}\left(2\Delta kx + 2\phi_2 \,\middle|\, \exp\left\{-\frac{1}{2}B_{22}\right\}\right) - 1\right]$$

$$\times \left[1 + \frac{\theta_{1,1}\left(3\Delta kx + \phi_1 + 2\phi_2 \,\middle|\, \exp\left\{-\frac{1}{2}(B_{11} + 2B_{12} + B_{22})\right\}\right) + \theta_{1,-1}\left(\Delta kx - \phi_1 \,\middle|\, \exp\left\{-\frac{1}{2}(B_{11} - 2B_{12} + B_{22})\right\}\right) - 2}{\theta_{1,0}\left(\Delta kx + \phi_1 \,\middle|\, \exp\left\{-\frac{1}{2}B_{11}\right\}\right) + \theta_{0,1}\left(2\Delta kx + 2\phi_2 \,\middle|\, \exp\left\{-\frac{1}{2}B_{22}\right\}\right) - 1}\right],$$

$$\theta_{MN}(x|q) = \theta_{1,0}\left(\Delta kx + \phi_1 \,\middle|\, \exp\left\{-\frac{1}{2}B_{11}\right\}\right)\theta_{0,1}\left(2\Delta kx + 2\phi_2 \,\middle|\, \exp\left\{-\frac{1}{2}B_{22}\right\}\right)$$

$$\times \left[1 + \frac{\theta_{1,1}\left(3\Delta kx + \phi_1 + 2\phi_2 \,\middle|\, \exp\left\{-\frac{1}{2}(B_{11} + 2B_{12} + B_{22})\right\}\right) + \theta_{1,-1}\left(\Delta kx - \phi_1 \,\middle|\, \exp\left\{-\frac{1}{2}(B_{11} - 2B_{12} + B_{22})\right\}\right) - 2}{\theta_{1,0}\left(\Delta kx + \phi_1 \,\middle|\, \exp\left\{-\frac{1}{2}B_{11}\right\}\right) + \theta_{0,1}\left(2\Delta kx + 2\phi_2 \,\middle|\, \exp\left\{-\frac{1}{2}B_{22}\right\}\right) - 1}\right]$$

The last step comes about because the sum of single-degree-of-freedom thetas reduces to the product when the off-diagonal terms are zero. The advantage of the present formulation over that which I have previously considered is that we now express everything in terms of single-degree-of-freedom thetas, that is, the nonlinear interaction terms are *explicit in terms of single-degree-of-freedom thetas*.

Now, we can write a new one-degree-of-freedom "theta function" which is normalized by removing "1" from it:

$$\hat{\theta}(x|q) = \frac{1}{2}[\theta(x|q) - 1] \quad \text{Normalized one-dimensional theta function.}$$

Hence, it is clear that the full N-dimensional theta function is then given by

$$\theta_{MN}(x, 0) = 1 + 2 \sum_{\mathbf{m}'=-\mathbf{m}}^{0^-} \text{Cos}[\hat{\theta}_{\mathbf{m}}(x, q_{\mathbf{m}})]$$

$$\text{Cos}[\hat{\theta}_{\mathbf{m}}(x|q_{\mathbf{m}})] = \text{Cos}\left\{ \frac{1}{2}[\theta_{\mathbf{m}}(x|q_{\mathbf{m}}) - 1] \right\} = \sum_{n=1}^{\infty} q_{\mathbf{m}}^{n^2} \cos[n(\mathbf{m}\cdot\mathbf{k}x + \mathbf{m}\cdot\boldsymbol{\phi})]$$

where the nome is given by

$$q_{\mathbf{m}} = \exp\left\{ -\frac{1}{2}\mathbf{m}^{\mathrm{T}}\mathbf{B}\mathbf{m} \right\}.$$

Note that the Cos function (beginning with a *capital letter* "C") is here defined to be an infinite series of cosines, and is *not* the usual cosine function of trigonometry. This form of the Riemann theta function in N dimensions is quite nice: It says that the N-dimensional theta function can be written as the linear superposition of an infinite number of normalized one-degree-of-freedom theta functions.

Now, you have all the information necessary to write a theta function program which sums over a large number of one-degree-of-freedom theta functions.

Appendix I: Various Notations for Theta Functions

The following notational forms of the theta function are all equivalent. The purpose of this appendix is to document some of these forms in order to provide a reference for the many applications of these important functions.

Exponential Forms

Scalar form in N dimensions:

$$\theta_N(X_1, X_2, \ldots, X_N) = \sum_{m_1=-\infty}^{\infty} \sum_{m_2=-\infty}^{\infty} \cdots \sum_{m_N=-\infty}^{\infty} \exp\left[i \sum_{k=1}^{N} m_k X_k - \frac{1}{2} \sum_{j=1}^{N} \sum_{k=1}^{N} m_j m_k B_{jk} \right].$$

Vector form:

$$\theta_N(\mathbf{X}) = \sum_{\mathbf{m}=-\infty}^{\infty} \exp\left\{ i\mathbf{m} \cdot \mathbf{X} - \frac{1}{2} \mathbf{m} \cdot \mathbf{Bm} \right\}.$$

Dynamical form:

$$\theta_N(x, t) = \sum_{m_1=-\infty}^{\infty} \sum_{m_2=-\infty}^{\infty} \cdots \sum_{m_N=-\infty}^{\infty} \exp\left\{ -\frac{1}{2} \sum_{i=1}^{N} \sum_{j=1}^{N} m_i m_j B_{ij} \right\}$$

$$\times \exp\left\{ i\left(\sum_{i=1}^{N} m_i k_i \right) x - i\left(\sum_{i=1}^{N} m_i \omega_i \right) t + i \sum_{i=1}^{N} m_i \phi_i \right\}.$$

Linear Fourier transform:

$$\theta_N(x, 0) = \sum_{n=1}^{\infty} \theta_n e^{ik_n x}, \quad \theta_n(t) = \sum_{\substack{\text{sum over subset} \\ \text{of } l \text{ for which } I_l = n}} q_l e^{i\Phi_l},$$

$$\theta_N(x, t) = \sum_{n=1}^{\infty} \theta_n(t) e^{ik_n x}, \quad \theta_n(t) = \sum_{\substack{\text{sum over subset} \\ \text{of } l \text{ for which } I_l = n}} q_l e^{-i(\Omega_l t - \Phi_l)},$$

$$\theta_N(x, 0) = \theta_0 + 2 \sum_{n=1}^{\infty} a_n \cos(k_n x) + b_n \sin(k_n x),$$

$$a_n = \sum_{\substack{\text{sum } l \text{ over subset} \\ \text{of } l \text{ for which } I_l = n}} q_l \cos(\Phi_l), \quad 1 \leq n \leq \infty,$$

$$b_n = - \sum_{\substack{\text{sum } l \text{ over subset} \\ \text{of } l \text{ for which } I_l = n}} q_l \sin(\Phi_l), \quad 1 \leq n \leq \infty.$$

Cosine Forms

Scalar form in N dimensions:

$$\theta_N(X_1, X_2, \ldots, X_N) = \sum_{m_1=-\infty}^{\infty} \sum_{m_2=-\infty}^{\infty} \cdots \sum_{m_N=-\infty}^{\infty}$$

$$\exp\left\{-\frac{1}{2}\sum_{j=1}^{N}\sum_{k=1}^{N} m_j m_k B_{jk}\right\} \cos\left[\sum_{k=1}^{N} m_k X_k\right].$$

Vector form:

$$\theta_N(\mathbf{X}) = \sum_{\mathbf{m}=-\infty}^{\infty} \exp\left\{-\frac{1}{2}\mathbf{m}\cdot\mathbf{Bm}\right\} \cos(\mathbf{m}\cdot\mathbf{X}),$$

$$\theta_N(\mathbf{X}) = 1 + 2 \sum_{\mathbf{m}=-\infty}^{0^-} \exp\left\{-\frac{1}{2}\mathbf{m}\cdot\mathbf{Bm}\right\} \cos(\mathbf{m}\cdot\mathbf{X}).$$

Dynamical form:
Vector:

$$\theta_N(x, t) = \sum_{\mathbf{m}=-\infty}^{\infty} \exp\left\{-\frac{1}{2}\mathbf{m}\cdot\mathbf{Bm}\right\} \cos\left[\mathbf{m}\cdot(\mathbf{K}x - \mathbf{\Omega}t + \mathbf{\Phi})\right].$$

Scalar:

$$\theta_N(x, t) = 1 + 2 \sum_{m_1=-\infty}^{0} \sum_{m_2=-\infty}^{0} \cdots \sum_{m_N=-\infty}^{-1} \exp\left\{-\frac{1}{2}\sum_{i=1}^{N}\sum_{j=1}^{N} m_i m_j B_{ij}\right\}$$

$$\times \cos\left[\left(\sum_{i=1}^{N} m_i K_i\right)x - \left(\sum_{i=1}^{N} m_i \omega_i\right)t + \left(\sum_{i=1}^{N} m_i \phi_i\right)\right].$$

Linear Fourier transform:

$$\theta_N(x, t) = \sum_{n=1}^{\infty} \theta_n(t)\exp(ik_n x),$$

$$\theta_n(t) = \sum_{\substack{\text{sum over subset} \\ \text{of } l \text{ for which } I_l = n}} q_l \exp\{-i(\Omega_l t - \Phi_l)\}.$$

Appendix II: Partial Sums of Theta Functions

The following notational forms of the theta function partial sums are all equivalent. The purpose of this appendix is to document some of these forms in order to provide a reference for the many applications of these important functions.

Exponential Forms

Scalar form in N dimensions:

$$\theta_{MN}(X_1, X_2, \ldots, X_N) = \sum_{m_1=-M_1}^{M_1} \sum_{m_2=-M_2}^{M_2} \cdots \sum_{m_N=-M_N}^{M_N}$$

$$\times \exp\left[i\sum_{k=1}^{N} m_k X_k - \frac{1}{2}\sum_{j=1}^{N}\sum_{k=1}^{N} m_j m_k B_{jk} \right].$$

Vector form:

$$\theta_{MN}(\mathbf{X}) = \sum_{\mathbf{m}=-\mathbf{M}}^{\mathbf{M}} \exp\left\{ i\mathbf{m}\cdot\mathbf{X} - \frac{1}{2}\mathbf{m}\cdot\mathbf{Bm} \right\}.$$

Dynamical form:

$$\theta_{MN}(x, t) = \sum_{m_1=-M_1}^{M_1} \sum_{m_2=-M_2}^{M_2} \cdots \sum_{m_N=-M_N}^{M_N} \exp\left\{ -\frac{1}{2}\sum_{i=1}^{N}\sum_{j=1}^{N} m_i m_j B_{ij} \right\}$$

$$\times \exp\left\{ i\left(\sum_{i=1}^{N} m_i k_i\right)x - i\left(\sum_{i=1}^{N} m_i \omega_i\right)t + i\sum_{i=1}^{N} m_i \phi_i \right\}.$$

Cosine Forms

Scalar form in N dimensions:

$$\theta_{MN}(X_1, X_2, \ldots, X_N) = \sum_{m_1=-M_1}^{M_1} \sum_{m_2=-M_2}^{M_2} \cdots \sum_{m_N=-M_N}^{M_N}$$

$$\times \exp\left\{ -\frac{1}{2}\sum_{j=1}^{N}\sum_{k=1}^{N} m_j m_k B_{jk} \right\} \cos\left[\sum_{k=1}^{N} m_k X_k \right].$$

Vector form ($\mathbf{m} = \{m_1, m_2, \ldots, m_N\}$):

$$\theta_{MN}(\mathbf{X}) = \sum_{\mathbf{m}=-\mathbf{M}}^{\mathbf{M}} \exp\left\{ -\frac{1}{2}\mathbf{m}\cdot\mathbf{Bm} \right\} \cos(\mathbf{m}\cdot\mathbf{X}),$$

$$\theta_{MN}(\mathbf{X}) = 1 + 2 \sum_{\mathbf{m}=-\mathbf{M}}^{0^-} \exp\left\{ -\frac{1}{2}\mathbf{m}\cdot\mathbf{Bm} \right\} \cos(\mathbf{m}\cdot\mathbf{X}).$$

Dynamical form:
Vector:

$$\theta_{MN}(x, t) = \sum_{\mathbf{m}=-\mathbf{M}}^{\mathbf{M}} \exp\left\{ -\frac{1}{2}\mathbf{m}\cdot\mathbf{Bm} \right\} \cos[\mathbf{m}\cdot(\mathbf{K}x - \mathbf{\Omega}t + \mathbf{\Phi})].$$

Scalar:

$$\theta_{MN}(x, t) = 1 + 2 \sum_{m_1=-M_1}^{0} \sum_{m_2=-M_2}^{0} \cdots \sum_{m_N=-M_N}^{-1} \exp\left\{ -\frac{1}{2}\sum_{i=1}^{N}\sum_{j=1}^{N} m_i m_j B_{ij} \right\}$$

$$\times \cos\left[\left(\sum_{i=1}^{N} m_i k_i\right)x - \left(\sum_{i=1}^{N} m_i \omega_i\right)t + \sum_{i=1}^{N} m_i \phi_i \right].$$

Appendix III: Fourier Series of Theta Functions at $t = 0$

The Fourier series for the theta function at $(x, t = 0)$:

$$\theta_N(x, 0) = \sum_{n=-N/2}^{N/2} \theta_n e^{i(k_n x + \phi_n)} = \theta_0 + 2\sum_{n=1}^{N/2} \theta_n \cos(k_n x + \phi_n)$$

$$= a_0 + 2\sum_{n=1}^{N/2} a_n \cos(k_n x) + b_n \sin(k_n x),$$

where

$$\theta_n = a_n + ib_n,$$

$$c_n = \sqrt{a_n^2 + b_n^2} \quad \text{and} \quad \tan\phi_n = -b_n/a_n,$$

$$a_n = d_n \cos\phi_n \quad \text{and} \quad b_n = -d_n \sin\phi_n.$$

The expression for $\theta_N(x, t)$ is indistinguishable from an ordinary Fourier series. One ordinarily would think of the coefficients of the Fourier series as being evaluated in terms of certain Fourier integrals. In terms of inverse scattering transform variables for the KdV equation, the Fourier coefficients are instead given by the appropriate coefficient series:

$$\theta_n = \sum_{\substack{\text{for each } n \text{ sum over subset} \\ \text{of } l \text{ for which } I_l = n}} q_l e^{i\Phi_l}, \quad 0 \le n \le N/2,$$

$$a_n = \sum_{\substack{\text{for each } n \text{ sum over subset} \\ \text{of } l \text{ for which } I_l = n}} q_l \cos(\Phi_l), \quad 0 \le n \le N/2,$$

$$b_n = - \sum_{\substack{\text{for each } n \text{ sum over subset} \\ \text{of } l \text{ for which } I_l = n}} q_l \sin(\Phi_l), \quad 0 \le n \le N/2,$$

where

$$q_l = \exp\left\{\frac{1}{2}\mathbf{m}^l \cdot \mathbf{Bm}^l\right\}, \, \Phi_l = \mathbf{m}_l \cdot \boldsymbol{\phi}, \quad I_l = \sum_{j=1}^{N} j m_j^l.$$

Appendix IV: Fourier Series of Theta Functions at Time t

The Fourier series for the theta function at (x, t):

$$\theta_N(x, t) = \sum_{n=-N/2}^{N/2} \theta_n(t) e^{i(k_n x + \phi_n)} = \theta_0 + 2 \sum_{n=1}^{N/2} \theta_n(t) \cos(k_n x + \phi_n)$$

$$= c_0 + 2 \sum_{n=1}^{N/2} a_n(t) \cos(k_n x) + b_n(t) \sin(k_n x),$$

where

$$\theta_n(t) = a_n(t) + i b_n(t),$$

$$d_n = \sqrt{a_n^2 + b_n^2} \quad \text{and} \quad \tan(\omega_n t - \phi_n) = b_n / a_n,$$

$$a_n(t) = d_n \cos(\omega_n t - \phi_n) \quad \text{and} \quad b_n(t) = d_n \sin(\omega_n t - \phi_n).$$

One ordinarily, for the solution of a *linear* wave equation, takes $\theta_n(t) = d_n e^{-i\omega_n t}$, where the coefficients in the Fourier series are given by particular Fourier integrals. For $\theta_N(x,\ t)$ to solve the Hirota-KdV equation, however, the time-varying Fourier coefficients must instead be given by the appropriate coefficient series:

$$\theta_n(t) = \sum_{\substack{\text{for each } n \text{ sum over subset} \\ \text{of } l \text{ for which } I_l = n}} q_l e^{-i(\Omega_l t - \Phi_l)}, \quad 0 \leq n \leq N/2,$$

$$a_n(t) = \sum_{\substack{\text{for each } n \text{ sum over subset} \\ \text{of } l \text{ for which } I_l = n}} q_l \cos(\Omega_l t - \Phi_l), \quad 0 \leq n \leq N/2,$$

$$b_n(t) = \sum_{\substack{\text{for each } n \text{ sum over subset} \\ \text{of } l \text{ for which } I_l = n}} q_l \sin(\Omega_l t - \Phi_l), \quad 0 \leq n \leq N/2.$$

9 Riemann Theta Functions as Ordinary Fourier Series

9.1 Introduction

One of the most important properties of Riemann theta functions for applications (primarily to data analysis and to numerical modeling) is that the theta function can be written as an ordinary *Fourier series with time-varying coefficients* as discussed in Chapters 7 and 8. This is a useful result from many points of view: (1) the "esoteric" Riemann theta function used normally by pure and applied mathematicians and theoretical physicists to solve difficult theoretical problems is now reduced to an ordinary Fourier series, which is relatively well known and easy to use for most investigators; (2) the software for computing operations with Fourier series (the FFT and other operations) is thus now available for applications with theta functions; and (3) analogical and digital computations available for decades for computations using Fourier transforms (think telephone communications) can also be used for computation of theta functions.

As mentioned elsewhere (Chapters 21, 22, and 32–34), we can therefore program solutions of nonlinear partial differential equations using theta functions, but most of the calculations, aside from a "preprocessor" step, can be computed using the ordinary fast Fourier transform, resulting in huge savings in computer time.

This can be seen by a simple example. In absence of the above results, i.e. that theta functions are just ordinary Fourier series, it would at first seem reasonable to program the Riemann theta function, $\theta(x, y, t | \mathbf{B}, \boldsymbol{\phi})$, as a single subroutine and then to call it for a certain number of values of x, y, and t. We call the theta function routine written in this way a kind of "brute force" class of algorithms, which are discussed in Chapter 20. Consider a wave simulation of 1024×1024 intervals in the x, y plane and 1000 steps in time t. Then the total computer time would be $1024 \times 1024 \times 1000 \sim 10^9$ times the execution time for a single theta function subroutine call. The execution time for a single call to a "brute force" theta function routine is proportional to $(2M + 1)^N$, where $(-M, M)$ are the lower and upper limits in a single summation in the theta function and N is the number of summations (number of degrees of freedom). For example, when $M = 4$, $N = 30$ we have $\sim 10^{30}$ operations in a single subroutine call. Clearly, the "brute force" algorithm leads to astronomical amounts

Doi: 10.1016/S0074-6142(10)97009-5

of computer time; not only is a single subroutine call far beyond the ability of modern-day computers, but we also need to call the subroutine 10^9 times!

An alternative approach would be to use the methods discussed in this chapter and in Chapters 7, 8, 32, and 33. We begin with the theta function written explicitly in two dimensions:

$$\theta(x, y, t|\mathbf{B}, \boldsymbol{\phi}) = \sum_{m_1=-M}^{M} \sum_{m_2=-M}^{M} \cdots \sum_{m_N=-M}^{M} \exp\left\{ -\frac{1}{2}\sum_{m=1}^{N}\sum_{n=1}^{N} m_m m_n B_{mn} \right\}$$
$$\exp\left\{ i\sum_{n=1}^{N} m_n \kappa_n x + i\sum_{n=1}^{N} m_n \lambda_n y - i\sum_{n=1}^{N} m_n \omega_n t + i\sum_{n=1}^{N} m_n \phi_n \right\}.$$

(9.0a)

I then convert the theta function to an ordinary Fourier series with time-varying coefficients:

$$\theta(x, y, t|\mathbf{B}, \boldsymbol{\phi}) = \sum_{m=-N_x/2}^{N_x/2} \sum_{n=-N_y/2}^{N_y/2} \theta_{mn}(t) e^{ik_m x + il_n y}$$

(9.0b)

where the *commensurable wavenumbers* are

$$k_m = \frac{2\pi m}{L_x}, \quad l_n = \frac{2\pi n}{L_y}.$$

The *Fourier coefficients* are given by

$$\theta_{mn}(t) = \sum_{\{j\in\mathbb{Z}: I_j=m, J_j=n\}} q_j e^{-i\Omega_j t + i\Phi_j}$$

(9.0c)

and L_x and L_y are the dimensions of a periodic box in which the spatial domain is computed. The integer wavenumbers in the summation Equation (9.0a) are given by

$$I_j = \frac{L_x}{2\pi}\sum_{n=1}^{N} m_n^j \kappa_n, \quad J_j = \frac{L_y}{2\pi}\sum_{n=1}^{N} m_n^j \lambda_n.$$

(9.0d)

The first equation (9.0b) is the *ordinary Fourier series for the theta function* (with time-varying coefficients $\theta_{mn}(t)$) and the second equation (9.0c) gives the *formula for the time-varying coefficients*. The derivations are given in Chapters 8 and 32. Here, I use group theoretic notation for the summation in Equation (9.0c), but the meaning is the same as in formula (8.52), that is, one computes the integer wavenumbers (9.0d) and when they are equal to m or n one adds a term to the summation (9.0c).

For a moment, let us assume that we know how to compute the formula for the time-varying Fourier coefficients, $\theta_{mn}(t)$, via execution of a *preprocessor* subroutine which executes over all m, n ($-N_x/2 \leq m < N_x/2$, $-N_y/2 \leq n < N_y/2$). Once the preprocessor operation is complete we have the time-varying coefficients, $\theta_{mn}(t)$, and we then need only compute the theta function $\theta(x, y, t | \mathbf{B}, \boldsymbol{\phi})$ by a two-dimensional FFT for the N_t values of time we desire, for example to make a movie of the surface elevation with N_t frames. And that is the algorithm for computing the theta function used herein: a preprocessor step, and N_t 2D FFTs! The *linear* solution of a *linear* wave equation requires N_t 2D FFTs, so we have reduced the nonlinear problem to this same form, but with the addition of a preprocessor. The result is a fast computation of the theta function. Of course, summing the preprocessor step over the N-*ellipsoid* (Chapter 22) after an appropriate *modular transformation* (Chapter 8) makes the operation "superfast" or "hyperfast."

Now, the preprocessor step turns out to use roughly the same amount of computer time as the call of the theta function routine discussed above for one set of input parameters (say, x_1, y_1, t_1), but the preprocessor is called only once instead of $1024 \times 1024 \times 1000 \sim 10^9$ times. Therefore, application of the results of this chapter provides a rough savings in computer time of about 10^9. It therefore seems reasonable to place these results in a separate chapter in which the main ideas are elaborated on. A fast algorithm for the preprocessor is discussed in detail in Chapter 22 where the details of the summation over an N-*ellipsoid* in lattice space are given.

9.2 Theoretical Considerations

We now restrict ourselves to one spatial dimension to simplify the exposition and graphics; the two-dimensional case follows in a similar fashion (Chapter 32). In one spatial dimension, the theta function can be written as an ordinary Fourier series with time variable coefficients (see discussion in Chapters 8 and 32):

$$\theta(x, t) = |\theta_n(x, t)| e^{i\varphi(x,t)} = \sum_{n=-\infty}^{\infty} \theta_n(t) e^{ik_n x}, \quad k_n = \frac{2\pi n}{L} \tag{9.1}$$

$$\theta_n(t) = \sum_{\{j \in \mathbb{Z}: I_j = n\}} q_j e^{-i\Omega_j t + i\Phi_j}, \quad -J \leq j \leq J \tag{9.2}$$

where as usual for periodic boundary conditions we have

$$I_j = \frac{L}{2\pi} \sum_{n=1}^{N} m_n^j k_n = \sum_{n=1}^{N} n m_n^j \tag{9.3}$$

or in vector notation:

$$I_{\mathbf{m}} = \frac{L}{2\pi} \mathbf{m} \cdot \mathbf{k} = \mathbf{m} \cdot \mathbf{n}, \quad \mathbf{n} = [1, 2, 3, \ldots, N]. \tag{9.4}$$

Also,

$$\mathbf{k} = [k_1, k_2, \ldots, k_N], \quad k_n = \frac{2\pi n}{L}.$$

Nota Bene. The result (9.1) is important for a number of reasons. First, for those unfamiliar with theta functions, one gets over the idea that thetas are hard and therefore should be avoided at all costs. The result (9.1) says that theta functions can be expressed as linear Fourier series with time-varying coefficients. This means that we can think of theta functions in exactly the same way as we do about applications of Fourier series to nonlinear problems, i.e. that the coefficients must change as a function of time to solve a nonlinear wave equation (Chapter 32). The result (9.1) means that theta functions, in spite of their aura as being too hard to use in applications, are really very simple mathematical objects with which we are already familiar. Normally, one thinks of the time-varying coefficients as being governed by ordinary differential equations. But for theta functions the coefficients are analytic formulas of the type (9.2)! Notice that Equation (9.2) is itself a Fourier series over time with incommensurable frequencies. In reality, theta functions are just as easy to address as ordinary Fourier series. They just require getting used to.

We can also write Equation (9.2) in vector notation:

$$\theta_n(t) = \sum_{\{\mathbf{m}\in\mathbb{Z}:I_{\mathbf{m}}=n\}} q_{\mathbf{m}} e^{-i\Omega_{\mathbf{m}}t + i\Phi_{\mathbf{m}}} \tag{9.5}$$

In summing the theta function, we need to sum over all $\mathbf{m}$ vectors and this is indeed the same sum in Equation (9.2).

Now, it is convenient to write the time-varying coefficients in the following alternative form:

$$\theta_n(t) = \Theta_n(t)e^{-i\omega_n t} = [e^{i\omega_n t}\theta_n(t)]e^{-i\omega_n t}, \quad \Theta_n(t) \equiv \theta_n(t)e^{i\omega_n t}, \tag{9.6}$$

so that

$$\theta(x, t) = |\theta_n(x, t)|e^{i\varphi(x, t)} = \sum_{n=-\infty}^{\infty} \Theta_n(t)e^{ik_n x - i\omega_n t} \tag{9.7}$$

We have in Equation (9.7) the linear Fourier transform of the theta function that has coefficients that vary in time. These coefficients, $\Theta_n(t) = \theta_n(t)e^{i\omega_n t}$, can be interpreted as a *complex modulation* $\theta_n(t)$ of the sinusoidal carrier wave $e^{i\omega_n t}$.

We further see that as a wave train is allowed to decrease in amplitude, and thus to become linear, we are left with the $\Theta_n(t)$ which must now be *time independent* (because there is no nonlinearity):

$$\Theta_n(t) = \Theta_n$$

and hence

$$\theta(x,t) = \sum_{n=-\infty}^{\infty} \Theta_n e^{ik_n x - i\omega_n t}, \tag{9.8}$$

which is the obvious linear limit of the theta function, where $\omega_n = \omega_n(k_n)$ is the linear dispersion relation. Therefore, to understand the nonlinear behavior of the theta function it is worthwhile graphing properties of $\Theta_n(t)$. If $\Theta_n(t)$ does not vary in time at all then we have a linear system. If $\Theta_n(t)$ has small, slowly varying time dependence then we have a weakly nonlinear system. Large variations in time describe a system that is strongly nonlinear.

Now notice that

$$|\theta_n(t)| = |\Theta_n(t)|.$$

This means that we can graph the modulus $|\theta_n(t)| = |\Theta_n(t)|$ to search for time dependence. This can be seen in Figure 9.7 below, generated in a numerical example in Section 9.3 below, where a sine wave in the linear Fourier spectrum ($n = 20$) is seen to be modulated by nonlinear effects. Thus, we have the interpretation of Equation (9.2), where $\theta_n(t) = |\theta_n(t)|e^{i\phi_n(t)}$ is written in complex form and where $|\theta_n(t)| = |\Theta_n(t)|$ is the modulus of the carrier wave $e^{i\omega_n t}$; $|\theta_n(t)|$ has time variation only if nonlinearity plays a role in the problem. Here, $\phi_n(t)$ is the time-varying phase. This particular case consists of a slow modulation of a fast oscillation, an approximation often used for the derivation of the nonlinear Schrödinger equation from the KdV equation (Zakharov and Kuznetsov, 1986). For lower wavenumbers, the modulation may not be slow with respect to the carrier. Indeed in the presence of solitons (for low wavenumber), the separation of long and short scales is not distinct. A complete numerical example is now given for the KdV equation to provide means for interpreting the significance of the results in this chapter.

9.3   A Numerical Example for the KdV Equation

In this section, I have used the KP equation to simulate results for the KdV equation (see Chapter 32). This is done by setting the y wavenumbers equal to zero. The water depth is taken to be $h = 8$ m. The input parameters leading to the computation of the Riemann spectrum are shown in the appendix. I now give several graphical results that are used to familiarize the reader with theta functions for the simulation given. First note that, since we are simulating the KdV equation, we in effect are using the linear dispersion relation, $\omega_n = c_0 k_n - \beta k_n^3$, shown in Figure 9.1; the solid line is the leading order part and the dotted line contains the cubic correction. For the particular simulation shown herein we have the spectrum given in Figure 9.2, which is just a Pierson-Moskowitz-type spectrum used often in physical oceanography to describe what is known as an "equilibrium spectrum." The initial conditions for this simulation are shown in Figure 9.3. This figure illustrates the use of the KP equation as a model for shallow-water nonlinear wave motion, but here applied to unidirectional KdV motion. This is done by setting all the y wavenumbers to zero, $k_y = 0$.

In Figure 9.4, I give the time evolution of the linear Fourier spectrum for the theta function. Figure 9.5 shows contours of this same spectrum as a function of frequency and time and Figure 9.6 gives the spectrum at time zero. The time evolution on one of the Fourier components ($n = 20$) is shown in Figure 9.7 and is graphed out to the recurrence time, $t \sim 2000$ s. This latter result illustrates how the linear Fourier coefficients of the theta function vary in time to solve the KdV equation. Of course, for inputs that are fully linear we would

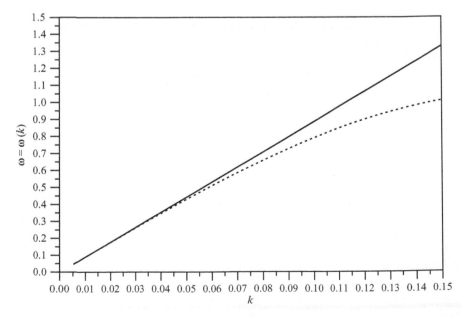

Figure 9.1 Dispersion relation and wavenumber range in the simulations.

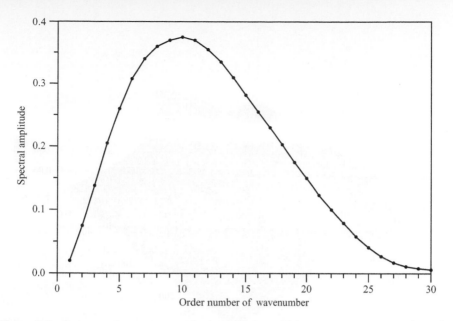

Figure 9.2 Spectrum input to numerical simulations. This spectrum corresponds to the diagonal elements in the period matrix and is typical of an ocean wave (Pierson-Moskowitz) spectrum, that is, it has exponential decay on the left of the peak and power law behavior to the right.

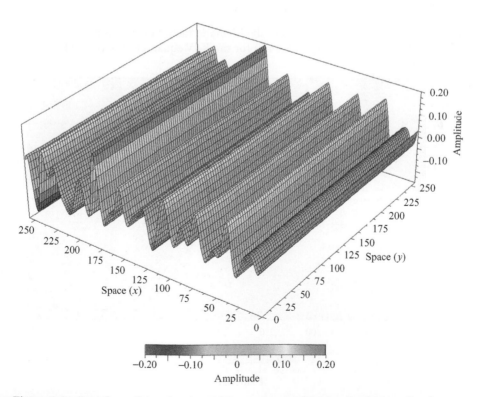

Figure 9.3 Initial condition for the KdV equation used in the simulations for this chapter. Note that the wave train is unidirectional as required for the KdV equation (this result is an output from a numerical simulation using the KP equation, Chapter 32). (See color plate).

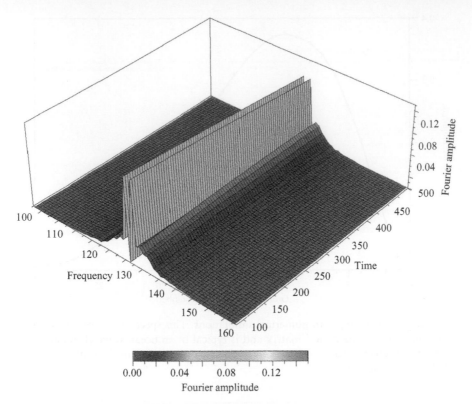

Figure 9.4   Linear Fourier spectrum of the Riemann theta function as a function of frequency and time for the simulation run for the KdV equation. (See color plate).

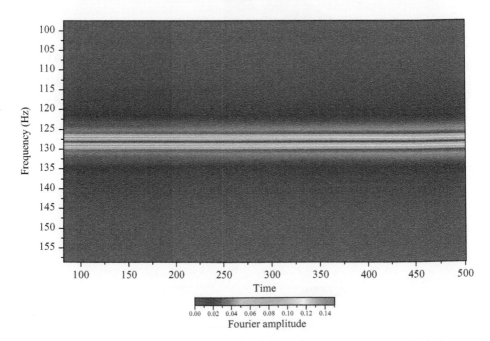

Figure 9.5   Contours of linear Fourier spectrum of the Riemann theta function as a function of frequency and time for the simulation run for the KdV equation. (See color plate).

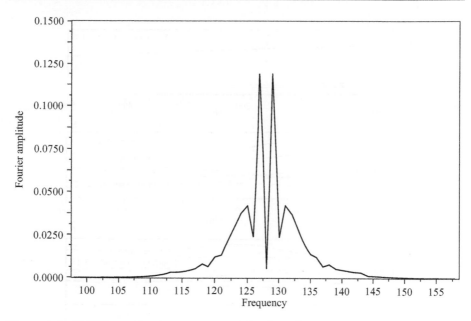

Figure 9.6   KdV linear Fourier spectrum of the theta function at time zero.

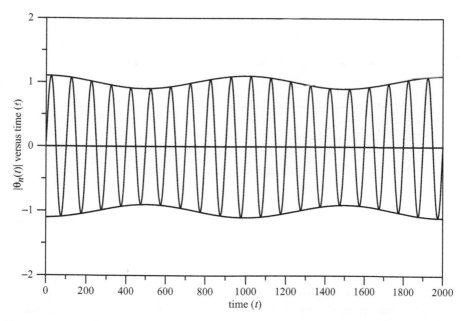

Figure 9.7   Graph of modulus $|\theta_n(t)|$ $(n = 20)$ as a function of time t for a linear Fourier component of the Riemann theta function (9.7). Variation of $|\theta_n(t)|$ with time is an indication of nonlinear dynamics in the theta function. For $|\theta_n(t)| = $ const, there is no nonlinearity and the ordinary linear Fourier transform (9.8) describes the dynamics.

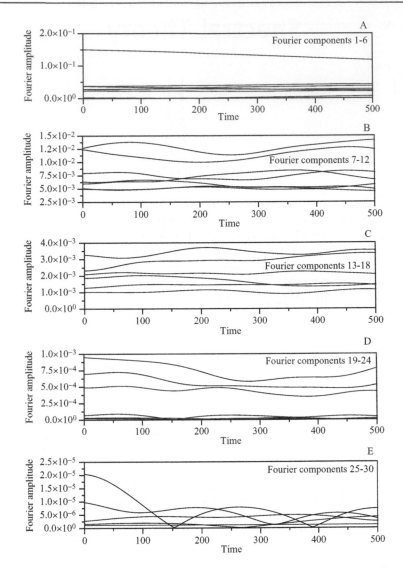

Figure 9.8 Time variation of linear Fourier spectral component amplitudes for KdV equation.

find no modulation of the theta function Fourier coefficient, $|\theta_n(t)|$. Indeed, it is the modulation in Figure 9.7 that gives us the nonlinearity. In Figure 9.8, I show the time evolution of all 30 of the linear Fourier component amplitudes of the theta function.

In the theta function formulation, we have seen that the individual Fourier modes in the theta functions have the following wavenumbers, frequencies, and phases:

$$K_j = \mathbf{m}_j \cdot \mathbf{k} = \sum_{n=1}^{N} m_n^j k_n,$$

$$\Omega_j = \mathbf{m}_j \cdot \boldsymbol{\omega} = \sum_{n=1}^{N} m_n^j \omega_n,$$

$$\Phi_j = \mathbf{m}_j \cdot \boldsymbol{\phi} = \sum_{n=1}^{N} m_n^j \phi_n.$$

Let us now look at some of the statistical properties of these important parameters in the theta function for the present simulation. The first result is shown in Figure 9.9 where the frequencies Ω_j are compared to the wavenumbers K_j, that is, we have points on the "nonlinear dispersion relation" for the KdV equation. We see that the wavenumbers are commensurable while the frequencies are not. In Figure 9.10 are the amplitudes q_j as a function of wavenumber K_j. Notice that at each of the commensurable wavenumbers K_j we have *many* q_j amplitudes. Summing the amplitudes at a particular wavenumber is the essence of Equation (9.5)! In this way, we obtain the ordinary linear Fourier amplitudes of Equation (9.5). Finally, in Figure 9.11 are the amplitudes as a function of frequency. Lack of commensurability means we have the simple diagram for the wavenumbers in Figure 9.11.

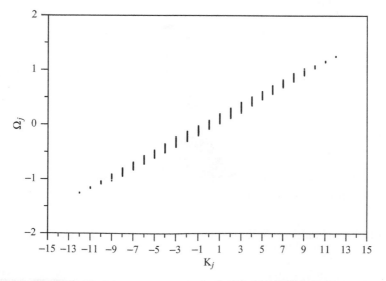

Figure 9.9 Ω_j versus K_j in theta function solution of the KdV equation.

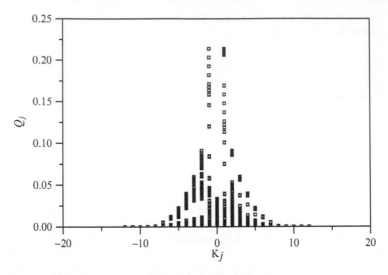

Figure 9.10   Amplitudes of the nomes, q_j, versus the wavenumbers, K_j.

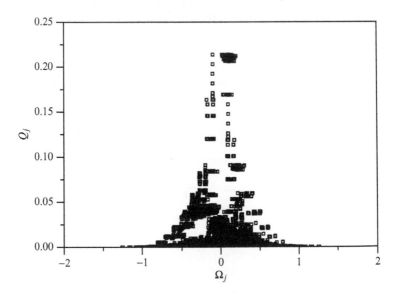

Figure 9.11   Amplitudes of the nomes, q_j, versus the frequencies, Ω_j.

Nota Bene. In the present example, there are 20,096 terms in the theta function summation. Attempts to graph other aspects of the properties of theta functions are difficult because of the large number of terms. Of course in typical data analysis and modeling applications, the number of terms may be much larger. To see best how to interpret Equations (9.1) and (9.2), let us look at Figure 9.10. Note that I am graphing the amplitudes of the theta function, q_j, versus the commensurable wavenumbers, K_j. Notice that for each of the commensurable wavenumbers there are *many* values of the q_j. To get the linear Fourier coefficients return to Equation (9.2) and set time $t = 0$; for simplicity, let the phases be zero, so that Equation (9.2) becomes

$$\theta_n(0) = \sum_{\{j \in \mathbb{Z} : I_j = n\}} q_j, \quad -J \le j \le J.$$

Now compare this formula with Figure 9.10. This formula says that if we sum all of the q_j in a column corresponding to a particular K_j (and take $j = n$), then we get the linear Fourier coefficient $\theta_n(0)$! So, by "collapsing" all the q_j onto the linear Fourier modes, we get the linear Fourier transform for the theta function. This paragraph thus contains the essence of Equations (9.1) and (9.2) and provides the user with the information needed to code these results.

Appendix: Theta Function Run with KP Program

The input parameters for the KdV simulation given in this chapter are: $N_{dof} = 30$ m (the Riemann matrix is therefore 30×30), $h = 8$ m, for an x-y grid of 256×256 points. Table 9.1 shows parameter values for the simulations. Here, n is the number of the individual degrees of freedom ($n = 1$, $2, \ldots, 30$), a_n is the amplitude of a cnoidal wave (see Chapter 8 for computation of Riemann matrix diagonal elements from the cnoidal wave amplitudes), k_x and k_y are the wavenumbers (since we are dealing with the KdV equation, all $k_y = 0$), $\theta_n = 0$ are the angles with the k_x-axis, f_n are the frequencies, ϕ_n are the phases, $|k_n| = (k_x^2 + k_y^2)^{1/2}$ are the moduli of the wavenumbers, L_n are the wavelengths, and T_n are the periods of the numeral simulation components. The method of Schottky uniformization (Charter 15) was used to compute the Riemann matrix and other parameters of the Riemann spectrum. Note that we have taken the arbitrary phases to be zero. Note further that in Table 32.2, we have defined $\kappa_n = k_x$ and $\lambda_n = k_y$.

Table 9.1 Parameters for the KP Simulation

n	a_n (m)	k_x (m^{-1})	k_x (m^{-1})	θ_n	f_n (Hz)	ϕ_n	$|k_n|$ (m^{-1})	L_n (m)	T_n (s)
1	0.00800	0.00524	0.00000	0.00000	0.00738	0.00000	0.00524	1200.00	135.45
2	0.03000	0.01047	0.00000	0.00000	0.01476	0.00000	0.01047	599.99	67.73
3	0.05520	0.01571	0.00000	0.00000	0.02215	0.00000	0.01571	399.99	45.15
4	0.08200	0.02094	0.00000	0.00000	0.02953	0.00000	0.02094	300.00	33.86
5	0.10400	0.02618	0.00000	0.00000	0.03691	0.00000	0.02618	240.00	27.09
6	0.12320	0.03142	0.00000	0.00000	0.04429	0.00000	0.03142	200.00	22.57
7	0.13600	0.03665	0.00000	0.00000	0.05168	0.00000	0.03665	171.42	19.35
8	0.14400	0.04189	0.00000	0.00000	0.05906	0.00000	0.04189	150.00	16.93
9	0.14800	0.04712	0.00000	0.00000	0.06644	0.00000	0.04712	133.33	15.05
10	0.15000	0.05236	0.00000	0.00000	0.07382	0.00000	0.05236	120.00	13.54
11	0.14800	0.05760	0.00000	0.00000	0.08121	0.00000	0.05760	109.09	12.31
12	0.14200	0.06283	0.00000	0.00000	0.08859	0.00000	0.06283	100.00	11.28
13	0.13400	0.06807	0.00000	0.00000	0.09597	0.00000	0.06807	92.30	10.41
14	0.12400	0.07330	0.00000	0.00000	0.10335	0.00000	0.07330	85.71	9.67
15	0.11280	0.07854	0.00000	0.00000	0.11074	0.00000	0.07854	80.00	9.03
16	0.10200	0.08378	0.00000	0.00000	0.11812	0.00000	0.08378	75.00	8.46
17	0.09200	0.08901	0.00000	0.00000	0.12550	0.00000	0.08901	70.58	7.96
18	0.08120	0.09425	0.00000	0.00000	0.13288	0.00000	0.09425	66.66	7.52
19	0.07000	0.09948	0.00000	0.00000	0.14027	0.00000	0.09948	63.15	7.12
20	0.06000	0.10472	0.00000	0.00000	0.14765	0.00000	0.10472	59.99	6.77
21	0.04920	0.10996	0.00000	0.00000	0.15503	0.00000	0.10996	57.14	6.45
22	0.04000	0.11519	0.00000	0.00000	0.16241	0.00000	0.11519	54.54	6.15
23	0.03160	0.12043	0.00000	0.00000	0.16980	0.00000	0.12043	52.17	5.88
24	0.02320	0.12566	0.00000	0.00000	0.17718	0.00000	0.12566	49.99	5.64
25	0.01640	0.13090	0.00000	0.00000	0.18456	0.00000	0.13090	47.99	5.41
26	0.01080	0.13614	0.00000	0.00000	0.19194	0.00000	0.13614	46.15	5.20
27	0.00680	0.14137	0.00000	0.00000	0.19933	0.00000	0.14137	44.44	5.01
28	0.00440	0.14661	0.00000	0.00000	0.20671	0.00000	0.14661	42.85	4.83
29	0.00320	0.15184	0.00000	0.00000	0.21409	0.00000	0.15184	41.37	4.67
30	0.00240	0.15708	0.00000	0.00000	0.22147	0.00000	0.15708	39.99	4.51

Part Four

Nonlinear Shallow-Water Spectral Theory

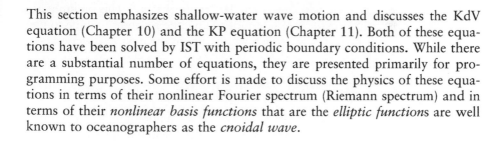

This section emphasizes shallow-water wave motion and discusses the KdV equation (Chapter 10) and the KP equation (Chapter 11). Both of these equations have been solved by IST with periodic boundary conditions. While there are a substantial number of equations, they are presented primarily for programming purposes. Some effort is made to discuss the physics of these equations in terms of their nonlinear Fourier spectrum (Riemann spectrum) and in terms of their *nonlinear basis functions* that are the *elliptic functions* are well known to oceanographers as the *cnoidal wave*.

10 The Periodic Korteweg-DeVries Equation

10.1 Introduction

The inverse scattering transform (IST) for the *periodic* Korteweg-deVries (KdV) equation is discussed from a numerical perspective. Two approaches are given for numerically evolving the equation in space and in time for an N degree-of-freedom wave train. Both the *hyperelliptic* and *θ-function representations* of the KdV equation are discussed. *Periodic boundary conditions* are assumed.

The KdV equation describes small-but-finite-amplitude, long-wave motion,

$$\eta_t + c_0\eta_x + \alpha\eta\eta_x + \beta\eta_{xxx} = 0, \tag{10.1}$$

which governs the space-time evolution of the nonlinear wave field, $\eta(x, t)$, here assumed to be spatially periodic, $\eta(x, t) = \eta(x + L, t)$, for $0 \leq x \leq L$, L the period. The coefficients of Equation (10.1) are constant parameters and have values that depend upon the physical application.

The KdV equation was the first of the soliton equations (Zabusky and Kruskal, 1965) that is now known to be integrable by IST. Both infinite line $(-\infty < x < \infty)$ and periodic boundary conditions $(0 \leq x \leq L)$ have been studied. The present chapter discusses the prerequisites for the space-time numerical integration of solutions to KdV using the structure of the *periodic* inverse scattering transform. Both the hyperelliptic function and the θ-function representations are considered. The actual numerical model is given in Chapter 32.

10.2 Linear Fourier Series Solution to the Linearized KdV Equation

The solution to the *linearized* KdV equation (set $\alpha = 0$ in Equation (10.1)):

$$\eta_t + c_0\eta_x + \beta\eta_{xxx} = 0 \tag{10.2}$$

is given by an ordinary Fourier series:

$$\eta(x, t) = \sum_{j=0}^{N-1} C_j \cos\left(k_j x - \omega_j t + \phi_j\right) \tag{10.3}$$

where the commensurable wavenumbers are given by

$$k_j = 2\pi j/L \tag{10.4}$$

and the associated frequencies are governed by the cubic dispersion relation

$$\omega_j = c_0 k_j - \beta k_j^3 \tag{10.5}$$

The Fourier transform of a wave train $\eta(x, t)$ consists of the set of Fourier amplitudes and phases $\{C_j, \phi_j\}$, for $0 \leq j \leq N - 1$. The mean of Equation (10.3) and of all spatiotemporal solutions of linear and nonlinear wave motion discussed here are assumed to have zero mean (Osborne and Bergamasco, 1986). Equation (10.3), for uniformly distributed random phases, has been used extensively for generating *linear* random functions (Osborne, 1982). Linear approaches of this type motivate the study of *nonlinear* random function solutions to KdV using the periodic inverse scattering transform (Osborne, 1993c) (see Chapter 1 for an example). A review of linear Fourier analysis is given by Champeney, 1973. The fundamental reference is of course Fourier (1955). See also Titchmarsh, 1937; Lighthill, 1959; Tolstov, 1962; Sneddon, 1995.

10.3   The Hyperelliptic Function Solution to KdV

The general spectral solution to periodic KdV (10.1) is written as a *linear superposition of nonlinearly interacting, nonlinear waves (hyperelliptic functions)*, $\mu_j (x, t)$ (see Belokolos et al., 1994, for a list of references):

$$\lambda\eta(x,t) = -E_1 + \sum_{j=1}^{N}[2\mu_j(x,t) - E_{2j} - E_{2j+1}] \tag{10.6}$$

$\lambda = \alpha/6\beta$ and the E_i $(1 \leq i \leq 2N + 1)$ are constant eigenvalues derived from Floquet theory for the time-independent Schrödinger equation (see Chapter 17). Equation (10.6) reduces to a linear Fourier series (10.3) in the limit of small-amplitude motion (Osborne and Bergamasco, 1985) (where the Fourier amplitudes and phases are written in terms of scattering transform variables); hence Equation (10.6) may be interpreted as a *nonlinear Fourier series*. The spatial evolution of the μ_j is governed by the following system of coupled, nonlinear, ordinary differential equations (ODEs):

$$\frac{d\mu_j}{dx} = \frac{2i\sigma_j R^{1/2}(\mu_j)}{\displaystyle\prod_{\substack{k=1 \\ j \neq k}}^{N} (\mu_j - \mu_k)} \tag{10.7}$$

where $0 \leq j \leq N$ and

$$R(\mu_j) = \prod_{k=1}^{2N+1} (\mu_j - E_k) \tag{10.8}$$

The $\sigma_j = \pm 1$ are the signs of the square root of the function $R(\mu_j)$. The $\mu_j(x, t)$ live on two-sheeted Riemann surfaces; the *branch points* connecting them are called "band edges" E_{2j} and E_{2j+1}. The μ_j lie in the intervals (E_{2j}, E_{2j+1}) and oscillate exactly j times between these limits as x is varied on $(0, L)$. When a μ_j reaches a band edge the σ_j changes sign and the motion moves to the adjacent Riemann sheet. The temporal evolution of the μ_j is described by:

$$\frac{d\mu_j}{dt} = 2[-\lambda\eta(x,t) + 2\mu_j]\mu_j' \qquad (10.9)$$

where $\mu_j' = d\mu_j/dx$ is given by Equation (10.7) and $\lambda\eta(x, t)$ is given by Equation (10.6). The space (10.7) and time (10.9) ODEs evolve the $\mu_j(x, t)$ and the non-linear Fourier series (10.6) constructs general spectral solutions to periodic KdV. It is often convenient to consider the Cauchy problem, for which $\eta(x, 0)$ is specified so that the $\mu_j(x, 0)$ are easily computed (without the aid of Equation (10.7), but instead with a technique called *base point iteration* (Osborne, 1994), Chapter 17). Then Equation (10.9) is used to compute the time evolution of the $\mu_j(x, t)$ and Equation (10.6) constructs the solution $\eta(x, t)$ of the KdV equation. Chapter 17 gives an example of the construction of a solution of the KdV equation using hyperelliptic functions.

10.4   The θ-Function Solution to the KdV Equation

In addition to the hyperelliptic function representation (10.6), the general solution to the KdV equation may be written in terms of a θ-*function* formulation:

$$\lambda\eta(x,t) = 2\frac{\partial^2}{\partial x^2}\ln\theta_N(X_1, X_2, \ldots X_N), \quad 1 \le j \le N \qquad (10.10)$$

where

$$\theta_N(X_1, X_2, \ldots X_N) = \sum_{m_1=-\infty}^{\infty}\sum_{m_2=-\infty}^{\infty}\cdots\sum_{m_N=-\infty}^{\infty}$$
$$\exp\left[i\sum_{k=1}^{N}m_kX_k - \frac{1}{2}\sum_{j=1}^{N}\sum_{k=1}^{N}m_jB_{jk}m_k\right] \qquad (10.11)$$

Here N is the number of degrees of freedom in a particular solution to the KdV equation. $X_n = X_n(x, t)$ is discussed with regard to (10.21) below. The θ-function is 2π periodic in each of the N phases

$$\theta[(X_1 + 2\pi), (X_2 + 2\pi), \ldots, (X_N + 2\pi)] = \theta[X_1, X_2, \ldots, X_N] \qquad (10.12)$$

Normalized *holomorphic differentials* on the Riemann surface Γ are defined by

$$d\Omega_n(E) = \sum_{m=1}^{N}C_{nm}\frac{E^{m-1}dE}{R^{1/2}(E)} \qquad (10.13)$$

where $R(E)$ is given by Equation (10.8) and the following normalization condition is assumed to hold:

$$\oint_{\alpha_j} d\Omega_n(E) = 2\pi i \delta_{nj} \tag{10.14}$$

These are the "α_j-cycles" over the open bands (E_{2j}, E_{2j+1}) in the spectrum. Inserting Equation (10.13) into Equation (10.14) gives:

$$\sum_{m=1}^{N} C_{nm} \oint_{\alpha_j} \frac{E^{m-1} dE}{R^{1/2}(E)} = 2\pi i \, \delta_{nj} \tag{10.15}$$

or

$$\sum_{m=1}^{N} C_{nm} J_{mj} = 2\pi i \, \delta_{nj}, \quad J_{mj} = \oint_{\alpha_j} \frac{E^{m-1} dE}{R^{1/2}(E)} \tag{10.16}$$

and in matrix notation

$$\mathbf{CJ} = 2\pi i \, \mathbf{1} \tag{10.17}$$

where $\mathbf{1}$ is the unit matrix, so that finally

$$\mathbf{C} = 2\pi i \mathbf{J}^{-1} \tag{10.18}$$

The normalization coefficients in Equation (10.13) are then given by:

$$\mathbf{C} = \{C_{jm}\} = 2\pi i \mathbf{J}^{-1} = 2\pi i \{J_{jm}\}^{-1} = 2\pi i \left\{ \oint_{\alpha_j} \frac{E^{m-1} dE}{\sqrt{\prod_{k=1}^{2N+1} (E - E_k)}} \right\}^{-1} \tag{10.19}$$

Elements of the J_{jm} matrix, corresponding to the α_j-cycles, may then be reduced to the following simple form:

$$J_{jm} = 2 \int_{E_{2j}}^{E_{2j+1}} \frac{E^{m-1} \, dE}{\sqrt{\prod_{k=1}^{2N+1} (E - E_k)}} \tag{10.20}$$

The *generalized phases* X_j of the θ-function (10.11) are then found by the following Abelian integrals

$$X_j(P_1, P_2, \ldots, P_j) = -i \sum_{m=1}^{N} \int_{E_{2m}}^{P_m(x,t)} d\Omega_j(E) = K_j x - \omega_j t + \phi_j \tag{10.21}$$

where $\omega_j = c_0 K_j - \beta v_j$ (see Equation (10.22) below) and $P_m(x, t) = [\mu_m(x, t), \sigma_m]$. Equation (10.21) is in effect a *linearization of the μ_j flow*, that is, integration over the holomorphic differentials Equation (10.13) from the lower band edge E_{2j} to the hyperelliptic functions $\mu_j(x, t)$ leads to the linear θ-function inverse problem for KdV. Equations (10.13) and (10.21) are an *Abel transform pair*. Generally speaking the phase of the hyperelliptic functions ϕ_j (10.21) depends upon the main spectrum (E_j) and the space-time evolution of the auxiliary spectrum $[\mu_j(x, t), \sigma_j]$, $1 \le j \le N$. This is one of the miracles of the periodic inverse scattering transform: the μ_j flow has been linearized by (10.21).

One finds then the following expressions for the K_j, v_j

$$K_j = 2C_{N,j}, \quad v_j = 8C_{N-1,j} + 4C_{N,j} \sum_{i=1}^{2N+1} E_i \tag{10.22}$$

Note that both K_j and v_j (and the ω_j) are real constants since the C_{jm} and the E_k are real constants. For the specific case of one degree of freedom the usual non-linear dispersion law (Whitham, 1973) is obtained. For many degrees of freedom, while the K_j are commensurable, the ω_j are generally incommensurable.

To obtain the phases ϕ_j set $x = 0$, $t = 0$ in Equation (10.21) to get:

$$\phi_j = -i \sum_{m=1}^{N} \int_{E_{2m}}^{P_m(0,0)} d\Omega_j(E) - i \sum_{m=1}^{N} C_{jm} \int_{E_{2m}}^{\mu_m(0,0)} \frac{E^{m-1} dE}{R^{1/2}(E)} = -i \sum_{m=1}^{N} C_{jm} \Phi_m \tag{10.23}$$

In matrix notation, $\boldsymbol{\phi} = -i\mathbf{C}\boldsymbol{\Phi}$ ($\mathbf{C}$ is given by Equation (10.19), where $\boldsymbol{\phi} = \{\phi_m\}$, $\boldsymbol{\Phi} = \{\Phi_m\}$, and:

$$\Phi_m = \int_{E_{2m}}^{\mu_m(0,0)} \frac{E^{m-1} dE}{R^{1/2}(E)} \tag{10.24}$$

and $P_j(0, 0) = [\mu_j(0, 0), \sigma_j]$. It is therefore clear that the constant phases ϕ_j of the hyperelliptic functions depend upon the *starting values of the hyperelliptic functions $\mu_j(0,0)$ and the Riemann sheet indices σ_j*. Note that these are the necessary initial conditions for integrating the space ODEs (10.2)–(10.4).

The *period matrix* in Equation (10.11) is given by:

$$B_{nj} = \oint_{\beta_j} d\Omega_n(E) = \sum_{m=1}^{N} C_{nm} \oint_{\beta_j} \frac{E^{m-1} dE}{R^{1/2}(E)} \tag{10.25}$$

$$B_{nj} = \sum_{m=1}^{N} C_{nm} \left[2 \int_{E_1}^{E_{2j}} \frac{E^{m-1} dE}{R^{1/2}(E)} \right] = \sum_{m=1}^{N} C_{nm} A_{mj} \tag{10.26}$$

and in matrix notation

$$\mathbf{B} = \mathbf{CA} = 2\pi i \, \mathbf{J}^{-1}\mathbf{A} \tag{10.27}$$

where

$$A_{mj} = 2 \int_{E_1}^{E_{2j}} \frac{E^{m-1} dE}{R^{1/2}(E)} \tag{10.28}$$

where the integrals are over the "β-cycles." Note that for N degrees of freedom the indices range over $1 \leq j \leq N$ and $1 \leq m \leq N$.

The determination of the main spectrum (E_i, $1 \leq i \leq 2N + 1$) and the auxiliary spectrum ($\mu_j (0,0)$, $\sigma_j = \pm 1$, $1 \leq j \leq N$) is referred to as the *direct scattering problem*. The determination of the hyperelliptic functions $\mu_j (x, t)$ by the solution of the nonlinear ODEs (10.7)–(10.8) and the construction of solutions of the KdV equation by Equation (10.6) constitutes the *inverse scattering problem* in the hyperelliptic function representation. See details and numerical examples in Chapter 17.

Alternatively one may construct solutions to KdV by Equations (10.10), (10.11) in the θ-function formulation. A *particular* θ-solution to KdV (10.10), $\eta(x, t)$, based upon a *particular* spectrum for the direct problem, is of course the same as that for the corresponding μ-function solution (10.6). The beauty and elegance of the θ-function formulation as a linearization of the μ-flow is of course a triumph of modern mathematics. Each of the two inverse methods offers particular advantages/disadvantages for numerical computations of the theta functions.

10.5   Special Cases of Solutions to the KdV Equation to Using θ-Functions

I now consider a number of cases that shed light on the physical significance of the θ-functions that aid in the numerical examples to follow. It is useful to rewrite Equation (10.11) in the following way:

$$\theta_N(X_1, X_2, \ldots X_N) = \sum_{m_1=-\infty}^{\infty} \exp\left[im_1 X_1 + \frac{1}{2}m_1^2 B_{11}\right] \sum_{m_2=-\infty}^{\infty} \exp\left[im_2 X_2 + \frac{1}{2}m_2^2 B_{22}\right] \cdots$$

$$\times \sum_{m_n=-\infty}^{\infty} \exp\left[im_N X_N + \frac{1}{2}m_N^2 B_{NN}\right] \exp\left[\frac{1}{2}\sum_{j=1}^{N}\sum_{\substack{k=1\\k \neq j}}^{N} m_j B_{jk} m_k\right] \tag{10.29}$$

This expression proves useful when considering the nonlinear interactions.

10.5.1 One Degree of Freedom

To gain some insight into the above formulation, consider Equation (10.29) for a single degree of freedom, $N = 1$. Hence only a single sum is considered for which $X_1 = X$, $B_{11} = b$:

$$\theta(X) = \sum_{n=-\infty}^{\infty} \exp\left[inX + \frac{1}{2}bn^2\right] \tag{10.30}$$

Since the real part of the period matrix must be negative definite, we set:

$$b = -b_o + 2\pi i$$

where the number b_o is real and positive. The factor of 2π will be justified in the following way. Note that the imaginary part of the exponent in Equation (10.30) is just

$$\exp[i(nX + \pi n^2)] = \exp[in(X + \pi n)] = \exp[in(X + \pi)]$$

so the resultant wave is phase shifted by π (one half the period of 2π). Set $q = \exp(-b_o/2)$ (the *nome*) and use $e^{-i\pi n^2} = (-1)^n$ to find:

$$\theta_4(x) = \theta_4(x, q) = 1 + 2\sum_{n=1}^{\infty} (-1)^n q^{n^2} \cos(nkx) \tag{10.31}$$

where we have identified the series with the Jacobi θ-function, θ_4, (Whittaker and Watson (1902)) and set $X = kx$, $k = 2\pi/L$. The related Jacobian function $\theta_3(x)$ has the form:

$$\theta_3(x) = \theta_3(x, q) = 1 + 2\sum_{n=1}^{\infty} q^{n^2} \cos(nkx)$$

The following variable definitions hold (Abramowitz and Stegun, 1964; Magnus, et al., 1966):

$$b_o = 2\pi K'/K \tag{10.32}$$

where

$$K(m) = K = \int_0^1 [(1 - t^2)(1 - mt^2)]^{-1/2} dt = \int_0^{\pi/2} \frac{d\theta}{[1 - m\sin^2\theta]^{1/2}} \tag{10.33}$$

$$K'(m) = K(m_1) = \int_0^1 [(1 - t^2)(1 - m_1 t^2)]^{-1/2} dt = \int_0^{\pi/2} \frac{d\theta}{[1 - m_1\sin^2\theta]^{1/2}} \tag{10.34}$$

with $m + m_1 = 1$ and $K(m) = K'(m_1) = K'(1 - m)$. Thus the *single degree-of-freedom* θ-function (10.30) is the usual Jacobi θ_4-function (10.31). The solution to KdV is then

$$\lambda \eta(x, t) = 2\partial_{xx} \ln \theta_4(x, q) \tag{10.35}$$

This can be explicitly computed from:

$$\ln \theta_4(x, q) = \ln \gamma - 2 \sum_{n=1}^{\infty} \frac{q^n}{1 - q^{2n}} \frac{\cos(nkx)}{n} \tag{10.36}$$

so that

$$\lambda \eta(x, t) = 2 \frac{\partial^2}{\partial x^2} \ln \theta_4(x, q) = 4k^2 \sum_{n=1}^{\infty} \frac{nq^n}{1 - q^{2n}} \cos(nkx) \tag{10.37}$$

This latter expression is the usual cnoidal wave solution to KdV. The following result, for q small, also holds

$$\lambda \eta(x, t) \cong 4k^2 \sum_{n=1}^{\infty} nq^n \cos(nkx) \tag{10.38}$$

which, provided we interpret q as proportional to the *Ursell number*, is just the *approximate* cnoidal wave found above.

Equation (10.37) is easily linearized for small amplitudes by letting $\ln(1 + x) \sim x$ and summing only over $m = 1$ (or over $-1, 0, 1$ in Equation (10.30) so that $\eta(x, t) \sim -4k^2 e^{-b/2} \cos kx$, a simple cosine. The linearization of Equation (10.29) for N degrees of freedom is discussed below.

To ascertain whether Equation (10.37) is really the cnoidal wave solution to KdV consider the Jacobi zeta function (Abramowitz and Stegun, 1964):

$$Z(u \mid m) = \frac{\pi}{2K} \frac{\theta_4'\left(\frac{\pi u}{2K}\right)}{\theta_4\left(\frac{\pi u}{2K}\right)} = E(u \mid m) - uE(m)/K(m) \tag{10.39}$$

so that the solution to KdV is given by

$$\lambda \eta(x) = \frac{d}{du} Z(u \mid m) = \frac{d}{du} E(u \mid m) - E(m)/K(m) \tag{10.40}$$

for

$$E(u \mid m) = \int_0^x (1 - t^2)^{-1/2} (1 - mt^2)^{1/2} dt = \int_0^\phi (1 - \sin^2\alpha \sin^2\theta)^{1/2} d\theta$$
$$= m_1 u + m \int_0^u cn^2 w \, dw \tag{10.41}$$

where

$$m = \sin^2\alpha, \quad \cos\phi = cnu \tag{10.42}$$

Finally

$$\lambda\eta(x) = m_1 + mcn^2u - E(m)/K(m) \tag{10.43}$$

We therefore see that the single degree-of-freedom solution to periodic KdV is a cnoidal wave. In the infinite period limit, it is easy to show that

$$Z(u \mid 1) = \tanh u \tag{10.44}$$

for which the solution to KdV (10.40) is the soliton:

$$\lambda\eta(x) = \frac{\mathrm{d}}{\mathrm{d}u} Z(u \mid 1) = \mathrm{sech}^2 u \tag{10.45}$$

Note further that due to the fact that

$$\lim_{m\to 1} K' = \frac{\pi}{2}, \quad \lim_{m\to 1} K = \frac{1}{2}\ln\frac{16}{1-m} \tag{10.46}$$

means that

$$\lim_{m\to 1} q(m) = \lim_{m\to 1} \exp\left[-\pi\frac{K'}{K}\right] = \exp\left[-\frac{\pi^2}{\ln[16/(1-m)]}\right] \tag{10.47}$$

so that as *m approaches* 1 we have the argument of the exponent becoming small and the nome slowly approaches 1. This occurs in the soliton limit of the single degree-of-freedom solution to KdV.

An alternative approach to solving the cnoidal wave solution to the KdV equation is to use the hyperelliptic function ODEs for $N = 1$. To this end, we get from Equations (10.7) and (10.8) (setting $E_1 = 0$ without loss of generality):

$$\left(\frac{\mathrm{d}\mu}{\mathrm{d}x}\right)^2 = 4(\mu - E_1)(E_2 - \mu)(\mu - E_3) \tag{10.48}$$

or

$$\frac{\mathrm{d}\mu}{\mathrm{d}x} = 2\sqrt{(\mu - E_1)(E_2 - \mu)(\mu - E_3)} \tag{10.49}$$

Solve this for dx and integrate over $(0, L)$ to get the period (twice the integral over the band edges, (E_2, E_3)):

$$L_1 = \int_{E_2}^{E_3} \frac{dE}{\sqrt{(E - E_1)(E - E_2)(E_3 - E)}} \tag{10.50}$$

Note that we have integrated from one open band edge to the other. We associate L_1 with the wavenumber through the relation: $K_1 = 2\pi/L_1$.

These same expressions may be found from the θ-function formulation from Equation (10.19)

$$C_{11} = \pi \left[\int_{E_2}^{E_3} \frac{dE}{\sqrt{(E - E_1)(E - E_2)(E_3 - E)}} \right]^{-1} = \frac{\pi}{L_1} = \frac{K_1}{2} \tag{10.51}$$

which returns the relation $K_1 = 2C_{11}$ as given by Equation (10.22). This confirms the notation for the wavenumbers as found by the α_j-cycle integrals (10.15). The explicit expression for K_1 is given by:

$$K_1 = \frac{2\pi}{\int_{E_2}^{E_3} \frac{dE}{\sqrt{(E-E_1)(E-E_2)(E_3-E)}}} = \frac{2\pi}{K(m)} \tag{10.52}$$

The period matrix B_{nj} for a single degree of freedom can be computed by Equation (10.28). We find

$$B_{11} = C_{11}A_{11} = 2C_{11} \int_{E_1}^{E_2} \frac{dE}{R^{1/2}(E)} = K_1 \int_{E_1}^{E_2} \frac{dE}{R^{1/2}(E)} \tag{10.53}$$

and finally

$$B_{11} = \frac{2\pi \int_{E_1}^{E_2} \frac{dE}{\sqrt{(E-E_1)(E_2-E)(E_3-E)}}}{\int_{E_2}^{E_3} \frac{dE}{\sqrt{(E-E_1)(E-E_2)(E_3-E)}}} = 2\pi \frac{F\left(\frac{\pi}{2} \mid m_1\right)}{F\left(\frac{\pi}{2} \mid m\right)} = 2\pi \frac{K'(m)}{K(m)} \tag{10.54}$$

where the last step was made with Equations (17.4.62) and (17.4.68) of Abramowitz and Stegun (1964) together with Equations (17.3.2) and (17.3.6). Equation (10.54) is just Equation (10.32). Note that in these expressions the following definitions are used:

$$m = \frac{E_3 - E_2}{E_3 - E_1}$$

$$m_1 = \frac{E_2 - E_1}{E_3 - E_1} = 1 - m \tag{10.55}$$

These results bring out anther form of the elliptic integrals which is more natural for work with the periodic inverse scattering transform:

$$K(m) = \frac{1}{2}\sqrt{E_3 - E_1} \int_{E_2}^{E_3} \frac{dE}{\sqrt{(E - E_1)(E - E_2)(E_3 - E)}}$$

$$K'(m) = \frac{1}{2}\sqrt{E_3 - E_1} \int_{E_1}^{E_2} \frac{dE}{\sqrt{(E - E_1)(E_2 - E)(E_3 - E)}}$$

(10.56)

Note that the first of Equation (10.56) is easily transformed into Equation (10.33) by

$$E = E_3 - (E_3 - E_2)t^2, \quad dE = -2(E_3 - E_2)t \, dt$$

for which the polynomial has the form

$$P(E) = (E - E_1)(E - E_2)(E_3 - E) = (E_3 - E_1)(E_3 - E_2)^2 t^2 (1 - t^2)(1 - mt^2)$$

for

$$m = \frac{E_3 - E_2}{E_3 - E_1}$$

(10.57)

Furthermore the second of Equation (10.56) is transformed into Equation (10.34) by

$$E = E_1 + (E_2 - E_1)t^2, \quad dE = 2(E_2 - E_1)t \, dt$$

with

$$P(E) = (E - E_1)(E_2 - E)(E_3 - E) = (E_3 - E_1)(E_2 - E_1)^2 t^2 (1 - t^2)(1 - m_1 t^2)$$

for

$$m_1 = \frac{E_2 - E_1}{E_3 - E_1}, \quad m + m_1 = 1$$

(10.58)

The first of Equation (10.56) can be seen to be a measure of the open bandwidth, while the second of Equation (10.56) is a measure of the associated gap width. The transformations from Equations (10.56) to (10.33), (10.34), suggests a way to remove the singular behavior near the limits in numerical calculations for the more general case of N degrees of freedom (see Chapter 19).

Nota Bene: The wavenumbers chosen in $X_n = K_n x$ in the argument of the exponentials are just the usual commensurable wavenumbers of linear Fourier analysis; it is the B_{nn} in the nome which govern the *shape* of the wave because of the intimate relationship with the modulus m. Therefore, we choose the period L on which a wave has wavenumber $k = 2\pi/L$; if one wants the wave to be a soliton then the nome must be chosen appropriately, that is, with $m \sim 1$. What happens when one has two solitons of differing amplitudes and phases? Each has the wavenumber $k = 2\pi/L$, since only one *oscillation* appears on the period L. We give them both the same wavenumber, but *different* **B** matrix elements to give us the interaction between two solitons. This suggests that the solitons exist one to the period, but that in the radiation solutions the nonlinear Fourier components can be *soliton trains*, that is, cnoidal waves which oscillate many times in a period, but for which each peak is graphically indistinguishable from a soliton.

10.5.2   On the Possibility of Multiple, Noninteracting Cnoidal Waves

Can multiple cnoidal waves be constructed which do not interact with one another? Do solutions of the KdV equation exist which do not have the interaction contribution? The answer is of course "no" as we see below. For no interactions to occur the off-diagonal terms in the period matrix must be zero in Equation (10.29) and this leads to the following expression for the "solution" of the KdV equation

$$\eta(x,t) = 2 \sum_{n=1}^{N} \eta_n cn^2 \{(K(m_n)/\pi)[k_n(x - C_n t)] \,|\, m\} \tag{10.59}$$

As we now see this condition cannot happen in general and can happen only in the linear limit. How do the properties of these cnoidal waves depend on the band structure in the direct problem? Recall that the period matrix has the following form:

$$\mathbf{B} = \mathbf{CA}$$

where

$$\mathbf{C} = 2\pi i \, \mathbf{J}^{-1}$$

and

$$J_{jm} = 2 \int_{E_{2j}}^{E_{2j+1}} \frac{E^{m-1} dE}{R^{1/2}(E)}$$

$$A_{mj} = 2 \int_{E_1}^{E_{2j}} \frac{E^{m-1} dE}{R^{1/2}(E)}$$

Suppose **C** and **A** are two-by-two, then the **B** matrix is given by:

$$\mathbf{B} = \begin{bmatrix} C_{11}A_{11} + C_{12}A_{21} & C_{11}A_{12} + C_{12}A_{22} \\ C_{21}A_{11} + C_{22}A_{21} & C_{21}A_{12} + C_{22}A_{22} \end{bmatrix} \tag{10.60}$$

For the two degree-of-freedom case, the explicit results are given by:

$$\mathbf{C} = 2\pi i \mathbf{J}^{-1} = \frac{2\pi i}{\det \mathbf{J}} \begin{bmatrix} J_{22} & -J_{12} \\ -J_{21} & J_{11} \end{bmatrix}$$

for $\det \mathbf{J} = J_{11}J_{22} - J_{12}J_{21}$ and then

$$\mathbf{B} = \mathbf{CA} = \frac{2\pi i}{\det \mathbf{J}} \begin{bmatrix} J_{22} & -J_{12} \\ -J_{21} & J_{11} \end{bmatrix} \begin{bmatrix} A_{11} & A_{12} \\ A_{21} & A_{22} \end{bmatrix}$$

$$= \frac{2\pi i}{\det \mathbf{J}} \begin{bmatrix} J_{22}A_{11} - J_{12}A_{21} & J_{22}A_{12} - J_{12}A_{22} \\ J_{11}A_{21} - J_{21}A_{11} & J_{11}A_{22} - J_{21}A_{12} \end{bmatrix} \tag{10.61}$$

We see that the off-diagonal terms cannot be zero in general. This point is further revisited in the chapter on Schottky uniformization, Chapter 15, and in Chapter 32 in which Schottky uniformization is used to construct the off-diagonal elements of the period matrix.

10.5.3 The Linear Fourier Limit

The linear limit to solutions of the KdV equation is obtained by considering the spectral amplitudes to be small (B_{nn} large) so that only the leading order first harmonic contributes to (10.11), (10.29) and:

$$\eta(x,t) = 2\partial_{xx} \ln \theta \simeq 2 \frac{\partial^2}{\partial x^2} \left\{ 1 + 2\exp\left[\frac{1}{2}B_{11}\right] \cos[X_1] + 2\exp\left[\frac{1}{2}B_{22}\right] \right.$$
$$\left. \times \cos[X_2] + \cdots + 2\exp\left[\frac{1}{2}B_{NN}\right] \cos[X_N] \right\} \tag{10.62}$$

This expression reduces to the linear Fourier transform (10.3) provided that we set for the coefficients

$$C_n = -4k_n^2 \exp\left[\frac{1}{2}B_{nn}\right] \tag{10.63}$$

and

$$X_n = n\Delta kx \tag{10.64}$$

where the B_{nn} are negative definite, $B_{nn} = -b_n$, for b_n positive. The linear Fourier phases in terms of scattering transform variables are given elsewhere (Osborne and Bergamasco, 1985).

10.5.4   The Soliton and the N-Soliton Limits

The soliton limit of a single mode may be easily found by the Poisson sum formula (Bellman (1951)):

$$\sum_{m=-\infty}^{\infty} \exp[m^2 s + 2miz] = \left(-\frac{\pi}{s}\right)^{1/2} \sum_{m=-\infty}^{\infty} \exp\left[\frac{(z - m\pi)}{s}\right] \tag{10.65}$$

The left side of this expression is an ordinary Fourier series. The right side is a Poisson summation. That the two are equal is a remarkable property of one degree-of-freedom θ-functions. See Chapter 8 for details.

10.5.5   Physical Selection of the Basis Cycles

A number of different definitions for the beta basis cycles can be considered. I discuss two physically relevant bases in this section.

10.5.5.1   Oscillation Basis

One takes the loop integrals to minus infinity from the lower edge of each open band. This is appropriate for having commensurable wavenumbers, so that each cnoidal wave in the spectrum oscillates exactly an integer number of times within the spatial period of the wave train. The wavenumbers in the oscillation basis are $k_n = 2\pi n/L$, $n = 0,1,2,\ldots,N$, exactly as for the linear Fourier analysis of periodic wave trains. See the Chapter 14 on the numerics of the loop integrals.

10.5.5.2   Soliton Basis

The beta cycle loop integrals are taken relative to the *reference level* (Chapter 17, Osborne and Bergamasco (1985)), looping to the left past the gap of each open band. This means the cycles are defined relative to the solitons as they appear in the hyperelliptic function representation. The important aspect of this definition of the loop integrals is that *there is only one oscillation per period* for each of the soliton degrees of freedom in this basis. The soliton wavenumbers are $k_n = 2\pi/L$, $n = 1, 2,\ldots, N$. As discussed in Chapter 8 one makes a simple modular transformation to transform a particular Riemann spectrum from the oscillation basis to the soliton basis or *vice versa*.

10.6   Exact and Approximate Solutions to the KdV Equation for Specific Cases

I now consider a number of cases that shed light on the physical interpretation of the θ-functions. It is useful to rewrite Equation (10.11) in the following way:

$$\theta_N(X_1, X_2, \ldots, X_N) = \sum_{m_1=-\infty}^{\infty} \exp\left[im_1 X_1 + \frac{1}{2}m_1^2 B_{11}\right] \sum_{m_2=-\infty}^{\infty} \exp\left[im_2 X_2 + \frac{1}{2}m_2^2 B_{22}\right] \cdots$$

$$\times \sum_{m_N=-\infty}^{\infty} \exp\left[im_N X_N + \frac{1}{2}m_N^2 B_{NN}\right] \exp\left[\frac{1}{2}\sum_{j=1}^{N}\sum_{\substack{k=1 \\ k \neq j}}^{N} m_j m_k B_{jk}\right]$$

(10.66)

This expression proves useful in the special cases given below.

10.6.1   A Single Cnoidal Wave

To gain some insight into the θ-function formulation, let us revisit Equation (10.66) for *a single degree of freedom*, $N = 1$. Hence only a single sum is considered for which $\eta_1 = \eta$, $B_{11} = -b$, for b a real constant:

$$\theta(X) = \sum_{n=-\infty}^{\infty} \exp\left[inX - \frac{1}{2}bn^2\right]$$

(10.67)

This is the classical function often referred to as the Jacobian θ-function, θ_3, which is normally written:

$$\theta_3(X) \equiv \theta_3(x, q) = 1 + 2\sum_{n=1}^{\infty} q^{n^2} \cos(nkx)$$

(10.68)

where $X = kx$, $k = 2\pi/L$, $q = \exp[-b/2]$. Here I have set $t = \phi = 0$ to simplify Equation (10.68) and the formulas which follow. To introduce the time t and an arbitrary phase ϕ into the formulation it is only necessary to make the simple replacement: $kx \rightarrow kx - \omega t + \phi$. As discussed elsewhere $b = 2\pi K'/K$ so that:

$$q = e^{-\pi K'/K}$$

(10.69)

where K, K' are elliptic integrals:

$$K(m) = K = \int_0^1 [(1-t^2)(1-mt^2)]^{-1/2}dt = \int_0^{\pi/2} \frac{d\theta}{[1-m\sin^2\theta]^{1/2}}$$

(10.70)

$$K'(m) = K(m_1) = \int_0^1 [(1-t^2)(1-m_1t^2)]^{-1/2} dt = \int_0^{\pi/2} \frac{d\theta}{[1 - m_1 \sin^2\theta]^{1/2}} \quad (10.71)$$

with $m + m_1 = 1$ and $K(m) = K'(m_1) = K'(1-m)$.

The *single degree-of-freedom θ-function* (10.67) may be written in terms of the Jacobi θ_3-function (10.68) or in terms of the function θ_4, since $\theta_3(kx + \pi, q) = \theta_4 (kx,q)$. The solution to the KdV equation is

$$\lambda\eta(x,0) = 2\frac{\partial^2}{\partial x^2} \ln \theta_3(x,q) \quad (10.72)$$

This can be explicitly computed from Abramowitz and Stegun (1964):

$$\ln \theta_3(x,q) = \ln \gamma - 2\sum_{n=1}^{\infty} \frac{(-1)^n q^n}{1 - q^{2n}} \frac{\cos(nkx)}{n}$$

so that

$$2\frac{\partial}{\partial x} \ln \theta_3 = \frac{2\pi\eta_0}{mK(m)k} [E(v \mid m) - (1-m)v]$$

where $E(v \mid m)$ is the elliptic integral of the second kind (Abramowitz and Stegun, 1964) and

$$v = \frac{K(m)}{\pi} [k(x - Ct)]$$

Finally

$$\lambda\eta(x,t) = 2\frac{\partial^2}{\partial x^2} \ln \theta_3(x,q) = 4k^2 \sum_{n=1}^{\infty} \frac{(-1)^n nq^n}{1 - q^{2n}} \cos(nkx) \quad (10.73)$$

which is the series expansion for the *cnoidal wave solution* to the KdV equation:

$$\eta(x,t) = 2\eta_0 cn^2\{(K(m)/\pi)[k(x - Ct)] \mid m\} \quad (10.74)$$

where the modulus m and the phase speed C have been defined in Equations (10.60) and (10.61) above. The following approximate result, for q small, also follows from Equation (10.73)

$$\lambda\eta(x,t) \cong 4k^2 \sum_{n=1}^{\infty} (-1)^n nq^n \cos(nkx) \quad (10.75)$$

There is an alternative way to consider the single degree-of-freedom case, which uses the product form for the θ-function (Whittaker and Watson, 1902):

$$\theta_3 = Q \prod_{n=1}^{\infty} [1 + 2q^{2n-1} \cos kx + q^{4n-2}], \quad Q = \prod_{n=1}^{\infty} (1 - q^{2n}) \tag{10.76}$$

The logarithm of this is

$$\ln \theta_3 = \ln Q + \sum_{n=1}^{\infty} \ln [1 + 2q^{2n-1} \cos kx + q^{4n-2}]$$

so that

$$u(x,0) = 2 \frac{\partial^2}{\partial x^2} \ln \theta_3 = -4k^2 \sum_{n=1}^{\infty} q^{2n-1} \left[\frac{(1 + q^{4n-2}) \cos kx + 2q^{2n-1}}{(1 + 2q^{2n-1} \cos kx + q^{4n-2})^2} \right] \tag{10.77}$$

which to leading order gives

$$u(x,0) = -4k^2 q \left[\frac{(1 + q^2) \cos kx + 2q}{(1 + 2q \cos kx + q^2)^2} \right] - 4k^2 q^3 \left[\frac{(1 + q^6) \cos kx + 2q^3}{(1 + 2q^3 \cos kx + q^6)^2} \right] - \cdots \tag{10.78}$$

Now suppose that we rewrite Equation (10.77) in terms of a fundamental wavelet

$$W(q) \equiv -4k^2 q \left[\frac{(1 + q^2) \cos kx + 2q}{(1 + 2q \cos kx + q^2)^2} \right] \tag{10.79}$$

then

$$u(x,0) = \sum_{n=1}^{\infty} W(\varepsilon_n), \quad \varepsilon_n = q^{2n-1} \tag{10.80}$$

Thus the general one degree-of-freedom solution to the KdV equation is the wavelet W with q as the parameter, plus the wavelet W with q^3 as the parameter, and so on. It is emphasized that Equations (10.77) and (10.80) are equivalent alternative expressions for the cnoidal wave (10.72).

10.6.2 Multiple, Noninteracting Cnoidal Waves

Now consider once again the case of N-cnoidal wave trains that do *not* undergo nonlinear interactions with each other. This occurs when the off-diagonal terms in the period matrix are neglected in Equation (10.66) so that one has:

$$
\theta_N(X_1, X_2, \ldots, X_N) = \sum_{m_1=-\infty}^{\infty} \exp\left[i m_1 X_1 + \frac{1}{2} m_1^2 B_{11}\right] \sum_{m_2=-\infty}^{\infty} \exp\left[i m_2 X_2 + \frac{1}{2} m_2^2 B_{22}\right] \cdots
$$
$$
\times \sum_{m_N=-\infty}^{\infty} \exp\left[i m_N X_N + \frac{1}{2} m_N^2 B_{NN}\right]
$$

(10.81)

It then follows that:

$$
\eta(x,t) = \frac{2}{\lambda} \frac{\partial^2}{\partial x^2} \ln \theta_N = 2 \frac{\partial^2}{\partial x^2} \left[\ln \sum_{m_1=-\infty}^{\infty} \exp\left[i m_1 X_1 + \frac{1}{2} m_1^2 B_{11}\right] \right.
$$
$$
\left. + \ln \sum_{m_2=-\infty}^{\infty} \exp\left[i m_2 X_2 + \frac{1}{2} m_2^2 B_{22}\right] + \cdots + \ln \sum_{m_N=-\infty}^{\infty} \exp\left[i m_N X_N + \frac{1}{2} m_N^2 B_{NN}\right] \right]
$$

(10.82)

Each sum in the latter expression corresponds to a cnoidal wave (see Equations (10.68), (10.72), and (10.74)) so that Equation (10.82) represents a linear superposition of N-cnoidal waves:

$$
\eta(x,t) = 2 \sum_{n=1}^{N} \eta_n cn^2\{(K(m_n)/\pi)[k_n(x - C_n t)] \mid m\}
$$

(10.83)

where the moduli m_n of each squared elliptic function is given by

$$
m_n K^2(m_n) = \frac{3\pi^2 \eta_n}{2k_n^2 h^3} = 4\pi^2 U_n, \quad U_n = \frac{3\eta_n}{8k_n^2 h^3}
$$

(10.84)

and the k_n are the wavenumbers. U_n is the spectral Ursell number. The nonlinear phase speeds C_n are

$$
C_n = c_0\{1 + 2\eta_n/h - 2k_n^2 h^2 K^2(m_n)/3\pi^2\}
$$

(10.85)

and U_n may be also called the *Ursell number* of the nth degree of freedom. One has, therefore, for the *diagonal form of the period matrix*, $B_{ij} = B_{ii} \delta_{ij}$, a linear superposition of cnoidal waves that do *not* interact with each other (10.83). As seen in the following section, the diagonal part of the period matrix (10.83) provides a fundamental contribution to the general solution of the KdV equation.

10.6.3 Cnoidal Waves with Interactions

It is possible to rewrite the complete θ-function theory in a form that emphasizes that the solutions to the KdV equation can be written *in terms of cnoidal waves plus their mutual nonlinear interactions* (Osborne, 1995a,b). To do this I change notation to vector form

$$\theta_N(\mathbf{X}) = \sum_{\mathbf{m} \in \mathbb{Z}} \exp\left(i\,\mathbf{m} \cdot \mathbf{X} + \frac{1}{2}\mathbf{m} \cdot \mathbf{Bm}\right) \tag{10.86}$$

such that the vectors have the following components: $\mathbf{m} = [m_1, m_2, \ldots, m_N]$ and $\mathbf{X} = [X_1, X_2, \ldots, X_N]$. Separating the period matrix $\mathbf{B}$ into diagonal ($\mathbf{D}$) and off-diagonal ($\mathbf{O}$) parts

$$\mathbf{B} = \mathbf{D} + \mathbf{O} \tag{10.87}$$

one can write

$$\theta_N(\mathbf{X}) = \sum_{\mathbf{m}} \exp\left(i\,\mathbf{m} \cdot \mathbf{X} + \frac{1}{2}\mathbf{m} \cdot \mathbf{Dm}\right)$$
$$+ \left\{ \sum_{\mathbf{m}} \exp\left(i\,\mathbf{m} \cdot \mathbf{X} + \frac{1}{2}\mathbf{m} \cdot \mathbf{Bm}\right) - \sum_{\mathbf{m}} \exp\left(i\,\mathbf{m} \cdot \mathbf{X} + \frac{1}{2}\mathbf{m} \cdot \mathbf{Dm}\right) \right\}$$

$$\theta_N(\mathbf{X}) = \sum_{\mathbf{m}} \exp\left(i\,\mathbf{m} \cdot \mathbf{X} + \frac{1}{2}\mathbf{m} \cdot \mathbf{Dm}\right)$$
$$+ \left\{ \sum_{\mathbf{m}} \left[\exp\left(\frac{1}{2}\mathbf{m} \cdot \mathbf{Om}\right) - 1 \right] \exp\left(i\,\mathbf{m} \cdot \mathbf{X} + \frac{1}{2}\mathbf{m} \cdot \mathbf{Dm}\right) \right\}$$

$$\theta_N(\mathbf{X}) = \sum_{\mathbf{m}} \exp\left(i\,\mathbf{m} \cdot \mathbf{X} + \frac{1}{2}\mathbf{m} \cdot \mathbf{Dm}\right)$$
$$\times \left\{ 1 + \frac{\sum_{\mathbf{m}} \left[\exp\left(\frac{1}{2}\mathbf{m} \cdot \mathbf{Om}\right) - 1 \right] \exp\left(i\,\mathbf{m} \cdot \mathbf{X} + \frac{1}{2}\mathbf{m} \cdot \mathbf{Dm}\right)}{\sum_{\mathbf{m}} \exp\left(i\,\mathbf{m} \cdot \mathbf{X} + \frac{1}{2}\mathbf{m} \cdot \mathbf{Dm}\right)} \right\} \tag{10.88}$$

Then the solution to the KdV equation can be written:

$$u(x,t) = 2\partial_{xx} \ln \sum_{\mathbf{m}} \exp\left(i\,\mathbf{m} \cdot \mathbf{X} + \frac{1}{2}\mathbf{m} \cdot \mathbf{Dm}\right) + u_{\text{int}}(\mathbf{X}) \tag{10.89}$$

where

$$u_{\text{int}}(x,t) = 2\partial_{xx} \ln \left\{ 1 + \frac{\sum_{\mathbf{m}} \left[\exp\left(\frac{1}{2}\mathbf{m} \cdot \mathbf{O\,m}\right) - 1 \right] \exp\left(i\,\mathbf{m} \cdot \mathbf{X} + \frac{1}{2}\mathbf{m} \cdot \mathbf{D\,m}\right)}{\sum_{\mathbf{m}} \exp\left(i\,\mathbf{m} \cdot \mathbf{X} + \frac{1}{2}\mathbf{m} \cdot \mathbf{Dm}\right)} \right\}$$
$$\tag{10.90}$$

The above formulation (10.88) can also be written in the following notation:

$$\theta_N(\mathbf{X}) = F(\mathbf{X})\left\{1 + \frac{F(\mathbf{X}, G)}{F(\mathbf{X})}\right\} \tag{10.91}$$

where

$$F(\mathbf{X}, G) = \sum_m G \exp\left(i\mathbf{m}\cdot\mathbf{X} + \frac{1}{2}\mathbf{m}\cdot\mathbf{D}\mathbf{m}\right), \quad F(\mathbf{X}) \equiv F(\mathbf{X}, 1) = \sum_{n=1}^N \theta_n \tag{10.92}$$

$$G = \left[\exp\left(\frac{1}{2}\mathbf{m}\cdot\mathbf{O}\mathbf{m}\right) - 1\right] \tag{10.93}$$

Here the θ_n in Equation (10.92) are the ordinary Jacobian θ-functions θ_3, one for each of the N degrees of freedom. Therefore, the solution to KdV is

$$u(x, t) = 2\partial_{xx} \ln F(\mathbf{X}) + u_{\text{int}}(\mathbf{X}) \tag{10.94}$$

where

$$u_{\text{int}}(x, t) = 2\partial_{xx} \ln\left\{1 + \frac{F(\mathbf{X}, G)}{F(\mathbf{X})}\right\} \tag{10.95}$$

The latter expressions (10.94), (10.95) may be interpreted as the general periodic solution to the KdV equation for N degrees of freedom written as *the linear superposition of N cnoidal waves plus interactions among the cnoidal waves*. To verify this latter statement assume that the off-diagonal terms of the period matrix are zero so that $G = 0$, and Equations (10.94) and (10.95) become:

$$u(x, t) = 2\partial_{xx} \ln F(\mathbf{X}) \tag{10.96}$$

which is equivalent to Equation (10.82) so that:

$$\eta(x, t) = 2\sum_{n=1}^N \eta_n cn^2\{(K(m_n)/\pi)[k_n(x - C_n t)]|m\} \tag{10.97}$$

The complete spectral solution to KdV (10.89) is then given by the linear superposition of cnoidal waves plus nonlinear interactions (10.95):

$$\eta(x, t) = 2\sum_{n=1}^N \eta_n cn^2\{(K(m_n)/\pi)[k_n(x - C_n t)]|m\} + u_{\text{int}}(x, t) \tag{10.98}$$

This latter expression is completely general and equivalent to the θ-function solution to KdV given by Equations (10.10) and (10.11).

When the off-diagonal elements of the periodic matrix **B** *are sufficiently small* then an approximate form for the interactions is given by the following expression (which follows from Equation (10.90)):

$$
u_{\text{int}}(x,t) \cong 2\frac{\partial^2}{\partial x^2}\ln\left\{1+\frac{1}{2}\frac{\sum\limits_{\mathbf{m}}(\mathbf{m}\cdot\mathbf{Om})\exp\left(i\,\mathbf{m}\cdot\mathbf{X}+\frac{1}{2}\mathbf{m}\cdot\mathbf{Dm}\right)}{\sum\limits_{\mathbf{m}}\exp\left(i\,\mathbf{m}\cdot\mathbf{X}+\frac{1}{2}\mathbf{m}\cdot\mathbf{Dm}\right)}\right\} \tag{10.99}
$$

While this latter expression is a tempting diversion, it is worth noting that Equation (10.99) is not a very precise approximation to the wave motion when the spectrum is dominated by solitons; in this case the interaction terms are large and cannot be treated as perturbations. On the other hand when the moduli are sufficiently small ($m_n \ll 1$) for the cnoidal waves used to construct Equation (10.98), the second term inside the brackets {.} of Equation (10.99) may be viewed as a small perturbation with respect to 1.

10.6.4 Approximate Solutions to KdV for Partial Theta Sums

To study nonlinear interactions with the θ-function solution to KdV it is instructive to look at *particular partial sums*, that is, by making a specific choice for the lower and upper limits in the sums of Equation (10.67) (rather than ∞); then the following Mth partial sum may be considered:

$$
\theta_{MN}(X_1, X_2, \ldots, X_N) = \sum_{m_1=-M}^{M}\sum_{m_2=-M}^{M}\cdots\sum_{m_N=-M}^{M}
$$
$$
\times \exp\left[i\sum_{k=1}^{N}m_k X_k + \frac{1}{2}\sum_{j=1}^{N}\sum_{k=1}^{N}m_j m_k B_{jk}\cdot\right] \tag{10.100}
$$

Note that the θ-function is now subscripted with M (which defines a partial sum for a particular approximate solution to the KdV equation) and N (the number of degrees-of-freedom or the number of cnoidal waves in the IST spectrum). One could generalize Equation (10.100) to have different limits for each of the nested sums; this idea is given a practical basis in the numerical examples discussed below.

Now consider the specific case for the partial sum of two degrees of freedom, $N = 2$ (two approximate cnoidal waves plus nonlinear interactions):

$$
\theta_{M2} = \sum_{m_1=-M}^{M}\sum_{m_2=-M}^{M}C(m_1, m_2)e^{im_1 X_1 + im_2 X_2} \tag{10.101}
$$

where

$$
C(m_1, m_2) = q^{m_1^2}p^{m_2^2}r^{m_1 m_2} \tag{10.102}
$$

and

$$q = e^{-\frac{1}{2}B_{11}}, \quad p = e^{-\frac{1}{2}B_{22}}, \quad r = e^{-B_{12}} \tag{10.103}$$

When $M = 1$, the following partial sum arises

$$\theta_{12} = 1 + 2q \cos X_1 + 2p \cos X_2 + 2qpr^{-1} \cos(X_1 - X_2) \\ + 2qpr \cos(X_1 + X_2) \tag{10.104}$$

and for $M = 2$, the partial sum at next order is given by

$$\theta_{22} = \theta_{12} + 2q^4 \cos 2X_1 + 2p^4 \cos 2X_2 + 2q^4 p^4 r^{-4} \cos 2(X_1 - X_2) \\ + 2qp^4 r^2 \cos(X_1 + 2X_2) + 2qp^4 r^{-2} \cos(X_1 - 2X_2) \tag{10.105}$$

while for $M = 3$:

$$\theta_{32} = \theta_{12} + \theta_{22} + 2q^9 \cos 3X_1 + 2p^9 \cos 3X_2 + 2q^9 p^9 r^9 \cos(3X_1 + 3X_2) \\ + 2q^9 p^4 r^6 \cos(3X_1 + 2X_2) + 2q^9 pr^3 \cos(3X_1 + X_2) \\ + 2q^9 pr^{-3} \cos(3X_1 - X_2) + 2q^9 p^4 r^6 \cos(3X_1 + 2X_2) \\ + 2q^9 p^9 r^{-9} \cos(3X_1 - 3X_2) + 2q^4 p^9 r^6 \cos(2X_1 + 3X_2) \\ + 2q^4 p^9 r^{-6} \cos(2X_1 - 3X_2) + 2qp^9 r^3 \cos(X_1 + 3X_2) \\ + 2qp^9 r^{-3} \cos(X_1 - 3X_2)$$

The latter expression in complete form (using Equations (10.104) and (10.105)) is:

$$\theta_{32} = 1 + 2q \cos X_1 + 2p \cos X_2 + 2q^4 \cos 2X_1 + 2p^4 \cos 2X_2 + 2q^9 \cos 3X_1 \\ + 2p^9 \cos 3X_2 + 2qpr^{-1} \cos(X_1 - X_2) + 2qpr \cos(X_1 + X_2) \\ + 2q^4 p^4 r^4 \cos 2(X_1 + X_2) + 2q^4 pr^2 \cos(2X_1 + X_2) \\ + 2q^4 pr^{-2} \cos(2X_1 - X_2) + 2q^4 p^4 r^{-4} \cos 2(X_1 - X_2) \\ + 2qp^4 r^2 \cos(X_1 + 2X_2) + 2qp^4 r^{-2} \cos(X_1 - 2X_2) \\ + 2q^9 p^9 r^9 \cos(3X_1 + 3X_2) + 2q^9 p^4 r^6 \cos(3X_1 + 2X_2) \\ + 2q^9 pr^3 \cos(3X_1 + X_2) + 2q^9 pr^{-3} \cos(3X_1 - X_2) \\ + 2q^9 p^4 r^6 \cos(3X_1 + 2X_2) + 2q^9 p^9 r^{-9} \cos(3X_1 - 3X_2) \\ + 2q^4 p^9 r^6 \cos(2X_1 + 3X_2) + 2q^4 p^9 r^{-6} \cos(2X_1 - 3X_2) \\ + 2qp^9 r^3 \cos(X_1 + 3X_2) + 2qp^9 r^{-3} \cos(X_1 - 3X_2)$$

Hence by increasing M one finds higher order partial-sum approximations of the cnoidal waves, including their mutual nonlinear interaction terms. Note that by going to higher order one also finds all possible combinations of the sums and differences of the wavenumbers ($X_n \sim k_n x$) in the IST spectrum.

This result should come as no surprise to those who have experience in computing bi and tri-spectra of nonlinear wave trains; in the periodic IST, however, one is able to compute *all* N-spectra of the wave train, provided of course that one has the patience to compute the appropriate partial sums.

In the above expressions, it is clear that by setting $r = 1$ (equivalent to setting he off-diagonal terms to zero) then the interactions are excluded and the θ-functions reduce identically to the factored forms:

$$\theta_{12} = (1 + 2q \cos X_1)(1 + 2p \cos X_2)$$

$$\theta_{22} = (1 + 2q \cos X_1 + 2q^4 \cos 2X_1)(1 + 2p \cos X_2 + 2p^4 \cos 2X_2) \quad (10.106)$$

$$\theta_{32} = (1 + 2q \cos X_1 + 2q^4 \cos 2X_1 + 2q^9 \cos 3X_1)$$
$$\times (1 + 2p \cos X_2 + 2p^4 \cos 2X_2 + 2p^9 \cos 3X_2)$$

In these latter expressions there are no nonlinear interactions between the two components m_1, m_2, for which one has successive approximations of the linear superposition of two cnoidal waves. Nonlinear interactions are taken into account only when $B_{12} \neq 0$ such that $r \neq 1$, as discussed above.

Another partial sum that will be useful below for the study of three degrees of freedom, $N = 3$, is given by

$$\theta_{M3} = \sum_{m_1=-M}^{M} \sum_{m_2=-M}^{M} \sum_{m_3=-M}^{M} C(m_1, m_2, m_3) e^{im_1 X_1 + im_2 X_2 + im_3 X_3} \quad (10.107)$$

The coefficients have the following form:

$$C(m_1, m_2, m_3) = q_1^{m_1^2} q_2^{m_2^2} q_3^{m_3^2} r_{12}^{m_1 m_2} r_{13}^{m_1 m_3} r_{23}^{m_2 m_3} \quad (10.108)$$

where

$$q_1 = e^{-\frac{1}{2}B_{11}}, \quad q_2 = e^{-\frac{1}{2}B_{22}}, \quad q_3 = e^{-\frac{1}{2}B_{33}}$$
$$\quad (10.109)$$
$$r_{12} = e^{-B_{12}}, \quad r_{13} = e^{-B_{13}}, \quad r_{23} = e^{-B_{23}}$$

For the specific case $M = 1$:

$$\begin{aligned}
\theta_{13} = 1 &+ 2q_1 \cos X_1 + 2q_2 \cos X_2 + 2q_3 \cos X_3 \\
&+ 2q_1 q_2 r_{12} \cos (X_1 + X_2) + 2q_1 q_2 r_{12}^{-1} \cos (X_1 - X_2) \\
&+ 2q_1 q_3 r_{13} \cos (X_1 + X_3) + 2q_1 q_3 r_{13}^{-1} \cos (X_1 - X_3) \\
&+ 2q_2 q_3 r_{23} \cos (X_2 + X_3) + 2q_2 q_3 r_{23}^{-1} \cos (X_2 - X_3) \\
&+ 2q_1 q_2 q_3 r_{12} r_{13} r_{23} \cos (X_1 + X_2 + X_3) \\
&+ 2q_1 q_2 q_3 r_{12} r_{13}^{-1} r_{23}^{-1} \cos (X_1 + X_2 - X_3) \\
&+ 2q_1 q_2 q_3 r_{12}^{-1} r_{13} r_{23}^{-1} \cos (X_1 - X_2 + X_3) \\
&+ 2q_1 q_2 q_3 r_{12}^{-1} r_{13}^{-1} r_{23} \cos (X_1 - X_2 - X_3)
\end{aligned} \quad (10.110)$$

In the absence of interactions one has $r_{12} = 1$, $r_{13} = 1$, and $r_{23} = 1$ so that

$$\theta_{13} = (1 + 2q_1 \cos X_1)(1 + 2q_2 \cos X_2)(1 + 2q_3 \cos X_3) \qquad (10.111)$$

This latter expression leads to a simple linear superposition of three small amplitude ($M = 1$) cnoidal waves.

10.6.5   Linear Limit of KdV Solutions

The linear limit of the θ-function formulation occurs when the wave amplitudes are so small that the cnoidal waves are reduced to sine waves and the nonlinear interaction terms no longer contribute. In this case, the period matrix is essentially diagonal with large negative elements. Thus the nomes $q_n = \exp[-B_{nn}/2]$ are small and each degree of freedom is a simple sine wave: $\lambda\eta \sim -qk^2 \cos(kx - \omega t + \phi)$. The following linear Fourier series then holds as an approximate small-amplitude solution to KdV:

$$\lambda\eta \cong \sum_{n=1}^{N} c_n \cos\left(k_n x - \omega_n t + \phi_n\right), \quad c_n = -q_n k_n^2$$

Because of the small amplitudes the frequencies are governed essentially by the linear dispersion relation, $\omega = c_o k - \beta k^3$, and the phases are arbitrary. See Osborne and Bergamasco (1985) for a discussion of this limit in terms of inverse scattering transform variables in the hyperelliptic function representation.

10.6.6   Approximate Solutions to KdV for Specific Cases

In this section the symbol θ refers to a θ-function, while X refers to a phase.

10.6.6.1   Case for One Degree of Freedom

Consider the case for which one sums the θ-function, for a single degree of freedom, over $M = -1, 0, 1$ in Equation (10.67) (or over $n = 1$ in Equation (10.68)) to find:

$$\theta_3 \cong \theta_{11} = 1 + 2q \cos X \qquad (10.112)$$

where $X = kx - \omega t + \phi$ and q is given by Equation (10.69). This latter expression is the simplest approximation for the θ-function. Applying Equation (10.72)

$$\lambda\eta(x, t) \cong 2\frac{\partial^2}{\partial x^2} \ln \theta_{11}(x, t, q) = -4qk^2 \left[\frac{\cos X + 2q}{(1 + 2q \cos X)^2}\right] \qquad (10.113)$$

Using a simple, consistent approximation (see Equation (10.84))

$$q = e^{-b/2} \cong U/(1 + U^2)$$

for U the Ursell number, one finds:

$$\eta(x,t) \cong -\frac{\eta_o}{1+U^2}\left[\frac{\cos X + 2q}{(1 + 2q\cos X)^2}\right] \cong -\eta_o\left[\frac{(1+U^2)\cos X + 2U}{(1+U^2 + 2U\cos X)^2}\right] \quad (10.114)$$

The amplitude η_o has the explicit form

$$\eta_o \equiv A/\lambda = 8Uk^2 h^3/3 \tag{10.115}$$

with $A = 4qk^2\,(1 + U^2) = 4Uk^2$. It follows that the Ursell number has the following expression:

$$U = 3\eta_o/8k^2 h^3 \tag{10.116}$$

Note that to obtain the *small-amplitude cnoidal wave* (10.114), the θ-function has been summed *not* over $(-\infty, \infty)$ as generally required by the theory, but *instead* over the smallest possible number of terms $(-1,0,1)$. How good is this seemingly trivial approximation? Comparison of Equation (10.114) and the cnoidal wave (10.74) for $m = 0.969$ shows that the two functions are graphically indistinguishable. Equation (10.114) is a physically significant approximation to the cnoidal wave (nominally valid for the approximate range of the modulus $0 \leq m < 0.9$). It is thus clear that, even though θ_{11} Equation (10.112) is only a simple cosine function, the corresponding approximate solution of KdV Equation (10.114) is quite nonlinear. Equation (10.75) is the exact Fourier series expansion for the approximate cnoidal wave (10.114).

Now recall the product formula (10.77) discussed above. Note that the leading order term for the solution to KdV, for $n = 1$, is given by Equation (10.78):

$$u(x, 0) = -4k^2 q\left[\frac{(1 + q^2)\cos kx + 2q}{(1 + 2q\,\cos kx + q^2)^2}\right] \tag{10.117}$$

which is, for sufficiently small $q \sim U$, the same as Equation (10.114) above. Thus to leading order in q the product formula (10.77) and the sum formula (10.68) give the same approximation for the single degree-of-freedom solution to KdV. However, for larger values of q the convergence of Equation (10.117) to a cnoidal wave is superior to Equation (10.114).

10.6.6.2 Case for Two Degrees of Freedom

Now consider the case for $N = 2$, $M = 1$ in which *two* of the approximate cnoidal waves (10.114) interact nonlinearly in the important regime $0 \leq m < 0.9$.

An approximate two degree-of-freedom solution to the KdV equation then arises
for which nonlinear interactions are explicitly computed. For this case, the following form for the θ-function arises:

$$\theta_{12} = 1 + 2q_1 \cos X_1 + 2q_2 \cos X_2 + 2q_1 q_2 r^{-1} \cos (X_1 - X_2)$$
$$+ 2q_1 q_2 r \cos (X_1 + X_2) \tag{10.118}$$

where

$$q_1 = e^{-\frac{1}{2}B_{11}}, \quad q_2 = e^{-\frac{1}{2}B_{22}}, \quad r = e^{-B_{12}} \tag{10.119}$$

and $X_1 = k_1 x - \omega_1 t + \phi_1$, $X_2 = k_2 x - \omega_2 t + \phi_2$.

In the particular case for which nonlinear interactions are small, $B_{12} \simeq 0$,
then $r \simeq 1$ and:

$$\theta_{12} \cong (1 + 2q_1 \cos X_1)(1 + 2q_2 \cos X_2) \tag{10.120}$$

This suggests that to include nonlinear interactions one might more generally
write:

$$\theta_{12} = (1 + 2q_1 \cos X_1)(1 + 2q_2 \cos X_2) + \varepsilon \tag{10.121}$$

where

$$\varepsilon = 2q_1 q_2 r^{-1} \cos (X_1 - X_2) + 2q_1 q_2 r \cos (X_1 + X_2) - 4q_1 q_2 \cos X_1 \cos X_2 \tag{10.122}$$

It will be seen, however, that ε is not necessarily a small correction, except
when the wave moduli are sufficiently small. Equations (10.121) and
(10.122) are identical to Equation (10.118). It follows that:

$$\theta_{12} = f_{12} + \varepsilon = f_{12}(1 + \varepsilon / f_{12}) \tag{10.123}$$

for which

$$\ln \theta_{12} = \ln(f_{12} + \varepsilon) = \ln f_{12} + \ln(1 + \varepsilon / f_{12}) \tag{10.124}$$

where $f_{12} = (1 + 2q1 \cos X_1)(1 + 2q_2 \cos X_2)$. These latter results ensure that
the approximate two degree-of-freedom solution to KdV has the simple form:

$$\lambda \eta(x,t) = 2\frac{\partial^2}{\partial x^2} \ln \theta_{12}(x,t) = 2\frac{\partial^2}{\partial x^2} [\ln(1 + 2q_1 \cos X_1)$$
$$+ \ln(1 + 2q_2 \cos X_2) + \ln(1 + \varepsilon / f_{12})] \tag{10.125}$$

Therefore

$$\eta(x,t) = -\eta_1 \left[\frac{(1+U_1^2)\cos X_1 + 2U_1}{(1+U_1^2 + 2U_1\cos X_1)^2} \right] - \eta_2 \left[\frac{(1+U_2^2)\cos X_2 + 2U_2}{(1+U_2^2 + 2U_2\cos X_2)^2} \right]$$

$$+ \eta_{int}(x,t)$$

(10.126)

where

$$\eta_{int}(x,t) = \frac{2}{\lambda}\frac{\partial^2}{\partial x^2}\ln\theta_{int}(x,t)$$

(10.127)

and

$$\theta_{int}(x,t) = (1+\varepsilon/f_{12}) = \{1 + 4q_1q_2[g(r)s(X_1)s(X_2) + f(r)c(X_1)c(X_2)]\}$$

(10.128)

Here

$$c(X_n) = \frac{\cos X_n}{1 + 2q_n\cos X_n}, \quad s(X_n) = \frac{\sin X_n}{1 + 2q_n\cos X_n}$$

(10.129)

for $n = 1, 2$. The following notation has been used

$$f(r) = \frac{r^{-1}+r}{2} - 1, \quad g(r) = \frac{r^{-1}-r}{2}$$

(10.130)

When $r = 1$ then $f(r) = g(r) = 0$ and no interactions occur. Inserting Equation (10.128) into Equation (10.127) one finds:

$$\eta_{int}(x,t) = \frac{2}{\lambda}\frac{[a(c_1''c_2 + 2c_1'c_2' + c_1c_2'') + b(s_1''s_2 + 2s_1's_2' + s_1s_2'')]}{1 + ac_1c_2 + bs_1s_2}$$

$$- \frac{2}{\lambda}\left[\frac{a(c_1'c_2 + c_1c_2') + b(s_1's_2 + s_1s_2')}{1 + ac_1c_2 + bs_1s_2} \right]^2$$

(10.131)

where primes denote differentiation with respect to x and

$$a = 4q_1q_2f(r), \quad b = 4q_1q_2g(r)$$

$$c_n \equiv c(X_n), \quad s_n \equiv s(X_n)$$

(10.132)

Hence

$$s'(X_n) = \frac{k_n(\cos X_n + 2q_n)}{(1 + 2q_n\cos X_n)^2} = \frac{1}{4q_nk_n}u_n(x,t)$$

$$s''(X_n) = \frac{1}{4q_n k_n} u'(x,t), \quad u_n(x,t) = 4q_n k_n^2 \left[\frac{\cos X_n + 2q_n}{(1 + 2q_n \cos X_n)^2} \right]$$

$$c'(X_n) = -\frac{k_n \sin X_n}{(1 + 2q_n \cos X_n)^2}$$

$$c''(X_n) = -\frac{k_n^2 \cos X_n(1 - 2q_n \cos X_n) + 4q_n k_n^2}{(1 + 2q_n \cos X_n)^3}$$

Note that the following relations hold approximately for *small interactions*, $B_{12} \ll 1$:

$$f(r) \simeq \frac{1}{2} B_{12}^2, \quad g(r) \simeq B_{12}$$

Then the interaction term can be written in the simple form (use $\ln(1+x) \sim x$):

$$\eta_{\text{int}}(x,t) \cong \frac{8}{\lambda} q_1 q_2 B_{12} \frac{\partial^2}{\partial x^2} \left\{ \frac{\sin X_1 \sin X_2}{(1 + 2q_1 \cos X_1)(1 + 2q_2 \cos X_2)} \right\}$$

$$\cong \frac{8}{\lambda} q_1 q_2 B_{12} \frac{\partial^2}{\partial x^2} [s(X_1)s(X_2)]$$

$$\cong \frac{8}{\lambda} q_1 q_2 B_{12} [s''(X_1)s(X_2) + 2s'(X_1)s'(X_2) + s(X_1)s''(X_2)]$$

$$(10.133)$$

The latter expression for small interactions holds only when the two degrees of freedom have moduli that are substantially less than one.

10.6.6.3   Case for Three Degrees of Freedom

Consider the case for $N = 3$, $M = 1$ in which *three* of the approximate cnoidal waves (10.114) interact nonlinearly. One has the following form for the θ-function:

$$
\begin{aligned}
\theta_{13} = 1 &+ 2q_1 \cos X_1 + 2q_2 \cos X_2 + 2q_3 \cos X_3 \\
&+ 2q_1 q_2 r_{12} \cos(X_1 + X_2) + 2q_1 q_2 r_{12}^{-1} \cos(X_1 - X_2) \\
&+ 2q_1 q_3 r_{13} \cos(X_1 + X_3) + 2q_1 q_3 r_{13}^{-1} \cos(X_1 - X_3) \\
&+ 2q_2 q_3 r_{23} \cos(X_2 + X_3) + 2q_2 q_3 r_{23}^{-1} \cos(X_2 - X_3) \\
&+ 2q_1 q_2 q_3 r_{12} r_{13} r_{23} \cos(X_1 + X_2 + X_3) \\
&+ 2q_1 q_2 q_3 r_{12} r_{13}^{-1} r_{23}^{-1} \cos(X_1 + X_2 - X_3) \\
&+ 2q_1 q_2 q_3 r_{12}^{-1} r_{13} r_{23}^{-1} \cos(X_1 - X_2 + X_3) \\
&+ 2q_1 q_2 q_3 r_{12}^{-1} r_{13}^{-1} r_{23} \cos(X_1 - X_2 - X_3)
\end{aligned}
$$

$$(10.134)$$

where

$$q_1 = e^{-\frac{1}{2}B_{11}}, \quad q_2 = e^{-\frac{1}{2}B_{22}}, \quad q_3 = e^{-\frac{1}{2}B_{33}}, \quad r_{12} = e^{-B_{12}}, \quad r_{13} = e^{-B_{13}}, \quad r_{23} = e^{-B_{23}}$$

$$(10.135)$$

and $X_1 = k_1 x - \omega_1 t + \phi_1$, $X_2 = k_2 x - \omega_2 t + \phi_2$, and $X_3 = k_3 x - \omega_3 t + \phi_3$.
   In the particular case for which nonlinear interactions are sufficiently small, $B_{12} \simeq 0$, $B_{13} \simeq 0$, and $B_{23} \simeq 0$, so that then $r_{12} \simeq 1$, $r_{13} \simeq 1$, and $r_{23} \simeq 1$:

$$\theta_{13} \cong (1 + 2q_1 \cos X_1)(1 + 2q_2 \cos X_2)(1 + 2q_3 \cos X_3) \qquad (10.136)$$

This observation therefore suggests that, for Equation (10.134), one may write the following expression for the θ-function partial sum that includes nonlinear interactions:

$$\theta_{13} = (1 + 2q_1 \cos X_1)(1 + 2q_2 \cos X_2)(1 + 2q_3 \cos X_3) + \varepsilon \qquad (10.137)$$

where

$$\begin{aligned}
\varepsilon = {}& 2\,q_1 q_2 r_{12} \cos (X_1 + X_2) + 2q_1 q_2 r_{12}^{-1} \cos (X_1 - X_2) \\
&+ 2q_1 q_3 r_{13} \cos (X_1 + X_3) + 2q_1 q_3 r_{13}^{-1} \cos (X_1 - X_3) \\
&+ 2q_2 q_3 r_{23} \cos (X_2 + X_3) + 2q_2 q_3 r_{23}^{-1} \cos (X_2 - X_3) \\
&+ 2q_1 q_2 q_3 r_{12} r_{13} r_{23} \cos (X_1 + X_2 + X_3) \\
&+ 2q_1 q_2 q_3 r_{12} r_{13}^{-1} r_{23}^{-1} \cos (X_1 + X_2 - X_3) \\
&+ 2q_1 q_2 q_3 r_{12}^{-1} r_{13} r_{23}^{-1} \cos (X_1 - X_2 + X_3) \\
&+ 2q_1 q_2 q_3 r_{12}^{-1} r_{13}^{-1} r_{23} \cos (X_1 - X_2 - X_3) \\
&- 4q_1 q_2 \cos X_1 \cos X_2 - 4q_1 q_3 \cos X_1 \cos X_3 - 4q_2 q_3 \cos X_2 \cos X_3 \\
&- 8q_1 q_2 q_3 \cos X_1 \cos X_2 \cos X_3
\end{aligned}$$

$$(10.138)$$

Again ε is not necessarily a small correction, except when $r_{ij} \simeq 1$. Equations (10.137) and (10.138) are identical to Equation (10.134). Equation (10.137) may be written:

$$\theta_{13} = f_{13} + \varepsilon = f_{13}(1 + \varepsilon/f_{13}) \qquad (10.139)$$

for which

$$\ln \theta_{13} = \ln(f_{13} + \varepsilon) = \ln f_{13} + \ln(1 + \varepsilon/f_{13}) \qquad (10.140)$$

where $f_{13} = (1 + 2q_1 \cos \theta_1)(1 + 2q_2 \cos \theta_2)(1 + 2q_3 \cos \theta_3)$. Finally the approximate three degree-of-freedom solution to KdV has the simple form:

$$\lambda\eta(x,t) = 2\frac{\partial^2}{\partial x^2}\ln\theta_{13}(x,t) = 2\frac{\partial^2}{\partial x^2}[\ln(1+2q_1\cos X_1) + \ln(1+2q_2\cos X_2)$$
$$+ \ln(1+2q_3\cos X_3) + \ln(1+\varepsilon/f_{13})]$$

$$(10.141)$$

Therefore

$$\eta(x,t) = -\eta_1\left[\frac{(1+U_1^2)\cos X_1 + 2U_1}{(1+U_1^2+2U_1\cos X_1)^2}\right] - \eta_2\left[\frac{(1+U_2^2)\cos X_2 + 2U_2}{(1+U_2^2+2U_2\cos X_2)^2}\right]$$
$$- \eta_3\left[\frac{(1+U_3^2)\cos X_3 + 2U_3}{(1+U_3^2+2U_3\cos X_3)^2}\right] + \eta_{int}(x,t)$$

$$(10.142)$$

which consists of three small-amplitude cnoidal waves plus nonlinear interactions. Here

$$\eta_{int}(x,t) = \frac{2}{\lambda}\frac{\partial^2}{\partial x^2}\ln\theta_{int}(x,t) \qquad (10.143)$$

for:

$$\theta_{int}(x,t) = (1 + \varepsilon/f_{13}) \qquad (10.144)$$

For brevity I do not give the complete calculation. Numerical implementation of the three degree-of-freedom case has been accomplished directly using Equation (10.142) with Equations (10.143), (10.144), and (10.138).

10.6.7 The Single Cnoidal Wave Solution to the KdV Equation

The simplest θ-function is $\theta_3(x)$, which is written:

$$\theta(x) = 1 + 2\sum_{n=1}^{\infty} q^{n^2}\cos[2\pi nx/L] \qquad (10.145)$$

where q is the *nome*, written in terms of the elliptic integrals as (1050.01) of Byrd and Friedman (p. 315, 1971):

$$q = e^{-\pi K'(m)/K(m)}, \quad L = 2K(m)$$

Here $K'(m) = K(m-1)$, where $K(m)$ is the elliptic integral of the first kind. We also use the notation

$$q = e^{-\frac{1}{2}B}, \quad B = -2\ln q$$

and B is the one-by-one "period matrix."
    The solution to the KdV equation is given by

$$\eta(x,0) = \frac{2}{\lambda}\partial_{xx}\ln\theta(x), \quad \lambda = \frac{3}{2h^3} \quad \text{(KdV nonlinearity parameter)}$$

First note that (Byrd and Friedman, 1971 (1050.02), p. 316):

$$\ln\theta(x) = \ln\gamma - 2\sum_{n=1}^{\infty}\frac{(-1)^n q^n}{n(1-q^{2n})}\cos\left(2\pi nx/L\right) \tag{10.146}$$

Here γ is Euler's number, $\gamma \sim 0.5772156649$. Formulas in Abramowitz and Stegun (1964; 16.29.3) and Whittaker and Watson (1902; p. 489, 12) give

$$\partial_x\ln\theta(x) = \frac{\theta'(x)}{\theta(x)} = \frac{4\pi}{L}\sum_{n=1}^{\infty}\frac{(-1)^n q^n}{1-q^{2n}}\sin\left[2\pi nx/L\right] \tag{10.147}$$

Then the cnoidal wave solution to KdV is:

$$\eta(x,0) = \frac{2}{\lambda}\partial_{xx}\ln\theta(x) = \frac{16\pi^2}{\lambda L^2}\sum_{n=1}^{\infty}(-1)^n\frac{nq^n}{1-q^{2n}}\cos\left[2\pi nx/L\right] \tag{10.148}$$

Now we are used to using the cnoidal wave with *modulus*, m, rather than the *nome*, q. Thus, the relationship between the two is established by first computing the elliptic integral, K, from q (Abramowitz and Stegun, 1964, 17.3.22):

$$K(q) = \frac{\pi}{2} + 2\pi\sum_{n=1}^{\infty}\frac{q^n}{1+q^{2n}} \tag{10.149}$$

Then the modulus m has the form (Abramowitz and Stegun, 1964; (16.38.7)):

$$m(q) = \frac{4\pi^2 q}{K^2(q)}\left(\sum_{n=0}^{\infty}q^{n(n+1)}\right)^4 \tag{10.150}$$

The above formulas work provided we choose B, compute q, and then evaluate the series for the various functions desired. But, what if we wish to select m and then compute q from m?

To this end we also know from Abramowitz and Stegun (1964; 17.3.17) that

$$q = \exp\left(-\pi\frac{K(1-m)}{K(m)}\right)$$

This is not very useful from a practical point of view. Instead we compute

$$\varepsilon = \frac{1}{2}\left(\frac{1-(1-m)^{1/4}}{1+(1-m)^{1/4}}\right)$$

and then

$$q = \varepsilon + 2\varepsilon^5 + 15\varepsilon^9 + 150\varepsilon^{13} + 1707\varepsilon^{17} + \cdots$$

(see Whittaker and Watson, 1902, p. 486 and Magnus et al., 1966, p. 378). This result allows us to select m and then to compute the nome q. From this we can compute $B = -2\ln q$.

10.6.8   The Ursell Number

A useful parameter is the *Ursell number* whose space-like form is:

$$U = \frac{3a}{4k^2h^3} = \frac{3aL^2}{16\pi^2h^3} = \frac{3}{16\pi^2}\left(\frac{a}{h}\right)\left(\frac{L}{h}\right)^2 \tag{10.151}$$

and whose time-like form is

$$U = \frac{3ac_o^2T^2}{16\pi^2h^3} = \frac{3}{16\pi^2}\left(\frac{a}{h}\right)\left(\frac{c_oT}{h}\right)^2 \tag{10.152}$$

The Ursell number is connected to the modulus m by the relation given in Section 10.6.9. See Figure 10.1 of Chapter 2.

10.6.9   The Cnoidal Wave as a Classical Elliptic Function
and Its Ursell Number

In the classical elliptic function notation, the cnoidal wave is given by:

$$\eta(x,t) = 2\eta_c cn^2\{[K(m)/\pi](kx - \omega_c t)\,|\,m\} - \bar{\eta}$$

The cnoidal wave amplitude, η_c, is related to the modulus, m, by:

$$mK^2(m) = (3\pi^2/2k^2h^3)\eta_c$$

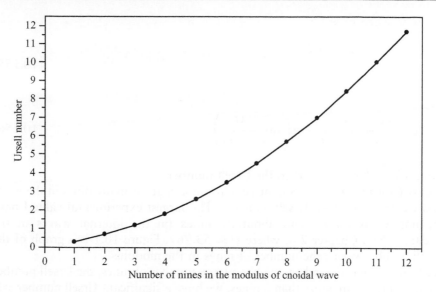

Figure 10.1 Ursell number as a function of the number of nines in the cnoidal wave modulus.

If we define the Ursell number to be

$$U = \frac{3\eta_c}{4k^2h^3}$$

Then we have the relationship between the Ursell, U, number and the modulus, m:

$$mK^2(m) = 2\pi^2(3\eta_c/4k^2h^3) = 2\pi^2 U \tag{10.153}$$

$$U = \frac{mK^2(m)}{2\pi^2} \tag{10.154}$$

When m is small we have:

$$K \approx \frac{\pi}{2}$$

so that:

$$U \approx m/8$$

Likewise when $m \approx 1$:

$$K = \ln \left(\frac{16}{1-m} \right)^{1/2} \tag{10.155}$$

$$U = \frac{mK^2(m)}{2\pi^2} \cong \frac{1}{2\pi^2} m \left(\ln \sqrt{\frac{16}{1-m}} \right)^2 \tag{10.156}$$

Table 10.1 relates m ($\sim$ 1) to the Ursell number.

$U \approx 1$ for $m \approx 0.9978$. So, the conclusion is that for many nines in the modulus we have very high Ursell numbers. The largest experimental value I have encountered is for m with about 27 nines (in the internal waves in the Andaman Sea, Chapter 25), where $U \approx 53.767$. Figure 10.1 is a graph of the Ursell number versus the number of nines in the modulus.

The message is that when one has a large number of nines, the Ursell number is large. If there are more than 3 nines, we have a significant Ursell number $\sim$1. If there are 16 nines, we have about the maximum possible value in double precision, which is an Ursell number of about 20. The way to interpret Figure 10.1 is to assume we have a single cnoidal wave component in the IST spectrum with modulus m. Then we have associated with it an Ursell number, given in the graph. This approach is quite nice, since we compute the Ursell number, which includes the amplitude of the component, its wavenumber (or frequency) and the depth, all in the single parameter, U.

Table 10.1 Values of the Elliptic Modulus m Versus the Ursell Number U

m	U
0.9	0.294
0.99	0.682
0.999	1.186
0.999 9	1.818
0.999 99	2.585
0.999 999	3.485
0.999 999 9	4.520
0.999 999 99	5.689
0.999 999 999	6.992
0.999 999 999 9	8.429
0.999 999 999 99	10.001
0.999 999 999 999	11.707
0.999 999 999 999 999 9	19.770
0.999 999 999 999 999 999 999 999	53.767

10.6.10 An Example Problem with 10 Degrees of Freedom

An example of a 10 degree-of-freedom solution of the KdV equation is shown in Figure 10.2. The 10 cnoidal waves are shown together with their summation. The nonlinear interactions are given and are summed to the 10 cnoidal waves to give a synthesized solution of the KdV equation. In this case, the nonlinear interactions are seen to be quite large and cannot be viewed as a simple small-amplitude perturbation of the summed cnoidal waves.

10.6.11 Relationship of Cnoidal Wave Parameters to the Parameter q

By now it should be clear that the parameters q, m, U, K, and B are all related to one another, uniquely. Each of these parameters can be expressed as one of the others.

For example

$$m = m(q)$$

$$m(q) = \frac{4\pi^2 q}{K^2(q)} \left(\sum_{n=0}^{\infty} q^{n(n+1)} \right)^4 \tag{10.157}$$

$$K = K(q)$$

$$K(q) = \frac{\pi}{2} + 2\pi \sum_{n=1}^{\infty} \frac{q^n}{1 + q^{2n}} \tag{10.158}$$

$$U = U(m)$$

$$U = \frac{mK^2(m)}{2\pi^2} \tag{10.159}$$

$$q = q(m)$$

$$q = \varepsilon + 2\varepsilon^5 + 15\varepsilon^9 + 150\varepsilon^{13} + 1707\varepsilon^{17} + \cdots,$$

$$\varepsilon = \frac{1}{2} \left(\frac{1 - (1-m)^{1/4}}{1 + (1-m)^{1/4}} \right) \tag{10.160}$$

$$B = B(q)$$

$$B = -2 \ln q \tag{10.161}$$

At the end of this chapter is a table relating various parameters of the theta and elliptic functions.

...

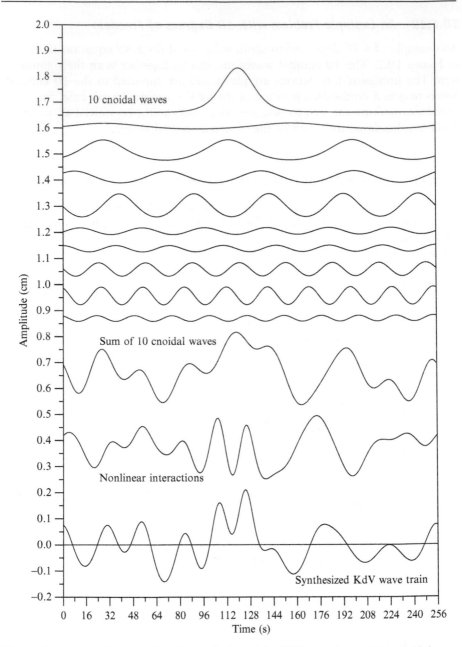

Figure 10.2 A 10 degree-of-freedom solution of the KdV equation and its cnoidal wave decomposition.

10.6.12 Wave Amplitudes and Heights for Each Degree of Freedom of KdV

The wave amplitudes and heights of each of the degrees of freedom in physical space, in the absence of interactions with the other components, are easily computed. First, recall that for a single degree of freedom n the associated cnoidal wave is given by:

$$\lambda \eta_n(x, 0) = 2 \frac{\partial^2}{\partial x^2} \ln \theta_3(x, q_n) = 4k_n^2 \sum_{m=1}^{\infty} \frac{m(-1)^m q_n^m}{1 - q_n^{2m}} \cos(mk_n x) \quad (10.162)$$

where

$$q_n = \exp\left[-\frac{1}{2} B_{nn}\right] \quad (10.163)$$

Note that $\eta_n(\pi/k_n, 0)$ is the maximum value of the wave and $\eta_n(0, 0)$ is the minimum. The *amplitude of a single degree of freedom* therefore is:

$$\eta_{\max} = \eta_n(\pi/k_n, 0) = 4k_n^2 \sum_{m=1}^{\infty} \frac{m q_n^m}{1 - q_n^{2m}} \quad (10.164)$$

Then the *height of a single degree of freedom* is given by

$$\lambda H_n = \lambda[\eta_n(\pi/k_n, 0) - \eta_n(0, 0)] = 4k_n^2 \sum_{m=1}^{\infty} \frac{m q_n^m}{1 - q_n^{2m}} (1 - (-1)^m) \quad (10.165)$$

so that

$$H_n = \frac{8k_n^2}{\lambda} \sum_{m=1, 3, 5 \ldots}^{\infty} \frac{m q_n^m}{1 - q_n^{2m}} \quad (10.166)$$

The number of terms required for convergence in this formula *is not the same number to be taken in the θ-function sum* (10.62). Generally, speaking the number of terms in Equation (10.66) is much less than the number of terms required in either Equation (10.162) or (10.164) to obtain a similar precision.

As demonstrated above, for $M = 1$, one obtains a small-amplitude cnoidal wave which works quite well to leading order in q or U (Equations (10.114) and (10.117)). One can easily estimate the height of one of these small-amplitude cnoidal waves. Recall that, approximately

$$\lambda \eta_n(x, t) \cong -A_n \left[\frac{(1 + U_n^2) \cos(k_n x - \omega_n t) + 2U_n}{[1 + U_n^2 + 2U_n \cos(k_n x - \omega_n t)]^2} \right] \quad (10.167)$$

for $U_n = 3\eta_n/8k_n^2 h^3$. Then

$$\eta_{\max} = \eta_n(\pi/k_n, 0) = \frac{\eta_n}{(1 - U_n)^2}, \quad \eta_n = A_n/\lambda = 8U_n k_n^2 h^3/3 \qquad (10.168)$$

$$\eta_{\min} = \eta_n(0, 0) = -\frac{\eta_n}{(1 + U_n)^2}$$

It follows that $H_n = \eta_n (\pi/k_n, 0) - \eta_n (0,0)$:

$$H_n = 2\eta_n \left[\frac{1 + U_n^2}{(1 - U_n^2)^2} \right] \qquad (10.169)$$

This expression is a leading order approximation for the height of a single cnoidal wave (10.164). For very small Ursell numbers, the solution is a sine wave and the latter formula (10.166) reduces to $H_n = 2\eta_n$, that is, twice the amplitude of the sine wave.

Another estimate of the amplitude of a single degree-of-freedom solution to KdV can be obtained by use of the product formulas (10.76) and (10.77). In analogy with the formulas obtained above one finds:

$$\eta_{\max} = \eta_n(\pi/k_n, 0) = -4k_n^2 \sum_{m=1}^{\infty} \frac{q_n^{2m-1}}{(1 - q_n^{2m-1})^2}$$

$$\eta_{\min} = \eta_n(0, 0) = 4k_n^2 \sum_{m=1}^{\infty} \frac{q_n^{2m-1}}{(1 + q_n^{2m-1})^2} \qquad (10.170)$$

This gives the amplitude of a cnoidal wave of nome for a particular modulus m_n or Ursell number U_n, with wavenumber k_n in water of depth h. It follows that:

$$H_n = (\eta_{\max} - \eta_{\min})/\lambda = \frac{8k_n^2}{\lambda} \sum_{m=1}^{\infty} q_n^{2m-1} \left[\frac{1 + q_n^{4m-1}}{(1 - q_n^{4m-2})^2} \right] \qquad (10.171)$$

The $n = 1$ term in this series corresponds to the approximation (10.166) for the height of the fundamental wavelet (10.79) when $q = e^{-b/2} \cong U/(1 + U^2) \simeq U$. We see that to leading order, for small q_n, we have

$$\frac{H_n}{h} \simeq \frac{16k_n^2 h^2}{3} q_n$$

See Table 10.2 for a quick lookup guide to the important nonlinear parameters for a cnoidal wave.

Table 10.2 Table of Parameters for One Degree-of-Freedom θ-Function

i	m	q	B_{11}	$K(q)$
1	0.1022051404D + 00	0.6737946999D − 02	0.1000000000D + 02	0.1613417366D + 01
2	0.1071548570D + 00	0.7083408929D − 02	0.9900000000D + 01	0.1615617971D + 01
3	0.1123290342D + 00	0.7446583071D − 02	0.9800000000D + 01	0.1617933020D + 01
4	0.1177362894D + 00	0.7828377549D − 02	0.9700000000D + 01	0.1620368553D + 01
5	0.1233854076D + 00	0.8229747049D − 02	0.9600000000D + 01	0.1622930934D + 01
6	0.1292853254D + 00	0.8651695203D − 02	0.9500000000D + 01	0.1625626875D + 01
7	0.1354451130D + 00	0.9095277102D − 02	0.9400000000D + 01	0.1628463453D + 01
8	0.1418739520D + 00	0.9561601931D − 02	0.9300000000D + 01	0.1631448132D + 01
9	0.1485811116D + 00	0.1005183574D − 01	0.9200000000D + 01	0.1634588788D + 01
10	0.1555759198D + 00	0.1056720438D − 01	0.9100000000D + 01	0.1637893727D + 01
11	0.1628677322D + 00	0.1110899654D − 01	0.9000000000D + 01	0.1641371715D + 01
12	0.1704658958D + 00	0.1167856697D − 01	0.8900000000D + 01	0.1645032004D + 01
13	0.1783797086D + 00	0.1227733990D − 01	0.8800000000D + 01	0.1648884359D + 01
14	0.1866183755D + 00	0.1290681258D − 01	0.8700000000D + 01	0.1652939090D + 01
15	0.1951909581D + 00	0.1356855901D − 01	0.8600000000D + 01	0.1657207087D + 01
16	0.2041063201D + 00	0.1426423391D − 01	0.8500000000D + 01	0.1661699849D + 01
17	0.2133730672D + 00	0.1499557682D − 01	0.8400000000D + 01	0.1666429525D + 01
18	0.2229994809D + 00	0.1576441648D − 01	0.8300000000D + 01	0.1671408954D + 01
19	0.2329934473D + 00	0.1657267540D − 01	0.8200000000D + 01	0.1676651706D + 01
20	0.2433623786D + 00	0.1742237464D − 01	0.8100000000D + 01	0.1682172127D + 01
21	0.2541131295D + 00	0.1831563889D − 01	0.8000000000D + 01	0.1687985387D + 01
22	0.2652519065D + 00	0.1925470178D − 01	0.7900000000D + 01	0.1694107533D + 01
23	0.2767841714D + 00	0.2024191145D − 01	0.7800000000D + 01	0.1700555546D + 01
24	0.2887145378D + 00	0.2127973644D − 01	0.7700000000D + 01	0.1707347394D + 01
25	0.3010466620D + 00	0.2237077186D − 01	0.7600000000D + 01	0.1714502105D + 01
26	0.3137831278D + 00	0.2351774586D − 01	0.7500000000D + 01	0.1722039826D + 01
27	0.3269253251D + 00	0.2472352647D − 01	0.7400000000D + 01	0.1729981903D + 01
28	0.3404733243D + 00	0.2599112878D − 01	0.7300000000D + 01	0.1738350957D + 01
29	0.3544257455D + 00	0.2732372245D − 01	0.7200000000D + 01	0.1747170969D + 01
30	0.3687796237D + 00	0.2872463965D − 01	0.7100000000D + 01	0.1756467371D + 01
31	0.3835302726D + 00	0.3019738342D − 01	0.7000000000D + 01	0.1766267146D + 01
32	0.3986711451D + 00	0.3174563638D − 01	0.6900000000D + 01	0.1776598932D + 01
33	0.4141936952D + 00	0.3337326996D − 01	0.6800000000D + 01	0.1787493136D + 01
34	0.4300872405D + 00	0.3508435410D − 01	0.6700000000D + 01	0.1798982061D + 01
35	0.4463388289D + 00	0.3688316740D − 01	0.6600000000D + 01	0.1811100032D + 01
36	0.4629331101D + 00	0.3877420783D − 01	0.6500000000D + 01	0.1823883550D + 01
37	0.4798522158D + 00	0.4076220398D − 01	0.6400000000D + 01	0.1837371441D + 01
38	0.4970756510D + 00	0.4285212687D − 01	0.6300000000D + 01	0.1851605027D + 01
39	0.5145801979D + 00	0.4504920239D − 01	0.6200000000D + 01	0.1866628312D + 01
40	0.5323398381D + 00	0.4735892439D − 01	0.6100000000D + 01	0.1882488179D + 01
41	0.5503256944D + 00	0.4978706837D − 01	0.6000000000D + 01	0.1899234612D + 01
42	0.5685059971D + 00	0.5233970595D − 01	0.5900000000D + 01	0.1916920926D + 01
43	0.5868460780D + 00	0.5502322006D − 01	0.5800000000D + 01	0.1935604033D + 01
44	0.6053083965D + 00	0.5784432087D − 01	0.5700000000D + 01	0.1955344717D + 01

Continued

Table 10.2 Table of Parameters for One Degree-of-Freedom θ-Function

i	m	q	B_{11}	$K(q)$
45	0.6238526009D + 00	0.6081006263D − 01	0.5600000000D + 01	0.1976207950D + 01
46	0.6424356294D + 00	0.6392786121D − 01	0.5500000000D + 01	0.1998263226D + 01
47	0.6610118547D + 00	0.6720551274D − 01	0.5400000000D + 01	0.2021584937D + 01
48	0.6795332746D + 00	0.7065121306D − 01	0.5300000000D + 01	0.2046252777D + 01
49	0.6979497523D + 00	0.7427357821D − 01	0.5200000000D + 01	0.2072352202D + 01
50	0.7162093086D + 00	0.7808166600D − 01	0.5100000000D + 01	0.2099974921D + 01
51	0.7342584678D + 00	0.8208499862D − 01	0.5000000000D + 01	0.2129219455D + 01
52	0.7520426573D + 00	0.8629358650D − 01	0.4900000000D + 01	0.2160191748D + 01
53	0.7695066621D + 00	0.9071795329D − 01	0.4800000000D + 01	0.2193005858D + 01
54	0.7865951306D + 00	0.9536916222D − 01	0.4700000000D + 01	0.2227784716D + 01
55	0.8032531310D + 00	0.1002588437D + 00	0.4600000000D + 01	0.2264660987D + 01
56	0.8194267521D + 00	0.1053992246D + 00	0.4500000000D + 01	0.2303778039D + 01
57	0.8350637428D + 00	0.1108031584D + 00	0.4400000000D + 01	0.2345291031D + 01
58	0.8501141831D + 00	0.1164841578D + 00	0.4300000000D + 01	0.2389368151D + 01
59	0.8645311764D + 00	0.1224564283D + 00	0.4200000000D + 01	0.2436192028D + 01
60	0.8782715508D + 00	0.1287349036D + 00	0.4100000000D + 01	0.2485961338D + 01
61	0.8912965579D + 00	0.1353352832D + 00	0.4000000000D + 01	0.2538892658D + 01
62	0.9035725522D + 00	0.1422740716D + 00	0.3900000000D + 01	0.2595222604D + 01
63	0.9150716364D + 00	0.1495686192D + 00	0.3800000000D + 01	0.2655210307D + 01
64	0.9257722538D + 00	0.1572371663D + 00	0.3700000000D + 01	0.2719140300D + 01
65	0.9356597094D + 00	0.1652988882D + 00	0.3600000000D + 01	0.2787325908D + 01
66	0.9447266017D + 00	0.1737739435D + 00	0.3500000000D + 01	0.2860113228D + 01
67	0.9529731444D + 00	0.1826835241D + 00	0.3400000000D + 01	0.2937885857D + 01
68	0.9604073617D + 00	0.1920499086D + 00	0.3300000000D + 01	0.3021070507D + 01
69	0.9670451371D + 00	0.2018965180D + 00	0.3200000000D + 01	0.3110143738D + 01
70	0.9729101031D + 00	0.2122479738D + 00	0.3100000000D + 01	0.3205640048D + 01
71	0.9780333563D + 00	0.2231301601D + 00	0.3000000000D + 01	0.3308161676D + 01
72	0.9824529898D + 00	0.2345702881D + 00	0.2900000000D + 01	0.3418390526D + 01
73	0.9862134378D + 00	0.2465969639D + 00	0.2800000000D + 01	0.3537102761D + 01
74	0.9893646313D + 00	0.2592402606D + 00	0.2700000000D + 01	0.3665186770D + 01
75	0.9919609741D + 00	0.2725317930D + 00	0.2600000000D + 01	0.3803665424D + 01
76	0.9940601531D + 00	0.2865047969D + 00	0.2500000000D + 01	0.3953723823D + 01
77	0.9957218065D + 00	0.3011942119D + 00	0.2400000000D + 01	0.4116744125D + 01
78	0.9970060891D + 00	0.3166367694D + 00	0.2300000000D + 01	0.4294349585D + 01
79	0.9979721794D + 00	0.3328710837D + 00	0.2200000000D + 01	0.4488460710D + 01
80	0.9986767924D + 00	0.3499377491D + 00	0.2100000000D + 01	0.4701367484D + 01
81	0.9991727714D + 00	0.3678794412D + 00	0.2000000000D + 01	0.4935823228D + 01
82	0.9995078439D + 00	0.3867410235D + 00	0.1900000000D + 01	0.5195167939D + 01
83	0.9997236318D + 00	0.4065696597D + 00	0.1800000000D + 01	0.5483492455D + 01
84	0.9998550009D + 00	0.4274149319D + 00	0.1700000000D + 01	0.5805860118D + 01
85	0.9999298203D + 00	0.4493289641D + 00	0.1600000000D + 01	0.6168610981D + 01
86	0.9999691662D + 00	0.4723665527D + 00	0.1500000000D + 01	0.6579786988D + 01
87	0.9999879549D + 00	0.4965853038D + 00	0.1400000000D + 01	0.7049738658D + 01
88	0.9999959282D + 00	0.5220457768D + 00	0.1300000000D + 01	0.7592011114D + 01

Continued

Table 10.2 Table of Parameters for One Degree-of-Freedom θ-Function

i	m	q	B_{11}	$K(q)$
89	0.9999988512D + 00	0.5488116361D + 00	0.1200000000D + 01	0.8224672696D + 01
90	0.9999997425D + 00	0.5769498104D + 00	0.1100000000D + 01	0.8972368215D + 01
91	0.9999999572D + 00	0.6065306597D + 00	0.1000000000D + 01	0.9869604507D + 01
92	0.9999999952D + 00	0.6376281516D + 00	0.9000000000D + 00	0.1096622713D + 02
93	0.9999999997D + 00	0.6703200460D + 00	0.8000000000D + 00	0.1233700550D + 02
94	0.1000000000D + 01	0.7046880897D + 00	0.7000000000D + 00	0.1409943486D + 02
95	0.1000000000D + 01	0.7408182207D + 00	0.6000000000D + 00	0.1644934067D + 02
96	0.1000000000D + 01	0.7788007831D + 00	0.5000000000D + 00	0.1973920880D + 02
97	0.1000000000D + 01	0.8187307531D + 00	0.4000000000D + 00	0.2467401100D + 02
98	0.1000000000D + 01	0.8607079764D + 00	0.3000000000D + 00	0.3289868134D + 02
99	0.1000000000D + 01	0.9048374180D + 00	0.2000000000D + 00	0.4934802201D + 02
100	0.1000000000D + 01	0.9512294245D + 00	0.1000000000D + 00	0.9869604401D + 02

11 The Periodic Kadomtsev-Petviashvili Equation

11.1 Introduction

The KP equation (Kadomtsev and Petviashvili, 1970) as studied by soliton physicists and mathematicians has the following scaled form:

$$(u_t + 6uu_x + u_{xxx})_x + \sigma^2 u_{yy} = 0 \tag{11.1}$$

where for $\sigma^2 = +1$ ($\sigma = 1$) the equation is called KP II and for $\sigma^2 = -1$ ($\sigma = i$) it is referred to as KP I. The book by Ablowitz and Clarkson (1991) discusses the integration of the KP equation on the infinite plane using the DBAR technique. The seminal work of Krichever integrates the KP equation for periodic/quasiperiodic boundary conditions (Krichever, 1976, 1977a,b, 1988, 1989, 1992). The thesis of Deconinck (1998) is a wonderful place to get started on the periodic/quasiperiodic KP equation. See also Deconinck and Segur (1998), Deconinck and van Hoeij (2001), Deconinck et al. (2004). The book by Belokolos et al. (1994) gives an overview of the algebro-geometric approach to the solution of nonlinear, integrable wave equations. The book by Baker (1897) is an excellent reference on many of the mathematical methods adopted for the solution of the periodic/quasiperiodic problem and has a broad and detailed explanation for the algebro-geometric background useful for the study of the KdV and KP equations.

The physical form of the equation for water waves is given by:

$$\frac{\partial}{\partial x}\left[\eta_t + c_o\eta_x + \alpha\eta\eta_x + \beta\left(1 - \frac{3T}{\rho g h^2}\right)\eta_{xxx}\right] + \frac{c_o}{2}\eta_{yy} = 0 \tag{11.2}$$

where $\eta(x, y, t)$ is the surface elevation, T is the surface tension, g the acceleration of gravity, h the depth, and ρ the density. The constant coefficients are given by: $c_o = \sqrt{gh}$, $\beta = c_o h^2/6$, and $\alpha = 3c_o/2h$.

Equation (11.2) reduces to Equation (11.1) under the transformation:

$$u = \frac{\alpha}{6\beta}\eta, \quad T = \beta^* t, \quad X = x - c_o t, \quad Y = ay \tag{11.3}$$

Doi: 10.1016/S0074-6142(10)97011-3

Where

$$\beta^* = \beta\left(1 - \frac{3T}{\rho g h^2}\right), \quad a^2 = \frac{2\beta^*}{c_o}\operatorname{sgn}\beta^*, \quad \sigma^2 = \operatorname{sgn}\beta^* \tag{11.4}$$

For shallow water waves (depths greater than $\sim$0.5 cm, see below), the wave motion is described by the KP II equation where the relative effects of surface tension are assumed small:

$$\frac{\partial}{\partial x}[\eta_t + c_o\eta_x + \alpha\eta\eta_x + \beta\eta_{xxx}] + \frac{c_o}{2}\eta_{yy} = 0, \text{ KP II} \tag{11.5}$$

Here of course $\sigma^2 = +1$, $\sigma = 1$. This is easily seen to reduce to the KdV equation when the variation in the y coordinate is small.

When the depth is shallower still, less than $\sim$0.5 cm ($h < \sqrt{3T/\rho g}$), the surface tension dominates and we have

$$\frac{\partial}{\partial x}\left[\eta_t + c_o\eta_x + \alpha\eta\eta_x - \left(\frac{c_o T}{18\rho g} - \beta\right)\eta_{xxx}\right] + \frac{c_o}{2}\eta_{yy} = 0, \text{ KPI} \tag{11.6}$$

A simple change of variables reduces this equation to KP I (for which $\sigma^2 = -1$, $\sigma = i$).

What is the depth at which one goes from KP I to KP II? This is given by, at 20 °C:

$$h = \sqrt{\frac{3T}{\rho g}} = \sqrt{\frac{3(72.75\,\text{dynes/cm})}{1\text{gm/cm}^3 \times 981\text{cm/s}^2}} = 0.472\,\text{cm} \tag{11.7}$$

This means that the KP I equation is valid for depths less than about a half centimeter. *This perspective supports our use of KP II for describing at leading-order directionally spread ocean surface waves.* Carrying shallow water waves to still higher order one finds the 2 + 1 Gardner equation (see Chapters 33 and 34).

11.2 Overview of Periodic Inverse Scattering

The *Kadomtsev and Petviashvili (KP) equation* in dimensional units will be used in the following form:

$$\eta_t + c_o\eta_x + \alpha\eta\eta_x + \beta\eta_{xxx} + \gamma\partial_x^{-1}\eta_{yy} = 0 \tag{11.8}$$

where $c_0 = \sqrt{gh}$, $\alpha = 3c_0/2h$, $\beta = c_0h^2/6$, and $\gamma = c_0/2$. Where $\eta(x, y, t)$ is the wave amplitude as a function of the two spatial variables, x, y and time, t. The subscripts correspond to derivatives with respect to space and time. The symbol ∂_x^{-1} is the antiderivative. The KP equation (11.8) (this is the KP II equation) is a natural two-space dimensional extension of the KdV equation.

The solution of the KP equation for periodic/quasiperiodic boundary conditions is due to the elegant work of Krichever (1976, 1977a,b, 1988, 1989, 1992). See Ablowitz and Clarkson (1991) for a complete overview of the physics and mathematics of the KP equation for infinite plane boundary conditions, that is, for which the solution $\eta(x, y, t)$ vanishes sufficiently fast in the limit:

$$\lim_{\substack{x \to \infty \\ y \to \infty}} \eta(x, y, t) \to 0$$

The periodic KP solutions include *directional spreading* in the wave field due to the linear spreading term $\gamma \partial_x^{-1} \eta_{yy}$. The periodic/quasiperiodic solution of KP is given by:

$$\eta(x, t) = \frac{2}{\lambda} \partial_{xx} \ln\theta(x, y, t | \mathbf{B}, \boldsymbol{\phi}) \tag{11.9}$$

Here, the generalized Fourier series has the same form as in one dimension, but the phase has the *two dimensional* expression:

$$\mathbf{X}(x, y, t) = \boldsymbol{\kappa}x + \boldsymbol{\lambda}y - \boldsymbol{\omega}t + \boldsymbol{\phi} \tag{11.10}$$

The spatial term $\boldsymbol{\kappa}x$ (corresponding to the dominant wave direction) is joined by the lateral spatial term $\boldsymbol{\lambda}y$ (perpendicular to the dominant direction), which is how wave spreading is taken into account.

The theta function has the explicit form:

$$\theta(x, t) = \sum_{m_1=-\infty}^{\infty} \sum_{m_2=-\infty}^{\infty} \cdots \sum_{m_N=-\infty}^{\infty}$$

$$\exp\left[-\frac{1}{2}\mathbf{m} \cdot \mathbf{Bm} + i\mathbf{m} \cdot \boldsymbol{\kappa}x + i\mathbf{m} \cdot \boldsymbol{\lambda}y - i\mathbf{m} \cdot \boldsymbol{\omega}t + i\mathbf{m} \cdot \boldsymbol{\phi} \right] \tag{11.11}$$

I now discuss the work of Bobenko (Belokolos et al., 1994) on determination of the Riemann spectrum via Schottky uniformization.

11.3   Computation of the Spectral Parameters in Terms of Schottky Uniformization

The fundamental problem is to select the Riemann spectrum such that one has a solution of the KP equation. Algebraic geometric loop integrals (Krichever, 1988), *Schottky uniformization* (Bobenko and Bordag, 1989) and the approach of Nakamura (1980) and Boyd (1984a,b,c) are the methods most commonly used for this purpose. Naturally one expects that all of these methods satisfy the *Novikov conjecture*, which states that to ensure physical solutions of the KP equation one must have a Riemann spectrum that corresponds to an underlying Riemann surface. The conjecture is even stronger for it states that out of all the possible finite genus theta functions only those associated with a compact Riemann surface solve the KP equation. The theoretical state of affairs is much less certain and I suspect that there are still some very interesting and important mathematical issues to resolve in this regard. For example, Shiota (1986) has formally proved the Novikov conjecture, but can we say for sure that Schottky uniformization and the Nakamura-Boyd approaches also obey the conjecture? That they provide procedures for determining a Riemann spectrum that lies on a compact Riemann surface? The Schottky approach is on somewhat firm ground because it is based upon holomorphic differentials normalized in a particular oscillation basis that is a uniformization of a Riemann surface; the holomorphic differentials are particular Poincaré series over a Schottky group which can represent any compact Riemann surface of genus N (see Appendix III of Chapter 15). Furthermore, the approach can be mapped to the Floquet spectrum for the KdV equation. The basis of the work herein lies with *numerical verifications* which suggest that all three methods obey the Novikov conjecture, that is, loop integrals, Schottky uniformization, and Nakamura-Boyd all give the same answers and all can, therefore, be used to ensure that numerical simulations or data analysis lie on an associated Riemann surface.

Here, I use the *Schottky uniformization of Riemann surfaces* to get access to all smooth, nonsingular, periodic solutions of the KP equation (Schottky, 1888; Baker, 1897; Krichever, 1988; Shiota, 1986; Bobenko and Bordag, 1989; Burnside, 1892; Belokolos et al., 1994). Chapter 32 gives a detailed overview of the numerical methods.

For Equations (11.9) and (11.11) to be a solution to the KP equation, one must first compute the appropriate wavenumbers, frequencies, and Riemann matrix. Formally speaking, one can select the diagonal elements of the Riemann matrix B_{nn} to give the cnoidal wave solutions that one desires. The wavenumbers are fixed for numerical modeling purposes to be commensurable, that is, $\kappa_n = 2\pi n/L_x$, $\lambda_n = 2\pi n/L_y$. And the phases ϕ_n can be arbitrarily chosen. Thus, the off-diagonal elements of the Riemann matrix, B_{mn}, and the frequencies, ω_n must be computed in such a way that Equations (11.9) and (11.11) solve KP on a Riemann surface. Schottky uniformization is the method of choice to compute these latter parameters, although the Nakamura-Boyd approach is also considered below.

The Schottky procedure provides *Poincaré series* for B_{mn}, ω_n as a function of the so-called *uniformization parameters* A_n, ρ_n, $n = 1, 2 \cdots, N$ where N (the genus) is the number of degrees of freedom (cnoidal waves) in the spectrum. To give a physical interpretation of these parameters (see the discussion in the following sections), it is enough to remember that at leading order the ρ_n (real numbers) are related to the diagonal elements of the Riemann matrix (or the amplitudes of the cnoidal waves) and the A_n (complex numbers) are related to the wavenumbers κ_n, λ_n, $n = 1, 2, \cdots, N$ (real numbers); the Poincaré series specify the wavenumbers, frequencies, and Riemann matrix in terms of the Schottky parameters.

11.3.1 Linear Fractional Transformation

Details and derivations for the Schottky uniformization procedure are due to Bobenko (see Belokolos et al., 1994, for cited references). In what follows we require the linear fractional (Möbius) transformation σ, which has the form:

$$\sigma z = \frac{\alpha z + \beta}{\gamma z + \delta} \tag{11.12}$$

whose constants are given in terms of the uniformization parameters

$$\alpha = \frac{1}{2}\left(\frac{1+\rho}{\sqrt{\rho}}\right), \quad \beta = -\frac{A}{2}\left(\frac{1-\rho}{\sqrt{\rho}}\right) \tag{11.13}$$

$$\gamma = -\frac{1}{2A}\left(\frac{1-\rho}{\sqrt{\rho}}\right), \quad \delta = \frac{1}{2}\left(\frac{1+\rho}{\sqrt{\rho}}\right) \tag{11.14}$$

The corresponding matrix operator for σ is given by

$$\tilde{\sigma} = \begin{bmatrix} \frac{1}{2}\left(\frac{1+\rho}{\sqrt{\rho}}\right) & -\frac{A}{2}\left(\frac{1-\rho}{\sqrt{\rho}}\right) \\ -\frac{1}{2A}\left(\frac{1-\rho}{\sqrt{\rho}}\right) & \frac{1}{2}\left(\frac{1+\rho}{\sqrt{\rho}}\right) \end{bmatrix} \tag{11.15}$$

with inverse

$$\sigma^{-1}z = \frac{\delta z - \beta}{-\gamma z + \alpha} \tag{11.16}$$

$$
\tilde{\sigma}^{-1} = \begin{bmatrix} \dfrac{1}{2}\left(\dfrac{1+\rho}{\sqrt{\rho}}\right) & \dfrac{A}{2}\left(\dfrac{1-\rho}{\sqrt{\rho}}\right) \\[3ex] \dfrac{1}{2A}\left(\dfrac{1-\rho}{\sqrt{\rho}}\right) & \dfrac{1}{2}\left(\dfrac{1+\rho}{\sqrt{\rho}}\right) \end{bmatrix}
$$

(11.17)

The group elements $\sigma_o = I$, σ_n, and σ_n^{-1} constitute the (Schottky) group over which the Poincaré series are summed.

11.3.2   Theta Function Spectrum as Poincaré Series of Schottky Parameters

For a general introduction to Poincaré series see Baker (1897), and for specific applications to KP see Bobenko and Bordag (1989) and Belokolos et al. (1994). You will be quite amazed and pleased to find many of the important formulas below in Baker's classic book from over a century ago; Bobenko's work has provided us with a great service in linking much of this formulation, via Schottky (1888), to modern theories of nonlinear integrable wave equations. Of course, it was Krichever (1988) who integrated KP II and his introduction to the latest version of Baker's book is absolutely marvelous in the perspective that it gives.

Here the Schottky problem is revisited in the small-amplitude, oscillatory limit. The soliton limit is discussed in detail by Bobenko (Belokolos et al., 1994). Here, I emphasize the oscillatory limit because it provides numerical advantages in the use of the ordinary Fourier transform for the theta function as discussed in Chapters 9 and 32. The diagonal elements of the Riemann matrix have the following Poincaré series:

$$
B_{nn} = \ln \mu_n + \sum_{\sigma \in G_n \backslash G / G_n, \, \sigma \neq I} \ln\left[\frac{(A_n^* - \sigma A_n^*)(A_n - \sigma A_n)}{(A_n^* - \sigma A_n)(A_n - \sigma A_n^*)}\right]
$$

(11.18)

The off-diagonal elements have the form:

$$
B_{mn} = \sum_{\sigma \in G_m \backslash G / G_n} \ln\left[\frac{(A_m^* - \sigma A_n^*)(A_m - \sigma A_n)}{(A_m^* - \sigma A_n)(A_m - \sigma A_n^*)}\right], \quad m \neq n
$$

(11.19)

In the last equation, I have brought out the term for $\sigma_o = I$. Thus the identity has been removed from the group theoretic summation; note that this latter summation in Equation (11.19) now excludes the identity term. The wavenumbers have the form:

$$
\kappa_n = \sum_{\sigma \in G / G_n} (\sigma A_n - \sigma A_n^*)
$$

(11.20)

$$\lambda_n = h \sum_{\sigma \in G/G_n} \left[(\sigma A_n)^2 - (\sigma A_n^*)^2 \right] \tag{11.21}$$

The frequency is

$$\omega_n = c_0 k_0 - 4\beta \sum_{\sigma \in G/G_n} \left[(\sigma A_n)^3 - (\sigma A_n^*)^3 \right] \tag{11.22}$$

Equations (11.18)–(11.22) have been written in dimensional form. Therefore, given a set of Schottky parameters (ρ_n, A_n) one can compute all of the parameters of the theta function, namely, B_{mn}, κ_n, λ_n, ω_n for an arbitrary set of phases ϕ_n. All of the Poincaré series given above are summed over the (Schottky) group whose elements are $\sigma_0 = I$, σ_n, and σ_n^{-1}.

In Chapter 32, I give the leading order behavior for the off-diagonal elements of the period matrix and the frequency for the KP II equation:

$$B_{mn} \cong \ln \left[\frac{(\kappa_m - \kappa_n)^2 + \left(\dfrac{\kappa_n \lambda_m - \kappa_m \lambda_n}{\kappa_m \kappa_n h} \right)^2}{(\kappa_m + \kappa_n)^2 + \left(\dfrac{\kappa_n \lambda_m - \kappa_m \lambda_n}{\kappa_m \kappa_n h} \right)^2} \right], \quad m \neq n \tag{11.23}$$

$$\omega_n \simeq c_0 k - \beta \kappa_n^3 + \frac{c_0}{2} \frac{\lambda_n^2}{\kappa_n} \tag{11.24}$$

To clearly understand the set theory notation and the algebra of Poincaré series see, for example, Baker (1897). Chapter 32 gives a detailed exposition of the use of the KP equation as a numerical model of water wave dynamics. In particular, the numerical implementation of the Schottky method is explained in some detail, together with physical motivation for the selection of the Schottky parameters in an oceanographic context.

11.4 The Nakamura-Boyd Approach for Determining the Riemann Spectrum

An alternative method for determining the Riemann spectrum is to use the approach developed by Nakamura (1980) and Boyd (1984a,b,c). Their procedure is briskly outlined for the KP equation. One first substitutes the transformation equation (11.9) into the KP equation (11.8) in order to obtain the Hirota-KP equation:

$$\theta \theta_{xt} - \theta_x \theta_t + c_0 \left(\theta \theta_{xx} - \theta_x^2 \right) + \beta \left(3\theta_{xx}^2 - 4\theta_x \theta_{xxx} + \theta \theta_{xxxx} \right)$$
$$+ \gamma \left(\theta \theta_{yy} - \theta_y^2 \right) + \frac{\alpha c}{12\beta} \theta^2 = 0 \tag{11.25}$$

This expression can be put into Hirota operator notation to give:

$$\left(D_x D_t + c_o D_x^2 + \beta D_{xxxx} + \gamma D_y^2 + \frac{\alpha c}{12\beta} \right) \theta \cdot \theta = 0 \tag{11.26}$$

We use the usual definition of the Hirota operator:

$$D_x^n a \cdot b \equiv \left(\frac{\partial}{\partial x} - \frac{\partial}{\partial y} \right)^2 a(x) b(y) \big|_{y=x} = \frac{\partial^n}{\partial y^n} a(x+y) b(x-y) \big|_{y=0} \tag{11.27}$$

so that

$$\begin{aligned}
D_x D_t \theta \cdot \theta &= 2(\theta \theta_{xt} - \theta_x \theta_t) \\
D_x^2 \theta \cdot \theta &= 2(\theta \theta_{xx} - \theta_x^2) \\
D_x^4 \theta \cdot \theta &= 2(3\theta_{xx}^2 - 4\theta_x \theta_{xxx} + \theta \theta_{xxxx}) \\
D_y^2 \theta \cdot \theta &= 2(\theta \theta_{yy} - \theta_y^2)
\end{aligned} \tag{11.28}$$

By introducing theta functions with characteristics (Nakamura, 1981; Boyd, 1983) (see Chapter 16) one arrives at a set of nonlinear equations written in terms of the period matrix, wavenumbers, and frequencies:

$$\sum_{m_1=-\infty}^{\infty} \sum_{m_2=-\infty}^{\infty} \cdots \sum_{m_N=-\infty}^{\infty} \left\{ \left[2\sum_{j=1}^{N} (m_j - \mu_j/2)\kappa_j \right] \left[2\sum_{j=1}^{N} (m_j - \mu_j/2)\omega_j \right] \right.$$

$$-c_o \left[2\sum_{j=1}^{N} (m_j - \mu_j/2)\kappa_j \right]^2 + \beta \left[2\sum_{j=1}^{N} (m_j - \mu_j/2)\kappa_j \right]^4 + \gamma \left[2\sum_{j=1}^{N} (m_j - \mu_j/2)\lambda_j \right]^2$$

$$\left. -\frac{\alpha c}{12\beta} \right\} e^{-\sum_{j=1}^{N}\sum_{k=1}^{N} (m_j - \mu_j/2)(m_k - \mu_k/2)B_{jk}} = 0$$

$$\tag{11.29}$$

In the above equation, the characteristics μ_j take on the values 0, 1; this leads to a set of 2^N equations for determining the parameters of the Riemann spectrum. In the present procedure, one assumes particular values for the diagonal elements of the period matrix and the Wavenumbers, here taken to be commensurable so that $\kappa_n = 2\pi n/L_x$, $\lambda_n = 2\pi n/L_y$. One then seeks to determine the off-diagonal elements of the period matrix and the frequencies so that Equation (11.9) gives a solution of the KP equation. It is not hard to show from Equation (11.27) that to leading order we obtain Equations (11.23) and (11.24) for the off-diagonal elements of the period matrix and the frequencies, thus coinciding with the leading order terms in the Schottky procedure (see details in Chapter 32). These are of course the starting values in an iterative procedure given in Chapter 16 to determine the off-diagonal elements of the period matrix B_{mn} and frequencies ω_n given the diagonal elements of the period matrix B_{nn} and the commensurable Wavenumbers κ_n, λ_n.

Extensive details about the KP equation and its use in numerical modeling and its implication for data analysis are discussed in Chapter 32.

Part Five

Nonlinear Deep-Water Spectral Theory

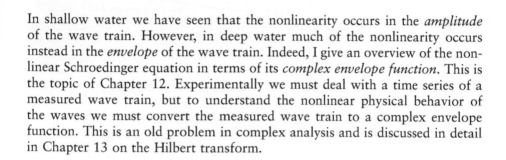

In shallow water we have seen that the nonlinearity occurs in the *amplitude* of the wave train. However, in deep water much of the nonlinearity occurs instead in the *envelope* of the wave train. Indeed, I give an overview of the nonlinear Schroedinger equation in terms of its *complex envelope function*. This is the topic of Chapter 12. Experimentally we must deal with a time series of a measured wave train, but to understand the nonlinear physical behavior of the waves we must convert the measured wave train to a complex envelope function. This is an old problem in complex analysis and is discussed in detail in Chapter 13 on the Hilbert transform.

Doi: 10.1016/S0074-6142(10)97044-7

12 The Periodic Nonlinear Schrödinger Equation

12.1 Introduction

This chapter summarizes some of the main results for the IST of the NLS equation with periodic boundary conditions (Kotljarov and Its, 1976; Tracy, 1984; Tracy and Chen, 1988; see also Sulem and Sulem, 1999):

$$iu_t + u_{xx} + 2|u|^2 u = 0, \quad u(x + L, t) = u(x, t) \tag{12.1a}$$

Additional indispensable references are those of Ablowitz and coworkers (see Ablowitz et al., 2004, and references cited therein). A number of the known "breather" or "rogue wave" solutions are given by Akmediev and coworkers (see Akhmediev et al., 2009 and cited references; a complete list of references by these authors is given below). For nonlinear effects beyond the order of the nonlinear Schrödinger equation, including chaotic dynamics, see the work of Schober and Calini and coworkers (Calini et al., 1996; Calini and Schober, 2002; Islas and Schober, 2005; Schober, 2006) and Trulsen and Dysthe (1996, 1997a,b). Additional useful references are Tracy et al., 1984; Tracy et al., 1987. Studies on the *statistics of rogue waves* are left to a sequel.

12.2 The Nonlinear Schrödinger Equation

The nonlinear Schrödinger equation, scaled to represent physical units, is given by Yuen and Lake (1982) and cited references (see also the original work of Zakharov (1967)):

$$i(\psi_t + C_g\psi_x) + \mu\psi_{xx} + v|\psi|^2\psi = 0, \quad \psi(x + L, t) = \psi(x, t) \tag{12.1b}$$

For reasons made clear below I call this the *space NLS* (sNLS) *equation*. The constant coefficients, for infinite water depth, are given by:

$$C_g = \frac{1}{2}\frac{\omega_o}{k_o} = \frac{1}{2}\frac{L_o}{T_o}, \quad \mu = -\frac{\omega_o}{8k_o^2}, \quad v = -\frac{1}{2}\omega_o k_o^2 \tag{12.2}$$

The constant

Doi: 10.1016/S0074-6142(10)97012-5

$$\rho = \sqrt{\frac{v}{2\mu}} = \sqrt{2}k_o^2 \qquad (12.3)$$

is important in the inverse scattering transform formulation, where it serves as a kind of nonlinearity parameter. The subscript "o" refers to the *carrier wave* that is modulated by the function $\psi(x, t)$: ω_o is the carrier wave frequency, k_o is the wavenumber, while L_o, T_o are the carrier wave length and period, respectively ($k_o = 2\pi/L_o$, $\omega_o = 2\pi/T_o = 2\pi f_o$). The sNLS equation describes the space/time dynamics of the *complex envelope function*, $\psi(x, t)$, of a deep-water wave train which propagates in the $+x$ direction as a function of time, t. The equation solves the *Cauchy problem*, that is, given the complex envelope at some initial time $t = 0$, $\psi(x, 0)$, Equation (12.1b) evolve the dynamics for all space and time, $\psi(x, t)$. The formulation of the sNLS equation is appropriate for the *space series analysis* of data whose assumed behavior is approximated by $\psi(x, 0)$. Space series measurements are most appropriately obtained by *remote sensing technology*.

The *sea surface elevation*, $\eta(x, t)$, is computed from the complex envelope function, $\psi(x, t)$, in the following way:

$$\eta(x, t) = \psi(x, t)e^{i(k_o x - \omega_o t)} + \text{c.c.} \qquad (12.4)$$

where "c.c." denotes complex conjugate. Thus the *carrier wave* $e^{i(k_o x - \omega_o t)}$ is modulated by the complex envelope function, $\psi(x, t)$, as governed by the NLS equation (12.1b) for *periodic boundary conditions*. Particular initial conditions, $\eta(x, 0)$ and $\psi(x, 0)$, are assumed for physical and numerical modeling purposes.

12.2.1 The "Time" NLS Equation and Its Relation to Physical Experiments

At leading order in nonlinearity in Equation (12.1b) we have

$$\psi_t + C_g\psi_x \cong 0$$

so that $\psi_x \cong -\psi_t/C_g$ and $\psi_{xx} \cong \psi_{tt}/C_g^2$. When these are used in the higher order terms in Equation (12.1b) we obtain the *time NLS equation* (tNLS):

$$i(\psi_x + C_g'\psi_t) + \mu'\psi_{tt} + v'|\psi|^2\psi = 0, \quad \psi(x,t) = \psi(x, t + T) \qquad (12.5)$$

where

$$C_g' = \frac{1}{C_g}, \quad \mu' = \frac{\mu}{C_g^3}, \quad v' = \frac{v}{C_g} \qquad (12.6)$$

The inverse scattering transform parameter has the form:

$$\rho' = \sqrt{\frac{v'}{2\mu'}} = C_g\sqrt{\frac{v}{2\mu}} = C_g\rho \qquad (12.7)$$

Solutions to tNLS (Equation (12.5)) are related to solutions of sNLS (Equation (12.1b)) by the simple transformation

$$x \to t, \quad t \to x, \quad \rho \to \rho', \quad C_g \to C_g', \quad v \to v', \quad \mu \to \mu' \tag{12.8}$$

Thus the space (Equation (12.1)) and time (Equation (12.5)) NLS equations are related by a simple change of variables and parameters (Equation (12.8)). Physically the tNLS Equation (12.5) solves a *boundary value problem*: Given the boundary value, $\psi(0, t)$, the space/time dynamics of Equation (12.5) determine the solutions over all space and time, $\psi(x, t)$. Equation (12.5) is thus suitable for the *time series analysis* of measured wave trains whose assumed behavior is approximated by $\psi(0, t)$. Time series observations are typically obtained by *in situ instrumentation* located at the spatial position $x = 0$.

12.2.2 A Scaled Form of the NLS Equation

The simple transformation

$$u = \rho\psi, \quad T = \mu t, \quad X = x - C_g t \tag{12.9}$$

allows Equation (12.1b) to be put into a simpler, scaled form suitable for work with periodic IST (Equation (12.1a)). One finds the scaled sNLS:

$$iu_T + u_{XX} + 2|u|^2 u = 0 \tag{12.10}$$

Note that the field $u(x, t)$ is that "seen" by the periodic inverse scattering transform as discussed below. It consists of the physical field $\psi(x, t)$ multiplied by ρ, thus providing the motivation for emphasizing the important parameter ρ (the nonlinear parameter, $\rho = \sqrt{2}k_o^2$ in deep water) in the context of nonlinear Fourier analysis. This observation is essential when applying IST to nonlinearly Fourier-analyzed data and for hyperfast numerical modeling. I first discuss some of the important physical results and then go on to address the IST for NLS.

12.2.3 Small-Amplitude Modulations of the NLS Equation

Yuen and Lake (1982) studied the NLS equation intensely, together with a number of other wave equations, to improve the understanding of instabilities in deep-water wave trains. Their work focused, in part, on numerical solutions of the NLS equation with periodic boundary conditions. They typically considered a small-amplitude modulated sine (a carrier) wave of the form:

$$\eta(x,t) = a_o[1 + \varepsilon \cos(Kx - \Omega t)] \cos[k_o x - \omega_o(1 + k_o^2 a_o^2/2)t] \tag{12.11}$$

The carrier wave has amplitude a_o, wavenumber k_o, and frequency ω_o. The small modulation amplitude is ε. Note the Stokes wave correction to the carrier frequency, $k_o^2 a_o^2/2$, often referred to as a *frequency shift*. Indeed Equation (12.11) for $\varepsilon = 0$ is just the leading order Stokes wave. Here K is the modulation wavenumber and Ω is the modulation frequency. The small-amplitude modulation for *small times* to the sNLS equation is given by:

$$\psi(x,t) = \left\{ a_o + \varepsilon_+ \exp\left(i\Omega t + iK\left(x - \frac{\omega_o}{2k_o}t \right) \right) \right.$$

$$\left. + \varepsilon_- \exp\left(-i\Omega t - iK\left(x - \frac{\omega_o}{2k_o}t \right) \right) \right\} \exp\left(\frac{1}{2} i\omega_o k_o^2 a_o^2 t \right)$$

Inserting this last equation into Equation (12.1b) and linearizing gives the *modulation dispersion relation*:

$$\Omega^2 = \frac{\omega_o^2}{8k_o^2} \left(\frac{K^2}{8k_o^2} - k_o^2 a_o^2 \right) K^2 \tag{12.12}$$

This expression shows that an *initial wave train with a small-amplitude modulation* is *unstable* to the modulation provided that the modulation wavenumber K lies in the range

$$0 < K < 2\sqrt{2}k_o^2 a_o$$

This is because Equation (12.12), in this range, gives a frequency which is imaginary so that $e^{i\Omega t} \sim e^{\Omega_I t}$ grows exponentially for small time (Ω_I is the imaginary part of the frequency). The wave train is *stable* if it lies outside this range because the frequency is real so that $e^{i\Omega t}$ is purely oscillatory. Thus the modulated wave train $\eta(x, t)$ can undergo exponential growth for *early time* provided we choose K in the interval $0 < K < 2\sqrt{2}k_o^2 a_o$. However, the essence of the early Yuen and Lake work was the experimental and numerical study of the *long-time evolution* of $\eta(x, t)$ and $\psi(x, t)$ for many unstable cases. They saw that the unstable mode solutions of the NLS equation did not grow exponentially for all time, but instead grew to a maximum value and then decreased in amplitude for later times and most of them repeated this oscillation periodically over time (Fermi-Pasta-Ulam recurrence). The *oscillatory unstable modes* of this type are often referred to as *breathers*. Indeed, the IST procedure discussed here provides the analytic means for determining the *long-time* evolution of solutions of the NLS equation (see Chapter 24 on the *nonlinear instability analysis for the NLS equation*) and the results are discussed later.

12.3   Representation of the IST Spectrum in the Lambda Plane

This section on the IST spectrum in the so-called *lambda plane* anticipates mathematics in later sections that might be skipped on a first reading. However, much of the physics for the interpretation and understanding of the nonlinear wave dynamics of the NLS equation is given in this section. The lambda plane is the complex plane where the IST spectrum for the NLS equation lives (the spectrum arises from the *Zakharov-Shabat eigenvalue problem* for periodic boundary conditions, see Chapter 18). The lambda plane has two axes corresponding to the real and imaginary parts of an eigenvalue: $\lambda = \lambda_R + i\lambda_I$ (see Figure 12.1 and note that we are dealing with the scaled NLS equation (12.1a)). The λ_I axis can be physically characterized as a *spectral amplitude* while the λ_R axis corresponds to the *wavenumbers* for the sNLS equation and to *frequencies* for the tNLS equation. In the linear, small-amplitude limit, the lambda plane analysis becomes the *linear Fourier spectrum* with which we are all familiar. Indeed all graphs in the lambda plane given here are prepared so that this limit is transparent, exact, and in the units of the ordinary linear Fourier transform, that is, meters or centimeters for the spectral amplitudes, 1/m or 1/cm for wavenumbers and Hz for frequencies.

For the simple problems discussed in this chapter, the lambda plane contains the simple *carrier eigenvalues*, $\lambda = \pm iA$, where A is the carrier amplitude in scaled coordinates; the value of the wavenumber on the horizontal axis below $\lambda = iA$ lies at the peak of the (ordinary linear Fourier) spectrum. Also shown in Figure 12.1 are the *points of simple spectrum* (simple eigenvalues) that are given the symbols ×. To have a *degree of freedom* two of these eigenvalues must be connected by a *spine*, a curve that connects the two simple eigenvalues. When the spine crosses the real axis the degree of freedom corresponds to a simple *Stokes wave* and the associated Riemann matrix has a single diagonal element corresponding to this degree of freedom (Chapter 24). When the spine does not cross the real axis the degree of freedom corresponds to an unstable (rogue wave) mode and the Riemann matrix has a 2 × 2 submatrix centered on the diagonal; the *unstable mode degree of freedom* thus corresponds to a

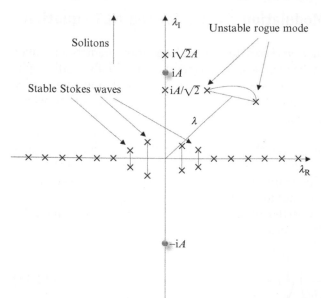

Figure 12.1 The lambda plane as a graph of the IST spectrum of the NLS equation. The lambda plane is symmetric about the real axis and so not all of the spectrum is shown here in the lower half plane. For this reason, a graph of the lambda plane need show only the upper half plane.

2×2 *Riemann matrix*. In Chapter 24, I discuss procedures for computing the Riemann spectrum (period matrix, wavenumbers, frequencies, and phases) from the spectrum in the lambda plane.

It is worthwhile anticipating some of the results of this chapter, in particular those of Section 12.4. In Figure 12.1, one can see the spectra of three types of rogue wave shown in Figures 12.6–12.8. On the imaginary axis of Figure 12.1 there is an $\times$ at $iA/\sqrt{2}$, corresponding to the NLS solution of Figure 12.6; the single $\times$ is in reality two superposed $\times$s corresponding to a double point which is the homoclinic solution to the NLS equation given by Equation (12.25). At the point iA there is another double point corresponding to Figure 12.7 that is given by Equation (12.26). Another mode occurs at $i\sqrt{2}A$, corresponding to Figure 12.8; the single $\times$ is in reality two superposed $\times$s corresponding to a double point which is the solution to the NLS equation given by Equation (12.27). Also in Figure 12.1 are several *Stokes waves* that correspond to two points of simple spectrum, one above and the other below the real axis, connected by a line or "spine" (across the real axis) whose mathematical definition is given below. It should be clear at this point that the Stokes waves are very similar to the linear Fourier modes with which we are all familiar. An infinite class of solutions of the NLS equation is described in Figure 12.1. Indeed many of the kinds of nonlinear behavior that occur in nonlinear waves are given here: (1) some waves are stable *Stokes wave trains* and (2) other waves can be *unstable modes* (nonlinear wave packets that "breathe" in space and time) that, under certain circumstances, can be "rogue waves." This preliminary discussion of the lambda plane is an attempt to give the reader a prelude to the mathematical details given below, which we discuss in the next few sections.

12.4 Overview of Modulation Theory for the NLS Equation

The *small-amplitude modulation theory* for the NLS equation predicts a number of interesting features about the nonlinear propagation of an initially small-amplitude modulation. Figure 12.2 shows a small-amplitude modulation of a carrier wave (for a discussion see Chapter 13 on the Hilbert transform, a method which allows us to compute the complex envelope from an oscillatory wave train). At a later time this small modulation develops into an unstable wave packet as shown in Figure 12.3. In the present case, the maximum amplitude is about 2.6 times the carrier height, $a_o = 1$. Note that many of the results discussed below are derived in Chapter 24 on the *nonlinear instability analysis* of the NLS equation. The equations in this section refer to the dimensional NLS Equation (12.1b).

One of the important properties of an unstable wave packet is the modulational frequency known as the *growth rate*:

$$\Omega = i\omega_o k_o^2 a_o^2 \left(\frac{K}{2\sqrt{2}k_o^2 a_o}\right) \sqrt{1 - \left(\frac{K}{2\sqrt{2}k_o^2 a_o}\right)^2} \tag{12.13}$$

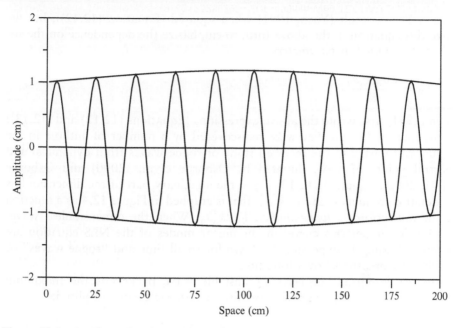

Figure 12.2 Small-amplitude initial modulation of a carrier wave.

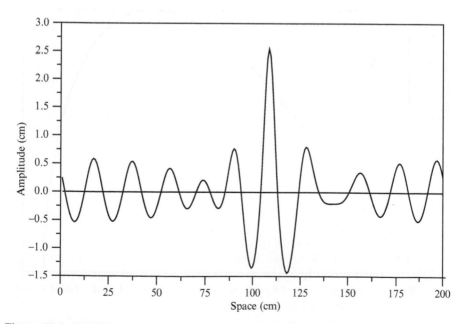

Figure 12.3 Small-amplitude initial modulation of Figure 12.2 has grown into an unstable wave packet.

This is Yuen's result (Yuen, 1991) when squared (Equation (12.12)), but we leave this equation in the above form to emphasize the dependence on the so-called *Benjamin-Feir parameter*:

$$I_{BF} = \frac{2\sqrt{2}k_o^2 a_o}{K}$$

Yuen (1991) first wrote the above expressions (Equations (12.12) and (12.13)) in the mid-1970s and they were rediscovered by a number of authors in the 1990s with regard to the renewed study of rogue waves from dynamical and spectral points of view, initiated by Osborne et al. (2000) and Osborne (2001, 2002). Equation (12.13) is just the imaginary part of the dimensionless modulation frequency $(2\Omega/\omega_o k_o^2 a_o^2)$ that is graphed in Figure 12.4 as a function of the *dimensionless wavenumber*, $K/2k_o^2 a_o$. When the dimensionless wavenumber lies under this curve, the nonlinear modes of the NLS equation are unstable, leading to exponential growth for small time and "rogue waves" or "oscillatory breathers" over long times.

Another important property of unstable wave packets is the maximum amplitude of the unstable packet with respect to the carrier amplitude:

$$\frac{A_{max}}{a_o} = 1 + 2\frac{\lambda_I}{a_o} = 1 + 2\sqrt{1 - \left(\frac{K}{2\sqrt{2}k_o^2 a_o}\right)^2} \qquad (12.14)$$

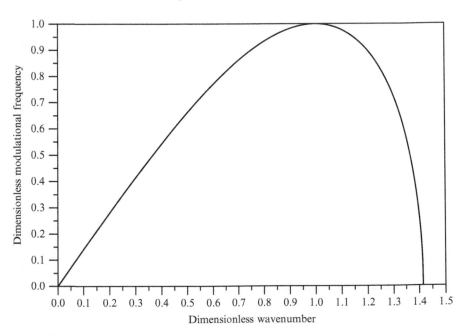

Figure 12.4   Instability diagram for small-amplitude modulations for the NLS equation.

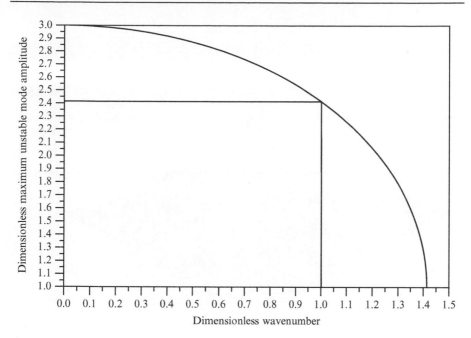

Figure 12.5 Maximum amplitude of an unstable wave packet as a function of dimensionless wavenumber. The maximum amplitude for the maximum growth rate (see Figure 12.4) is shown. The maximum amplitude is ~2.4142 (see space/time evolution examples in Figures 12.3 and 12.6).

This function is graphed in Figure 12.5. We see that the maximum growth rate of Figure 12.4 (dimensionless wavenumber 1) is associated with an unstable wave packet that reaches a height of $1 + \sqrt{2} \sim 2.41$ times the carrier height. Smaller modulation wavenumbers are necessary to get larger packet amplitudes (up to a maximum of three times the carrier height for *small-amplitude modulations*), although they will take longer to reach their maximum height because the growth rate is smaller (see discussion below on Equation (12.17)). However, the maximum height of 3 occurs only for very small wavenumbers; this case corresponds to modulation wavelengths that span many carrier oscillations. Chapter 24 shows the derivation of Equation (12.14) and extends it beyond the factor of 3; indeed waves higher than three times the carrier height are easily found and characterized as a *large-amplitude modulation*. An example is given later (Figure 12.8; Equation (12.27)).

We can also compute the imaginary part of the IST eigenvalue (see below and Chapter 24) by the simple relation (in the examples given below the eigenvalues lie on the imaginary axis in the lambda plane):

$$\frac{\lambda_{\mathrm{I}}}{a_{\mathrm{o}}} = \sqrt{1 - \left(\frac{K}{2\sqrt{2}k_{\mathrm{o}}^2 a_{\mathrm{o}}}\right)^2} \qquad (12.15)$$

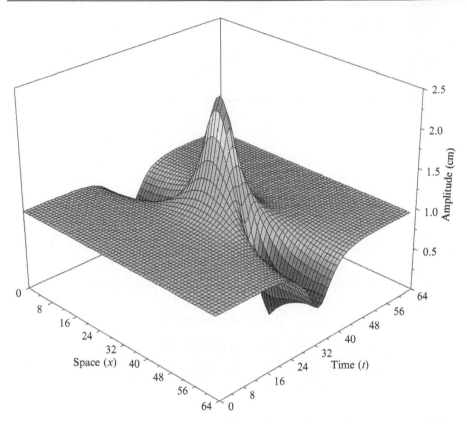

Figure 12.6 Modulus $|u(x, t)|$ of unstable wave packet that lies below the carrier in the complex lambda plane with spectrum: $\{A, 0, 0, 0, A/\sqrt{2}\}$. The initial condition at time $t = 0$ is seen to be a small-amplitude modulation. (See color plate).

As before we see that when the eigenvalue is near, but below the carrier in the lambda plane, the ratio $\lambda_I/a_o \sim 1$ and hence the modulation wavenumber must be small and the modulation wavelength must be long. *For long-modulation wavelengths, the rogue waves become higher and greater in number, but rise up slowly to their maximum amplitudes.*

Another parameter that is very useful is the actual *time to appearance of an unstable mode*, beginning with its initial small-amplitude modulation. This is computed by noting that:

$$\varepsilon e^{\gamma t} \sim O(1)$$

where ε is the initial modulation amplitude and γ is the imaginary part of the IST frequency (Equation (12.16)):

$$\gamma = \omega_o k_o^2 a_o^2 \left(\frac{K}{2\sqrt{2}k_o^2 a_o}\right) \sqrt{1 - \left(\frac{K}{2\sqrt{2}k_o^2 a_o}\right)^2} \tag{12.16}$$

This gives the approximate *rise time* from the initial modulation to the maximum of an unstable mode:

$$T_{BF} \approx |\ln \varepsilon|/\gamma \qquad (12.17)$$

This is a very useful formula for determining how long it takes for a small-amplitude modulation to develop into a rogue wave.

Likewise we may ask if *dissipation* is important in wave propagation problems where instabilities are present. Will a rogue mode come up before the waves are dissipated? For example, for linear, dissipated waves we have the approximation:

$$\varepsilon e^{-\delta t} \sim O(1)$$

so that the waves are dissipated in time approximately by:

$$T_{diss} \approx |\ln \varepsilon|/\delta \qquad (12.18)$$

Modulationally unstable packets can be expected to occur in the *dissipated NLS equation*

$$i(\psi_t + C_g \psi_x) + \mu \psi_{xx} + v|\psi|^2 \psi + i\delta \psi = 0$$

if $T_{BF} \ll T_{diss}$, which is typically true in the ocean where $T_{BF} \sim$ minutes and $T_{diss} \sim$ tens of hours. A simple transformation of coordinates $\psi(x,\,t) \rightarrow c^{-\delta t}\psi(x,\,t)$ gives:

$$i(\psi_t + C_g \psi_x) + \mu \psi_{xx} + v e^{-2\delta t}|\psi|^2 \psi = 0$$

Thus if dissipation is much slower than the BF rise time ($T_{BF} \ll T_{diss}$), then the unstable rogue modes may "breathe" or oscillate up to their maximum heights and back down again many times during the characteristic dissipation decay time, δ. The competition between the appearance of unstable modes and dissipation is studied in wave tank experiments in Chapter 30. Segur et al. (2005) have studied the particular case where $T_{BF} \sim T_{diss}$, which occurs for small-amplitude waves, ~ 2 mm; in this case, the dissipation dominates the dynamics and the Benjamin-Fier instability is stabilized.

The periodic inverse scattering theory (Kotljarov and Its, 1976; Tracy, 1984; Tracy and Chen, 1988) tells us that unstable wave packets associated with small-amplitude modulations exist when $\rho a_o L > n\pi$ where $\rho = \sqrt{2}k_o^2$ and $L > 2\pi/K$ where K is the modulation wavenumber, a_o is the carrier amplitude, and k_o is the carrier wavenumber. Here n is an integer ($n = 1,\ 2,\ \ldots$) that counts the number of unstable wave packets in a wave train. Note that the expression $\rho a_o L$ is just the dimensional area of a box of height a_o and length

L scaled by the nonlinearity parameter ρ. This provides a useful definition of nonlinearity in terms of a kind of *Benjamin-Feir (BF) parameter*:

$$I_{BF} = \frac{\rho a_o L}{\pi} = \frac{2\sqrt{2}k_o^2 a_o}{K} > n \tag{12.19}$$

Note that I_{BF} is just the inverse of the dimensionless wavenumber $K/2\sqrt{2}k_o^2 a$ used in Equation (12.12). Thus Equation (12.19) might also be called the *Yuen-Lake parameter* (after all they wrote it down in the mid-1970s) or *Tracy parameter* (Tracy, 1984) (whose thesis provides the proof of the assertion that $\rho a_o L > n\pi$ must be true for unstable modes to occur) or the *parameter for the modulational instability*. The BF parameter increases with *increasing steepness*, $k_o a_o$, and *decreasing bandwidth*, K/k_o. Another useful form for Equation (12.19) is obtained by noting that the number of carrier oscillations, N_x, in a space series below the modulation envelope can be written

$$N_x = \frac{k_o}{K} = \frac{L}{L_o} \tag{12.20}$$

so that the BF parameter is then:

$$I_{BF} = 2\sqrt{2}N_x k_o a_o > n \tag{12.21}$$

Thus the nonlinearity is increased by increasing the steepness, $k_o a_o$, and/or the number of carrier oscillations, N_x, under the modulation envelope. The bandwidth of the spectrum is $K/k_o = 1/N_x$.

We see that I_{BF} is the same parameter that appears in the growth rate (Equation (12.13)) and the maximum amplitude (Equation (12.14)) of an unstable packet, which we now rewrite:

$$\Omega = i\omega_o k_o^2 a_o^2 \frac{\sqrt{I_{BF}^2 - 1}}{I_{BF}^2} \tag{12.22}$$

$$\frac{a_{max}}{a_o} = 1 + 2\frac{\sqrt{I_{BF}^2 - 1}}{I_{BF}} \tag{12.23}$$

Thus two of the most useful results for estimating unstable wave packet behavior can be written in terms of the Benjamin-Feir parameter.

It is also clear that an unstable wave packet (a nonlinear mode in the spectrum) has the imaginary part of the centroid of the two points of a simple spectrum (Equation (12.15)) that is also a function of the Benjamin-Feir parameter (see also Chapter 24):

$$\lambda_I = a_o \frac{\sqrt{I_{BF}^2 - 1}}{I_{BF}} \tag{12.24}$$

where the inverse is given by:

$$I_{BF} = \frac{a_o}{\sqrt{a_o^2 - \lambda_I^2}}$$

Thus the spectral parameter λ_I of the IST of an unstable wave packet in the (spectral) lambda plane also defines the BF parameter.

12.5   Analytical Formulas for Unstable Wave Packets

A large number of examples of unstable wave packets are known (see, e.g., Akhmediev et al., 1985, 1987, 1990; Akhmediev, 1986, 2001; Akhmediev and Ankiewicz, 1997; Osborne et al., 2000; and cited references). We consider three cases given by Equations (12.25) (IST spectral parameters $\{A, 0, 0, 0, A/\sqrt{2}\}$ see Figure 12.2, Chapter 24 for this notation), (12.26) ($\{A, 0, 0, 0, A\}$), and (12.27) ($\{A, 0, 0, 0, \sqrt{2}A\}$) below. The first case lies on the imaginary axis below the carrier in the lambda plane, the second case lies directly on the carrier, and the third case lies above the carrier (more details on the lambda plane are given in Chapter 24). The first case considered has the following solution to the NLS equation (12.1a):

$$u(x,t) = A\left[\frac{\cos\left[\sqrt{2}Ax\right]\operatorname{sech}[2\Lambda^2 t] + i\sqrt{2}\,\tanh[2A^2 t]}{\sqrt{2} - \cos\left[\sqrt{2}Ax\right]\operatorname{sech}[2A^2 t]}\right]e^{2iA^2 t} \qquad (12.25)$$

The imaginary part of the eigenvalue is $\lambda_I = iA/\sqrt{2}$ ($\lambda_R = 0$) (which is below the carrier eigenvalue iA in the lambda plane, Figure 12.1) and the maximum packet amplitude is then given by

$$\frac{u_{max}}{A} = 2\frac{|\lambda_I|}{A} + 1 \cong 2.414$$

This case is typical of previous studies of the Benjamin-Feir instability, that is, a small-amplitude modulation in the far past evolves into an extreme wave event in the present (Yuen, 1991). As shown in Figure 12.6, the small modulation is not easily visible at early times even though it is present, that is, it appears to be a broad flat plane over all x for early time t. Exponential growth is seen to lead to a large amplitude of ~2.41 times the carrier amplitude and then the wave decreases in amplitude as the modulation effectively disappears for large times. This solution to NLS (Equation (12.25)) is periodic in x and decays exponentially for large past and future times; it can be viewed as a single nonlinear mode (a single nonlinear Fourier component) of NLS with a 2×2 Riemann matrix (Chapter 24 gives details of this calculation). This rogue wave rises up to its full height only once in its lifetime! For the rest of its lifetime the wave train is indistinguishable from a sine wave.

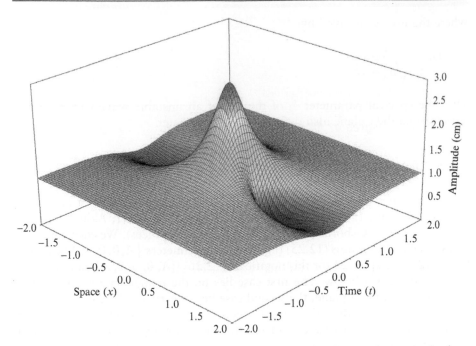

Figure 12.7 Modulus $|u(x, t)|$ of unstable wave packet that lies on the carrier in the complex lambda plane with spectrum: $\{A, 0, 0, 0, A\}$. The initial condition at time $t = -2$ is seen to be a small-amplitude modulation. (See color plate).

The second case (which lies directly on the carrier eigenvalue in the lambda plane) is shown in Figure 12.7. It has the exact solution given by

$$u(x,t) = A\left[1 - \frac{4(1 + 4iAt)}{1 + 16A^4t^2 + 4A^2x^2}\right]e^{2iA^2t} \tag{12.26}$$

Here the imaginary part of the eigenvalue is $\lambda_I = iA$ ($\lambda_R = 0$) and thus the maximum wave height is given by:

$$\frac{u_{max}}{A} = 2\frac{|\lambda_I|}{A} + 1 = 3$$

From Equation (12.26) we see that this solution to NLS is characterized by an algebraic decay for large x and t. In the spirit of the periodic inverse scattering transform Equation (12.26) is a nonlinear Fourier component in the theory.

The third case (above the carrier) is shown in Figure 12.8:

$$u(x,t) = A\left[1 + \frac{2\left(\cos\left[4\sqrt{2}A^2t\right] + i\sqrt{2}\sin\left[4\sqrt{2}A^2t\right]\right)}{\cos\left[4\sqrt{2}A^2t\right] + \sqrt{2}\cosh[2Ax]}\right]e^{2iA^2t} \tag{12.27}$$

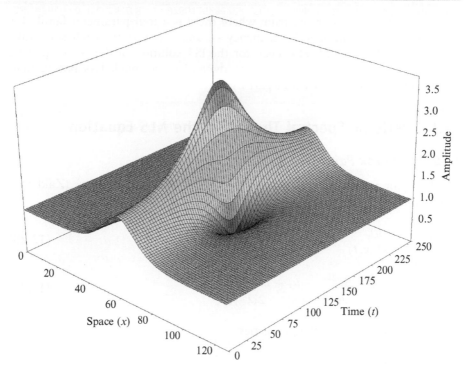

Figure 12.8 Modulus $|u(x, t)|$ of unstable wave packet that lies above the carrier in the complex lambda plane with spectrum: $\{A, 0, 0, 0, \sqrt{2}A\}$. The initial condition at time $t = 0$ is seen to be a large-amplitude modulation. (See color plate).

The eigenvalue is given by $\lambda_I = i\sqrt{2}A$ ($\lambda_R = 0$) so that the maximum height has the value

$$\frac{u_{max}}{A} = 2\frac{|\lambda_I|}{A} + 1 = 2\sqrt{2} + 1 \cong 3.828$$

This case lies *above the carrier* and is *no longer a small-amplitude modulation* for times far in the past. Indeed, the solution is periodic in t and exponentially decaying in x. Note that for small time in Figure 12.8 the spatial variation in the solution is a *large-amplitude modulation* (see Ma, 1979; Osborne, 2001, 2002; and cited references). This behavior is characteristic of spectral components with centroid above the carrier in the lambda plane. This kind of solution, ubiquitous in the nonlinear instability analysis of the NLS equation, is *not* accessible to *linear instability analysis*. Nonlinear instability analysis as discussed in Chapter 24 provides a systematic solution to this problem.

At this point it seems clear that there are an infinite number of solutions to the NLS equation, each corresponding to particular values for the parameters in the spectrum $\{A, \varepsilon, \theta, \lambda_{Rc}, \lambda_{Ic}\}$ (see Figure 12.1 in Chapter 24). This is also true of the linear Fourier spectrum where there is a four-parameter family for amplitude A, wavenumber k, frequency ω, and phase ϕ for each sine wave component: $\{A, k, \omega, \phi\}$. However, for the IST solution of the NLS equation the basis functions and the space/time dynamics are much less boring than simple sine waves, as verified in part by Figures 12.6–12.8.

12.6 Periodic Spectral Theory for the NLS Equation

12.6.1 The Lax Pair

The Lax pair for the NLS equation is given by the two operators (Zakharov and Shabat, 1972):

$$L \equiv \begin{bmatrix} i\partial_x & u(x,t) \\ -u^*(x,t) & -i\partial_x \end{bmatrix} \tag{12.28}$$

$$A \equiv \begin{bmatrix} i|u|^2 - 2i\lambda^2 & -u_x + 2i\lambda u \\ -u_x^* + 2i\lambda u^* & -i|u|^2 + 2i\lambda^2 \end{bmatrix} \tag{12.29}$$

The two operators have the properties

$$L\phi = \lambda\phi \tag{12.30}$$

$$\phi_t = A\phi \tag{12.31}$$

where ϕ is a two component eigenfunction

$$\phi = \begin{pmatrix} \phi_1 \\ \phi_2 \end{pmatrix} \tag{12.32}$$

Take the time derivative of Equation (12.30) and assume $\lambda_t = 0$ to get

$$L_t\phi + L\phi_t = \lambda\phi_t \tag{12.33}$$

Use Equation (12.31) (with $\lambda A\phi = A(\lambda\phi) = AL\phi$) in Equation (12.33) to find a condition that is necessary for Equations (12.30) and (12.31) to be *compatible*

$$L_t + [L, A] = 0 \tag{12.34}$$

where $[L, A] = LA - AL$ is the commutator of L and A. Equation (12.34) is a *nonlinear evolution equation* if L and A are chosen correctly; for Equations (12.28) and (12.29) we easily find the NLS equation (12.1a). Given L, Lax (1968) shows how to construct A so that Equation (12.34) gives a nontrivial evolution equation. In the present case, Equation (12.34) is the NLS equation. Notice that by abuse of notation I am using the symbol L to denote the Lax operator and the period of the wave train; the meaning is clear from the context.

12.6.2 The Spectra Eigenvalue Problem and Floquet Analysis

Now consider the eigenvalue problem first found by Zakharov and Shabat (1972) (Equation (12.30)), now written in component form:

$$
\begin{bmatrix} i\partial_x & u(x,t) \\ -u^*(x,t) & -i\partial_x \end{bmatrix} \begin{bmatrix} \phi_1 \\ \phi_2 \end{bmatrix} = \lambda \begin{bmatrix} \phi_1 \\ \phi_2 \end{bmatrix}
\tag{12.35}
$$

and solve it using the *Cauchy initial condition* $u(x, 0)$. The mathematical procedure is called Floquet analysis. Since we are studying the periodic problem for Equation (12.35) we have the boundary condition

$$
u(x + L, 0) = u(x, 0)
$$

Choose an arbitrary base point $x = x_o$ and introduce two independent solutions of Equation (12.35) which are assumed to have the following values at $x = x_o$ (the so-called (c, s) basis):

$$
\phi(x_o, x_o; \lambda) = \begin{pmatrix} 1 \\ 0 \end{pmatrix}, \quad \tilde{\phi}(x_o, x_o; \lambda) = \begin{pmatrix} 0 \\ 1 \end{pmatrix}
\tag{12.36}
$$

The solution matrix of Equation (12.35) is then

$$
\mathbf{\Phi}(x, x_o; \lambda) = \begin{bmatrix} \phi_1(x, x_o; \lambda) & \tilde{\phi}_1(x, x_o; \lambda) \\ \phi_2(x, x_o; \lambda) & \tilde{\phi}_2(x, x_o; \lambda) \end{bmatrix}
\tag{12.37}
$$

which satisfies Equation (12.35)

$$
L\mathbf{\Phi} = \lambda\mathbf{\Phi}, \quad \text{with the boundary condition } \mathbf{\Phi}(x_o, x_o; \lambda) = \begin{bmatrix} 1 & 0 \\ 0 & 1 \end{bmatrix}
\tag{12.38}
$$

The Wronskian of any two solutions is defined as $W(f, g) = f_1 g_2 - g_1 f_2$. Hence

$$
W(\phi, \tilde{\phi}) = \det(\mathbf{\Phi})
\tag{12.39}
$$

Using Equation (12.35) one finds that $\partial_x W = 0$ which means $\det(\mathbf{\Phi}(x)) = \det(\mathbf{\Phi}(x_o)) = 1$.

Using Equation (12.38) we can find $\mathbf{\Phi}(x)$ for any x. In particular, we can find the solution for $x_o + L$:

$$
\mathbf{\Phi}(x_o + L, x_o; \lambda) = \begin{bmatrix} \phi_1(x_o + L, x_o; \lambda) & \tilde{\phi}_1(x_o + L, x_o; \lambda) \\ \phi_2(x_o + L, x_o; \lambda) & \tilde{\phi}_2(x_o + L, x_o; \lambda) \end{bmatrix}
\tag{12.40}
$$

This is the *monodromy matrix* that can be written as

$$
\mathbf{M}(x_o; \lambda) \equiv \mathbf{\Phi}(x_o + L, x_o; \lambda)
$$

If we change the base point from x_o to x_1, $x_o \rightarrow x_1$, then the new functions

$$\phi(x, x_1; \lambda) \quad \text{and} \quad \tilde{\phi}(x, x_1; \lambda) \tag{12.41}$$

are simple linear combinations of the original functions $\phi(x, x_o; \lambda)$, $\tilde{\phi}(x, x_o; \lambda)$. This means that the spatial translation $x_o \rightarrow x_1$ results in a change of basis and therefore $M(x_1; \lambda)$ is related to $M(x_o; \lambda)$ by a similarity transformation

$$M(x_1; \lambda) = SM(x_o; \lambda)S^{-1} \tag{12.42}$$

The trace and determinant of the monodromy matrix are preserved under similarity transformations and therefore

$$[TrM](x_o, \lambda) = [TrM](\lambda) \equiv \Delta(\lambda), \quad \det M = 1 \tag{12.43}$$

The function $\Delta(\lambda)$ is called the *Floquet discriminant*; it is a function *only* of λ and it is fundamental for the determination of the *spectral properties* of the operator L. To see how this happens we construct the Bloch (or Floquet) solutions of Equation (12.35), $\psi(x; \lambda)$. The Bloch eigenfunctions have the property

$$\psi(x + L) = e^{ip(\lambda)}\psi(x; \lambda) \tag{12.44}$$

where $p(\lambda)$ is referred to as the *Floquet exponent* or the *quasi-momentum*. Floquet theory tells us (Magnus and Winkler, 1966) that for every λ (which is a complex parameter) there exists a Bloch eigenfunction. It is clear that these functions can be expressed as a linear combination of the standard solutions ϕ and $\tilde{\phi}$:

$$\psi(x; \lambda) = A\phi(x; \lambda) + B\tilde{\phi}(x; \lambda) \tag{12.45}$$

where A and B are complex constants. From Equation (12.36) we have

$$\psi(x_o; \lambda) = A\phi(x_o; \lambda) + B\tilde{\phi}(x_o; \lambda) = \begin{bmatrix} A \\ B \end{bmatrix} \tag{12.46}$$

The fundamental relation we seek to solve is

$$\psi(x_o + L; \lambda) = m(\lambda)\psi(x_o; \lambda) \tag{12.47}$$

where $m(\lambda)$ is the *Floquet multiplier*. From Equation (12.45) we can write

$$\begin{bmatrix} \psi_1(x_o + L; \lambda) \\ \psi_2(x_o + L; \lambda) \end{bmatrix} = \begin{bmatrix} \phi_1(x_o + L; \lambda) & \tilde{\phi}_1(x_o + L; \lambda) \\ \phi_2(x_o + L; \lambda) & \tilde{\phi}_2(x_o + L; \lambda) \end{bmatrix} \begin{bmatrix} A \\ B \end{bmatrix}$$

Applying this latter expression together with Equation (12.46) gives an *eigenvalue problem for the monodromy matrix*:

$$
\mathbf{M}\begin{bmatrix} A \\ B \end{bmatrix} = m(\lambda)\begin{bmatrix} A \\ B \end{bmatrix}
\tag{12.48}
$$

This last equation is fundamental in Floquet analysis. It implies that we can compute the Bloch eigenfunctions and the related Floquet multipliers (or exponents) by finding the *eigenvectors* and *eigenvalues* of the *monodromy matrix*. The eigenvalues are given by

$$
\det(\mathbf{M} - m) = m^2 - (\mathrm{Tr}\mathbf{M})m + \det \mathbf{M} = m^2 - \Delta(\lambda)m + 1
$$

Equation (12.48) has nontrivial solutions if and only if $\det(\mathbf{M} - m) = 0$. This gives:

$$
m^{\pm}(\lambda) = \frac{\Delta(\lambda) \pm (\Delta^2 - 4)^{1/2}}{2}
\tag{12.49}
$$

The scattering (spectral) problem (Equation (12.35)), for the *defocusing* NLS equation, is *self-adjoint* and it is sufficient to study the properties of $\Delta(\lambda)$ and $m^{\pm}(\lambda)$ along the *real and imaginary axes* (as is done for the KdV equation). However, in the present case, for the *focusing* (deep water) NLS equation, the spectral problem is *not self-adjoint* and we must therefore study $\Delta(\lambda)$ and $m^{\pm}(\lambda)$ *in the complex λ plane*.

The following results from the Floquet analysis are in the references (Magnus and Winkler, 1979; Ma and Ablowitz, 1981; Forest and McLaughlin, 1982; Tracy, 1984; Tracy and Chen, 1988). The discriminant, $\Delta(\lambda)$, a function of the complex parameter λ, is analytic in the complex plane. Therefore $\Delta(\lambda)$ *is real* (e.g., $\mathrm{Im}[\Delta(\lambda)] = 0$) *along one-dimensional curves in the complex λ plane*. The entire *real λ axis* is one of these curves.

Along *spectral curves* of this type there are three distinct domains: (I) $\Delta^2(\lambda) < 4$, (II) $\Delta^2(\lambda) = 4$, and (III) $\Delta^2(\lambda) > 4$. A discussion of the IST spectrum in these three domains follows.

Region I is referred to as the "band of stability" because when $\Delta^2(\lambda) < 4$ the Floquet multiplier $m^{\pm}(\lambda)$ (12.49) is a complex number with modulus one, $|m^{\pm}(\lambda)| = 1$. This implies that the Bloch eigenfunctions are *stable under a spatial translation*. In this context, the entire real λ axis is a band of stability. All other stable bands in the complex λ plane are called *spines*. This latter concept is useful in understanding the physics of spectral components of nonlinear wave solutions of the NLS equation and relates to the discussion above with regard to Figure 12.1. See also Chapter 19.

Region II consists of discrete points in the λ plane where the Floquet discriminant $\Delta(\lambda) = \pm 2$; in this case $m^{\pm}(\lambda) = \pm 1$ and the Bloch eigenfunctions are either *periodic* or *antiperiodic*. This set of eigenvalues in the λ domain is called the *main spectrum of the periodic NLS equation*, $\{\lambda_j\}$, $j = 1, 2, \ldots, N$.

Region III has $m^{\pm}(\lambda)$ real for which $|m^{\pm}(\lambda)| \neq 1$. This implies that the Bloch eigenfunctions are *unstable* to spatial translations along the x axis.

The remainder of the complex λ plane (where $\text{Im}[\Delta(\lambda)] \neq 0$) is *unstable*. Because $\Delta(\lambda)$ is real for real λ we have

$$[\Delta(\lambda)]^* = \Delta(\lambda^*)$$

Here the asterisk means complex conjugation. Consequently, if λ lies in a band of stability or is an eigenvalue in the main spectrum, then so too is λ^*. Thus the *main spectrum* is *symmetric about the real axis under conjugation*.

For *all* of the regions of the complex λ plane, *except where* $\Delta = \pm 2$, the two eigenvalues $m^{\pm}(\lambda)$ of the monodromy matrix $\mathbf{M}$ are distinct. This implies that the two eigenvectors will be independent and there are therefore two Bloch eigenfunctions

$$\begin{aligned}
\psi^+(x + L) &= m^+ \psi^+(x) \\
\psi^-(x + L) &= m^- \psi^-(x)
\end{aligned} \tag{12.50}$$

On the other hand, when $\Delta(\lambda) = \pm 2$, the two eigenvalues are no longer distinct and *two* independent Bloch eigenfunctions can no longer exist.

Definition of degenerate and nondegenerate spectra: If λ_j is an eigenvalue in the main spectrum which has *two* independent Bloch eigenfunctions ($\Delta(\lambda) \neq \pm 2$), then λ_j is said to be a *degenerate* eigenvalue. On the other hand, if λ_j is an eigenvalue in the main spectrum which has only one Bloch eigenfunction ($\Delta(\lambda) = \pm 2$) then λ_j is a *nondegenerate* eigenvalue.

This definition is useful for understanding when the NLS equation has unstable solutions. See also Tracy and Chen (1988) who discuss the fact that the existence and location of degeneracies in the main spectrum determine whether a solution is *unstable*.

The *spines* of stable λ *can terminate only at nondegenerate eigenvalues*. Any Cauchy initial condition $u(x, 0)$ that has no degenerate eigenvalue is referred to as *generic*. When degeneracies are present in the spectrum of $u(x, 0)$, it is *nongeneric* in the sense that if we add a small perturbation to $u(x, 0)(u(x, 0) \rightarrow u(x, 0) + \varepsilon\phi(x))$ then the degeneracy will in general be broken.

It can be shown that the discriminant $\Delta(\lambda)$ is invariant in space and time. Hence so too is the main spectrum; the $\{\lambda_j\}$ are constants of the motion for the NLS equation.

Throughout I have considered only *finite band potentials*, that is, those for which there are a finite number of nondegenerate eigenvalues $\{\lambda_j\}$, $j = 1, 2, \ldots$, where N is generally the number of *degrees of freedom*. This is true for the applications given here, that is, for the analysis of *discrete* space or time series and for numerical modeling.

12.7 Overview of the Spectrum and Hyperelliptic Functions

I assume the form of the NLS equation given by Equation (12.1a). The *spectral eigenvalue problem* (Equation (12.35)) is given by:

$$
\begin{aligned}
i\psi_{1x} + iu\psi_2 &= \lambda\psi_1 \\
-i\psi_{2x} + iu^*\psi_1 &= \lambda\psi_2
\end{aligned}
\tag{12.51}
$$

or in matrix form:

$$
i\begin{bmatrix} 1 & u \\ u^* & -1 \end{bmatrix}\begin{bmatrix} \psi_{1x} \\ \psi_{2x} \end{bmatrix} = \lambda\begin{bmatrix} \psi_1 \\ \psi_2 \end{bmatrix}
$$

It is this problem that provides the main and auxiliary spectra of the NLS equation, as will now be described.

12.7.1 The IST Spectrum

The solution to NLS is constructed in the following manner using the inverse scattering transform developed by Kotljarov and Its (1976), Tracy (1984), and Tracy and Chen (1988):

(1) The *main spectrum* consists of the complex constants λ_k, $k = 1, 2, \ldots, 2N$.
(2) The *auxiliary spectrum* consisting of the "μ variables" μ_j and their Riemann sheet indices $\sigma_j:(\mu_j,\sigma_j)$, $j = 1, 2, \ldots, N-1$. The μ_j are functions of space and time: $\mu_j(x, t)$. These dynamical variables are paired with their complex conjugates, $\mu_j^*(x, t)$, and are the "nonlinear modes" for the NLS equation in the so-called μ-function representation. The Riemann sheet indices take on the values $\sigma_j = \pm 1$, as with the KdV equation.

The *auxiliary variables*, $\mu_j(x, t)$, have the following equations of motion in space and time:

$$
\mu_{jx} = \frac{-2i\sigma_j\sqrt{\prod\limits_{k=1}^{2N}(\mu_j - \lambda_k)}}{\prod\limits_{\substack{m=1 \\ m\neq j}}^{N-1}(\mu_j - \lambda_k)}, \quad
\mu_{jt} = -2\left(\sum\limits_{\substack{m=1 \\ m\neq j}}^{N-1}\mu_m - \frac{1}{2}\sum\limits_{k=1}^{2N}\lambda_k\right)\mu_{jx}
\tag{12.52}
$$

Then the space/time evolution of the nonlinear Schrödinger equation is found by the *trace formula*:

$$
\partial_x \ln u(x, t) = 2i\left(\sum\limits_{j=1}^{N-1}\mu_j - \frac{1}{2}\sum\limits_{k=1}^{2N}\lambda_k\right)
\tag{12.53}
$$

(see Appendix for practical details). Hence the solution, as written so far, remarkably resembles that for the KdV equation except that the λ_j and μ_j variables are *complex* and the left-hand side of the trace formula (Equation (12.53)) is the *derivative of the logarithm of the solution.*

However, there is another difference between constructing solutions of the NLS and KdV equations. This is because a *constraint relation* (a reality condition) (Tracy, 1984) must be satisfied in order that the above formulae satisfy NLS. To introduce the constraint consider the following *polynomials* for the *squared eigenfunctions g, h,* and *f*:

$$g(x,t;\lambda) = iu(x,t) \prod_{j=1}^{N-1} (\lambda - \mu_j(x,t)) \tag{12.54}$$

$$h(x,t;\lambda) = iu^*(x,t) \prod_{j=1}^{N-1} (\lambda - \mu_j^*(x,t)) \tag{12.55}$$

$$P(\lambda) = \prod_{k=1}^{2N} (\lambda - \lambda_k) \tag{12.56}$$

The first two polynomials depend upon the auxiliary spectrum eigenvalues $\mu_j(x, t)$ and $\mu_j^*(x, t)$, while the third depends on the main spectrum eigenvalues λ_k. Here of course λ_k is the spectral parameter, complex in the present case, which has units of *wavenumber.* Now use these three functions to define a new function:

$$f(x,t;\lambda) = \sqrt{P+gh} \quad \text{or} \quad f^2 - gh = P(\lambda) \tag{12.57}$$

Here then is the *constraint relation*:

The above method for solving the μ-function ordinary differential Equation (12.52) plus the trace formula (Equation (12.53)), yield a solution to NLS *only if* $f(x,t;\lambda) = \sqrt{P+gh}$ is a *finite order polynomial.*

Now that we know $f(x,t;\lambda) = \sqrt{P+gh}$ must be a finite-order polynomial, it is interesting to see exactly the form of the requisite polynomial:

$$f(x,t;\lambda) = \prod_{j=1}^{N} (\lambda - \gamma_j(x,t)) \tag{12.58}$$

where the $\gamma_j(x, t)$ are the nonlinear modes of the *radiation stress* contribution to a nonlinear wave train (defined to be $|u|^2$) (Tracy et al., 1991):

$$|u|^2 = \frac{1}{2} \sum_{k=1}^{2N} \lambda_k^2 - \sum_{j=1}^{N} \gamma_j^2 \tag{12.59}$$

This latter expression simply means that the modulus squared of the complex envelope solution of the NLS equation has itself a spectral decomposition in terms of the squared variables $\gamma_j^2 = \gamma_j^2(x,t)$.

12.7.2 Generating Solutions to the NLS Equation

Now we discuss a procedure for constructing a solution to NLS using (a) the μ-function ordinary differential equations, (b) the trace formula, and (c) the constraint relation. Chapter 24 discusses the solutions in terms of Riemann theta functions.

To generate a solution to NLS:

(1) Choose a set of $2N$ complex constants λ_k.
(2) Construct the polynomial f by choosing a set of N complex numbers f_k such that:

$$f = \sum_{k=0}^{N} f_k \lambda^k \tag{12.60}$$

(3) Solve for the zeros and leading order coefficients of

$$f^2 - P = gh = -uu^* \prod_{j=1}^{N-1} (\lambda - \mu_j(x,t)) \prod_{j=1}^{N-1} (\lambda - \mu_j^*(x,t)) \tag{12.61}$$

to obtain $uu^*(0, 0)$ and the auxiliary spectrum eigenvalues $\mu_j(x, t)$ and $\mu_j^*(x, t)$.

(4) Find the Riemann sheet indices (± 1) from:

$$\sigma_k = \left. \frac{f(\lambda)}{\sqrt{P(\lambda)}} \right|_{\lambda = \mu_k} \tag{12.62}$$

(5) Integrate the μ-function ODEs Equation (12.52) to find the spatial and temporal evolution of the $\mu_j(x, t)$. Then the trace formula (Equation (12.53)) gives the space-time evolution of the NLS equation.

12.7.3 Applications to the Cauchy Problem: Space and Time Series Analysis

In order to apply this method to the analysis of data we need to extend the above approach to allow us to compute the *main and auxiliary spectra* from the function $u(x, 0)$ as related to the sNLS equation (12.1b) or the function $u(0, t)$ as related to the tNLS equation (12.5). In what follows we consider the analysis of the spatial function $u(x, 0)$ for space series analysis. Time series analysis follows from a similar analysis of $u(0, t)$. Since only a simple redefinition of the variables is necessary (Equation (12.8)) between space and time, the same computer program suffices for both space and time series analysis.

One way to determine the spectrum is to develop procedures for extracting the spectral data from the *monodromy matrix*. The monodromy matrix may be written in the following form:

$$\mathbf{T}(x_0, \lambda) = \begin{pmatrix} a & b^* \\ b & a^* \end{pmatrix} \tag{12.63}$$

The *squared eigenfunctions* can be written in terms of these matrix elements:

$$f(x_o, \lambda) = -ia_I(x_o, \lambda) = i \operatorname{Im}(T_{11})$$
$$g(x_o, \lambda) = b^*(x_o, \lambda) = T_{12}^*$$
$$h(x_o, \lambda) = -b(x_o, \lambda) = -T_{21}$$

(12.64)

Here

$$a_I(x_o, \lambda) = \frac{1}{2}i[a^*(x_o, \lambda) - a(x_o, \lambda)]$$

(12.65)

It should be noted that the latter expression is *not* the imaginary part of $a(x_o, t)$, written $\operatorname{Im}[a(x_o, t)]$; instead $a_I(x_o, \lambda)$ is in general a complex number. Generally $a_I(x_o, t) = \operatorname{Im}[a(x_o, t)]$ *only* for λ real. The monodromy matrix can be written in terms of the squared eigenfunctions:

$$\mathbf{T}(x_o, \lambda) = \begin{pmatrix} a_R(x_o, \lambda) - f(x_o, \lambda) & g(x_o, \lambda) \\ -h(x_o, \lambda) & a_R(x_o, \lambda) + f(x_o, \lambda) \end{pmatrix}$$

(12.66)

Where we have:

$$a_R(x_o, \lambda) = \frac{1}{2}i[a(x_o, \lambda) + a^*(x_o, \lambda^*)]$$

(12.67)

where, once again, the subscript does not refer to the real part of the function; a_R is instead a complex number. Naturally, we also have

$$a_R^2 + a_I^2 = aa^*$$

(12.68)

The recipe for extracting the *NLS spectrum* is then given in the following sections.

12.7.4   The Main Spectrum

The trace of the monodromy matrix determines the *main spectrum eigenvalues*:

$$\frac{1}{2}\operatorname{Tr}\mathbf{T} = \frac{1}{2}(T_{11} + T_{22}) = a_R(x_o, \lambda) = \pm 1$$

(12.69)

gives the complex eigenvalues λ_k, $k = 1, 2, \ldots, 2N$

12.7.5   The Auxiliary Spectrum of the $\mu_j(x, 0)$

The auxiliary spectrum eigenvalues are found by:

$$T_{12} = g(x_o, \lambda) = 0$$
gives the complex eigenvalues $\mu_j(x_o, t = 0)$, $j = 1, 2, \ldots, N$

(12.70)

To pick up both the μ_j and their complex conjugates μ_j^* one can of course use $T_{12}T_{21} = 0$ to instead determine the spectrum. These are the Kotljarov-Its-Tracy (KIT) μ-function modes that can be *inverted* and *linearized* by theta functions using the general algebraic geometrical procedure for solving the Jacobian inverse problem. Note that Equation (12.53) constructs the potential $u(x, t)$ from the hyperelliptic functions $\mu_j(x,t)$.

12.7.6 The Auxiliary Spectrum of the Riemann Sheet Indices σ_j

The *Riemann sheet indices* are found from:

$$\sigma_k = \left.\frac{f(\lambda)}{\sqrt{P(\lambda)}}\right|_{\lambda=\mu_k} = \left.\frac{i\,\mathrm{Im}(T_{11})}{\sqrt{T_{12}^* T_{21} - \mathrm{Im}^2(T_{11})}}\right|_{\lambda=\mu_k} \tag{12.71}$$

12.7.7 The Auxiliary Spectrum of the $\gamma_j(x, 0)$

The *radiation stress modes* are solutions of:

$$\frac{1}{2}(T_{22} - T_{11}) = f(x_0, \lambda) = 0$$

that give the eigenvalues
$$\gamma_j(x_0, t = 0) \quad \text{for} \quad 0 \leq x_0 \leq L, \quad j = 1, 2, \ldots, N$$

$$\tag{12.72}$$

The superposition law (Equation (12.59)) allows a nonlinear spectral construction of the radiation stress in terms of the γ_j modes.

Nota Bene: The expression $a_R(x_0, \lambda)$ is *not* the real part of $a(x_0, \lambda)$ except when λ is real. Generally speaking the main spectrum eigenvalue values λ_k lie somewhere in the complex plane (see Tracy, 1984 for details). Note further that T_{22} needs to be computed separately and independently of T_{11}. Likewise, note that a_I is *not* Im $a(x_0, \lambda)$ in general.

Nota Bene: The theta-function solutions of the NLS equation are discussed in Chapter 24.

Appendix—Interpretation of the Hyperelliptic Function Superposition Law

Recall the linear superposition law:

$$\partial_x \ln u(x,t) = 2i\left(\sum_{j=1}^{N-1} \mu_j - \frac{1}{2}\sum_{k=1}^{2N} \lambda_k\right)$$

Let us integrate this over the spatial variable to get

$$\ln u(x,t) = 2i\left(\sum_{j=1}^{N-1}\int_0^x \mu_j(x,t)dx - \frac{1}{2}x\sum_{k=1}^{2N}\lambda_k\right)$$

$$= 2i\sum_{j=1}^{N-1}\int_0^x \mu_j(x,t)dx - iKx; \quad K = \sum_{k=1}^{2N}\lambda_k$$

Here the symbol K has been used to represent a "wavenumber." Also set

$$\mu_j'(x,t) = 2\int_0^x \mu_j(x,t)dx$$

so that

$$\ln u(x,t) = i\sum_{j=1}^{N-1}\mu_j'(x,t) - iKx$$

Take the exponential of this to get

$$u(x,t) = \exp\left(-iKx + i\sum_{j=1}^{N-1}\mu_j'(x,t)\right)$$

Now suppose the modes are *small in amplitude*. Then this expression can be written as:

$$u(x,t) = \exp\left(-iKx + i\sum_{j=1}^{N-1}\mu_j'(x,t)\right) \cong e^{-iKx}\left[1 + i\sum_{j=1}^{N-1}\mu_j'(x,t) + \cdots\right]$$

This looks like the classical superposition law, that is, where the wave field $u(x, t)$ consists of a linear superposition of the modes. Note that the modes are themselves complex and hence so is $u(x, t)$.

The general superposition law can be written and interpreted in the compact shorthand notation:

$$u(x,t) = e^{-iKx}\left[1 + i\sum_{j=1}^{N-1} \mu_j'(x,t)\right] + u_{\text{int}}(x,t)$$

where $u_{\text{int}}(x,t)$ are the interactions among the $u_j'(x,t)$ modes. Thus we have, essentially, a linear superposition of the hyperelliptic $u_j'(x,t)$ modes plus nonlinear interactions.

13 The Hilbert Transform

13.1 Introduction

We have seen in previous chapters that certain classes of *nonlinear wave equations* are best described in terms of a *complex envelope function*, $\psi(x, t)$ or $\psi(x, y, t)$. Examples of wave equations that admit solutions of this class are the *nonlinear Schrödinger (NLS) equation* and generalizations. Recall the NLS equation in $1 + 1$ dimensions:

$$i\left(\psi_t + \frac{\omega_o}{2k_o}\psi_x\right) - \frac{\omega_o}{8k_o^2}\psi_{xx} - \frac{\omega_o k_o^2}{2}|\psi|^2\psi = 0 \tag{13.1}$$

and in $2 + 1$ dimensions:

$$i\left(\psi_t + \frac{\omega_o}{2k_o}\psi_x\right) - \frac{\omega_o}{8k_o^2}\psi_{xx} + \frac{\omega_o}{4k_o^2}\psi_{yy} - \frac{\omega_o k_o^2}{2}|\psi|^2\psi = 0 \tag{13.2}$$

The solutions to these equations are analytic functions of a complex variable, $\psi(x, t) = \psi_R(x, t) + i\psi_I(x, t)$, where $\psi_R(x, t)$ is the real part of the function and $\psi_I(x, t)$ is the imaginary part.

The complex envelope function, $\psi(x, t)$, is related to the free surface elevation, $\eta(x, t)$, at leading order in nonlinearity, by the simple relation:

$$\eta(x, t) = \text{Re}\{\psi(x, t)\, e^{i(k_o x - \omega_o t)}\} \tag{13.3}$$

Here the dispersion relation, $\omega_o = \omega_o(k_o)$, is that for deep-water waves, that is, $\omega_o^2 = gk_o$, where g is the acceleration of gravity and h is the water depth. One often assumes in this case that the *space series*, $\eta(x) \equiv \eta(x, 0)$, or *time series*, $\eta(t) \equiv \eta(0, t)$, are *narrow-banded processes*, that is, their Fourier spectra (in terms of wavenumber, k, or frequency, ω) are *narrowly distributed* about the central (carrier) wavenumber, k_o, or frequency, ω_o.

There are essentially two major results described in this chapter which make the Hilbert transform, to be defined below, useful in the study of nonlinear waves: (1) The Hilbert transform provides a means to construct the complex envelope $\psi(x, t)$ from the free surface elevation $\eta(t) \equiv \eta(0, t)$ in time series analysis applications. (2) The Hilbert transform provides a means to compute the imaginary part of the solution $\psi_I(x, t)$ from the real part of the solution $\psi_R(x, t)$.

© 2010 Elsevier Inc. All rights reserved.
Doi: 10.1016/S0074-6142(10)97013-7

This chapter focuses on a generalized way to treat narrow-banded processes from mathematical and physical points of view using the *Hilbert transform*. We also learn that the methods are not necessarily limited to narrow-banded processes, but work perfectly well for *broad-banded spectra* as well. The Hilbert transform is named after one of the greatest mathematicians of the twentieth century, David Hilbert (1862-1943), who, in his studies of integral equations, was the first to derive what is now referred to as the Hilbert transform pair. The Hilbert transform and its properties were developed mainly by Titchmarsh (1925, 1928, 1930) and Hardy (1924).

When working with theoretical or experimental wave trains it is often found convenient to consider the *complex surface elevation* that may be expressed in the following forms:

$$\Xi(x,t) = \eta(x,t) + i\tilde{\eta}(x,t) = \psi(x,t)\,e^{i(k_0 x - \omega_0 t)} = [\psi_R(x,t) + \psi_I(x,t)]\,e^{i(k_0 x - \omega_0 t)}$$

$$= A(x,t)\,e^{i\theta(x,t)+i(k_0 x - \omega_0 t)} = A(x,t)\,e^{i\phi(x,t)}$$

$$(13.4)$$

Here $\eta(x,t)$, $\tilde{\eta}(x,t)$ are the real and complex parts of the complex surface elevation. Also $\psi(x,t) = A(x,t)\,e^{i\theta(x,t)}$ is the *complex envelope function* of the wave train and $\phi(x,t) = (k_0 x - \omega_0 t) + \theta(x,t)$.

The *real envelope* of the wave train is $A(x,t)$ and the *real phase* is $\theta(x,t)$. Theoretically and experimentally we think of $\eta(x,t)$, $\tilde{\eta}(x,t)$ as describing the space-time dynamics of the free surface elevation and $\psi(x,t) = A(x,t)\,e^{i\theta(x,t)}$ as describing the space-time dynamics of the complex envelope function. These two functions are related via the expression (13.4).

In a purely experimental context one measures a time series of the surface elevation $\eta(0,t)$ and seeks to determine a time series for the complex envelope $\psi(0,t) = A(0,t)\,e^{i\theta(0,t)}$. According to Equation (13.4), first we need to compute $\tilde{\eta}(0,t)$. As will be discussed below this is done by the Hilbert transform: $\tilde{\eta}(0,t) = H[\eta(0,t)]$, where the symbol $H[\cdot]$ signifies the Hilbert transform operation. When this operation is complete the complex envelope is given by $\psi(x,t) = [\eta(x,t) + i\tilde{\eta}(x,t)]\,e^{-i(k_0 x - \omega_0 t)}$.

Let us discuss the consequences of this formulation. First assume that we have the expressions for the real and imaginary parts $\eta(x,t)$, $\tilde{\eta}(x,t)$ of the complex surface elevation, $\Xi(x,t) = \eta(x,t) + i\tilde{\eta}(x,t)$. Then it follows that

$$\eta(x,t) = \text{Re}\{A(x,t)e^{i\phi(x,t)}\} = A(x,t)\cos\phi(x,t) = A(x,t)\cos[k_0 x - \omega_0 t + \theta(x,t)]$$

$$\tilde{\eta}(x,t) = \text{Im}\{A(x,t)e^{i\phi(x,t)}\} = A(x,t)\sin\phi(x,t) = A(x,t)\sin[k_0 x - \omega_0 t + \theta(x,t)]$$

$$(13.5)$$

Thus, we see that the surface elevation $\eta(x,t)$ is interpreted as a simple cosine with (slowly varying) amplitude, $A(x,t)$, and phase given by that for the carrier, $k_0 x - \omega_0 t$, plus a slowly varying oscillating part, $\theta(x,t)$. The imaginary contribution to the surface elevation, $\tilde{\eta}(x,t)$, is the same, but with the cosine function replaced by a sine function, that is, there is a simple phase shift of $\pi/2$ between the real and imaginary parts of the complex surface elevation.

Space series are seen to take the form $\Xi(x,0) = \eta(x,0) + i\tilde{\eta}(x,0)$, or in simpler notation:

$$\Xi(x) = A(x)e^{i\phi(x)} = A(x)e^{ik_ox + i\theta(x,\,0)} = \eta(x) + i\tilde{\eta}(x)$$
$$= A(x)\cos[k_ox + \theta(x,0)] + iA(x)\sin[k_ox + \theta(x,0)] \tag{13.6}$$

Time series $(\Xi(0,t) = \eta(0,t) + i\tilde{\eta}(0,t))$ have the form:

$$\Xi(t) = A(t)e^{i\phi(t)} = A(x)e^{-i\omega_ox + i\theta(x,\,0)} = \eta(x) + i\tilde{\eta}(x)$$
$$= A(x)\cos[\omega_ot - \theta(0,t)] - iA(t)\sin[\omega_ot - \theta(0,t)] \tag{13.7}$$

From the above relations, we see that the space and time series for A, ϕ (the real envelope and real phase) have the explicit forms:

$$A(x) = \sqrt{\eta^2(x) + \tilde{\eta}^2(x)}, \quad A(t) = \sqrt{\eta^2(t) + \tilde{\eta}^2(t)} \tag{13.8}$$

$$\phi(x) = \arctan\frac{\tilde{\eta}(x)}{\eta(x)}, \quad \phi(t) = \arctan\frac{\tilde{\eta}(t)}{\eta(t)} \tag{13.9}$$

The *Hilbert transformed phase*, $\theta(x, t)$, *after removal of the carrier wave*, $(k_ox - \omega_ot)$, is given by: $\theta(x, t) = \phi(x, t) - (k_ox - \omega_ot)$.

To best understand what all this means let us consider the following specific cases.

Example 1—The Simplest Case: Constant Amplitude and Phase

Consider the case when the amplitude of the modulation and the phase are constants, $A(x, t) = a$, $\theta(x, t) = \theta_o$. Then we have

$$\Xi(x,t) = a\,e^{i(k_ox - \omega_ot + \theta_o)} = \eta(x,t) + i\tilde{\eta}(x,t)$$

This means that

$$\eta(x,t) = a\cos(k_ox - \omega_ot + \theta_o)$$

$$\tilde{\eta}(x,t) = a\sin(k_ox - \omega_ot + \theta_o)$$

A simple cosine wave for the surface elevation has as its auxiliary surface elevation a simple sine wave. Thus, if $\eta(x, t)$ is a cosine wave, then $\tilde{\eta}(x,t)$ is phase shifted from the latter by $\pi/2$, a generic property of the Hilbert transform. Clearly, $A(x,t) = \sqrt{\eta^2(x,t) + \tilde{\eta}^2(x,t)} = a$ and $\tan\theta(x,t) = \tilde{\eta}(x,t)/\eta(x,t) = \tan(k_ox - \omega_ot + \theta_o) = \tan\phi$. The modulation phase is $\theta_o = \phi - (k_ox - \omega_ot)$; this latter expression returns the input phase after "removing the carrier wave."

Example 2—The Carrier Wave Solution of the NLS Equation

Suppose the complex envelope function is given by $\psi(x,t) = a\,e^{-i\omega_o k_o^2 a^2 t/2}$, that is, the unmodulated carrier solution of the dimensional NLS equation (Yuen and Lake, 1982). Then the real envelope is given by $A(x, t) = a$, while the phase has the form $\theta(x,t) = \omega_o k_o^2 a^2 t/2$. Then the surface elevation is given by $\eta(x,t) = a\cos[k_o x - \omega_o(1 + k_o^2 a^2/2)t]$ and the auxiliary surface elevation is $\tilde{\eta}(x,t) = a\sin[k_o x - \omega_o(1 + k_o^2 a^2/2)t]$. This is just the *leading order Stokes wave* with the nonlinear frequency correction $\omega_o k_o^2 a^2 t/2$, as seen in an earlier chapter. The space series are $\eta(x,0) = a\cos[k_o x]$ and $\tilde{\eta}(x,0) = a\sin[k_o x]$. The time series are $\eta(0,t) = a\cos[\omega_o(1 + k_o^2 a^2/2)t]$ and $\tilde{\eta}(0,t) = a\sin[\omega_o(1 + k_o^2 a^2/2)t]$. We have the envelope $A(x,t) = \sqrt{\eta^2(x,t) + \tilde{\eta}^2(x,t)} = a$ and phase $\tan\phi(x,t) = \tilde{\eta}(x,t)/\eta(x,t) = \tan(k_o x - \omega_o(1 + k_o^2 a^2/2)t)$. The modulation phase is $\phi(x,t) - (k_o x - \omega_o t) = \omega_o k_o^2 a^2 t/2$; this latter expression returns the input phase after "removing the carrier wave, $k_o x - \omega_o t$."

Example 3—The General Case

Consider the case when the amplitude of the modulation, $a(x, t)$, and the phase, $\theta(x, t)$ are arbitrary. Then we have

$$\Xi(x,t) = a(x,t)e^{i[k_o x - \omega_o t + \phi(x,t)]} = \eta(x,t) + i\tilde{\eta}(x,t) =$$

This means that

$$\eta(x,t) = a(x,t)\,\cos(k_o x - \omega_o t + \theta(x,t))$$
$$\tilde{\eta}(x,t) = a(x,t)\,\sin(k_o x - \omega_o t + \theta(x,t))$$

These are just simple amplitude/phase-modulated wave trains. We have the real envelope function $A(x,t) = \sqrt{\eta^2(x,t) + \tilde{\eta}^2(x,t)} = a(x,t)$ and the phase $\tan\phi(x,t) = \tilde{\eta}(x,t)/\eta(x,t) = \tan(k_o x - \omega_o t + \theta(x,t))$. The modulation phase is $\phi(x, t) - (k_o x - \omega_o t) = \theta(x, t)$, after a correction for the carrier wave, $k_o x - \omega_o t$.

13.2 The Hilbert Transform

We now confine most of the discussion to time series. From the point of view of an experiment, we measure a time series for the surface elevation, $\eta(t)$. Can we also determine the auxiliary surface elevation, $\tilde{\eta}(t)$, associated with $\eta(t)$?

Clearly, we can find the complex time series $\Xi(t)$, whose real component is $\eta(t)$, provided we also know $\tilde{\eta}(t)$. The formulas in the last section express relationships between $\eta(t), \tilde{\eta}(t)$ and $A(x, t)$, $\theta(x, t)$ but in general we find in the discussion below that $\eta(t)$ will be available (from time series measurements), but $\tilde{\eta}(t)$ will not. This is because the function $\tilde{\eta}(t)$ is not uniquely determined by the above formulas.

How do we determine the time series $\tilde{\eta}(t)$ from $\eta(x, t)$? To answer this question, we use results from a well-known theorem in the theory of functions of a complex variable. Simply stated, the function $\tilde{\eta}(t)$, and hence the complex variable $\Xi(t)$ will be uniquely determined by the requirement that the latter be an analytic function. In this case, we are faced with computing an analytic function $\Xi(t) = \eta(t) + i\tilde{\eta}(t)$ from values of the function $\eta(t)$ on the real axis. The solution to this problem is given by the *Schwarz integral* or the *Hilbert transformation*:

$$\tilde{\eta}(t) \equiv \mathbf{H}[\eta(t)] = \frac{1}{\pi} \int_{-\infty}^{\infty} \frac{\eta(\tau)d\tau}{t - \tau} \tag{13.10}$$

The inverse Hilbert transform has the form:

$$\eta(t) \equiv \mathbf{H}^{-1}[\tilde{\eta}(t)] = -\frac{1}{\pi} \int \frac{\tilde{\eta}(\tau)d\tau}{t - \tau} \tag{13.11}$$

Thus, the Hilbert transform uniquely carries a time series $\eta(t)$ over to another time series $\tilde{\eta}(t)$ that is in a very real sense "orthogonal" to the original series (see more details below). This operation contrasts to the Fourier transform which carries the original signal from the time domain to the frequency domain or vice versa.

An alternative perspective is that the Hilbert transform can be used to determine properties of the imaginary part of a solution to the NLS equations or extensions, that is, given $\psi_R(x, t)$ compute $\psi_I(x, t)$ by

$$\psi_I(x,t) = \mathbf{H}[\psi_R(x,t)] \tag{13.12}$$

The Hilbert transform provides a means to compute the imaginary part of the solution $\psi_I(x, t)$ from the real part of the solution $\psi_R(x, t)$.

Detailed properties of the Hilbert transform are given in Beckmann (1967) and Bendat and Piersol (1986). We outline a few of these properties below.

13.2.1 Properties of the Hilbert Transform

Let us now look at some of the important properties of the Hilbert transform, which for the time series $\eta(t)$ has the form

$$\tilde{\eta}(t) \equiv \mathbf{H}[\eta(t)] = \frac{1}{\pi} \int_{-\infty}^{\infty} \frac{\eta(\tau)d\tau}{\tau - t} \tag{13.13}$$

(1) *Envelope property*

On what basis do we call the function $A(t) = \sqrt{\eta^2(t) + \tilde{\eta}^2(t)}$ the *real envelope* of the wave train, $\eta(t)$? First note that $A(t) \geq |\eta(t)|$. This occurs because $\eta(t), \tilde{\eta}(t)$ are "orthogonal coordinates" and $A(t)$ is the "diagonal." Let us think momentarily in terms of time series. Since $A^2(t) = \eta^2(t) + \tilde{\eta}^2(t)$, by taking the derivative with respect to time t, it is easy to show that

$$A\frac{\mathrm{d}A}{\mathrm{d}t} = \eta\frac{\mathrm{d}\eta}{\mathrm{d}t} + \tilde{\eta}\frac{\mathrm{d}\tilde{\eta}}{\mathrm{d}t}$$

and hence that all points where $A(t) = \eta(t)$ have the property

$$\frac{\mathrm{d}A}{\mathrm{d}t} = \frac{\mathrm{d}\eta}{\mathrm{d}t}$$

Hence the curve $A(t)$ is never crossed by $\eta(t)$; indeed $\eta(t)$ lies beneath $A(t)$ and the two functions have the same time derivative as $\eta(t)$ rises from below to touch $A(t)$. Hence $A(t)$ is truly an "envelope" of $\eta(t)$. In the same way, $A(t)$ is also an envelope of $\tilde{\eta}(t)$.

(2) *Linear property*

$$\mathbf{H}[a\eta(t) + b\phi(t)] = a\tilde{\eta}(t) + b\tilde{\phi}(t) \tag{13.14}$$

for any time series $\eta(t)$, $\phi(t)$ and any constants a, b.

(3) *Shift property*

$$\mathbf{H}[\eta(t - a)] = \tilde{\eta}(t - a) \tag{13.15}$$

The Hilbert transform of a shifted (or lagged time series) is equivalent to a lagged, Hilbert transformed time series.

(4) *Hilbert transform of the Hilbert transform*

$$\mathbf{H}\{\mathbf{H}[\eta(t)]\} = \mathbf{H}[\tilde{\eta}(t)] = -\eta(t) \tag{13.16}$$

The application of two successive Hilbert transforms returns the negative of the original time series.

(5) *Inverse Hilbert transform*

$$\eta(t) \equiv \mathbf{H}^{-1}[\tilde{\eta}(t)] = -\frac{1}{\pi}\int\frac{\tilde{\eta}(\tau)\mathrm{d}\tau}{\tau - t} \tag{13.17}$$

Thus, $\eta(t)$ is the convolution of $\tilde{\eta}(t)$ with $(-1/\pi t)$. In terms of the Fourier transform $\eta(t)$ has the form:

$$\eta(t) = \mathbf{F}^{-1}[(\mathrm{i}\,\mathrm{sgn}f)\tilde{\mathbf{F}}(f)], \quad \text{for} \quad \tilde{\mathbf{F}}(f) = \mathbf{F}[\tilde{\eta}(t)] \tag{13.18}$$

(6) *Even and odd function properties*

If $\eta(t)$ is an even (odd) function of time t, then $\tilde{\eta}(t)$ is an odd (even) function of t:

$$\eta(t) \text{ even } \leftrightarrow \tilde{\eta}(t) \text{ odd}$$
$$\eta(t) \text{ odd } \leftrightarrow \tilde{\eta}(t) \text{ even}$$

(13.19)

(7) *Similarity property*

$$\mathbf{H}[\eta(at)] = \tilde{\eta}(at)$$

(13.20)

The Hilbert transform is invariant to rescaled time t.

(8) *Energy (or variance) property*

$$\int_{-\infty}^{\infty} \eta^2(t)\mathrm{d}t = \int_{-\infty}^{\infty} \tilde{\eta}^2(t)\mathrm{d}t$$

(13.21)

This result easily follows from Parseval's theorem:

$$\int_{-\infty}^{\infty} \eta^2(t)\mathrm{d}t = \int_{-\infty}^{\infty} |F(f)|^2\mathrm{d}f$$

(13.22)

Likewise

$$\int_{-\infty}^{\infty} \tilde{\eta}^2(t)\mathrm{d}t = \int_{-\infty}^{\infty} |\tilde{F}(f)|^2\mathrm{d}f$$

Combining these results with the fact that

$$|\tilde{F}(f)|^2 = |F(f)|^2$$

gives the above variance property.

(9) *Orthogonality property*

$$\int_{-\infty}^{\infty} \eta(t)\tilde{\eta}(t)\mathrm{d}t = 0$$

(13.23)

This result is a consequence of the Parseval theorem since

$$\int_{-\infty}^{\infty} \eta(t)\tilde{\eta}(t)\mathrm{d}t = \int_{-\infty}^{\infty} F^*(f)\tilde{F}(f)\mathrm{d}f$$

It then follows that

$$F^*(f)\tilde{F}(f) = (-\mathrm{i}\,\mathrm{sgn}\,f)|F(f)|^2$$

The latter expression is an odd function of f and hence the above right-hand integral is zero.

(10) *Modulation property*

One of the most important properties of the Hilbert transform results from the fact that

$$\mathbf{H}[\eta(t)\cos(\omega_o t)] = \eta(t)\sin(\omega_o t) \qquad \omega_o = 2\pi f_o \tag{13.24}$$

This result holds true provided that $\eta(t)$ is a time series that has a Fourier transform $\mathbf{F}(f)$ which is bandwidth limited, that is

$$\mathbf{F}(f) = \begin{cases} \mathbf{F}(f) & f > -f_o \\ 0 & \text{otherwise} \end{cases}$$

The following inverse relation also holds:

$$\mathbf{H}[\eta(t)\sin(\omega_o t)] = -\eta(t)\cos(\omega_o t)$$

(11) *Convolution property*

The following convolution property holds:

$$\mathbf{H}[\eta(t) * \phi(t)] = \tilde{\eta}(t) * \phi(t) = \eta(t) * \tilde{\phi}(t) \tag{13.25}$$

This can be seen by noting that

$$\mathbf{F}[\eta(t) * \phi(t)] = \mathbf{F}(f)\Phi(f)$$

where $\Phi(f)$ is the Fourier transform of $\phi(t)$. Furthermore,

$$[(-i\operatorname{sgn} f)\mathbf{F}(f)]\Phi(f) = \tilde{\mathbf{F}}(f)\Phi(f) = \mathbf{F}(f)[(-i\operatorname{sgn} f)\Phi(f)] = \mathbf{F}(f)\tilde{\Phi}(f)$$

(12) *Lack of commutation of the Fourier and Hilbert transforms*

The following relation holds:

$$\mathbf{F}\{\mathbf{H}[\eta(t)]\} \neq \mathbf{H}\{\mathbf{F}[\eta(t)]\} \tag{13.26}$$

Thus, the Fourier and Hilbert transforms do not commute.

Based upon what has been presented above, it may seem that this is all there is to do, that is, assume the surface elevation is a modulated cosine, the auxiliary surface elevation is a modulated sine and it all ends there. However, the main focus of this chapter is the development of practical procedures for space and time series analysis and the processing of nonlinear signals, both in the laboratory and in the ocean. We now proceed to look for practical structure in the Hilbert transform.

13.2.2 Numerical Procedure for Determining the Hilbert Transform

To compute the Hilbert transform of the surface elevation $\eta(t)$ we can use the so-called phase-shifting property. This is accomplished by a simple phase shifting $\pi/2$ filter. As seen above $\eta(t)$ is the convolution of $\tilde{\eta}(t)$ with $(-1/\pi t)$. In terms of the Fourier transform, $\eta(t)$ has the form (see Equation (13.18)):

$$\eta(t) = \mathbf{F}^{-1}[(\mathrm{i}\,\mathrm{sgn}\,f)\tilde{\mathrm{H}}(f)], \quad \text{for } \tilde{\mathrm{H}}(f) = \mathbf{F}[\tilde{\eta}(t)]$$

The inverse of this is just

$$\tilde{\mathrm{H}}(f) = (-\mathrm{i}\,\mathrm{sgn}\,f)\mathbf{F}[\eta(t)]$$

which gives

$$\tilde{\eta}(f) = \mathbf{F}^{-1}\{(-\mathrm{i}\,\mathrm{sgn}\,f)\mathbf{F}[\eta(t)]\}$$

Therefore, the Hilbert transform is obtained by the FFT. First one obtains the Fourier transform of $\eta(t)$, then applies the filter $-\mathrm{i}\,\mathrm{sgn}\,f$, and finally takes the inverse Fourier transform! This is the recommended numerical procedure.

13.2.3 Table of Simple Hilbert Transforms

Here is a table of simple Hilbert transforms:

$\eta(t)$	$\tilde{\eta}(t)$	$A(t)$
$\cos\omega_o t$	$\sin\omega_o t$	1
$\sin\omega_o t$	$-\cos\omega_o t$	1
$\dfrac{\sin t}{t}$	$\dfrac{1 - \cos t}{t}$	$\dfrac{\sin(t/2)}{t/2}$
$\dfrac{1}{1 + t^2}$	$\dfrac{t}{1 + t^2}$	$\left[\dfrac{1}{1 + t^2}\right]^{1/2}$
$e^{-c|t|}\cos\omega_o t$	$e^{-c|t|}\sin\omega_o t$	$e^{-c|t|}$

13.3 Narrow-Banded Processes

Although *any* time series, $\eta(t)$, has a well-defined Hilbert transform $\tilde{\eta}(t)$, the Hilbert transform is often viewed as being most useful for (although not limited to) a *narrow-banded process*. As noted above we set:

$$\theta_o(t) = \theta(t) - \mathrm{i}\omega_o t$$

where ω_o is the dominant or peak frequency of the narrow-banded spectrum. Thus, ω_o is the frequency of the carrier wave that has wavenumber, k_o. We therefore have

$$\eta(t) = A(t) \cos\theta_o(t) = A(t) \cos[\theta(t) - \omega_o t]$$
$$= A(t) \cos\theta(t) \cos\omega_o t + A(t) \sin\theta(t) \sin\omega_o t$$

Now set

$$C(t) = A(t) \cos\theta(t), \quad S(t) = A(t) \sin\theta(t)$$

and finally

$$\eta(t) = C(t) \cos\omega_o t + S(t) \sin\omega_o t \tag{13.27}$$

Clearly, for a narrow-banded process, $\eta(t)$, the functions $C(t)$ and $S(t)$ must be slowly varying in time with respect to the fast-time oscillation of the carrier wave, $\omega_o t$. It is clear that $A(t)$, $\theta(t)$ must also be slowly varying.

Electrical Engineering Analogy of Water Wave Dynamics

In the field of communications engineering (Beckmann, 1967), one speaks of an "amplitude modulated" wave train (AM radio) with a carrier wave of frequency, ω_o, whose amplitude, $A(t)$, and phase, $\theta(t)$, vary slowly in time with respect to the carrier frequency. On the radio one "dials" the carrier frequency, ω_o (the "radio station") and the "message" (the music or conversation) is contained in the modulation functions $A(t)$, $\theta(t)$. The message, that is the voice, is thus sent as a modulation of the carrier wave. A whole spectrum of sounds, music, etc. is encoded as a modulation by the radio station, sent out from the antenna as electromagnetic waves and then received by a radio antenna, and finally demodulated by radio circuitry and transmitted to a loud speaker. Thus, the ideas discussed in this chapter are inherent in the body of knowledge used in modern radio communications. Many of the ideas presented here are not novel to the communications engineer.

Fortunately, for all of us the modulation envelope of *electromagnetic waves* is *linear* and *stable*, that is, the modulation always remains under control by the circuitry, doing exactly what it was designed to do, that is, act as a tool for the communication of information. Thus, the fact that electromagnetic waves are *stable* as they propagate in a vacuum (or air) means they are wonderful for carrying messages. This means that the machinery of the Fourier transform, ideal for *linear* wave dynamics, is perfectly adequate for communications purposes.

On the other hand, *water waves are not ideal for carrying messages*, although in principle they could be used to do so, but nonlinear effects would have to be taken into account. For example, suppose one would want to communicate to a friend on the other side of a large body of water, say a lake (let us assume that the phones are down and that there is a heavy fog so that we cannot send him Morse code with, say, a flashlight). We establish ourselves on the pier in our backyard overlooking the lake. Undaunted, we construct a large paddle which we use to make a sine wave (a small-amplitude Stokes wave) that will be our "carrier" which will arrive at our friend's house on the other side of the lake. If we do not modulate the sine wave carrier our friend will receive a rather boring message, no signal at all, in the language of communications engineering, just the "hum" of the carrier (Stokes) wave. We decide the best way to encode information is the linear Fourier transform (radio engineers have already known this since the 1920s). So, we put our information into the modulation envelope using the Fourier transform and send the wave out across the lake to our friend on the other side. He is waiting and, because he is a good friend, he knows the carrier frequency and so tunes into it with his Fourier transformer (his "radio" is just a wave-measuring device connected to a "computer" to do the Fourier transform; he is a smart guy so we leave it up to him to design a mechanical or electronic Fourier transformer for this purpose, which is not very hard to do).

Under ideal conditions (not too much dissipation in the waves) one would think that this way to communicate, although somewhat novel and unusual, would be guaranteed to work, provided of course that all sources of noise, etc. could be taken into account. *But this is just not the case.* Generally speaking, water waves are *not* a very good way to communicate messages because they are both *nonlinear* and *unstable by the Benjamin-Fier (BF) mechanism*, described in detail herein. Deep-water waves are *nonlinear* and hence can grow *exponentially* in time. The modulation envelope just gets very big, that is, it does not remain small and controllable as the "side bands" in the radio problem, but becomes large and difficult to control. Now from this point of view the most likely outcome of our attempt to communicate is that the BF instability just makes some of the waves get big and then nonlinear effects take over and *destroy all information in the original modulated signal*, at least from the point of view of the linear Fourier transform. As we look further into this situation later in this book, we will see that the exploding, large-amplitude solutions are a kind of *rogue wave*. So, water waves are generally not very good for communication purposes; they are just too complicated, because of their nonlinear behavior, to be very practical to send a message to your friend "to come over for barbecue on Sunday afternoon." One would of course have to address the signal processing from the point of view of the inverse scattering transform (IST).

Let us continue our electromagnetic analogy for a moment. Suppose, as above, we use water waves to carry a message on a carrier water (Stokes) wave, but now we hook up the output to a loud speaker so we can hear the message itself (this is an advantage, since we do the Fourier transform in our ears and heads rather than in a computer). Now if water waves were perfect linear carriers of information you would understand

Continued

the message perfectly, just as Edison first did it on the telephone or others did on the phonograph and other recording instruments developed over a century ago. On the other hand, in the water wave problem, the nonlinear effects (the BF instability) will modify what we hear from the loud speaker. Generally speaking, we will hear the message with some small variations in amplitudes and frequencies (like playing an old fashion phonograph record too fast or too slow) *plus occasional large bursts of volume*. These will be the BF instabilities or *very loud volume enhancements*, right in the middle of your favorite aria! In fact, I discuss a theorem elsewhere in this volume which says that deep water wave trains can be decomposed into quasi-linear "music" (the "Gaussian" or "near Gaussian" sea state that is often assumed to govern ocean waves) plus loud bursts of very nonlinear "rogue waves."

Here is a simple example. Suppose a radio engineer wants to sound a simple tone on the radio. This means to modulate the carrier with a simple sine wave, that is, a single frequency, note or tone. This is of course a very boring radio program, not often broadcast, primarily because few advertisers are convinced that listeners will be attracted by a constant tone or note. What would happen if we use deep water waves to broadcast such a message? First, for early time we would "hear" a single tone, but soon the volume would begin to rapidly increase, by say a factor of 3-5, and then it would return to the original background tone; then the process may repeat itself or become even more complicated with other spikes in the signal. The presence of these *large bursts of noise* is just manifestations of the sporadic appearance (Fermi-Pasta-Ulam *recurrence*) of rogue waves in the deep-water wave field.

There are a couple of important messages that I would like to communicate to potential end users of this complex nonlinear wave dynamics. First, the synthesis and Fourier decomposition of these signals is not very easy and requires implementation of the IST as discussed in the rest of this book. Second, true communications using such signals is probably not very efficient. Without IST one virtually stands no chance of "decoding" the signals sent by nonlinear water waves.

13.4 Statistical Properties of Complex Time Series

Let us consider the possibility that the time series, $\eta(t)$, is a random function of time with a well-defined power spectrum. The Fourier representation, for an N-point time series discretized at Δt seconds, gives:

$$\eta(t) = \sum_{n=1}^{N} C_n \cos(\omega_n t - \phi_n) \tag{13.28}$$

where of course $t \to t_m = m \, \Delta t$, $m = 1, 2, \ldots, N$. Here the phases, ϕ_n, are often assumed to be uniformly distributed random numbers on the interval $(0, 2\pi)$. The power spectrum, $P(\omega_n)$, is related to the constant coefficients by the relation: $C_n = \sqrt{2P(\omega_n)\Delta\omega}$ ($\Delta\omega$ is the frequency interval under the power spectrum, $\Delta\omega = 2\pi \, \Delta f = 2\pi/T$, where T is the length of the time series in seconds; the Nyquist frequency has the value $f_N = 1/2\Delta t$). It is not hard to show that the auxiliary (Hilbert transformed) time series, $\tilde{\eta}(t)$, is given by

$$\tilde{\eta}(t) = \sum_{n=1}^{N} C_n \sin(\omega_n t - \phi_n) \tag{13.29}$$

This expression results because the Hilbert transform phase shifts the trigonometric functions by $\pi/2$. Since the above Fourier series between $\eta(t)$ and $\tilde{\eta}(t)$ differ only by a simple phase in the trigonometric functions, the power spectra for $\eta(t), \tilde{\eta}(t)$ are the same. It then follows that the variances of $\eta(t), \tilde{\eta}(t)$ are the same. It is furthermore an easy matter to remove the means of the two processes. Finally, it can be shown that $\eta(t), \tilde{\eta}(t)$, at the same moment of time, are uncorrelated with each other (Beckmann, 1967).

What are the statistical properties of $\eta(t), \tilde{\eta}(t)$? This is easy, since by the above linear Fourier superpositions, both $\eta(t), \tilde{\eta}(t)$ are Gaussian (by the central limit theorem) with the same power spectrum, $P(\omega)$, and both have a zero mean and variance, σ^2, determined by the area under the power spectrum:

$$\sigma^2 = 2 \int_0^{\omega_N} P(\omega) d\omega \tag{13.30}$$

The probability density function for $\eta(t), \tilde{\eta}(t)$ is given by

$$p(\eta) = \frac{1}{\sigma\sqrt{2\pi}} \exp\left[\frac{\eta^2}{2\sigma^2}\right], \quad p(\tilde{\eta}) = \frac{1}{\sigma\sqrt{2\pi}} \exp\left[\frac{\tilde{\eta}^2}{2\sigma^2}\right] \tag{13.31}$$

Clearly, since the average values have the definition

$$\langle \eta^n \rangle = \int \eta^n p(\eta) d\eta, \quad \langle \tilde{\eta}^n \rangle = \int \tilde{\eta}^n p(\tilde{\eta}) d\tilde{\eta} \tag{13.32}$$

we have

$$\langle \eta \rangle = \langle \tilde{\eta} \rangle = 0, \quad \langle \eta^2 \rangle = \langle \tilde{\eta}^2 \rangle = \sigma^2$$

To obtain the correlation between $\eta(t), \tilde{\eta}(t)$ we proceed as follows. Using the definition of the Hilbert transform we have

$$\langle \eta(t)\tilde{\eta}(t) \rangle = \frac{1}{\pi} \left\langle \eta(t) \int \frac{\eta(\tau)d\tau}{\tau - t} \right\rangle = \frac{1}{\pi} \int \frac{\langle \eta(t)\eta(\tau) \rangle d\tau}{\tau - t}$$

Now the correlation function between $\eta(t), \tilde{\eta}(t)$ is given by

$$\langle \eta(t)\tilde{\eta}(t+\tau)\rangle = B(\tau) = \lim_{T\to\infty} \frac{1}{T}\int_0^T \eta(t)\tilde{\eta}(t+\tau)dt \qquad (13.33)$$

In the correlation calculation above we have $\langle \eta(t)\tilde{\eta}(\tau)\rangle$ which is just $B(\tau-t)$. Therefore

$$\langle \eta(t)\tilde{\eta}(t)\rangle = \frac{1}{\pi}\int \frac{B(\tau-t)d\tau}{\tau-t} = \frac{1}{\pi}\int \frac{B(u)du}{u} = 0$$

This expression vanishes since $B(u)$ is even and the integrand is odd. We therefore conclude that $\eta(t), \tilde{\eta}(t)$ are uncorrelated at any moment in time.

Example 4—Statistical Properties of the Envelope and Phase

The statistical properties of the envelope, $A(t)$, and phase, $\theta(t)$, can be computed from the statistical properties of the surface elevation, $\eta(t)$, and auxiliary surface elevation, $\tilde{\eta}(t)$. The first step consists in answering the following question. Given the *joint probability density* $p(\eta,\tilde{\eta})$ of the random orthogonal coordinates $\eta(t), \tilde{\eta}(t)$, what is the probability density in polar coordinates A, θ? We know that

$$A = \sqrt{\eta^2 + \tilde{\eta}^2}, \quad \theta = \arctan(\tilde{\eta}/\eta)$$

with the inverse formulas

$$\eta = A\cos\theta, \quad \tilde{\eta} = A\sin\theta$$

The Jacobian of the transformation is given by

$$\frac{\partial(\eta,\tilde{\eta})}{\partial(A,\theta)} = \begin{vmatrix} \cos\theta & -A\sin\theta \\ \sin\theta & A\cos\theta \end{vmatrix} = A$$

which means that $d\eta d\tilde{\eta} = AdAd\theta$. We invoke the conservation of probability so that

$$p(\eta,\tilde{\eta})d\eta d\tilde{\eta} = p(A,\theta)dAd\theta$$

and we finally have

$$p(A,\theta) = Ap(\eta,\tilde{\eta}) = Ap(A\cos\theta, A\sin\theta)$$

As $\eta(t), \tilde{\eta}(t)$ vary on the interval $(-\infty, \infty)$, $A(t)$ varies on $(0, \infty)$ and $\theta(t)$ varies on $(C, C+2\pi)$, C an arbitrary constant (usually set to 0 or $-\pi$).

Now $\eta(t), \tilde{\eta}(t)$ have the joint Gaussian probability density

$$p(\eta, \tilde{\eta}) = \frac{1}{2\sigma^2} \exp\left[-\frac{\eta^2 + \tilde{\eta}^2}{2\sigma^2}\right]$$

We now derive the joint density function $p(A, \theta)$ and the one-dimensional densities $p(A)$, $p(\theta)$. We have

$$p(A, \theta) = A p(A \cos\theta, A \sin\theta) = \frac{A}{2\pi\sigma^2} e^{-A^2/2\sigma^2}$$

Then

$$p(A) = \int p(A, \theta) d\theta \quad \text{and} \quad p(\theta) = \int p(A, \theta) dA$$

for which

$$p(A) = \frac{2A}{\alpha} e^{-A^2/\alpha}, \quad p(\theta) = \frac{1}{2\pi}$$

The probability density for the envelope is Rayleigh and for the phases it is uniform. Of course Longuet-Higgins taught us this many decades ago.

13.5 Relations Between the Surface Elevation and the Complex Envelope Function

In this section we work with space series only. The equivalent time series formulas can be obtained by the simple transformation $x \rightarrow t$, $k_o \rightarrow -\omega_o$. We write the relationship between the spatial surface elevation, $\eta(x, 0) \rightarrow \eta(x)$, and the complex envelope, $\psi(x, 0) \rightarrow \psi(x)$, in the following form:

$$\Xi(x) = \Psi(x) e^{ik_o x} \tag{13.33}$$

where $\Xi(x) = \eta(x) + i\tilde{\eta}(x)$, $\Psi(x) = \psi_R(x) + i\psi_I(x)$, and $e^{ik_o x} = \cos(k_o x) + i \sin(k_o x)$. Notice that these expressions assume we have knowledge of the real and imaginary parts of the envelope function, $\psi_R(x), \psi_I(x)$. Furthermore, Equation (13.33) can be put in matrix form:

$$\begin{pmatrix} \eta(x) \\ \tilde{\eta}(x) \end{pmatrix} = \begin{pmatrix} \cos(k_o x) & -\sin(k_o x) \\ \sin(k_o x) & \cos(k_o x) \end{pmatrix} \begin{pmatrix} \psi_R(x) \\ \psi_I(x) \end{pmatrix} \tag{13.34}$$

Since the matrix is orthogonal the inverse is just the transpose:

$$\begin{pmatrix} \psi_R(x) \\ \psi_I(x) \end{pmatrix} = \begin{pmatrix} \cos(k_o x) & \sin(k_o x) \\ -\sin(k_o x) & \cos(k_o x) \end{pmatrix} \begin{pmatrix} \eta(x) \\ \tilde{\eta}(x) \end{pmatrix} \tag{13.35}$$

As discussed above, we naturally call $\eta(x)$ the *surface elevation* and $\tilde{\eta}(x)$ the *auxiliary surface elevation*. We have the relation

$$\tilde{\eta}(x) = \mathbf{H}[\eta(x)] = \mathbf{F}^{-1}\{(-i\,\mathrm{sgn}\,f)\mathbf{F}[\eta(t)]\} \qquad (13.36)$$

where $\mathbf{H}[\cdot]$ is the Hilbert transform. Since the two complex fields are related by an orthogonal transformation, the lengths of the vectors are invariant to a rotation, so that

$$A(x) = \sqrt{\eta^2(x) + \tilde{\eta}^2(x)} = \sqrt{\psi_R^2(x) + \psi_I^2(x)} \qquad (13.37)$$

Furthermore, since

$$\Psi(x) = \Xi(x)e^{-ik_ox} \qquad (13.38)$$

the Fourier transforms of the two complex fields are given by the *shifting property*:

$$F_\Psi(k) = F_\Xi(k - k_o) \qquad (13.39)$$

Thus, the complex modulation field, $\Psi(x)$, has a Fourier transform which is just the Fourier transform of the complex surface elevation, $\Xi(x)$, shifted to the *left* by exactly the carrier wavenumber, k_o. Examples will now be given.

Example 5—A Simple Real-Valued Modulation

Consider the following complex modulation with two sine wave components:

$$\Psi = a[1 + \varepsilon_1 \cos(Kx) + \varepsilon_2 \cos(2Kx)] \qquad (13.40)$$

Note that in this example the imaginary part is zero. The surface elevation is then, by Equation (13.34):

$$\eta(x) = a[1 + \varepsilon_1 \cos(Kx) + \varepsilon_2 \cos(2Kx)] \cos(k_ox)$$

It is easy to see that the carrier wave is given by $a \cos(k_ox)$. The real function, $1 + \varepsilon_1 \cos(Kx) + \varepsilon_2 \cos(2Kx)$, is just the (presumed small-amplitude) modulation of the carrier. And also by Equation (13.34) we have the auxiliary surface elevation:

$$\tilde{\eta}(x) = a[1 + \varepsilon_1 \cos(Kx) + \varepsilon_2 \cos(2Kx)] \sin(k_ox)$$

The modulus of the modulation envelope is then

$$A(x) = \sqrt{\eta^2(x) + \tilde{\eta}^2(x)} = \sqrt{\psi_R^2(x) + \psi_I^2(x)}$$
$$= a[1 + \varepsilon_1 \cos(Kx) + \varepsilon_2 \cos(2Kx)]$$

Consider the special case: $a = 1$, $\varepsilon_1 = 0.2$, $\varepsilon_2 = 0.1$, $L = 100(2\pi) = 628.318$, $L_o = L/10 = 62.8318$. Then $N = L/L_o = 10$, $k_o = 2\pi/L_o = 0.1$, $K = 2\pi/L = 0.01$. In this calculation I have used 256 points in the space series.

The surface wave elevation is given in Figure 13.1. Note that the horizontal x-axis has 256 points and has length $L = 100(2\pi) = 628.318$.

Figure 13.2 shows both the surface elevation given in the graph of Figure 13.1 and the auxiliary surface elevation (dotted line) and the modulus of the envelope function. It is easy to see how one forms an envelope from these two functions.

The Fourier spectrum of the free surface elevation is shown in Figure 13.3. Shown are the carrier and the two side bands.

The complex envelope is computed from the output $\tilde{\eta}(x)$ from the Hilbert transform. It has only the real part with two Fourier components; the imaginary part is zero as given by the input waveform. Since $\Psi(x) = \psi_R(x) + i\,\psi_I(x)$ was input in this way we have just verified that it is correct after Hilbert transforming the surface elevation. The spectrum of the envelope is shown in Figure 13.4. Note that it is shifted by the carrier frequency relative to the Fourier transform and appears at low frequency just as we intuitively expect.

Let us give an analytical explanation for these results. Here is the free surface elevation:

$$\eta(x) = a[1 + \varepsilon_1 \cos(Kx) + \varepsilon_2 \cos(2Kx)] \cos(k_o x)$$

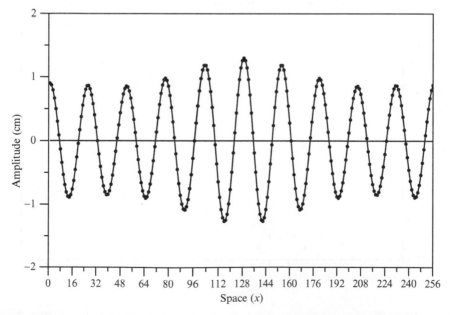

Figure 13.1 A simple modulation of a carrier wave. The length of the space series is 256 points, corresponding to a modulation wavelength of 628.318 cm.

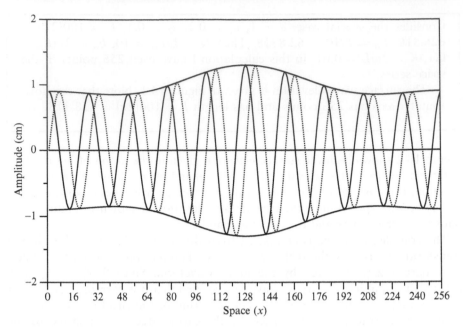

Figure 13.2 Simple modulation of the case in Figure 13.1 where the auxiliary surface elevation (dotted line) is also shown together with the modulus of the envelope function. The lengths of the space series are 256 points, corresponding to a modulation wavelength of 628.318 cm.

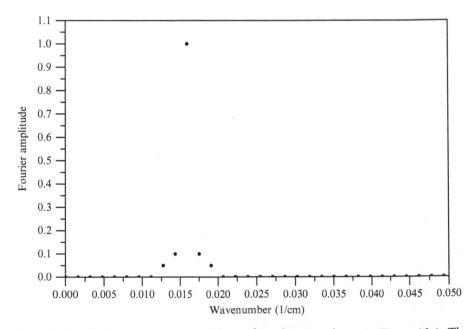

Figure 13.3 The Fourier spectrum of the surface elevation shown in Figure 13.1. The carrier wavenumber is $k_0 = 10/628.318$ cm $= 0.0159$ cm^{-1}.

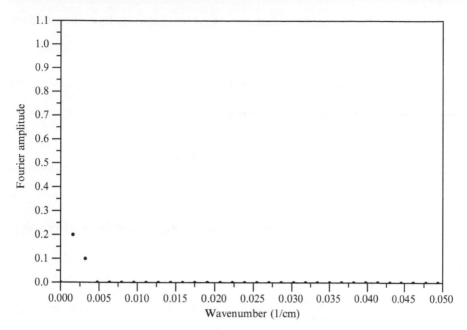

Figure 13.4 Fourier spectrum of envelope given in Figure 13.3.

Expand this and use

$$\cos(Kx)\cos(k_o x) = \frac{1}{2}\cos[(k_o - K)x] + \frac{1}{2}\cos[(k_o + K)x]$$

to get

$$\eta(x) = a\cos(k_o x) + a\varepsilon_1 \cos(Kx)\cos(k_o x) + a\varepsilon_2 \cos(2Kx)\cos(k_o x)$$

And finally

$$\eta(x) = a\cos(k_o x) + \frac{a\varepsilon_1}{2}\cos[(k_o - K)x] + \frac{a\varepsilon_1}{2}\cos[(k_o + K)x]$$

$$+ \frac{a\varepsilon_2}{2}\cos[(k_o - 2K)x] + \frac{a\varepsilon_2}{2}\cos[(k_o + 2K)x]$$

Note that the amplitudes of the sidebands, located at $k_o \pm K$ and $k_o \pm 2K$ have amplitudes exactly one-half the original Fourier amplitudes $(a\varepsilon_i/2)$ in the envelope function $a\varepsilon_i$. In this way we have the analytic form of the Fourier components shown in Figure 13.3.

Now let us go after the general relationships between the surface elevation, the auxiliary surface elevation, and their Hilbert transforms. Consider the following surface elevation:

$$\eta(x) = a\left[1 + \sum_{n=1}^{N}\varepsilon_n \cos(nKx - \phi_n)\right]\cos(k_o x) \qquad (13.41)$$

It is easy to see that the carrier is just $a \cos(k_o x)$. And also we have the auxiliary surface elevation:

$$\tilde{\eta}(x) = a \left[1 + \sum_{n=1}^{N} \varepsilon_n \cos(nKx - \phi_n) \right] \sin(k_o x) \qquad (13.42)$$

Now the surface elevation can be written:

$$\eta(x) = a \cos(k_o x) + \sum_{n=1}^{N} \varepsilon_n \cos(nKx - \phi_n) \cos(k_o x)$$

Use

$$\cos(nKx - \phi_n) \cos(k_o x) = \frac{1}{2} \cos[(k_o - nK)x + \phi_n] + \frac{1}{2} \cos[(k_o + nK)x - \phi_n]$$

Hence

$$\eta(x) = a \cos(k_o x) + \sum_{n=1}^{N} \frac{a\varepsilon_n}{2} \{ \cos[(k_o - nK)x + \phi_n] + \cos[(k_o + nK)x - \phi_n] \}$$

which can be written as

$$\eta(x) = \sum_{n=-N}^{N} C_n \cos[k_n x + \phi_n] \qquad (13.43a)$$

where

$$k_n = k_o + nK \qquad (13.43b)$$

$$C_n = \frac{a\varepsilon_n}{2} \qquad (13.43c)$$

$$\phi_{-n} = -\phi_n \qquad (13.43d)$$

This is just one of the standard forms for the Fourier decomposition for a surface wave train, here related directly to the equivalent wave train written as a modulation.

13.6 Fourier Representation of the Free Surface Elevation and the Complex Envelope Function

Let us consider a general form for the complex envelope that has the usual mathematical definition:

$$\psi(x) = \psi_R(x) + i\psi_I(x) = A(x)e^{i\phi(x)} \qquad (13.44)$$

where $A(x) = 0$ is the envelope and $\phi(x)$ is the phase. For many physical reasons I set

$$\psi_R(x) = a[1 + C(x)], \quad \psi_I(x) = aS(x)$$

Here "a" is the carrier wave amplitude and $C(x)$, $S(x)$ are "small-amplitude" modulation functions. Hence

$$\psi(x) = a[1 + C(x) + iS(x)] \tag{13.45}$$

When $C(x) = S(x) = 0$ there is no modulation; this justifies the "1" in this form for $\psi(x)$. For $S(x) = 0$ the modulation is real and determined by $C(x)$. For $C(x) = 0$ the modulation is determined by the imaginary component $S(x)$. For many applications the modulation is taken to be small and $C(x)$, $S(x)$ may be thought of as small oscillations (albeit with many Fourier components). The envelope and phase have the form:

$$A(x) = a\sqrt{[1 + C(x)]^2 + S^2(x)}$$
$$\tan \phi(x) = \frac{S(x)}{1 + C(x)} \tag{13.46}$$

The *free surface* and *auxiliary free surface* have the expressions (use (13.45)):

$$\eta(x) = a[1 + C(x)] \cos(k_o x) - aS(x) \sin(k_o x)$$
$$\tilde{\eta}(x) = a[1 + C(x)] \sin(k_o x) + aS(x) \cos(k_o x) \tag{13.47}$$

Case I: $S(x) = 0$ *(modulation only from real part of $\psi(x)$)*

$$\psi(x) = a[1 + C(x)]$$

$$A(x) = a[1 + C(x)]$$

$$\tan \phi(x) = 0$$

$$\eta(x) = a[1 + C(x)] \cos(k_o x)$$

$$\tilde{\eta}(x) = a[1 + C(x)] \sin(k_o x)$$

Case II: $C(x) = 0$) *(modulation only from imaginary part of $\psi(x)$)*

$$\psi(x) = a[1 + iS(x)]$$

$$A(x) = a\sqrt{1 + S^2(x)}$$

$$\tan \phi(x) = S(x)$$

$$\eta(x) = a\cos(k_\mathrm{o}x) - aS(x)\,\sin(k_\mathrm{o}x)$$

$$\tilde{\eta}(x) = a\sin(k_\mathrm{o}x) + aS(x)\,\cos(k_\mathrm{o}x)$$

Case III: $C(x) = S(x) = 0$ (*no modulation*)

$$\psi(x) = a$$

$$A(x) = a$$

$$\tan\phi(x) = 0$$

$$\eta(x) = a\cos(k_\mathrm{o}x)$$

$$\tilde{\eta}(x) = a\sin(k_\mathrm{o}x)$$

13.6.1 Fourier Representations

Now consider a *Fourier representation* for the complex envelope function:

$$\psi(x) = \sum_{n=0}^{N} a_n\cos(nKx) + b_n\sin(nKx) = a + \sum_{n=1}^{N} a_n\cos(nKx) + b_n\sin(nKx)$$

$$(13.48)$$

where the $n = 0$ term has the constant $a_\mathrm{o} = a$. Here N is the number of components. K is the modulation wavenumber that is related to the length of the periodic interval of the space series, $L = 2\pi/K$; clearly K is the smallest wavenumber in the spectrum, that is, the longest modulation. Now let both a_n and b_n be *complex numbers*:

$$a_n = c_n + id_n, \quad b_n = e_n + if_n$$

where c_n, d_n, e_n, and f_n are real. Thus

$$\psi(x) = a + \sum_{n=1}^{N} (c_n + id_n)\,\cos(nKx) + (e_n + if_n)\,\sin(nKx)$$

and

$$\psi(x) = a + \sum_{n=1}^{N} [c_n\cos(nKx) + e_n\sin(nKx)] + i\sum_{n=1}^{N} [d_n\cos(nKx) + f_n\sin(nKx)]$$

In this way, we have particular Fourier representations for the real and imaginary parts of the complex envelope:

$$\psi_\mathrm{R}(x) = a + \sum_{n=1}^{N} [c_n\cos(nKx) + e_n\sin(nKx)] \equiv a[1 + C(x)]$$

$$\psi_I(x) = \sum_{n=1}^{N} [d_n \, \cos(nKx) + f_n \, \sin(nKx)] \equiv aS(x)$$

This allows us to write ordinary Fourier series for $C(x)$ and $S(x)$:

$$C(x) = \frac{1}{a} \sum_{n=1}^{N} [c_n \, \cos(nKx) + e_n \, \sin(nKx)]$$

$$S(x) = \frac{1}{a} \sum_{n=1}^{N} [d_n \, \cos(nKx) + f_n \, \sin(nKx)] \tag{13.49}$$

It is clear that the Fourier series for $C(x)$ and $S(x)$ make explicit the oscillatory nature of the modulation; in the narrow-banded case, these must of course be small.

Now let us compute the Fourier series representation for the surface elevation using the above expressions for $C(x)$ and $S(x)$. From

$$\eta(x) = a[1 + C(x)] \, \cos(k_o x) - aS(x) \, \sin(k_o x)$$

we have

$$\eta(x) = a\left[1 + \frac{1}{a} \sum_{n=1}^{N} [c_n \, \cos(nKx) + e_n \, \sin(nKx)]\right] \cos(k_o x)$$

$$-a\frac{1}{a} \sum_{n=1}^{N} [d_n \, \cos(nKx) + f_n \, \sin(nKx)] \, \sin(k_o x)$$

$$\eta(x) = a\cos(k_o x) + \sum_{n=1}^{N} [c_n \, \cos(nKx) \, \cos(k_o x) + e_n \, \sin(nKx) \, \cos(k_o x)]$$

$$-\sum_{n=1}^{N} [d_n \, \cos(nKx) \, \sin(k_o x) + f_n \, \sin(nKx) \, \sin(k_o x)]$$

$$\tag{13.50}$$

Then use the following formulas:

$$\cos(nKx) \, \cos(k_o x) = \frac{1}{2} \cos[(k_o - nK)x] + \frac{1}{2} \cos[(k_o + nK)x]$$

$$\sin(nKx) \, \cos(k_o x) = \frac{1}{2} \sin[(k_o + nK)x] - \frac{1}{2} \sin[(k_o - nK)x]$$

$$\cos(nKx) \, \sin(k_o x) = \frac{1}{2} \sin[(k_o + nK)x] + \frac{1}{2} \sin[(k_o - nK)x]$$

$$\sin(nKx) \, \sin(k_o x) = \frac{1}{2} \cos[(k_o - nK)x] - \frac{1}{2} \cos[(k_o + nK)x]$$

Introduce the notation

$$\theta_n^\pm = (k_o \pm nK)x$$

and we have

$$\cos(nKx)\cos(k_o x) = \frac{1}{2}[\cos\theta_n^- + \cos\theta_n^+]$$

$$\sin(nKx)\cos(k_o x) = \frac{1}{2}[\sin\theta_n^+ - \sin\theta_n^-]$$

$$\cos(nKx)\sin(k_o x) = \frac{1}{2}[\sin\theta_n^+ + \sin\theta_n^-]$$

$$\sin(nKx)\sin(k_o x) = \frac{1}{2}[\cos\theta_n^- - \cos\theta_n^+]$$

Then

$$\eta(x) = a\cos(k_o x) + \sum_{n=1}^{N}\left[\frac{c_n}{2}(\cos\theta_n^- + \cos\theta_n^+) + \frac{e_n}{2}(\sin\theta_n^+ - \sin\theta_n^-)\right]$$

$$-\sum_{n=1}^{N}\left[\frac{d_n}{2}(\sin\theta_n^+ + \sin\theta_n^-) + \frac{f_n}{2}(\cos\theta_n^- - \cos\theta_n^+)\right]$$

$$\eta(x) = a\cos(k_o x) + \frac{1}{2}\sum_{n=1}^{N}[(c_n + f_n)\cos\theta_n^- + (c_n - f_n)\cos\theta_n^+$$

$$+(e_n - d_n)\sin\theta_n^+ - (e_n + d_n)\sin\theta_n^-]$$

$$\eta(x) = a\cos(k_o x) + \frac{1}{2}\sum_{n=1}^{N}\{(c_n + f_n)\cos[(k_o - nK)x] + (c_n - f_n)\cos[(k_o + nK)x]\}$$

$$+\frac{1}{2}\sum_{n=1}^{N}\{(e_n - d_n)\sin[(k_o + nK)x] - (e_n + d_n)\sin[(k_o - nK)x]\}$$

To best understand this expression let us look at the spectrum as shown in Figure 13.5, where

$$\eta(x) = \sum_{n=1}^{2N_{\mathrm{osc}}}[g_n\cos(k_n x) + h_n\sin(k_n x)]$$

$$g_n = \frac{c_n + f_n}{2}, \quad h_n = -\left(\frac{e_n + d_n}{2}\right), \quad \text{for } n < N_{\mathrm{osc}}$$

$$g_n = a, \quad h_n = 0, \quad \text{for } n = N_{\mathrm{osc}}$$

$$g_n = \frac{c_n - f_n}{2}, \quad h_n = \left(\frac{e_n - d_n}{2}\right), \quad \text{for } n > N_{\mathrm{osc}}$$

$$(13.51)$$

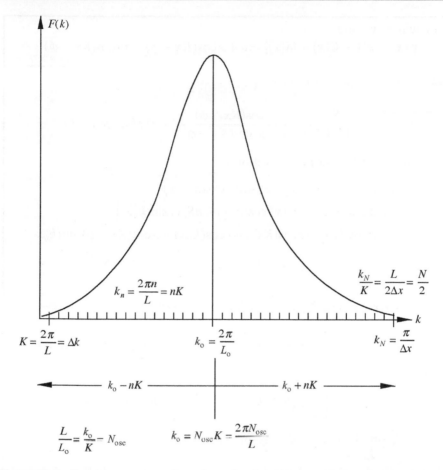

Figure 13.5 Fourier spectrum of envelope that has a many component spectrum. This is typical of ocean wave spectra. The Hilbert transform of the surface wave field shifts this spectrum to the left by k_o, but otherwise leaves the spectrum unchanged.

Example 6—General Form for a Single Fourier Mode in the Complex Envelope Function

Consider this general case when we have only one component in the modulation so that

$$C(x) = \frac{1}{a}[c\cos(Kx) + e\sin(Kx)] = \frac{1}{a}\varepsilon\cos(Kx - \phi), \quad c = \varepsilon\cos\phi, \quad e = \varepsilon\sin\phi$$

$$S(x) = \frac{1}{a}[d\cos(Kx) + f\sin(Kx)] = \frac{1}{a}\rho\cos(Kx - \phi), \quad d = \rho\cos\phi, \quad f = \rho\sin\phi$$

Continued

For which we find

$$\psi(x) = a[1 + C(x) + iS(x)] = a + \varepsilon \cos(Kx - \phi) + i\rho \cos(Kx - \phi)$$

$$A(x) = a\sqrt{[1 + C(x)]^2 + S^2(x)} = a\sqrt{[1 + \varepsilon' \cos(Kx - \phi)]^2 + \rho'^2 \cos^2(Kx - \phi)}$$

$$\tan \phi(x) = \frac{S(x)}{1 + C(x)} = \frac{\varepsilon' \cos(Kx - \phi)}{1 + \rho' \cos(Kx - \phi)}, \quad \varepsilon' = \varepsilon/a, \quad \rho' = \rho/a$$

Therefore, the free surface elevation is

$$\eta(x) = a[1 + C(x)] \cos(k_o x) - aS(x) \sin(k_o x)$$

$$= a\cos(k_o x) + aC(x) \cos(k_o x) - aS(x) \sin(k_o x)$$

$$= a\cos(k_o x) + \varepsilon\cos(Kx - \phi) \cos(k_o x) - \rho\cos(Kx - \phi) \sin(k_o x)$$

Now use

$$\cos(Kx - \phi) \cos(k_o x) = \frac{1}{2} \cos[(k_o - K)x + \phi] + \frac{1}{2} \cos[(k_o + K)x - \phi]$$

$$\sin(k_o x) \cos(Kx - \phi) = \frac{1}{2} \sin[(k_o + K)x - \phi] + \frac{1}{2} \sin[(k_o - K)x + \phi]$$

to get

$$\eta(x) = a\cos(k_o x) + \frac{\varepsilon}{2} \cos[(k_o - K)x + \phi] + \frac{\varepsilon}{2} \cos[(k_o + K)x - \phi]$$
$$- \frac{\rho}{2} \sin[(k_o + K)x - \phi] - \frac{\rho}{2} \sin[(k_o - K)x + \phi] \qquad (13.52)$$

The amplitudes of the sidebands at $k_o \pm K$ are exactly one-half the Fourier amplitudes $(a\varepsilon_i/2)$ in the envelope function $(a\varepsilon_i)$.

Example 7—Driving Unstable Modes with a Wave Maker in the Laboratory Environment

Consider the following example often used to drive the paddle of a wave maker:

$$\eta(x) = a\cos(k_o x) + b_+ \cos[(k_o + K)x - \phi_+] + b_- \cos[(k_o - K)x - \phi_-]$$

Now rearrange this to give coefficients multiplying $\cos(k_o x)$, $\sin(k_o x)$:

$$\eta(x) = [a + b_+ \cos(Kx - \phi_+) + b_- \cos(Kx + \phi_-)] \cos(k_o x)$$
$$- [b_+ \sin(Kx - \phi_+) - b_- \sin(Kx + \phi_-)] \sin(k_o x) \qquad (13.53)$$

But we know in general that

$$\eta(x) = a[1 + C(x)] \cos(k_o x) - aS(x) \sin(k_o x)$$

Hence it is clear that

$$C(x) = \frac{1}{a}[b_+ \cos(Kx - \phi_+) + b_- \cos(Kx + \phi_-)]$$

$$S(x) = \frac{1}{a}[b_+ \sin(Kx - \phi_+) - b_- \sin(Kx + \phi_-)]$$

This is the modulation that makes the most general paddle motion with two unequal side bands, that is, with different amplitudes and phases.

Here is the form of the surface elevation with the trigonometric functions expanded:

$$\eta(x) = [a + b_+ \cos\phi_+ \cos(Kx) + b_- \cos\phi_- \cos(Kx)$$
$$+ b_+ \sin\phi_+ \sin(Kx) - b_- \sin\phi_- \sin(Kx)] \cos(k_o x)$$
$$- [b_+ \cos\phi_+ \sin(Kx) - b_- \cos\phi_- \sin(Kx)$$
$$- b_+ \sin\phi_+ \cos(Kx) - b_- \sin\phi_- \cos(Kx)] \sin(k_o x)$$

This identifies all the coefficients in the Fourier series expansion for $C(x)$, $S(x)$.

$$C(x) = \frac{1}{a}[b_+ \cos\phi_+ \cos(Kx) + b_- \cos\phi_- \cos(Kx)$$
$$+ b_+ \sin\phi_+ \sin(Kx) - b_- \sin\phi_- \sin(Kx)]$$

$$S(x) = \frac{1}{a}[b_+ \cos\phi_+ \sin(Kx) - b_- \cos\phi_- \sin(Kx)$$
$$- b_+ \sin\phi_+ \cos(Kx) - b_- \sin\phi_- \cos(Kx)]$$

We need to check these against the most general coefficients. Simplifying

$$C(x) = \frac{1}{a}[(b_+ \cos\phi_+ + b_- \cos\phi_-) \cos(Kx) + (b_+ \sin\phi_+ - b_- \sin\phi_-) \sin(Kx)]$$

$$S(x) = \frac{1}{a}[(b_+ \cos\phi_+ - b_- \cos\phi_-) \sin(Kx) - (b_+ \sin\phi_+ + b_- \sin\phi_-) \cos(Kx)]$$

$$(13.54)$$

Here then are the Fourier coefficients of the complex modulation in terms of the Fourier parameters used to drive the paddle:

$$c = \frac{1}{a}(b_+ \cos\phi_+ + b_- \cos\phi_-)$$

Continued

$$e = \frac{1}{a}(b_+ \sin \phi_+ - b_- \sin \phi_-)$$

$$d = -\frac{1}{a}(b_+ \sin \phi_+ + b_- \sin \phi_-)$$

$$f = \frac{1}{a}(b_+ \cos \phi_+ - b_- \cos \phi_-)$$

Therefore, in terms of these Fourier modes we can write the modulation functions

$$C(x) = c \cos(Kx) + e \sin(Kx)$$

$$S(x) = f \sin(Kx) + d \cos(Kx)]$$

as we have seen above.

Thus, we have the transformation between the parameters (c, d, e, f) of the complex modulation function and the parameters $(b_+, b_-, \phi_+, \phi_-)$ of the paddle driving equation. The procedure for making waves in the laboratory is then quite simple: Decide the paddle parameters, compute the modulation parameters, and then use IST to get the kinds of unstable modes in the spectrum.

13.7 Initial Modulations for Certain Special Solutions of the NLS Equation

First let us consider the "Peregrine" equation which occurs at $\lambda = ia_o$ in the IST lambda plane:

$$u(x, t) = a \left[1 - \frac{4(1 + 4ia^2 t)}{1 + 16a^4 t^2 + 4a^2 x^2} \right] e^{2ia^2 t}$$

This can be written

$$u(x, t) = a \left[1 - \frac{4}{1 + 16a^4 t^2 + 4a^2 x^2} - \frac{16 i a^2 t}{1 + 16a^4 t^2 + 4a^2 x^2} \right] e^{2ia^2 t}$$

so that the modulation functions are

$$C(x, t) = -\frac{4}{1 + 16a^4 t^2 + 4a^2 x^2}$$

$$S(x,t) = \frac{16a^2t}{1 + 16a^4t^2 + 4a^2x^2}$$

Let us first dimensionalize these expressions. To this end set

$$ax \to \lambda a_0(x - C_gt) = \sqrt{2}k_0^2a_0(x - C_gt)$$

$$a^2t \to \lambda^2 a_0^2\mu t = \frac{k_0^2a_0^2\omega_0}{4}t$$

and we get dimensional forms for

$$C(x,t) = -\frac{4}{1 + (k_0^2a_0^2\omega_0t)^2 + (2\sqrt{2}k_0^2a_0(x - C_gt))^2}$$

$$S(x,t) = \frac{4k_0^2a_0^2\omega_0t}{1 + (k_0^2a_0^2\omega_0t)^2 + [2\sqrt{2}k_0^2a_0(x - C_gt)]^2}$$

Now let us construct the free surface elevation (keeping the temporal part, since it is not appropriate here to set $t=0$, since the state is never small):

$$\eta(x,t) = a[1 + C(x)]\cos(k_0x - \omega_0t) - aS(x)\sin(k_0x - \omega_0t)$$

and we find

$$\eta(x,t) = a\left[1 - \frac{4}{1 + (k_0^2a_0^2\omega_0t)^2 + [2\sqrt{2}k_0^2a_0(x - C_gt)]^2}\right]\cos(k_0x - \omega_0t)$$

$$- \left[a\frac{4k_0^2a_0^2\omega_0t}{1 + (k_0^2a_0^2\omega_0t)^2 + [2\sqrt{2}k_0^2a_0(x - C_gt)]^2}\right]\sin(k_0x - \omega_0t)$$

$$(13.55)$$

This latter expression is what should be used to move a wave paddle to produce the Peregrine solitary wave! Of course we can carry out this same approach for any of the other special solutions of the NLS equation.

Part Six

Theoretical Computation of the Riemann Spectrum

We have stated that to understand a particular, integrable nonlinear wave equation we need: (1) a Hirota transformation, (2) the Riemann spectrum, and (3) the theta function. This section is dedicated to three methods for determining the Riemann spectrum for a particular nonlinear wave equation. These are (1) algebraic-geometric loop integrals (Chapter 14), (2) Schottky uniformization (Chapter 15), and (3) the approach of Nakamura and Boyd (Chapter 16). Methods (1) and (2) arise from the field of algebraic geometry, the third is purely algebraic and is my preferred way to "quickly" determine the Riemann spectrum for a particular nonlinear wave equation. The methods from algebraic geometry also play an important role in this book.

Some perspective is in order here. You can probably rely on someone else to get the Hirota transformation for you (see Hirota, 2004). Now one seeks the Riemann spectrum. The linear Fourier spectrum of a linear wave equation consists of the amplitudes and phases for a particular data set or for a particular Cauchy initial value problem. All the Fourier components have the same mathematical form, that is, they are sine waves. In the nonlinear problem the (Riemann) spectrum has Fourier modes that may vary drastically from sine waves. Furthermore, the *random phase approximation* may well work, but there may also be *constraints* on the phases that require *phase locking* (think of the Stokes wave or the unstable mode solution of the nonlinear Schroedinger (NLS) equation). Therefore, while *any Fourier spectrum for a linear wave equation solves the equation* for the linear problem, only a *particular Riemann spectrum solves the nonlinear wave equation*.

A simple way to think of the problem is this. Suppose you want to arbitrarily pick a solution of a nonlinear wave equation by choosing the Riemann spectrum. First you must choose the diagonal elements of the Riemann matrix. Then the off-diagonal elements and the nonlinear frequencies (and possibly also the phases) must be computed so that the solution *lies on a Riemann surface*; this is the so-called Novikov conjecture. Each of the three chapters of this section provides different ways to compute the off-diagonal elements given the diagonal elements (and other parameters) so that the Novikov conjecture is met. Of course all three methods must give the same answer and, therefore, which method you might want to use is a matter of perspective or taste. Finally, there may appear, as with the NLS equation, nonlinear modes that require additional structure in the Riemann matrix. For example, in NLS equation you can find a two-by-two submatrix, centered on the diagonal, which corresponds to a single "unstable mode," often referred to as a "rogue wave" in a measured wave train. In this case you will also find that the phases can be very particular (Chapters 12, 18, 24, 29). The occurrence of a two-by-two submatrix also occurs for vortex dynamics (Chapter 27). An amazing thing about the formulation is that you will always find, after consideration for any coherent structures present, that you can still apply the random phase approximation to the nonlinear modes, a useful result for many oceanographic applications.

14 Algebraic-Geometric Loop Integrals

14.1 Introduction

The aim of this chapter is to discuss the mathematical formulation for computing the wavenumber, frequency, phase, and period matrix of a solution to the KdV equation beginning with the Floquet spectrum of the Schrödinger eigenvalue problem (Dubrovin and Novikov, 1976) (Chapter 17). The formulation is in terms of the so-called *loop integrals* of algebraic geometry. This chapter emphasizes results that will aid in the development of a program for computing the loop integrals. The actual description of the computer program is given in Chapter 19. The references for this chapter are Dubrovin and Novikov (1975a,b), Dubrovin et al. (1976), and Dubrovin (1981); see also Baker (1897), Ablowitz and Segur (1981) and Belokolos et al. (1994).

14.2 The Theta-Function Solutions to the KdV Equation

The general solution to the KdV equation may be written in terms of the Riemann *theta function*

$$\lambda \eta(x, t) = 2 \frac{\partial^2}{\partial x^2} \ln \theta_N(X_1, X_2, \ldots, X_N) \tag{14.1}$$

where

$$\theta_N(X_1, X_2, \ldots, X_N) = \sum_{m_1=-\infty}^{\infty} \sum_{m_2=-\infty}^{\infty} \cdots \sum_{m_N=-\infty}^{\infty} \exp\left[i \sum_{k=1}^{N} m_k X_k + \frac{1}{2} \sum_{j=1}^{N} \sum_{k=1}^{N} m_j m_k B_{jk} \right] \tag{14.2}$$

The theta function is 2π periodic in each of the N phases X_j

$$\theta[(X_1 + 2\pi), (X_2 + 2\pi), \ldots, (X_N + 2\pi)] = \theta[X_1, X_2, \ldots, X_N] \tag{14.3}$$

Here N is the number of degrees of freedom in a particular solution to the KdV equation. $\mathbf{B} = \{B_{ij}\}$ is the period or Riemann matrix. The vector $\mathbf{m} = [m_1, m_2, \ldots, m_N]$ is a vector of the integers, $\mathbf{m} \in \mathbb{Z}$. Here the B_{ij} are negative definite (see discussion below); this fact, insures convergence of Equation (14.2). The phases in Equation (14.2) are given by:

$$X_j = k_j x - \omega_j t + \phi_j, \quad 1 \leq j \leq N \tag{14.4}$$

Here k_j are the wavenumbers, ω_j are the frequencies, and the ϕ_j are the phases.

This section summarizes the determination of the wavenumbers k_j, frequencies ω_j, phases ϕ_j, and the period or interaction matrix $\mathbf{B}$ in the theta-function solution to KdV equations (14.1)–(14.4) (see Chapter 10). The formulation discussed here is the N dimensional generalization of the classical Jacobian elliptic integrals (Abramowitz and Stegun, 1964). Actual derivation of the loop integrals and other aspects of the theory for the solution of nonlinear wave equations are discussed in Baker (1897), Dubrovin and Novikov (1975a,b), Dubrovin, Matveev, and Novikov (1976), Ablowitz and Segur (1981), Dubrovin (1981), and Belokolos et al. (1994).

14.2.1   Holomorphic Differentials

Normalized *holomorphic differentials* on the Riemann surface Γ are first introduced:

$$d\Omega_m(E) = \sum_{k=1}^{N} C_{km} \frac{E^{k-1} dE}{R^{1/2}(E)} \tag{14.5}$$

where $R(E)$ is given by

$$R(E) = \prod_{n=1}^{2N+1} (E - E_n) \tag{14.6}$$

Note that the square root of $R(E)$ may be written as:

$$R^{1/2}(E) \rightarrow \sigma(E)\delta(E)|R(E)|^{1/2} \tag{14.7}$$

where $\sigma(E) = \pm 1$ is the sign of the square root of $R(E)$ and $\delta(E) = -i$ or 1 depending upon the sign of the polynomial $R(E)$. The σ arise from the main spectrum (they have unique values inside each open band) and the δ are found by looking at the behavior of the $R(E)$ in a particular band. It is worth establishing some nomenclature. A *band* lies between two successive eigenvalues of the main spectrum. An *open band* (or *forbidden band*) corresponds to intervals on E for which the half-trace of the monodromy matrix is greater (or less) than 1. A *gap* lies between the open bands (also known as *allowed bands*) where the half-trace is less than 1.

The following *normalization condition* is assumed to hold:

$$\oint_{\alpha_j} d\Omega_m(E) = 2\pi i \delta_{jm} \tag{14.8}$$

These are the "α_j-cycles" or contour integrals around the open bands (E_{2j}, E_{2j+1}) in the Floquet spectrum. Combining Equations (14.8) and (14.5) gives:

$$\oint_{\alpha_j} d\Omega_m(E) = \sum_{k=1}^{N} C_{km} \oint_{\alpha_j} \frac{E^{k-1} dE}{R^{1/2}(E)} = 2\pi i \delta_{jm}$$

so that

$$\oint_{\alpha_j} d\Omega_m(E) = \sum_{k=1}^{N} J_{jk} C_{km} = 2\pi i \delta_{jm}$$

$$J_{jk} = \oint_{\alpha_j} \frac{E^{k-1} dE}{R^{1/2}(E)} \tag{14.9}$$

This latter can be written in matrix notation

$$\mathbf{JC} = 2\pi i \mathbf{1}, \quad \mathbf{C} = 2\pi i \mathbf{J}^{-1} \tag{14.10}$$

The normalization coefficients C_{nm} in Equation (14.10) are then given by:

$$C_{jk} = 2\pi i \left[\oint_{\alpha_j} \frac{E^{k-1} dE}{R^{1/2}(E)} \right]^{-1}$$

We now need to evaluate the contour integrals. We use particular contours about the open bands (E_{2j}, E_{2j+1}). The contours (Figure 14.1) are taken in the counterclockwise direction:

$$\oint_{\alpha_j} \frac{E^{k-1} dE}{R^{1/2}(E)} = \int_{\substack{\text{right} \\ \text{circle}}} \frac{E^{k-1} dE}{R^{1/2}(E)} + \int_{\substack{\text{upper} \\ \text{line}}} \frac{E^{k-1} dE}{R^{1/2}(E)} + \int_{\substack{\text{left} \\ \text{circle}}} \frac{E^{k-1} dE}{R^{1/2}(E)} + \int_{\substack{\text{lower} \\ \text{line}}} \frac{E^{k-1} dE}{R^{1/2}(E)}$$

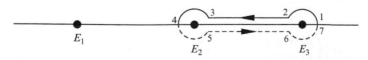

Figure 14.1 Alpha cycle for a one degree-of-freedom solution to the KdV equation.

Now, the integrals about the circles go to zero as their radius goes to zero. That leaves us with the upper and lower lines. These are written explicitly:

$$\oint_{\alpha_j} \frac{E^{k-1}dE}{R^{1/2}(E)} = \int_{\substack{E_{2j+1} \\ \text{upper}}}^{E_{2j}} \frac{E^{k-1}dE}{R^{1/2}(E)} + \int_{\substack{E_{2j} \\ \text{lower}}}^{E_{2j+1}} \frac{E^{k-1}dE}{R^{1/2}(E)}$$

$$= -\int_{\substack{E_{2j} \\ \text{upper}}}^{E_{2j+1}} \frac{E^{k-1}dE}{R^{1/2}(E)} + \int_{\substack{E_{2j} \\ \text{lower}}}^{E_{2j+1}} \frac{E^{k-1}dE}{R^{1/2}(E)}$$

To evaluate these, we recognize that (E_{2j}, E_{2j+1}) are branch points and that a branch cut connects them. Thus, one expects a phase difference in the integrand above and below the branch cut. This arises because:

$$R(E) = |R(E)|e^{i\phi}$$

$$R^{-1/2}(E) = |R(E)|^{-1/2}e^{-i\phi/2} = \sigma^{-1}(E)^{-1}\delta(E)|R(E)|^{-1/2}$$

where $\sigma(E) = \pm 1$ is the sign of the square root of $R(E)$ and $\delta(E) = -i$ or 1 depending upon the sign of the polynomial $R(E)$. The factors σ, δ are related to the phase in the square root of the polynomial:

$$e^{i\phi/2} = \sigma(E)\delta(E)$$

For both upper and lower lines we find $\delta = -i$. On the upper line $\sigma = 1$ while on the lower line $\sigma = -1$. Explicitly on the *upper line* we have ($e^{i\phi/2} = \sigma(E)\delta(E) = -i$):

$$R^{1/2}(E) \rightarrow -i|R(E)|^{1/2}$$

and on the *lower line* ($e^{i\phi/2} = \sigma(E)\delta(E) = i$):

$$R^{1/2}(E) \rightarrow i|R(E)|^{1/2}$$

and we find:

$$\oint_{\alpha_j} \frac{E^{k-1}dE}{R^{1/2}(E)} = -2i\int_{E_{2j}}^{E_{2j+1}} \frac{E^{k-1}dE}{|R(E)|^{1/2}} .$$

So that finally:

$$C_{jk} = 2\pi i \left[\oint_{\alpha_j} \frac{E^{k-1}dE}{R^{1/2}(E)}\right]^{-1} = -\pi \left[\int_{E_{2j}}^{E_{2j+1}} \frac{E^{k-1}dE}{|R(E)|^{1/2}}\right]^{-1} \tag{14.11}$$

Note that each loop integral has been reduced to an ordinary definite integral across an open band in the Floquet spectrum. Since the integrals are positive definite the matrix C is *negative definite*, ensuring that the B matrix is also negative definite, which insures convergence of the theta functions.

A perhaps more generic form for Equation (14.11) is given by:

$$C = 2\pi K^{-1}$$

$$C_{jk} = 2\pi \left[-2 \int_{E_{2j}}^{E_{2j+1}} \frac{E^{k-1} dE}{|R^{1/2}(E)|} \right]^{-1} \tag{14.12a}$$

$$K_{jk} = -2 \int_{E_{2j}}^{E_{2j+1}} \frac{E^{k-1} dE}{|R^{1/2}(E)|} \tag{14.12b}$$

These latter expressions look like the discussion in Abramowitz and Stegun (1964) of the one degree-of-freedom case, where the function $K(m)$ has been replaced by the matrix K. Equations (14.12a) and (14.12b) are the starting points for programming the loop integrals (Chapter 19).

Here is an interlude, based upon a single α cycle. The following results verify what I showed above, but they also give an additional perspective that is useful for programming this problem. First note that we can write the polynomial in radial form:

$$R(E) = |R(E)| e^{i\phi}$$

What we are instead interested in is

$$\frac{1}{R^{1/2}(E)} = \frac{1}{|R(E)|^{1/2} e^{i\phi/2}}$$

where

$$e^{i\phi/2} = \sigma(E)\delta(E) = \cos(\phi/2) + i \sin(\phi/2)$$

It is clear that we will find $\delta(E) = 1$ or i and that $\sigma(E) = \pm 1$. To elaborate on this, consider the one degree-of-freedom case where we have three eigenvalues so that

$$R(E) = (E - E_1)(E - E_2)(E - E_3)$$

Now we have the three ordered eigenvalues E_1, E_2, E_3. In the gap (E_1, E_2) we have

$$R(E) = (E - E_1)(E_2 - E)(E_3 - E)$$

so that

in the *gap* (E_1, E_2) we have $R^{1/2}(E) = \sigma |R(E)|^{1/2}$, $\quad \delta = 1$

In the open band (E_2, E_3) we have

$$R(E) = (E - E_1)(E - E_2)(E_3 - E) = -(E - E_1)(E - E_2)(E - E_3)$$

so that

in the *open band* (E_2, E_3) we have $R^{1/2}(E) = i\sigma |R(E)|^{1/2}$, $\quad \delta = i$

We have still to compute the σ. Based on the polar form given above

---

On the upper part of the gap contour $\sigma\delta = 1$ for $\phi = 0$ and therefore $\sigma = 1$
On the upper part of the open band contour $\sigma\delta = i$ for $\phi = \pi$ and therefore $\sigma = 1$
and
On the lower part of the gap contour $\sigma\delta = -1$ for $\phi = \pi$ and therefore $\sigma = -1$
On the lower part of the open band contour $\sigma\delta = -i$ for $\phi = 2\pi$ and therefore
$\quad \sigma = -1$

This says that the Riemann sheet index is 1 above and -1 below the branch cut. Now consider a typical α cycle about the open band:

The aim now is to examine the phase as we go from points 1-7 (see Figure 14.1) around the closed contour of the open band (E_2, E_3) for the square root of the polynomial, $R^{-1/2}$ *(E)*. At point 1 the phase is 0 and we have $\sigma = 1$, so that we are on the first of the 2 Riemann sheets. At point 2, the phase is $\pi/2$ and still $\sigma = 1$. At point 3, the phase is the same as at point 2. At point 4, we pick up another $\pi/2$ and we now have a total phase of π. At this angle (plus some small ε), we have $\sigma = -1$ and we are on the second of the 2 Riemann sheets. At point 5, we have gained another phase jump of $\pi/2$ and arrive at a total phase of 0 (minus some small ε) and we still (thanks to the fact that the angle lies between π, 2π) have $\sigma = -1$. The fact that σ changes signs from above to below the branch cut means we get a factor of two times the usual integral from one branch point to the other.

Finally, we get no contribution from the circles around each of the two branch points, but we get twice the integral from one branch point to the other. This is consistent with the calculation made at the first part of this section.

14.2.2 Phases of the Theta Functions

The *phases* X_j of the theta function in Equation (14.2) are found by the following Abelian integrals:

$$X_m(P_1, P_2, \ldots, P_m) = i \sum_{j=1}^{N} \int_{E_{2j}}^{P_j(x,t)} d\Omega_m(E) = k_j x - \omega_j t + \phi_j \qquad (14.13)$$

where ω_j is given by the second of Equation (14.14) below and $P_m(x, t) = [\mu_m(x, t), \sigma_m]$ for $1 \leq m \leq j$. Equation (14.13) may be interpreted as a *linearization of the hyperelliptic function representation of the flow*, that is, integration over the holomorphic differentials (Equation (14.5)) from the lower band edge E_{2j} to the hyperelliptic functions $\mu_j(x, t)$ in effect linearizes the μ_j (intrinsically nonlinear functions which provide the solution to KdV through a linear superposition law (see Section 10.3)) (Dubrovin and Novikov, 1975a,b). This leads to the linear theta-function inverse problem for the KdV equation. Equations (14.2) and (14.13) are an *Abel transform pair*. Generally speaking the phase of the hyperelliptic functions X_j (Equation (14.13)) depends upon the main spectrum (E_i, $1 \leq i \leq 2N + 1$) and the space-time evolution of the auxiliary spectrum $[\mu_j(x, t), \sigma_j]$, $1 \leq j \leq N$.

It then follows that the wavenumbers k_j, and frequencies ω_j are given by

$$k_j = 2C_{N,j} \qquad (14.14)$$

$$\omega_j = 8C_{N-1,j} + 4C_{N,j} \sum_{i=1}^{2N+1} E_i \qquad (14.15)$$

Both k_j and ω_j are real constants since the C_{jm} and the E_k are real constants. The usual dispersion law for a single degree of freedom may easily be obtained. The k_j are *commensurable wavenumbers* in the cycle integral basis considered here, while the frequencies ω_j are generally *incommensurable*.

The phases ϕ_j are found by using Equation (14.5) in Equation (14.13) and then setting $x = 0$, $t = 0$ to get:

$$\phi_m = -i \sum_{j=1}^{N} \int_{E_{2j}}^{P_j(0,0)} d\Omega_m(E) = -i \sum_{j=1}^{N} \sum_{k=1}^{N} C_{km} \int_{E_{2j}}^{P_j(0,0)} \frac{E^{k-1} dE}{R^{1/2}(E)} = -i \sum_{j=1}^{N} \sum_{k=1}^{N} F_{jk} C_{km}$$

$$(14.16)$$

where

$$F_{jk} = \int_{E_{2j}}^{P_j(0,0)} \frac{E^{k-1} dE}{R^{1/2}(E)} \qquad (14.17)$$

and C_{mj} is given by Equation (14.12) and the phases have the form:

$$\phi_m = \sum_{j=1}^{N} E_{jm}, \quad \mathbf{E} = -i\,\mathbf{FC}, \quad \mathbf{C} = 2\pi\mathbf{K}^{-1} \tag{14.18}$$

Thus, since $P_m(0,0) = [\mu_m(0,0), \sigma_m]$, the *constant phases* ϕ_j of the hyperelliptic functions depend upon the *starting values of the hyperelliptic functions* $\mu_j(0,0)$ *and the Riemann sheet indices* σ_j.

Equation (14.17) can also be estimated by:

$$F_{jk} = \frac{2\sigma_j}{\delta_j} \int_{E_{2j}}^{\mu_j(0,0)} \frac{E^{j-1}dE}{|R(E)|^{1/2}} = -2i\sigma_j \int_{E_{2j}}^{\mu_j(0,0)} \frac{E^{j-1}dE}{|R(E)|^{1/2}} \tag{14.19}$$

where in the latter $1/\delta_j = -i$.

14.2.3   The Period Matrix

The *period matrix* is given by:

$$B_{jk} = B_{kj} = \oint_{\beta_k} d\Omega_j(E) = \sum_{i=1}^{N} C_{ij} \oint_{\beta_k} \frac{E^{i-1}dE}{R^{1/2}(E)} \tag{14.20}$$

where the integrals are over the "β-cycles" of the theory. In vector notation we have

$$\mathbf{B} = \mathbf{K}'\mathbf{C} \tag{14.21}$$

A notation more familiar can be introduced by setting

$$\mathbf{C} = 2\pi\mathbf{K}^{-1} \tag{14.22}$$

The period matrix then has the form

$$\mathbf{B} = 2\pi\mathbf{K}'\mathbf{K}^{-1} \tag{14.23}$$

where

$$K'_{ki} = \oint_{\beta_k} \frac{E^{i-1}dE}{R^{1/2}(E)} \tag{14.24}$$

Now I discuss the β_k integrals: Here the branch cuts are made in the gaps of the main spectrum (see Figure 14.1). First note

$$K'_{ki} = \oint_{\beta_k} \frac{E^{i-1}dE}{R^{1/2}(E)} = \oint_{\text{gap1}} \frac{E^{i-1}dE}{R^{1/2}(E)} + \oint_{\text{band1}} \frac{E^{i-1}dE}{R^{1/2}(E)} + \oint_{\text{gap2}} \frac{E^{i-1}dE}{R^{1/2}(E)} + \oint_{\text{band2}} \frac{E^{i-1}dE}{R^{1/2}(E)}$$

(14.25)

Use the fact that

$$R^{1/2}(E) = \sigma(E)\delta(E)|R(E)|^{1/2}$$

(14.26)

and find that $\sigma_j = +1$ on the upper line, $\sigma_j = -1$ on the lower line (note that $\sigma_j = 1/\sigma_j$) and $\delta_j = 1$:

$$\oint_{\text{gap1}} \frac{E^{i-1}dE}{R^{1/2}(E)} = \frac{2\sigma}{\delta} \int_{E_1}^{E_2} \frac{E^{i-1}dE}{|R(E)|^{1/2}} = 2\int_{E_1}^{E_2} \frac{E^{i-1}dE}{|R(E)|^{1/2}}$$

(14.27a)

where I used $\sigma = \delta = 1$. Hence

$$\oint_{\text{gap2}} \frac{E^{i-1}dE}{R^{1/2}(E)} = \frac{2\sigma}{\delta} \int_{E_3}^{E_4} \frac{E^{i-1}dE}{|R(E)|^{1/2}} = 2\int_{E_3}^{E_4} \frac{E^{i-1}dE}{|R(E)|^{1/2}}$$

(14.27b)

$$\oint_{\text{band1}} \frac{E^{i-1}dE}{R^{1/2}(E)} = -2\sigma i \int_{E_2}^{E_3} \frac{E^{i-1}dE}{|R(E)|^{1/2}} = 0$$

(14.27c)

$$\oint_{\text{band2}} \frac{E^{i-1}dE}{R^{1/2}(E)} = 0$$

(14.27d)

where the last two integrals result because the real part of the integrand is zero. Therefore, the matrix $\mathbf{K}'$ is found from the sum of the gap integrals.

14.2.4 One Degree of Freedom

Consider a single degree of freedom, $N = 1$. For the theta function set $X_1 = X$, $B_{11} = b$:

$$\theta(X) = \sum_{n=-\infty}^{\infty} \exp\left[inX + \frac{1}{2}bn^2\right]$$

(14.28)

Let $q = \exp(-b/2)$ be the *nome* so that:

$$\theta_3(X) = \theta_3(x, q) = 1 + 2\sum_{n=1}^{\infty} q^{n^2} \cos(nkx)$$

(14.29)

where $X = kx$, $k = 2\pi/L$.

14.2.5   Notation for Classical Jacobian Integrals

The standard notation for the classical Jacobian integrals is (Abramowitz and Stegun, 1964):

$$b = 2\pi K'/K, \quad q = e^{-b/2} = e^{-\pi K'/K} \tag{14.30}$$

where

$$K(m) = K = \int_0^1 [(1 - t^2)(1 - mt^2)]^{-1/2} dt = \int_0^{\pi/2} \frac{d\theta}{[1 - m \sin^2 \theta]^{1/2}} \tag{14.31}$$

$$K'(m) = K(m_1) = \int_0^1 [(1 - t^2)(1 - m_1 t^2)]^{-1/2} dt = \int_0^{\pi/2} \frac{d\theta}{[1 - m_1 \sin^2 \theta]^{1/2}} \tag{14.32}$$

with $m + m_1 = 1$ and $K(m) = K'(m_1) = K'(1 - m)$. The trigometric forms of the above integrals arise by setting $t = \sin\theta$.

The solution to the KdV equation is then

$$\lambda\eta(x, t) = 2 \frac{\partial^2}{\partial x^2} \ln \theta_3(x, q) \tag{14.33}$$

so that

$$\lambda\eta(x, t) = 4k^2 \sum_{n=1}^{\infty} \frac{nq^n}{1 - q^{2n}} \cos(nkx) \tag{14.34}$$

This is the usual travelling (cnoidal) wave solution to the KdV equation.

14.2.6   Notation for the Theta-Function Formulation

These same expressions may be found from the theta-function formulation (Equations (14.5) and (14.6)). For the one degree-of-freedom case we have the polynomial $P(E) = (E - E_1)(E - E_2)(E_3 - E)$ for which:

$$C_{11} = \frac{2\pi}{K_{11}} = \pi \left[\int_{E_2}^{E_3} \frac{dE}{\sqrt{(E - E_1)(E - E_2)(E_3 - E)}} \right]^{-1} \tag{14.35}$$

The *wavenumber* is given by $k_1 = 4\pi/K_{11}$:

$$k_1 = \frac{2\pi}{\int_{E_2}^{E_3} \frac{dE}{\sqrt{(E - E_1)(E - E_2)(E_3 - E)}}} = \Delta k \tag{14.36}$$

The right-hand side of Equation (14.36) derives from the fact that the commensurable wavenumbers are given by

$$k_n = n\Delta k, \quad \Delta k = \frac{2\pi}{L}$$

Therefore

$$L = \int_{E_2}^{E_3} \frac{dE}{\sqrt{(E - E_1)(E - E_2)(E_3 - E)}}$$

is the *spatial period* of the wave train. Now by the first part of Equation (14.7), $k_j = 2C_{N,j}$, we have

$$C_{N,j} = \frac{k_j}{2} = \frac{j\pi}{L}$$

This latter expression provides numerical control of the elements of the matrices C, J, K. In one dimension

$$C_{11} = \frac{k_1}{2} = \frac{\pi}{L}$$

while in two dimensions

$$C_{21} = \frac{k_1}{2} = \frac{\pi}{L} \Rightarrow k_1 = \frac{2\pi}{L}$$

$$C_{22} = \frac{k_2}{2} = \frac{2\pi}{L} \Rightarrow k_2 = \frac{4\pi}{L}$$

So we have commensurable wavenumbers by construction for a periodic wave train. Here I have used $C = 2\pi K^{-1}$. Of course to have commensurable wavenumbers you must have the correct main spectrum, otherwise the wave train is quasi-periodic.

Now compute the *frequency* ω_1:

$$\omega_1 = 2(E_1 + E_2 + E_3)\sqrt{E_3 - E_1} = ck$$

where

$$c = 2(E_1 + E_2 + E_3)$$

$$k = \sqrt{E_3 - E_1}$$

For one degree of freedom, we have for the phase of the wave:

$$\phi_1 = -i\Phi_{11}C_{11}$$

$$C_{11} = \frac{\pi}{L}$$

$$\Phi_{11} = \sigma_1\delta_1 \int_{E_2}^{\mu_1(0,0)} \frac{dE}{R^{1/2}(E)}$$

and

$$\phi_1 = -i\frac{\sigma_1\delta_1\pi}{L} \int_{E_2}^{\mu_1(0,0)} \frac{dE}{R^{1/2}(E)}$$

To evaluate this further note that

$$R = (E - E_1)(E - E_2)(E - E_3) = -(E - E_1)(E - E_2)(E_3 - E)$$

Therefore

$$R^{1/2}(E) = i|R(E)|^{1/2}$$

So that $\delta_1=-i$ and

$$\phi_1 = -\frac{\sigma_1\pi}{L} \int_{E_2}^{\mu_1(0,0)} \frac{dE}{|R(E)|^{1/2}}$$

The minus sign means we could set $kx - \omega t + \phi \to kx - \omega t - \phi$ and then use the plus sign in the above equation. It is clear that the following results hold:

$$\phi_1 = -\frac{\sigma_1\pi}{L}L = -\sigma_1\pi, \quad \text{for } \mu_1(0,0) = E_3$$

$$\phi_1 = 0, \quad \text{for } \mu_1(0,0) = E_2$$

$$\phi_1 = -\sigma_1\frac{\pi}{2}, \quad \text{for } \mu_1(0,0) = (E_2 + E_3)/2$$

The interaction matrix B_{ij} from Equations (14.22)–(14.24) is:

$$B_{11} = 2\pi K'_{11}K_{11}^{-1} = \frac{4\pi}{K_{11}} \int_{E_1}^{E_2} \frac{dE}{R^{1/2}(E)} \qquad (14.37)$$

and using Equation (14.8)

$$B_{11} = \frac{2\pi \int_{E_1}^{E_2} \frac{dE}{\sqrt{(E-E_1)(E_2-E)(E_3-E)}}}{\int_{E_2}^{E_3} \frac{dE}{\sqrt{(E-E_1)(E-E_2)(E_3-E)}}} = 2\pi \frac{K'_{11}}{K_{11}} = 2\pi \frac{K'(m)}{K(m)} \tag{14.38}$$

for $K_{11} = K$, $K'_{11} = K'$. Note that:

$$m = \frac{E_3 - E_2}{E_3 - E_1}$$
$$m_1 = \frac{E_2 - E_1}{E_3 - E_1} = 1 - m \tag{14.39}$$

These results bring out another form of the elliptic integrals which is more natural for work with the periodic scattering transform:

$$K_{11} = K(m) = \frac{1}{2}\sqrt{E_3 - E_1} \int_{E_2}^{E_3} \frac{dE}{\sqrt{(E - E_1)(E - E_2)(E_3 - E)}}$$
$$K'_{11} = K'(m) = \frac{1}{2}\sqrt{E_3 - E_1} \int_{E_1}^{E_2} \frac{dE}{\sqrt{(E - E_1)(E_2 - E)(E_3 - E)}} \tag{14.40}$$

Note that the factor $\frac{1}{2}\sqrt{E_3 - E_1}$ before the integrals is not normally retained in Equations (14.12b) and (14.24) since it cancels out in the ratio K'_{11}/K_{11} (Equation (14.23)). The first part of Equation (14.40) is easily transformed into Equation (14.31) by

$$E = E_3 - (E_3 - E_2)t^2, \quad dE = -2(E_3 - E_2)t\, dt \tag{14.41}$$

for which the polynomial has the form

$$P(E) = (E - E_1)(E - E_2)(E_3 - E) = (E_3 - E_1)(E_3 - E_2)^2 t^2 (1 - t^2)(1 - mt^2) \tag{14.42}$$

Further substitution of $t = \sin\theta$ leads to the second part of Equation (14.31). Then the second part of Equation (14.40) is transformed into Equation (14.32) by

$$E = E_1 + (E_2 - E_1)t^2, \quad dE = 2(E_2 - E_1)t\, dt \tag{14.43}$$

with

$$P(E) = (E - E_1)(E_2 - E)(E_3 - E) = (E_3 - E_1)(E_2 - E_1)^2 t^2 (1 - t^2)(1 - m_1 t^2) \tag{14.44}$$

14.3 On the Possibility of "Interactionless" Potentials for the Two Degree-of-Freedom Case

Let us now consider the possibility of finding a solution for the KdV equation without nonlinear interactions. To this end suppose $C = 2\pi K^{-1}$ and K' are two-by-two real matrices, so that the C matrix is given by:

$$C = 2\pi K^{-1} = \frac{2\pi}{\det K} \begin{bmatrix} K_{22} & -K_{12} \\ -K_{21} & K_{11} \end{bmatrix} \tag{14.45}$$

with $\det K = K_{11}K_{22} - K_{12}K_{21}$. Then:

$$\begin{aligned} B = 2\pi K'K^{-1} &= \frac{2\pi}{\det K} \begin{bmatrix} K'_{11} & K'_{12} \\ K'_{21} & K'_{22} \end{bmatrix} \begin{bmatrix} K_{22} & -K_{12} \\ -K_{21} & K_{11} \end{bmatrix} \\ &= \frac{2\pi}{\det K} \begin{bmatrix} K'_{11}K_{22} - K'_{12}K_{21} & K'_{12}K_{11} - K'_{11}K_{12} \\ K'_{21}K_{22} - K'_{22}K_{21} & K'_{22}K_{11} - K'_{21}K_{12} \end{bmatrix} \end{aligned} \tag{14.46}$$

Since the period matrix is symmetric

$$B_{12} = B_{21} \tag{14.47}$$

then it is clear that

$$K'_{12}K_{11} - K'_{11}K_{12} = K'_{21}K_{22} - K'_{22}K_{21} \tag{14.48}$$

It is now worthwhile analyzing the two degree-of-freedom case. We have the polynomial:

$$R(E) = (E - E_1)(E - E_2)(E - E_3)(E - E_4)(E - E_5)$$

- In the first open band (2, 3):

$$R(E) = -(E - E_1)(E - E_2)(E_3 - E)(E_4 - E)(E_5 - E)$$

Therefore $\delta = -i$.

- In the second open band (4, 5):

$$R(E) = -(E - E_1)(E - E_2)(E_3 - E)(E_4 - E)(E_5 - E)$$

Therefore $\delta = -i$.

- In the first gap (1, 2):

$$R(E) = (E - E_1)(E_2 - E)(E_3 - E)(E_4 - E)(E_5 - E)$$

Therefore $\delta = 1$.

- In the second gap $(3, 4)$:

$$R(E) = (E - E_1)(E - E_2)(E - E_3)(E_4 - E)(E_5 - E)$$

Therefore $\delta = 1$.

The **K** matrix elements are given specifically by:

$$K_{11} = 2 \int_{E_2}^{E_3} \frac{dE}{|R(E)|^{1/2}}, \quad K_{12} = 2 \int_{E_4}^{E_5} \frac{dE}{|R(E)|^{1/2}}$$

$$K_{21} = 2 \int_{E_2}^{E_3} \frac{EdE}{|R(E)|^{1/2}}, \quad K_{22} = 2 \int_{E_4}^{E_5} \frac{EdE}{|R(E)|^{1/2}} \tag{14.49}$$

While the **K′** matrix elements are:

$$K'_{11} = 2 \int_{E_1}^{E_2} \frac{dE}{R^{1/2}(E)}, \quad K'_{12} = 2 \int_{E_1}^{E_4} \frac{dE}{R^{1/2}(E)}$$

$$K'_{21} = 2 \int_{E_1}^{E_2} \frac{EdE}{R^{1/2}(E)}, \quad K'_{22} = 2 \int_{E_1}^{E_4} \frac{EdE}{R^{1/2}(E)} \tag{14.50}$$

The phases have the form:

$$\Phi_1 = \int_{E_2}^{\mu_1(0,0)} \frac{dE}{R^{1/2}(E)}, \quad \Phi_2 = \int_{E_4}^{\mu_2(0,0)} \frac{EdE}{R^{1/2}(E)} \tag{14.51}$$

In the above expressions, we use the fifth-order polynomial:

$$R(E) = (E - E_1)(E - E_2)(E - E_3)(E - E_4)(E - E_5) \tag{14.52}$$

In the open band (E_2, E_3), we have:

$$R(E) = -(E - E_1)(E - E_2)(E_3 - E)(E_4 - E)(E_5 - E)$$

and in (E_4, E_5)

$$R(E) = -(E - E_1)(E - E_2)(E - E_3)(E - E_4)(E_5 - E)$$

Therefore in Equation (14.32):

$$R^{1/2}(E) = i|R(E)|^{1/2}$$

Figure 14.2 shows a typical Floquet discriminant. In the two degree-of-freedom case, the eigenvalues are degenerate after E_5.

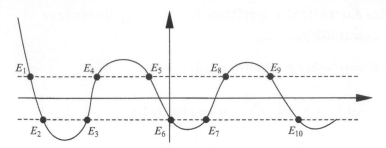

Figure 14.2 Floquet diagram for a typical solution of the KdV equation.

If the nonlinear interactions are discarded in both **K**, **K**′ (i.e., if both matrices are rendered diagonal) we have:

$$
\mathbf{B} = \mathbf{D}^* = 2\pi \mathbf{K}'\mathbf{K}^{-1} = \frac{2\pi}{\det \mathbf{K}}
\begin{bmatrix} K'_{11} & 0 \\ 0 & K'_{22} \end{bmatrix}
\begin{bmatrix} K_{22} & 0 \\ 0 & K_{11} \end{bmatrix}
$$

$$
= \frac{2\pi}{K_{11}K_{22}}
\begin{bmatrix} K'_{11}K_{22} & 0 \\ 0 & K'_{22}K_{11} \end{bmatrix}
= 2\pi
\begin{bmatrix} K'_{11}/K_{11} & 0 \\ 0 & K'_{22}/K_{22} \end{bmatrix}
\tag{14.53}
$$

This then yields the classical sum of cnoidal waves, that is, the theta function reduces to a sum of two cnoidal waves where each has a nome of the form:

$$
q_n = e^{2\pi K'_{nn}/K_{nn}}, \quad 1 \le n \le 2
$$

However, it is not possible to arbitrarily set the off-diagonal elements to zero as done here because the resultant Riemann matrix does not correspond to a solution of the KdV equation. Only in the linear limit do the off-diagonal terms make a small enough contribution that they can be neglected in the formulation.

14.4   Numerical Computation of the Riemann Spectrum

If we introduce the integral

$$
G_{ji} = 2 \int_{E_j}^{E_{j+1}} \frac{E^{i-1} dE}{R^{1/2}(E)}
\tag{14.54}
$$

Then the parameters in the theta function can be computed from the following relations:

$$
\mathbf{G} \Rightarrow G_{ij} = \{G_{ji}\}^{\mathrm{T}}
$$

$$
\mathbf{K} \Rightarrow K_{ij} = (-1)^{\alpha_j} G_{2i,j}, \quad \alpha_j = j/2, \quad \text{for } j \text{ even } \alpha_j = (j+1)/2, \quad \text{for } j \text{ odd}
$$

$$\mathbf{K}' \Rightarrow K'_{ij} = \sum_{k=1}^{2i-1} G_{2k,j}$$

$$\mathbf{B} = 2\pi \mathbf{K}'\mathbf{K}^{-1}$$

$$\mathbf{C} = 2\pi \mathbf{K}^{-1}$$

$$\mathbf{k} \Rightarrow k_j = 2|K_{N,j}^{-1}|$$

$$\boldsymbol{\omega} \Rightarrow \omega_j = c_0 k_j - \beta v_j$$

$$v_j = 8C_{N-1,j} + 4C_{N,j} \sum_{i=1}^{2N+1} E_i$$

$$\phi_m = \sum_{j=1}^{N} E_{jm}, \quad \mathbf{E} = -2\pi i\,\mathbf{F}\mathbf{K}^{-1}$$

where

$$F_{jk} = \int_{E_{2j}}^{P_j(0,0)} \frac{E^{k-1}dE}{R^{1/2}(E)}, \quad P_m(0,0) = [\mu_m(0,0), \sigma_m]$$

These relations form the basis for the *loop-integral program* discussed in Chapter 19. Note that the *inputs* to this program accept the *outputs* of the program that analyzes the *Floquet problem* for the *Schrödinger eigenvalue problem* for *periodic boundary conditions* discussed in Chapters 10 and 17.

Appendix: Summary of Formulas for the Loop Integrals of the KdV Equation

The general solution to the KdV equation may be written in terms of the *theta function* representation

$$\eta(x,t) = \frac{2}{\lambda}\frac{\partial^2}{\partial x^2} \ln \theta_N(X_1, X_2, \ldots, X_N) \tag{14.A.1}$$

where

$$\theta_N(X_1, X_2, \ldots, X_N) = \sum_{m_1=-\infty}^{\infty} \sum_{m_2=-\infty}^{\infty} \cdots \sum_{m_N=-\infty}^{\infty} \exp\left[i\sum_{k=1}^{N} m_k X_k - \frac{1}{2}\sum_{j=1}^{N}\sum_{k=1}^{N} m_j m_k B_{jk}\right] \tag{14.A.2}$$

Here N is the number of degrees of freedom in a particular solution to the KdV equation. $\mathbf{B} = \{B_{ij}\}$ is the period or interaction matrix. The vector $\mathbf{m} = [m_1, m_2, \ldots, m_N]$ is a vector of the integers. Here the B_{ij} are assumed to be positive definite; this fact, together with the minus sign in front of the double sum in the exponential insures convergence of Equation (14.A.2).

The *phases* X_j of the theta function (Equation (14.A.2)) are given by:

$$X_j = k_j x - \omega_j t + \phi_j, \quad 1 \le j \le N \tag{14.A.3}$$

where

$$\cdot \omega_j = c_o k_j - \beta v_j \tag{14.A.4}$$

To compute the k_j, ω_j and ϕ_j we need the following matrix of coefficients

$$\mathbf{C} = 2\pi i \, \mathbf{J}^{-1} = 2\pi \mathbf{K}^{-1} \tag{14.A.5}$$

Elements of the matrices $\mathbf{J} = \{J_{ji}\}$ and $\mathbf{K} = \{K_{ji}\}$, for $\mathbf{J} = i \, \mathbf{K}$ are given by:

$$J_{ji} = 2 \int_{E_{2j}}^{E_{2j+1}} \frac{E^{i-1} dE}{R^{1/2}(E)}, \quad K_{ji} = 2 \int_{E_{2j}}^{E_{2j+1}} \frac{E^{i-1} dE}{|R(E)|^{1/2}}, \quad 1 \le i \le N, \quad 1 \le j \le N \tag{14.A.6}$$

The notation for the matrix $\mathbf{K}$ is consistent with the elliptic integral $K(m)$ used in the definition of one degree-of-freedom theta functions (Abramowitz and Stegun, 1964). These are integrated over an open band and hence have information about the individual hyperelliptic functions, μ_j. Here, the Nth-order polynomial $R(E)$ is given by

$$R(E) = \prod_{k=1}^{2N+1} (E - E_k)$$

One then finds the following expressions for the k_j, ω_j

$$k_j = 2C_{N,j} \tag{14.A.7}$$

$$\omega_j = c_o k_j - \beta v_j, \quad v_j = 8C_{N-1,j} + 4C_{N,j} \sum_{i=1}^{2N+1} E_i \tag{14.A.8}$$

To obtain the *phases* ϕ_j:

$$\phi_m = \sum_{j=1}^{N} E_{jm}, \quad \mathbf{E} = -i \, \mathbf{FC}, \quad \mathbf{C} = 2\pi \mathbf{K}^{-1} \tag{14.A.9}$$

where

$$F_{jk} = \int_{E_{2j}}^{P_j(0,0)} \frac{E^{k-1}dE}{R^{1/2}(E)}, \quad P_m(0,0) = [\mu_m(0,0), \sigma_m]$$

The *vector of phase integrals* $\Phi = \{\Phi_j\}$ is given by:

$$\Phi_j = \sigma_j \int_{E_{2j}}^{\mu_j(0,0)} \frac{E^{j-1}dE}{R^{1/2}(E)} \tag{14.A.10}$$

Note that the phases explicitly contain information about the starting values of the hyperelliptic functions, $\mu_j(0,0)$.

The *period matrix* or *interaction matrix* has the form:

$$\mathbf{B} = 2\pi \quad \underbrace{\mathbf{K}'} \quad \underbrace{\mathbf{K}^{-1}} \tag{14.A.11}$$

Contains information   Contains information
among the μ_j about each μ_j

where

$$K'_{ji} = 2 \int_{E_1}^{E_{2j}} \frac{E^{i-1}dE}{R^{1/2}(E)} \tag{14.A.12}$$

Note that for N degrees of freedom the indices range over $1 \leq i \leq N$ and $1 \leq j \leq N$. Further note that the index i indicates the wavenumber $k_i = i\Delta k$, $\Delta k = 2\pi/L$, where L is the spatial period of the wave train and the index j is the number of a particular open band.

15 Schottky Uniformization

15.1 Introduction

The focus of this chapter is to make some simple observations about *Poincaré series* as a method for computing the *Riemann Spectrum* or *theta function parameters* in the *Schottky domain* and their *numerical implementation*. Relevant references are Schottky (1887), Burnside (1892a,b), Baker (1897), Bobenko (1987), Bobenko and Bordag (1987), Bobenko and Kubensky (1987), Bobenko and Bordag (1989), and Belokolos et al. (1994), Mumford et al. (2002). This chapter focuses on useful information for *programming* the Riemann spectrum for the KdV equation using the Schottky uniformization approach.

15.2 IST Spectral Domain

Suppose we have the *Floquet diagram* and *Schrödinger Eigenvalues* in the *E-domain* (i.e., the *IST spectral domain*, taken specifically for KdV evolution, see Chapters 10, 14, 16, and 17) as shown in Figure 15.1.

Note that the open bands (E_2, E_3) and (E_4, E_5) define the two hyperelliptic function "degrees of freedom" for this particular case. Hence the *degenerate bands* (E_6, E_7), (E_8, E_9), (E_{10}, E_{11}), etc., make no contribution to the dynamics. We, therefore, have a system that has the five eigenvalues: $(E_1, E_2, E_3, E_4, E_5)$. Figure 17.1 of Chapter 17 provides a complete overview of the spectral problem for the KdV equation and will help orient the reader to Figure 15.1.

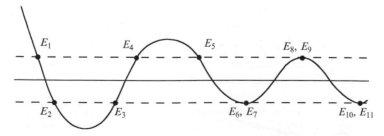

Figure 15.1 Floquet diagram in the spectral domain E-plane for evolution described by the KdV equation.

Now, let us exam what happens when we use *Schottky uniformization* to *reparameterize* the spectrum. There are two sets of basis cycles, the first is the *oscillation basis* and the second is the *soliton basis* (Chapters 8 and 10).

15.3 Linear Oscillation Basis

The focus of this section is to discuss Schottky uniformization in the oscillation basis. The idea is the following: Given the IST eigenvalues $(E_1, E_2, E_3, E_4, E_5)$ (for two degrees of freedom) associate the open bands (E_2, E_3), (E_4, E_5) with each of the *degrees of freedom* (the *hyperelliptic functions* live inside these bands as seen in Figure 17.1). Then make a transformation (a mapping from the *IST E-plane* to the *Schottky z-plane*, which is a *Poincaré series*) to the new variable pairs (again associated with each open band) (A_1, ρ_1), (A_2, ρ_2). Once these variables are computed, we can compute via other Poincaré series the 2×2 *period matrix* and *two wavenumbers, frequencies and phases*. Why go to all this trouble to compute the parameters of the *Riemann theta functions* using the Schottky procedure? Because the Schottky method avoids many of the difficulties in the computation of the loop integrals, that is, there are no singularities or near singularities as they occur in the loop integrals and the computational speed is much improved.

15.3.1 An Overview of Schottky Uniformization in the Oscillation Basis

In Figure 15.2, upper panel, is a schematic of the *IST E-domain* showing the five eigenvalues and the appropriate *cycles* for the linear oscillation (sine wave) basis (see Fig. 15.1 and Chapter 17).

The *spectral E-plane* is the place where the *IST Floquet problem* is computed (upper panel). The *Schottky z-plane* is the place where *Schottky uniformization* is carried out (lower panel). The *mapping* between the IST E-plane and the

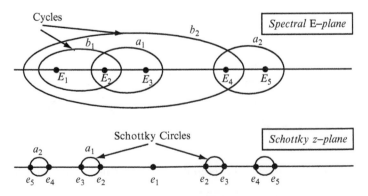

Figure 15.2 The IST E-plane is mapped to the Schottky z-plane.

Schottky z-plane is a *Poincaré series* (see Section 5.2.4 for details) whose leading order approximation is given by:

$$E \sim z^2 + \text{higher order terms in } z \tag{15.1}$$

This is a leading order expression that takes us from the Schottky z-plane to the IST E-plane. The Poincaré series on the right-hand side of this expression (higher order terms) is not easily invertible, but inversion is desirable since what we need in the space/time series analysis of data is an algorithm to take us from E to z. For now, we use $E \sim z^2$ to get a first look at Schottky uniformization in the small-amplitude, near-linear limit of KdV evolution.

The physical implication of $E \sim z^2$ is, since $E \sim (k/2)^2$ (k is the wavenumber) for the radiation modes of IST (Osborne and Bergamasco, 1986), then the Schottky variable, $z \sim e_n \sim k/2$, is a wavenumber. As we will see below A_n has units of wavenumber also and ρ_n is dimensionless. In the small-amplitude limit considered in this section, we will eventually assume $\rho_n \ll 1$.

This means that the *Schottky images* e_n (in the Schottky z-plane) of the *IST eigenvalues* E_n (in the IST E-plane) are related approximately by $E_n - E_1 \sim e_n^2$ (note that the arbitrary shift by E_1; this is legal since *any arbitrary shift of the Floquet eigenvalues* leaves the KdV spectrum invariant (Osborne and Bergamasco, 1986)).

15.3.2 The Schottky Circles and Parameters

Also note the quadratic equation for the so-called *Schottky circles* in Figure 15.2

$$(z - A_n)^2 = \rho_n (z + A_n)^2 \tag{15.2}$$

has important information relating the circles and the Schottky parameters. To see this note that the solutions to this equation, $z_\pm$, correspond to the images (e_n, e_{n+1}) of the IST eigenvalues (E_n, E_{n+1}) in the Schottky z-plane. Thus we have that the Schottky circles cross the real axis approximately at the following positions in the *Schottky z-plane*, $e_n \sim \sqrt{E_n - E_1}$. Expand the quadratic equation (15.2) and use the usual solution to the equation, to get

$$z_\pm = e_\pm = A_n \left(\frac{1 + \rho_n}{1 - \rho_n} \right) \pm 2 A_n \frac{\sqrt{\rho_n}}{1 - \rho_n} = A_n \frac{(1 \pm \sqrt{\rho_n})^2}{1 - \rho_n} = A_n \left(\frac{1 \mp \sqrt{\rho_n}}{1 \pm \sqrt{\rho_n}} \right) \tag{15.3}$$

or

$$e_n = e_- = A_n \frac{(1 - \sqrt{\rho_n})^2}{1 - \rho_n} = A_n \left(\frac{1 - \sqrt{\rho_n}}{1 + \sqrt{\rho_n}} \right) \tag{15.4}$$

and

$$e_{n+1} = e_+ = A_n \frac{(1 + \sqrt{\rho_n})^2}{1 - \rho_n} = A_n \left(\frac{1 + \sqrt{\rho_n}}{1 - \sqrt{\rho_n}} \right) \tag{15.5}$$

On this basis, we are able to conclude that the *radius of the Schottky circle* is:

$$r_n = \frac{1}{2}(e_{n+1} - e_n) = 2A_n \frac{\sqrt{\rho_n}}{1 - \rho_n} \quad \text{Has units of wavenumber} \tag{15.6}$$

and the *center* is:

$$c_n = \frac{1}{2}(e_n + e_{n+1}) = A_n \left(\frac{1 + \rho_n}{1 - \rho_n} \right) \quad \text{Has units of wavenumber} \tag{15.7}$$

Inversely

$$A_n = c_n \left(\frac{1 - \rho_n}{1 + \rho_n} \right) \quad \text{Has units of wavenumber} \tag{15.8}$$

Note that the radius of the Schottky circle is a measure of the amplitudes of the degrees of freedom (related in a nontrivial way to the amplitudes of the hyperelliptic functions). In the near-linear limit, for which $\rho_n \ll 1$, we have

$$r_n \cong 2A_n \sqrt{\rho_n}(1 + \rho_n)$$
$$c_n \cong A_n(1 + 2\rho_n) \tag{15.9}$$

The *equation for the Schottky circles* of center c_n and radius r_n is:

$$(z - c_n)(z^* - c_n) = r_n^2 \tag{15.10}$$

From Equations (15.4) and (15.5) we find the interpretation of A_n as the *geometric mean of the images*:

$$A_n = \sqrt{e_n e_{n+1}} \tag{15.11}$$

Recalling the formula (15.7) for the center, c_n, we see that the center lies to the right of the geometric average of the images, A_n (Equation (15.11)). The importance of the parameters A_n is that they lie near the center of the Schottky circles and, indeed, the A_n are the *fixed points* of the *linear fractional transformation*, σ_n, given below. The value of the parameter ρ_n in terms of the Schottky images is then given by:

$$\rho_n = \left[\frac{\sqrt{e_n} - \sqrt{e_{n+1}}}{\sqrt{e_n} + \sqrt{e_{n+1}}} \right]^2 \tag{15.12}$$

The importance of the parameters ρ_n is that they are related to the diagonal elements of the period matrix approximately by $\rho_n \approx e^{-2\pi B_{nn}}$ as seen more precisely below.

In summary, the Schottky images (e_n, e_{n+1}) (of the IST eigenvalues (E_n, E_{n+1}) via the approximate relation $e_n \sim \sqrt{E_n - E_1}$ or, more importantly, via the full Poincaré series discussed below) map to the Schottky parameters (A_n, ρ_n) by:

$$A_n = \sqrt{e_n e_{n+1}} = c_n \left(\frac{1 - \rho_n}{1 + \rho_n}\right), \quad \rho_n = \left[\frac{\sqrt{e_n} - \sqrt{e_{n+1}}}{\sqrt{e_n} + \sqrt{e_{n+1}}}\right]^2 = \frac{c_n - A_n}{c_n + A_n} \tag{15.13}$$

and inversely by

$$e_n = A_n \frac{1 - \sqrt{\rho_n}}{1 + \sqrt{\rho_n}}, \quad e_{n+1} = A_n \frac{1 + \sqrt{\rho_n}}{1 - \sqrt{\rho_n}} \tag{15.14}$$

Our perspective at this point is quite simple: Given the E_n as numerical eigenvalues of the Schrödinger eigenvalue problem for a particular time series, compute the Schottky images e_n and hence the (A_n, ρ_n) which then allow computation of the parameters in the $\theta(x, t)$ function, namely B_{mn}, k_n, and ω_n.

A summary of results for the *small-amplitude limit* in the *oscillation basis* is given in Appendix I.

15.3.3   Linear Fractional Transformations

Define the *linear fractional transformation* (see, e.g., Ablowitz and Fokas, 1997; an additional very useful reference for this chapter is Needham (1997)) by the following map:

$$\sigma z = \frac{\alpha z + \beta}{\gamma z + \delta} \tag{15.15}$$

where σ is an operator which operates on z by the rule given on the right-hand side of the equation. The constants α, β, γ, δ are determined in terms of the Schottky parameter pair (A, ρ) (which is associated with a particular Schottky circle defined by the IST *eigenvalues* (E_n, E_{n+1}) that are transformed by the *Poincaré series* to the *Schottky image pair* (e_n, e_{n+1}), yielding the *Schottky parameter pair* (A_n, ρ_n). Here are the elements of the linear fractional transformation:

$$\alpha = \frac{1}{2}\left(\frac{1 + \rho}{\sqrt{\rho}}\right), \quad \beta = -\frac{A}{2}\left(\frac{1 - \rho}{\sqrt{\rho}}\right)$$

$$\tag{15.16}$$

$$\gamma = -\frac{1}{2A}\left(\frac{1 - \rho}{\sqrt{\rho}}\right), \quad \delta = \frac{1}{2}\left(\frac{1 + \rho}{\sqrt{\rho}}\right)$$

These elements form a *matrix* for σ which is denoted $\tilde{\sigma}$:

$$\tilde{\sigma} = \begin{bmatrix} \dfrac{1}{2}\left(\dfrac{1+\rho}{\sqrt{\rho}}\right) & -\dfrac{A}{2}\left(\dfrac{1-\rho}{\sqrt{\rho}}\right) \\[4mm] -\dfrac{1}{2A}\left(\dfrac{1-\rho}{\sqrt{\rho}}\right) & \dfrac{1}{2}\left(\dfrac{1+\rho}{\sqrt{\rho}}\right) \end{bmatrix} \tag{15.17}$$

Note that

$$\lim_{\rho \to 0} \tilde{\sigma} \sim \frac{1}{2\sqrt{\rho}} \begin{bmatrix} 1 & -A \\[2mm] -\dfrac{1}{A} & 1 \end{bmatrix} \text{ corresponding to } B_{nn} \to \infty \text{ in linear limit}$$

$$\tag{15.18}$$

$$\lim_{\rho \to 1} \tilde{\sigma} = \begin{bmatrix} 1 & 0 \\ 0 & 1 \end{bmatrix} \text{ corresponding to } B_{nn} \to 0 \text{ in soliton limit} \tag{15.19}$$

Note further that $\sigma = 1$ (the identity) and $\tilde{\sigma} = \tilde{1}$ (the identity matrix) for the particular value $\rho = 1$ corresponds to the maximum value of ρ ($0 \leq \rho \leq 1$), which occurs in the soliton limit. In contrast, small values of ρ (linear oscillation, sine wave case) have large matrix elements.

Note that the solution of

$$z = \frac{\alpha z + \beta}{\gamma z + \delta}$$

gives the *fixed points* $z = \pm A_n$. To demonstrate this one finds the quadratic equation $\gamma z^2 + (\delta - \alpha)z - \beta = 0$ and uses the fact that $\delta - \alpha = 0$, so $z = \pm\sqrt{\beta/\gamma} = \pm A$.

The *inverse linear fractional transformation* has the form:

$$\sigma^{-1}z = \frac{\delta z - \beta}{-\gamma z + \alpha} \tag{15.20}$$

where again the constants α, β, γ, δ are defined in terms of the Schottky parameter pair (A, ρ) by Equation (15.16). The inverse linear fractional transformation in matrix form is:

$$\tilde{\sigma}^{-1} = \begin{bmatrix} \dfrac{1}{2}\left(\dfrac{1+\rho}{\sqrt{\rho}}\right) & \dfrac{A}{2}\left(\dfrac{1-\rho}{\sqrt{\rho}}\right) \\[4mm] \dfrac{1}{2A}\left(\dfrac{1-\rho}{\sqrt{\rho}}\right) & \dfrac{1}{2}\left(\dfrac{1+\rho}{\sqrt{\rho}}\right) \end{bmatrix} \tag{15.21}$$

It is clear that $\sigma\sigma^{-1} = 1$ and $\tilde{\sigma}\tilde{\sigma}^{-1} = \tilde{1}$. In the application of these trans-formations the following notation is useful: $\sigma = \sigma(A, \rho)$. Each *IST degree of freedom* is associated with a particular (A_n, ρ_n) pair and we therefore see that the Schottky group generators are $\sigma_n = \sigma(A_n, \rho_n)$.

15.3.4 *Poincaré Series Relating the IST E-plane to the Schottky z-plane*

The relevant Poincaré series is (Belokolos et al., 1994)

$$E(z) - E_1 = \sum_{\sigma \in G}^{N} [(\sigma z)^2 - (\sigma 0)^2] \tag{15.22}$$

As discussed above, the σ are *linear fractional transformations*, G is the *group* of these transformations, and N is the *number of degrees of freedom* (often referred to as "genus N"). There are N *generators*, σ_n, $n = 1, 2, \ldots, N$, in the group, one for each degree of freedom, $\sigma_n = \sigma(A_n, \rho_n)$. To form the entire group, one needs to add also the *identity transformation*, $\sigma_o = 1$, and the *inverse transformations*, $\sigma_n^{-1}, n = 1, 2, \ldots, N$. There are thus $2N + 1$ *group elements* in the Schottky group. The above Poincaré series can be rewritten in the form (evaluate the first term for the identity, $\sigma_o = 1$, for which $\sigma_o z = z$, so that $\sigma_o 0 = 0$):

$$E(z) - E_1 = z^2 + \sum_{\sigma \in G, \, \sigma \notin I}^{N} \left[(\sigma z)^2 - (\sigma 0)^2 \right] \tag{15.23a}$$

Ignoring the Poincaré sum (presumably small in the small-amplitude limit of the theory) gives the approximation used above for the oscillatory basis: $E(z) - E_1 \sim z^2$.

Now in the Schottky z-plane we have the eigenvalues as a result of an *involution* (see Belokolos et al., 1994 for details): e_n, $n = 1, 2, \ldots, N$. The above Poincaré series evaluated at the e_n is then (this is the *most important conformal mapping in this chapter*):

$$E_n - E_1 = e_n^2 + \sum_{\sigma \in G, \, \sigma \notin I}^{N} \left[(\sigma e_n)^2 - (\sigma 0)^2 \right] \tag{15.23b}$$

where we have used $E_n = E(e_n)$ which are the IST eigenvalues (see Chapters 10, 14 and 17). Now for the applications herein we have the obvious need, given the E_n, to compute the e_n. The above Poincaré series, however, only allows computation of the E_n given the e_n, which is the *opposite* of what we require. This means that we need to develop a capability for numerically computing *inverse Poincaré series* (see below).

Nota Bene: Solving the Cauchy Problem for the KdV Equation. First, we assume some form for the initial condition for KdV, $\eta(x, 0)$, and then one solves the Floquet problem for the Schrödinger eigenvalue problem to obtain the *main spectrum* $(E_n, n = 1, 2, \ldots, 2N + 1)$ (Chapter 17). We then invert Equation (15.24) numerically (see Chapter 32 for both the KdV and KP equations) to obtain the Schottky images (e_n). From the images one computes the Schottky parameters (A_n, ρ_n) and finally by the Poincaré series discussed below (Equations (15.24), (15.25), (15.28), and (15.29)) one computes the Riemann spectrum $(B_{mn}, k_n, \text{ and } \omega_n, m, n = 1, 2, \ldots N)$. We have of course chosen arbitrary values for the diagonal elements of the Riemann matrix, B_{nn}, and the phases, ϕ_n; periodic boundary conditions are also assumed. The Riemann spectrum is then inserted into the Riemann theta function and the second derivative of the logarithm of the latter is then evaluated at all space x and time t to obtain the solution of the KdV equation.

15.3.5 Poincaré Series for the Period Matrix

The *period matrix in the oscillation basis* has the following form (See pp. 348, 356 of Baker, 1897): Here are the *diagonal elements*:

$$B_{mm} = \log \rho_m + \sum_{\sigma \in G_m \backslash G / G_m, \, \sigma \neq I} \ln \left(\frac{A_m - \sigma A_m}{A_m - \sigma(-A_m)} \right)^2 \tag{15.24}$$

The *off-diagonal elements* are

$$B_{mn} = \sum_{\substack{\sigma \in G_n \backslash G / G_m \\ m \neq n}} \ln \left(\frac{A_n - \sigma A_m}{A_n - \sigma(-A_m)} \right)^2 \tag{15.25}$$

which can be simplified to read (by separating out the identity, $\sigma = 1$)

$$B_{mn} = \ln \left(\frac{A_n - A_m}{A_n + A_m} \right)^2 + \sum_{\sigma \in G_n \backslash G / G_m, \, \sigma \notin I} \ln \left(\frac{A_n - \sigma A_m}{A_n - \sigma(-A_m)} \right)^2, m \neq n \tag{15.26}$$

The summation convention (see the box below) has the following definitions. $\sigma \in G_n \backslash G / G_n, \sigma \notin I$ means take the group products and decimate them on the right by G_n and then on the left again by G_n; exclude the identity. Likewise $\sigma \in G_n \backslash G / G_m$ means take the group products and decimate them on the right by G_m and then on the left again by G_n.

Neglecting the Poincaré series in the above expressions for the period matrix gives the small-amplitude approximations as discussed in Appendix I:

$$B_{mm} \approx \log \rho_m$$
$$B_{mn} \approx \ln \left(\frac{A_n - A_m}{A_n + A_m} \right)^2 \tag{15.27}$$

15.3.6 Poincaré Series for the wavenumbers and Frequencies

The *wavenumbers* have the following Poincaré series

$$k_n = \sum_{\sigma \in G/G_n} (\sigma A_n - \sigma(-A_n)) = 2A_n + \sum_{\sigma \in G/G_n, \sigma \neq I} (\sigma A_n - \sigma(-A_n)) \tag{15.28}$$

and the *frequencies* have the following form

$$\omega_n = \sum_{\sigma \in G/G_n} ((\sigma A_n)^3 - (\sigma(-A_n))^3) = 2A_n^3 + \sum_{\sigma \in G/G_n, \sigma \neq I} ((\sigma A_n)^3 - (\sigma(-A_n))^3)$$

$$\tag{15.29}$$

The summation convention $\sigma \in G/G_n$ means decimate the group on the right by G_n.

By excluding the (presumed small) Poincaré series for the linear limit we find to leading order

$$k_n \approx 2A_n$$

$$\omega_n \approx 2A_n^3$$

15.3.7 How to Sum the Poincaré Series

I now discuss somewhat pedantically how to sum the Poincaré series. The exposition follows Baker (1897) with more modern notation.

Here is how to sum the Poincaré series (Equation (15.23)): The Schottky group consists of the particular elements $\sigma_o = 1, \sigma_n, \sigma_n^{-1}$ where the subscript ranges over all the degrees of freedom, $n = 1, 2, \ldots, N$. The series is summed in a *particular order*. In the Poincaré series the σ have a *product form*: $\sigma = \sigma_i^{\pm} \sigma_j^{\pm} \ldots \sigma_k^{\pm}$, which is made clear below.

Continued

First, make the *identity transformation* $\sigma_o = 1$ $(\sigma_o z = (z + 0)/(0.z + 1) = z)$. Evaluate the term $(\sigma_o z)^2 - (\sigma_o 0)^2$.

Second, make the $2N$ substitutions whose products contain one factor, σ_n or σ_n^{-1}. Evaluate the terms for which $(\sigma_n^{\pm} z)^2 - (\sigma_n^{\pm} 0)^2$. Note that $\sigma^- = \sigma^{-1}$ and that $\sigma^+ = \sigma^{+1} = \sigma$.

Third, make the $2N(2N - 1)$ substitutions whose products are in *pairs* of the form $\sigma_i \sigma_j, \sigma_i \sigma_j^{-1}, \sigma_i^{-1} \sigma_j, \sigma_i^{-1} \sigma_j^{-1}$, in which the paired substitutions are *not* to include mutual inverses, that is, exclude terms of the form $\sigma_i \sigma_i^{-1}, \sigma_i^{-1} \sigma_i$. Evaluate the relevant terms in the Poincaré series for which $(\sigma_i^{\pm} \sigma_j^{\pm} z)^2 - (\sigma_i^{\pm} \sigma_j^{\pm} 0)^2$.

Fourth, make the $2N(2N - 1)^2$ substitutions whose products contain *three factors*, $\sigma_i^{\pm} \sigma_j^{\pm} \sigma_k^{\pm}$. Evaluate the relevant terms for which $(\sigma_i^{\pm} \sigma_j^{\pm} \sigma_k^{\pm} z)^2 - (\sigma_i^{\pm} \sigma_j^{\pm} \sigma_k^{\pm} 0)^2$.

Generally, one seeks higher order terms including p factors for which there are $2N(2N - 1)^{p-1}$ members. And so on.

Terms of p factors are said to be of pth order. The identity transformation is of zeroth order, single σ_n are of first order, pairs $\sigma_i \sigma_j$ are of second order, triples $\sigma_i \sigma_j \sigma_k$ are of third order, and so on. The final step is to sum all the terms, for all the computed orders, in the Poincaré series.

It is convenient to extract the identity transformation from the Poincaré series (15.22):

$$E(z) - E_1 = z^2 + \sum_{\sigma \in G, \, \sigma \neq 1}^{N} [(\sigma z)^2 - (\sigma 0)^2]$$

where I have used $\sigma_o 0 = 0$. This establishes the relation $E(z) - E_1 \sim z^2$ discussed above, which is true of course only when the other terms in the Poincaré series can be neglected. This form of the Poincaré series is what we will use below in the linear oscillation basis.

On the basis of the above discussion we can abstractly write the Poincaré series:

$$E(z) - E_1 = z^2 + \sum_{i=1}^{2N}[(\sigma_i^{\pm} z)^2 - (\sigma_i^{\pm} 0)^2] + \sum_{i=1}^{2N(2N-1)} \sum_{j=1}^{2N(2N-1)} (\sigma_i^{\pm} \sigma_j^{\pm} z)^2 - (\sigma_i^{\pm} \sigma_j^{\pm} 0)^2$$

$$+ \sum_{i=1}^{2N(2N-1)^2} \sum_{j=1}^{2N(2N-1)^2} \sum_{k=1}^{2N(2N-1)^2} (\sigma_i^{\pm} \sigma_j^{\pm} \sigma_k^{\pm} z)^2 - (\sigma_i^{\pm} \sigma_j^{\pm} \sigma_k^{\pm} 0)^2 + \cdots$$

One of course excludes the terms for which inverses appear in pairs, since they have no effect on the summation other than to reduce the result to lower order, which has already been computed earlier and therefore can be neglected.

15.3.8   One Degree of Freedom

This example consists of *one mode in the Riemann matrix* (which is 1×1) and is often referred to as the "genus 1" case, $N=1$. The *IST main spectrum* is the triple of points (E_1, E_2, E_3). This implies a mapping to the images in the Schottky plane: (e_1, e_2, e_3). The Schottky parameters then are (A_1, ρ_1). The goal is to compute the relevant group and to sum the Poincaré series for this case. Since there is only one degree of freedom, the group elements are $(\sigma_0, \sigma_1, \sigma_1^-)$, that is, the identity and the linear fractional transformation and its inverse. Since we have already evaluated the influence of the identity on the Poincaré series we now focus on the remaining terms.

First discuss the single product terms for the group elements (σ_1, σ_1^-). The only option we have is to apply σ_1 and its inverse σ_1^-. Figure 15.3 gives a schematic of the one degree-of-freedom case.

Here are the cases to consider ($N=1$), up to third order:

$\sigma_0 = 1$ Only one term leading to $E(z) - E_1 \approx z^2$
                                        in the Poincaré series

σ_1, σ_1^- $2N$ terms $= 2$ terms in the Poincaré series

$\left.\begin{array}{l}\sigma_1\sigma_1 \\ \sigma_1\sigma_1^- \\ \sigma_1^-\sigma_1 \\ \sigma_1^-\sigma_1^-\end{array}\right\}$ *Inverses* $2N(2N-1) = 2$ terms in the Poincaré series

$\left.\begin{array}{l}\sigma_1\sigma_1\sigma_1 \\ \sigma_1\sigma_1^-\sigma_1 \\ \sigma_1^-\sigma_1\sigma_1 \\ \sigma_1^-\sigma_1^-\sigma_1^-\end{array}\right\}$ *Inverses* $2N(2N-1)^2 = 2$ terms in the Poincaré series

In conclusion, a sum over *all group elements* to third order implies

(1) Include the zeroth term for the identity: $\sigma_0 = 1$.
(2) Include the first-order term for the single products: σ_1, σ_1^{-1}.
(3) Include the second term for the double products: $\sigma_1\sigma_1 = \sigma_1^2, \sigma_1^-\sigma_1^- = \sigma_1^{-2}$.
(4) Include the third term for the triple products: $\sigma_1\sigma_1\sigma_1 = \sigma_1^3, \sigma_1^-\sigma_1^-\sigma_1^- = \sigma_1^{-3}$.

Note that I have deleted all the pairs with products of a linear fractional transformation and its inverse as shown in the table above. This leads to the

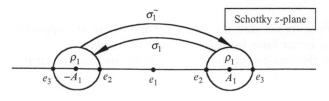

Figure 15.3 Schottky z-plane for a one degree-of-freedom system. Shown are the Schottky circles and the linear fractional transformation and its inverse.

following Poincaré series (Equation (15.23)) for the one degree-of-freedom case.

$$E(z) - E_1 = z^2 + \left[(\sigma_1 z)^2 - (\sigma_1 0)^2\right] + \left[(\sigma_1^- z)^2 - (\sigma_1^- 0)^2\right]$$

$$+ \left[(\sigma_1 \sigma_1 z)^2 - (\sigma_1 \sigma_1 0)^2\right] + \left[(\sigma_1^- \sigma_1^- z)^2 - (\sigma_1^- \sigma_1^- 0)^2\right]$$

$$+ \left[(\sigma_1 \sigma_1 \sigma_1 z)^2 - (\sigma_1 \sigma_1 \sigma_1 0)^2\right] + \left[(\sigma_1^- \sigma_1^- \sigma_1^- z)^2 - (\sigma_1^- \sigma_1^- \sigma_1^- 0)^2\right] + \cdots$$

Thus this case is really simple.

Can we invert this series? Suppose that we are near the linear limit, $E(z) - E_1 \approx z^2$. Since the eigenvalues are what we want we can write this as:

$$z^2 = E(z) - E_1 - [(\sigma_1 z)^2 - (\sigma_1 0)^2] - [(\sigma_1^- z)^2 - (\sigma_1^- 0)^2]$$

$$- [(\sigma_1 \sigma_1 z)^2 - (\sigma_1 \sigma_1 0)^2] - [(\sigma_1^- \sigma_1^- z)^2 - (\sigma_1^- \sigma_1^- 0)^2]$$

$$- [(\sigma_1 \sigma_1 \sigma_1 z)^2 - (\sigma_1 \sigma_1 \sigma_1 0)^2] - [(\sigma_1^- \sigma_1^- \sigma_1^- z)^2 - (\sigma_1^- \sigma_1^- \sigma_1^- 0)^2] - \cdots$$

Thus the first iteration is

$$z^2 = E(z) - E_1 \quad \text{or} \quad z = \sqrt{E(z) - E_1}$$

This latter value of z can then be used to iterate on the solution. Of course this procedure can only be used when the bracketed terms are small. A numerical iteration process is used for all computations herein (see more details in Chapter 32).

Now look at the other IST parameters and their Poincaré series. We need to provide decimated values: Decimate from the right according to $\sigma \in G/G_1$, leaves only the identity term!

Consider the Poincaré series for the wavenumbers. Only the identity, $\sigma_o = 1$, remains and this gives

$$k_n = \sigma_o A_n - \sigma_o(-A_n) = 2A_n$$

and the frequencies resulting only from the identity are

$$\omega_n = (\sigma_o A_n)^3 - (\sigma_o(-A_n))^3 = 2A_n^3$$

These are amazingly simple results and are exact for n = 1 or are approximately correct in the near linear limit for many degrees of freedom.

Now the period matrix also has a contribution only from the identity for the B_{11} term:

$$B_{nn} = \frac{1}{2\pi} \log \rho_n$$

These results are to be interpreted in terms of a very simple fact: When there are no interactions with other components (which occurs for the one degree-of-freedom case or in the near linear limit for many degrees of freedom), all terms in the Poincaré series vanish except for the first term, for the identity σ_o.

Note on Checkout of Numerical Algorithm

For one degree of freedom the wavenumber, k_1, frequency, ω_1, and period matrix element, B_{11}, do not interact with any other degrees of freedom and their Poincaré series have no contribution in the summation (other than the identity). Hence in the one dof case, *the only Poincaré series contribution comes from the IST plane to Schottky plane mapping*. The Poincaré series truncate exactly at the identity. Thus the e_ns can be computed directly from the loop integrals (the normal procedure which has been programed) by the formulas given above and the Poincaré series can be then be checked out exactly. A checkout procedure is to compute k_1, ω_1, and B_{11} from the cycle integrals. We then compute A_1, ρ_1 by

$$\rho_1 = e^{-2\pi B_{11}} \quad \text{or} \quad \sqrt{\rho_1} = e^{-\pi B_{11}}$$

$$A_1 = 2\pi k_1/2 = \pi k_1$$

The e_n are then given by

$$e_1 = 0$$

$$e_2 = A_1 \left(\frac{1 - \sqrt{\rho_1}}{1 + \sqrt{\rho_1}} \right)$$

$$e_3 = A_1 \left(\frac{1 + \sqrt{\rho_1}}{1 - \sqrt{\rho_1}} \right)$$

This gives *exact values of the e_n* from which the *inverse Poincaré series* can be *checked out to all orders* for the *one dof case*. This case has the advantage of providing exact e_n values, resulting from the previously computed E_n of the IST plane. In this way one can completely check out the Poincaré series (mapping) from the Schottky z-plane to the IST E-plane (Equation (15.23)).

15.3.9   Two Degrees of Freedom

This example consists of two modes in the Riemann matrix (which is 2×2) and is often referred to as the "genus 2" case, $N = 2$. The IST main spectrum is the quintuple of points $(E_1, E_2, E_3, E_4, E_5)$. This implies a need to map $(E_1, E_2, E_3, E_4, E_5)$ to the images in the Schottky plane: $(e_1, e_2, e_3, e_4, e_5)$. The Schottky parameters then are $(A_1, \rho_1, A_2, \rho_2)$ as warrants a two dof system. The goal is to compute the relevant group and to sum the Poincaré series for this case. Since there are only two degrees of freedom, the group

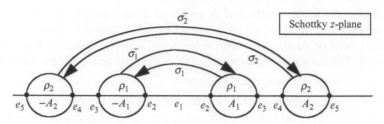

Figure 15.4 Schottky z-plane for two degree-of-freedom system. Shown are the Schottky circles and the linear fractional transformation and its inverse.

elements are $(\sigma_o, \sigma_1, \sigma_1^-, \sigma_2, \sigma_2^-)$, that is, the identity and the linear fractional transformation and its inverse for both degrees of freedom. Since we have already evaluated the influence of the identity on the Poincaré series we now focus on the remaining terms (Fig. 15.4).

Physically summing over $\sigma \in G/G_n$, which means "decimate from the right by G_n," implies that we are excluding "self-interactions" and are including only interactions with other degrees of freedom (in the oscillation basis these are the "Schottky circles").

Here are the *relevant Group Products* to consider for two degrees of freedom up to third order:

1. $\sigma_o = 1$ — Only 1 term leading to z^2 in the Poincaré series (here $\sigma^- = \sigma^{-1}$, etc.)

1. σ_1
2. σ_1^- — 2N terms = 4 terms in the Poincaré series
3. σ_2
4. σ_2^-

1. $\sigma_2^{-1}\sigma_2^{-1}$
2. $\sigma_2^{-1}\sigma_1^{-1}$
3. $\sigma_2^{-1}\sigma_1$
4. $\sigma_1^{-1}\sigma_2^{-1}$
5. $\sigma_1^{-1}\sigma_1^{-1}$
6. $\sigma_1^{-1}\sigma_2$ — 2N(2N − 1) = 12 terms in the Poincaré series (inverse pairs already omitted)
7. $\sigma_1\sigma_2^{-1}$
8. $\sigma_1\sigma_1$
9. $\sigma_1\sigma_2$
10. $\sigma_2\sigma_1^{-1}$
11. $\sigma_2\sigma_1$
12. $\sigma_2\sigma_2$

1. $\sigma_2^{-1}\sigma_2^{-1}\sigma_2^{-1}$
2. $\sigma_2^{-1}\sigma_2^{-1}\sigma_1^{-1}$
3. $\sigma_2^{-1}\sigma_2^{-1}\sigma_1$
4. $\sigma_2^{-1}\sigma_1^{-1}\sigma_2^{-1}$
5. $\sigma_2^{-1}\sigma_1^{-1}\sigma_1^{-1}$
6. $\sigma_2^{-1}\sigma_1^{-1}\sigma_2$

7. $\sigma_2^{-1}\sigma_1\sigma_2^{-1}$ $2N(2N-1)^2 = 36$ terms in the Poincaré series (inverse pairs already
8. $\sigma_2^{-1}\sigma_2^{-1}\sigma_1^{-1}$ omitted, not all terms included)
9. $\sigma_2^{-1}\sigma_1\sigma_1$
10. $\sigma_1^{-1}\sigma_2^{-1}\sigma_2^{-1}$

First we compute the single terms $(\sigma_1, \sigma_1^-, \sigma_2, \sigma_2^-)$, that is we apply σ_1 and its inverse σ_1^-, and σ_2 and its inverse σ_2^-. At higher order I have deleted all the pairs with products of a linear fractional transformation with its inverse. This leads to the following leading order Poincaré series for the two degree-of-freedom case.

$$E(z) - E_1 = z^2 + [(\sigma_1 z)^2 - (\sigma_1 0)^2] + [(\sigma_1^- z)^2 - (\sigma_1^- 0)^2]$$
$$+ [(\sigma_2 z)^2 - (\sigma_2 0)^2] + [(\sigma_2^- z)^2 - (\sigma_2^- 0)^2] + \cdots$$

I have omitted all terms after the first order terms, but these are easily applied. The wavenumbers have the following Poincaré series

$$k_n = 2A_n + \sum_{\sigma \in G/G_n,\, \sigma \neq I} (\sigma A_n - \sigma(-A_n))$$

So we need to decimate from the right by G_n. For the case $n=1$ we have

$$k_1 = 2A_1 + \sum_{\sigma \in G/G_1,\, \sigma \neq I} (\sigma A_1 - \sigma(-A_1))$$

This leads to the following decimation

1. σ_2
2. σ_2^- two terms in the Poincaré series
1. $\sigma_2^{-1}\sigma_2^{-1}$
4. $\sigma_1^{-1}\sigma_2^{-1}$
6. $\sigma_1^{-1}\sigma_2$ six terms in the Poincaré series (inverse pairs already omitted)
7. $\sigma_1\sigma_2^{-1}$
9. $\sigma_1\sigma_2$
13. $\sigma_2\sigma_2$
1. $\sigma_2^{-1}\sigma_2^{-1}\sigma_2^{-1}$
4. $\sigma_2^{-1}\sigma_1^{-1}\sigma_2^{-1}$
6. $\sigma_2^{-1}\sigma_1^{-1}\sigma_2$ 5+ terms in the Poincaré series
7. $\sigma_2^{-1}\sigma_1\sigma_2^{-1}$
9. $\sigma_1^{-1}\sigma_2^{-1}\sigma_2^{-1}$

The leading order terms in the series for the wavenumbers are then given by

$$k_1 = 2A_1 + \sigma_2 A_1 - \sigma_2(-A_1) + \sigma_2^- A_1 - \sigma_2^-(-A_1) + \cdots$$

$$k_2 = 2A_2 + \sigma_1 A_2 - \sigma_1(-A_2) + \sigma_1^- A_2 - \sigma_1^-(-A_2) + \cdots$$

The frequencies have the following form

$$\omega_n = 2A_n^3 + \sum_{\sigma \in G/G_n,\, \sigma \neq I} ((\sigma A_n)^3 - (\sigma(-A_n))^3)$$

For $n = 1$

$$\omega_1 = 2A_1^3 + (\sigma_2 A_1)^3 - (\sigma_2(-A_1))^3 + (\sigma_2^- A_1)^3 - (\sigma_2^-(-A_1))^3 + \cdots$$

$$\omega_2 = 2A_2^3 + (\sigma_1 A_2)^3 - (\sigma_1(-A_2))^3 + (\sigma_1^- A_2)^3 - (\sigma_1^-(-A_2))^3 + \cdots$$

These results give some idea about the behavior of the leading order corrections in the Poincaré series.

Now look at the period matrix, first the *diagonal terms*

$$B_{nn} = \log \rho_n + \sum_{\sigma \in G_n \backslash G/G_n,\, \sigma \neq I} \ln\left(\frac{A_n - \sigma A_n}{A_n - \sigma(-A_n)}\right)^2$$

The first diagonal element is:

$$B_{11} = \log \rho_1 + \sum_{\sigma \in G_1 \backslash G/G_1,\, \sigma \neq I} \ln\left(\frac{A_1 - \sigma A_1}{A_1 - \sigma(-A_1)}\right)^2$$

For the decimated group elements: $\sigma_2, \sigma_2^-, \sigma_2^{-1}\sigma_2^{-1}, \sigma_2\sigma_2$, and we get

$$B_{11} = \log \rho_1 + \ln\left(\frac{A_1 - \sigma_2 A_1}{A_1 - \sigma_2(-A_1)}\right)^2 + \ln\left(\frac{A_1 - \sigma_2^- A_1}{A_1 - \sigma_2^-(-A_1)}\right)^2$$

$$+ \ln\left(\frac{A_1 - \sigma_2\sigma_2 A_1}{A_1 - \sigma_2\sigma_2(-A_1)}\right)^2 + \ln\left(\frac{A_1 - \sigma_2^-\sigma_2^- A_1}{A_1 - \sigma_2^-\sigma_2^-(-A_1)}\right)^2 + \cdots$$

For the second diagonal element we have

$$B_{22} = \log \rho_1 + \sum_{\sigma \in G_2 \backslash G/G_2,\, \sigma \neq I} \ln\left(\frac{A_2 - \sigma A_2}{A_2 - \sigma(-A_2)}\right)^2$$

for which the decimated group elements are: $\sigma_1, \sigma_1^-, \sigma_1^{-1}\sigma_1^{-1}, \sigma_1\sigma_1$,

$$B_{22} = \log \rho_2 + \ln\left(\frac{A_1 - \sigma_1 A_2}{A_2 - \sigma_1(-A_2)}\right)^2 + \ln\left(\frac{A_2 - \sigma_1^- A_2}{A_2 - \sigma_1^-(-A_2)}\right)^2$$

$$+ \ln\left(\frac{A_2 - \sigma_1\sigma_1 A_2}{A_2 - \sigma_1\sigma_1(-A_2)}\right)^2 + \ln\left(\frac{A_2 - \sigma_1^-\sigma_1^- A_2}{A_2 - \sigma_1^-\sigma_1^-(-A_2)}\right)^2 + \cdots$$

Then the *off-diagonal terms* are given by

$$B_{mn} = \ln\left(\frac{A_n - A_m}{A_n + A_m}\right)^2 + \sum_{\substack{\sigma \in G_n \backslash G/G_m,\, \sigma \notin I \\ m \neq n}} \ln\left(\frac{A_n - \sigma A_m}{A_n - \sigma(-A_m)}\right)^2$$

For the first off-diagonal term

$$B_{12} = \ln\left(\frac{A_2 - A_1}{A_2 + A_1}\right)^2 + \sum_{\sigma \in G_2 \backslash G/G_1,\, \sigma \notin I} \ln\left(\frac{A_2 - \sigma A_1}{A_2 - \sigma(-A_1)}\right)^2$$

The decimated group elements are (none occur at the single product level): $\sigma_1^{-1}\sigma_2^{-1}, \sigma_1^{-1}\sigma_2, \sigma_1\sigma_2^{-1}, \sigma_1\sigma_2$. We get

$$B_{12} = \ln\left(\frac{A_2 - A_1}{A_2 + A_1}\right)^2 + \ln\left(\frac{A_2 - \sigma_1^{-1}\sigma_2^{-1}A_1}{A_2 - \sigma_1^{-1}\sigma_2^{-1}(-A_1)}\right)^2 + \ln\left(\frac{A_2 - \sigma_1^{-1}\sigma_2 A_1}{A_2 - \sigma_1^{-1}\sigma_2(-A_1)}\right)^2$$

$$+ \ln\left(\frac{A_2 - \sigma_1\sigma_2^{-1}A_1}{A_2 - \sigma_1\sigma_2^{-1}(-A_1)}\right)^2 + \ln\left(\frac{A_2 - \sigma_1\sigma_2 A_1}{A_2 - \sigma_1\sigma_2(-A_1)}\right)^2 + \cdots$$

and for the second off-diagonal term:

$$B_{21} = \ln\left(\frac{A_1 - A_2}{A_1 + A_2}\right)^2 + \sum_{\sigma \in G_1 \backslash G/G_2,\, \sigma \notin I} \ln\left(\frac{A_1 - \sigma A_2}{A_1 - \sigma(-A_2)}\right)^2$$

The decimated group elements are (none occur at the single product level): $\sigma_2^{-1}\sigma_1^{-1}, \sigma_2^{-1}\sigma_1, \sigma_2\sigma_1^{-1}, \sigma_2\sigma_1$. Get

$$B_{12} = \ln\left(\frac{A_2 - A_1}{A_2 + A_1}\right)^2 + \ln\left(\frac{A_2 - \sigma_2^{-1}\sigma_1^{-1}A_1}{A_2 - \sigma_2^{-1}\sigma_1^{-1}(-A_1)}\right)^2 + \ln\left(\frac{A_2 - \sigma_2^{-1}\sigma_1 A_1}{A_2 - \sigma_2^{-1}\sigma_1(-A_1)}\right)^2$$

$$+ \ln\left(\frac{A_2 - \sigma_2\sigma_1^{-1}A_1}{A_2 - \sigma_2\sigma_1^{-1}(-A_1)}\right)^2 + \ln\left(\frac{A_2 - \sigma_2\sigma_1 A_1}{A_2 - \sigma_2\sigma_1(-A_1)}\right)^2 + \cdots$$

Note that we have $B_{12} = B_{21}$ as required.

The reader by now is familiar with Poincaré series and should have no problem in programming them. Additional details for the KdV and KP equations are discussed in Chapter 32.

Appendix I: Schottky Uniformization in the Small-Amplitude Limit of the Oscillation Basis

Here is the algorithm as it is now employed. Shown only is the *small ampli-tude, oscillatory limit* of the formulation.

Compute the Images of the Floquet Eigenvalues in the Schottky Domain

$$e_i = (E_i - E_1)^{1/2}, \quad 1 \leq i \leq 2N+1$$

where N is the number of degrees of freedom (number of sine waves). This includes a simple shift to put the *first* Floquet eigenvalue at the origin, which of course does not change the spectrum.

Compute Schottky Parameters

$A_i = \sqrt{e_{2i}e_{2i+1}}$ (geometric mean of images in an open band, $\mathbf{A} = [A_1, A_2, \ldots, A_N]$)

$C_i = (e_{2i} + e_{2i+1})/2$ (arithmetic mean of images in an open band)

$\rho_i = \dfrac{1}{4}\left[\dfrac{\sqrt{e_{2i}} - \sqrt{e_{2i+1}}}{\sqrt{e_{2i}} + \sqrt{e_{2i+1}}}\right]^2$ (determines diagonal elements of period matrix)

Period Matrix in Oscillatory Basis

$D_{ii} = \ln \rho_i$ (diagonal elements)

$D_{ij} = \ln\left[\dfrac{A_j - A_i}{A_j + A_i}\right]^2$ (off-diagonal elements)

$\mathbf{B}_{\mathrm{osc}} = -\mathbf{D}/2\pi$ (period matrix in oscillatory basis, $\mathbf{D} = \{D_{ij}\}, \mathbf{B}_{\mathrm{osc}} = \{B_{\mathrm{osc},ij}\}$)

Period Matrix in Soliton Basis by Modular Transformation

$\mathbf{B}_{\mathrm{sol}} = \mathbf{A}_{\mathrm{o}}\mathbf{B}_{\mathrm{osc}}\mathbf{A}_{\mathrm{o}}^{\mathrm{T}}$ (period matrix in soliton representation, $\mathbf{B}_{\mathrm{osc}} = \{B_{\mathrm{osc},ij}\}, \mathbf{B}_{\mathrm{sol}} = \{B_{\mathrm{sol},ij}\}$)

$\mathbf{A}_{\mathrm{o}} = \begin{pmatrix} 1 & 0 & 0 & 0 \\ -1 & 1 & 0 & 0 \\ 0 & -1 & 1 & 0 \\ 0 & 0 & -1 & 1 \end{pmatrix}$ (matrix in modular transformation)

Wavenumbers in Oscillatory Basis

$\mathbf{K}_{\text{osc}} = \mathbf{A}/\pi$ 　　　　(Wavenumbers in oscillatory basis, $\mathbf{A} = [A_1, A_2 \ldots A_N]$, $\mathbf{K} = [K_1, K_2 \ldots K_N]$)

Wavenumbers in Soliton Basis by Modular Transformation

$\mathbf{K}_{\text{sol}} = \mathbf{A}_\text{o}\mathbf{K}_{\text{osc}}$ 　　　　(wavenumbers in *soliton basis*, $\mathbf{K} = [K_1, K_2, \ldots, K_N]$)

Appendix II: Schottky Uniformization in the Large-Amplitude Limit of the Soliton Basis

Here is the leading order algorithm in the soliton basis. Shown only is the *soliton limit of the formulation*.

Compute the Images of the Floquet Eigenvalues in the Schottky Domain

$$e_i = |E_i - E_{2N+1}|^{1/2}, \quad 1 \leq i \leq 2N + 1$$

where N is the number of degrees of freedom (number of solitons). This is just a simple shift to put the *last Floquet eigenvalue* at the origin, followed by an absolute value.

Compute Schottky Parameters

$A_i = \sqrt{e_{2i-1}e_{2i}}$ 　　　　(geometric mean of images in a *gap*, $\mathbf{A} = [A_1, A_2, \ldots, A_N]$)

$C_i = (e_{2i-1} + e_{2i})/2$ 　　　　(arithematic mean of images in a *gap*)

$\rho_i = \dfrac{1}{4}\left[\dfrac{\sqrt{e_{2i}} - \sqrt{e_{2i-1}}}{\sqrt{e_{2i}} + \sqrt{e_{2i-1}}}\right]^2$ 　　　　(determines diagonal elements of period matrix)

Period Matrix in Soliton Basis

$D_{ii} = \ln \rho_i$ 　　　　(diagonal elements)

$D_{ij} = \ln\left[\dfrac{A_j - A_i}{A_j + A_i}\right]^2$ 　　　　(off-diagonal elements)

$\mathbf{B}_{\text{sol}} = (2\pi)^2\mathbf{D}^{-1}$ 　　　　(period matrix in soliton basis, $\mathbf{D} = \{D_{ij}\}, \mathbf{B}_{\text{sol}} = \{B_{\text{sol},ij}\}$)

Period Matrix in Oscillatory Basis

$$\mathbf{B}_{\text{osc}} = \mathbf{A}_o^{-1}\mathbf{B}_{\text{sol}}\left(\mathbf{A}_o^{T}\right)^{-1}$$

(period matrix in oscillatory basis via a modular transformation; inverse is

$$\mathbf{B}_{\text{sol}} = \mathbf{A}_o\mathbf{B}_{\text{osc}}\mathbf{A}_o^{T}, \mathbf{B}_{\text{osc}} = \{B_{\text{osc},ij}\})$$

$$\mathbf{A}_o = \begin{pmatrix} 1 & 0 & 0 & 0 \\ -1 & 1 & 0 & 0 \\ 0 & -1 & 1 & 0 \\ 0 & 0 & -1 & 1 \end{pmatrix}$$

(matrix in modular transformation)

Wavenumbers in Soliton Basis

$$\mathbf{K}_{\text{sol}} = 2\mathbf{A}\mathbf{D}^{-1} = \frac{1}{2\pi^2}\mathbf{A}\mathbf{B}_{\text{sol}}^{-1}$$

(wavenumbers in *soliton basis*, $\mathbf{A} = [A_1, A_2, \ldots, A_N], \mathbf{K} = [K_1, K_2, \ldots, K_N]$)

Wavenumbers in Oscillatory Basis

$$\mathbf{K}_{\text{osc}} = \mathbf{A}_o^{-1}\mathbf{K}_{\text{sol}}$$

(wavesnumbers in *oscillation basis*; inverse is $\mathbf{K}_{\text{sol}} = \mathbf{A}_o\mathbf{K}_{\text{osc}}$)

$$\mathbf{A}_o^{-1} = \begin{pmatrix} 1 & 0 & 0 & 0 \\ 1 & 1 & 0 & 0 \\ 0 & 1 & 1 & 0 \\ 0 & 0 & 1 & 1 \end{pmatrix}$$

(inverse modular transformation matrix)

Appendix III: Poincaré Series from the Holomorphic Differentials

The focus of this Appendix is to derive the Poincaré series used in the numerical computation of the theta function parameters (Belokolos et al., 1994).

The Oscillation Basis of Dubrovin and Novikov

The holomorphic differentials on the Riemann surface are introduced:

$$\Omega_m(E) = \sum_{k=1}^{N} C_{km}\frac{E^{k-1}\mathrm{d}E}{R^{1/2}(E)}$$

where $R(E)$ is a polynomial given by (this is the Riemann surface for the hyperelliptic case)

$$R(E) = \prod_{n=1}^{2N+1} (E - E_n)$$

and the C_{km} are normalization coefficients defined by

$$\oint_{\alpha_j} \Omega_m(E) = 2\pi i \delta_{jm}$$

To find the C_{km} insert the holomorphic differentials into this latter expression and get

$$\oint_{\alpha_j} \Omega_m(E) = \sum_{k=1}^{N} C_{km} \oint_{\alpha_j} \frac{E^{k-1}dE}{R^{1/2}(E)} = 2\pi i \delta_{jm}$$

$$\oint_{\alpha_j} \Omega_m(E) = \sum_{k=1}^{N} J_{jk} C_{km} = 2\pi i \delta_{jm}$$

where

$$J_{jk} = \oint_{\alpha_j} \frac{E^{k-1}dE}{R^{1/2}(E)}$$

In matrix notation

$$\mathbf{J}\mathbf{C} = 2\pi i \mathbf{1}$$

$$\mathbf{C} = 2\pi i \mathbf{J}^{-1}$$

or

$$C_{jk} = 2\pi i \left[\oint_{\alpha_j} \frac{E^{k-1}dE}{R^{1/2}(E)} \right]^{-1}$$

The *phases* X_m of the θ-function (Equation (6.29)) are found by the following Abelian integrals

$$X_m(P_1, P_2, \ldots, P_m) = -i \sum_{j-1}^{N} \int_{E_{2j}}^{P_j(0,0)} \Omega_m(E) = K_m x - \omega_m t + \phi_m$$

where $P_j(x,t) = [\mu_j(x,t),\sigma_j]$ for $1 \leq j \leq N$. This leads, upon substitution of the holomorphic differentials for the wavenumbers, frequencies, and phases:

$$K_m = 2C_{N,m}$$

$$\omega_m = 8C_{N-1,m} + 4C_{N,m} \sum_{k=1}^{2N+1} E_k$$

$$\phi_m = -i \sum_{j=1}^{N} \oint_{E_{2j}}^{P_j(0,0)} \Omega_m(E)$$

Note that the phases are new information and depend on the phases of the hyperelliptic functions at $x = 0$, $t = 0$. Specifically the phases have the form:

$$\phi_m = \sum_{j=1}^{N} E_{jm}, \quad \mathbf{E} = \mathbf{FC}$$

where

$$F_{jk} = i \oint_{E_{2j}}^{P_j(0,0)} \frac{E^{k-1} dE}{R^{1/2}(E)}$$

The *period matrix* is given by

$$B_{jk} = B_{kj} = \oint_{\beta_k} \Omega_j(E)$$

which leads to the period matrix

$$\mathbf{B} = \mathbf{AC}$$

where

$$A_{ki} = \oint_{\beta_k} \frac{E^{i-1} dE}{R^{1/2}(E)}$$

The Oscillation Basis in the Schottky Domain Due to Bobenko

The *holomorphic differentials* in the *Schottky plane* are given by:

$$\Omega_m(E) = \sum_{\sigma \in \Gamma / G_n} \left(\frac{1}{z - \sigma(-A_n)} - \frac{1}{z - \sigma A_n} \right) dz$$

The *period matrix* has the definition

$$B_{mn} = \int_z^{\sigma_n z} \Omega_m$$

The normalization conditions are

$$\int_{a_n} \omega_m = 2\pi i \sigma_{mn}$$

This leads to expressions for the period matrix:

$$B_{mn} = \log \rho_n + \sum_{\sigma \in G_n \backslash G / G_m, \, \sigma \neq I} \log\{-A_n, A_n, \sigma(-A_m), \sigma A_m\}$$

$$B_{mn} = \sum_{\sigma \in G_n \backslash G / G_m} \log\{-A_n, A_n, \sigma(-A_m), \sigma A_m\}, m \neq n$$

where the curly brackets indicate the *cross ratio*:

$$\{z_1, z_2, z_3, z_4\} = \frac{(z_1 - z_3)(z_2 - z_4)}{(z_1 - z_4)(z_2 - z_3)}$$

The final forms given by Bobenko (refined somewhat from those in Baker (1897) are:

$$B_{nn} = \log \rho_n + \sum_{\sigma \in G_n \backslash G / G_n, \, \sigma \neq I} \ln \left(\frac{A_n - \sigma A_n}{A_n - \sigma(-A_n)} \right)^2$$

$$B_{mn} = \sum_{\substack{\sigma \in G_n \backslash G / G_m \\ m \neq n}} \ln \left(\frac{A_n - \sigma A_m}{A_n - \sigma(-A_m)} \right)^2$$

The solution of KPII is then of the form:

$$u(x, y, z) = 2\frac{\partial^2}{\partial x^2}\log\theta(\kappa x + \lambda y + \omega t + D) + 2c$$

We have

$$\kappa_n = f_n(0), \quad \lambda_n = \frac{d}{dp}f_n(p)\bigg|_{p=0}, \quad \omega_n = \frac{1}{2}\frac{d^2}{dp^2}f_n(p)\bigg|_{p=0}$$

where $f(p)$ is defined below, and we show that:

$$\kappa_n = \sum_{\sigma \in G/G_n}(\sigma A_n - \sigma(-A_n))$$

$$\lambda_n = \sum_{\sigma \in G/G_n}((\sigma A_n)^2 - (\sigma(-A_n))^2)$$

$$\omega_n = \sum_{\sigma \in G/G_n}((\sigma A_n)^3 - (\sigma(-A_n))^3)$$

Now let us evaluate the last three Poincaré series. The holomorphic differentials are given by:

$$\Omega_m(E) = \sum_{\sigma \in G/G_n}\left(\frac{1}{z - \sigma(-A_n)} - \frac{1}{z - \sigma A_n}\right)dz$$

Make the transformation $z \to 1/p$ and get

$$\Omega_m = f_n(p)dp = \sum_{\sigma \in G/G_n}\left(\frac{1}{\frac{1}{p} - \sigma(-A_n)} - \frac{1}{\frac{1}{p} - \sigma A_n}\right)\frac{dz}{dp}dp, \quad \text{for} \quad \frac{dz}{dp} = -\frac{dp}{p^2}$$

and

$$\Omega_m = f_n(p)dp = -\sum_{\sigma \in G/G_n}\left(\frac{1}{\frac{1}{p} - \sigma(-A_n)} - \frac{1}{\frac{1}{p} - \sigma A_n}\right)\frac{dp}{p^2}$$

for which we find

$$f_n(p) = \sum_{\sigma \in G/G_n}\frac{(\sigma A_n - \sigma(-A_n))}{(1 - p\sigma(-A_n))(1 - p\sigma A_n)}$$

So that

$$f_n(0) = \sum_{\sigma \in G/G_n} (\sigma A_n - \sigma(-A_n))$$

Then

$$\frac{\mathrm{d}}{\mathrm{d}p} f_n(p) = \sum_{\sigma \in G/G_n} \frac{(\sigma A_n - \sigma(-A_n))(\sigma(-A_n) - \sigma A_n(2p\sigma(-A_n) - 1))}{(1 - p\sigma A_n)^2 (1 - p\sigma(-A_n))^2}$$

and then

$$\left.\frac{\mathrm{d}}{\mathrm{d}p} f_n(p)\right|_{p=0} = \sum_{\sigma \in G/G_n} ((\sigma A_n)^2 - (\sigma(-A_n))^2)$$

Likewise we find

$$\left.\frac{1}{2}\frac{\mathrm{d}^2}{\mathrm{d}p^2} f_n(p)\right|_{p=0} = \sum_{\sigma \in G/G_n} ((\sigma A_n)^3 - (\sigma(-A_n))^3)$$

and the fundamental Poincaré series arise as anticipated.

Appendix IV: One Degree-of-Freedom Schottky z-Plane to IST E-Plane Poincaré Series

For those interested in developing a program for testing the Schottky method for the one degree-of-freedom case I give a few analytical results which may prove useful.

We have the Poincaré series:

$$E(z) - E_1 = z^2 + [(\sigma_1 z)^2 - (\sigma_1 0)^2] + [(\sigma_1^- z)^2 - (\sigma_1^- 0)^2]$$

$$+ [(\sigma_1 \sigma_1 z)^2 - (\sigma_1 \sigma_1 0)^2] + [(\sigma_1^- \sigma_1^- z)^2 - (\sigma_1^- \sigma_1^- 0)^2]$$

$$+ [(\sigma_1 \sigma_1 \sigma_1 z)^2 - (\sigma_1 \sigma_1 \sigma_1 0)^2] + [(\sigma_1^- \sigma_1^- \sigma_1^- z)^2 - (\sigma_1^- \sigma_1^- \sigma_1^- 0)^2] + \dots$$

For the one dof case we have (there are three IST eigenvalues so that n = 1, 2, 3)

$$E_n - E_1 = e_n^2 + [(\sigma_1 e_n)^2 - (\sigma_1 0)^2] + [(\sigma_1^- e_n)^2 - (\sigma_1^- 0)^2]$$

$$+ [(\sigma_1 \sigma_1 e_n)^2 - (\sigma_1 \sigma_1 0)^2] + [(\sigma_1^- \sigma_1^- e_n)^2 - (\sigma_1^- \sigma_1^- 0)^2]$$

$$+ [(\sigma_1 \sigma_1 \sigma_1 e_n)^2 - (\sigma_1 \sigma_1 \sigma_1 0)^2] + [(\sigma_1^- \sigma_1^- \sigma_1^- e_n)^2 - (\sigma_1^- \sigma_1^- \sigma_1^- 0)^2] + \dots$$

For $n = 1$ we have $e_n = 0$ so that we get the expected result

$$E_n - E_1 = 0$$

For $n = 2$ we have

$$E_2 - E_1 = e_2^2 + [(\sigma_1 e_2)^2 - (\sigma_1 0)^2] + [(\sigma_1^- e_2)^2 - (\sigma_1^- 0)^2]$$
$$+ [(\sigma_1 \sigma_1 e_2)^2 - (\sigma_1 \sigma_1 0)^2] + [(\sigma_1^- \sigma_1^- e_2)^2 - (\sigma_1^- \sigma_1^- 0)^2]$$
$$+ [(\sigma_1 \sigma_1 \sigma_1 e_2)^2 - (\sigma_1 \sigma_1 \sigma_1 0)^2] + [(\sigma_1^- \sigma_1^- \sigma_1^- e_2)^2 - (\sigma_1^- \sigma_1^- \sigma_1^- 0)^2] + \cdots$$

For $n = 3$ we have

$$E_3 - E_1 = e_3^2 + \left[(\sigma_1 e_3)^2 - (\sigma_1 0)^2\right] + \left[(\sigma_1^- e_3)^2 - (\sigma_1^- 0)^2\right]$$
$$+ \left[(\sigma_1 \sigma_1 e_3)^2 - (\sigma_1 \sigma_1 0)^2\right] + \left[(\sigma_1^- \sigma_1^- e_3)^2 - (\sigma_1^- \sigma_1^- 0)^2\right]$$
$$+ \left[(\sigma_1 \sigma_1 \sigma_1 e_3)^2 - (\sigma_1 \sigma_1 \sigma_1 0)^2\right] + \left[(\sigma_1^- \sigma_1^- \sigma_1^- e_3)^2 - (\sigma_1^- \sigma_1^- \sigma_1^- 0)^2\right] + \cdots$$

where

$$\tilde{\sigma} = \begin{bmatrix} \dfrac{1}{2}\left(\dfrac{1+\rho}{\sqrt{\rho}}\right) & -\dfrac{A}{2}\left(\dfrac{1-\rho}{\sqrt{\rho}}\right) \\[3mm] -\dfrac{1}{2A}\left(\dfrac{1-\rho}{\sqrt{\rho}}\right) & \dfrac{1}{2}\left(\dfrac{1+\rho}{\sqrt{\rho}}\right) \end{bmatrix}$$

and its inverse are

$$\tilde{\sigma}^{-1} = \begin{bmatrix} \dfrac{1}{2}\left(\dfrac{1+\rho}{\sqrt{\rho}}\right) & \dfrac{A}{2}\left(\dfrac{1-\rho}{\sqrt{\rho}}\right) \\[3mm] \dfrac{1}{2A}\left(\dfrac{1-\rho}{\sqrt{\rho}}\right) & \dfrac{1}{2}\left(\dfrac{1+\rho}{\sqrt{\rho}}\right) \end{bmatrix}$$

The following results are found to hold:

$$\sigma_1(0) = -A\left(\dfrac{1-\rho}{1+\rho}\right)$$

$$\sigma_1^2(0) = \sigma_1\sigma_1(0) = -A\left(\dfrac{1-\rho^2}{1+\rho^2}\right)$$

So that

$$\sigma_1^n(0) = -A\left(\frac{1-\rho^n}{1+\rho^n}\right)$$

Likewise

$$\sigma_1^{-n}(0) = A\left(\frac{1-\rho^n}{1+\rho^n}\right) = -\sigma_1^n(0)$$

and

$$\sigma_1^n(e_2) = -A\left(\frac{1-\rho^{n-1/2}}{1+\rho^{n-1/2}}\right). \quad \text{Note:}\ \sigma_1^1(e_2) = -A\left(\frac{1-\rho^{1/2}}{1+\rho^{1/2}}\right) = -e_2$$

$$\sigma_1^{-n}(e_2) = A\left(\frac{1-\rho^{n+1/2}}{1+\rho^{n+1/2}}\right)$$

Finally

$$\sigma_1^n(e_3) = -A\left(\frac{1+\rho^{n-1/2}}{1-\rho^{n-1/2}}\right). \quad \text{Note:}\ \sigma_1^1(e_3) = -A\left(\frac{1+\rho^{1/2}}{1-\rho^{1/2}}\right) = -e_3$$

$$\sigma_1^{-n}(e_3) = A\left(\frac{1+\rho^{n+1/2}}{1-\rho^{n+1/2}}\right)$$

For $n = 2$ we have

$$\begin{aligned}
E_2 - E_1 = e_2^2 &+ \left[\left(-A\left(\frac{1-\rho^{1/2}}{1+\rho^{1/2}}\right)\right)^2 - \left(-A\left(\frac{1-\rho}{1+\rho}\right)\right)^2\right] \\
&+ \left[\left(A\left(\frac{1-\rho^{3/2}}{1+\rho^{3/2}}\right)\right)^2 - \left(A\left(\frac{1-\rho}{1+\rho}\right)\right)^2\right] \\
&+ \left[\left(-A\left(\frac{1-\rho^{3/2}}{1+\rho^{3/2}}\right)\right)^2 - \left(-A\left(\frac{1-\rho^2}{1+\rho^2}\right)\right)^2\right] \\
&+ \left[\left(A\left(\frac{1-\rho^{5/2}}{1+\rho^{5/2}}\right)\right)^2 - \left(A\left(\frac{1-\rho^2}{1+\rho^2}\right)\right)^2\right] \\
&+ \left[\left(-A\left(\frac{1-\rho^{5/2}}{1+\rho^{5/2}}\right)\right)^2 - \left(-A\left(\frac{1-\rho^3}{1+\rho^3}\right)\right)^2\right] \\
&+ \left[\left(A\left(\frac{1-\rho^{7/2}}{1+\rho^{7/2}}\right)\right)^2 - \left(A\left(\frac{1-\rho^3}{1+\rho^3}\right)\right)^2\right] + \cdots
\end{aligned}$$

then

$$E_2 - E_1 = e_2^2 + A^2 \left\{ \left[\left(\frac{1 - \rho^{1/2}}{1 + \rho^{1/2}}\right)^2 - \left(\frac{1 - \rho}{1 + \rho}\right)^2 \right] + \left[\left(\frac{1 - \rho^{3/2}}{1 + \rho^{3/2}}\right)^2 - \left(\frac{1 - \rho}{1 + \rho}\right)^2 \right] \right.$$

$$+ \left[\left(\frac{1 - \rho^{3/2}}{1 + \rho^{3/2}}\right)^2 - \left(\frac{1 - \rho^2}{1 + \rho^2}\right)^2 \right] + \left[\left(\frac{1 - \rho^{5/2}}{1 + \rho^{5/2}}\right)^2 - \left(\frac{1 - \rho^2}{1 + \rho^2}\right)^2 \right]$$

$$\left. + \left[\left(\frac{1 - \rho^{5/2}}{1 + \rho^{5/2}}\right)^2 - \left(\frac{1 - \rho^3}{1 + \rho^3}\right)^2 \right] + \left[\left(\frac{1 - \rho^{7/2}}{1 + \rho^{7/2}}\right)^2 - \left(\frac{1 - \rho^3}{1 + \rho^3}\right)^2 \right] + \cdots \right\}$$

Recall that

$$e_2 = A\left(\frac{1 - \sqrt{\rho}}{1 + \sqrt{\rho}}\right)$$

The first two terms in the above series combine, and the other terms give pairs so that we have:

$$E_2 - E_1 = 2A^2 \sum_{n=0}^{\infty} \left[\left(\frac{1 - \rho^{n+1/2}}{1 + \rho^{n+1/2}}\right)^2 - \left(\frac{1 - \rho^n}{1 + \rho^n}\right)^2 \right]$$

Evaluating explicitly the first term gives:

$$E_2 - E_1 = 2e_2^2 + 2A^2 \sum_{n=1}^{\infty} \left[\left(\frac{1 - \rho^{n+1/2}}{1 + \rho^{n+1/2}}\right)^2 - \left(\frac{1 - \rho^n}{1 + \rho^n}\right)^2 \right]$$

This is an exact expression.

Now the next case is for e_3. From the above we have:

$$E_3 - E_1 = e_3^2 + [(\sigma_1 e_3)^2 - (\sigma_1 0)^2] + [(\sigma_1^- e_3)^2 - (\sigma_1^- 0)^2]$$

$$+ [(\sigma_1 \sigma_1 e_3)^2 - (\sigma_1 \sigma_1 0)^2] + [(\sigma_1^- \sigma_1^- e_3)^2 - (\sigma_1^- \sigma_1^- 0)^2]$$

$$+ [(\sigma_1 \sigma_1 \sigma_1 e_3)^2 - (\sigma_1 \sigma_1 \sigma_1 0)^2] + [(\sigma_1^- \sigma_1^- \sigma_1^- e_3)^2 - (\sigma_1^- \sigma_1^- \sigma_1^- 0)^2] + \cdots$$

where

$$\sigma_1^n e_3 = -A\left(\frac{1 + \rho^{n-1/2}}{1 - \rho^{n-1/2}}\right). \quad \text{Note: } \sigma_1^{-n} e_3 = A\left(\frac{1 + \rho^{n+1/2}}{1 - \rho^{n+1/2}}\right) = -e_3$$

$$\sigma_1^{-n} e_3 = A\left(\frac{1 + \rho^{n+1/2}}{1 - \rho^{n+1/2}}\right)$$

and

$$\sigma_1^n 0 = -A\left(\frac{1-\rho^n}{1+\rho^n}\right)$$

Likewise

$$\sigma_1^{-n}(0) = A\left(\frac{1-\rho^n}{1+\rho^n}\right)$$

Have

$$E_3 - E_1 = e_3^2 + \left\{ [(\sigma_1 e_3)^2 - (\sigma_1 0)^2] + [(\sigma_1^- e_3)^2 - (\sigma_1^- 0)^2] + [(\sigma_1^2 e_3)^2 - (\sigma_1^2 0)^2] \right.$$
$$\left. + [(\sigma_1^{-2} e_3)^2 - (\sigma_1^{-2} 0)^2] + [(\sigma_1^3 e_3)^2 - (\sigma_1^3 0)^2] + [(\sigma_1^{-3} e_3)^2 - (\sigma_1^{-3} 0)^2] + \ldots \right\}$$

So that

$$E_3 - E_1 = e_3^2 + \left\{ \left[\left(A\left(\frac{1+\rho^{1/2}}{1-\rho^{1/2}}\right) \right)^2 - \left(A\left(\frac{1-\rho}{1+\rho}\right) \right)^2 \right] \right.$$

$$+ \left[\left(A\left(\frac{1+\rho^{3/2}}{1-\rho^{3/2}}\right) \right)^2 - \left(A\left(\frac{1-\rho}{1+\rho}\right) \right)^2 \right]$$

$$+ \left[\left(A\left(\frac{1+\rho^{3/2}}{1-\rho^{3/2}}\right) \right)^2 - \left(A\left(\frac{1-\rho^2}{1+\rho^2}\right) \right)^2 \right]$$

$$+ \left[\left(A\left(\frac{1+\rho^{5/2}}{1-\rho^{5/2}}\right) \right)^2 - \left(A\left(\frac{1-\rho^2}{1+\rho^2}\right) \right)^2 \right]$$

$$+ \left[\left(A\left(\frac{1+\rho^{5/2}}{1-\rho^{5/2}}\right) \right)^2 - \left(A\left(\frac{1-\rho^3}{1+\rho^3}\right) \right)^2 \right]$$

$$\left. + \left[\left(A\left(\frac{1+\rho^{7/2}}{1-\rho^{7/2}}\right) \right)^2 - \left(A\left(\frac{1-\rho^3}{1+\rho^3}\right) \right)^2 \right] + \ldots \right\}$$

$$E_3 - E_1 = 2e_3^2 + 2A^2 \sum_{n=1}^{\infty} \left[\left(\frac{1+\rho^{n+1/2}}{1-\rho^{n+1/2}}\right)^2 - \left(\frac{1-\rho^n}{1+\rho^n}\right)^2 \right]$$

or

$$E_3 - E_1 = 2A^2 \sum_{n=0}^{\infty} \left[\left(\frac{1+\rho^{n+1/2}}{1-\rho^{n+1/2}}\right)^2 - \left(\frac{1-\rho^n}{1+\rho^n}\right)^2 \right]$$

These are exact expressions.

Now let us return to the expression for $n = 2$ we have

$$E_2 - E_1 = e_2^2 + [(\sigma_1 e_2)^2 - (\sigma_1 0)^2] + [(\sigma_1^- e_2)^2 - (\sigma_1^- 0)^2]$$
$$+ [(\sigma_1 \sigma_1 e_2)^2 - (\sigma_1 \sigma_1 0)^2] + [(\sigma_1^- \sigma_1^- e_2)^2 - (\sigma_1^- \sigma_1^- 0)^2]$$
$$+ [(\sigma_1 \sigma_1 \sigma_1 e_2)^2 - (\sigma_1 \sigma_1 \sigma_1 0)^2] + [(\sigma_1^- \sigma_1^- \sigma_1^- e_2)^2 - (\sigma_1^- \sigma_1^- \sigma_1^- 0)^2] + \ldots$$

Now use the facts that

$$\sigma_1^{-n} 0 = -\sigma_1^n 0, \quad \sigma_1^{n+1} e_2 = \sigma_1^{-n} e_2$$

This can be written

$$E_2 - E_1 = 2e_2^2 + 2[(\sigma_1^2 e_2)^2 - (\sigma_1 0)^2] + 2[(\sigma_1^3 e_2)^2 - (\sigma_1^2 0)^2]$$
$$+ 2[(\sigma_1^4 e_2)^2 - (\sigma_1^3 0)^2] + 2[(\sigma_1^5 e_2)^2 - (\sigma_1^4 0)^2] + \ldots$$

so that, finally

$$E_2 - E_1 = 2e_2^2 + 2\sum_{n=1}^{\infty}[(\sigma_1^{n+1} e_2)^2 - (\sigma_1^n 0)^2]$$

Inserting the above results we get to the series already discussed above:

$$E_2 - E_1 = 2e_2^2 + 2A^2 \sum_{n=1}^{\infty} \left[\left(\frac{1 - \rho^{n+1/2}}{1 + \rho^{n+1/2}}\right)^2 - \left(\frac{1 - \rho^n}{1 + \rho^n}\right)^2 \right]$$

For $n = 3$ we have

$$E_3 - E_1 = 2e_3^2 + 2[(\sigma_1^2 e_3)^2 - (\sigma_1 0)^2] + 2[(\sigma_1^3 e_3)^2 - (\sigma_1^2 0)^2]$$
$$+ 2[(\sigma_1^4 e_3)^2 - (\sigma_1^3 0)^2] + 2[(\sigma_1^5 e_3)^2 - (\sigma_1^4 0)^2] + \ldots$$

or

$$E_3 - E_1 = 2e_3^2 + 2\sum_{n=1}^{\infty}[(\sigma_1^{n+1} e_3)^2 - (\sigma_1^n 0)^2]$$

and finally

$$E_3 - E_1 = 2e_3^2 + 2A^2 \sum_{n=1}^{\infty} \left[\left(\frac{1 + \rho^{n+1/2}}{1 - \rho^{n+1/2}}\right)^2 - \left(\frac{1 - \rho^n}{1 + \rho^n}\right)^2 \right]$$

16 Nakamura-Boyd Approach

16.1 Introduction

This chapter discusses how to determine the *Riemann spectrum* of the Korteweg-deVries (KdV) equation with periodic boundary conditions using the method of Nakamura and Boyd. The KdV equation has the form:

$$\eta_t + c_o\eta_x + \alpha\eta\eta_x + \beta\eta_{xxx} = 0, \quad \eta(x,t) = \eta(x+L,t) \tag{16.1}$$

where $c_o = \sqrt{gh}$, $\alpha = 3c_o/2h$, and $\beta = c_oh^2/6$; h is the depth and g is the acceleration of gravity. The Riemann spectrum is of course the natural, nonlinear generalization of the ordinary, linear Fourier spectrum for the KdV equation. This chapter deals with certain aspects of the determination of the Riemann spectrum for numerical modeling and data analysis purposes.

The Riemann spectrum for the KdV equation consists of the Riemann matrix, frequencies and phases. The Riemann matrix diagonal elements specify the *nonlinear modes* (for the KdV equation these are cnoidal waves) and the off-diagonal terms provide the *nonlinear interactions* among the nonlinear modes. The frequencies are corrected for nonlinear interactions in a natural way. Any set of Riemann phases solves the KdV equation. A particular set of phases solves the Cauchy problem. This chapter uses the approach of Nakamura (1980), Nakamura and Matsuno (1980), Hirota and Ito (1981), and Boyd (1984a,b,c) to determine the Riemann spectrum and *nonlinear dispersion relation* for the KdV equation. We have two kinds of problems that are of interest:

(1) Given the Cauchy initial condition, $\eta(x, 0)$, determine the corresponding Riemann spectrum using the methods of Chapters 14, 17, and 19. Additionally, one can also use the methods of Chapter 23 (nonlinear adiabatic annealing) together with those of this chapter.

(2) Otherwise one can *choose the desired diagonal elements* of the Riemann matrix and then use the Nakamura-Boyd procedure (the subject of this chapter) to get the off-diagonal terms and the nonlinear dispersion relations for each of the degrees of freedom. For the KdV equation the phases can take arbitrary values. The phases can also be chosen to be random numbers, as is often assumed in the study of ocean waves. A set of zero phases places the largest possible wave for the Riemann spectrum at $x = 0$ (for a space series) or at $t = 0$ (for a time series). This technique is also well known for use with the linear Fourier transform.

Doi: 10.1016/S0074-6142(10)97016-2

16.2   The Hirota Direct Method for the KdV Equation with Periodic Boundary Conditions

The method of approach is the so-called *direct method* or *physical effectiviza-tion method* (Nakamura, 1980; Nakamura and Matsuno, 1980; Hirota and Ito, 1982; Boyd, 1994a,b,c). This method has been found to be quite useful in the *time series analysis* and *numerical modeling* of nonlinear wave trains, see Chapters 32–34. The direct method contrasts to the elegant *algebro-geometric method* (Belokolos et al., 1994) founded on *algebraic geometry* and discussed in Chapters 10, 12, 14, 17–19, and 32. Algebraic geometry provides two methods for determining the Riemann spectrum: (1) the loop integrals (Chapter 14) and (2) Schottky uniformization (Chapter 15).

Now consider the following Hirota dependent variable transformation to the dimensional KdV equation (16.1):

$$\eta(x,t) = \frac{2}{\lambda} \partial_{xx} \ln \theta(x,t), \quad \lambda = \frac{\alpha}{6\beta} \tag{16.2}$$

First make the substitution $\eta = w_x$ and integrate once in x to find:

$$w_t + c_0 w_x + \frac{\alpha}{2} w_x^2 + \beta w_{xxx} + c = 0 \tag{16.3}$$

In the search for soliton solutions, where infinite-line boundary conditions hold, it is natural to take $c = 0$. However, we are concerned with periodic boundary conditions in the present chapter and therefore c must be kept finite. Now make the final substitution $w = 2\partial_x \ln \theta$ and obtain for the terms in Equation (16.3):

$$\lambda \frac{\theta^2}{2} w_t = \theta\theta_{xt} - \theta_x\theta_t \tag{16.4}$$

$$\lambda \frac{\theta^2}{2} c_0 w_x = c_0(\theta\theta_{xx} - \theta_x^2) \tag{16.5}$$

$$\lambda \frac{\theta^2}{2} \frac{\alpha}{2} w_x = \frac{6\beta(\theta_x^2 - \theta\theta_{xx})^2}{\theta^2} \tag{16.6}$$

$$\lambda \frac{\theta^2}{2} \beta w_{xxx} = \beta \left(\frac{12\theta_x^2\theta_{xx}}{\theta} - \frac{6\theta_x^4}{\theta^2} - 3\theta_{xx}^2 - 4\theta_x\theta_{xxx} + \theta\theta_{xxxx} \right) \tag{16.7}$$

Combining Equations (16.4)–(16.7) we have the following *bilinear form*:

$$\theta\theta_{xt} - \theta_x\theta_t + c_0(\theta\theta_{xx} - \theta_x^2) + \beta(3\theta_{xx}^2 - 4\theta_x\theta_{xxx} + \theta\theta_{xxxx}) + \frac{\alpha c}{12\beta}\theta^2 = 0 \tag{16.8}$$

Notice that in the determination of Equation (16.8) the higher order nonlinear and dispersive terms partially cancel upon adding Equations (16.6) and (16.7) and we are left with a bilinear form. This is important, because without this cancellation we would be faced with solving a quadrilinear form instead of the bilinear form (Equation (16.8)). This amazing simplification is characteristic of integrable wave equations, here physically interpreted as a balance between nonlinearity and dispersion. Using the Hirota bilinear operator notation, Equation (16.8) takes the form:

$$(D_x D_t + c_o D_x^2 + \beta D_x^4 + \rho)\theta \cdot \theta = 0 \qquad (16.9)$$

where

$$\rho = \frac{\alpha c}{12\beta} \qquad (16.10)$$

and where we have used

$$D_x D_t \theta \cdot \theta = 2(\theta \theta_{xt} - \theta_x \theta_t)$$
$$D_x^2 \theta \cdot \theta = 2(\theta \theta_{xx} - \theta_x^2) \qquad (16.11)$$
$$D_x^2 \theta \cdot \theta = 2(\theta \theta_{xxxx} - 4\theta_x \theta_{xxx} + 3\theta_{xx}^2)$$

Note that the shorthand symbol $\theta \cdot \theta$ just means θ^2. Chapters 4 and 6 describe the Hirota operator notation (Hirota, 2004). Note that Equation (16.9) has the *linear* dispersion relation for the KdV equation. This consideration often motivates early explorations of the possible integrability of nonlinear wave equations.

Nota Bene: The bilinear form does not depend explicitly on the nonlinear coefficient, α, but only the linear phase speed c_o, the dispersion coefficient β, and the normalized integration constant ρ. *Therefore, the theta functions do not have the nonlinear coefficient in their bilinear form representation.* Indeed, the nonlinear coefficient, α, enters in the formulation only through the transformation $(\alpha/12\beta)\eta(x,t) = \partial_{xx} \ln \theta(x,t)$; α rescales the physical amplitude of the waves to include nonlinearity. Thus, wave motion associated with large α "scales up" the theta function to give *smaller* diagonal elements; this is the physical basis for obtaining large nonlinearity.

Nota Bene: To provide some physical insight to the bilinear form (Equation (16.8)) we note that, in analogy with the linear problem, we can associate the derivative operations with the wavenumber and frequency:

$$D_x \leftrightarrow ik, \quad D_t \leftrightarrow -i\omega$$

When these expressions are inserted into the bilinear form we have the dispersion relation

$$\omega = c_o k - \beta k^3 + \frac{\rho}{k}$$

The dispersion relation for KdV derived rigorously below is given by

---

Continued

$$\omega = c_0 k - \beta k^3 + \frac{9 c_0 a^2}{16 b^4} k^{-1}$$

which means that

$$\rho = \frac{9 c_0 a^2}{16 b^4}$$

for a single *Stokes wave solution* of the KdV equation. This identifies the integration constant ρ with the *nonlinear correction* (proportional to the small wave amplitude a^2) to the linear dispersion relation. It is clear physically that ρ must be kept finite in any complete formulation of the solutions of the periodic/quasi-periodic KdV equation. The advantage of the Nakamura-Boyd method is that we can carry the Stokes wave solution and dispersion relation out to infinite order. Furthermore we can work the problem for N interacting Stokes waves to all orders.

The dependent variable transformation (Equation (16.2)) has "almost linearized" the KdV equation by giving us the bilinear form (Equation (16.8)). While the bilinear form appears more complex than the KdV equation itself, it is important to notice that (a) Equation (16.8) is homogenous (each term in the bilinear form consists of the product of two functions from the set $\theta, \theta_t, \theta_x, \theta_{xx}, \theta_{xxx}, \theta_{xxxx}$) and (b) the actual solution of the bilinear form is linear (in terms of Riemann theta functions).

Now normalize the KdV equation

$$u = \lambda \eta, \quad x \to x - c_0 t, \quad t \to \beta t \tag{16.12}$$

to get

$$u_t + 6 u u_x + u_{xxx} = 0 \tag{16.13}$$

Then the transformation

$$u(x, t) = 2[\ln \theta(x, t)]_{xx} \tag{16.14}$$

reduces KdV to the *bilinear form*:

$$\theta \theta_{xt} - \theta_x \theta_t + \theta \theta_{xxxx} - 4 \theta_x \theta_{xxx} + 3 \theta_{xx}^2 + \rho \theta^2 = 0 \tag{16.15}$$

Here ρ is again a constant of integration. The equation can be written in *Hirota operator form* as

$$[D_x(D_t + D_x^3) + \rho]\theta \cdot \theta = 0 \tag{16.16}$$

This normalized form is that normally used in theoretical work.

16.3 Theta Functions with Characteristics

The Nakamura-Boyd approach allows one to compute the Riemann spectrum using a modification of theta functions called "theta functions with characteristics" (Baker, 1897). Write the *theta function with characteristics* in *vector form*:

$$\theta(\mathbf{X}, \mathbf{a}, \mathbf{b} \mid \mathbf{B}) = \sum_{\mathbf{m}=-\infty}^{\infty} e^{-\frac{1}{2}(\mathbf{m}+\mathbf{a})\cdot\mathbf{B}(\mathbf{m}+\mathbf{a})} \, e^{i(\mathbf{m}+\mathbf{a})\cdot(\mathbf{X}+\mathbf{b})} \tag{16.17}$$

where $\mathbf{m} = [m_1, m_2, \ldots, m_N]$ are integers, $\mathbf{a} = [a_1, a_2, \ldots, a_N]$ and $\mathbf{b} = [b_1, b_2, \ldots, b_N]$ are arbitrary constants (typically half-integers for many of the applications given herein, but can also be constants in general), $\mathbf{X} = [X_1, X_2, \ldots, X_N]$ where

$$X_j = k_j x - \omega_j t + \phi_j, \quad j = 1, 2, \ldots, N \tag{16.18}$$

The scalar form of the theta function with characteristics is given by

$$\theta(\mathbf{X}, \mathbf{a}, \mathbf{b} \mid \mathbf{B}) = \sum_{m_1=-\infty}^{\infty} \sum_{m_2=-\infty}^{\infty} \cdots \sum_{m_N=-\infty}^{\infty} \exp\left[-\frac{1}{2} \sum_{j=1}^{N} \sum_{k=1}^{N} (m_j + a_j)(m_k + a_k) B_{jk} \right]$$
$$\times \exp\left[i \sum_{j=1}^{N} (m_j + a_j)(X_j + b_j) \right] \tag{16.19}$$

Note that by setting $\mathbf{a} = \mathbf{b} = 0$ we get the usual theta function without characteristics, that is

$$\theta(\mathbf{X} \mid \mathbf{B}, \boldsymbol{\delta}) = \theta(\mathbf{X}, 0, 0 \mid \mathbf{B}, \boldsymbol{\delta}) \tag{16.20}$$

Notice that $\theta(\mathbf{X}, 0, \mathbf{b} \mid \mathbf{B})$ can be a pure space $(k_j x + b_j = k_j (x - x_j); x_j = -b_j/k_j)$ or time translation $(-\omega_j t + b_j = -\omega_j(t - t_j); t_j = -b_j/\omega_j)$ of the theta function.

Theta functions with the form $\theta(\mathbf{X}, \mathbf{a}, \mathbf{b} = 0 \mid \mathbf{B}, \boldsymbol{\delta})$ will be seen to be useful for determining the Riemann spectrum or constant parameters B_{ij} $(i \neq j)$, ω_i, ρ: off-diagonal elements of the period matrix, frequencies, and integration constant) necessary for *numerical wave modeling*. Theta functions with the form $\theta(\mathbf{X}, \mathbf{a} = 0, \mathbf{b} \mid \mathbf{B}, \boldsymbol{\delta})$ will be found to be useful in the *method of surrogates*, a common approach used to study the *statistics of nonlinear wave motion*. An Appendix discusses several important properties of theta functions with characteristics.

16.4   Solution of the KdV Equation for the Theta Function with Characteristics

To better understand theta functions with characteristics we now show that the solution to the KdV equation can be computed with theta functions with characteristics. We have the following

> **Theorem:** The KdV equation can be solved using the theta function with characteristics for arbitrary value of the characteristic, $\mathbf{a}$. Formally,
>
> $$u(x,t) = 2\partial_{xx} \ln \theta(\mathbf{X}, \mathbf{a}, 0 \mid \mathbf{B}, \boldsymbol{\delta}) = 2\partial_{xx} \ln \theta(x, t \mid \mathbf{B}, \boldsymbol{\delta} - \boldsymbol{\alpha})$$
>
> for $\boldsymbol{\alpha} = \mathbf{Ba}$. Thus, for each choice of the vector $\mathbf{a}$ we get a solution of the KdV equation which differs from the case $\mathbf{a} = 0$ by the phase shift $\boldsymbol{\delta} \to \boldsymbol{\delta} - \boldsymbol{\alpha}$. For all choices of the vector $\mathbf{a}$ we have a unique solution of the KdV equation with the original period matrix, $\mathbf{B}$, wavenumbers, $\mathbf{k}$, and frequencies, $\boldsymbol{\omega}$, but with phases depending on $\mathbf{a}$ given by $\boldsymbol{\delta} - \mathbf{Ba}$. Since the phases in the theta function are arbitrary, any set of phases will solve the KdV equation.

Proof: To see how this happens notice that

$$\exp\left[i\sum_{j=1}^{N}(m_j + a_j)X_j\right] = \exp\left(i\sum_{j=1}^{N}m_j X_j\right)\exp\left(i\sum_{j=1}^{N}a_j X_j\right)$$

This result will let us move the second exponential out from the nested summation in Equation (16.19). Furthermore, it is easily shown that

$$\exp\left[-\frac{1}{2}\sum_{j=1}^{N}\sum_{k=1}^{N}(m_j + a_j)(m_k + a_k)B_{jk}\right]$$
$$= \exp\left(-\frac{1}{2}\sum_{j=1}^{N}\sum_{k=1}^{N}m_j m_k B_{jk}\right)\exp\left(-\sum_{j=1}^{N}m_j \alpha_j\right)\exp\left(-\frac{1}{2}\sum_{j=1}^{N}\sum_{k=1}^{N}a_j a_k B_{jk}\right)$$

where

$$\alpha_j = \sum_{k=1}^{N}a_k B_{jk} = \sum_{k=1}^{N}a_k B_{kj}$$

since $B_{jk} = B_{kj}$ due to the symmetry property of the Riemann matrix. Inserting these results into the theta function with characteristics, and get

$$\theta(\mathbf{X}, \mathbf{a}, 0 | \mathbf{B}, \boldsymbol{\delta}) = \underbrace{\exp\left(-\frac{1}{2}\sum_{j=1}^{N}\sum_{k=1}^{N} a_j a_k B_{jk}\right)}_{\boxed{1}} \underbrace{\exp\left(i\sum_{j=1}^{N} a_j X_j\right)}_{\boxed{2}} \theta(\mathbf{X}, 0, -\boldsymbol{\alpha} | \mathbf{B}, \boldsymbol{\delta})$$

Here the numbered symbols indicate: $\boxed{1}$ The *exponential terms* in the theta function which contain the period matrix and characteristic vector, a_j, and $\boxed{2}$ a *dynamical phase term* which is computed from the phase, X_j, and the characteristic vector, a_j. We then have the solution to the KdV equation:

$$u(x, t) = 2\partial_{xx} \ln \theta(\mathbf{X}, \mathbf{a}, 0 \mid \mathbf{B}, \boldsymbol{\delta})$$

$$= 2\partial_{xx}\left(-\frac{1}{2}\sum_{j=1}^{N}\sum_{k=1}^{N} a_j a_k B_{jk} + i\sum_{j=1}^{N} a_j X_j + \ln \theta(\mathbf{X}, 0, -\boldsymbol{\alpha} \mid \mathbf{B}, \boldsymbol{\delta})\right)$$

Note that the spatial derivative of the first term on the right of the above equation is zero, and the second derivative of the second term is also zero. We thus arrive at a solution of the KdV equation that we can write in the original notation without characteristics:

$$u(x, t) = 2\partial_{xx} \ln \theta(x, t \mid \mathbf{B}, \boldsymbol{\delta} - \boldsymbol{\alpha})$$

As the result of the theorem we see that each choice of the characteristic vector, $\mathbf{a}$, gives us a new solution to the KdV equation which is changed from the original only by a simple translation of the phase vector.

Alternatively, *setting the first characteristic to zero* gives

$$\theta(\mathbf{X}, 0, \mathbf{b} \mid \mathbf{B}) = \sum_{m_1=-\infty}^{\infty}\sum_{m_2=-\infty}^{\infty} \cdots \sum_{m_N=-\infty}^{\infty} \exp\left(-\frac{1}{2}\sum_{j=1}^{N}\sum_{k=1}^{N} m_j m_k B_{jk}\right)\exp\left(i\sum_{j=1}^{N} m_j(X_j + b_j)\right)$$

This expression results in a solution of the KdV equation with a *new set of (possibly) random phases* given by $\phi_j + b_j$, *equivalent to what dynamical systems researchers call the method of surrogates, an approach used to test the statistical invariance of time series of nonlinear dynamical motions*. The difference here is that we are using Riemann theta functions instead of ordinary Fourier series to build surrogate space or time series. The results of this paragraph, while seemingly trivial, are important for the study of the *statistical properties of nonlinear dynamical motions*. Additional discussion is given in the Appendix.

Notice that the above results show that theta functions with characteristics also solve the NLS equation, since the additional factors in front of the theta function cancel out when we take the ratio of two theta functions, leaving only an arbitrary phase in the solution to the equation.

16.5 Determination of Theta-Function Parameters

This section addresses how to compute the Riemann spectrum necessary to *model waves* using the Riemann theta functions (see Chapters 10 and 32). This step of the Nakamura-Boyd approach parallels the substitution of the linear Fourier transform into a linear PDE to determine the linear dispersion relation. The nonlinear problem contains several additional features not included in the linear problem. What are these features? Here is a summary: (1) In the linear problem we select the Fourier amplitudes and phases. In the nonlinear problem we select the *diagonal elements of the period matrix* and the *phases*. (2) In the linear problem we obtain the linear dispersion relation by direct substitution of the linear Fourier transform into the linear PDE. In the nonlinear case we first use the Hirota direct method to obtain the bilinear form. The theta function is then substituted into the bilinear form to obtain the *off-diagonal elements of the period matrix*, *the integration constant*, and the *nonlinear dispersion relation*. This exercise provides us with the necessary parameters for simulating the wave motion using Riemann theta functions. Let us see how this procedure plays out.

The theta functions with characteristics, $\theta(\mathbf{X} \mid \mathbf{B}, \boldsymbol{\delta}) \Rightarrow \theta(\mathbf{X}, \mathbf{a}, 0 \mid \mathbf{B} \, \boldsymbol{\delta})$, are useful for determining and computing *particular nonlinear equations for the parameters B_{mn} ($m \neq n$), ω_n, ρ*, given the diagonal elements of the Riemann matrix, B_{nn}, the wavenumbers, k_n, and the phases, ϕ_n. The results are valid for any choice of the diagonal elements, B_{nn}, and the phases, ϕ_n. As a practical consideration we assume periodic boundary conditions so that the resultant wavenumbers are commensurable:

$$k_n = \frac{2\pi n}{L}$$

where L is the period of the wave train.

Specifically, inserting the theta function with characteristics into the bilinear form we get a set of 2^N equations (for N degrees of freedom) for determining the parameters (wavenumbers, frequencies, period matrix, and ρ) in the theta-function solution of the KdV equation. As a practical consideration we use the half-integers for the characteristics, $\mathbf{a} = -\boldsymbol{\mu}/2$ or $a_j = -\mu_j/2$ for $j = 1$, $2, \ldots, N$. The resultant nonlinear equations for B_{ij} ($i \neq j$), ω_i, ρ, given the B_{ii}, k_i, and ϕ_i are given by

$$\sum_{m_1=-\infty}^{\infty} \sum_{m_2=-\infty}^{\infty} \cdots \sum_{m_N=-\infty}^{\infty} \left\{ \left[2\sum_{j=1}^{N}(m_j - \mu_j/2)k_j \right]\left[2\sum_{j=1}^{N}(m_j - \mu_j/2)\omega_j \right] \right.$$

$$\left. + \left[2\sum_{j=1}^{N}(m_j - \mu_j/2)k_j \right]^4 + \rho \right\} \exp\left[-\sum_{j=1}^{N}\sum_{k=1}^{N}(m_j - \mu_j/2)(m_k - \mu_k/2)B_{jk} \right] = 0$$

$$(16.21a)$$

This is the most important equation in this chapter. The equation is nonlinear and serves to compute the Riemann spectrum for a particular solution of the KdV equation. In this expression we have

$$\boldsymbol{\mu} = \{\mu_1, \mu_2, \ldots, \mu_N\} \quad \text{where } \mu_j = 0, 1$$

Note that the sign in front of the term

$$\left[\sum_{j=1}^{N}(m_j - \mu_j/2)k_j \right]\left[\sum_{j=1}^{N}(m_j - \mu_j/2)\omega_j \right]$$

is positive. In Hirota and Ito (1981) it is negative because of the sign choice in the ω_j term, which is taken to be positive in that work. Note, further, that $B_{ij} \rightarrow 2B_{ij}$ in the formulation. Additionally, as discussed in more detail below, $k_i \rightarrow 2k_i$, $\omega_i \rightarrow 2\omega_i$.

The above nonlinear equations (16.21a) can be written for the *dimensional, physical form of the KdV equation*:

$$\sum_{m_1=-\infty}^{\infty} \sum_{m_2=-\infty}^{\infty} \cdots \sum_{m_N=-\infty}^{\infty} \left\{ \left[2\sum_{j=1}^{N}(m_j - \mu_j/2)k_j \right]\left[2\sum_{j=1}^{N}(m_j - \mu_j/2)\omega_j \right] \right.$$

$$\left. - c_0 \left[2\sum_{j=1}^{N}(m_j - \mu_j/2)k_j \right]^2 + \beta \left[2\sum_{j=1}^{N}(m_j - \mu_j/2)k_j \right]^4 - \frac{\alpha\rho}{12\beta} \right\} \quad (16.21b)$$

$$\times \exp\left[-\sum_{j=1}^{N}\sum_{k=1}^{N}(m_j - \mu_j/2)(m_k - \mu_k/2)B_{jk} \right] = 0$$

where again we take the components of the vector $\boldsymbol{\mu} = [\mu_1, \mu_2, \ldots, \mu_N]$ to be 0 or 1.

Nota Bene: This dimensional form of the nonlinear Riemann spectra equations (16.21b) includes the physical constants of the KdV equation, namely, the linear phase speed, c_0, and the linear dispersion constant, β. It is a natural memory aid to view Equation (16.21b) as being "quasi-linear" since it lacks the nonlinear term with the constant α. Note further that the terms get their signs from $\eta(x,t) \sim e^{ikx - i\omega t}$. Thus, heuristically we see that: $(ik)(-i\omega) + c_0(ik)^2 + \beta(ik)^4 = k\omega - c_0k^2 + \beta k^4 = 0$, hence $\omega = c_0k - \beta k^3$, which is the correct linear dispersion relation.

16.6 Linearized Form for Riemann Spectrum for the KdV Equation

We can look at the linear form of Equation (16.21b) by setting the off-diagonal terms to be zero and setting the diagonal elements to be large and hence the q's to be small. It is then a simple task to show that *all the degrees of freedom* have the linear dispersion relation of the KdV equation, $\omega_n = c_o k_n - \beta k_n^3$, for large values of the diagonal elements in the Riemann matrix. This is of course the physical linear limit that must occur in any theory that determines the Riemann spectrum.

16.7 Strategy for Determining Solutions of Nonlinear Equations

Now let us discuss the strategy for obtaining the parameters (integration constant ρ, period matrix B_{ij}, and nonlinear dispersion relation $\omega_j = \omega_j\,(k_j)$) from Equation (16.21b). First we assume that we are dealing with periodic boundary conditions so that $k_n = 2\pi n/L$ for L the spatial period. Hence we are assuming the values of the wavenumbers are fixed and proportional to the integers. Next assume that the diagonal elements of the period matrix are known or supplied, B_{jj}. Likewise assume the phases, ϕ_j, are known or supplied (note that the phases do not appear in Equation (16.21b)). *Then we must seek the off-diagonal elements, B_{ij} ($i \neq j$), there are $N(N+1)/2 - N = N(N-1)/2$ of them), the frequencies, ω_j (there are N of them), and the constant, ρ (1 of these) from the nonlinear equations (16.21b).* The total number of parameters is then given by: $1 + N + N(N-1)/2 = 1 + N(N+1)/2$.

Example: For a two degree-of-freedom case, $N = 2$, the number of equations is $2^N = 4$ and the number of parameters is $1 + N(N+1)/2 = 4$. Thus, for this simple case, the number of equations and the number of parameters are the same, 4.

Nota Bene: The wavenumbers and diagonal elements of the period matrix are assumed given. We then compute the frequencies ω_n, off-diagonal elements B_{mn}, and the parameter ρ from Equation (16.21b). This means we are assuming that we have the cnoidal waves in the spectrum as described by their amplitudes A_n (corresponding to the diagonal elements of the Riemann matrix, in the small amplitude approximation $A_n \sim k_n^2 \exp\left[-B_{nn}/2\right]$; the exact expression is given by Equation (10.164)), for modulus m_n, wavenumbers k_n, and phases ϕ_n; this is equivalent to inputting the diagonal elements B_{mm}, wavenumbers k_n, and phases ϕ_n. A flowchart of the algorithm is given in Figure 16.1.

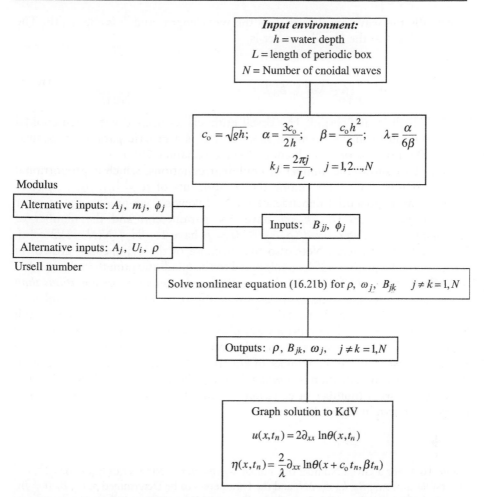

Figure 16.1 Flowchart of the algorithm for computing the Riemann spectrum and periodic solutions of the KdV equation. One defines the water depth, length of the periodic box in which solutions are to be computed and the number of cnoidal waves desired. Additionally, the diagonal elements of the Riemann matrix and the phases are specified. The program computes the parameter ρ, the off-diagonal elements of the Riemann matrix, B_{jk}, $j \neq k = 1, N$, and the frequencies, ω_j. Then the solution to KdV is computed for a specified time interval $(0, T)$ using the associated Riemann theta function with the determined Riemann spectrum.

A useful parameter in the computation of solutions of the KdV and KP equations is the *Ursell number*, whose space-like form is

$$U = \frac{3a}{4k^2 h^3} = \frac{3aL^2}{16\pi^2 h^3} = \frac{3}{16\pi^2}\left(\frac{a}{h}\right)\left(\frac{L}{h}\right)^2 \qquad (16.22)$$

where a is the wave amplitude, L is the wavelength, and h is the depth. The time-like form of the Ursell number is

$$U = \frac{3ac_0^2 T^2}{16\pi^2 h^3} = \frac{3}{16\pi^2}\left(\frac{a}{h}\right)\left(\frac{c_0 T}{h}\right)^2 \tag{16.23}$$

where T is the wave period. The Ursell numbers associated with each cnoidal wave component in the Riemann spectrum; U is a generic parameter, equivalent to the modulus m of each cnoidal wave (Chapter 10).

Table 16.1 shows the number of nonlinear equations, which is proportional to 2^N, obtained by simply setting the components of $\boldsymbol{\mu} = [\mu_1, \mu_2, \ldots, \mu_N]$ to 0 and 1, an exponential dependence on N. Likewise, we present the number of *given parameters* (B_{jj}, k_j, ϕ_j have 3N parameters) and the number of *parameters to be determined* (ρ, ω_j, B_{ij} ($i \neq j$) have $1 + N + N(N - 1)/2 = 1 + N(N +1)/2$ parameters). Note also that because of the exponential growth in the number of equations (2^N), as compared to the quadratic growth in the number of unknowns ($1 + N(N + 1)/2$), the generic case has *many more equations than unknowns*. Thus, one could easily, as a first approach, decide on the number of unknown parameters and then take the same number of equations from the full list. Of course the method must always provide the same answers, independent of *which equations* are selected from the total 2^N. This suggests that *recursion relations* must truncate the number of equations beyond $1 + N(N + 1)/2$ when the nonlinear wave equation is *completely integrable* (Nakamura, 1980). A graph of the two curves (number of equations vs. number of parameters to be determined) is shown in Figure 16.2.

Table 16.1 For a Given Number of Cnoidal Waves N, the Number of Equations 2^N, the Given Parameters, B_{jj}, k_j, ϕ_j: 3N, and the Parameters to be Determined ρ, ω_j, B_{ij} ($i \neq j$): $1 + N(N + 1)/2$ are Tabularized (See Figure 16.2 for a Graphical Comparison)

Number of Equations		Given Parameters: B_{jj}, k_j, ϕ_j	Parameters to Find: ρ, ω_j, B_{ij} ($i \neq j$)
N	2^N	3N	$1 + N(N + 1)/2$
1	2	3	2
2	4	6	4
3	8	9	7
4	16	12	11
5	32	15	16
6	64	18	22
7	132	21	29
8	264	24	37
9	512	27	46
10	1024	30	56

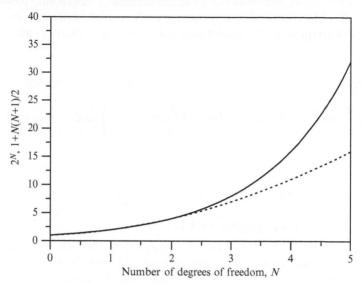

Figure 16.2 Graph of 2^N (number of equations, upper curve) versus $1 + N(N + 1)/2$ (the number of parameters to be determined, lower curve).

Typically, in the computer program I use up to 12 nines in the modulus. See Sections 10.6.10 through 10.6.12 in Chapter 10 for a numerical example and a table of values of elliptic function parameters.

16.8 One Degree-of-Freedom Riemann Spectrum and Solution of the KdV Equation

For one degree of freedom we have two equations ($2^1 = 2$) and two unknowns ($1 + 1(1 + 1)/2 = 2$). The given parameters are B_{11}, k_1, ϕ_1 and the unknowns are ρ, ω_1. Here are the equations from Equation (16.21b):

$$\sum_{m_1=-\infty}^{\infty} \left\{ [2(m_1 - \mu_1/2)k_1][2(m_1 - \mu_1/2)\omega_1] - c_0[2(m_1 - \mu_1/2)k_1]^2 \right.$$
$$\left. + \beta[2(m_1 - \mu_1/2)k_1]^4 - \frac{\alpha\rho}{12\beta} \right\} e^{-(m_1-\mu_1/2)^2 B_{11}} = 0$$

or

$$\sum_{m_1=-\infty}^{\infty} \left\{ [2(m_1 - \mu_1/2)k_1][2(m_1 - \mu_1/2)\omega_1] - c_0[2(m_1 - \mu_1/2)k_1]^2 \right.$$

$$\left. + \beta[2(m_1 - \mu_1/2)k_1]^4 - \frac{\alpha\rho}{12\beta} \right\} q^{2(m_1-\mu_1/2)^2} = 0 \tag{16.24}$$

for $q = e^{-\frac{1}{2}B_{11}}$. Here is another form that may be useful (written as coefficients of powers of frequency and wavenumber, $k_1 = k$, $\omega_1 = \omega$, $\mu_1 = \mu$):

$$\sum_{m=-\infty}^{\infty} \left\{ 4(m - \mu/2)^2 q^{(m-\mu/2)^2} k\omega - c_0[2(m - \mu/2)k]^2 q^{(m-\mu/2)^2} \right.$$
$$\left. + 16\beta(m - \mu/2)^4 q^{(m-\mu/2)^2} k^4 - \frac{\alpha\rho}{12\beta} q^{2(m-\mu/2)^2} \right\} = 0$$

which can be written

$$A_\mu k\omega + B_\mu \rho' = c_0 k^2 C_\mu - \beta k^4 D_\mu, \quad \rho' = -\frac{\alpha\rho}{12\beta}$$

$$A_\mu = 4 \sum_{m=-\infty}^{\infty} (m - \mu/2)^2 q^{2(m-\mu/2)^2}$$

$$B_\mu = \sum_{m=-\infty}^{\infty} q^{2(m-\mu/2)^2}$$

$$C_\mu = 4 \sum_{m=-\infty}^{\infty} (m - \mu/2)^2 q^{2(m-\mu/2)^2}$$

$$D_\mu = 16 \sum_{m=-\infty}^{\infty} (m - \mu/2)^4 q^{2(m-\mu/2)^2}$$

Notice that $A_\mu = C_\mu$, so the equations become

$$A_\mu(k\omega - c_0 k^2) + B_\mu \rho' = -\beta k^4 C_\mu \qquad (16.25)$$

$$A_\mu = 4 \sum_{m=-\infty}^{\infty} (m - \mu/2)^2 q^{2(m-\mu/2)^2}$$

$$B_\mu = \sum_{m=-\infty}^{\infty} q^{2(m-\mu/2)^2} \qquad (16.26)$$

$$C_\mu = 16 \sum_{m=-\infty}^{\infty} (m - \mu/2)^4 q^{2(m-\mu/2)^2}$$

The coefficients in the above equations can be viewed as *theta constants* with characteristics. Here we take $\mu = 0,1$ so that the above equations become

$$A_0(k\omega - c_0 k^2) + B_0 \rho' = -\beta k^4 C_0$$
$$A_1(k\omega - c_0 k^2) + B_1 \rho' = -\beta k^4 C_1$$

or in matrix form:

$$\begin{bmatrix} A_0 & B_0 \\ A_1 & B_1 \end{bmatrix} \begin{bmatrix} k\omega - c_0 k^2 \\ \rho' \end{bmatrix} = -\beta k^4 \begin{bmatrix} C_0 \\ C_1 \end{bmatrix}$$

These are linear equations in two unknowns ρ, ω, for the givens B_{11}, k, ϕ (the phase is arbitrary and does not enter into the computations). The coefficients are given by

$$
A_0 = 4 \sum_{m=-\infty}^{\infty} m^2 q^{2m^2} \cong 8q^2 + 32q^8 + \cdots
$$

$$
B_0 = \sum_{m=-\infty}^{\infty} q^{2m^2} \cong 1 + 2q^2 + 2q^8 + \cdots
$$

$$
C_0 = 16 \sum_{m=-\infty}^{\infty} m^4 q^{2m^2} \cong 32q^2 + 512q^8 + \cdots
$$

$$
A_1 = 4 \sum_{m=-\infty}^{\infty} (m - 1/2)^2 q^{2(m-1/2)^2} \cong 2\sqrt{q} + 18q^{9/2} + 25q^{25/2} + \cdots
$$

$$
B_1 = \sum_{m=-\infty}^{\infty} q^{2(m-1/2)^2} \cong 2\sqrt{q} + 2q^{9/2} + q^{25/2} + \cdots
$$

$$
C_1 = 16 \sum_{m=-\infty}^{\infty} (m - 1/2)^4 q^{2(m-1/2)^2} \cong 2\sqrt{q} + 162q^{9/2} + 625q^{25/2} + \cdots
$$

$$(16.27)$$

Here I have given the first few terms of the polynomials in the nome, q.
 The solution in matrix form is given by

$$
\begin{bmatrix} k\omega - c_o k^2 \\ \rho' \end{bmatrix} = -\frac{\beta k^4}{A_0 B_1 - B_0 A_1} \begin{bmatrix} B_1 & -B_0 \\ -A_1 & A_0 \end{bmatrix} \begin{bmatrix} C_0 \\ C_1 \end{bmatrix} \tag{16.28a}
$$

Then

$$
\omega = c_o k - \beta k^3 \left[\frac{B_1 C_0 - B_0 C_1}{A_0 B_1 - B_0 A_1} \right] = c_o k - \beta k^3 \left[\frac{1 - 30q^2 + 81q^4 + 130q^6 + \cdots}{1 - 6q^2 + 9q^4 + 10q^6 + \cdots} \right]
$$

$$(16.28b)$$

$$
\rho' = -k^4 \left[\frac{A_0 C_1 - A_1 C_0}{A_0 B_1 - B_0 A_1} \right] = k^4 \left[\frac{24q^2 - 360q^6 + \cdots}{1 - 6q^2 + 9q^4 + 10q^6 + \cdots} \right]
$$

The frequency depends generally on the period matrix term B_{11}, or its nome $q = \exp(-B_{11}/2)$.
 Now recall the linear dispersion relation for KdV in dimensional form: $\omega = c_o k - \beta k^3$. Hence we have

$$
\omega = c_o k - \beta k^3 [1 - 24q^2 - 72q^4 - 96q^6 - 168q^8 + \cdots] \tag{16.29}
$$

$$
\rho' = \frac{\alpha \rho}{12\beta} = k^4 [24q^2 + 144q^4 + 288q^6 + 672q^8 + \cdots]
$$

Finally

$$\rho = \frac{12\beta}{\alpha} k^4 [24q^2 + 144q^4 + 288q^6 + 672q^8 + \cdots] \qquad (16.30)$$

Let us look at the results out to second order in nonlinearity, that is, out to q^2:

$$\omega = c_o k - \beta k^3 + 24\beta k^3 q^2 + \cdots$$

$$\rho = \frac{288\beta}{\alpha} k^4 q^2 + \cdots$$

We need no longer pursue the parameter ρ, since it has no further use in constructing solutions of the KdV equation. Now, we know (from the Stokes series solution of KdV, Chapter 10) that

$$q = \frac{3a}{8h^3 k^2}, \quad \lambda = \frac{3}{2h^3}$$

This will be derived rigorously below. Using this later expression in the dispersion relation given above leads to the result (agreeing with Equation (13.119) in Whitham, 1974):

$$\frac{\omega}{c_o k} = 1 - \frac{h^2 k^2}{6} + \frac{9a^2}{16h^4 k^2} + \cdots \qquad (16.31)$$

Now let us compute the solution of KdV in terms of the one degree-of-freedom Riemann theta function:

$$\theta(x, t) = 1 + 2 \sum_{n=1}^{\infty} q^{n^2} \cos [n(kx - \omega t)]$$

where $q = \exp[-(1/2)B_{11}]$, the frequency is given by (16.28b) and $k = 2\pi/L$.
From Chapter 10 we have the Fourier series expansion of the cnoidal wave:

$$\eta(x, t) = \frac{2}{\lambda} \partial_{xx} \ln \theta(x, t) = \frac{4k^2}{\lambda} \sum_{n=1}^{\infty} \frac{nq^n}{1 - q^{2n}} \cos [n(kx - \omega t)] \qquad (16.32)$$

This is the *exact solution* with dispersion relation (16.29), but it is also interesting to compare this to the "classical" Stokes wave. To this end write the approximation (assuming $q \ll 1$):

$$\eta(x, t) = \frac{2}{\lambda} \partial_{xx} \ln \theta(x, t) \cong \frac{4k^2}{\lambda} \sum_{n=1}^{\infty} nq^n \cos [n(kx - \omega t)] \qquad (16.33)$$

for which

$$\eta(x, t) = \frac{4k^2}{\lambda} \{q \cos [kx - \omega t] + 2q^2 \cos [2(kx - \omega t)] + 3q^3 \cos [3(kx - \omega t)] + \cdots\}$$

Now assume that the leading order term $\cos[kx - \omega t]$ should have an amplitude a:

$$a = \frac{4k^2 q}{\lambda}, \quad \text{where } \lambda = \frac{3}{2h^3}$$

Therefore

$$q = \frac{\lambda a}{4k^2} = \frac{3a}{8h^3 k^2} \tag{16.34}$$

which verifies the expression we used above in the derivation of the dispersion relation (16.31). Now evaluate the coefficients in the above Fourier series to get

$$\eta(x,t) = a\cos[kx - \omega t] + \frac{3a^2}{4k^2 h^3}\cos[2(kx - \omega t)] + \frac{27a^3}{64k^4 h^6}\cos[3(kx - \omega t)] + \cdots \tag{16.35}$$

which agrees with Whitham's derivation of the *Stokes wave* from the KdV equation (Whitham, 1974), Equation (13.118), by the *classical multiscale expansion technique*. The dispersion relation is given by (16.31) as derived by the method of Nakamura and Boyd.

An alternative approximate form of the Stokes wave is to truncate the theta function at first order in the nome:

$$\theta(x,t) \cong 1 - 2q\cos(kx - \omega t)$$

and this gives

$$\eta(x,t) = \frac{2}{\lambda}\partial_{xx}\ln\theta(x,t) \cong \frac{4qk^2}{\lambda}\left[\frac{\cos(kx - \omega t) - 2q}{(1 - 2q\cos(kx - \omega t))^2}\right]$$

Upon expanding this expression in a Fourier series we have to leading order in q the "classical" Stokes wave (Equations (16.31) and (16.35))! This completes the derivation of the one degree-of-freedom solution of the KdV equation using the method of Nakamura and Boyd to find the Riemann spectrum.

16.9 Two Degree of Freedom Riemann Spectrum and Solution of the KdV Equation

We now take a look at two degrees of freedom, $N = 2$, so that Equation (16.21b) becomes

$$\sum_{m_1=-\infty}^{\infty} \sum_{m_2=-\infty}^{\infty} \left\{ \left[2\sum_{j=1}^{2}(m_j - \mu_j/2)k_j \right] \left[2\sum_{j=1}^{2}(m_j - \mu_j/2)\omega_j \right] - c_0 \left[2\sum_{j=1}^{2}(m_j - \mu_j/2)k_j \right]^2 \right.$$
$$\left. + \beta \left[2\sum_{j=1}^{2}(m_j - \mu_j/2)k_j \right]^4 - \frac{\alpha\rho}{12\beta} \right\} \exp\left[-\sum_{j=1}^{2}\sum_{k=1}^{2}(m_j - \mu_j/2)(m_k - \mu_k/2)B_{jk} \right] = 0$$

This last result can also be put into the form:

$$\sum_{m_1=-\infty}^{\infty} \sum_{m_2=-\infty}^{\infty} \left\{ 4[(m_1 - \mu_1/2)k_1 + (m_2 - \mu_2/2)k_2][(m_1 - \mu_1/2)\omega_1 + (m_2 - \mu_2/2)\omega_2] \right.$$
$$\left. - 4c_0[(m_1 - \mu_1/2)k_1 + (m_2 - \mu_2/2)k_2]^2 + 16\beta[(m_1 - \mu_1/2)k_1 + (m_2 - \mu_2/2)k_2]^4 + \rho \right\}$$
$$\times e^{-(m_1-\mu_1/2)^2 B_{11} - (m_2-\mu_2/2)^2 B_{22} - 2(m_1-\mu_1/2)(m_2-\mu_2/2)B_{12}} = 0$$

Note that here and below I drop the prime on the symbol ρ. Write this in terms of the nomes:

$$\sum_{m_1=-\infty}^{\infty} \sum_{m_2=-\infty}^{\infty} \left\{ 4[(m_1 - \mu_1/2)k_1 + (m_2 - \mu_2/2)k_2][(m_1 - \mu_1/2)\omega_1 + (m_2 - \mu_2/2)\omega_2] \right.$$
$$\left. - 4c_0[(m_1 - \mu_1/2)k_1 + (m_2 - \mu_2/2)k_2]^2 + 16\beta[(m_1 - \mu_1/2)k_1 + (m_2 - \mu_2/2)k_2]^4 + \rho \right\}$$
$$\times q_1^{2(m_1-\mu_1/2)^2} q_2^{2(m_2-\mu_2/2)^2} q_{12}^{2(m_1-\mu_1/2)(m_2-\mu_2/2)} = 0$$

$$q_1 = q_1(B_{11}) = e^{-\frac{1}{2}B_{11}}, \quad q_2 = q_2(B_{22}) = e^{-\frac{1}{2}B_{22}}, \quad q_{12} = q_{12}(B_{12}) = e^{-B_{12}}$$

Now let

$$k_n = nk, \quad k \equiv \Delta k = 2\pi/L$$

which holds true for a periodic wave train of length L, as we assume here.

From this equation we have given parameters $k \equiv \Delta k = 2\pi / L$, q_1, q_2 and seek ω_1, ω_2, ρ, q_{12}. Then we write

$$A^1_{\mu_1\mu_2}(q_{12})\omega_1 + A^2_{\mu_1\mu_2}(q_{12})\omega_2 + B_{\mu_1\mu_2}(q_{12})\rho = c_o k^2 C_{\mu_1\mu_2} - \beta k^4 D_{\mu_1\mu_2}(q_{12})$$

$$A^1_{\mu_1\mu_2} = 4 \sum_{m_1=-\infty}^{\infty} \sum_{m_2=-\infty}^{\infty} (m_1 - \mu_1/2)[(m_1 - \mu_1/2)k_1 + (m_2 - \mu_2/2)k_2]q_1^{2(m_1-\mu_1/2)^2}$$

$$\times q_2^{2(m_2-\mu_2/2)^2} q_{12}^{2(m_1-\mu_1/2)(m_2-\mu_2/2)}$$

$$A^2_{\mu_1\mu_2} = 4 \sum_{m_1=-\infty}^{\infty} \sum_{m_2=-\infty}^{\infty} (m_2 - \mu_2/2)[(m_1 - \mu_1/2)k_1 + (m_2 - \mu_2/2)k_2]q_1^{2(m_1-\mu_1/2)^2}$$

$$\times q_2^{2(m_2-\mu_2/2)^2} q_{12}^{2(m_1-\mu_1/2)(m_2-\mu_2/2)}$$

$$B_{\mu_1\mu_2} = \sum_{m_1=-\infty}^{\infty} \sum_{m_2=-\infty}^{\infty} q_1^{2(m_1-\mu_1/2)^2} q_2^{2(m_2-\mu_2/2)^2} q_{12}^{2(m_1-\mu_1/2)(m_2-\mu_2/2)}$$

$$C_{\mu_1\mu_2} = 4 \sum_{m_1=-\infty}^{\infty} \sum_{m_2=-\infty}^{\infty} [(m_1 - \mu_1/2)k_1 + (m_2 - \mu_2/2)k_2]^2 q_1^{2(m_1-\mu_1/2)^2} q_2^{2(m_2-\mu_2/2)^2}$$

$$\times q_{12}^{2(m_1-\mu_1/2)(m_2-\mu_2/2)}$$

$$D_{\mu_1\mu_2} = 16 \sum_{m_1=-\infty}^{\infty} \sum_{m_2=-\infty}^{\infty} [(m_1 - \mu_1/2)k_1 + (m_2 - \mu_2/2)k_2]^4 q_1^{2(m_1-\mu_1/2)^2} q_2^{2(m_2-\mu_2/2)^2}$$

$$\times q_{12}^{2(m_1-\mu_1/2)(m_2-\mu_2/2)}$$

$$(16.36)$$

Nota Bene: The "4" and the "16" come about because the wavenumbers are doubled when forming the nonlinear equations for the Riemann spectrum: $k \to 2k$, together with the period matrix, $B \to 2B$. In the final analysis, the wavenumbers in (16.36) k, k^2, k^4 can all be replaced by "1" and then k_1, $k_2,\ldots$, k_N are then interpreted as the usual wavenumbers. The reason that I keep for now k, k^2, k^4 in (16.36) is that it makes the physics transparent on the right-hand side. Think of the above coefficients as polynomials in q_1, q_2 for fixed q_{12}. The *linear limit* in the first equation above shows that: All the coefficients become unit matrices, $\rho = 0$, the off-diagonal terms in the period matrix become insignificant relative to the diagonal elements and we get $\omega_1 = c_o k_1 - \beta k_1^3$ and $\omega_2 = c_o k_2 - \beta k_2^3$ for the dispersion relations.

The coefficients A, B, C, D are series in q_1, q_2, and q_{12}. Here μ_1, μ_2 take on values of 0, 1:

$$
\begin{array}{cc}
0 & 0 \\
0 & 1 \\
1 & 0 \\
1 & 1
\end{array}
$$

Thus, there are $2^N = 2^2 = 4$ equations. The unknowns are ω_1, ω_2, ρ, q_{12}, also 4 in number.

Now let us look at the equations again with simplified notation. Let $a_j = m_j - \mu_j/2$ and get

$$A^1_{\mu_1\mu_2}(q_{12})\omega_1 + A^2_{\mu_1\mu_2}(q_{12})\omega_2 + B_{\mu_1\mu_2}(q_{12})\rho = c_o k^2 C_{\mu_1\mu_2}(q_{12}) - \beta k^4 D_{\mu_1\mu_2}(q_{12})$$

$$A^1_{\mu_1\mu_2} = 4 \sum_{m_1=-\infty}^{\infty} \sum_{m_2=-\infty}^{\infty} a_1[a_1 k_1 + a_2 k_2] q_1^{2a_1^2} q_2^{2a_2^2} q_{12}^{2a_1 a_2}, \quad a_j = m_j - \mu_j/2$$

$$A^2_{\mu_1\mu_2} = 4 \sum_{m_1=-\infty}^{\infty} \sum_{m_2=-\infty}^{\infty} a_2[a_1 k_1 + a_2 k_2] q_1^{2a_1^2} q_2^{2a_2^2} q_{12}^{2a_1 a_2}$$

$$B_{\mu_1\mu_2} = \sum_{m_1=-\infty}^{\infty} \sum_{m_2=-\infty}^{\infty} q_1^{2a_1^2} q_2^{2a_2^2} q_{12}^{2a_1 a_2}$$

$$C_{\mu_1\mu_2} = 4 \sum_{m_1=-\infty}^{\infty} \sum_{m_2=-\infty}^{\infty} [a_1 k_1 + a_2 k_2]^2 q_1^{2a_1^2} q_2^{2a_2^2} q_{12}^{2a_1 a_2}$$

$$D_{\mu_1\mu_2} = 16 \sum_{m_1=-\infty}^{\infty} \sum_{m_2=-\infty}^{\infty} [a_1 k_1 + a_2 k_2]^4 q_1^{2a_1^2} q_2^{2a_2^2} q_{12}^{2a_1 a_2}$$

$$(16.37)$$

Now let us write the four nonlinear equations explicitly:

$$A^1_{00}(q_{12})\omega_1 + A^2_{00}(q_{12})\omega_2 + B_{00}(q_{12})\rho = c_o k^2 C_{00}(q_{12}) - \beta k^4 D_{00}(q_{12})$$

$$A^1_{01}(q_{12})\omega_1 + A^2_{01}(q_{12})\omega_2 + B_{01}(q_{12})\rho = c_o k^2 C_{01}(q_{12}) - \beta k^4 D_{01}(q_{12})$$

$$A^1_{10}(q_{12})\omega_1 + A^2_{10}(q_{12})\omega_2 + B_{10}(q_{12})\rho = c_o k^2 C_{10}(q_{12}) - \beta k^4 D_{10}(q_{12})$$

$$A^1_{11}(q_{12})\omega_1 + A^2_{11}(q_{12})\omega_2 + B_{11}(q_{12})\rho = c_o k^2 C_{11}(q_{12}) - \beta k^4 D_{11}(q_{12})$$

$$(16.38)$$

We can write the equations with $k_n = nk$, $k = 2\pi / L$:

$$A^1_{\mu_1\mu_2}(q_{12})k\omega_1 + A^2_{\mu_1\mu_2}(q_{12})k\omega_2 + B_{\mu_1\mu_2}(q_{12})\rho = c_0 k^2 C_{\mu_1\mu_2}(q_{12}) - \beta k^4 D_{\mu_1\mu_2}(q_{12})$$

$$A^1_{\mu_1\mu_2} = 4 \sum_{m_1=-\infty}^{\infty} \sum_{m_2=-\infty}^{\infty} a_1[a_1 + 2a_2] q_1^{2a_1^2} q_2^{2a_2^2} q_{12}^{2a_1 a_2}, \quad a_j = m_j - \mu_j/2$$

$$A^2_{\mu_1\mu_2} = 4 \sum_{m_1=-\infty}^{\infty} \sum_{m_2=-\infty}^{\infty} a_2[a_1 + 2a_2] q_1^{2a_1^2} q_2^{2a_2^2} q_{12}^{2a_1 a_2}$$

$$B_{\mu_1\mu_2} = \sum_{m_1=-\infty}^{\infty} \sum_{m_2=-\infty}^{\infty} q_1^{2a_1^2} q_2^{2a_2^2} q_{12}^{2a_1 a_2}$$

$$C_{\mu_1\mu_2} = 4 \sum_{m_1=-\infty}^{\infty} \sum_{m_2=-\infty}^{\infty} [a_1 + 2a_2]^2 q_1^{2a_1^2} q_2^{2a_2^2} q_{12}^{2a_1 a_2} w$$

$$D_{\mu_1\mu_2} = 16 \sum_{m_1=-\infty}^{\infty} \sum_{m_2=-\infty}^{\infty} [a_1 + 2a_2]^4 q_1^{2a_1^2} q_2^{2a_2^2} q_{12}^{2a_1 a_2}$$

$$(16.39)$$

In Section 16.11 I discuss an iterative procedure to solve the above two degrees-of-freedom case.

16.10 *N* Degree of Freedom Riemann Spectrum and Solution of the KdV Equation

The nonlinear equations (16.21a) that we seek to solve for the Riemann spectrum are:

$$\sum_{m_1=-\infty}^{\infty} \sum_{m_2=-\infty}^{\infty} \cdots \sum_{m_N=-\infty}^{\infty} \left\{ \left[2\sum_{j=1}^{N}(m_j - \mu_j/2)k_j \right] \left[2\sum_{j=1}^{N}(m_j - \mu_j/2)\omega_j \right] \right.$$

$$\left. + \left[2\sum_{j=1}^{N}(m_j - \mu_j/2)k_j \right]^4 + \rho \right\} \times \exp\left[-\sum_{j=1}^{N}\sum_{k=1}^{N}(m_j - \mu_j/2)(m_k - \mu_k/2)B_{jk} \right] = 0$$

It is convenient to write these equations in several forms, depending on the desired goal, whether it be physical interpretation or analytical or numerical computation of the Riemann spectrum.

16.10.1 Form Number 1

For the coefficients let us write:

$$
A_n = \sum_{m_1=-\infty}^{\infty} \sum_{m_2=-\infty}^{\infty} \cdots \sum_{m_N=-\infty}^{\infty} (m_n - \mu_n/2) \left[4\Delta k D^{\mu_1,\mu_2,\dots,\mu_N}_{m_1,m_2,\dots,m_N} D^{\mu_1,\mu_2,\dots,\mu_N}_{m_1,m_2,\dots,m_N} \sum_{j=1}^{N} (m_j - \mu_j/2)j \right]
$$

where

$$
Q^{\mu_1,\mu_2,\dots,\mu_N}_{m_1,m_2,\dots,m_N}(B_{ii}) = \prod_{i=1}^{N} q_{ii}^{2(m_i - \mu_i/2)^2}
$$

$$
P^{\mu_1,\mu_2,\dots,\mu_N}_{m_1,m_2,\dots,m_N}(B_{ij}) = \prod_{j=1}^{N} \prod_{k=1,\,k>j}^{N} q_{jk}^{2(m_j - \mu_j/2)(m_k - \mu_k/2)}
$$

Now the original equations take the form:

$$
\sum_{m_1=-\infty}^{\infty} \sum_{m_2=-\infty}^{\infty} \cdots \sum_{m_N=-\infty}^{\infty} \left\{ \left[2\sum_{j=1}^{N} (m_j - \mu_j/2)k_j \right] \left[2\sum_{j=1}^{N} (m_j - \mu_j/2)\omega_j \right] \right.
$$

$$
\left. + \left[2\sum_{j=1}^{N} (m_j - \mu_j/2)k_j \right]^4 + \rho \right\} Q^{\mu_1,\mu_2,\dots,\mu_N}_{m_1,m_2,\dots,m_N}(B_{ii}) P^{\mu_1,\mu_2,\dots,\mu_N}_{m_1,m_2,\dots,m_N}(B_{ij}) = 0
$$

which have been written in terms of the diagonal and off-diagonal parts of the period matrix. We now have the desired form:

$$
\sum_{n=1}^{N} A_{n,\mu}[\mathbf{B}]\omega_n + B_{\mu}[\mathbf{B}]\rho = C_{\mu}[\mathbf{B}], \quad \boldsymbol{\mu} = [\mu_1, \mu_2, \dots, \mu_N],
$$

(16.40)

$$
\text{where } \mu_j = 0, 1, \quad j = 1, 2, \dots, N
$$

for which the coefficients are

$$A_{n,\boldsymbol{\mu}}[\mathbf{B}] = \sum_{m_1=-\infty}^{\infty} \sum_{m_2=-\infty}^{\infty} \cdots \sum_{m_N=-\infty}^{\infty} (m_n - \mu_n/2)$$

$$\times \left[4\Delta k Q^{\mu_1,\mu_2,...,\mu_N}_{m_1,m_2,...,m_N} D^{\mu_1,\mu_2,...,\mu_N}_{m_1,m_2,...,m_N} \sum_{j=1}^{N} (m_j - \mu_j/2)j \right]$$

$$B_{\boldsymbol{\mu}}[\mathbf{B}] = \sum_{m_1=-\infty}^{\infty} \sum_{m_2=-\infty}^{\infty} \cdots \sum_{m_N=-\infty}^{\infty} Q^{\mu_1,\mu_2,...,\mu_N}_{m_1,m_2,...,m_N} D^{\mu_1,\mu_2,...,\mu_N}_{m_1,m_2,...,m_N}$$

$$C_{\boldsymbol{\mu}}[\mathbf{B}] = -16\Delta k^4 \sum_{m_1=-\infty}^{\infty} \sum_{m_2=-\infty}^{\infty} \cdots \sum_{m_N=-\infty}^{\infty} \qquad\qquad (16.41)$$

$$\times \left\{ \left[\sum_{j=1}^{N} (m_j - \mu_j/2)j \right]^4 Q^{\mu_1,\mu_2,...,\mu_N}_{m_1,m_2,...,m_N} D^{\mu_1,\mu_2,...,\mu_N}_{m_1,m_2,...,m_N} \right\}$$

$$Q^{\mu_1,\mu_2,...,\mu_N}_{m_1,m_2,...,m_N}(B_{ii}) = \prod_{i=1}^{N} q_{ii}^{2(m_i-\mu_i/2)^2}$$

$$P^{\mu_1,\mu_2,...,\mu_N}_{m_1,m_2,...,m_N}(B_{ij}) = \prod_{j=1}^{N} \prod_{k=1,\,k>j}^{N} q_{jk}^{2(m_j-\mu_j/2)(m_k-\mu_k/2)}$$

16.10.2   Form Number 2

We can modify the form of the coefficients so that we have:

$$A_{n,\boldsymbol{\mu}}[\mathbf{B}] = 4 \sum_{m_1=-\infty}^{\infty} \sum_{m_2=-\infty}^{\infty} \cdots \sum_{m_N=-\infty}^{\infty} (m_n - \mu_n/2) \left(\sum_{j=1}^{N} (m_j - \mu_j/2)k_j \right)$$

$$\times \exp\left[-\sum_{j=1}^{N}\sum_{k=1}^{N} (m_j - \mu_j/2)(m_k - \mu_k/2)B_{jk} \right]$$

$$B_{\boldsymbol{\mu}}[\mathbf{B}] = \sum_{m_1=-\infty}^{\infty} \sum_{m_2=-\infty}^{\infty} \cdots \sum_{m_N=-\infty}^{\infty} \exp\left[-\sum_{j=1}^{N}\sum_{k=1}^{N} (m_j - \mu_j/2)(m_k - \mu_k/2)B_{jk} \right]$$

$$C_{\boldsymbol{\mu}}[\mathbf{B}] = -16 \sum_{m_1=-\infty}^{\infty} \sum_{m_2=-\infty}^{\infty} \cdots \sum_{m_N=-\infty}^{\infty} \left(\sum_{j=1}^{N} (m_j - \mu_j/2)k_j \right)^4$$

$$\times \exp\left[-\sum_{j=1}^{N}\sum_{k=1}^{N} (m_j - \mu_j/2)(m_k - \mu_k/2)B_{jk} \right]$$

$$(16.42)$$

> *Nota Bene:* The diagonal terms of the period matrix are assumed to be given, B_{ii} ($i = 1, 2, \ldots, N$), along with the phases, $\phi_1, \phi_2, \ldots, \phi_N$ at selected wavenumbers, $k_1, k_2, \ldots, k_N = 2\pi n/L$. Then the above equations are to be solved for the frequency components $\omega_1, \omega_2, \ldots, \omega_N$, the constant, ρ, and the off-diagonal terms of the period matrix, B_{ij} ($i \neq j$).

16.10.3   Form Number 3

Another useful form can be derived from the original equations if we set $a_j = m_j - \mu_j/2$ to find

$$
\sum_{m_1=-\infty}^{\infty} \sum_{m_2=-\infty}^{\infty} \cdots \sum_{m_N=-\infty}^{\infty} \left[\left(2\sum_{j=1}^{N} a_j k_j \right) \left(2\sum_{j=1}^{N} a_j \omega_j \right) + \left(2\sum_{j=1}^{N} a_j k_j \right)^4 + \rho \right]
$$
$$
\times \exp\left(-\sum_{j=1}^{N} \sum_{k=1}^{N} a_j a_k B_{jk} \right) = 0
$$

(16.43)

This is a compact way to write the equations. They can also be written in the following form:

$$
\sum_{n=1}^{N} A_{n,\mu}[\mathbf{B}] \omega_n + B_\mu[\mathbf{B}] \rho = C_\mu[\mathbf{B}], \quad \mu = [\mu_1, \mu_2, \ldots, \mu_N],
$$

(16.44)

$$
\text{where } \mu_j = 0, 1, \quad j = 1, 2, \ldots, N
$$

where the coefficients are given by

$$
A_{n,\mu}[\mathbf{B}] = 4 \sum_{m_1=-\infty}^{\infty} \sum_{m_2=-\infty}^{\infty} \cdots \sum_{m_N=-\infty}^{\infty} a_n \left(\sum_{j=1}^{N} a_j k_j \right) \exp\left(-\sum_{j=1}^{N} \sum_{k=1}^{N} a_j a_k B_{jk} \right)
$$

$$
B_\mu[\mathbf{B}] = \sum_{m_1=-\infty}^{\infty} \sum_{m_2=-\infty}^{\infty} \cdots \sum_{m_N=-\infty}^{\infty} \exp\left(-\sum_{j=1}^{N} \sum_{k=1}^{N} a_j a_k B_{jk} \right)
$$

$$
C_\mu[\mathbf{B}] = -16 \sum_{m_1=-\infty}^{\infty} \sum_{m_2=-\infty}^{\infty} \cdots \sum_{m_N=-\infty}^{\infty} \left(\sum_{j=1}^{N} a_j k_j \right)^4 \exp\left(-\sum_{j=1}^{N} \sum_{k=1}^{N} a_j a_k B_{jk} \right)
$$

$$
\text{where } a_j = (m_j - \mu_j/2)
$$

(16.45)

This last form of the equations seems to be quite useful for *programming*. Here is the procedure: (1) Select the number of modes N. (2) Compute 2^N, which is generally larger than or equal to the number of parameters, $1 + N(N + 1)/2$. (3) Truncate the number of equations at $1 + N(N + 1)/2$. (4) Particularly for large N, we can take a second or third set of equations to test for *approximate* or *complete integrability*.

16.11 Numerical Algorithm for Solving Nonlinear Equations

The equations we want to solve are

$$
\sum_{m_1=-\infty}^{\infty} \sum_{m_2=-\infty}^{\infty} \cdots \sum_{m_N=-\infty}^{\infty} \left\{ \left[2\sum_{j=1}^{N}(m_j - \mu_j/2)k_j \right] \left[2\sum_{j=1}^{N}(m_j - \mu_j/2)\omega_j \right] \right.
$$
$$
\left. + \left[2\sum_{j=1}^{N}(m_j - \mu_j/2)k_j \right]^4 + \rho \right\} \exp\left[-\sum_{j=1}^{N}\sum_{k=1}^{N}(m_j - \mu_j/2)(m_k - \mu_k/2)B_{jk} \right] = 0
$$

given the wavenumbers

$$
k_j = 2\pi j/L
$$

and the diagonal elements of the period matrix and the phases

$$
B_{ii}, \phi_i, \quad i = 1, 2, \ldots, N
$$

We then compute the following frequencies, constant, and off-diagonal elements of the period matrix:

$$
\omega_i, \rho, B_{ij}, \quad i \neq j = 1, 2, \ldots, N
$$

The number of equations is 2^N and the number of parameters is $1 + N(N + 1)/2$. To this end write the above system of equations as

$$
F_n(\omega_1, \omega_2, \ldots, \omega_N, \rho, B_{12}, \ldots; B_{11}, B_{22}, \ldots, B_{NN}, \phi_1, \phi_2, \ldots, \phi_N) = 0,
$$
$$
n = 1, 2, \ldots, 1 + N(N + 1)/2
$$

Or in vector form have

$$
\mathbf{F}(\boldsymbol{\omega}, \rho, \mathbf{O}; \mathbf{D}, \boldsymbol{\phi}) = 0
$$

Here $\boldsymbol{\omega}$ is the frequency vector, ρ is the integration constant, $\mathbf{O}$ is the vector of off-diagonal terms in the period matrix. The given parameters are the diagonal elements of the period matrix, $\mathbf{D}$, and vector of phases, $\boldsymbol{\phi}$. Now let $\mathbf{x} = [x_1,$

$x_2, \ldots, x_N] = [\boldsymbol{\omega}, \rho, \mathbf{O}]$ and the nonlinear equations for the Riemann spectrum become:

$$F(x) = 0 \tag{16.46}$$

whose solution can be written as a simple expansion:

$$\mathbf{F}(\mathbf{x} + \delta\mathbf{x}) = \mathbf{F}(\mathbf{x}) + \mathbf{J} \cdot \delta\mathbf{x} + O(\delta\mathbf{x}^2), \quad \mathbf{J} \Rightarrow \{J_{ij}\} = \frac{\partial F_i}{\partial x_j}$$

Here $\mathbf{J}$ is the Jacobian matrix. Now suppose that the iteration happens to be perfect and that we render

$$\mathbf{F}(\mathbf{x} + \delta\mathbf{x}) = 0$$

This means that to leading order

$$\mathbf{J} \cdot \delta\mathbf{x} = -\mathbf{F}(\mathbf{x})$$

formally

$$\delta\mathbf{x} = -\mathbf{J}^{-1}\mathbf{F}(\mathbf{x}) \tag{16.47}$$

This linear set of equations can be solved by *LU* decomposition. Finally, the new vector we seek is computed by

$$\mathbf{x}_{\text{new}} = \mathbf{x}_{\text{old}} + \delta\mathbf{x} \tag{16.48}$$

This procedure can be iterated when the equations are nonlinear to obtain the desired accuracy.

Consider the following system of linear equations ($\mathbf{Ax} = \mathbf{b}$) that we write:

$$\mathbf{F}(\mathbf{x}) = \mathbf{Ax} - \mathbf{b} = 0$$

One generally seeks the vector of solutions, $\mathbf{x}$, given the matrix, $\mathbf{A}$, and vector, $\mathbf{b}$. Provided that $\mathbf{A}$ is not singular (implying that it has an inverse) the exact solution is

$$\mathbf{x} = \mathbf{A}^{-1}\mathbf{b}$$

For the linear system the number of iterations is only one for which the Jacobian is $\mathbf{J} = \mathbf{A}$ (we will see that the approach works equally well when the system of equations is nonlinear, but iterated).

Given an initial estimate for $\mathbf{x}$ which we call $\mathbf{x}_0$. Then the next iterate on the solution is computed by (the Taylor series expansion):

$$\mathbf{F}(\mathbf{x}_0 + \delta\mathbf{x}) = \mathbf{F}(\mathbf{x}_0) + \mathbf{J}\cdot\delta\mathbf{x}$$

The focus herein is to use this method iteratively when the systems of equations, $\mathbf{F}(\mathbf{x}) = 0$, is nonlinear, that is, when $\mathbf{F}(\mathbf{x}) \neq \mathbf{A}\mathbf{x} - \mathbf{b}$.

16.12 Solving Systems of Two-Dimensional Nonlinear Equations

In Section 16.8 we found the following four equations for four unknowns ω_1, ω_2, ρ, q_{12}:

$$
\begin{aligned}
F_1(\omega_1, \omega_2, \rho, q_{12}) &= A_{00}^1(q_{12})\omega_1 + A_{00}^2(q_{12})\omega_2 + B_{00}(q_{12})\rho - c_o k^2 C_{00}(q_{12}) \\
&\quad + \beta k^4 D_{00}(q_{12}) = 0 \\
F_2(\omega_1, \omega_2, \rho, q_{12}) &= A_{01}^1(q_{12})\omega_1 + A_{01}^2(q_{12})\omega_2 + B_{01}(q_{12})\rho - c_o k^2 C_{01}(q_{12}) \\
&\quad + \beta k^4 D_{01}(q_{12}) = 0 \\
F_3(\omega_1, \omega_2, \rho, q_{12}) &= A_{10}^1(q_{12})\omega_1 + A_{10}^2(q_{12})\omega_2 + B_{10}(q_{12})\rho - c_o k^2 C_{10}(q_{12}) \\
&\quad + \beta k^4 D_{10}(q_{12}) = 0 \\
F_4(\omega_1, \omega_2, \rho, q_{12}) &= A_{11}^1(q_{12})\omega_1 + A_{11}^2(q_{12})\omega_2 + B_{11}(q_{12})\rho - c_o k^2 C_{11}(q_{12}) \\
&\quad + \beta k^4 D_{11}(q_{12}) = 0
\end{aligned}
$$

$$(16.49)$$

where by abuse of notation the $B_{ij}(q_{12})$ are not elements of the period matrix but are the coefficients given by (16.45).

In the above equations I have associated the equation numbers with the μ-pair values:

$$F_1 \Rightarrow 0,0, \quad F_2 \Rightarrow 0,1, \quad F_3 \Rightarrow 1,0, \quad F_4 \Rightarrow 1,1$$

Now in the present problem (for two degrees of freedom or two cnoidal waves in the Riemann spectrum) we have the unknowns, which we write in vector notation

$$
\mathbf{x} = \begin{bmatrix} \omega_1 \\ \omega_2 \\ \rho \\ q_{12} \end{bmatrix}
$$

The inputs are the wavenumbers k_1, k_2, and nomes q_1, q_2 (the phases do not enter in the nonlinear equations and their solution). Here

$$q_1 = e^{-\frac{1}{2}B_{11}}, \quad q_2 = e^{-\frac{1}{2}B_{22}}, \quad q_{12} = e^{-B_{12}}$$

Here are the equations and coefficients:

$$A^1_{\mu_1\mu_2}(q_{12})\omega_1 + A^2_{\mu_1\mu_2}(q_{12})\omega_2 + B_{\mu_1\mu_2}(q_{12})\rho = c_0 k^2 C_{\mu_1\mu_2}(q_{12}) - \beta k^4 D_{\mu_1\mu_2}(q_{12})$$

$$A^1_{\mu_1\mu_2} = 4 \sum_{m_1=-\infty}^{\infty} \sum_{m_2=-\infty}^{\infty} (m_1 - \mu_1/2)[(m_1 - \mu_1/2)k_1 + (m_2 - \mu_2/2)k_2]$$

$$\times q_1^{2(m_1-\mu_1/2)^2} q_2^{2(m_2-\mu_2/2)^2} q_{12}^{2(m_1-\mu_1/2)(m_2-\mu_2/2)}$$

$$A^2_{\mu_1\mu_2} = 4 \sum_{m_1=-\infty}^{\infty} \sum_{m_2=-\infty}^{\infty} (m_2 - \mu_2/2)[(m_1 - \mu_1/2)k_1 + (m_2 - \mu_2/2)k_2]$$

$$\times q_1^{2(m_1-\mu_1/2)^2} q_2^{2(m_2-\mu_2/2)^2} q_{12}^{2(m_1-\mu_1/2)(m_2-\mu_2/2)}$$

$$B_{\mu_1\mu_2} = \sum_{m_1=-\infty}^{\infty} \sum_{m_2=-\infty}^{\infty} q_1^{2(m_1-\mu_1/2)^2} q_2^{2(m_2-\mu_2/2)^2} q_{12}^{2(m_1-\mu_1/2)(m_2-\mu_2/2)}$$

$$C_{\mu_1\mu_2} = 4 \sum_{m_1=-\infty}^{\infty} \sum_{m_2=-\infty}^{\infty} [(m_1 - \mu_1/2)k_1 + (m_2 - \mu_2/2)k_2]^2$$

$$\times q_1^{2(m_1-\mu_1/2)^2} q_2^{2(m_2-\mu_2/2)^2} q_{12}^{2(m_1-\mu_1/2)(m_2-\mu_2/2)}$$

$$D_{\mu_1\mu_2} = 16 \sum_{m_1=-\infty}^{\infty} \sum_{m_2=-\infty}^{\infty} [(m_1 - \mu_1/2)k_1 + (m_2 - \mu_2/2)k_2]^4$$

$$\times q_1^{2(m_1-\mu_1/2)^2} q_2^{2(m_2-\mu_2/2)^2} q_{12}^{2(m_1-\mu_1/2)(m_2-\mu_2/2)}$$

$$(16.50)$$

A leading order expansion of the coefficients is given below for the values of μ_1, μ_2 as mentioned above. *These can be seen to be polynomials in q_1, q_2, q_{12}. The results have been obtained by summing m_1, m_2 over $(-1,0,1)$. The polynomials to sixth order, $\sim \theta(q^6)$, are*

$$\mu_1, \mu_2 \Rightarrow 0,0$$

$$A^1_{00} = 8k_1 q_1^2 - 8(k_2 - k_1)\frac{q_1^2 q_2^2}{q_{12}^2} + 8(k_1 + k_2)q_1^2 q_2^2 q_{12}^2 + \cdots$$

$$A^2_{00} = 8k_2 q_2^2 + 8(k_2 - k_1)\frac{q_1^2 q_2^2}{q_{12}^2} + 8(k_1 + k_2)q_1^2 q_2^2 q_{12}^2 + \cdots$$

$$B_{00} = 1 + 2(q_1^2 + q_2^2) + \frac{2q_1^2 q_2^2}{q_{12}^2} + 2q_1^2 q_2^2 q_{12}^2 + \cdots$$

$$C_{00} = 8(k_1^2 q_1^2 + k_2^2 q_2^2) - \frac{8(k_2 - k_1)^2 q_1^2 q_2^2}{q_{12}^2} + 8(k_1 + k_2)^2 q_1^2 q_2^2 q_{12}^2 + \cdots$$

$$D_{00} = 32(k_1^4 q_1^2 + k_2^4 q_2^2) - \frac{32(k_2 - k_1)^4 q_1^2 q_2^2}{q_{12}^2} + 32(k_1 + k_2)^4 q_1^2 q_2^2 q_{12}^2 + \cdots$$

$$(16.51)$$

$$\mu_1, \mu_2 \Rightarrow 0, 1$$

$$A^1_{01} = 2(2k_1 + 3k_2)q_1^2 q_2^{9/2} q_{12}^3 + \frac{2(2k_1 - 3k_2)q_1^2 q_2^{9/2}}{q_{12}^3} + 4(2k_1 + k_2)q_1^2\sqrt{q_2}\sqrt{q_2 q_{12}} + \frac{4(2k_1 - k_2)q_1^2\sqrt{q_2}}{q_{12}} + \cdots$$

$$A^2_{01} = 3(2k_1 + 3k_2)q_1^2 q_2^{9/2} q_2 \, q_{12}^3 + 2k_2\sqrt{q_2} + 9k_2 q_2^{9/2} + \frac{3(3k_2 - 2k_1)q_1^2 q_2^{9/2}}{q_{12}^3} - \frac{2(2k_1 - k_2)q_1^2\sqrt{q_2}}{q_{12}} + 2(2k_1 + k_2)q_1^2\sqrt{q_2 q_{12}} + \cdots$$

$$B_{01} = 2\sqrt{q_2} + q_2^{9/2} + \frac{q_1^2 q_2^{9/2}}{q_{12}^3} + \frac{2q_1^2\sqrt{q_2}}{q_{12}} + 2q_1^2\sqrt{q_2}q_{12} + q_1^2 q_2^{9/2} q_{12}^3 + \cdots$$

$$C_{01} = 2k_2^2\sqrt{q_2} + \frac{q_2^{9/2}[(2k_1 - 3k_2)^2 q_1^2 + 9k_2^2 q_{12}^3]}{q_{12}^3} + \frac{2q_1^2\sqrt{q_2}[(k_2 - 2k_1)^2 + (2k_1 + k_2)^2 q_{12}^2]}{q_{12}} + (2k_1 + 3k_2)^2 q_1^2 q_2^{9/2} q_{12}^3 + \cdots$$

$$D_{01} = \sqrt{q_2}\left\{ k_2^4(2 + 81q_2^4) + \frac{q_1^2[(2k_1 - 3k_2)^4 q_2^4 + 2(k_2 - 2k_1)^4 q_{12}^2]}{q_{12}^3} + q_1^2 q_{12}[2(2k_1 + k_2)^4 + (2k_1 + 3k_2)^4 q_2^4 q_{12}^2] \right\} + \cdots$$

$$(16.52)$$

$$\mu_1, \mu_2 \rightrightarrows 1, 0$$

$$A_{10}^1 = 2k_1\sqrt{q_1} + \frac{3q_2^{9/2}[(3k_1 - 2k_2)]q_2^2 + 3k_1q_{12}^3}{q_{12}^3} + \frac{2q_2^2\sqrt{q_1}[k_1 - 2k_2 + (k_1 + 2k_2)q_{12}^2]}{q_{12}} + 3(3k_1 + 2k_2)q_1^{9/2}\,q_2^2q_{12}^3 + \cdots$$

$$A_{10}^2 = 2(3k_1 - 2k_2)q_1^{9/2}q_2^2q_{12}^3 + \frac{2(2k_2 - 3k_1)q_1^{9/2}\,q_2^2}{q_{12}^3} + 4(k_1 + 2k_2)\sqrt{q_1q_2^2}q_{12} - \frac{4(k_1 - 2k_2)\sqrt{q_1q_2^2}}{q_{12}} + \cdots$$

$$B_{10} = 2\sqrt{q_1} + q_1^{9/2} + \frac{\sqrt{q_1}q_2^2(q_1^4 + 2q_{12}^2)}{q_{12}^3} + \sqrt{q_1}q_2^2q_{12}(2 + q_1^4q_{12}^2) + \cdots$$

$$C_{10} = 2k_1^2\sqrt{q_1} + 9k_1^2q_1^{9/2} + \frac{(3k_1 - 2k_2)^2q_1^{9/2}\,q_2^2}{q_{12}^3} + \frac{2(k_1 - 2k_2)^2\sqrt{q_1q_2^2}}{q_{12}} + 2(k_1 + 2k_2)^2\sqrt{q_1q_2^2}q_{12} + (3k_1 + 2k_2)^2q_1^{9/2}\,q_2^2q_{12}^3 + \cdots$$

$$D_{10} = \sqrt{q_1}\left\{ k_1^4(2 + 81q_1^4) + \frac{q_2^2[(3k_1 - 2k_2)^4q_1^4 + 2(k_1 - 2k_2)^4q_{12}^2]}{q_{12}^3} + q_2^2q_{12}[2(k_1 + 2k_2)^4 + (3k_1 + 2k_2)^4q_1^4q_{12}^2] + \cdots \right\} + \cdots$$

$$(16.53)$$

$\mu_1, \mu_2 \Rightarrow 1, 1$

$$A^1_{11} = \frac{\sqrt{q_1 q_2}\left\{2q_{12}[k_1 - k_2 + q_{12}(k_1+k_2)] + q_2^4[k_1 - 3k_2 + q_{12}^3(k_1+3k_2)] + 3q_1^4[3k_1 - k_2 + q_{12}^3(3k_1+k_2)] + 3(k_1+k_2)q_1^2 q_{12}^6\right\}}{q_{12}^{3/2}} + \cdots$$

$$A^2_{11} = \frac{\sqrt{q_1 q_2}\left\{2q_{12}[k_2 - k_1 + q_{12}(k_1+k_2)] + 3q_2^4[3k_2 - k_1 + q_{12}^3(k_1+3k_2)] + q_1^4[k_2 - 3k_1 + q_{12}^3(3k_1+k_2)] + 9(k_1+k_2)q_1^2 q_{12}^6\right\}}{q_{12}^{3/2}} + \cdots$$

$$B_{11} = 2\sqrt{q_2} + q_2^{9/2} + \frac{q_1^2 q_2^{9/2}}{q_{12}^3} + \frac{2q_1^2\sqrt{q_2}}{q_{12}} + 2q_1^2\sqrt{q_2}\,q_{12} + q_1^2 q_2^{9/2} q_{12}^3 + \cdots$$

$$C_{11} = \frac{\sqrt{q_1 q_2}\left\{2q_{12}[(k_1-k_2)^2 + q_{12}(k_1+k_2)^2] + q_2^4[(k_1-3k_2)^2 + q_{12}^3(k_1+3k_2)^2] + q_1^4[(3k_1-k_2)^2 + q_{12}^3(3k_1+k_2)^2] + 9(k_1+k_2)^2 q_2^4 q_{12}^6\right\}}{q_{12}^{3/2}} + \cdots$$

$$D_{11} = \frac{\sqrt{q_1 q_2}\left\{2q_{12}[(k_1-k_2)^4 + q_{12}(k_1+k_2)^4] + q_2^4[(k_1-3k_2)^4 + q_{12}^3(k_1+3k_2)^4] + q_1^4[(3k_1-k_2)^4 + q_{12}^3(3k_1+k_2)^4] + 81(k_1+k_2)^4 q_2^4 q_{12}^6\right\}}{q_{12}^{3/2}} + \cdots$$

(16.54)

The first column of Jacobian elements is:

$$J_{11} = \frac{\partial F_1}{\partial \omega_1} = A_{00}^1, \quad J_{21} = \frac{\partial F_2}{\partial \omega_1} = A_{01}^1, \quad J_{31} = \frac{\partial F_3}{\partial \omega_1} = A_{10}^1, \quad J_{41} = \frac{\partial F_4}{\partial \omega_1} = A_{11}^1$$

The second column of Jacobian elements is

$$J_{12} = \frac{\partial F_1}{\partial \omega_2} = A_{00}^2, \quad J_{22} = \frac{\partial F_2}{\partial \omega_2} = A_{01}^2, \quad J_{32} = \frac{\partial F_3}{\partial \omega_2} = A_{10}^2, \quad J_{42} = \frac{\partial F_4}{\partial \omega_2} = A_{11}^2$$

The third column of Jacobian elements is

$$J_{13} = \frac{\partial F_1}{\partial \rho} = B_{00}, \quad J_{23} = \frac{\partial F_2}{\partial \rho} = B_{01}, \quad J_{33} = \frac{\partial F_3}{\partial \rho} = B_{10}, \quad J_{43} = \frac{\partial F_4}{\partial \rho} = B_{11}$$

The forth column of Jacobian elements (the only column not trivial) is

$$J_{14} = \frac{\partial F_1}{\partial q_{12}}, \quad J_{24} = \frac{\partial F_2}{\partial q_{12}}, \quad J_{34} = \frac{\partial F_3}{\partial q_{12}}, \quad J_{44} = \frac{\partial F_4}{\partial q_{12}}$$

Then the equations we seek to solve are

$$\begin{bmatrix} A_{00}^1(q_{12}) & A_{00}^2(q_{12}) & B_{00}(q_{12}) & J_{14}(\omega_1, \omega_2, \rho, q_{12}) \\ A_{01}^1(q_{12}) & A_{01}^2(q_{12}) & B_{01}(q_{12}) & J_{24}(\omega_1, \omega_2, \rho, q_{12}) \\ A_{10}^1(q_{12}) & A_{10}^2(q_{12}) & B_{10}(q_{12}) & J_{34}(\omega_1, \omega_2, \rho, q_{12}) \\ A_{11}^1(q_{12}) & A_{11}^2(q_{12}) & B_{11}(q_{12}) & J_{44}(\omega_1, \omega_2, \rho, q_{12}) \end{bmatrix} \begin{bmatrix} \delta\omega_1 \\ \delta\omega_2 \\ \delta\rho \\ \delta q_{12} \end{bmatrix} = - \begin{bmatrix} F_1(\omega_1, \omega_2, \rho, q_{12}) \\ F_2(\omega_1, \omega_2, \rho, q_{12}) \\ F_3(\omega_1, \omega_2, \rho, q_{12}) \\ F_4(\omega_1, \omega_2, \rho, q_{12}) \end{bmatrix}$$

So that

$$\begin{bmatrix} \delta\omega_1 \\ \delta\omega_2 \\ \delta\rho \\ \delta q_{12} \end{bmatrix} = - \begin{bmatrix} A_{00}^1(q_{12}) & A_{00}^2(q_{12}) & B_{00}(q_{12}) & J_{14}(\omega_1, \omega_2, \rho, q_{12}) \\ A_{01}^1(q_{12}) & A_{01}^2(q_{12}) & B_{01}(q_{12}) & J_{24}(\omega_1, \omega_2, \rho, q_{12}) \\ A_{10}^1(q_{12}) & A_{10}^2(q_{12}) & B_{10}(q_{12}) & J_{34}(\omega_1, \omega_2, \rho, q_{12}) \\ A_{11}^1(q_{12}) & A_{11}^2(q_{12}) & B_{11}(q_{12}) & J_{44}(\omega_1, \omega_2, \rho, q_{12}) \end{bmatrix}^{-1} \begin{bmatrix} F_1(\omega_1, \omega_2, \rho, q_{12}) \\ F_2(\omega_1, \omega_2, \rho, q_{12}) \\ F_3(\omega_1, \omega_2, \rho, q_{12}) \\ F_4(\omega_1, \omega_2, \rho, q_{12}) \end{bmatrix}$$

$$(16.55)$$

with

$$\begin{bmatrix} \omega_1 \\ \omega_2 \\ \rho \\ q_{12} \end{bmatrix}_{new} = \begin{bmatrix} \omega_1 \\ \omega_2 \\ \rho \\ q_{12} \end{bmatrix} + \begin{bmatrix} \delta\omega_1 \\ \delta\omega_2 \\ \delta\rho \\ \delta q_{12} \end{bmatrix} \qquad (16.56)$$

Then take the appropriate derivatives from the function F:

$$F_{\mu_1\mu_2}(\omega_1,\omega_2,\rho,B_{12}) = \sum_{m_1=-\infty}^{\infty} \sum_{m_2=-\infty}^{\infty} \Big\{ 4[(m_1-\mu_1/2)k_1 + (m_2-\mu_2/2)k_2]$$

$$\times [(m_1-\mu_1/2)\omega_1 + (m_2-\mu_2/2)\omega_2] - 4c_o[(m_1-\mu_1/2)k_1 + (m_2-\mu_2/2)k_2]^2$$

$$+ 16\beta[(m_1-\mu_1/2)k_1 + (m_2-\mu_2/2)k_2]^4 + \rho \Big\}$$

$$q_1^{2(m_1-\mu_1/2)^2} q_2^{2(m_2-\mu_2/2)^2} q_{12}^{2(m_1-\mu_1/2)(m_2-\mu_2/2)} = 0$$

$$\frac{\partial F_{\mu_1\mu_2}(\omega_1,\omega_2,\rho,B_{12})}{\partial \omega_1} = \sum_{m_1=-\infty}^{\infty} \sum_{m_2=-\infty}^{\infty} \Big\{ 4(m_1-\mu_1/2)[(m_1-\mu_1/2)k_1$$

$$+ (m_2-\mu_2/2)k_2] \Big\} \times q_1^{2(m_1-\mu_1/2)^2} q_2^{2(m_2-\mu_2/2)^2} q_{12}^{2(m_1-\mu_1/2)(m_2-\mu_2/2)} = 0$$

$$\frac{\partial F_{\mu_1\mu_2}(\omega_1,\omega_2,\rho,B_{12})}{\partial \omega_2} = \sum_{m_1=-\infty}^{\infty} \sum_{m_2=-\infty}^{\infty} \Big\{ 4(m_2-\mu_2/2)[(m_1-\mu_1/2)k_1$$

$$+ (m_2-\mu_2/2)k_2] \Big\} \times q_1^{2(m_1-\mu_1/2)^2} q_2^{2(m_2-\mu_2/2)^2} q_{12}^{2(m_1-\mu_1/2)(m_2-\mu_2/2)} = 0$$

$$\frac{\partial F_{\mu_1\mu_2}(\omega_1,\omega_2,\rho,B_{12})}{\partial \rho} = \sum_{m_1=-\infty}^{\infty} \sum_{m_2=-\infty}^{\infty} q_1^{2(m_1-\mu_1/2)^2} q_2^{2(m_2-\mu_2/2)^2} q_{12}^{2(m_1-\mu_1/2)(m_2-\mu_2/2)}$$

$$\frac{\partial F_{\mu_1\mu_2}(\omega_1,\omega_2,\rho,B_{12})}{\partial q_{12}} = 2q_{12}^{-1} \sum_{m_1=-\infty}^{\infty} \sum_{m_2=-\infty}^{\infty} (m_1-\mu_1/2)(m_2-\mu_2/2)$$

$$\times \Big\{ 4[(m_1-\mu_1/2)k_1 + (m_2-\mu_2/2)k_2] \times [(m_1-\mu_1/2)\omega_1 + (m_2-\mu_2/2)\omega_2]$$

$$- 4c_o[(m_1-\mu_1/2)k_1 + (m_2-\mu_2/2)k_2]^2 + 16\beta[(m_1-\mu_1/2)k_1 + (m_2-\mu_2/2)k_2]^4 + \rho \Big\}$$

$$\times q_1^{2(m_1-\mu_1/2)^2} q_2^{2(m_2-\mu_2/2)^2} q_{12}^{2(m_1-\mu_1/2)(m_2-\mu_2/2)} = 0$$

We now seek:

$$\begin{bmatrix} \delta\omega_1 \\ \delta\omega_2 \\ \delta\rho \\ \delta q_{12} \end{bmatrix} = - \begin{bmatrix} \partial_{\omega_1}F_{00} & \partial_{\omega_2}F_{00} & \partial_{\rho}F_{00} & \partial_{q_{12}}F_{00} \\ \partial_{\omega_1}F_{01} & \partial_{\omega_2}F_{01} & \partial_{\rho}F_{01} & \partial_{q_{12}}F_{01} \\ \partial_{\omega_1}F_{10} & \partial_{\omega_2}F_{10} & \partial_{\rho}F_{10} & \partial_{q_{12}}F_{10} \\ \partial_{\omega_1}F_{11} & \partial_{\omega_2}F_{11} & \partial_{\rho}F_{11} & \partial_{q_{12}}F_{11} \end{bmatrix}^{-1} \begin{bmatrix} F_1(\omega_1,\omega_2,\rho,q_{12}) \\ F_2(\omega_1,\omega_2,\rho,q_{12}) \\ F_3(\omega_1,\omega_2,\rho,q_{12}) \\ F_4(\omega_1,\omega_2,\rho,q_{12}) \end{bmatrix}$$

$$(16.57)$$

To get the linear Fourier limit we assume $q_1 \ll 1$, $q_2 \ll 1$. In this limit the initial values of the parameters we seek have the form:

$$
\mathbf{x}_o = \begin{bmatrix} \omega_1 \\ \omega_2 \\ \rho \\ q_{12} \end{bmatrix} = \begin{bmatrix} c_o k_1 - \beta k_1^3 \\ c_o k_1 - \beta k_1^3 \\ 0 \\ \left(\dfrac{k_1 - k_2}{k_1 + k_2} \right)^2 \end{bmatrix}
\tag{16.58}
$$

which agrees with the leading order estimate from Schottky uniformization (Chapter 15):

$$
q_{12} = e^{-B_{12}} \simeq \left(\frac{k_1 - k_2}{k_1 + k_2} \right)^2 \quad \text{or} \quad B_{12} \simeq -\ln\left(\frac{k_1 - k_2}{k_1 + k_2} \right)^2
$$

After iterating the procedure we get

$$
\mathbf{x}_1 = \begin{bmatrix} \omega_1 \\ \omega_2 \\ \rho \\ q_{12} \end{bmatrix} = \begin{bmatrix} c_o k_1 - \beta k_1^3 (1 - 24 q_1^2) \\ c_o k_1 - \beta k_1^3 (1 - 24 q_2^2) \\ 0 + 12(k_1^4 q_1^2 + k_2^4 q_2^2) \\ \left(\dfrac{k_1 - k_2}{k_1 + k_2} \right)^2 + 32 k_1 k_2 [(k_1^2 q_1^2 + k_2^2 q_2^2)/(k_1 + k_2)^4 \end{bmatrix}
\tag{16.59}
$$

These results are the leading order nonlinear corrections to the initial estimates. A computer program would iterate again on these results until convergence. Of course (16.59) is the Riemann spectrum computed to second order.

Appendix: Theta Functions with Characteristics

Write the *theta function with characteristics in vector form*:

$$
\theta(\mathbf{X}; \mathbf{a}, \mathbf{b} \mid \mathbf{B}) = \sum_{\mathbf{m}=-\infty}^{\infty} e^{-\frac{1}{2}(\mathbf{m}+\mathbf{a}) \cdot \mathbf{B}(\mathbf{m}+\mathbf{a})} \, e^{i(\mathbf{m}+\mathbf{a}) \cdot (\mathbf{X}+\mathbf{b})}
\tag{16.A.1}
$$

where $\mathbf{m} = [m_1, m_2, \ldots, m_N]$ are integers, $\mathbf{a} = [a_1, a_2, \ldots, a_N]$ and $\mathbf{b} = [b_1, b_2, \ldots, b_N]$ are arbitrary constants (generally half-integers), $\mathbf{X} = [X_1, X_2, \ldots, X_N]$ where

$$
X_j = k_j x - \omega_j t + \phi_j, \quad j = 1, 2, \ldots, N
\tag{16.A.2}
$$

The *scalar form of the theta function with characteristics* is given by

$$\theta(\mathbf{X}, \mathbf{a}, \mathbf{b} \mid \mathbf{B}) = \sum_{m_1=-\infty}^{\infty} \sum_{m_2=-\infty}^{\infty} \cdots \sum_{m_N=-\infty}^{\infty} \exp\left[-\frac{1}{2}\sum_{j=1}^{N}\sum_{k=1}^{N}(m_j + a_j)(m_k + a_k)B_{jk}\right]$$

$$\times \exp\left[i\sum_{j=1}^{N}(m_j + a_j)(X_j + b_j)\right]$$

$$(16.A.3)$$

Note that by setting $\mathbf{a} = \mathbf{b} = 0$ we get the usual theta function without characteristics, that is

$$\theta(\mathbf{X} \mid \mathbf{B}) = \theta(\mathbf{X}, 0, 0 \mid \mathbf{B}) \tag{16.A.4}$$

Notice that $\theta(\mathbf{X}, 0, \mathbf{b} \mid \mathbf{B})$ is a *pure space or time translation of the theta function*. This becomes

$$\theta(\mathbf{X}, 0, \mathbf{b} \mid \mathbf{B}) = \sum_{m_1=-\infty}^{\infty} \sum_{m_2=-\infty}^{\infty} \cdots \sum_{m_N=-\infty}^{\infty} \exp\left(-\frac{1}{2}\sum_{j=1}^{N}\sum_{k=1}^{N}m_j m_k B_{jk}\right)$$

$$\times \exp\left[i\sum_{j=1}^{N}m_j(X_j + b_j)\right] \tag{16.A.5}$$

Now write

$$X_j + b_j = k_j x - \omega_j t + \phi_j + b_j$$

Let $b_j = k_j L$, for L a constant spatial coherence length or spatial lag distance (it is not the period of the wave train in this appendix), and get

$$X_j + b_j = k_j x - \omega_j t + \phi_j + k_j L = k_j(x + L) - \omega_j t + \phi_j$$

and

$$\theta(\mathbf{X}, 0, \mathbf{k}L \mid \mathbf{B}) = \sum_{m_1=-\infty}^{\infty} \sum_{m_2=-\infty}^{\infty} \cdots \sum_{m_N=-\infty}^{\infty} \exp\left(-\frac{1}{2}\sum_{j=1}^{N}\sum_{k=1}^{N}m_j m_k B_{jk}\right)$$

$$\times \exp\left[i\sum_{j=1}^{N}m_j k_j(x + L) - i\sum_{j=1}^{N}m_j \omega_j t + i\sum_{j=1}^{N}m_j \phi_j\right] \tag{16.A.6}$$

Therefore

$$\theta(x, t; 0, \mathbf{k}L \mid \mathbf{B}) = \theta(x + L, t; 0, 0 \mid \mathbf{B})$$

Therefore, a *spatially translated theta function with zero characteristics* $(0,0)$ is equivalent to a *theta function with particular characteristics* $(0, kL)$.

Let $b_j = \omega_j \tau$, for τ a constant time lag, and get

$$X_j + b_j = k_j x - \omega_j t + \phi_j + \omega_j \tau = k_j x - \omega_j (t + \tau) + \phi_j$$

and

$$\theta(\mathbf{X}, 0, \boldsymbol{\omega}\tau \mid \mathbf{B}) = \sum_{m_1=-\infty}^{\infty} \sum_{m_2=-\infty}^{\infty} \cdots \sum_{m_N=-\infty}^{\infty} \exp\left(-\frac{1}{2}\sum_{j=1}^{N}\sum_{k=1}^{N} m_j m_k B_{jk}\right)$$
$$\times \exp\left[i\sum_{j=1}^{N} m_j k_j x - i\sum_{j=1}^{N} m_j \omega_j (t+\tau) + i\sum_{j=1}^{N} m_j \phi_j\right] \quad (16.A.7)$$

Therefore

$$\theta(x, t; 0, \boldsymbol{\omega}\tau \mid \mathbf{B}) = \theta(x, t+\tau; 0, 0 \mid \mathbf{B})$$

Therefore, a *temporally shifted theta function with zero characteristics* $(0,0)$ is equivalent to a *theta function with particular characteristics* $(0,\boldsymbol{\omega}\tau)$.

Nota Bene: The spatial and temporal lags just discussed in terms of characteristics are what I refer to as being *experimentally accessible*. Just by lagging a measured time series we can access a theta function with characteristics. This is because a lagged theta function (whether viewed as a function of space or time) can be interpreted in terms of a theta function with characteristics, as seen above, a solution of the KdV equation! This also means that the method of surrogates can be applied to theta functions to enhance the data analysis procedures in Chapter 23.

Another choice is to let the characteristics be phases, that is, $b_j = \Phi_j$ and get

$$\theta(\mathbf{X}, 0, \boldsymbol{\Phi} \mid \mathbf{B}) = \sum_{m_1=-\infty}^{\infty} \sum_{m_2=-\infty}^{\infty} \cdots \sum_{m_N=-\infty}^{\infty} \exp\left(-\frac{1}{2}\sum_{j=1}^{N}\sum_{k=1}^{N} m_j m_k B_{jk}\right)$$
$$\times \exp\left[i\sum_{j=1}^{N} m_j k_j x - i\sum_{j=1}^{N} m_j \omega_j t + i\sum_{j=1}^{N} m_j (\phi_j + \Phi_j)\right] \quad (16.A.8)$$

This choice just means we can change the phase by a constant, or jumble each phase as desired. For example, we could use $\phi_j = 0$ and then make a random choice of phases just by changing the second characteristic. Therefore, the theta function $\theta(x,t;0,\boldsymbol{\Phi} \mid \mathbf{B}, \boldsymbol{\phi} = 0) = \theta(x,t; 0,0 \mid \mathbf{B}, \boldsymbol{\Phi}) = \theta(x,t \mid \mathbf{B}, \boldsymbol{\Phi})$ is a *statistical realization* of a particular theta function. Generally, $\theta(x,t;0,\Phi \mid \mathbf{B}, \boldsymbol{\phi}) = \theta(x,t; 0,0 \mid \mathbf{B}, \boldsymbol{\phi} + \boldsymbol{\Phi})$.

Now let us go to the first characteristic and exam it in detail. To do this we need to compute:

$$(\mathbf{m} + \mathbf{a}) \cdot \mathbf{B}(\mathbf{m} + \mathbf{a}) = (\mathbf{m} + \mathbf{a}) \cdot \mathbf{B}(\mathbf{m} + \mathbf{a}) = (\mathbf{m} + \mathbf{a}) \cdot (\mathbf{Bm} + \mathbf{Ba})$$
$$= \mathbf{m} \cdot \mathbf{Bm} + \mathbf{m} \cdot \mathbf{Ba} + \mathbf{a} \cdot \mathbf{Bm} + \mathbf{a} \cdot \mathbf{Ba}$$

or in scalar form:

$$(\mathbf{m} + \mathbf{a}) \cdot \mathbf{B}(\mathbf{m} + \mathbf{a}) = \sum_{i=1}^{N}\sum_{j=1}^{N} m_i m_j B_{ij} + \sum_{i=1}^{N}\sum_{j=1}^{N} a_i a_j B_{ij} + \sum_{i=1}^{N}\sum_{j=1}^{N} m_i a_j B_{ij}$$
$$+ \sum_{i=1}^{N}\sum_{j=1}^{N} a_i m_j B_{ij}$$

The last two terms may be written:

$$(\mathbf{m} + \mathbf{a}) \cdot \mathbf{B}(\mathbf{m} + \mathbf{a}) = \sum_{i=1}^{N}\sum_{j=1}^{N} m_i m_j B_{ij} + \sum_{i=1}^{N}\sum_{j=1}^{N} a_i a_j B_{ij} + \sum_{i=1}^{N} m_i \sum_{j=1}^{N} a_j B_{ij}$$
$$+ \sum_{j=1}^{N} m_j \sum_{i=1}^{N} a_i B_{ij}$$

Interchange the summation indices in the last term, to give

$$(\mathbf{m} + \mathbf{a}) \cdot \mathbf{B}(\mathbf{m} + \mathbf{a}) = \sum_{i=1}^{N}\sum_{j=1}^{N} m_i m_j B_{ij} + \sum_{i=1}^{N}\sum_{j=1}^{N} a_i a_j B_{ij} + \sum_{i=1}^{N} m_i \sum_{j=1}^{N} a_j B_{ij}$$
$$+ \sum_{i=1}^{N} m_i \sum_{j=1}^{N} a_j B_{ji}$$

But $B_{ij} = B_{ji}$ so that

$$(\mathbf{m} + \mathbf{a}) \cdot \mathbf{B}(\mathbf{m} + \mathbf{a}) = \sum_{i=1}^{N}\sum_{j=1}^{N} m_i m_j B_{ij} + \sum_{i=1}^{N}\sum_{j=1}^{N} a_i a_j B_{ij} + 2 \sum_{i=1}^{N} m_i \sum_{j=1}^{N} a_j B_{ij}$$

This leads to the form for theta functions with characteristics discussed earlier in this Chapter.

Part Seven

Nonlinear Numerical and Time Series Analysis Algorithms

To analyze or assimilate data and construct nonlinear models one needs a number of numerical algorithms. This Section of the book gives several approaches that allow the user to apply the inverse scattering transform to a variety of physical situations.

It is often convenient to compute the *direct scattering transform* (DST) via the spectral eigenvalue problem (Chapters 17, 18). This approach can be applied to the analysis of data, including internal waves and surface waves, topics of Part 8. Since the DST provides the spectral eigenvalues one can immediately compute the loop integrals to give the Riemann spectrum (Chapters 14 and 19).

Of course the most important of the numerical methods are the approaches that allow one to compute the Riemann theta functions (Chapters 20–23). Chapter 20 provides an overview of several "brute-force" approaches for computing theta functions. This chapter is useful for developing codes that are not fast, but nevertheless are easy to program and to check out. Codes of this type are also useful for providing diagnostics that can aid in initiating strategies to develop new fast algorithms, or at least to improve one's understanding of how the theta function works. Once checked out these programs also serve as controls during the development of "fast" theta function algorithms. I discuss the discrete theta function as a fast algorithm in Chapter 21. This approach serves to compare and contrast the discrete theta function with the discrete Fourier transform and numerous examples are given. In contrast to the brute-force algorithms, which sum the theta function over the n-cube in lattice space, Chapter 22 discusses an explicit algorithm that allows one instead to sum over an n-sphere or n-ellipsoid. This approach allows one to compute theta functions several orders of magnitude faster than the brute-force algorithms.

Doi: 10.1016/S0074-6142(10)97046-0

Finally, in Chapter 23 I discuss how to determine the Riemann spectrum from a data set or from a numerical simulation. The algorithm begins with a measurement of the sea surface and computes the Riemann matrix and phases. The approach is purely numerical and depends on an iterative method that I call "nonlinear adiabatic annealing on a Riemann surface." This approach does not use the eigenvalue part of the Lax pair, but instead "extracts" the Riemann spectrum from data as a kind of least-squares fit of the data. The algorithm must of course ensure that we remain on a Riemann surface in order to obey the Novikov conjecture, which when translated into physical terms means "the physics lies on Riemann surfaces."

17 Automatic Algorithm for the Spectral Eigenvalue Problem for the KdV Equation

17.1 Introduction

The inverse scattering transform (IST) for the *periodic* Korteweg-deVries (KdV) equation in the μ-function representation is described in this chapter and numerical formulations are given (1) for determining the *direct scattering transform spectrum* of an input discrete space or time series and (2) for reconstructing the wave train from the spectrum via the *inverse scattering problem* in the *hyperelliptic-function linear superposition law*. The advantage of the present method for solving the KdV eigenvalue problem is that the numerical computations are *automatic*, that is, one is guaranteed that the algorithm will search out and find *all* the nonlinear modes (to within the input numerical precision) of a given arbitrary, N degree-of-freedom wave train. Therefore, one never need be concerned that certain zeroes (eigenvalues of the direct spectral problem) may be missed in the search process. Thus the algorithm may be viewed as "worry free" in the sense that the user need not be an expert in periodic IST theory to analyze nonlinear wave data using the nonlinear Fourier approach given here. The algorithms are most appropriate for the *time series analysis* of measured and computed data. One is thus numerically able to analyze an input time series with M discrete points: (1) to construct the IST spectrum, (2) to determine the $N = M/2$ hyperelliptic-function oscillation modes, (3) to reconstruct the input wave train by a linear superposition law, and (4) to nonlinearly filter the input wave train. I discuss the numerical details of the algorithm and give a number of numerical examples. A reduced version of this chapter is given in Osborne (1994).

17.2 Formulation of the Problem

The Korteweg-deVries equation is the classical prototypical partial differential equation for describing small-but-finite amplitude, long waves in shallow water (Korteweg and deVries, 1895; Whitham, 1974; Miles, 1980):

$$\eta_t + c_o \eta_x + \alpha \eta \eta_x + \beta \eta_{xxx} = 0 \tag{17.1}$$

The surface elevation $\eta(x, t)$ varies nonlinearly as a function of space x and time t. The constant coefficients are given by $c_0 = \sqrt{gh}$, $\alpha = 3c_0/2h$, and $\beta = c_0 h^2/6$; g is the acceleration of gravity. Subscripts in Equation (17.1) refer to partial derivatives with respect to x and t; h is the water depth. The linearized KdV equation (set $\alpha=0$ in Equation (17.1)) has the associated dispersion relation $\omega_0 = c_0 k - \beta k^3$ and c_0 is the linear, dispersionless phase speed. Herein it is assumed that (17.1) is governed by periodic boundary conditions, $\eta(x, t) = \eta(x+L, t)$ for L the period. As discussed elsewhere in this monograph, many other applications of the KdV equation are known. These include internal waves (Benny, 1966), Rossby waves (Benny, 1966; Maxworthy and Redekopp, 1976), plasma waves (Karpman, 1975), and bores (Peregrine, 1966).

It is useful in the present context, to think of the KdV equation as a *source of convenient, nonlinear mathematical basis functions* useful for the analysis of time series. Recognizing the importance of this perspective is an important step in understanding the use of IST as a way to improve understanding of nonlinear data in the form of time series.

The presence of the nonlinear term $\alpha\eta\eta_x$ historically raised unique challenges in the determination of solutions to Equation (17.1). Without this term the general *spectral solution* to Equation (17.1) can be easily found using the Fourier transform, and in this sense the linear problem is trivial. From the time of its discovery in 1895 (Korteweg and deVries, 1895), 72 years passed before the general spectral solution to (17.1) was found for *infinite-line boundary conditions* (Gardner et al., 1967). This technique has since been christened the IST (Novikov et al., 1984; Ablowitz and Segur, 1981; Dodd et al., 1982; Newell, 1985; Degasperis, 1991; Ablowitz and Clarkson, 1991). The work of Gardner et al. (1967) evolved from the precise numerical experiments of Zabusky and Kruskal (1965), who formulated the concept of the soliton as a stable "particle" in the nonlinear dynamics of KdV. They were further able to establish the presence of a pair-wise *phase shift* during soliton interactions with each other. Thus soliton physics was born by studying the mathematical and physical structure of the KdV equation.

The nonlinear Fourier structure of the KdV equation, as given by the IST with *periodic boundary conditions*, has also attracted considerable interest (Dubrovin and Novikov, 1975a,b; Dubrovin et al., 1976; Flaschka and McLaughlin, 1976; McKean and Trubowitz, 1976). IST allows solutions to the periodic KdV equation to be constructed by a linear superposition of the so-called hyperelliptic (μ-function) nonlinear oscillation modes. The hyperelliptic modes are generalizations of the sine waves of the associated Fourier series solution to the *linearized* problem. While an alternative formulation of the inverse problem for periodic IST exists in terms of the θ-function representation (see Chapters 18 and 19), I focus on the μ-function representation in this chapter. As discussed in detail below the μ-function representation provides a numerical alternative to the θ-function representation for time series analysis purposes.

The present work builds on previous progress in the numerical implementation of the IST (Taha and Ablowitz, 1984a,b; Bishop and Lomdahl, 1986; Bishop et al., 1986; Terrones et al., 1990; Flesch et al., 1991; Osborne et al., 1991; McLaughlin and Schober, 1992). Special emphasis is given herein on *time series analysis methods*. These contrast to *numerical methods* in that the latter focus on precise computation of specific results, while time series analysis methods generally focus on a rather complete determination of *all* relevant spectral information in both the direct and inverse problems, primarily for a particular *unique discretization of the spectral problem*. Furthermore time series analysis methods generally focus on the behavior of nonlinear wave motion as a function of time; the spatial evolution is not a necessary ingredient. Of course, in the time series analysis of data one is concerned with the amount of computer time required, for example, any algorithm must complete the computation in a "reasonable" amount of time. A number of previous efforts with regard to time series analysis are given in the literature (Osborne, 1983, 1991a,b; Osborne and Bergamasco, 1985, 1986).

Here I focus (1) on a particular discretization of the KdV direct and inverse spectral problems and (2) on an approach, based upon a variable step algorithm, for the *automatic control* of the numerical computations. The latter method provides a way to determine *all* of the discrete eigenvalues (zeros) in the spectral problem, to within the input numerical precision. If in the search process certain eigenvalues are missed, the algorithm automatically returns to find them. In this sense, the computer code is "worry free," that is, the user need not be an expert in inverse scattering theory to use the approach in the analysis of data. For those interested in the coding of the methods discussed herein, I give a rather detailed numerical analysis of the properties of the periodic IST, particularly from the point of view of the "oscillation theorems" relating to the behavior of certain elements of the monodromy matrix as a function of space (or time) and squared-wavenumber ($E = k^2$). These results lead to procedures for the automatic control of the algorithms given herein.

The most common experimental situation is to record data as a function of time at a single spatial location. The reasons are often economical. The measurement of time series requires a single wave staff, resistance gauge or pressure recorder; the measurement of space series requires remote sensing capability; it is clear that satellites are more expensive than *in situ* instruments. One is thus motivated to develop procedures for determining the scattering transform of a *time series*, $\eta(0, t)$. To this end one may employ the *time-like* KdV equation (tKdV) (Karpman, 1975; Ablowitz and Segur, 1981):

$$\eta_x + c_0'\eta_t + \alpha'\eta\eta_t + \beta'\eta_{ttt} = 0 \tag{17.2}$$

where $c_0' = 1/c_0$, $\alpha' = -\alpha/c_0$, and $\beta' = -\beta/c_0^4 = -h^2/6c_0^3$; (17.2) has the linearized dispersion relation $k = \omega/c_0 + (\beta/c_0^4)\omega^3$. TKdV solves a *boundary value problem*: given the temporal evolution $\eta(0,t)$ at a fixed spatial location $x = 0$, Equation (17.2) determines the wave motion over all space as a function of time $\eta(x,t)$.

Periodic boundary conditions ($\eta(x, t) = \eta(x, t+T)$) are assumed herein to be consistent with linear Fourier algorithms (discrete and fast Fourier transforms). These issues are discussed in detail later and elsewhere (Osborne, 1991a,b).

The present chapter is organized as follows. Section 17.3 gives an overview of periodic IST for KdV in the μ-representation. Section 17.4 gives certain facts about periodic IST useful for physical understanding and numerical implementation. Section 17.5 describes a numerical discretization procedure for the construction of the IST spectrum and for determination of the hyperelliptic inverse problem. Section 17.6 extends the approach, applying results from certain "oscillation theorems" of periodic IST theory and extensions thereof. A method is then developed which *automatically controls* the search for and determination of the eigenvalues in the spectrum. Finally in Section 17.7 I discuss the application of the algorithm as a tool for the *space or time series analysis of data* and give a numerical example in which a wave train with $N = 128$ degrees of freedom is analyzed. I furthermore discuss periodic IST as a tool for the *nonlinear filtering of data*. Using this procedure one is able to surgically remove and isolate selected IST spectral components and to subsequently reconstruct these in configuration space, that is, to graph the soliton components in the absence of the radiation spectrum or *vice versa*.

17.3   Periodic IST for the KdV Equation in the μ-Function Representation

The general spectral solution to the periodic KdV equation (17.1) may be written as a *linear superposition of nonlinearly interacting, nonlinear waves that are referred to as hyperelliptic functions*, $\mu_j(x; x_0, t)$:

$$\lambda\eta(x, t) = -E_1 + \sum_{j=1}^{N}[2\mu_j(x; x_0, t) - E_{2j} - E_{2j+1}] \qquad (17.3)$$

Here x_0 is referred to as a base point and will be discussed in detail below. It is helpful to refer to Figure 17.1 for the discussion that follows. Here $\lambda = \alpha/6\beta$ for a space series, $\eta(x, 0)$, and $\lambda = \alpha c_0^2/6\beta$ for a time series, $\eta(0, t)$, and the E_{2j} are constant eigenvalues to be defined later. Equation (17.3) is the first of the so-called *trace formulae* for the KdV equation (Dubrovin and Novikov, 1974; Flaschka and McLaughlin, 1976; McKean and Trubowitz, 1976) and may be thought of as a *linear superposition law*. Since Equation (17.3) reduces to an ordinary linear Fourier series in the limit of small amplitude wave motion (Osborne and Bergamasco, 1985), we refer to (17.3) as a *nonlinear Fourier series*. The hyperelliptic functions $\mu_j(x; x_0, t)$ are the *nonlinear oscillation modes* of periodic KdV, that is, they are the "sine waves" of the nonlinear Fourier series for KdV (Osborne and Bergamasco, 1986). The $\mu_j(x; x_0, t)$ are however generally quite nonsinusoidal in shape and implicitly contain the nonlinear dynamics of the

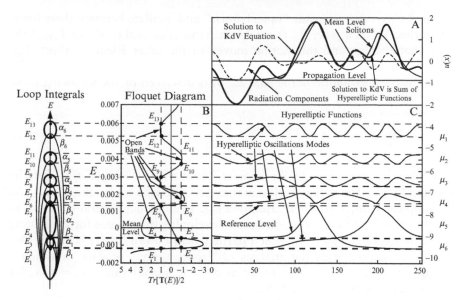

Figure 17.1 Typical spectral representation of a particular solution of the KdV equation. Shown in (A) is a solution to the KdV equation. In (B) is the Floquet discriminant (the trace of the monodromy matrix as a function of $E = k^2$) that has exactly six open bands (degrees of freedom) in the spectrum. Shown in (C) are the six hyperelliptic-function oscillation modes. The linear superposition of these six modes gives the exact solution to the KdV equation shown in (A).

solitons and radiation solutions of periodic KdV. The $\mu_j(x; x_o, t)$ evolve in space x, for a fixed value of time $t = 0$, according to the following xsystem of coupled, nonlinear, ordinary differential equations (ODEs):

$$\frac{d\mu_j}{dx} = \frac{2i\sigma_j R^{1/2}(\mu_j)}{\prod\limits_{\substack{k=1 \\ j \neq k}}^{N} (\mu_j - \mu_k)} \tag{17.4}$$

and

$$R(\mu_j) = \prod_{k=1}^{2N+1} (\mu_j - E_k) \tag{17.5}$$

The $\sigma_j = \pm 1$ specify the signs of the square root of the function $R(\mu_j)$. The nonlinear functions $\mu_j(x; x_o, t)$ live on two-sheeted Riemann surfaces, each of which is specified by $\sigma_j = +1$ or -1. The branch points connecting the surfaces are called the "band edges" E_{2j} and E_{2j+1}. The $\mu_j(x; x_o, t)$ lie in the intervals

$E_{2j} \leq \mu_j \leq E_{2j+1}$, that is, inside "open bands," and oscillate between these limits as x varies. When a $\mu_j(x; x_o, t)$ reaches a band edge (either E_{2j} or E_{2j+1}) the index changes sign and the motion moves to the other Riemann sheet. Figure 17.1 graphically illustrates many of these ideas.

The temporal evolution of the $\mu_j(x; x_o, 0)$ is described by the following set of ODEs:

$$\frac{d\mu_j}{dt} = 2[-u(x,t) + 2\mu_j]\mu_j' \tag{17.6}$$

where $\mu_j' = d\mu_j/dx$ is given by Equation (17.4) and $u(x, t) = \lambda\eta(x, t)$ is given by Equation (17.3). The space (17.4) and time (17.6) ODEs evolve the $\mu_j(x, t)$ and the *nonlinear Fourier series* (17.3) allows one to construct general solutions to KdV. In what follows we concentrate on developing procedures for computing the oscillation modes $\mu_j(x; x_o, 0)$ at a particular instant of time, $t = 0$. Thus we are here concerned with the analysis of "space series" of data in which a measured signal, $\eta(x_n, 0)$, is recorded at discrete spatial values $x_n = n\Delta x$, for Δx a constant discretization interval.

I now discuss *time series* analysis. To this end we may think of a measured signal, $\eta(0, t_n)$, which is recorded at discrete temporal intervals, $t_n = n\Delta t$, where Δt is the sampling interval. It is important however in what follows to remember that x and t are in a sense *interchangeable*; this property provides for the possibility of analyzing both space and time series by the methods described below. Herein I often speak of the analysis of space series (governed by sKdV Equation (17.1)) but alternatively, with the following changes of variable, one can just as easily analyze time series (governed by tKdV Equation (17.2)):

$$x \rightarrow t, \quad t \rightarrow x, \quad \lambda \rightarrow c_o^2\lambda, \quad k \rightarrow \omega \tag{17.7}$$

One interchanges the space and time coordinates and modifies the nonlinearity parameter λ. In the spectral domain, the transformation is equally simple, that is, the wavenumber k becomes the frequency ω. Apart from the rescaling of the nonlinearity parameter $(\lambda \rightarrow c_o^2\lambda)$ these results are exactly parallel to linear Fourier analysis (Bendat and Piersol, 1971).

Generally speaking we refer to the construction of the main spectrum $(E_k, \ 1 \leq k \leq 2N+1)$ and the auxiliary spectrum $(\mu_j(x_o; x_o, 0), \ \sigma_j = \pm 1, \ 1 \leq j \leq N)$ as the *direct scattering problem* (see details in Sections 17.4 and 17.5). The computation of the hyperelliptic functions $\mu_j(x, t)$ by the solution of the nonlinear ODEs (17.4)–(17.6) and the construction of solutions of the KdV equation by the trace formula (17.3) constitutes the *inverse scattering problem* (Section 17.5). Herein I (a) discuss numerical procedures for obtaining the direct scattering transform and (b) show that direct numerical integration of Equations (17.4)–(17.6) (as discussed in detail in Osborne and Segre (1990)) can be replaced by a much simpler, more precise and faster algorithm for the specific cases of space or time series analysis.

17.4   The Spectral Structure of Periodic IST

The *direct spectral problem* for KdV is the Schrödinger eigenvalue problem:

$$\psi_{xx} + [\lambda\eta(x) + k^2]\psi = 0 \qquad (k^2 = E) \tag{17.8}$$

where $\eta(x) = \eta(x,0)$ is the solution to the KdV equation (17.1) at the arbitrary time $t = 0$; k is the wavenumber, and E the "energy." Here periodic boundary conditions are assumed so that $\eta(x, t) = (x + L, t)$ for L the spatial period of the wave train.

For solving the *periodic scattering problem* (17.8), one normally begins by choosing a basis of eigenfunctions $\phi(x; x_o, k)$ such that: $\phi(x_o; x_o, k) = 1$, $\phi_x(x_o; x_o, k) = ik$, $\phi^*(x_o; x_o, k) = 1$, $\phi_x^*(x_o; x_o, k) = -ik$ where $*$ indicates complex conjugate, the subscript x refers to a derivative and x_o is an arbitrary *base point* in the interval $0 \le x \le L$. To satisfy the periodic boundary conditions for the wave train $\eta(x, 0)$, one assumes that the eigenfunctions one period to the right, $\phi(x+L; x_o, k)$, are linear combinations of the solutions $\phi(x; x_o, k)$. Introduce the matrix of independent eigenfunctions

$$\Phi(x; x_o, k) = \begin{pmatrix} \phi & \phi_x \\ \phi^* & \phi_x^* \end{pmatrix} \tag{17.9}$$

To satisfy the periodic boundary conditions for the wave train $\eta(x, 0)$, one assumes that the eigenfunctions one period to the right $\phi(x+L; x_o, k)$ are linear combinations of the solutions $\phi(x; x_o, k)$. Introducing the matrix of independent eigenfunctions (with determinant equal to $-2ik$)

$$\Phi(x + L; x_o, k) = S(x_o, k)\Phi(x; x_o, k) \tag{17.10}$$

one writes

$$S(x_o, k) = \begin{pmatrix} a & b \\ b^* & a^* \end{pmatrix} \tag{17.11}$$

$S(x_o, k)$ is the *monodromy matrix* where a and b are complex numbers. Thus the monodromy matrix by definition carries the solutions of Equation (17.8) one period from the point x to the point $x + L$.

The *main spectrum* consists of the set of real constants $\{E_k\}$ ($1 \le k \le 2N + 1$, where N is the integer *number of degrees of freedom* of a particular solution to KdV); the E_k are defined as solutions to the relation:

$$|a_R(E)| = 1 \tag{17.12}$$

(where the subscript "R" means "take the real part of a").

The *auxiliary spectrum* $\{\mu_j(x; x_o, t)\}$ $(1 \le j \le N)$ is given as the solution to:

$$(a_I + b_I)(E) = 0 \tag{17.13}$$

("I" denotes "imaginary part of a or b"). The $\{E_k\}$ are the eigenvalues corresponding to Bloch eigenfunctions which are either *periodic* or *antiperiodic* on the period L (a periodic function has the property $f(x) = f(x + L)$ and anti-periodic means $f(x) = -f(x + L)$). The "signs" of the spectrum are given by the *Riemann sheet indices*

$$\sigma_j = \text{sgn}[b_R(k)]_{\mu_j} \tag{17.14}$$

The auxiliary spectrum $\{\mu_j; \sigma_j\}$ may be viewed as the source of phase information in the hyperelliptic-function representation of KdV, that is the $\{\mu_j; \sigma_j\}$ may be suitably combined to give the phases of the $\mu_j(x, t)$ (Osborne, 1993e).

The spectrum $\{E_k; \mu_j, \sigma_j\}$ constitutes the *direct scattering transform* of a wave train with N degrees of freedom; $1 \le k \le 2N + 1$; $1 \le j \le N$. This is the fundamental information that is to be computed from a space or time series by the numerical methods given below.

For numerical purposes it is appropriate to consider a different basis set (c, s) (Flaschka and McLaughlin, 1976) (the symbol "c" might be read as "cosine" and the "s" as "sine," but this identification is not rigorous, just informative):

$$\begin{pmatrix} c(x_o) & c'(x_o) \\ s(x_o) & s'(x_o) \end{pmatrix} = \begin{pmatrix} 1 & 0 \\ 0 & 1 \end{pmatrix} \tag{17.15}$$

That (c, s) is a basis is verified by the fact that the wronskian $W(c, s) = 1$. The matrix α carries the solution from point x to $x + L$:

$$\begin{pmatrix} c(x + L) & c'(x + L) \\ s(x + L) & s'(x + L) \end{pmatrix} = \begin{pmatrix} \alpha_{11} & \alpha_{12} \\ \alpha_{21} & \alpha_{22} \end{pmatrix} \begin{pmatrix} c(x) & c'(x) \\ s(x) & s'(x) \end{pmatrix} \tag{17.16}$$

where $\boldsymbol{\alpha}$ is determined from the S matrix by the *similarity transformation*:

$$\alpha = Q^{-1}SQ, \qquad Q = \begin{pmatrix} 1 & ik \\ 1 & -ik \end{pmatrix} \tag{17.17}$$

Therefore, $\boldsymbol{\alpha}$ is the monodromy matrix in the (c, s) representation.

With reasoning similar to that for the basis $\boldsymbol{\Phi}$ (Equation (17.10)) the main spectrum in the (c, s) basis consists of eigenvalues E_k that correspond to the Bloch eigenfunctions for a particular period L. The auxiliary spectrum is defined as the eigenvalues for which the eigenfunctions $s(x)$ have the fixed boundary conditions $s(x_o + L) = s(x_o)$. To this end we have the specific spectral definitions (Osborne and Bergamasco, 1985, 1986):

Main spectrum $\{E_k\}$: $\frac{1}{2}(\alpha_{11} + \alpha_{22}) = \frac{1}{2}(S_{11} + S_{22}) = \pm 1$ (17.18)

Auxiliary spectrum $\{\mu_j\}$: $\alpha_{21} = -\frac{i}{2k}(S_{11} + S_{12} - S_{21} - S_{22}) = 0$ (17.19)

Riemann Sheet Indices $\{\sigma_j\}$:

$$\sigma_j = \{\text{sgn}[\alpha_{11}(E) - \alpha_{22}(E)]_{E=\mu_j}\} = \{\text{sgn}[S_{12}(E) + S_{21}(E)]_{E=\mu_j}\}$$ (17.20)

Equations (17.18)–(17.20) constitute the solution of the direct scattering problem. We now discuss briefly the *inverse scattering problem* for the special case when $t = 0$. This of course implies that we seek the $\mu_j(x, 0)$ for all x on $(0, L)$, $1 \leq j \leq N$. The solution to KdV at $t = 0$, $\eta(x,0)$, may then be computed by the trace formula (17.3). We would therefore like to compute the $\mu_j(x, 0)$ *not just for a single, arbitrary base point* $x = x_o$, *but for all base points in the interval* $(0 \leq x \leq L)$. The potential function $\eta(x, 0)$ can then be reconstructed by the linear superposition law (17.3) or nonlinear filtering can be accomplished by considering the sum (17.3) over a selected wavenumber range. Formally speaking the $\mu_j(x, 0)$ are solutions of the nonlinear, coupled ODEs (17.4) (since the time ODEs (17.6) are excluded in the case for which $t = 0$). Is there an alternative to the difficult numerical integration of the ill-conditioned ODEs (17.4) for computing the $\mu_j(x, 0)$ by the procedure discussed in Osborne and Segre (1990)?

The answer is in the affirmative. First note that the IST spectrum as represented by Equations (17.18)–(17.20) corresponds to the particular base point $x_o = 0$; one thus obtains the main and auxiliary spectra at $x_o = 0$: $\{E_k\}$, $\{\mu_j(0, 0)\}$. To obtain the $\mu_j(x, 0)$ for all x, we seek the auxiliary spectrum at a nearby point $x_o = \Delta x \ll L$, $\mu_j(\Delta x, 0)$, and then at $x_o = 2\Delta x$, $\mu_j(2\Delta x, 0)$, etc. To carry out this procedure, it is enough to notice that by considering the wave train $\eta(x, 0)$ to be on the *associated periodic interval* $(\Delta x \leq x \leq L + \Delta x)$ we find $\mu_j(\Delta x, 0)$ by application of Equation (17.19). Then for $\eta(x,0)$ on $(2\Delta x \leq x \leq L + 2\Delta x)$ Equation (17.19) yields $\mu_j(2\Delta x, 0)$, etc. In this way we can determine the *hyperelliptic functions* $\mu_j(x, 0)$ for all x by iterating the *direct problem* (17.18)–(17.20) for the potential $\eta(x, 0)$ on successive intervals $[x_o, x_o + L] = [n\Delta x, (n + M - 1)\Delta x]$, where $L = (M-1)\Delta x$ is the period, $x_o = n\Delta x$ is an arbitrary base point, and n is an integer on $0 \leq n \leq M-1$. Use of the linear superposition law (17.3) then allows the original wave train to be reconstructed. Therefore, instead of numerically solving the spatial evolution of the ODEs (17.4), we simply repeat the direct scattering problem for each desired spatial point in the function $\mu_j(x,0)$. This is a large improvement over the numerical problems encountered in the solution to Equation (17.4). It can be shown that the iterative method for obtaining the $\mu_j(x,0)$ via the direct problem is equivalent to a *similarity transformation* which carries the monodromy matrix from the spatial point $n\Delta x$ to $(n + 1)\Delta x$. Details of this calculation are given in Section 17.5.

17.5   A Numerical Discretization

17.5.1   Formulation

The numerical search for the scattering eigenvalues $\{E_k; \mu_j; \sigma_j\}$ ($1 \le k \le 2N + 1$; $1 \le j \le N$) suggests that knowledge of the derivatives of the matrix α_{ij} with respect to the energy E may be important. This is because Newton's numerical root-finding algorithm is often applied to find the solutions to Equations (17.18) and (17.19). Clearly, this process is improved if exact expressions for the derivatives of the matrix elements with respect to E are known. For this reason, the following method has been developed for obtaining the evolution of the eigenfunction ψ as a function of x and E for a particular wave train $\eta(x, 0)$; this method includes exact expressions for the required derivatives.

The scattering equations are the Schrödinger eigenvalue problem (17.8) and its derivative with respect to E:

$$\begin{aligned} \psi_{xx} &= -q\psi \\ \psi_{xxE} &= -q\psi_E - \psi \end{aligned} \qquad (17.21)$$

Here the subscripts refer to differentiation with respect to x and E; $q(x) = \lambda\eta(x) + E$. The scattering problem may be written in the following matrix form:

$$\Psi_x = \mathbf{A}\Psi \qquad (17.22)$$

where $\Psi(x, E)$ is understood to be a four-vector with components $(\psi, \psi_x, \psi_E, \psi_{xE})$ and $\mathbf{A}$ is the four-by-four matrix

$$\mathbf{A} = \begin{pmatrix} 0 & 1 & 0 & 0 \\ -q & 0 & 0 & 0 \\ 0 & 0 & 0 & 1 \\ -1 & 0 & -q & 0 \end{pmatrix} \qquad (17.23)$$

The four-vector field $\Psi(x + \Delta x)$, for Δx small, may be expanded in a Taylor series around x:

$$\Psi(x + \Delta x, E) = \Psi(x, E) + \frac{\partial\Psi(x, E)}{\partial x}\Delta x + \frac{1}{2}\frac{\partial^2\Psi(x, E)}{\partial x^2}\Delta x^2 + \ldots = \mathbf{H}(x, E)\Psi(x, E) \qquad (17.24)$$

The matrix $\mathbf{H}(x, E)$, which translates the field Ψ from x to $x + \Delta x$ at the eigenvalue E, is the exponential of the matrix $\mathbf{A}$

$$\mathbf{H}(x, E) = e^{\Delta x \mathbf{A}} \qquad (17.25)$$

The following explicit results then follow:

$$
\begin{pmatrix}
\psi(x + \Delta x) \\
\psi_x(x + \Delta x) \\
\psi_E(x + \Delta x) \\
\psi_{xE}(x + \Delta x)
\end{pmatrix}
= \mathbf{H}
\begin{pmatrix}
\psi(x) \\
\psi_x(x) \\
\psi_E(x) \\
\psi_{xE}(x)
\end{pmatrix}
\tag{17.26}
$$

where

$$
\mathbf{H} = \begin{pmatrix} \mathbf{T} & 0 \\ \mathbf{T}_E & \mathbf{T} \end{pmatrix}
$$

and

$$
\mathbf{T} = \begin{pmatrix} \cos(\kappa\Delta x) & \dfrac{\sin(\kappa\Delta x)}{\kappa} \\ -\kappa\sin(\kappa\Delta x) & \cos(\kappa\Delta x) \end{pmatrix}
\tag{17.27}
$$

And

$$
\mathbf{T}_E = \frac{\partial \mathbf{T}}{\partial E} = \begin{pmatrix} -\dfrac{\Delta x \sin(\kappa\Delta x)}{2\kappa} & \dfrac{\Delta x \cos(\kappa\Delta x)}{2\kappa^2} - \dfrac{\sin(\kappa\Delta x)}{2\kappa^3} \\ -\dfrac{\Delta x \cos(\kappa\Delta x)}{2} - \dfrac{\sin(\kappa\Delta x)}{2\kappa} & -\dfrac{\Delta x \sin(\kappa\Delta x)}{2\kappa} \end{pmatrix}
\tag{17.28}
$$

for $\kappa = \sqrt{q}$. While κ may be either real or imaginary, the matrices $\mathbf{T}$, $\mathbf{T}_E$ are always real with determinant 1. This property is exploited in the numerical algorithm below.

In the above formulation (17.24)–(17.28), the waveform $\eta(x, 0)$ is a continuous function. However, as in previous numerical problems of this type (Osborne, 1991a,b), I assume the wave train $\eta(x, 0)$ has the form of a piecewise constant function with M partitions on the periodic interval $(0, L)$, where the discretization interval is $\Delta x = L/M$. Each partition has constant wave amplitude η_n, $0 \leq n \leq M-1$, associated with a discrete value of x: $x_n = n\Delta x$. The four-by-four scattering matrix $\mathbf{M}$ therefore follows from iterating (17.26):

$$
\mathbf{M} = \prod_{n=M-1}^{0} \mathbf{T}(\eta_n, \Delta x)
\tag{17.29}
$$

where, due to the structure of the $\mathbf{H}$ matrix, the following properties hold: $M_{13} = M_{14} = M_{23} = M_{24} = 0$ and $M_{33} = M_{11}$, $M_{44} = M_{22}$, $M_{34} = M_{12}$, $M_{43} = M_{21}$.

The initial conditions of the numerically convenient basis set (c, s) (17.16) at the base point x_o are given by:

$$\begin{pmatrix} c(x_o) \\ c'(x_o) \\ c_E(x_o) \\ c'_E(x_o) \end{pmatrix} = \begin{pmatrix} 1 \\ 0 \\ 0 \\ 0 \end{pmatrix}, \quad \begin{pmatrix} s(x_o) \\ s'(x_o) \\ s_E(x_o) \\ s'_E(x_o) \end{pmatrix} = \begin{pmatrix} 0 \\ 1 \\ 0 \\ 0 \end{pmatrix} \tag{17.30}$$

From the definition of the matrix α_{ij} (17.17) we therefore have:

$$\{\alpha_{ij}\} = \begin{pmatrix} c(x+L) & c'(x+L) \\ s(x+L) & s'(x+L) \end{pmatrix} \begin{pmatrix} c(x) & c'(x) \\ s(x) & s'(x) \end{pmatrix}^{-1} \tag{17.31}$$

Thus at x_o we find

$$\frac{1}{2}(\alpha_{11} + \alpha_{22}) = \frac{1}{2}(M_{11} + M_{22}) \tag{17.32}$$

$$\alpha_{21} = M_{12} \tag{17.33}$$

while the derivatives are given by

$$\frac{\partial}{\partial E}\frac{1}{2}(\alpha_{11} + \alpha_{22}) = \frac{1}{2}(M_{31} + M_{42}) \tag{17.34}$$

$$\frac{\partial \alpha_{21}}{\partial E} = M_{32} \tag{17.35}$$

It is worthwhile remarking that the algorithm (17.24) (with Equations (17.27) and (17.28) defining the matrix **H**) is equivalent numerically to that in Osborne (1991a,b) if the derivative matrix T_E is excluded from the calculations. The main advantage of the present approach is that it can be implemented as a real computer code, rather than complex, and the present algorithm is consequently about four times faster than that in Osborne (1991b).

17.5.2   Implementation of the Numerical Algorithm

Because $\kappa = \sqrt{\lambda \eta(x; 0) + k^2}$ can be either real or imaginary, but not complex, the matrix **H** is always real. This result allows implementation of an algorithm that is entirely real. As a consequence the following relations have been used in the program:

$$T_{11} = T_{22} = \begin{cases} \cos(\kappa' \Delta x), & \text{if } \kappa^2 \geq 0 \\ \cosh(\kappa' \Delta x), & \text{if } \kappa^2 < 0 \end{cases} \tag{17.36}$$

$$T_{12} = \begin{cases} \dfrac{\sin(\kappa' \Delta x)}{\kappa'}, & \text{if } \kappa^2 \geq 0 \\[2mm] \dfrac{\sinh(\kappa' \Delta x)}{\kappa'}, & \text{if } \kappa^2 < 0 \end{cases} \tag{17.37}$$

$$T_{21} = \begin{cases} -\kappa' \sin(\kappa' \Delta x), & \text{if } \kappa^2 \geq 0 \\ \kappa' \sinh(\kappa' \Delta x), & \text{if } \kappa^2 < 0 \end{cases} \tag{17.38}$$

where

$$\kappa' = \sqrt{|\lambda\eta + k^2|} = \sqrt{|\kappa^2|} \tag{17.39}$$

and analogously for the matrix $\mathbf{T}_E$.

17.5.3 Reconstruction of Hyperelliptic Functions and Periodic Solutions to the KdV Equation

The reconstruction of an input space or time series from the spectrum by (17.3) is carried out by computing the auxiliary spectra $\mu_j(x_o = x_n, 0)$ for $0 \leq n \leq M-1$ for which the M different base points have the values $x_o = x_o$, $x_1, x_2, \ldots x_M - 1$. This is done by computing M different monodromy matrices (17.29) which differ from each other by a shift Δx in the wave train $\eta_n = \eta(x_n, 0)$. Thus the direct scattering problem is repeated M times to construct the $\mu_j(x_n, 0)$ for all x_n, $0 \leq x_n \leq L$, $0 \leq n \leq M-1$. This process is referred to as *base point iteration* and arises theoretically from the following similarity transformation which is easily seen from Equation (17.29):

$$\mathbf{M}(x_{n+1}, E) = \mathbf{H}(\eta_n, E)\mathbf{M}(x_n, E)\mathbf{H}(\eta_n, E)^{-1} \tag{17.40}$$

This expression relates the matrix $\mathbf{M}(x_{n+1}, E)$ at a point x_{n+1} to the previously computed matrix $\mathbf{M}(x_n, E)$ at x_n for a particular value of the eigenvalue E. Values of the auxiliary spectra $\{\mu_j(x_n, 0)\}$ for each x_n are computed from the associated matrices $\mathbf{M}(x_n, E)$. For numerical convenience we use base point iteration, rather than Equation (17.40), because $\mathbf{M}(x_n, E)$ can be evaluated at *arbitrary* values of E in the iterative search for a zero. This approach contrasts to Equation (17.40) which computes the monodromy matrix at different base points for the *same* value of E. Using the values of $\mu_j(x_n, 0)$ obtained by base point iteration allows reconstruction of the wave train amplitude $\eta(x_n, 0)$ via a discrete version of (17.3):

$$\lambda\eta(x_n, 0) = -E_1 + \sum_{j-1}^{N}[2\mu_j(x_n, 0) - E_{2j} - E_{2j+1}] \tag{17.41}$$

for $n = 0, 1, 2, \ldots M-1$. Osborne and Bergamasco (1985) have shown that the number of degrees of freedom $N = M/2$ for an arbitrary discrete wave train of M points. Equation (17.41) is a finite-term nonlinear generalization of Fourier series for the discrete wave train $\eta(x_n, 0)$. As indicated by the notation, each nonlinear oscillation mode $\mu_j(x_n, 0)$ depends upon the wavenumber of the mode, theoretically given by $k_j = 2\pi j/L$, a result identical to that for the linear Fourier transform.

A summary of IST for the KdV equation is given in Figure 17.1. The Floquet discriminant, $(\Delta(E) = (M_{11} + M_{22})/2)$, is shown on the left in panel (Figure 17.1B); Δ is graphed horizontally while the energy $E = k^2$ is graphed vertically; the discrete eigenvalues E_k are shown at the intersections of $\Delta(E)$ with ± 1. The open bands are indicated by the shaded regions which also highlight the hyperelliptic function basis states $\mu_j(x_n, 0)$ (Figure 17.1C), here graphed as a function of the spatial coordinate x. Note that the $\mu_j(x_n, 0)$ oscillate between the edges of their respective open bands as x is varied. The linear superposition of the hyperelliptic functions by Equation (17.41) gives the solution to the KdV equation shown on the upper right (Figure 17.1A). The algorithms presented herein begin with the discrete potential $\eta(x_n, 0)$ and generate the Floquet discriminant $\Delta(E)$, the discrete eigenvalues E_k $(1 \leq k \leq 2N + 1)$, the hyperelliptic functions $\mu_j(x_n, 0)$ $(1 \leq j \leq N; \ 0 \leq n \leq M-1)$ and their linear superposition (i.e., reconstruction of the input potential $\eta(x_n, 0)$ by Equation (17.41)).

17.6   Automatic Numerical IST Algorithm

The purpose of an *automatic* IST algorithm is to find the exact number of degrees of freedom, N, in the input space or time series, and to subsequently find *all* the eigenvalues in the main and auxiliary spectra, *independent* of any initial choice for the resolution in the squared-wavenumber domain, $E = k^2$. To this end I give two numerical procedures which have proved to be equally effective, although one of these has certain advantages in the requisite coding of the method and, to a certain extent, this preferred approach is somewhat esthetically more pleasing than its alternative.

In the numerical construction of the IST spectrum, it is natural to first compute the main spectrum $\{E_k; 1 \leq k \leq 2N + 1\}$ and then the auxiliary spectrum $\{\mu_j, \sigma_j; 1 \leq j \leq N\}$ (Osborne and Bergamasco, 1986; Osborne, 1991a). In the calculation of the main spectrum, one normally chooses a particular resolution in the squared-wavenumber $(E = k^2)$ domain, $\Delta E = (E_{max} - E_{min})/N_E$, where E_{max} and E_{min} are the maximum and the minimum values of E and N_E is the desired number of discrete values of E. Then the Floquet discriminant, $\Delta(E) = (M_{11} + M_{22})/2$, is computed at each $E = E_i$, $1 \leq i \leq N_E$, for $\Delta E = E_{i+1} - E_i$. Of course there is always the possibility of missing an eigenvalue in the main spectrum (17.18, 17.19). In the interval $E_{min} \leq E \leq E_{max}$, we know that in principle there are N degrees of freedom, but it is not always clear *how many there are and what their spatial density might be*. It is convenient to chose E_{min} to be so far to the left that no eigenvalues exist for $E<E_{min}$. One way to ensure this is to set $E_{min} = -\lambda(\eta_{max} - \bar{\eta})$, where η_{max} is the maximum value in the time series and $\bar{\eta}$ is its mean (Osborne, 1991a). With this restriction the parameter E_{max} controls the number of degrees of freedom whose scattering transform spectrum one seeks. It can be shown that when $E_{max} = E_{Ny} = (k_{Ny}/2)^2$, $k_{Ny} = 2\pi/\Delta x$ (the Nyquist cutoff wavenumber) then $N = M/2$, where M is the number of points in the input space or time series (Osborne and Bergamasco, 1985). Generally speaking for $E_{max}<E_{Ny}$, $N<M/2$. Computation of the IST spectrum and the subsequent construction of the wave train for the case $E_{max}<E_{Ny}$ constitutes a *low pass filter* (see discussion below) with $E_{max} = (k_c/2)^2$, where k_c is the chosen upper wavenumber cutoff in the spectrum.

Previous approaches have attempted to avoid the problem of missing eigenvalues, particularly when they may be very closely spaced, by devising complex codes for finding N maxima and minima in the trace of the monodromy matrix (Osborne, 1991a). However, in pathological cases where, say, a large number of eigenvalues lie in some small interval, such algorithms are difficult to code, nontrivial to check out, tend to use inordinate amounts of computer time and may even fail for very closely space eigenvalues (Osborne, 1991a). I propose here an alternative method that, within the precision of a particular machine, seems to always work.

The approach proposed herein derives from the Schrödinger eigenvalue problem together with particular boundary conditions:

$$\psi_{xx} + [u(x,0) + E_{max}]\psi = 0 \qquad (17.42)$$

for

$$\psi(x_0; x_0, E_{max}) = 1, \quad \psi_x(x_0; x_0, E_{max}) = 0$$

The number of zeros in the wave function $\psi(x; x_0, E_{max})$ is the number of degrees of freedom in the system in the energy interval $-\infty<E \leq E_{max}$. This idea derives from the usual determination of the number of solitons for KdV with infinite-line boundary conditions, for which E_{max} is set identically to zero (Segur, 1973). For the above problem (17.42), there are exactly N zeros in the interval $-\infty<E \leq E_{max}$. In the algorithm that follows we normally set $x_0 = 0$, so that the spatial interval of interest is $0 \leq x \leq L$. According to (17.23-17.28) we have:

$$\begin{pmatrix} \psi(x_n; x_0, E) \\ \psi_x(x_n; x_0, E) \end{pmatrix} = \begin{pmatrix} M_{11}(x_n, E) & M_{12}(x_n, E) \\ M_{21}(x_n, E) & M_{22}(x_n, E) \end{pmatrix} \begin{pmatrix} \psi(x_0; x_0, E) \\ \psi_x(x_0; x_0, E) \end{pmatrix} \qquad (17.43)$$

which with the boundary conditions in Equation (17.42) gives:

$$\psi(x_n; x_0, E) = M_{11}(x_n, E), \quad \psi_x(x_n; x_0, E) = M_{21}(x_n, E) \qquad (17.44)$$

Of course (17.43, 44) have, for an M-point discrete wave train (where $N<M/2$), the form:

$$\begin{pmatrix} \psi(x_o + L; x_o, E) \\ \psi_x(x_o + L; x_o, E) \end{pmatrix} = \begin{pmatrix} M_{11}(x_o + L, E) & M_{12}(x_o + L, E) \\ M_{21}(x_o + L, E) & M_{22}(x_o + L, E) \end{pmatrix} \begin{pmatrix} \psi(x_o; x_o, E) \\ \psi_x(x_o; x_o, E) \end{pmatrix}$$

(17.45)

for M given by Equation (17.29). Equation (17.44) suggests that the number of degrees of freedom of a particular input wave train may be found by judicious scrutiny of the M_{11} element of the monodromy matrix. Generally speaking M_{11} depends both upon the spatial variable x_n and on the parameter E. Equation (17.43) is computed by iterating on the following recursion relations for the **T** matrix:

$$\mathbf{T}(x_{n+1}, E) = \begin{pmatrix} \cos \kappa_n \Delta x & \dfrac{\sin \kappa_n \Delta x}{\kappa_n} \\ -\kappa_n \sin \kappa_n \Delta x & \cos \kappa_n \Delta x \end{pmatrix} \mathbf{T}(x_n, E)$$

(17.46)

These in effect carry the Schrödinger eigenfunctions from point x_n to x_{n+1} and, after all discrete points in the wave train are iterated on, beginning with $\mathbf{T}(x_o, E)$, one arrives at the monodromy matrix one period to the right, $\mathbf{M}(x_o + L, E)$. Here the wavenumber κ_n is given by

$$\kappa_n(x_n, E) = \sqrt{u(x_n) + E}$$

(17.47)

and $\Delta x = (x_{n+1} - x_n)$.

At this point I introduce the function $S(x, E)$ which is related to the number of times that the monodromy matrix element $M_{11}(x, E)$ changes sign between $x = x_o = 0$ and L. The solutions

$$M_{11}(x_o + L, E) = 0$$

(17.48)

are referred to as E_j^* $(1 \le j \le N)$ so that strictly speaking it is convenient to require

$$S(x_o + L, E_j^*) = j, \quad 1 \le j \le N$$

(17.49)

Thus the function S has monotonically increasing integer values as E increases so that $S(x_o + L, E_j^*) = j$ up to $S(x_o + L, E_N^*) = N$, where N has a maximum value equal to one half the number of points M in the space or time series, that is, $E_{M/2}^* = (k_{Ny}/2)^2$ so that $S(x_o + L, E_{M/2}^*) = M/2$. The function $S(x, E)$ is given the following precise definition:

$$S(x_n, E) = \sum_{i=0}^{n} |\mathrm{sgn}(M_{11}(x_{i+1}, E)) - \mathrm{sgn}(M_{11}(x_i, E))|/2$$

(17.50)

Only when $M_{11}(x_n, E)$ makes a sign change between successive values of x_n and x_{n+1} (or E_i and E_{i+1}) is an integer contribution made to the sum in Equation (17.50). Therefore $S(x_n, E)$ is monotonically increasing in both x_n and E and is piecewise constant, experiencing only unit jumps at the zeros of M_{11} in the x and E domains.

The functions $M_{11}(x, E)$ and $S(x, E)$ are shown in Figure 17.2 for a particular example problem: a sine wave potential (as considered by Zabusky and Kruskal (1965)) with $M = 300$, $\lambda = 0.012$, $N_E = 1000$, $E_{min} = -0.02$, and $E_{max} = 0.06$ (Osborne and Bergamasco, 1986). In Figure 17.2A, $M_{11}(x, E_{max})$ and $S(x, E_{max})$ are graphed as a function of x $(0 \leq x \leq 300)$. Note that as M_{11} oscillates as a function of x, S increases monotonically in integer steps up to $N = 23$ degrees of freedom for the selected value of E_{max}. Each stepwise increase in S occurs precisely at a zero of M_{11}, $M_{11}(x, E = E_{max}) = 0$. Another example of the x dependence of M_{11} and S is given in Figure 17.2B for $E = E_{max}/2 = 0.03$; here $N = 16$ for the selected energy value $E = E_{max}$.

In Figure 17.3 I show the E dependence of the "oscillation function" $M_{11}(x_o + L, E)$ and the "accounting function" $S(x_o + L, E)$. Note that M_{11} is graphed linearly as a function of E when $|M_{11}| < 1$ and logarithmically outside this range; the vertical scale in the figure indicates powers of ten. Each oscillation in M_{11} corresponds to a *degree of freedom* in the solution to KdV. Each zero of M_{11} in effect counts a degree of freedom in the E domain and causes a stepwise, integer increase in S; note that S increases up to $N = 23$ as E approaches E_{max}.

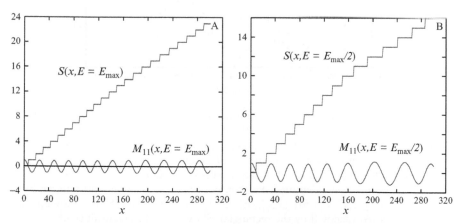

Figure 17.2 The Zabusky and Kruskal initial sine wave is used to compute (A) the monodromy matrix element $M_{11}(x, E)$ versus x for the parameter value $E = E_{max} = 0.06$: M_{11} is seen to oscillate many times (~ 12) on the interval $0 \leq x \leq 300$. The function $S(x, E = E_{max})$ (Equation (17.50)) is a piecewise constant monotonically increasing function that has exactly 23 steps (degrees of freedom) in the system. In (B) are shown $M_{11}(x, E)$ and $S(x, E)$ for half the previous value: $E = E_{max}/2 = 0.03$.

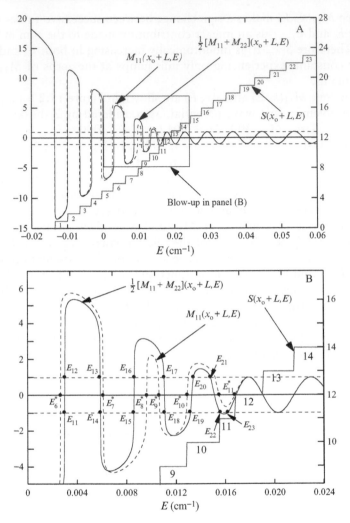

Figure 17.3 The Zabusky and Kruskal initial sine wave is used to compute (A) the trace of the monodromy matrix $(M_{11} + M_{22})(x_{M-1}, E)/2$, the oscillation function $M_{11}(x_{M-1}, E)$, and the accounting function $S(x_{M-1}, E)$. Note that the piecewise, monotonically increasing behavior of $S(x_{M-1}, E)$ coincides identically with the zero crossings of $M_{11}(x_{M-1}, E)$. All of the degrees of freedom in the selected E-domain range are easily identified. Further, note that the maxima and minima in $(M_{11} + M_{22})(x_{M-1}, E)/2$ and $M_{11}(x_{M-1}, E)$ nearly correspond. In (B) is a blowup of the region in (A) (denoted by the rectangle). Shown are the eigenvalues of $(M_{11} + M_{22})(x_{M-1}, E)/2 = \pm 1$ (the "main spectrum," E_j) and $M_{11}(x_{M-1}, E) = 0$ (the oscillation function eigenvalues, E_j^*).

It is worth pointing out that the main advantage of the present approach is that the function $S(x_{M-1},E)$

$$S(x_{M-1}, E) = \sum_{i=0}^{M-1} |\text{sgn}(M_{11}(x_{i+1}, E)) - \text{sgn}(M_{11}(x_i, E))|/2 \qquad (17.51)$$

does not depend on any chosen resolution in the E *domain.* Here $x_{M-1} = x_o + L = L$. One selects a particular value of $E = E_{max}$ and determines uniquely the number of the degrees of freedom in the spectrum in the range $E_{min} \le E \le E_{max}$. For the chosen resolution one can then determine S for all values of E as graphed in Figure 17.3; in the event that any degrees of freedom are missing in the search process (i.e., the sequence 1,2,3...N is interrupted) then it is quite straightforward to search for intermediate values of S by the *interval bisection technique* as discussed below.

To better understand the present approach, I now give a rather extensive look at the behavior of the following functions of the monodromy matrix elements in the E domain: (1) the element $M_{11}(x_o + L, E)$ which counts the numbers of degrees of freedom in the system, $S(x_o + L, E_{max})$, (2) the trace of the monodromy matrix, $(M_{11} + M_{22})/2$, which determines the "main spectrum" eigenvalues, E_k, and (3) the element M_{12}, which determines the auxiliary spectrum eigenvalues, $\mu_j(0,0)$. I consider the above example problem of the ZK initial sine wave and graph all three functions as a function of E for $x = x_o + L$. The results illustrate and extend the properties of these functions as established by the so-called "oscillation theorems" (Ablowitz and Segur, 1981) which in particular describe the oscillatory behavior of the trace of the monodromy matrix and the element M_{12}. I give rather detailed graphs to illustrate the important aspects of the behavior of these functions, which need to be clearly understood for developing efficient numerical codes.

In Figure 17.3A, I compare the functions $M_{11}(x_o + L, E)$, $(M_{11} + M_{22})(x_o + L, E)/2$ and $S(x_o + L, E)$. The trace of the monodromy matrix determines the "main spectrum" (E_k) as found from the zeros of $(E_{11} + M_{22})/2 = \pm 1$. Note further that the oscillation function M_{11} tracks rather nicely with the trace of the monodromy matrix. The maxima and minima, while they do not coincide, are rather close together in the E domain. The function S is also seen to be rather well behaved, that is, it increases monotonically by 1 whenever M_{11} crosses zero. S is evaluated numerically by Equation (17.50) for a selected value of $E = E'(E_{min} \le E' \le E_{max})$ and thus effectively provides a value of the number of degrees of freedom in the range $-\infty < E \le E'$. Furthermore, the *density* of the number of degrees of freedom as a function of E can also be easily determined in this way. Consequently, one always knows the *local resolution in* E required to determine the eigenvalues in the spectrum. To aid the reader in the description of the numerical codes (see details below) a blowup of the rectangular region of Figure 17.3A is shown in Figure 17.3B; I explicitly denote the important eigenvalues in the figure for $M_{11}(E) = 0$ (denoted by E_j^*) and for $(E_{11} + M_{22})/2 = \pm 1$ (the main spectrum E_k).

In Figure 17.4A the functions M_{11}, M_{12}, and S are contrasted. This figure provides perspective on the behavior of M_{12} (which generates the auxiliary spectrum, μ_j) relative to the oscillation function M_{11} (which generates the E_j^*) and the accounting function S. Here we see that M_{12} does not track with M_{11} as well as the latter follows the trace of the monodromy matrix (Figure 17.3). Nevertheless, for some localized region in the E domain, one finds that the maxima and minima of M_{12} are not too far from those of M_{11}. The rectangle in the center of Figure 17.4A has been exploded in Figure 17.4B to provide details not visible in the former figure. Furthermore, the specific eigenvalues of M_{11} (the E_j^*) and those of M_{12} (the hyperelliptic-function auxiliary spectrum values, μ_j) are also shown. These figures provide excellent perspective for coding the algorithm given herein.

Figure 17.4 The Zabusky and Kruskal initial sine wave is used to compute (A) the elements of the monodromy matrix $M_{11}(x_{M-1}, E)$, $M_{12}(x_{M-1}, E)$, and the accounting function $S(x_{M-1}, E)$. Note that the maxima and minima in M_{11} only roughly correspond. In (B) is a blowup of the region in (A) (denoted by the rectangle). Shown are the eigenvalues of $M_{12}(x_{M-1}, E)$ (the "auxiliary spectrum," μ_j) and $M_{11}(x_{M-1}, E)$ (the oscillation function, E_j^*).

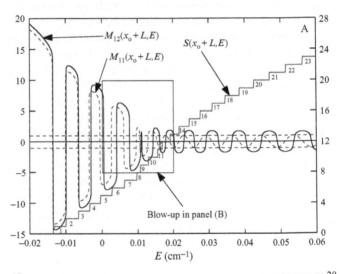

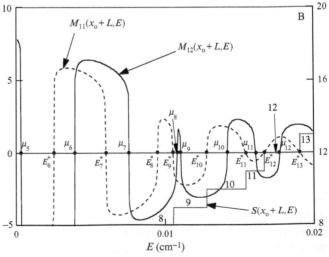

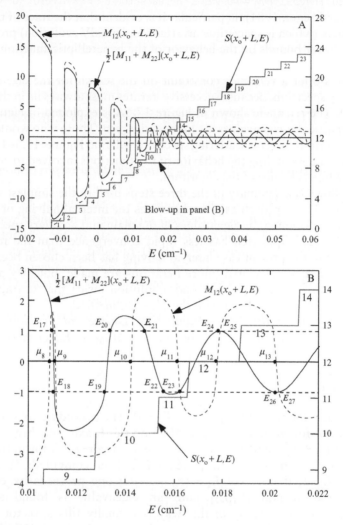

Figure 17.5 The Zabusky and Kruskal initial sine wave is used to compute (A) the trace of the monodromy matrix $(M_{11}+M_{22})(x_{M-1},E)/2$, the element $M_{12}(x_{M-1}, E)$, and the accounting function $S(x_{M-1}, E)$. Further note that the maxima and minima in $(M_{11}+M_{22})(x_{M-1},E)/2$ and $M_{12}(x_{M-1},E)$ roughly correspond. In (B) is a blowup of the region in (A) denoted by the rectangle. Shown are the eigenvalues of $(M_{11}+M_{22})(x_{M-1}, E)/2=\pm1$ (the "main spectrum," E_j) and $M_{12}(x_{M-1}, E)=0$ (the "auxiliary spectrum," μ_j).

Finally in Figure 17.5A I contrast the trace of the monodromy matrix and the monodromy matrix element M_{12}. Thus the essential ingredients of the *direct scattering problem* of the IST are graphed in Figure 17.5. Also shown in Figure 17.5A is the accounting function S. One clearly sees some of the most important results of the oscillation theorems of the Schrödinger eigenvalue problem in this figure

(Ablowitz and Segur, 1981); in particular it is evident that the trace of the monodromy matrix (which is space/time invariant under KdV evolution) provides the upper and lower bounds on the behavior of the hyperelliptic-function values μ_j which must always lie in the "open bands" of the theory ($E_{2j} \leq \mu_j \leq E_{2j+1}$). These results offer a very nice constraint on the search for the eigenvalues of the auxiliary spectrum, because successive iterates must always lie in the interval (E_{2j}, E_{2j+1}). The rectangle shown in Figure 17.5A is exploded in Figure 17.5B, where the particular eigenvalues are more easily seen. Those who would attempt to code the periodic inverse scattering spectrum will find Figures 17.3–17.5 a welcome aid in understanding the behavior of the functions involved to ensure efficient, error-free behavior of the algorithm.

What follows is a summary of the three steps used in the simplest version of the program. The approach as given here uses the interval halving or bisection method for determining the eigenvalues in the main and auxiliary spectra (Press et al., 1992). The approach has less rapid convergence than the method of Newton, but in the present case interval halving has been chosen because derivatives need not be calculated (more iterations are made, but they are faster) and because one is *guaranteed* to find a solution inside the starting interval, while this is not necessarily so with the Newton method.

Step I: Determine the Exact Number of Degrees of Freedom. Between $E_{min} = -\lambda(\eta_{max} - \bar{\eta})$ and E_{max} there are $N = S(x_{M-1}, E_{max})$ eigenvalues E_j^* such that

$$S(x_{M-1}, E_j^*) = j, \quad 1 \leq j \leq N \tag{17.52}$$

The E_j^* are the solutions of $M_{11}(x_{M-1}, E) = 0$. The procedure followed is that for every j *not* found between 1 and N along the E axis, one begins with the eigenvalues $E_1 = E_{j-1}^*$ and $E_2 = E_{j+1}^*$ for which S is already known; the interval $[E_1, E_2]$ is divided and $S(x_{M-1}, (E_1 + E_2)/2)$ is computed: If $S > j$, set $E_2 = (E_1 + E_2)/2$, if not take $E_1 = (E_1 + E_2)/2$; one proceeds by continuing to divide the interval $[E_1, E_2]$ until an eigenvalue is found such that $S(x_{M-1}, E_j^*) = j$. By iterating in this way one finally fills a vector of eigenvalues E_j^*, $1 \leq j \leq N$. This process is complete when Equation (17.52) is satisfied for all j, that is, the sequence $1, 2, 3, \ldots, N-1$ is uninterrupted. The program is infinitely nested so that all eigenvalues are eventually found within the chosen numerical precision ε. This approach provides a simple means for establishing *all* the degrees of freedom of a system and for localizing them in the E-domain.

Step II: The Auxiliary Spectrum. Here one computes the $N-1$ hyperelliptic-function starting values $\mu_j(0,0)$. Figure 17.4 suggests that one can localize the μ_j between the successive eigenvalues E_j^*. According to the figure, there are either zero, one, or two auxiliary eigenvalues (perhaps more in other cases) in an interval (E_j^*, E_{j+1}^*). The search is made by first determining the sign of M_{12} at E_j^*, $\text{sgn}[M_{12}(E_j^*)]$: If the sign is $(-1)^j$ this implies that $E_j^* < \mu_j$, otherwise $E_j^* > \mu_j$. Two eigenvalues E_1 and E_2 are searched for until the signs of

$M_{12}(E_1)$ and $M_{12}(E_2)$ are opposite. At this point, the bisection search for the eigenvalues begins. If

$$\text{sgn}[M_{12}(E_j^*)] = (-1)^j \tag{17.53}$$

set $E_2 = E_j^*$, if not set $E_1 = E_j^*$. The interval is then halved and the checks are made again. The algorithm continues to check the signs of the adjacent values of the M_{12} (at E_1, E_2) to be sure that they are opposite throughout the search process; this ensures that the bisection process will function. The iteration of the dividing process is exited from when $\mu_2 - \mu_1 < \varepsilon(\mu_1 + \mu_2)/2$, for $\varepsilon \sim 10^{-12}$. The Riemann sheet indices σ_j (17.20) are then computed for each of the μ_j.

Step III: The Main Spectrum. Here one computes the $2M + 1$ eigenvalues in the main spectrum E_k, $1 \leq k \leq 2N + 1$. Given the spectrum of the μ_j we know that their are exactly two associated main spectrum eigenvalues $(E_{2j + 1}, E_{2j + 2})$ in the interval $(\mu_j, \mu_{j + 1})$; one eigenvalue corresponds to $\text{TrM}/2 = +1$, the other to $\text{TrM}/2 = -1$. Thus the bisection search is quite simple since only one main spectrum eigenvalue of each type lies in the interval $(\mu_j, \mu_{j + 1})$. The search is made by first determining the sign of $\text{TrM}/2$ at $\mu_j(0,0)$, $\text{sgn}[\text{TrM}(\mu_j)/2]$: If the sign is $(-1)^j$, this implies that $\mu_j < E_j$, alternatively, $\mu_j < E_j$. Two eigenvalues E_1 and E_2 are searched for until the signs of $\text{TrM}(E_1)/2$ and $\text{TrM}(E_2)/2$ are opposite. At this point, the bisection search for the main spectrum eigenvalues is initiated. If

$$\text{sgn}[\text{TrM}(E_j^*)/2] = (-1)^j \tag{17.54}$$

is true set $E_2 = \mu_j$, if not set $E_1 = \mu_j$. The interval is then halved and the checks are made again. Once again the iteration of the dividing process is exited from when $E_2 - E_1 < \varepsilon(E_1 + E_2)/2$, for $\varepsilon \sim 10^{-12}$.

An alternative procedure to that just given would be to invert steps II and III. One would thus use the accounting eigenvalues E_j^* to localize the main spectrum E_k. Then the auxiliary spectrum eigenvalues μ_j are found from the bracketing E_k. Both procedures have been programmed and they give satisfactory results. Because the method outlined above is simpler to program (I also find it theoretically and esthetically more pleasing) I recommend it over the second approach. The bottom line is that both methods work just fine; their codes are of equal length and are a pleasure to use. Given an input space or time series with hundreds or thousands of discrete points, the program computes both the direct and inverse scattering problems, and reconstructs the input series with little difficulty. The direct scattering problem takes at most a few minutes on a modern workstation, while the inverse problem may take somewhat longer. The reason for the larger times for computing the hyperelliptic functions is that one is repeating the direct problem M times for an M-point time series. Since the direct problem requires M^2 steps, the total time for constructing the inverse problem is M^3.

Very often in the analysis of data one first determines which hyperelliptic functions he would like to filter from the input series (Osborne, 1991a). For example, to filter the solitons from the series it is enough to compute only the soliton hyperelliptic functions; these are normally much fewer than M. As a consequence the amount of computer time required is substantially reduced, as seen in the following example.

17.7 Example of the Analysis of a Many-Degree-of-Freedom Wave Train and Nonlinear Filtering

Here I consider the analysis of a complex wave train solution of the KdV equation, that is, one that has a many-degree-of-freedom spectrum with $N = 128$. The example I choose has the Cauchy initial condition $\eta(x, 0)$ which is given by a time series that is governed by a power-law power spectrum, $f^{-\alpha}$, with uniformly distributed random Fourier phases on the interval $(0, 2\pi)$. Such a time series, even though it is generated by a linear Fourier series, may be viewed as an *initial value state* of KdV and is here analyzed using the algorithm described above. The results are given in Figure 17.6. The input time series corresponds to $\alpha = 1.0$ and is shown in Figure 17.6A. The associated Floquet diagram is given in Figure 17.6B and has nine solitons in the spectrum (i.e., nine zero crossings of the Floquet discriminant to the left of the "reference level") (Osborne and Bergamasco, 1986). Both the hyperelliptic-function spectrum and the soliton index are given in Figure 17.6C. The hyperelliptic-function spectrum is a graph of the widths of the open bands, $a_j = (E_{2j+1} - E_{2j})$, as a function of wavenumber, $k_j = 2\pi j/L$. The "soliton index" is given by $m_j = (E_{2j+1} - E_{2j})/(E_{2j+1} - E_{2j-1})$; m_j is the "modulus" of the hyperelliptic functions and is nearly 1 for a soliton and substantially less than 1 for the radiation components; the "reference level" divides these two fundamental spectral types. The first ten hyperelliptic functions are given in Figure 17.6D. The radiation components occur for m_j to the right of the reference level (see Figure 17.6C); these correspond to the oscillation mode numbers 10-128 and are shown in Figure 17.6E. The solitons are found by first establishing the maximum value of $j = J$ for which $m_j \geq 0.99$. In the present case $j = 9$ so there are nine solitons that propagate on a reference level defined by $\eta_{\text{ref}} = -E_J/\lambda$ in real space (Osborne and Bergamasco, 1986).

Shown in Figure 17.6F the soliton components (the linear superposition of the hyperelliptic mode numbers 1-9) together with the radiation components (sum of modes 10-128). The reference level, η_{ref}, is also displayed. In Figure 17.7 I show how the input series may be reconstructed from the spectrum via the linear superposition law (17.3). I sum the modes 1-126 (rather than the total number of modes, 1-128) so that some difference can be graphically seen between the input space series and its reconstruction. Summing all the 128 modes results in a series that is accurate to within about 10^{-10} of the input series.

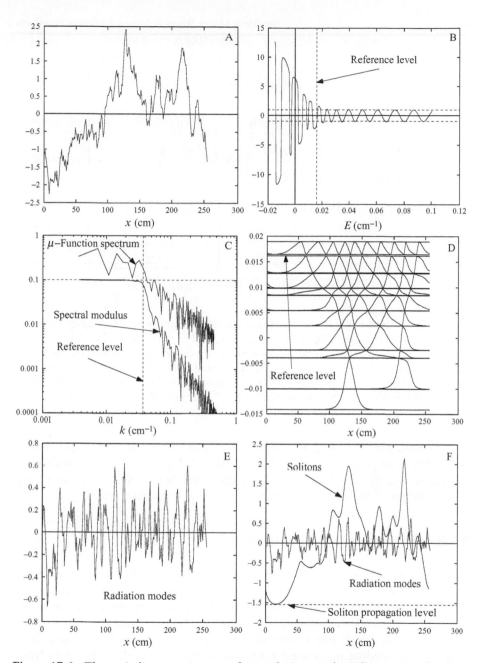

Figure 17.6 The periodic scattering transform solutions to the KdV equation for the particular numerical case of a many degree-of-freedom ($N = 128$) wave train (A) whose linear Fourier spectrum is a power law, $f^{-\alpha}$, $\alpha = 1$, and whose Fourier phases are random on $(0, 2\pi)$. The Floquet spectrum (one-half the trace of the monodromy matrix, $\mathrm{TrT}(E)/2$, graphed as a function of $E = k^2$) is shown in (B); nine solitons are found in the spectrum. The amplitudes of the hyperelliptic functions, $A_j = (E_{2j+1}-E_{2j})/2$, and the spectral index, $m_j = (E_{2j+1}-E_{2j})/(E_{2j+1}-E_{2j-1})$, are given in (C). The vertical scale of m_j has been shifted downward one decade for clarity. The first 10 hyperelliptic functions are shown in (D). The sum of the radiation modes (10-128) are shown in (E). In (F) the solitons (modes 1-9) and the radiation are shown on the same scale.

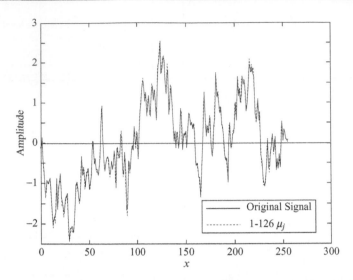

Figure 17.7 An example of the reconstruction of the input wave train using linear superposition of the hyperelliptic functions, $\mu_j(x,0)$. The first 126 of the 128 hyperelliptic functions have been summed; these are shown as a dotted line, while the original input signal is given as a solid line. The small differences are due to the fact that modes 127, 128 have not been included in the reconstructed wave train. Including these last two modes gives excellent agreement ($\sim 10^{-10}$) between the input and reconstructed wave trains.

It is emphasized that the determination of a large number of hyperelliptic modes is a highly nontrivial numerical task that is handled quite straightforwardly with the automatic numerical IST algorithm discussed herein. Computation of the results just given requires a few seconds on a modern workstation.

17.8 Summary and Conclusions

Numerical procedures are introduced to implement the IST of the periodic KdV equation (1) for solving the direct scattering problem of an input discrete waveform (potential) and (2) for computing the associated inverse scattering problem. The methods developed here are useful in time series analysis applications in which one is able to determine the spectrum of a measured, N degree-of-freedom input wave train and to reconstruct the wave train from the spectrum or to nonlinearly filter the data.

It is worth pointing out that the filtering approach considered in Figures 17.6 and 17.7 is the nonlinear analogue of the linear "perfect filter," the latter of which is most often implemented as the exact sum of specific terms in an ordinary Fourier series; typical applications are normally made in terms of low pass, high pass, and band pass filters. It is well known that the "perfect filter" is far from perfect in its behavior even in the linear problem, primarily due to the Gibbs phenomenon. It should not come as a great surprise that the nonlinear version of the "perfect filter" is, after all, not "perfect" either. What is required is a generalization of well-established linear filtering approaches to the nonlinear problem. This can be accomplished by modifying the form of Equation (17.29), for example, the filter itself may be incorporated into the "partition" matrices H. These results are beyond the scope of the present chapter. Additional nonlinear filtering methods are discussed in Chapter 28. The development of nonlinear filtering techniques offers a significant challenge to future research.

What does the present algorithm offer that has not been provided by other algorithms for the direct and ISTs given previously in the literature (Osborne, 1983, 1991a,b; Osborne and Bergamasco, 1986)? Here is a list of items:

(1) The construction of the monodromy matrix given herein consists of purely real numbers. This contrasts to other formulations that require complex numbers. As a consequence the computer code described here is about four times faster than fully complex formulations (Osborne, 1991b).

(2) The partition matrix, T, by a leading order perturbation expansion in $\kappa \Delta x \ll 1$, has the form:

$$T = \begin{pmatrix} 1 & 1 \\ -\kappa^2 \Delta x & 1 \end{pmatrix}$$

where $\kappa^2 = \lambda \eta(x_n, 0) + k^2$. This simple expression for the matrix T is related to the approach of Ablowitz and Ladik (1975, 1976a,b) who show that the AKNS class of nonlinear wave equations (Ablowitz et al., 1974, Ablowitz and Segur, 1981) may be discretized and integrated by the IST. While the numerical approach given herein is discrete, it is not however integrable by IST; nevertheless the method has excellent numerical properties (Osborne, 1993c). A future challenge will be the determination of other forms for the matrix T that may have even more desirable properties for numerical work, such as higher precision and a greater amenability to nonlinear filtering (Osborne, 1993b).

(3) The derivatives of the elements of the monodromy matrix with respect to the parameter E are analytic Equation (17.28). These provide a useful aid in applying the Newton iteration scheme for determination of the eigenvalues; the exact derivatives speed convergence of the method.

(4) Finally, the algorithm given herein is *automatic* in that it finds *all* the eigenvalues in the direct spectral problem and subsequently finds the hyperelliptic oscillation modes in the inverse scattering problem without requiring the user to have a deep knowledge of the periodic IST. Thus it seems that this is the first paper in which an algorithm has been presented that allows the investigator to do experimental soliton physics without having to first master the mathematics of IST. The algorithm should also enable the theoretician to explore details of periodic IST heretofore inaccessible, that is, in cases where the mathematical recipe may be so complex that physical understanding is not easily obtainable without numerical results.

(5) The IST algorithm uses computer time proportional to N^3 (N^2 for each of the N values of the hyperelliptic functions computed a single base point, $\mu_j(x_o, 0)$). This may be contrasted with the Gel'fand-Levitan-Marchenko integral equation for KdV on the infinite line (which is N^4) and with the θ-function inverse problem for periodic boundary conditions (which is discussed in detail in this monograph, see Chapters 20-23 and 32).

(6) The algorithm given in this chapter is in the hyperelliptic-function representation, and therefore has no numerical pathologies (i.e., exploding exponentials) in the soliton part of the spectrum as does the Gel'fand-Levitan-Marchenko equation. The requisite large roundoff errors and very large numbers (perhaps even exceeding the range of many computers) that occur in formulations of the latter type are *not* present in the hyperelliptic-function representation. Additional reading includes recent use of the present algorithm in the analysis of data (Zimmerman and Haarlemmer, 1999; Hawkins et al., 2008; Christov, 2009).

18 The Spectral Eigenvalue Problem for the NLS Equation

18.1 Introduction

This chapter discusses a numerical algorithm for solving the nonlinear Schrö-dinger equation *spectral eigenvalue problem*, an important ingredient for the *nonlinear spectral analysis of deep-water wave trains*. This algorithm has been used extensively in Chapter 12 where a discussion of the physics of the nonlinear Schrödinger equation is given in the context of its spectral eigenvalue problem in which unstable mode (rogue wave) dynamics appear in the λ plane. Chapter 30 gives a discussion on the use of the algorithm in the analysis of laboratory data.

18.2 Numerical Algorithm

The nonlinear Schrödinger equation as studied in this chapter has the following form:

$$iu_t - u_{xx} + 2\sigma|u|^2 u = 0 \tag{18.1}$$

where $\sigma = \pm 1$. The deep water (the so-called "focusing") case has $\sigma = -1$. The shallow water ("defocusing") case has $\sigma = 1$. I assume periodic boundary conditions so that $u(x, t) = u(x + L, t)$ for $0 \le x \le L$.

The eigenvalue or spectral problem for the NLS equation is given by:

$$\Psi_x = Q(\lambda)\Psi, \quad Q = \begin{pmatrix} -i\lambda & u \\ \sigma u^* & i\lambda \end{pmatrix} \tag{18.2}$$

Here λ is the time-independent complex spectral wavenumber. Some of the important details in the Floquet analysis of this problem are given in Chapter 12.

The numerical algorithm is designed to solve the Floquet problem for Equation (18.2) by discretizing the wave train "space series" $u(x, 0)$ (or $u(0, t)$ for a time series) into steps of piecewise constant values u_n at spatial points

Doi: 10.1016/S0074-6142(10)97018-6

$x_n = n\Delta x$ where $\Delta x = L/M$, $n = 1,2\ldots M$. Periodic boundary conditions are assumed so that $u(x,0) = u(x+L,0)$ and therefore $u_n = u_{n+M}$. The solution of the spectral eigenfunction $\Psi(x)$ in each interval Δx is then obtained by integrating the eigenvalue problem for a constant potential:

$$\Psi(x_n + \Delta x) = U(u_n, \Delta x)\Psi(x_n) \tag{18.3}$$

where $U(u_n, \Delta x)$ is the exponential of the trace-vanishing matrix $Q(\lambda)$:

$$U(u) = e^{\Delta x Q(\lambda)} = \exp\left[\Delta x \begin{pmatrix} -i\lambda & u \\ \sigma u* & i\lambda \end{pmatrix}\right]$$

$$= \begin{pmatrix} \cosh(k\Delta x) - \dfrac{i\lambda}{k}\sinh(k\Delta x) & \dfrac{u}{k}\sinh(k\Delta x) \\ \dfrac{\sigma u*}{k}\sinh(k\Delta x) & \cosh(k\Delta x) + \dfrac{i\lambda}{k}\sinh(k\Delta x) \end{pmatrix}$$

$$\tag{18.4}$$

Here $k^2 = \sigma|u|^2 - \lambda^2$ is constant inside an interval Δx. This approach is equivalent to the propagator method for solving the scattering problem when $u(x, 0)$ is constant in the interval Δx.

At this point it is convenient to introduce a four-component field consisting of Ψ and its derivative with respect to λ:

$$\Xi(x, \lambda) = \begin{pmatrix} \Psi \\ \Psi' \end{pmatrix} \tag{18.5}$$

where

$$\Psi' = \partial\Psi/\partial\lambda$$

It is clear that the field $\Xi(x,\lambda)$ has a recursion relation

$$\Xi(x_n + \Delta x) = T(u_n)\Xi(x_n) \tag{18.6}$$

where

$$T(u_n) = \begin{pmatrix} U(u_n) & 0 \\ U'(u_n) & U(u_n) \end{pmatrix} \tag{18.7}$$

is a four-by-four matrix and $U'(u_n) = \partial U(u_n)/\partial\lambda$ is given by the four elements:

$$U'_{11} = i\Delta x \frac{\lambda^2}{k^2} \cosh(k\Delta x) - \left(\lambda\Delta x + i + i\frac{\lambda^2}{k^2} \right) \frac{\sinh(k\Delta x)}{k}$$

$$U'_{12} = -\frac{u\lambda}{k^2} \left(\Delta x \cosh(k\Delta x) - \frac{\sinh(k\Delta x)}{k} \right)$$

$$U'_{21} = -\frac{\sigma u^* \lambda}{k^2} \left(\Delta x \cosh(k\Delta x) - \frac{\sinh(k\Delta x)}{k} \right)$$

$$U'_{22} = -i\Delta x \frac{\lambda^2}{k^2} \cosh(k\Delta x) - \left(\lambda\Delta x - i - i\frac{\lambda^2}{k^2} \right) \frac{\sinh(k\Delta x)}{k}$$

(18.8)

Discretizing the field $u(x, 0)$ into M steps gives:

$$\Xi(x_n) = \prod_{j=n-1}^{0} T(u_j)\Xi(x_0)$$

(18.9)

The *monodromy matrix of Floquet theory* is then given by

$$T(x_o, \lambda) = \prod_{j=M}^{0} U(u_j, \lambda)$$

(18.10)

The NLS *spectrum* is given in Section 18.3.

18.3 The NLS Spectrum

The NLS spectrum is that set of mathematical information that is derivable from the eigenvalue problem (18.2) (see details in Chapter 12) and which is necessary to compute the inverse problem, that is, the loop integrals and Riemann theta-function solutions as discussed in Chapter 24. Thus, the spectral information from Equation (18.2) allows us to compute the Riemann spectrum (Riemann matrix, wavenumbers, frequencies, and phases) that is then used in the theta-function solution of NLS.

18.3.1 The Main Spectrum

The trace of the monodromy matrix determines the *main spectrum eigenvalues*:

$$\frac{1}{2}\text{Tr}\mathbf{T} = \frac{1}{2}(T_{11} + T_{22}) = a_R(x_o, \lambda) = \pm 1$$

(18.11)

gives the complex eigenvalues λ_k, $k = 1, 2 \dots 2N$

18.3.2   The Auxiliary Spectrum of the $\mu_j(x,0)$

The auxiliary spectrum eigenvalues are found by:

$$T_{12} = g(x_o, \lambda) = 0 \ \text{gives the complex eigenvalues} \ \mu_j(x_o, t = 0), \ j = 1, 2, \ldots, N$$

$$(18.12)$$

To compute both the μ_j and their complex conjugates μ_j^* one can of course use $T_{12}T_{21} = 0$ to instead determine the spectrum. These are the Kotljarov-Its-Tracy (KIT) μ-function modes that can be *inverted* and *linearized* by theta functions using the general algebraic geometrical procedure for solving the Jacobian inverse problem. Note further that Equation (12.53) constructs the potential $u(x, t)$ from the hyperelliptic functions $\mu_j(x, t)$. The function $g(x_o, \lambda)$ is given by (12.54).

18.3.3   The Auxiliary Spectrum of the Riemann Sheet Indices σ_j

The *Riemann sheet indices* are found from:

$$\sigma_k = \frac{f(\lambda)}{\sqrt{P(\lambda)}}\bigg|_{\lambda=\mu_k} = \frac{i \, \text{Im}(T_{11})}{\sqrt{T_{12}^* T_{21} - \text{Im}^2(T_{11})}}\bigg|_{\lambda=\mu_k} \qquad (18.13)$$

These indices tell us on which Riemann sheet a particular hyperelliptic function lies.

18.3.4   The Auxiliary Spectrum of the $\gamma_j(x,0)$

The *radiation stress modes or the carrier envelope of the wave train* are solutions of:

$$\frac{1}{2}(T_{22} - T_{11}) = f(x_o, \lambda) = 0, \quad \text{which give the eigenvalues} \ \gamma_j(x_o, t = 0),$$

$$\text{for} \quad 0 \leq x_o \leq L, \quad j = 1, 2, \ldots, N$$

$$(18.14)$$

where the $\gamma_j(x, t)$ are the nonlinear modes of the *radiation stress* contribution to a nonlinear wave train (defined to be $|u|^2$) (Tracy et al., 1991):

$$|u|^2 = \frac{1}{2}\sum_{k=1}^{2N}\lambda_k^2 - \sum_{j=1}^{N}\gamma_j^2 \qquad (18.15)$$

Here $u(x,t)$ is the solution of NLS in Equation (18.2) form. The function $f(x_o, \lambda)$ is given by (12.58).

18.3.5 Spines in the Spectrum

The main spectrum provides the information necessary to compute the spectral quantity known as the *spines*. These are curves in the complex plane with values of λ which insure that the Bloch eigenfunctions are *stable*, that is, they do not blow up exponentially fast for certain values of the spatial variable, x (i.e., for arbitrary spatial translations of the Bloch eigenfunctions). The spines are defined by:

$$\mathrm{Im}[\mathrm{TrT}/2] = 0, \quad -2 \leq \mathrm{Re}[\mathrm{TrT}/2] \leq 2 \tag{18.16}$$

Additional analysis reveals that spines typically connect two (or even three, but this is a rarer case) points of simple spectrum. When two points of simple spectrum are connected by a spine, the combination of spectral information is called a *nonlinear mode*. There are two kinds of nonlinear mode: (1) When two points of spectrum are connected by a spine that crosses the real axis we have a *stable Stokes wave*. (2) When the two points of spectrum are connected by a spine that does not cross the real axis, we have an *unstable Stokes wave* or "rogue" mode in the spectrum.

18.4 Examples of Spectral Solutions of the NLS Equation

I will now give some examples of spectra of various types of wave trains.

18.4.1 Plane Waves

A first example is given in Figure 18.1 that corresponds to a plane (unmodulated, square) wave with amplitude $A = 1$ and wavelength $L = 2$. Since $AL = 2 < \pi$, this is a *stable* mode. The spines are seen to correspond to the entire real axis and the imaginary axis between the carrier eigenvalues.

Another example is shown in Figure 18.2 that is again a plane (unmodulated, square) wave with amplitude $A = 1$ and wavelength $L = \sqrt{2}\pi \cong 4.44\ldots$. Since $AL = 4.44\ldots > \pi$, this is an *unstable mode*. An empty (degenerate) double point appears below the carrier on the imaginary axis. Once again the spines are seen to correspond to the real axis and the imaginary axis between the carrier eigenvalues.

18.4.2 Small Modulations

Still another example is given in Figure 18.3 for a small-amplitude modulation with amplitude $A = 1$ and wavelength $L = \sqrt{2}\pi \cong 4.44\ldots$ and $\varepsilon = 10^{-5}$. Since $AL = 4.44\ldots > \pi$, this is an unstable mode. A small, near-degenerate double point appears below the carrier on the imaginary axis. The spines are seen to correspond to the real axis and the imaginary axis between the carrier eigenvalues. The value of ε is so small that it is not observable between the two closely

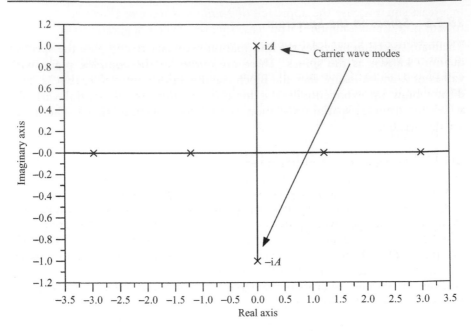

Figure 18.1   Lambda plane spectrum of a plane (unmodulated) carrier wave.

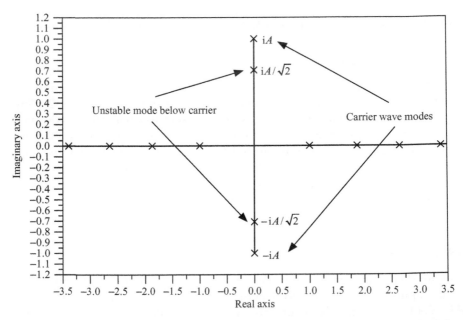

Figure 18.2   Lambda plane spectrum of a plane (unmodulated) carrier wave which has one (homoclinic) eigenmode.

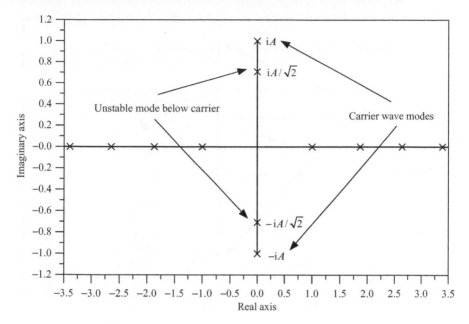

Figure 18.3 Lambda plane spectrum of a plane carrier wave which is modulated by an unstable, small-amplitude ($\varepsilon = 10^{-5}$) sine wave and which has one (homoclinic) unstable mode. The small value of the modulation amplitude means this graph is indistinguishable by eye from Figure 18.2.

spaced eigenvalues of the small-amplitude modulation. The small value of ε renders the two cases of Figures 18.2 and 18.3 indistinguishable by eye. This is because in Figure 18.3 the × at $iA/\sqrt{2}$ is in reality two ×s (two points of simple spectrum) spaced slightly apart ($\varepsilon \sim 10^{-5}$), but the distance is so small that we cannot see it. This is the typical case for the problem of "small-amplitude modulations," that is, the spectrum resembles an unmodulated box. This motivates the study of the spectrum of a "zero potential" in terms of an unmodulated box as discussed at length in Tracy (1984).

Now consider a case, shown in Figure 18.4, with a stable modulation on the real axis. The results were computed with a real perturbation of a plane wave, $1+\varepsilon \cos(Kx)$. Here I used $\varepsilon = 0.1$, $A = 1$, $L = 2$, $K = 2\pi/L$. Since $AL < \pi$, two bands open up on the real axis. Note the appearance of the spines between these two points of simple spectrum. The spines are seen to cross the real axis. Physically, these nonlinear modes are *stable Stokes waves*.

Now let us treat an unstable mode on the imaginary axis, Figure 18.5. The results were computed with a real perturbation of a plane wave, $1 + \varepsilon \cos (Kx)$, for which $\varepsilon = 0.1$, $A = 1$, $L = 4.44$, $K = 2\pi/L$. Since $AL > \pi$, two bands open up on the imaginary axis. In this case, a gap opens up in the spines. This type of nonlinear mode is often called a "gap" state, because the spine has been removed between two points of simple spectrum.

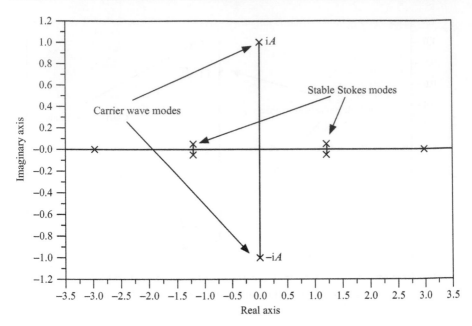

Figure 18.4 Lambda plane spectrum of a plane carrier wave which is modulated by a stable, small-amplitude sine wave $(1 + \varepsilon \cos(Kx))$, which has two stable eigenmodes.

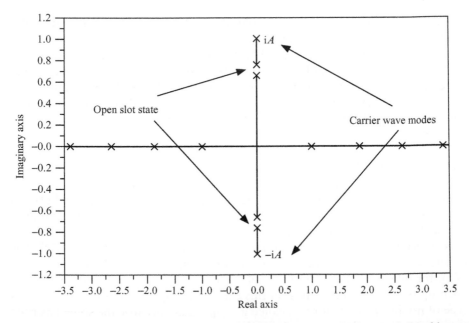

Figure 18.5 Lambda plane spectrum of a plane carrier wave which is modulated by an unstable, small-amplitude $(1 + \varepsilon \cos(Kx))$ sine wave, which has an unstable eigenmode.

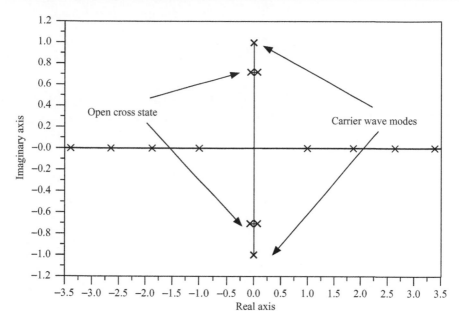

Figure 18.6 Lambda plane spectrum of a plane carrier wave which is modulated by an unstable, small-amplitude, imaginary $(1 + \varepsilon \cos(Kx),\ \varepsilon = 0.1i)$ sine wave, which has an unstable eigenmode (two points of simple spectrum connected by a spine which is perpendicular to the imaginary axis) called a "cross state."

Now let us treat a cross state unstable mode on the imaginary axis, see Figure 18.6. The results were computed with an imaginary perturbation of a plane wave, $1 + i\varepsilon \cos(Kx)$, for which $\varepsilon = 0.1$, $A = 1$, $L = 4.44$, $K = 2\pi/L$. Since $AL > \pi$, two bands open up on the imaginary axis, but they are "cross states." In this case, two bands open up horizontally, extending small spines to the right and the left of the imaginary axis.

In Chapter 24, I give some concrete examples of many types of unstable "rogue" modes using Riemann theta functions. Specifically given are a large class of solutions of the NLS equation for which the modulation amplitude is formally "small," $\varepsilon \ll 1$.

18.5 Summary

The above formulas provide a well-defined numerical algorithm for the study of the spectrum of the NLS equation. Several numerical examples are given in Chapter 12 where different kinds of nonlinear modes of the NLS equation are discussed in terms of their IST spectrum. Computations of the Riemann spectrum are discussed in Chapter 24. The actual computer code that computes the spectrum uses the results given in this chapter. Additional application of the NLS spectral algorithm is given in Chapter 30 where the approach is used to analyze measured laboratory time series.

19 Computation of Algebraic-Geometric Loop Integrals for the KdV Equation

19.1 Introduction

This chapter gives numerical methods for computing the Riemann spectrum by the method of loop integrals; in particular, the results for the KdV equation are given. Computation of the generic loop integral G_{ij} is the central topic of conversation and in particular how to remove singularities from the numerical computations due to particular configurations of the IST eigenvalues. Two kinds of singularities (or near singularities) exist in the loop integrals. The first occurs exactly at the band edges and these are trivially accounted for. The second is a near singularity that occurs when the nearest eigenvalues immediately to the right and left of the band-edge integration limits. The near singularities occur either in the small-amplitude limit (for sine waves, the near-degenerate case occurs when two open-band eigenvalues are nearly equal) or in the large amplitude soliton limit (when two adjacent gap eigenvalues are nearly equal).

19.2 Convenient Transformations

I now discuss correction of the singularities that occur at the band edges. The following integral can be used to construct the numerical values of the loop integrals considered here:

$$G_{ji} = 2 \int_{E_j}^{E_{j+1}} \frac{E^{i-1} dE}{R^{1/2}(E)} \tag{19.0}$$

Because the root of the polynomial $R(E)$ (see Equation (14.6)) can be either real or imaginary, the quantity G_{ji} can also be real or imaginary. Also the square root of the polynomial can be either positive or negative, depending on the Riemann sheet that the polynomial lies on. Therefore, the above integral might be modified at most by a phase factor $R^{1/2}(E) = \pm i|R(E)|^{1/2}$. It is of course important to remember in the evaluation of a particular loop integral over an open band that it is the $\text{Re}\{G_{ji}\}$ that contributes to the contour integrations; purely

Doi: 10.1016/S0074-6142(10)97019-8

imaginary values of G_{ii} yield zero for the loop integrals. The goal now is to construct a numerical procedure for computing Equation (19.1) below. We are guided by the fact that singularities or near singularities occur at or near the branch points of a particular band. Therefore, for numerical computations, it is necessary to remove these singularities in order to have a straightforward numerical computation.

Here is the form of Equation (19.0) which I use in the numerical computations:

$$G_{ji} = 2 \int_{E_j}^{E_{j+1}} \frac{E^{i-1}dE}{\left[\prod_{n=1}^{2N+1} |E - E_n| \right]^{1/2}} \tag{19.1}$$

19.2.1   First Transformation

Consider the following transformation:

$$E(\theta) = \left(\frac{E_{j+1} - E_j}{2} \right) \cos \theta + \left(\frac{E_{j+1} + E_j}{2} \right) \tag{19.2}$$

or

$$E(\theta) = A_j \cos \theta + B_j \tag{19.3}$$

where

$$A_j = \left(\frac{E_{j+1} - E_j}{2} \right), \quad B_j = \left(\frac{E_{j+1} + E_j}{2} \right) \tag{19.4}$$

Note that

$$dE = -A_j \sin \theta \, d\theta \tag{19.5}$$

We thus have the triangle as shown in Figure 19.1:
Then it is easy to see that

$$A_j^2 \sin^2\theta = A_j^2 - (E - B_j)^2 = -(E^2 - 2B_jE + B_j^2 - A_j^2) \tag{19.6}$$

Now compute $B_j^2 - A_j^2$:

$$B_j^2 = \frac{(E_{j+1} + E_j)^2}{4} = \frac{E_{j+1}^2 + 2E_jE_{j+1} + E_j^2}{4}$$

$$A_j^2 = \frac{(E_{j+1} - E_j)^2}{2} = \frac{E_{j+1}^2 - 2E_jE_{j+1} + E_j^2}{4} \tag{19.7}$$

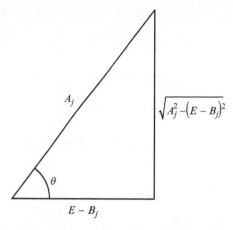

Figure 19.1 Triangle of parameters in first transformation.

Therefore

$$B_j^2 - A_j^2 = E_j E_{j+1} \tag{19.8}$$

and

$$A_j^2 \sin^2\theta = -(E^2 - 2B_j E + E_j E_{j+1}) = (E - E_j)(E_{j+1} - E) \tag{19.9}$$

Finally

$$A_j \sin\theta = \sqrt{(E - E_j)(E - E_{j+1})} \tag{19.10}$$

Now get for Equation (19.4):

$$dE = -\sqrt{(E - E_j)(E_{j+1} - E)}\,d\theta \tag{19.11}$$

So that Equation (19.1) becomes

$$G_{ji} = -2\int_{E_j}^{E_{j+1}} \frac{E(\theta)^{i-1}\sqrt{(E - E_j)(E_{j+1} - E)}\,d\theta}{\left[\prod_{n=1}^{2N+1} |E - E_n|\right]^{1/2}} \tag{19.12}$$

So that:

$$G_{ji} = -2\int_{E_j}^{E_{j+1}} \frac{E(\theta)^{i-1}\,d\theta}{\left[\substack{2N+1 \\ \prod \\ n=1 \\ n \neq j \\ n \neq j+1}} |E - E_n|\right]^{1/2}}$$

Now let us get the limits of the integral. When $E = E_j$ we have:

$$E_j = A_j \cos \theta + \frac{E_{j+1} + E_j}{2}$$

hence

$$\frac{E_{j+1} - E_j}{2} = -A_j = A_j \cos \theta \Rightarrow \cos \theta = -1 \Rightarrow \theta = \pi$$

When $E = E_{j+1}$ we have:

$$E_{j+1} = A_j \cos \theta + \frac{E_{j+1} + E_j}{2}$$

hence

$$\frac{E_{j+1} - E_j}{2} = A_j = A_j \cos \theta \Rightarrow \cos \theta = 1 \Rightarrow \theta = 0$$

The integral becomes:

$$G_{ji} = 2 \int_0^{\pi} \frac{E(\theta)^{i-1} \mathrm{d}\theta}{\left[\prod_{\substack{n=1 \\ n \neq j \\ n \neq j+1}}^{2N+1} |E(\theta) - E_n| \right]^{1/2}} \tag{19.13}$$

The signs of the **G** matrix need to be modified in the following way for programming purposes:

$$G_{ji} \rightarrow (-1)^{\alpha_j} G_{ji}$$

where

$$\alpha_j = j/2, \quad \text{for } j \text{ even}, \quad \alpha_j = (j+1)/2, \quad \text{for } j \text{ odd}$$

This convention generalizes the above results.

The relationship among the variables is illustrated in Figure 19.2.

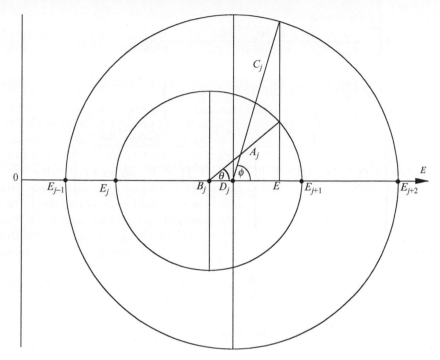

Figure 19.2 Relationship of transformation variables.

19.2.2 Second Transformation

Let us try the simple transformation $\theta = 2\phi$ to Equation (19.13) (Figure 19.2). This gives:

$$G_{ji} = 4 \int_0^{\pi/2} \frac{E(2\phi)^{i-1} d\phi}{\left[\displaystyle\prod_{\substack{n=1 \\ n \neq j \\ n \neq j+1}}^{2N+1} E(2\phi) - E_n \right]^{1/2}} = 4 \int_0^{\pi/2} \frac{E(2\phi)^{i-1} d\phi}{\left[\displaystyle\prod_{\substack{n=1 \\ n \neq j \\ n \neq j+1}}^{2N+1} \left| A_j \cos 2\phi + B_j - E_n \right| \right]^{1/2}}$$

$$G_{ji} = 4 \int_0^{\pi/2} \frac{E(2\phi)^{i-1} d\phi}{\left[\displaystyle\prod_{\substack{n=1 \\ n \neq j \\ n \neq j+1}}^{2N+1} \left| A_j (1 - 2\sin^2\phi) + B_j - E_n \right| \right]^{1/2}}$$

$$G_{ji} = 4 \int_0^{\pi/2} \frac{E(2\phi)^{i-1} d\phi}{\left[\displaystyle\prod_{\substack{n=1 \\ n \neq j \\ n \neq j+1}}^{2N+1} |A_j + B_j - E_n - 2A_j \sin^2\phi| \right]^{1/2}}$$

$$= \frac{4}{\left\{ \displaystyle\prod_{\substack{n=1 \\ n \neq j \\ n \neq j+1}}^{2N+1} |A_j + B_j - E_n| \right\}^{1/2}} \int_0^{\pi/2} \frac{E(2\phi)^{i-1} d\phi}{\left[\displaystyle\prod_{\substack{n=1 \\ n \neq j \\ n \neq j+1}}^{2N+1} [1 - m_{jn} \sin^2\phi] \right]^{1/2}}$$

where

$$m_{jn} = \frac{2A_j}{A_j + B_j - E_n} = \frac{E_{j+1} - E_j}{E_{j+1} - E_n}$$

Now look at the numerator of the integral above.

$$E(2\phi) = A_j \cos 2\phi + B_j - E_0, \quad E_0 = 0$$

$$E(2\phi) = (A_j + B_j - E_0)\left[1 - \frac{2A_j}{A_j + B_j - E_0} \sin^2\phi \right] = E_{j+1}(1 - m_{j0} \sin^2\phi)$$

$$m_{j0} = \frac{2A_j}{A_j + B_j - E_0} = \frac{E_{j+1} - E_j}{E_{j+1} - E_0}$$

Finally, we have the generalization of ordinary elliptic integrals:

$$G_{ji} = \frac{4(E_{j+1})^{i-1}}{\Delta} \int_0^{\pi/2} \frac{(1 - m_{j0} \sin^2\phi)^{i-1} d\phi}{\left[\displaystyle\prod_{\substack{n=1 \\ n \neq j \\ n \neq j+1}}^{2N+1} [1 - m_{jn} \sin^2\phi] \right]^{1/2}} \qquad (19.14)$$

$$m_{jn} = \frac{E_{j+1} - E_j}{E_{j+1} - E_n}, \quad \Delta = \left\{ \displaystyle\prod_{\substack{n=1 \\ n \neq j \\ n \neq j+1}}^{2N+1} |E_j - E_n| \right\}^{1/2}$$

The key question now is "Can we derive a suitable Landen transformation which allows the arithmetic-geometric mean (AGM) method to be applied to this problem for numerical computations?" See the discussion in Section 19.3.

19.2.3 A Final Transformation

Now, based upon later consideration, it is prudent to make a modification to the transformation used above so that it reads:

$$E(\theta) = (-1)^j A_j \cos \theta + B_j$$

This transformation, although a simple one, makes sure that the gap integrals look like the classical integrals K'. This will be seen in the single and double degree-of-freedom examples given below. Thus, the only change needed here is the factor $(-1)^j$ before A_j in all the computations. Here are the final results

$$G_{ji} = \frac{4(E_{j+1})^{i-1}}{\Delta} \int_0^{\pi/2} \frac{(1 - m_{j0} \sin^2 \phi)^{i-1} d\phi}{\left[\prod_{\substack{n=1 \\ n \neq j \\ n \neq j+1}}^{2N+1} [1 - m_{jn} \sin^2 \phi] \right]^{1/2}} \tag{19.15}$$

$$m_{jn} = \frac{2(-1)^j A_j}{(-1)^j A_j + B_j - E_n} = \frac{(-1)^j (E_{j+1} - E_j)}{\frac{[1+(-1)^j]E_{j+1} + [1-(-1)^j]E_j}{2} - E_n}$$

$$\Delta = \left\{ \prod_{\substack{n=1 \\ n \neq j \\ n \neq j+1}}^{2N+1} \left| (-1)^j A_j + B_j - E_n \right| \right\}^{1/2}$$

$$= \left\{ \prod_{\substack{n=1 \\ n \neq j \\ n \neq j+1}}^{2N+1} \left| \frac{[1+(-1)^j]E_{j+1} + [1-(-1)^j]E_j}{2} - E_n \right| \right\}^{1/2}$$

19.3   The Landen Transformation

The classical Landen transformation (Abramowitz and Stegun, 1964; Press et al., 1992) provides a fast, stable algorithm for computation of elliptic integrals. I provide pedantic results for the one and two degree-of-freedom cases because the results prove invaluable in the checkout of computer codes. Let us consider the following form for the above integral:

$$
G_{ji} = \frac{4(E_{j+1})^{i-1}}{\Delta} \int_0^{\pi/2} \frac{(\cos^2\phi + (1 - m_{j0})\sin^2\phi)^{i-1}\,d\phi}{\left[\displaystyle\prod_{\substack{n=1\\ n\neq j\\ n\neq j+1}}^{2N+1} [\cos^2\phi + (1 - m_{jn})\sin^2\phi]\right]^{1/2}}
\tag{19.16}
$$

Inserting appropriate parameters:

$$
G_{ji} = \frac{4(E_{j+1})^{i-1}}{\Delta} \int_0^{\pi/2} \frac{(a_{j0}^2 \cos^2\phi + b_{j0}^2 \sin^2\phi)^{i-1}\,d\phi}{\left[\displaystyle\prod_{\substack{n=1\\ n\neq j\\ n\neq j+1}}^{2N+1} [a_{jn}^2 \cos^2\phi + b_{jn}^2 \sin^2\phi]\right]^{1/2}},
\tag{19.17}
$$

$$
a_{jn} = 1, \quad b_{jn} = \sqrt{1 - m_{jn}}
$$

A final step for the Landen transformation is to iterate on the angle ϕ until $a_N = b_N$ and the integral becomes:

$$
G_{2i} = \frac{4(E_{j+1})^{i-1} a_{j0,N}^{2(i-1)}}{\Delta \displaystyle\prod_{\substack{n=1\\ n\neq j\\ n\neq j+1}}^{2N+1} a_{jn,N}} K_o \int_0^{\phi_N} d\phi = \frac{4K_o(E_{j+1})^{i-1} a_{j0,N}^{2(i-1)}}{\Delta \displaystyle\prod_{\substack{n=1\\ n\neq j\\ n\neq j+1}}^{2N+1} a_{jn,N}} \phi_N
\tag{19.18}
$$

where K_o is a numerical factor which accumulates for each iteration of the Landen transformation.

19.4   Search for an AGM Method for the Loop Integrals

The method of the AGM (Abramowitz and Stegun, 1964; Borwein and Borwein, 1987) has been historically important for the computation of special functions. To this end let us look at the loop integrals in the form

$$G_{ji} = 2 \int_{E_j}^{E_{j+1}} \frac{E^{i-1} dE}{R^{1/2}(E)}, \quad 1 \leq i \leq N, \quad 1 \leq j \leq 2N \tag{19.19}$$

or more specifically

$$G_{ji} = 2 \int_{E_j}^{E_{j+1}} \frac{E^{i-1} d\theta}{\left[\prod_{n=1}^{2N+1} (E - E_n)\right]^{1/2}}, \quad 1 \leq i \leq N, \quad 1 \leq j \leq 2N \tag{19.20}$$

19.4.1 One Degree-of-Freedom Case

Now write the loop integrals for the one-dimensional case:

$$C_{11} = \pi \left[\int_{E_2}^{E_3} \frac{dE}{\sqrt{(E - E_1)(E - E_2)(E_3 - E)}} \right]^{-1} = \frac{\pi}{L_1} = \frac{K_1}{2} \tag{19.21}$$

which returns the relation $K_1 = 2 \, C_{11}$. Thus we have an appropriate notation for the wavenumbers as found by the α_j cycle integrals. The explicit expression for K_1 is given by:

$$K_1 = \frac{2\pi}{\int_{E_2}^{E_3} \dfrac{dE}{\sqrt{(E - E_1)(E - E_2)(E_3 - E)}}} = \frac{2\pi}{K(m)} \tag{19.22}$$

The interaction matrix B_{nj} for a single degree of freedom can be computed:

$$B_{11} = C_{11}A_{11} = 2C_{11} \int_{E_1}^{E_2} \frac{dE}{R^{1/2}(E)} = K_1 \int_{E_1}^{E_2} \frac{dE}{R^{1/2}(E)} \tag{19.23}$$

and finally

$$B_{11} = \frac{2\pi \int_{E_1}^{E_2} \dfrac{dE}{\sqrt{(E - E_1)(E_2 - E)(E_3 - E)}}}{\int_{E_2}^{E_3} \dfrac{dE}{\sqrt{(E - E_1)(E - E_2)(E_3 - E)}}} = 2\pi \frac{F\left(\frac{\pi}{2}\big|m_1\right)}{F\left(\frac{\pi}{2}\big|m\right)} = 2\pi \frac{K'(m)}{K(m)} \tag{19.24}$$

where the last step was made with Equations (17.4.62) and (17.4.68) of Abramowitz and Stegun (1964) together with Equation (17.3.2) and (17.3.6). Note that in these expressions the following definitions are used:

$$m = \frac{E_3 - E_2}{E_3 - E_1}$$

$$m_1 = \frac{E_2 - E_1}{E_3 - E_1} = 1 - m \tag{19.25}$$

These results bring out another form of the elliptic integrals that is more natural for work with the periodic scattering transform:

$$K(m) = \frac{1}{2}\sqrt{E_3 - E_1} \int_{E_2}^{E_3} \frac{dE}{\sqrt{(E - E_1)(E - E_2)(E_3 - E)}}$$

$$K'(m) = \frac{1}{2}\sqrt{E_3 - E_1} \int_{E_1}^{E_2} \frac{dE}{\sqrt{(E - E_1)(E_2 - E)(E_3 - E)}}$$

(19.26)

The first of Equation (19.26) is easily transformed into the standard form by

$$E = E_3 - (E_3 - E_2)t^2 \tag{19.27}$$

for which the polynomial has the form

$$\begin{aligned} P(E) &= (E - E_1)(E - E_2)(E_3 - E) \\ &= (E_3 - E_1)(E_3 - E_2)^2 t^2 (1 - t^2)(1 - mt^2) \end{aligned} \tag{19.28}$$

for

$$m = \frac{E_3 - E_2}{E_3 - E_1} \tag{19.29}$$

Furthermore, the second of Equation (19.26) is transformed using

$$E = E_1 + (E_2 - E_1)t^2, \quad dE = 2(E_2 - E_1)t \, dt \tag{19.30}$$

with

$$\begin{aligned} P(E) &= (E - E_1)(E_2 - E)(E_3 - E) \\ &= (E_3 - E_1)(E_2 - E_1)^2 t^2 (1 - t^2)(1 - m_1 t^2) \end{aligned} \tag{19.31}$$

for

$$m_1 = \frac{E_2 - E_1}{E_3 - E_1}, \quad m + m_1 = 1 \tag{19.32}$$

The first of Equation (19.26) can be seen to be a measure of the open bandwidth, while the second of Equation (19.26) is a measure of the associated gap width. The transformations from Equation (19.26) to Equations (19.13) and (19.14) suggest a way to remove the singular behavior near the limits in numerical calculations for the more general case of N degrees of freedom.

It is useful to compare what follows to the classical elliptic integrals in the notation of Abramowitz and Stegun (1964):

$$K(m) = K = \int_0^1 [(1 - t^2)(1 - mt^2)]^{-1/2} dt = \int_0^{\pi/2} \frac{d\theta}{[1 - m\sin^2\theta]^{1/2}} \quad (19.33)$$

$$K'(m) = K(m_1) = \int_0^1 [(1 - t^2)(1 - m_1 t^2)]^{-1/2} dt = \int_0^{\pi/2} \frac{d\theta}{[1 - m_1 \sin^2\theta]^{1/2}}$$

$$(19.34)$$

with $m + m_1 = 1$ and $K(m) = K'(m_1) = K'(1 - m)$.

19.4.2 An Alternative Approach

Now let us take a look at a streamlined approach that improves the formulation. Consider the elliptic integrals in the following form (which is easily programmed):

$$K(m) = \int_{E_2}^{E_3} \frac{dE}{\sqrt{(E - E_1)(E - E_2)(E_3 - E)}}$$

$$K'(m) = \int_{E_1}^{E_2} \frac{dE}{\sqrt{(E - E_1)(E_2 - E)(E_3 - E)}}$$

$$(19.35)$$

Consider the transformation that I have used in the numerical work:

$$E(\theta) = \left(\frac{E_3 - E_2}{2}\right)\cos\theta + \left(\frac{E_3 + E_2}{2}\right) = A\cos\theta + B$$

Use this in K to get:

$$dE = -\sqrt{(E - E_2)(E_3 - E)}\, d\theta$$

and then we have

$$K = \int_0^\pi \frac{d\theta}{\sqrt{E(\theta) - E_1}} = \int_0^\pi \frac{d\theta}{\sqrt{A\cos\theta + B - E_1}}$$

The final step is to let $\theta = 2\phi$ so that:

$$K = 2\int_0^{\pi/2} \frac{d\phi}{\sqrt{E(2\phi) - E_1}} = 2\int_0^{\pi/2} \frac{d\phi}{\sqrt{A\cos 2\phi + B - E_1}}$$

Now use the formula

$$\cos 2\phi = 1 - 2\sin^2\phi$$

so that

$$K = 2\int_0^{\pi/2} \frac{d\phi}{\sqrt{E(2\phi) - E_1}} = 2\int_0^{\pi/2} \frac{d\phi}{\sqrt{A(1 - 2\sin^2\phi) + B - E_1}}$$

$$= 2\int_0^{\pi/2} \frac{d\phi}{\sqrt{A + B - E_1 - 2A\sin^2\phi}}$$

This has the correct form for the elliptic integral provided we factor out

$$A + B - E_1 = E_3 - E_1$$

Finally

$$K = 2\int_0^{\pi/2} \frac{d\phi}{\sqrt{E(2\phi) - E_1}} = \frac{2}{\sqrt{A + B - E_1}}\int_0^{\pi/2} \frac{d\phi}{\sqrt{1 - m\sin^2\theta}}$$

$$= \frac{2}{\sqrt{E_3 - E_1}}\int_0^{\pi/2} \frac{d\phi}{\sqrt{1 - m\sin^2\theta}}$$

or

$$K = 2\int_0^{\pi/2} \frac{d\phi}{\sqrt{E(2\phi) - E_1}} = \frac{2}{\sqrt{E_3 - E_1}}\int_0^{\pi/2} \frac{d\phi}{\sqrt{1 - m\sin^2\theta}}$$

where

$$m = \frac{2A}{A + B - E_1} = \frac{E_3 - E_2}{E_3 - E_1}$$

Now return to Equation (19.14) to get the primed elliptic integral:

$$K'(m) = \int_{E_1}^{E_2} \frac{dE}{\sqrt{(E - E_1)(E_2 - E)(E_3 - E)}}$$

The transformation that I have used in the numerical work is:

$$E(\theta) = \left(\frac{E_2 - E_1}{2}\right)\cos\theta + \left(\frac{E_2 + E_1}{2}\right) = A\cos\theta + B$$

Use this in K to get:

$$dE = -\sqrt{(E - E_1)(E_2 - E)}\,d\theta$$

and then we have

$$K = \int_0^\pi \frac{d\theta}{\sqrt{E_3 - E(\theta)}} = \int_0^\pi \frac{d\theta}{\sqrt{E_3 - B - A\cos\theta}}$$

The final step is to let $\theta = 2\phi$ so that:

$$K = 2\int_0^{\pi/2} \frac{d\phi}{\sqrt{E_3 - E(2\phi)}} = 2\int_0^{\pi/2} \frac{d\phi}{\sqrt{E_3 - B - A\cos 2\phi}}$$

Now use the formula

$$\cos 2\phi = 1 - 2\sin^2\phi$$

so that

$$K = 2\int_0^\pi \frac{d\phi}{\sqrt{E_3 - E(2\phi)}} = 2\int_0^{\pi/2} \frac{d\phi}{\sqrt{E_3 - B - A(1 - 2\sin^2\phi)}}$$

$$= 2\int_0^{\pi/2} \frac{d\phi}{\sqrt{E_3 - B - A + 2A\sin^2\phi}}$$

This has the correct form for the elliptic integral provided we factor out

$$E_3 - B - A = E_3 - E_1$$

Finally

$$K = 2\int_0^{\pi/2} \frac{d\phi}{\sqrt{E_3 - E_1 + 2A\sin^2\phi}} = 2\int_0^{\pi/2} \frac{d\phi}{\sqrt{E(2\phi) - E_1}}$$

$$= \frac{2}{\sqrt{E_3 - E_1}} \int_0^{\pi/2} \frac{d\phi}{\sqrt{1 - m_1\sin^2\phi}}$$

where

$$m_1 = 1 - m = \frac{2A}{E_3 - B - A} = 1 - \left(\frac{E_3 - E_2}{E_3 - E_1}\right) = \frac{E_2 - E_1}{E_3 - E_1}$$

The bottom line is that after applying the transformation to remove the singularities at the limits of the integral then it remains to apply the simple transformation $\theta = 2\phi$ and use the formula $\cos 2\phi = 1 - 2\sin^2\theta$ to put things into the standard elliptic integral form.

19.4.3   Two Degree-of-Freedom Case

For two degrees of freedom, the integral

$$G_{ji} = 2\int_{E_j}^{E_{j+1}} \frac{E^{i-1}d\theta}{\left[\prod\limits_{n=1}^{2N+1}(E - E_n)\right]^{1/2}}, \quad 1 \le i \le N, \ \ 1 \le j \le 2N \tag{19.36}$$

explicitly becomes

$$G_{11} = 2\int_{E_1}^{E_2} \frac{dE}{\left[\prod\limits_{n=1}^{5}(E - E_n)\right]^{1/2}}$$

$$= 2\int_{E_1}^{E_2} \frac{dE}{[(E - E_1)(E_2 - E)(E_3 - E)(E_4 - E)(E_5 - E)]^{1/2}}$$

$$G_{21} = 2\int_{E_2}^{E_3} \frac{dE}{\left[\prod\limits_{n=1}^{5}(E - E_n)\right]^{1/2}}$$

$$= 2\int_{E_2}^{E_3} \frac{dE}{[(E - E_1)(E - E_2)(E_3 - E)(E_4 - E)(E_5 - E)]^{1/2}}$$

$$G_{31} = 2\int_{E_3}^{E_4} \frac{d\theta}{\left[\prod\limits_{n=1}^{5}(E - E_n)\right]^{1/2}}$$

$$= 2\int_{E_3}^{E_4} \frac{d\theta}{[(E - E_1)(E - E_2)(E - E_3)(E_4 - E)(E_5 - E)]^{1/2}}$$

$$G_{41} = 2\int_{E_4}^{E_5} \frac{dE}{\left[\prod\limits_{n=1}^{5}(E - E_n)\right]^{1/2}}$$

$$= 2\int_{E_4}^{E_5} \frac{dE}{[(E - E_1)(E - E_2)(E - E_3)(E - E_4)(E_5 - E)]^{1/2}}$$

$$G_{12} = 2 \int_{E_1}^{E_2} \frac{E \, dE}{\left[\prod_{n=1}^{5} (E - E_n) \right]^{1/2}}$$

$$= 2 \int_{E_1}^{E_2} \frac{E \, dE}{[(E - E_1)(E_2 - E)(E_3 - E)(E_4 - E)(E_5 - E)]^{1/2}}$$

$$G_{22} = 2 \int_{E_2}^{E_3} \frac{E \, dE}{\left[\prod_{n=1}^{5} (E - E_n) \right]^{1/2}}$$

$$= 2 \int_{E_2}^{E_3} \frac{E \, dE}{[(E - E_1)(E - E_2)(E_3 - E)(E_4 - E)(E_5 - E)]^{1/2}}$$

$$G_{32} = 2 \int_{E_3}^{E_4} \frac{E \, dE}{\left[\prod_{n=1}^{5} (E - E_n) \right]^{1/2}}$$

$$= 2 \int_{E_3}^{E_4} \frac{E \, dE}{[(E - E_1)(E - E_2)(E - E_3)(E_4 - E)(E_5 - E)]^{1/2}}$$

$$G_{42} = 2 \int_{E_4}^{E_5} \frac{E \, dE}{\left[\prod_{n=1}^{5} (E - E_n) \right]^{1/2}}$$

$$= 2 \int_{E_4}^{E_5} \frac{E \, dE}{[(E - E_1)(E - E_2)(E - E_3)(E - E_4)(E_5 - E)]^{1/2}}$$

Now apply the transformation to remove the singularities from the integral

$$G_{2i} = 2 \int_{E_2}^{E_3} \frac{E^{i-1} dE}{\left[\prod_{n=1}^{5} (E - E_n) \right]^{1/2}}$$

$$= 2 \int_{E_2}^{E_3} \frac{E^{i-1} dE}{[(E - E_1)(E - E_2)(E_3 - E)(E_4 - E)(E_5 - E)]^{1/2}} \qquad (19.37)$$

Take

$$E = A \cos \theta + B$$

where

$$A = \frac{E_3 - E_2}{2}, \quad B = \frac{E_3 + E_2}{2}$$

so that

$$dE = -\sqrt{(E - E_2)(E_3 - E)}\, d\theta$$

Hence

$$G_{2i} = -2 \int_\pi^0 \frac{E^{i-1}(\theta)\, d\theta}{[(E(\theta) - E_1)(E_4 - E(\theta))(E_5 - E(\theta))]^{1/2}}$$

$$G_{2i} = -2 \int_\pi^0 \frac{E^{i-1}(\theta)\, d\theta}{[(A\cos\theta + B - E_1)(E_4 - A\cos\theta - B)(E_5 - A\cos\theta - B)]^{1/2}}$$

$$G_{2i} = -2 \int_\pi^0 \frac{E^{i-1}(\theta)\, d\theta}{[(A\cos\theta + B - E_1)(A\cos\theta + B - E_4)(A\cos\theta + B - E_5)]^{1/2}}$$

Now let $\theta \to 2\phi$

$$G_{2i} = 4 \int_0^{\pi/2} \frac{E^{i-1}(2\phi)\, d\phi}{[(A(1 - 2\sin^2\phi) + B - E_1)(A(1 - 2\sin^2\phi) + B - E_4)(A(1 - 2\sin^2\phi) + B - E_5)]^{1/2}}$$

$$G_{2i} = 4 \int_0^{\pi/2} \frac{E^{i-1}(2\phi)\, d\phi}{[(A + B - E_1 - 2A\sin^2\phi)(A + B - E_4 - 2A\sin^2\phi)(A + B - E_5 - 2A\sin^2\phi)]^{1/2}}$$

$$G_{2i} = \frac{4}{\Delta} \int_0^{\pi/2} \frac{E^{i-1}(2\phi)\, d\phi}{[(1 - m_1\sin^2\phi)(1 - m_4\sin^2\phi)(1 - m_5\sin^2\phi)]^{1/2}}$$

where

$$\Delta = \sqrt{(A + B - E_1)(A + B - E_4)(A + B - E_5)}$$

and

$$m_1 = \frac{2A}{A + B - E_1} = \frac{E_3 - E_2}{E_3 - E_1}$$

$$m_4 = \frac{2A}{A + B - E_4} = \frac{E_3 - E_2}{E_3 - E_4}$$

$$m_5 = \frac{2A}{A + B - E_5} = \frac{E_3 - E_2}{E_3 - E_5}$$

Now look at the numerator of the integral above.

$$E(2\phi) = A \cos 2\phi + B - E_0, \ E_0 = 0$$

$$E(2\phi) = (A + B - E_0)\left[1 - \frac{2A}{A + B - E_0} \sin^2\phi\right] = E_3(1 - m_0 \sin^2\phi)$$

$$m_0 = \frac{2A}{A + B - E_0} = \frac{E_3 - E_2}{E_3 - E_0}$$

Finally

$$E(2\phi) = E_3(1 - m_0 \sin^2\phi)$$

$$m_0 = \frac{E_3 - E_2}{E_3}$$

And at long last

$$G_{2i} = \frac{4E_3^{i-1}}{\Delta} \int_0^{\pi/2} \frac{(1 - m_0 \sin^2\phi)^{i-1} d\phi}{[(1 - m_1 \sin^2\phi)(1 - m_4 \sin^2\phi)(1 - m_5 \sin^2\phi)]^{1/2}}$$

where

$$\Delta = \sqrt{(E_3 - E_1)(E_3 - E_4)(E_3 - E_5)}$$

and

$$m_0 = \frac{E_3 - E_2}{E_3}, \quad m_1 = \frac{E_3 - E_2}{E_3 - E_1}$$

$$m_4 = \frac{E_3 - E_2}{E_3 - E_4}, \quad m_5 = \frac{E_3 - E_2}{E_3 - E_5}$$

Now replace all the "ones" in the above integral by sine-squared plus cosine-squared:

$$G_{2i} = \frac{4E_3^{i-1}}{\Delta} \int_0^{\pi/2} \frac{[\cos^2\phi + (1 - m_0)\sin^2\phi]^{i-1}d\phi}{[[\cos^2\phi + (1 - m_1)\sin^2\phi][\cos^2\phi + (1 - m_4)\sin^2\phi][\cos^2\phi + (1 - m_5)\sin^2\phi]]^{1/2}}$$

Then introduce the usual parameters that appeared in the one degree-of-freedom case:

$$G_{2i} = \frac{4E_3^{i-1}}{\Delta} \int_0^{\pi/2} \frac{[a_0^2\cos^2\phi + b_0^2\sin^2\phi]^{i-1}d\phi}{[[a_1^2\cos^2\phi + b_1^2\sin^2\phi][a_4^2\cos^2\phi + b_4^2\sin^2\phi][a_5^2\cos^2\phi + b_5^2\sin^2\phi]]^{1/2}}$$

where

$$a_n = 1$$

$$b_n = \sqrt{1 - m_n}$$

This is suggestive of a Landen transformation that iterates on the angle ϕ until $a_N = b_N$ and the integral becomes:

$$G_{2i} = \frac{4E_3^{i-1}a_{0N}^{i-1}}{\Delta a_{1N}a_{4N}a_{5N}} K \int_0^{\phi'} d\phi = \frac{4KE_3^{i-1}a_{0N}^{i-1}}{\Delta a_{1N}a_{4N}a_{5N}} \phi'$$

where K is a numerical factor which accumulates for each iteration of the Landen transformation.

19.5   Improving Loop Integral Behavior

Consider the polynomial

$$P(E) = (E - E_1)(E - E_2)(E - E_3)$$

Far to the left of the three eigenvalues each factor is negative, rendering the polynomial negative as $E \to -\infty$. Far to the right each factor is positive and the polynomial is therefore positive as $E \to \infty$.

The above example for three eigenvalues is also useful for all *odd* polynomials, that is, a polynomial that goes negative to the left and positive to the right and oscillates in between. An even polynomial goes positive both to the left and to the right. For the inverse scattering transform, the polynomials are always odd. Therefore, all polynomials go negative to the left and positive to the right.

Now consider what happens for the loop integrand $E^{i-1}/\sqrt{|P(E)|}$. To illustrate this, consider the case for two degrees-of-freedom, so that:

$$P(E) = (E - E_1)(E - E_2)(E - E_3)(E - E_4)(E - E_5)$$

The loop integrals for this case are shown in Figure 19.3 and Section 19.6 relates these to the Riemann spectrum. The integrand for the case with two degrees-of-freedom case is shown for $i = 1$ in Figure 19.4. The singularities are easily seen at each of the five eigenvalues. These singularities are easily removed by a simple transformation (19.2). For example, the singularities are removed in Figure 19.5 for the first open band E_2, E_3. As a consequence, in this

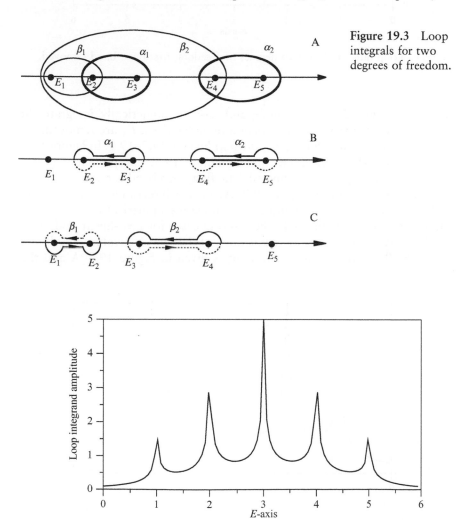

Figure 19.3 Loop integrals for two degrees of freedom.

Figure 19.4 Loop integrand for two degrees of freedom, no bands removed, $i = 1$.

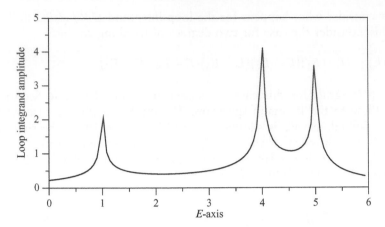

Figure 19.5 Loop integrand for two degrees of freedom, first open band removed, $i = 1$.

interval, the integrand is very simple and easy to numerically integrate. In Figure 19.6, the singularities in the second open band E_4, E_5 are removed.

The second case for which $i = 2$ is shown in Figures 19.7–19.9. In Figure 19.7 is the integrand, which resembles that in Figure 19.1, except that it grows slightly with increasing E as anticipated. The first open band is removed in Figure 19.8 and the second in Figure 19.9. Again the integrands over the band intervals are now quite smooth and easy to integrate numerically.

A different situation happens in the limiting cases for near-linear and near-soliton dynamics. The approximate positions of the five eigenvalues are shown in Figure 19.10 where the sine wave limit is given in Figure 19.10A and the

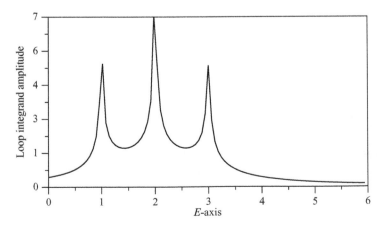

Figure 19.6 Loop integrand for two degrees of freedom, second open band removed, $i = 1$.

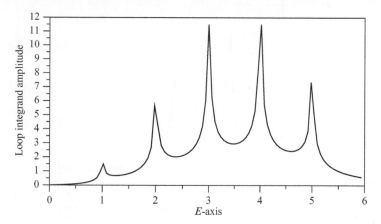

Figure 19.7 Loop integrand for two degrees of freedom, no open bands removed, $i = 2$.

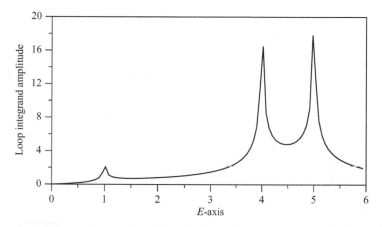

Figure 19.8 Loop integrand for two degrees of freedom, first open band removed, $i = 2$.

soliton limit is given in Figure 19.10B. First, consider the sine wave case, which is shown in Figure 19.11. Note that the first eigenvalue stands alone, while the second two are close together, as are the last two. In Figure 19.12, I have removed the first open band; note that the integrand is quite smooth in the open band (with positions 2, 2.2) and is simple to numerically integrate. Now in Figure 19.13, I remove the *gap* rather than the open band (2.2, 4.0); now the situation is not very good for numerical integration. Note that the integrand is no longer well behaved in (2.2, 4.0). Instead, the integrand becomes very large near the integration limits, which means that the numerical integrations require many more steps to for the integration than for a smoother integrand.

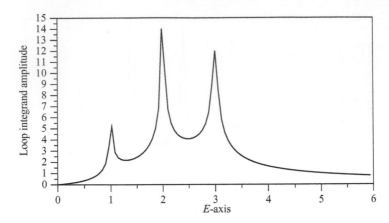

Figure 19.9   Loop integrand for two degrees of freedom, second open band removed, $i = 2$.

Figure 19.10   Linear (A) and Soliton (B) limits of the eigenvalue positions.

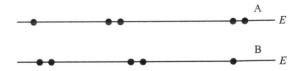

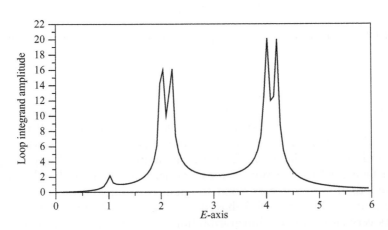

Figure 19.11   Loop integrand for two degrees of freedom, sine wave limit, no open bands removed, $i = 2$.

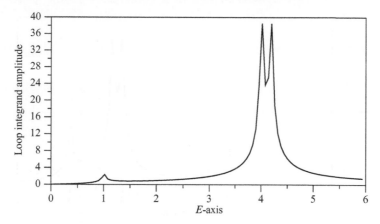

Figure 19.12 Loop integrand for two degrees of freedom, sine wave limit, first open band removed, $i = 2$.

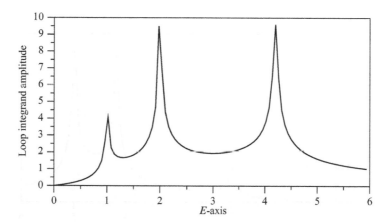

Figure 19.13 Loop integrand for two degrees of freedom, sine wave limit, second gap removed, $i = 2$.

This effect is much more difficult to handle the closer are the adjacent eigenvalues. In fact for, say 10^{-12} separation between adjacent "almost degenerate" eigenvalues, the computer time can go up by a factor of 1000 for the Romberg integrator in *Numerical Recipes* (Press et al., 1992). This example illustrates that one must remove the appropriate eigenvalues from the integrals before numerically integrating.

Finally, let me look at the soliton limit (Figure 19.10B). The integrand is graphed in Figure 19.14, where the closely spaced eigenvalues are easily discerned. In Figure 19.15, I remove the first open band (corresponding to the first

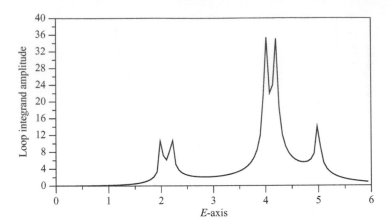

Figure 19.14 Loop integrand for two degrees of freedom, soliton limit, no open bands removed, $i = 2$.

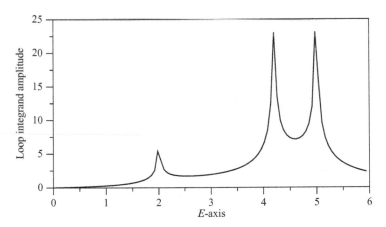

Figure 19.15 Loop integrand for two degrees of freedom, soliton limit, first open band removed, $i = 2$.

soliton in the spectrum). Note, however, that the adjacent eigenvalues (*outside* the open band) have near-singular behavior for the integrand over the interval of the open band. This is because of the close proximity of the first eigenvalue to the integration limit.

The issues just addressed for the sine wave and soliton limits (and experience with the numerical integrations for a wide variety of time series) suggest that some additional kind of transformation should be made for *both sides* of a band to improve the numerical convergence in general.

19.6   Constructing the Loop Integrals and Parameters of Periodic IST

The matrix G_{ji} (which is $2N \times N$) is the transpose of the matrix G_{ij} used in the computer program. The following results are compatible with this notation, $\mathbf{G} = \{G_{ij}\}, \mathbf{G}^{\mathrm{T}} = \{G_{ij}^{\mathrm{T}}\} = \{G_{ji}\}$.

The *hyperelliptic integral matrix of the first kind* is given by:

$$K_{ij} = G_{2i,j}^{\mathrm{T}}, \quad i = 1, \ldots, N, \quad j = 1, \ldots, N$$

The *hyperelliptic integral matrix of the second kind* is given by:

$$K'_{ij} = \sum_{k=1,3,5,\ldots}^{2i-1} G_{kj}$$

The *period matrix* $(N \times N)$ is given by

$$\mathbf{B} = 2\pi \mathbf{K}' \mathbf{K}^{-1}$$

where $\mathbf{K} = \{K_{ij}\}$ and $\mathbf{K}' = \{K'_{ij}\}$ are also $N \times N$.

The *wavenumbers* are given by

$$\kappa_i = 4\pi \left| K_{Ni}^{-1} \right|, \quad i = 1, \ldots, N$$

The *frequencies* are given by

$$\omega_i = c_o k_i + \beta v_i^3, \quad v_i^3 = 8 \left| K_{N-1,i}^{-1} \right| + 4 \left| K_{N,i}^{-1} \right| \sum_{k=1}^{2N+1} E_k$$

where $\kappa_i = 2\pi i / L$, L the period of the wave train.

Finally, the *phases* are given by:

$$\phi_i = 2\pi \sum_{m=1}^{N} \sigma_m E_{mi}, \quad \mathbf{E} = \mathbf{D}^{\mathrm{T}} \mathbf{K}^{-1}, \quad D_{ji} = 2 \int_{E_{2j}}^{\mu_j(0,0)} \frac{E^{i-1} dE}{|R(E)|^{1/2}}$$

where $\mathbf{D}^{\mathrm{T}} = \{D_{ij}\}$ and the σ_i are the Riemann sheet indices.

Given the above formulas for the parameters of the Riemann spectrum (the loop integrals) we can construct a periodic IST solution of the KdV equation in terms of an N dimensional θ-function:

$$\lambda \eta(x, t) = 2 \frac{\partial^2}{\partial x^2} \ln \theta_N(x, t)$$

where

$$\theta_N(x,t) = \sum_{m_1=-m_1}^{m_1} \sum_{m_2=-m_2}^{m_2} \cdots \sum_{m_N=-m_N}^{m_N}$$

$$\times \exp\left[i\sum_{k=1}^{N} m_k(\kappa_k x - \omega_k t + \phi_k) + \frac{1}{2}\sum_{j=1}^{N}\sum_{k=1}^{N} m_j m_k B_{jk}\right]$$

where the integers $[m_1, m_2, \ldots, m_N]$ are the numerical limits in the nested sums of the θ-function; methods for the computation of θ-functions are discussed in other chapters.

Figure 19.16 shows the matrix G_{ij}, $i = 1, \ldots, N$, $j = 1, 2, \ldots, N$ for a particular solution of the KdV equation. The Riemann spectrum ($\{B_{ij}, \kappa_i, \omega_i, \phi_i\}$, $i = 1, \ldots, N$, $j = 1, \ldots, N$) is extracted from this matrix using the results of this section. In the work of this chapter, the Riemann spectrum is computed for a particular

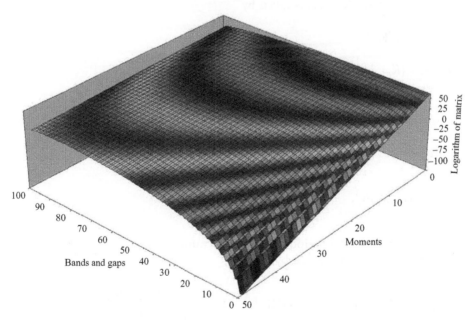

Figure 19.16 The logarithm of the matrix G_{ij}, $i = 1, 2, \ldots, N$, $j = 1, 2, \ldots, N$ for a particular case for the KdV equation. The i are referred to as "moments" in the figure and the j are referred to as "bands and gaps." The associated Riemann matrix is 50×50 for this case that is genus 50. Interpretation of this graph can be made using the results of Chapters 14 and 19 on loop integrals, Chapters 15 and 32 on Schottky uniformization, and Chapters 16 and 32 on the Nakamura-Boyd approach. In particular, the practiced eye will notice the power-law behavior of the off-diagonal elements of the Riemann matrix, an approximate, leading order result of all three approaches: $B_{ij} \simeq \ln[(\kappa_i - \kappa_j)/(\kappa_i + \kappa_j)]$. (See color plate).

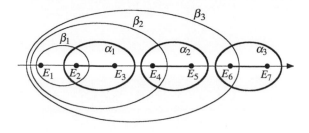

Figure 19.17 Cycles for oscillatory basis.

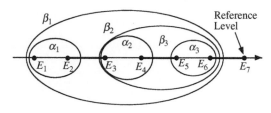

Figure 19.18 Cycles for soliton basis.

Cauchy initial conditions, that is, $\eta(x, 0)$. Then the *spectral problem* for the KdV equation (Chapter 17) is solved for the main spectrum E_j, $j = 1, 2, \dots, 2N + 1$ that is used to compute the matrix G_{ij}, and finally, from this the Riemann spectrum is then computed.

For aid in computing the loop integrals and for constructing the Riemann spectrum from the elements of the matrix G_{ij} I give the loop cycles for the three degree of freedom case for the sine wave case in Figure 19.17 and the soliton case in Figure 19.18.

20 Simple, Brute-Force Computation of Theta Functions and Beyond

20.1 Introduction

The focus of the present chapter is to address methods that are useful for the *brute-force numerical computation* of the *Riemann theta function*. These approaches provide insight into the computation of theta functions, *are easy to program*, but at the same time they are *slow to execute*. Additionally, I discuss the transition to a new method that is based upon "collapsing" the theta function onto the associated linear Fourier modes, thus allowing for *considerable speed improvement*. Finally, this chapter provides a variety of algorithms and notations for theta functions that are used elsewhere in this monograph.

The N-dimensional theta function has the form:

$$\theta_N(x,t) = \sum_{m_1=-\infty}^{\infty} \sum_{m_2=-\infty}^{\infty} \cdots \sum_{m_N=-\infty}^{\infty} \exp\left[i\sum_{j=1}^{N} m_j X_j - \frac{1}{2}\sum_{i=1}^{N}\sum_{j=1}^{N} m_i m_j B_{ij}\right],$$

(20.1)

where

$$X_j = k_j x - \omega_j t + \phi_j.$$

Computation of the parameters of the theta function $\{k_j, \omega_j, \phi_j, \mathbf{B}\}$ is discussed elsewhere (Chapters 14–16 and 23). Here, I assume these parameters to be known and seek to determine $\theta_N(x,t)$ for all x and t to some reasonable approximation. The minus sign in front of the double summation in the exponential of (20.1) means that the period matrix is assumed to be positive definite to ensure convergence of the theta function.

20.2 Brute-Force Method

We can consider the numerical evaluation of the theta functions in terms of a simple nested sum. The first step is to set limits on the infinite summations in Equation (20.1). We limit each sum to some finite value M_m:

$$\theta_N(x,t) = \sum_{m_1=-M_1}^{M_1} \sum_{m_2=-M_2}^{M_2} \cdots \sum_{m_N=-M_N}^{M_N} \exp\left[i \sum_{j=1}^{N} m_j X_j - \frac{1}{2} \sum_{i=1}^{N} \sum_{j=1}^{N} m_i m_j B_{ij}\right].$$

(20.2)

Some consideration must be made as to the selection of the M_m, in particular to ensure convergence of Equation (20.2) as discussed in detail in Chapter 22. How many exponential terms exp[...] must be evaluated in Equation (20.2)? This number has been evaluated in Chapter 8 to be

$$(2M+1)^N + 1,$$

where for the moment we take $M_m = M$. The number of complex double-precision operations in each term of Equation (20.2) grows exponentially with the number of degrees of freedom (N, the number of cnoidal waves, say, for the KdV equation). Let us suppose that $M = 4$ and $N = 1000$, then we have roughly $(2M+1)^N + 1 \sim 10^N = 10^{1000}$ exponentials in the partial theta sum (20.2)! On a gigaflop workstation we can do, say, 10^{10} operations in about an hour. On a teraflop computer we have about 10^{13} operations per hour. Even if we suppose that computers will become seven orders of magnitude faster in the near future, we are still talking about "only" 10^{20} operations per hour, vastly less than 10^{1000}, which would take more than 10^{970} Universal lifetimes to complete!

This simple example illustrates the enormous difficulty encountered in the practical computation of the Riemann theta function. Indeed the "brute-force method" considered in the present section provides only an initial look into these difficulties, which are discussed further below and in Chapters 8 and 22. Probably, the simplest FORTRAN code for the brute-force method would be to nest a few FORTRAN "do loops," with the exponential term computed within the inner loop. This approach is quite direct and provides a simple algorithm that is straightforward and simple to checkout, albeit for only a few degrees of freedom, that is, $N \sim 10$-20 for typical modern workstations. Note that the number of terms in the theta function is $\sim(2M+1)^N$, so that if we increase N by 1, say, then for $M = 4$ the algorithm is nine times slower! So if you are simulating $N = 20$ and need to go to $N = 25$ your brute-force algorithm will run $9^5 = 59,049$ times slower! This is the nature of exponential dependence on the number of degrees of freedom.

20.3   Vector Algorithm for the Theta Function

One of the difficulties with computing the theta functions is that for dynamical simulations one needs to compute them over a range of values in space, x, and time, t. If one were to treat the theta function as a subroutine, then one would need to call this subroutine for each value of space, x, and time, t. If we require

1000 spatial points and 1000 time points, then we would need 10^6 calls to a subroutine that would require (e.g., $M = 4$ and $N = 1000$) $\sim 10^{1000}$ operations! One way to avoid repeating many of the calculations in subsequent subroutine calls is to store the IST parameters in a *vector* and to recall them in subsequent calculations. This does indeed save large amounts of computer time, but results in a memory-bound code that requires $\sim 10^{1000}$ memory locations. Such huge amounts of memory are not possible with today's computers.

The vector procedure follows by noting that the theta function can be written in the following form:

$$\theta_N(x, t) = \sum_{j=-J}^{J} C_j e^{i(K_j x - \Omega_j t + \Phi_j)}, \quad C_j = e^{-(1/2)\mathbf{m}_j \cdot \mathbf{B} \mathbf{m}_j}, \tag{20.3}$$

where j is an ordering parameter associated with each vector $\mathbf{m} \equiv \mathbf{m}_j$ in the θ sum and $J = [(2M + 1)^N - 1]/2$. The operations in Equation (20.3) are carried out in such a way as to be equivalent to those in Equation (20.2). In addition to the C_j we have the following parameters to compute

$$K_j = \mathbf{m}_j \cdot \mathbf{k}, \quad \Omega_j = \mathbf{m}_j \cdot \boldsymbol{\omega}, \quad \Phi_j = \mathbf{m}_j \cdot \boldsymbol{\phi}. \tag{20.4}$$

For most applications the wavenumbers are assumed to be commensurable:

$$\mathbf{k} = [k_1, k_2, \ldots, k_N] = [1, 2, \ldots, N]\Delta k, \quad \Delta k = 2\pi/L,$$

where L is the length of the spatially periodic box in which the solution is assumed to evolve. Having commensurable wavenumbers means that the theta function is spatially periodic, $\theta(x, t) = \theta(x + L, t)$. Of course, the frequencies $\boldsymbol{\omega} = [\omega_1, \omega_2, \ldots, \omega_N]$ are *never* commensurable. Recall that in the linear limit for the KdV equation the dispersion relation is $\omega_j = c_0 k_j - \beta_j k_j^3$ and clearly commensurable frequencies cannot exist even in the linear limit. In these variables, the inverse scattering transform in the theta-function formulation given by Equations (20.3) and (20.4) resembles linear Fourier analysis, albeit with $(2M + 1)^N + 1$ terms.

How then do the wavenumbers K_j for periodic boundary conditions behave for the theta function given by Equation (20.3)? Clearly their formulation is different than those for the linear Fourier transform, that is,

$$K_j = \mathbf{m}_j \cdot \mathbf{k} = [m_1^j, m_2^j, \ldots, m_N^j] \cdot [1, 2, \ldots, N]\Delta k = \Delta k \sum_{n=1}^{N} nm_n^j. \tag{20.5}$$

Hence the theta-function wavenumbers, K_j, for the KdV equation are *integer multiples* I_j of Δk:

$$K_j = I_j \Delta k, \quad I_j = \sum_{n=1}^{N} nm_n^j \tag{20.6}$$

and are therefore *commensurable, often duplicated, and not ordered with the integers*. According to Equation (20.5), the wavenumbers in Equation (20.3) fall on the ordinary Fourier wavenumbers $k_n = 2\pi n/L$. In fact, an *infinite number of terms* in Equation (20.3) fall on each k_n. This leads to the idea of a histogram of occurrence for each k_n, as we see below.

As a numerical example, I graph I_j as a function of j in Figure 20.1 for $M = 2$, $N = 2$. The line denoted by the dots is precisely I_j versus j. Note that the integer pair (m_1, m_2) is shown near each dot in the plane. Note further that, in this simple case, I_j is just the inner product between (m_1, m_2) and $(1, 2)$ (the integer wavenumbers), that is, $I_j = m_1 + 2m_2$. The line denoted by the open boxes is the modulus $|m_1| + |m_2|$.

20.4 Theta Functions as Ordinary Fourier Series

Since all the wavenumbers in the θ sum fall on the k_n and since the wavenumbers are duplicated, we can write from Equation (20.3) (for the special case $t = 0$) (Osborne, 1995a,b) (see also Chapter 9):

$$\theta_N(x, 0) = \sum_{j=-J}^{J} C_j e^{i(K_j x + \Phi_j)} = \sum_{n=-\infty}^{\infty} \theta_n e^{ik_n x}, \qquad (20.7)$$

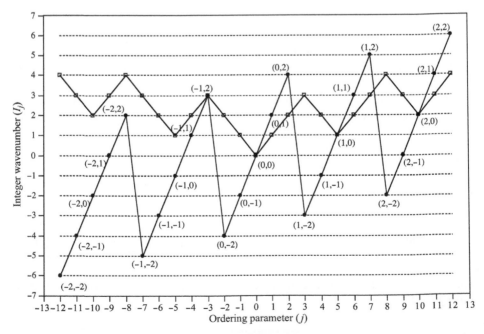

Figure 20.1 The integer wavenumber I_j versus j for the case $M = 2$, $N = 2$.

where

$$\theta_n = \sum_{\substack{\text{sum over subset} \\ \text{of } j \text{ for which } I_j = n}} C_j e^{i\Phi_j}, \quad -\infty \leq n \leq \infty. \tag{20.8}$$

It is worthwhile discussing the two distinct forms of Equation (20.7). The summation on the left is just a natural extension of the partial theta sum given by Equation (20.3). However, the second summation in Equation (20.7) is just an ordinary Fourier transform. The Fourier coefficients, θ_n, in Equation (20.7) are given by the series (20.8); the θ_n might be referred to as kinds of "theta constants." Theta constants are normally taken to be the constants that arise when we set $x = t = 0$ in the theta function. For Equation (20.1), we have

$$\theta_N(0,0) = \sum_{m_1=-\infty}^{\infty} \sum_{m_2=-\infty}^{\infty} \cdots \sum_{m_N=-\infty}^{\infty} \exp\left[i\sum_{j=1}^{N} m_j \phi_j - \frac{1}{2}\sum_{i=1}^{N}\sum_{j=1}^{N} m_i m_j B_{ij}\right] \tag{20.1'}$$

and for Equation (20.7) we get

$$\theta_N(0,0) = \sum_{j=-J}^{J} C_j e^{i\Phi_j} = \sum_{n=-\infty}^{\infty} \theta_n. \tag{20.7'}$$

We thus see that the "theta constants" given by Equation (20.8) are just subsets of the full theta constants given in Equation (20.7'), that is, each of the *ordinary linear Fourier coefficients* θ_n are just the sum of a *subset* of the terms in Equation (20.7') which *correspond to the wavenumbers* $k_n = 2\pi n/L$. One can thus use *set theory notation* to write Equation (20.8) (in this expression I have also included time dependence):

$$\theta_n(t) = \sum_{\{j\in\mathbb{Z}:I_j=n\}} q_j e^{-i\Omega_j t + i\Phi_j}, \quad -J \leq j \leq J \tag{20.8'}$$

Of course, Equations (20.8) and (20.8') are equivalent provide we set $t = 0$ in (20.8'). The notation in Equation (20.8') just means pick a value of n and sum over j; for each j compute I_j; if $I_j = n$ then add the term to the sum (20.8'). More details on the linear Fourier decomposition for theta functions (20.7), (20.8), or (20.8') are given in Chapter 9.

Figure 20.1 is useful for simultaneously understanding the *partial theta summation* (20.2), the *vector theta summation* (20.3), and the *ordinary linear Fourier summation of the theta function* (20.7). The case considered in the Figure 20.1 is for $M = 2$, $N = 2$, for which $J = [(2M+1)^N - 1]/2 = 12$; note that the ordering parameter j is shown on the abscissa of Figure 20.1.

Note the oscillating line segments punctuated by solid dots; these dots represent each of the terms in the partial theta summation (20.2); the dots are graphed on the ordinate as I_j (20.6). Next to each of these dots is a pair of numbers that represents the integers in the summation of the theta function: (m_1, m_2). Thus, the graph simultaneously represents the complex exponentials in Equation (20.2) or Equation (20.3) and makes clear that both methods of summation are equivalent.

Let us see how to compute the ordinary, linear Fourier coefficients of the theta function, θ_n, (20.8), (20.8') again using Figure 20.1. First, interpret the ordinate as representing the commensurable wavenumbers, $k_n = 2\pi n/L$. Note the horizontal dotted lines in Figure 20.1; these correspond to a particular wavenumber and intercept the function I_j as it oscillates as a function of j; each of the solid dots represents an exponential term in the partial theta summation (20.2). When the dotted line corresponding to a particular wavenumber, $k_n = 2\pi n/L$, intercepts I_j we have a contribution to the summation (20.8'). In general, several interceptions (say N_n of them) of k_n and I_j means there are N_n terms in the group theoretic summation in Equation (20.8'). This is how partial theta functions reduce to linear Fourier analysis! I call this process "collapsing the theta function onto it linear Fourier modes." Of course in this case we have commensurable wavenumbers and spatial periodicity in the theta functions. The collapse of theta-function modes on to linear Fourier modes has a great advantage for computing theta functions, but one must remember that after the collapsing process we have formally lost the *inverse scattering Riemann spectrum*. To recover this information from the linear Fourier modes, one must introduce the inverse problem (see Chapters 7, 8, 9, 11, 14–17, 19, and 23). More details on the collapsing process are given in Chapter 21.

Nota Bene. Now we see the motivation for adding "and beyond" to the title of this chapter. Introducing the idea of a *partial theta summation* (20.2) and its *vector representation* (20.3) has led us to the *ordinary Fourier representation* for theta functions (20.7), (20.8), and (20.8'). I will now show how to leap from a brute-force algorithm to a much faster algorithm using the ordinary Fourier representation for theta functions (20.8'). Consider the partial theta function (20.2) to represent a brute-force subroutine. To evaluate a space-time field we need to call the subroutine $N_x \times N_t$ times, where N_x is the number of discrete values on the spatial domain x and N_t is the number of values of the temporal variable t. Suppose $N_x = 1000$ and $N_t = 1000$. Then one must call the partial theta-function brute-force routine $N_x \times N_t = 10^6$ times. Instead of carrying out this brute-force procedure, just collapse the theta functions onto the linear Fourier modes via Equation (20.8'). Once you have the Fourier modes, θ_n, call the fast Fourier transform (FFT) over the spatial variable $N_t = 1000$ times instead of calling the theta-function routine $N_x \times N_t = 10^6$ times! The collapse process (20.8') to determine the Fourier coefficients θ_n takes about as much computer time as a single call to the brute-force theta-function routine. In consequence, we have an algorithm that is about $N_x \times N_t = 10^6$ time faster than the brute-force method! This is a good place to start to improve execution speeds for theta functions.

Now let us investigate other forms for the ordinary Fourier representation for the theta function:

$$
\begin{aligned}
\theta_N(x, 0) &= \sum_{n=-\infty}^{\infty} \theta_n e^{ik_n x} = a_0 + 2 \sum_{n=1}^{\infty} d_n \cos (k_n x + \phi_n) \\
&= a_0 + 2 \sum_{n=1}^{\infty} a_n \cos (k_n x) + b_n \sin (k_n x)
\end{aligned}
\tag{20.9}
$$

where we have used the fact that $\theta_{-n} = \theta_n^*$ (to ensure that $\theta_N(x, 0)$ is real for the KdV equation) so that

$$
\begin{aligned}
&d_n = d_{-n} \quad \text{and} \quad \phi_{-n} = -\phi_n, \\
&\theta_0 = d_0 \quad \text{and} \quad \theta_n = d_n e^{i\phi_n} = a_n - i b_n, \quad \theta_{-n} = \theta_n^* = d_n e^{-i\phi_n} = a_n + i b_n \\
&d_n = \sqrt{a_n^2 + b_n^2} \quad \text{and} \quad \tan \phi_n = -b_n / a_n, \\
&a_n = d_n \cos \phi_n \quad \text{and} \quad b_n = -d_n \sin \phi_n.
\end{aligned}
\tag{20.10}
$$

The coefficients are given by

$$
\theta_n = \sum_{\substack{\text{sum over subset} \\ \text{of } j \text{ for which } I_j = n}} C_j e^{i\Phi_j}, \quad -\infty \leq n \leq \infty
\tag{20.11}
$$

and

$$
a_n = \sum_{\substack{\text{sum over subset} \\ \text{of } j \text{ for which } I_j = n}} C_j \cos (\Phi_j), \quad -\infty \leq n \leq \infty,
$$

$$
b_n = -\sum_{\substack{\text{sum over subset} \\ \text{of } j \text{ for which } I_j = n}} C_j \sin (\Phi_j), \quad -\infty \leq n \leq \infty,
\tag{20.12}
$$

where

$$
C_j = e^{-(1/2) \mathbf{m}_j \cdot \mathbf{B} \mathbf{m}_j}, \quad \Phi_j = \mathbf{m}_j \cdot \boldsymbol{\phi}.
\tag{20.13}
$$

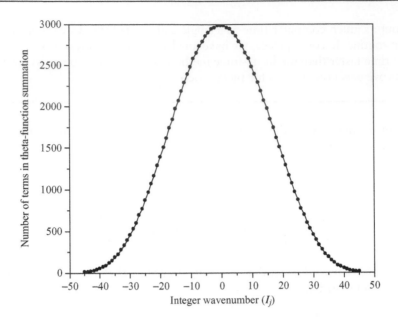

Figure 20.2   Histogram of occurrence of terms in the partial theta summation for $M = 3$, $N = 5$.

We have therefore reduced the theta function to an ordinary Fourier transform with coefficients a_n and b_n as given above in terms of inverse scattering transform variables. Equation (20.12) is the method of Osborne (1995a,b).

   Now let us look at a case that provides information about the statistical behavior of a theta summation. Consider $M = 3$, $N = 5$ for which

$$J_{max} = \mathbf{M} \cdot \mathbf{n} = [3,3,3,3,3] \cdot [1,2,3,4,5] = 45.$$

Here $n = [1,2,3,4,5]$ is the integer wavenumber vector.

A simulation has been run to construct the histogram seen in Figure 20.2, which appears to be approximated by a binomial distribution. Note that for low wavenumbers near zero there are about 3000 terms in each of the series (20.8′). Higher wavenumbers have many fewer terms ranging out to the values $J_{max} = \pm 45$ in the tail of the distribution. The total number of exponential terms in the theta function is the sum of the number of occurances in the histogram, $(2M + 1)^N = 16807$.

20.5   A Memory-Bound Brute-Force Method

We can develop another brute-force method based upon Equation (20.3) (Osborne, 1995a,b):

$$\theta_N(x, t) = \sum_{j=-J}^{J} C_j e^{i(K_j x - \Omega_j t + \Phi_j)}, \quad C_j = e^{-(1/2)\mathbf{m}_j \cdot \mathbf{Bm}_j}, \tag{20.14}$$

where, as before, $J = [(2M + 1)^N - 1]/2$. In this approach, one saves the wavenumbers, K_j, frequencies, Ω_j, phases, Φ_j, and amplitudes, C_j, for $t = 0$. Then, for each succeeding value of time, Equation (20.14) is executed without the need to recompute the C_j, K_j, Ω_j, and Φ_j. Thus, the algorithm is quite a lot faster than the algorithm of Section 20.3, but the amount of storage for the various vectors is quite prohibitive because the integer J is rather large for a large number of degrees of freedom.

20.6   Poisson Series for Theta Functions

As seen in Chapter 8, we can also write the theta functions as a sum of Gaussians. The mathematical construction is known as a Poisson or Gaussian series in N dimensions:

$$\theta(\mathbf{X}) = \sum_{\mathbf{m} \in \mathbb{Z}} \exp\left[-\frac{1}{2}\boldsymbol{\chi}^T \mathbf{F}\boldsymbol{\chi}\right],$$

where

$$\boldsymbol{\chi} = \mathbf{X} - \pi\mathbf{m}, \quad \mathbf{X} = [X_1, X_2, \ldots, X_N], \quad \mathbf{m} = [m_1, m_2, \ldots, m_N],$$

$$\chi_i = X_i - \pi m_i, \quad X_i = k_i x - \omega_i t + \phi_i, \quad \mathbf{F} = \mathbf{B}^{-1}.$$

The Poisson series can be programmed as a brute-force algorithm, that is, summed out to a sufficient number of terms as a partial sum. The series is rapidly convergent near the soliton limit, but only slowly convergent near the linear limit. In the small-amplitude, linear limit the slow convergence occurs because sine waves need to be constructed from a periodic sequence of Gaussians.

20.7   Decomposition of Space Series into Cnoidal Wave Modes

In Chapter 8 (Osborne, 1995a,b), I discuss how the solutions to the KdV equation can be written as a sum over cnoidal waves plus nonlinear interactions:

$$\eta(x, 0) = \sum_{n=1}^{N} \eta_n cn^2[k_n x + \phi_n | m_n] + \eta_{int}(x, 0).$$

A simple numerical example is shown in Figure 20.3, where I have taken $N = 10$ so that there are 10 cnoidal waves in the spectrum. The cnoidal waves are nonlinear and have a shape that varies from simple sine waves to Stokes

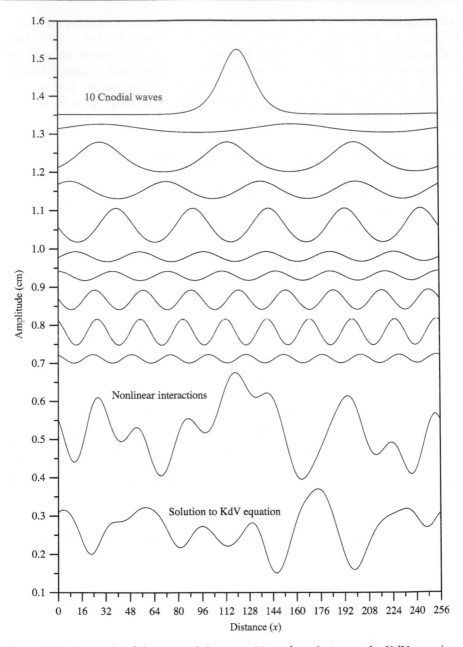

Figure 20.3 Example of the spectral decomposition of a solution to the KdV equation. Shown are the 10 cnoidal waves in the spectrum together with the nonlinear interactions and the actual waveform solution of the KdV equation.

waves to solitons, depending on the modulus m_n that can vary between 0 and 1. In the example shown we see that the cnoidal waves have all of these types of waveforms. Note also that the number of oscillations in each of the 10 modes is the same as in linear Fourier analysis, that is, n, the index of the spatial wavenumbers, $k_n = 2\pi n/L$, $n = 1, 2, \ldots, N$. The waves in this case are in the *oscillatory basis*. The nonlinear interactions are also shown. The sum of the cnoidal waves plus the nonlinear interactions gives the wave train $\eta(x, 0)$ of the KdV equation, that is, it contains the *Riemann matrix and phases* that were input to the program. A simple trick to estimate if nonlinearity is important for any of the cnoidal waves is to turn the figure upside down. Wave trains which are obviously up-down asymmetric are Stokes waves or solitons.

21 The Discrete Riemann Theta Function

21.1 Introduction

It is safe to say that most modern numerical applications of the linear Fourier transform have occurred due to the development of the *fast Fourier transform* (FFT), an approach which was first developed by Gauss and then rediscovered several times over the last couple of centuries by many researchers (Press et al., 1992). This approach begins with the so-called *discrete Fourier transform* (*DFT*). I discuss the DFT and its numerical implementation as the FFT. Many of these ideas are then applied to the development of the *discrete multidimensional Fourier transform* (DMFT). Finally, applications of the approach to the *discrete theta function* (DTF) are given. The DTF is the basis of many of the applications in this book.

21.2 Discrete Fourier Transform

It is instructive to, at first review, some of the vast body of work on the *discrete and fast Fourier transforms*. The results will be quite useful in subsequent sections on the discrete and fast theta-function. The FFT algorithm first begins with a definition of the *DFT* (Cooley and Tukey, 1967):

$$y_j = \sum_{k=0}^{N-1} x_k e^{-2\pi i j k/N}, \quad j = 0, 1, 2, \ldots, N-1 \tag{21.1}$$

The *inverse DFT* is then given by

$$x_k = \frac{1}{N} \sum_{j=0}^{N-1} y_j e^{2\pi i j k/N}, \quad k = 0, 1, 2, \ldots, N-1 \tag{21.2}$$

where $i = \sqrt{-1}$. These expressions may be written in matrix form:

$$y_j = \sum_{k=0}^{N-1} W_{jk} x_k, \quad W_{jk} = e^{-2\pi i j k/N} = (e^{-2\pi i/N})^{jk} = w^{jk} \tag{21.3}$$

where

$$w = e^{-2\pi i/N}$$

Thus, we can write the *DFT*:

$$Y = WX \tag{21.4}$$

where X, Y are column vectors of the input and outputs of the Fourier transform, respectively, and $W = \{W_{jk}\}$ is the matrix defined above.

The *inverse DFT* can formally be found by

$$X = W^{-1}Y \tag{21.5}$$

where W^{-1} is the inverse of the matrix W. To find the inverse note that the matrix W has the following property:

$$W^{*T}W = NI$$

where I is the unit matrix. This means that W, aside from the scale factor $\sqrt{N}$, is a *unitary matrix*, the latter of which commonly appears in the field of quantum mechanics. It is then clear that the inverse of W is given by

$$W^{-1} = \frac{1}{N}W^{*T} \tag{21.6}$$

This expression simplifies further by noticing that $W^T = W$, that is, W is also symmetric.

Several numerical examples are now considered.

Example 1: Consider $N = 2^1 = 2$:

$$\begin{bmatrix} y_0 \\ y_1 \end{bmatrix} = \begin{bmatrix} w^0 & w^0 \\ w^0 & w^1 \end{bmatrix} \begin{bmatrix} x_0 \\ x_1 \end{bmatrix} = \begin{bmatrix} 1 & 1 \\ 1 & w \end{bmatrix} \begin{bmatrix} x_0 \\ x_1 \end{bmatrix}$$

Example 2: Consider $N = 2^2 = 4$:

$$\begin{bmatrix} y_0 \\ y_1 \\ y_2 \\ y_3 \end{bmatrix} = \begin{bmatrix} w^0 & w^0 & w^0 & w^0 \\ w^0 & w^1 & w^2 & w^3 \\ w^0 & w^2 & w^4 & w^6 \\ w^0 & w^3 & w^6 & w^9 \end{bmatrix} \begin{bmatrix} x_0 \\ x_1 \\ x_2 \\ x_3 \end{bmatrix} = \begin{bmatrix} 1 & 1 & 1 & 1 \\ 1 & w & w^2 & w^3 \\ 1 & w^2 & 1 & w^2 \\ 1 & w^3 & w^2 & w \end{bmatrix} \begin{bmatrix} x_0 \\ x_1 \\ x_2 \\ x_3 \end{bmatrix}$$

where the following formula has been used for reducing the powers:

$$w^{jk} = w^{jk - N[jk/N]}$$

In this latter expression the square brackets mean "integer part of." Here is a simple example: For $N = 2^2 = 4$, consider the case $w^{(j=2)(k=3)} = w^6 = w^{6-4[6/4]} = w^2$, since the integer part of 6/4, [6/4], is 1.

Example 3: Consider $N = 2^3 = 8$. We have

$$
\begin{bmatrix} y_0 \\ y_1 \\ y_2 \\ y_3 \\ y_4 \\ y_5 \\ y_6 \\ y_7 \end{bmatrix}
=
\begin{bmatrix}
w^0 & w^0 & w^0 & w^0 & w^0 & w^0 & w^0 & w^0 \\
w^0 & w^1 & w^2 & w^3 & w^4 & w^5 & w^6 & w^7 \\
w^0 & w^2 & w^4 & w^6 & w^8 & w^{10} & w^{12} & w^{14} \\
w^0 & w^3 & w^6 & w^9 & w^{12} & w^{15} & w^{18} & w^{21} \\
w^0 & w^4 & w^8 & w^{12} & w^{16} & w^{20} & w^{24} & w^{28} \\
w^0 & w^5 & w^{10} & w^{15} & w^{20} & w^{25} & w^{30} & w^{35} \\
w^0 & w^6 & w^{12} & w^{18} & w^{24} & w^{30} & w^{36} & w^{42} \\
w^0 & w^7 & w^{14} & w^{21} & w^{28} & w^{35} & w^{42} & w^{49}
\end{bmatrix}
\begin{bmatrix} x_0 \\ x_1 \\ x_2 \\ x_3 \\ x_4 \\ x_5 \\ x_6 \\ x_7 \end{bmatrix}
$$

This reduces to

$$
=
\begin{bmatrix}
1 & 1 & 1 & 1 & 1 & 1 & 1 & 1 \\
1 & w^1 & w^2 & w^3 & w^4 & w^5 & w^6 & w^7 \\
1 & w^2 & w^4 & w^6 & 1 & w^2 & w^4 & w^6 \\
1 & w^3 & w^6 & w & w^4 & w^7 & w^2 & w^5 \\
1 & w^4 & 1 & w^4 & 1 & w^4 & 1 & w^4 \\
1 & w^5 & w^2 & w^7 & w^4 & w & w^6 & w^3 \\
1 & w^6 & w^4 & w^2 & 1 & w^6 & w^4 & w^2 \\
1 & w^7 & w^6 & w^5 & w^4 & w^3 & w^2 & w
\end{bmatrix}
\begin{bmatrix} x_0 \\ x_1 \\ x_2 \\ x_3 \\ x_4 \\ x_5 \\ x_6 \\ x_7 \end{bmatrix}
$$

Consider the particular matrix element corresponding to $j = 3$, $k = 3$ and in this case $w^{(j=3)(k=3)} = w^9 = w^{9-8[9/8]} = w^{9-8} = w$, since the integer part of 9/8, [9/8], is 1. From the above examples, it is clear that there is considerable redundancy in the computation of the Fourier transform.

This redundancy and other properties lead to a unique factorization of the matrix **W** that can be used to develop the *FFT* algorithm. We note that the matrix **W** can be written for the case $N = 2^n$ (when the length of a space or time series is a power of 2):

$$
\mathbf{W} = \left[\prod_{i=1}^{n} \mathbf{S}_i \right] \mathbf{P}
$$

where the $\mathbf{S}_i$ are *sparse matrices* and $\mathbf{P}$ is a *permutation matrix*. Below we develop formulas for computing these matrices, but first we give two concrete examples to illustrate the method.

Example 2 continued: Consider $N = 2^2 = 4$, where the sparse matrices and permutation matrix can be written:

$$
\mathbf{W} = S_1 S_2 P = \begin{bmatrix} 1 & 0 & 1 & 0 \\ 0 & 1 & 0 & 1 \\ w^0 & 0 & w^2 & 0 \\ 0 & w^0 & 0 & w^2 \end{bmatrix} \begin{bmatrix} 1 & 1 & 0 & 0 \\ w^0 & w^2 & 0 & 0 \\ 0 & 0 & 1 & 1 \\ 0 & 0 & w^1 & w^3 \end{bmatrix} \begin{bmatrix} 1 & 0 & 0 & 0 \\ 0 & 0 & 1 & 0 \\ 0 & 1 & 0 & 0 \\ 0 & 0 & 0 & 1 \end{bmatrix} = \begin{bmatrix} 1 & 1 & 1 & 1 \\ 1 & w & w^2 & w^3 \\ 1 & w^2 & 1 & w^2 \\ 1 & w^3 & w^2 & w \end{bmatrix}
$$

This verifies that decomposing the matrix **W** into sparse matrices and a permutation matrix does indeed give back the required **W** matrix, as required.

A direct calculation of the DFT gives

$$
Y = \mathbf{W}X = \begin{bmatrix} 1 & 1 & 1 & 1 \\ 1 & w & w^2 & w^3 \\ 1 & w^2 & 1 & w^2 \\ 1 & w^3 & w^2 & w \end{bmatrix} \begin{bmatrix} x_0 \\ x_1 \\ x_2 \\ x_3 \end{bmatrix} = \begin{bmatrix} w^0 x_0 + w^0 x_1 + w^0 x_2 + w^0 x_3 \\ w^0 x_0 + w^2 x_2 + w^1 x_1 + w^3 x_3 \\ w^0 x_0 + w^2 x_1 + w^0 x_2 + w^2 x_3 \\ w^0 x_0 + w^3 x_1 + w^2 x_2 + w x_3 \end{bmatrix}
$$

This matrix-vector multiplication requires N^2 multiply-add operations. Now let us count the number of operations required for the factorized case of the FFT in terms of the sparse and permutation matrices:

$$Y = \mathbf{W}X = S_1 S_2 P X$$

$$
= \begin{bmatrix} 1 & 0 & 1 & 0 \\ 0 & 1 & 0 & 1 \\ w^0 & 0 & w^2 & 0 \\ 0 & w^0 & 0 & w^2 \end{bmatrix} \begin{bmatrix} 1 & 1 & 0 & 0 \\ w^0 & w^2 & 0 & 0 \\ 0 & 0 & 1 & 1 \\ 0 & 0 & w^1 & w^3 \end{bmatrix} \begin{bmatrix} 1 & 0 & 0 & 0 \\ 0 & 0 & 1 & 0 \\ 0 & 1 & 0 & 0 \\ 0 & 0 & 0 & 1 \end{bmatrix} \begin{bmatrix} x_0 \\ x_1 \\ x_2 \\ x_3 \end{bmatrix}
$$

$$
Y = \mathbf{W}X = S_1 S_2 P X = \begin{bmatrix} 1 & 0 & 1 & 0 \\ 0 & 1 & 0 & 1 \\ w^0 & 0 & w^2 & 0 \\ 0 & w^0 & 0 & w^2 \end{bmatrix} \begin{bmatrix} 1 & 1 & 0 & 0 \\ w^0 & w^2 & 0 & 0 \\ 0 & 0 & 1 & 1 \\ 0 & 0 & w^1 & w^3 \end{bmatrix} \begin{bmatrix} x_0 \\ x_2 \\ x_1 \\ x_3 \end{bmatrix}
$$

$$
Y = \mathbf{W}X = S_1 S_2 P X = \begin{bmatrix} 1 & 0 & 1 & 0 \\ 0 & 1 & 0 & 1 \\ w^0 & 0 & w^2 & 0 \\ 0 & w^0 & 0 & w^2 \end{bmatrix} \begin{bmatrix} x_0 + x_2 \\ w^0 x_0 + w^2 x_2 \\ x_1 + x_3 \\ w^1 x_1 + w^3 x_3 \end{bmatrix}
$$

To compute the vector result on the above right-hand side required four multiplies and four adds. Call this latter vector **Z**:

$$
Z = S_2 P X = \begin{bmatrix} z_0 \\ z_1 \\ z_2 \\ z_3 \end{bmatrix} = \begin{bmatrix} x_0 + x_2 \\ w^0 x_0 + w^2 x_2 \\ x_1 + x_3 \\ w^1 x_1 + w^3 x_3 \end{bmatrix}
$$

Then we have

$$\mathbf{Y} = \mathbf{W}\mathbf{X} = \mathbf{S}_1\mathbf{S}_2\mathbf{P}\mathbf{X} = \begin{bmatrix} 1 & 0 & 1 & 0 \\ 0 & 1 & 0 & 1 \\ w^0 & 0 & w^2 & 0 \\ 0 & w^0 & 0 & w^2 \end{bmatrix} \begin{bmatrix} z_0 \\ z_1 \\ z_2 \\ z_3 \end{bmatrix} = \begin{bmatrix} z_0 + z_2 \\ z_1 + z_3 \\ w^0 z_0 + w^2 z_2 \\ w^0 z_1 + w^2 z_3 \end{bmatrix}$$

Thus, the second matrix multiply also takes four computer multiplies and four adds. The final result is

$$\mathbf{Y} = \mathbf{W}\mathbf{X} = \mathbf{S}_1\mathbf{S}_2\mathbf{P}\mathbf{X} = \begin{bmatrix} x_0 + x_2 + x_1 + x_3 \\ w^0 x_0 + w^2 x_2 + w^1 x_1 + w^3 x_3 \\ w^0(x_0 + x_2) + w^2(x_1 + x_3) \\ w^0(w^0 x_0 + w^2 x_2) + w^2(w^1 x_1 + w^3 x_3) \end{bmatrix}$$

This is in agreement with the DFT discussed above, as required.

Thus, apart from the permutation matrix that requires no multiplies or adds, the DFT calculation illustrated above in terms of sparse matrices, requires $N \ln_2 N = 4 \times 2 = 8$ ($N = 2^2$) multiply-and-add operations. This compares to $N^2 = 4 \times 4 = 16$ multiply-and-add operations for the full DFT computation (matrix times a vector). We thus have a factor $16/8 = 2$ improvement in computer time.

Example 3 continued: Consider $N = 2^3 = 8$, where the matrix $\mathbf{W}$ is given by

$$\mathbf{W} = \mathbf{S}_1\mathbf{S}_2\mathbf{S}_3\mathbf{P} =$$

$$\begin{bmatrix} 1 & 0 & 0 & 0 & 1 & 0 & 0 & 0 \\ 0 & 1 & 0 & 0 & 0 & 1 & 0 & 0 \\ 0 & 0 & 1 & 0 & 0 & 0 & 1 & 0 \\ 0 & 0 & 0 & 1 & 0 & 0 & 0 & 1 \\ w^0 & 0 & 0 & 0 & w^4 & 0 & 0 & 0 \\ 0 & w^0 & 0 & 0 & 0 & w^4 & 0 & 0 \\ 0 & 0 & w^0 & 0 & 0 & 0 & w^4 & 0 \\ 0 & 0 & 0 & w^0 & 0 & 0 & 0 & w^4 \end{bmatrix} \begin{bmatrix} 1 & 0 & 1 & 0 & 0 & 0 & 0 & 0 \\ 0 & 1 & 0 & 1 & 0 & 0 & 0 & 0 \\ w^0 & 0 & w^4 & 0 & 0 & 0 & 0 & 0 \\ 0 & w^0 & 0 & w^4 & 0 & 0 & 0 & 0 \\ 0 & 0 & 0 & 0 & 1 & 0 & 1 & 0 \\ 0 & 0 & 0 & 0 & 0 & 1 & 0 & 1 \\ 0 & 0 & 0 & 0 & w^2 & 0 & w^6 & 0 \\ 0 & 0 & 0 & 0 & 0 & w^2 & 0 & w^6 \end{bmatrix}$$

$$\times \begin{bmatrix} 1 & 1 & 0 & 0 & 0 & 0 & 0 & 0 \\ w^0 & w^4 & 0 & 0 & 0 & 0 & 0 & 0 \\ 0 & 0 & 1 & 1 & 0 & 0 & 0 & 0 \\ 0 & 0 & w^2 & w^6 & 0 & 0 & 0 & 0 \\ 0 & 0 & 0 & 0 & 1 & 1 & 0 & 0 \\ 0 & 0 & 0 & 0 & w^1 & w^5 & 0 & 0 \\ 0 & 0 & 0 & 0 & 0 & 0 & 1 & 1 \\ 0 & 0 & 0 & 0 & 0 & 0 & w^3 & w^7 \end{bmatrix} \begin{bmatrix} 1 & 0 & 0 & 0 & 0 & 0 & 0 & 0 \\ 0 & 0 & 0 & 0 & 1 & 0 & 0 & 0 \\ 0 & 0 & 1 & 0 & 0 & 0 & 0 & 0 \\ 0 & 0 & 0 & 0 & 0 & 0 & 1 & 0 \\ 0 & 1 & 0 & 0 & 0 & 0 & 0 & 0 \\ 0 & 0 & 0 & 0 & 0 & 1 & 0 & 0 \\ 0 & 0 & 0 & 1 & 0 & 0 & 0 & 0 \\ 0 & 0 & 0 & 0 & 0 & 0 & 0 & 1 \end{bmatrix}$$

which as expected reduces to the original matrix **W**:

$$
\mathbf{W} = S_1 S_2 S_3 P = \begin{bmatrix}
1 & 1 & 1 & 1 & 1 & 1 & 1 & 1 \\
1 & w^1 & w^2 & w^3 & w^4 & w^5 & w^6 & w^7 \\
1 & w^2 & w^4 & w^6 & 1 & w^2 & w^4 & w^6 \\
1 & w^3 & w^6 & w & w^4 & w^7 & w^2 & w^5 \\
1 & w^4 & 1 & w^4 & 1 & w^4 & 1 & w^4 \\
1 & w^5 & w^2 & w^7 & w^4 & w & w^6 & w^3 \\
1 & w^6 & w^4 & w^2 & 1 & w^6 & w^4 & w^2 \\
1 & w^7 & w^6 & w^5 & w^4 & w^3 & w^2 & w
\end{bmatrix}
$$

Now let us compute the FFT in terms of sparse and permutation matrices:

$$
Y = \mathbf{W} X = S_1 S_2 S_3 P X
$$

Consider the partial product $S_3 P X$:

$$
S_3 P X = \begin{bmatrix}
1 & 1 & 0 & 0 & 0 & 0 & 0 & 0 \\
w^0 & w^4 & 0 & 0 & 0 & 0 & 0 & 0 \\
0 & 0 & 1 & 1 & 0 & 0 & 0 & 0 \\
0 & 0 & w^2 & w^6 & 0 & 0 & 0 & 0 \\
0 & 0 & 0 & 0 & 1 & 1 & 0 & 0 \\
0 & 0 & 0 & 0 & w^1 & w^5 & 0 & 0 \\
0 & 0 & 0 & 0 & 0 & 0 & 1 & 1 \\
0 & 0 & 0 & 0 & 0 & 0 & w^3 & w^7
\end{bmatrix}
\begin{bmatrix}
1 & 0 & 0 & 0 & 0 & 0 & 0 & 0 \\
0 & 0 & 0 & 0 & 1 & 0 & 0 & 0 \\
0 & 0 & 1 & 0 & 0 & 0 & 0 & 0 \\
0 & 0 & 0 & 0 & 0 & 0 & 1 & 0 \\
0 & 1 & 0 & 0 & 0 & 0 & 0 & 0 \\
0 & 0 & 0 & 0 & 0 & 1 & 0 & 0 \\
0 & 0 & 0 & 1 & 0 & 0 & 0 & 0 \\
0 & 0 & 0 & 0 & 0 & 0 & 0 & 1
\end{bmatrix}
\begin{bmatrix}
x_0 \\ x_1 \\ x_2 \\ x_3 \\ x_4 \\ x_5 \\ x_6 \\ x_7
\end{bmatrix}
$$

$$
S_3 P X = \begin{bmatrix}
1 & 1 & 0 & 0 & 0 & 0 & 0 & 0 \\
w^0 & w^4 & 0 & 0 & 0 & 0 & 0 & 0 \\
0 & 0 & 1 & 1 & 0 & 0 & 0 & 0 \\
0 & 0 & w^2 & w^6 & 0 & 0 & 0 & 0 \\
0 & 0 & 0 & 0 & 1 & 1 & 0 & 0 \\
0 & 0 & 0 & 0 & w^1 & w^5 & 0 & 0 \\
0 & 0 & 0 & 0 & 0 & 0 & 1 & 1 \\
0 & 0 & 0 & 0 & 0 & 0 & w^3 & w^7
\end{bmatrix}
\begin{bmatrix}
x_0 \\ x_4 \\ x_2 \\ x_6 \\ x_1 \\ x_5 \\ x_3 \\ x_7
\end{bmatrix}
= \begin{bmatrix}
x_0 + x_4 \\
w^0 x_0 + w^4 x_4 \\
x_2 + x_6 \\
w^2 x_2 + w^6 x_6 \\
x_1 + x_5 \\
w^1 x_1 + w^5 x_5 \\
x_3 + x_7 \\
w^3 x_3 + w^7 x_7
\end{bmatrix}
$$

There are eight multiply-and-add operations due the matrix-vector multiply. Set the right-hand side to a vector **Z** and proceed to the next matrix multiply:

$$
S_2 S_3 P X = S_2 Z = \begin{bmatrix}
1 & 0 & 1 & 0 & 0 & 0 & 0 & 0 \\
0 & 1 & 0 & 1 & 0 & 0 & 0 & 0 \\
w^0 & 0 & w^4 & 0 & 0 & 0 & 0 & 0 \\
0 & w^0 & 0 & w^4 & 0 & 0 & 0 & 0 \\
0 & 0 & 0 & 0 & 1 & 0 & 1 & 0 \\
0 & 0 & 0 & 0 & 0 & 1 & 0 & 1 \\
0 & 0 & 0 & 0 & w^2 & 0 & w^6 & 0 \\
0 & 0 & 0 & 0 & 0 & w^2 & 0 & w^6
\end{bmatrix}
\begin{bmatrix}
z_0 \\ z_4 \\ z_2 \\ z_6 \\ z_1 \\ z_5 \\ z_3 \\ z_7
\end{bmatrix}
= \begin{bmatrix}
z_0 + z_2 \\
z_4 + z_6 \\
w^0 z_0 + w^4 z_2 \\
w^0 z_4 + w^4 z_6 \\
z_1 + z_3 \\
z_5 + z_7 \\
w^2 z_1 + w^6 z_3 \\
w^2 z_5 + w^6 z_7
\end{bmatrix}
$$

Once again there are eight multiply-add operations from the matrix multiply. Set the right-hand side to a vector $\mathbf{R}$ and compute:

$$S_1 S_2 S_3 PX = S_1 R = \begin{bmatrix} 1 & 0 & 0 & 0 & 1 & 0 & 0 & 0 \\ 0 & 1 & 0 & 0 & 0 & 1 & 0 & 0 \\ 0 & 0 & 1 & 0 & 0 & 0 & 1 & 0 \\ 0 & 0 & 0 & 1 & 0 & 0 & 0 & 1 \\ w^0 & 0 & 0 & 0 & w^4 & 0 & 0 & 0 \\ 0 & w^0 & 0 & 0 & 0 & w^4 & 0 & 0 \\ 0 & 0 & w^0 & 0 & 0 & 0 & w^4 & 0 \\ 0 & 0 & 0 & w^0 & 0 & 0 & 0 & w^4 \end{bmatrix} \begin{bmatrix} r_0 \\ r_1 \\ r_2 \\ r_3 \\ r_4 \\ r_5 \\ r_6 \\ r_7 \end{bmatrix} = \begin{bmatrix} r_0 + r_4 \\ r_1 + r_5 \\ r_2 + r_6 \\ r_3 + r_7 \\ w^0 r_0 + w^4 r_4 \\ w^0 r_1 + w^4 r_5 \\ w^0 r_2 + w^4 r_6 \\ w^0 r_3 + w^4 r_7 \end{bmatrix}$$

There are again eight multiply-add operations from the matrix-vector multiply. There are thus a total of three matrix-vector multiplies leading to $N \ln_2 N = 8 \times 3 = 24$ operations. This compares to $N^2 = 8 \times 8 = 64$ for the DFT. The FFT thus gives a factor of $64/24 \approx 2.7$ improvement in speed for the case $N = 2^3 = 8$.

We have thus come to the conclusion that by introducing a factorization of the DFT algorithm into n sparse matrices and a permutation matrix leads to an $N \ln N$ FFT algorithm. We have improved computational speed from about N^2 to $N \ln N$ multiply-and-add operations. A simple example is $N = 1024 = 2^{10}$ for which $N^2 = 1,048,576$ and $N \ln_2 N = 1024 \times 10 = 10,240$. Thus, for this case the FFT algorithm is $N^2 / N \ln_2 N \sim 102$ times faster than the DFT. For $N = 8192 = 2^{13}$ the FFT is ~ 630 times faster.

Many of the ideas just discussed will be exploited below to search for fast multidimensional Fourier transform and fast theta-function algorithms (FMFT, FTF).

21.3 The Multidimensional Fourier Transform

We have seen in Chapter 7 that the N-dimensional Fourier transform may be written in *vector notation*:

$$\Phi(\mathbf{X}) = \sum_{\mathbf{m}=-\infty}^{\infty} C_{\mathbf{m}} e^{i\mathbf{m}\cdot\mathbf{X}} \tag{21.7}$$

where the associated *inverse N-dimensional Fourier transform* has the expression:

$$C_{\mathbf{m}} = \frac{1}{(2\pi)^N} \int_0^{2\pi} e^{i\mathbf{m}\cdot\mathbf{X}} d\mathbf{X} \tag{21.8}$$

The integer vector is $\mathbf{m} = [m_1, m_2, \ldots, m_N]$ and the vector of dimensional is $\mathbf{X} = [X_1, X_2, \ldots, X_N]$. These are of course the defining equations for the multi-dimensional Fourier transform.

The computation of the multidimensional Fourier transform is in general very difficult and is most commonly addressed in one, two, and three dimensions (Press et al., 1992). Indeed the Riemann theta functions are much simpler because the Fourier coefficients depend on a matrix, not on a large-dimensional tensor (see Chapter 7 for a discussion).

21.4   The Theta Function

As pointed out numerous times in this monograph, the application of theta functions to the dynamics of nonlinear wave motion has a serious consequence: The inverse problem for the Riemann theta function requires knowledge of the space part of the Lax pair. Thus, the inverse problem is related to the solution of an eigenvalue problem for which the Floquet problem must be solved, together with the algebraic-geometric loop integrals for computing the Riemann spectrum. The simple inverse problem, valid for multidimensional Fourier series, is just not available for *dynamical* applications. This difficulty does not, however, prevent us from searching for numerical methods for determining the Riemann spectrum.

The Riemann theta function is given by

$$\theta(x) = \sum_{m=-\infty}^{\infty} e^{-\frac{1}{2}m \cdot Bm + im \cdot kx + im \cdot \phi}$$

which can be rewritten

$$\theta(x) = \sum_{m=-\infty}^{\infty} Q_m \, e^{im \cdot kx}, \quad Q_m = e^{-\frac{1}{2}m \cdot Bm + im \cdot \phi} = q_m \, e^{im \cdot \phi}, \quad q_m = e^{-\frac{1}{2}m \cdot Bm}$$

The summations are over all the positive and negative integers, that is, $m \in \mathbb{Z}$. Written as a *partial theta-function sum* the theta function can be expressed as

$$\theta(x) = \sum_{m=-M}^{M} Q_m \, e^{im \cdot kx} \tag{21.9}$$

For the moment we take

$$M = [M, M, \ldots, M]$$

for M an integer; this expression and those which follow can be generalized to arbitrary values for the limits of the N sums in the theta function. Here N is the number of degrees of freedom (number of cnoidal waves for the KdV equation or the Stokes modes for the nonlinear Schrödinger equation). The q_m are the *nomes*, **B** is the IST period or Riemann matrix, $\boldsymbol{\phi}$ are the IST *phases*, and the **k** are *wavenumbers*, assumed to be commensurable for most applications, $\mathbf{k} = 2\pi[1, 2, \ldots, N]/L$ and L the period of the wave train. This means that all results presented henceforth in this chapter assume *periodic boundary conditions*.

The scalar form of the Riemann theta function is given by

$$\theta(x) = \sum_{m_1=-\infty}^{\infty} \sum_{m_2=-\infty}^{\infty} \cdots \sum_{m_N=-\infty}^{\infty} \exp\left(-\frac{1}{2}\sum_{i=1}^{N}\sum_{j=1}^{N} m_i m_j B_{ij} + i\sum_{i=1}^{N} m_i k_i x + i\sum_{i=1}^{N} m_i \phi_i\right)$$

while the partial theta-function summation is

$$\theta(x) = \sum_{m_1=-M}^{M} \sum_{m_2=-M}^{M} \cdots \sum_{m_N=-M}^{M} \exp\left(-\frac{1}{2}\sum_{i=1}^{N}\sum_{j=1}^{N} m_i m_j B_{ij} + i\sum_{i=1}^{N} m_i k_i x + i\sum_{i=1}^{N} m_i \phi_i\right)$$

(21.10)

These latter expressions emphasize that the number of degrees of freedom, N, is also the number of nested summations in the theta function.

Now let us rewrite the theta function as a single sum over an integer j:

$$\theta(x) = \sum_{j=1}^{(2M+1)^N} Q_j\, e^{iK_j x}$$

(21.11)

where

$$Q_j = e^{-\frac{1}{2}\mathbf{m}_j \cdot \mathbf{B}\mathbf{m}_j + i\mathbf{m}_j \cdot \boldsymbol{\phi}} = \exp\left(-\frac{1}{2}\sum_{m=1}^{N}\sum_{n=1}^{N} m_m^j m_n^j B_{mn} + i\sum_{n=1}^{N} m_n^j \phi_n\right)$$

$$= q_j \exp\left(i\sum_{n=1}^{N} m_n^j \phi_n\right)$$

(21.12)

where

$$q_j = \exp\left(-\frac{1}{2}\sum_{m=1}^{N}\sum_{n=1}^{N} m_m^j m_n^j B_{mn}\right)$$

(21.13)

and the wavenumbers have the form $K_j = \mathbf{m} \cdot \mathbf{k}$. Here the notation $\mathbf{m}_j = [m_1^j, m_2^j, \ldots, m_N^j]$ has been adapted, that is, each wavenumber vector is associated with a particular value of index j. Since we have periodic boundary conditions in the x variable it is clear that

$$\mathbf{k} = [1, 2, \ldots, N]\Delta k = \frac{2\pi}{L}[1, 2, \ldots, N]$$

Here, we assume the wavenumbers to be commensurable:

$$k_n = \frac{2\pi n}{L}$$

This gives the important relation for the wavenumbers of the theta function for periodic boundary conditions:

$$K_j = \mathbf{m} \cdot \mathbf{k} = \frac{2\pi}{L} \sum_{n=1}^{N} m_n^j n = \frac{2\pi}{L} I_j, \quad I_j = \sum_{n=1}^{N} m_n^j n \tag{21.14}$$

Now the *partial theta-function summation* can also be written in slightly different notation:

$$\theta(x) = \sum_{j=0}^{M-1} Q_j \exp\left(i\frac{2\pi}{L} I_j x\right), \quad M = (2M+1)^N \tag{21.15}$$

Nota Bene: We could also write this in the form $-\infty, \infty \Rightarrow -J, J$:

$$\theta(x) = \sum_{j=-J}^{J} Q_j \exp\left(i\frac{2\pi}{L} I_j x\right), \quad J = \frac{1}{2}[(2M+1)^N - 1] \tag{21.16}$$

Here M is an odd number because $2M + 1$ is always odd. It is also worth noting that all of the forms for the theta function just mentioned are equivalent to the partial theta sum above. The reason for using the different notations is as a precursor for discussing the DFT in the next section.

21.5   The Discrete Theta Function

We are now ready to *discretize the spatial variable x* in the theta function. To this end write

$$x_k = k\Delta x = k\frac{L}{N}, \quad k = 0, 1, 2, \ldots, N-1 \tag{21.17}$$

where L is the length of the periodic box for x, $0 \le x \le L$. The box of length L has been divided into N small spatial intervals, $\Delta x = L/N$. Then we have the *DTF*:

$$\theta(x_k) = \theta_k = \sum_{j=0}^{M-1} Q_j \exp\left(i\frac{2\pi}{L} I_j x_k\right) = \sum_{j=0}^{M-1} Q_j \exp\left(i\frac{2\pi}{L} I_j k L / N\right)$$

$$= \sum_{j=0}^{M-1} Q_j \exp(i2\pi I_j k / N)$$

Finally

$$\theta(x_k) = \sum_{j=0}^{M-1} e^{2\pi i k I_j / N} Q_j \quad \text{(Discrete Theta Function)} \tag{21.18}$$

where k varies as $k = 0, 1, 2, \ldots, N - 1$ (here N is the number of points in the vector of values of the DTF, typically the length of a measured space or time series, say 1024 points) and

$$Q_j = e^{-\frac{1}{2}\mathbf{m}_j \cdot \mathbf{B}\mathbf{m}_j + i\mathbf{m}_j \cdot \boldsymbol{\phi}} = \exp\left(-\frac{1}{2} \sum_{m=1}^{N} \sum_{n=1}^{N} m_m^j m_n^j B_{mn} + i \sum_{n=1}^{N} m_n^j \phi_n\right)$$

$$= q_j \exp\left(i \sum_{n=1}^{N} m_n^j \phi_n\right) \tag{21.19}$$

$$q_j = \exp\left(-\frac{1}{2} \sum_{m=1}^{N} \sum_{n=1}^{N} m_m^j m_n^j B_{mn}\right)$$

Furthermore

$$I_j = \sum_{n=1}^{N} m_n^j n \tag{21.20}$$

An alternative form for the *DTF* is given by

$$\theta(x_k) = \sum_{j=-J}^{J} Q_j \, e^{2\pi i k I_j / N}, \quad J = [(2M + 1)^N - 1]/2 \tag{21.21}$$

Now some perspective is in order here. The Q_j are the multidimensional Fourier coefficients. There are $M = (2M + 1)^N$ of them, generally a very large number. The I_j are of course integers on the interval $-I_{max} \leq I_j \leq I_{max}$ where $I_{max} = M[1, 1, \ldots, 1] \cdot [1, 2, \ldots, N] = M \sum_{n=1}^{N} n = MN(N + 1)/2$; the I_j are rapidly varying functions of the integer j (see Chapters 7 and 8). We see that the DTF has indices (j, k) where j ranges over all the $M = (2M + 1)^N$ component sine waves and k ranges over all the spatial points. If one assumes that the Riemann matrix is diagonal and that the diagonal elements are large then the multidimensional Fourier coefficients q_j are small, $q_j \ll 1$. In this case the DTF reduces to the DFT. This occurs because in this case only those terms with the relation

$$I_j = \sum_{n=1}^{N} m_n^j n = j$$

significantly contribute to the partial theta sum, and thus the DTF reduces to the discrete Fourier transform.

Example: Suppose $M = N = 2$, then there are $M = (2M + 1)^N = 25$ sine-wave components and $J = [(2M + 1)^N - 1]/2 = 12$, so that the summation of the DTF is over $(0,24)$ $(j = 0, 1, 2, \ldots, M - 1)$, $(1, 25)$ $(j = 1, 2, \ldots, M)$, or $(-12,12)$ $(j = -J, -J + 1, \ldots, J-1, J)$, depending upon the summation notation used above.

Nota Bene: $m_n^j = -m_n^{-j}$ and $I_j = -I_{-j}$ both hold. Thus, $\mathbf{m}^j = -\mathbf{m}^{-j}$ also. Now let

$$w = e^{2\pi i/N}$$

The *DTF* can be written:

$$\theta_k = \theta(x_k) = \sum_{j=0}^{M-1} W_{kj} Q_j, \quad W_{kj} = w^{kI_j} \tag{21.22}$$

This says that the *theta-function vector* $\theta(x_k) = \theta_k$ of length N ($k = 0$, $1, 2, \ldots, N - 1$) is equal to a *matrix* $W_{kj} = w^{I_j k} = w^{kI_j}$ of dimension $M \times N$ $(j = -J, -J+1, \ldots, J-1, J, k = 0, 1, 2, \ldots, N-1)$ *times a vector of components*

$$Q_j = \exp\left[-\frac{1}{2} \sum_{m=1}^{N} \sum_{n=1}^{N} m_m^j m_n^j B_{mn} + i \sum_{n=1}^{N} m_n^j \phi_n \right]$$

of length $M = (2M+1)^N$. Here N is the number of points in the theta-function space or time series and M is the number of terms in the partial sum of the theta function.

The *matrix form* of these equations is

$$\boldsymbol{\theta} = \mathbf{W}\mathbf{Q} \tag{21.23}$$

where

$$\boldsymbol{\theta} = \{\theta_k, \quad k = 0, 1, 2, \ldots, N - 1\} = [\theta_0, \theta_1, \theta_2, \ldots, \theta_{N-1}]$$

$$\mathbf{W} = \{W_{kj}, \quad k = 0, 1, 2, \ldots, N - 1, \quad j = 0, 1, 2, \ldots, N - 1\}$$

$$\mathbf{Q} = \{Q_k, \quad k = 0, 1, 2, \ldots, N - 1\} = [Q_0, Q_1, Q_2, \ldots, Q_{N-1}]$$

In general the dimension, N, of the theta vector $\theta(x_k)$, $k = 0, 1, 2, \ldots, N - 1$, is much smaller than the dimension of the nome-phase vector, Q_j, $j = 0, 1, 2, \ldots, M-1$: $N \ll M$. This means that one can easily compute the vector

$\theta(x_k)$ from the vector Q_j. However, one cannot easily compute the vector Q_j from the $\theta(x_k)$. This is because the inverse problem, to compute the Q_j (and hence compute the period matrix and phases) from the $\theta(x_k)$, requires the solution to the above set of linear equations in the *unknown vector* Q_j. This cannot strictly be accomplished because there are fewer equations (for which there are N) than unknowns (for which there are M)! Only in the case when $N = M$ and the matrix W_{kj} is therefore square can we can speak of inverting Equation (21.23). Since $N \ll M$ we must presumably use the inverse scattering machinery (eigenvalue problem, loop integrals, Schottky uniformization, Nakamura-Boyd, etc.) to compute the parameters of the theta function. An alternative approach for determining the Riemann matrix and phases from $\theta(x_k)$, which I call *nonlinear adiabatic annealing*, is discussed in Chapter 23.

The other form for the DTF is given by

$$\theta(x_k) = \sum_{j=-J}^{J} Q_j\, e^{2\pi i k I_j/N} = \sum_{j=-J}^{J} W_{kj} Q_j, \quad W_{kj} = e^{2\pi i k I_j/N} = w^{kI_j}, \quad w = e^{2\pi i/N} \quad (21.24)$$

To get an idea of the brute-force computational effort that must be applied to compute a DTF I provide a convenient table for the total number of complex exponentials in the theta function for typical values of M and N: $M = (2M + 1)^N$. Recall that N is the number of nested sums in the partial theta summation and $\pm M$ are the limits in each sum. Table 21.1 is summarizes this relaltionship.

Nota Bene: The *DTF* given above does not necessarily imply the existence of an inverse problem. It is easily demonstrated that the vector form $\theta = WQ$ can be written:

$$\theta = WQ = W[Q_1 + Q_2 + Q_3 + \cdots + Q_R] \quad (21.25)$$

where the summation $WQ_1 + WQ_2 + WQ_3 + \cdots + WQ_R$ contains R ordinary Fourier transforms (the $Q_1, Q_2, \ldots$ and R must be computed from the details of the theta function). The matrix W is the same as the matrix from *ordinary linear Fourier analysis*: it is $N \times N$, that is, it is a square unitary matrix. Inverting the above equation to obtain $Q = W^{-1}\theta$ is thus not strictly possible. Equation (21.25) is a direct consequence of the fact that the theta function can be written as a linear Fourier series with coefficients that depend on a series of inverse scattering transform parameters (Chapter 9, Equations (9.1) and (9.2)). Rewriting Equation (9.2) as a sum of vectors for the $Q_1, Q_2, \ldots$ leads to Equation (21.25).

The equation $\theta = WQ = W[Q_1 + Q_2 + Q_3 + \ldots + Q_R]$ for computing the DTF offers two possibilities: (1) sum the appropriate Q's first and then take the FFT or (2) take the FFT R times. Item (1) is discussed in detail in Chapter 9, albeit from a different perspective than Equation (21.25).

Table 21.1 Number of Complex Exponentials in a Partial Theta-Function Summation for Typical Values of M and N

M/N	1	2	3	4	5	6	7	8	9	10
1	3	9	27	81	243	729	2187	6561	19,683	59,049
2	5	25	125	625	3125	15,625	78,125	390,625	1,953,125	9,765,625
3	7	49	343	2401	16,807	117,649	823,543	5,764,801	40,353,607	282,475,249
4	9	81	729	6561	59,049	531,441	4,782,969	43,046,721	387,420,489	3,486,784,401
5	11	121	1331	14,641	161,051	1,771,561	19,487,171	214,358,881	2,357,947,691	25,937,424,601

21.6   Determination of the Period Matrix and Phases from a Space/Time Series

Is there any other hope for using the theta functions directly for computing the theta-function parameters? Here are two possibilities. The first is to try and apply in a least square sense the approach of *Numerical Recipes* (Press et al., 1992) (p. 26, p. 57, and Chapter 15 on least squares). Their suggestion is to reduce the problem to a linear least-squares equation and then to solve it. The second is that the vector Q_j depends on far fewer parameters (period matrix elements and phases) so the Q_j do not all contain unique information so that a *reduced set* of the Q_j is best applied to compute the Riemann spectrum.

Let us consider the first approach. We have the equation for the DTF:

$$\boldsymbol{\theta} = \mathbf{W}\mathbf{Q} \tag{21.26}$$

which in the present context (the inverse problem) may be thought of as a set of linear equations in the vector $\mathbf{Q}$, which we seek given the matrix $\mathbf{W}$ and the vector $\boldsymbol{\theta}$. The technique to determine $\mathbf{Q}$ from $\mathbf{W}$ and $\boldsymbol{\theta}$ arises from the method of least squares. One multiplies Equation (21.26) by the transpose $\mathbf{W}^T$ to obtain the set of $M \times M$ equations:

$$[\mathbf{W}^T\mathbf{W}]\mathbf{Q} = \mathbf{W}^T\boldsymbol{\theta} \tag{21.27}$$

Thus, we have a square matrix $(\mathbf{W}^T\mathbf{W})$ times a vector $(\mathbf{Q})$ yielding a vector $(\mathbf{W}^T\boldsymbol{\theta})$. These are the *normal equations of least-squares linear analysis*. Note that this is a very large system M of equations. One can of course attempt to solve these equations for $\mathbf{Q}$ by the standard methods, including Gaussian elimination, LU decomposition, and singular value decomposition. However, we here follow a different path that arises because of the particular, unique properties of the matrix $\mathbf{W}$.

To this end take the *transpose* of the matrix $\mathbf{W}$ and its *complex conjugate* (*) (these operations are actually interchangeable) and apply the result to the left of the vector form of the theta function, $\boldsymbol{\theta} = \mathbf{W}\mathbf{Q}$:

$$\mathbf{W}^{T*}\boldsymbol{\theta} = \mathbf{W}^{T*}\mathbf{W}\mathbf{Q} = \mathbf{N}\mathbf{Q}$$

The last expression on the right is strictly notational for the moment and is motivated by the ordinary, linear Fourier transform, where the matrix $\mathbf{I}$ is the identity. We have yet to establish the form of $\mathbf{I}$ in the present case, but will do so below. Now divide by the number of points in the spatial domain, N, to give:

$$\frac{\mathbf{W}^{T*}}{N}\boldsymbol{\theta} = \frac{\mathbf{W}^{T*}\mathbf{W}}{N}\mathbf{Q} = \mathbf{I}\mathbf{Q}$$

or finally

$$\mathbf{I}\mathbf{Q} = \frac{\mathbf{W}^{T*}}{N}\boldsymbol{\theta}, \quad \mathbf{I} = \frac{\mathbf{W}^{T*}\mathbf{W}}{N} \tag{21.28}$$

For the ordinary, linear Fourier transform $\mathbf{I}$ is the identity matrix. For the DTF it differs from the identity matrix by the inclusion of a number of matrix elements which are "1" *off the diagonal*. Generally speaking $\mathbf{I} = \mathbf{W}^{T}{}^{*}\mathbf{W} / N$ is singular and it therefore requires some ingenuity to get information from Equation (21.28), which for purposes of this chapter will be called the *inverse theta function*.

Now return to the above equation and write:

$$\mathbf{IQ} = \frac{\mathbf{W}^{T*}}{N}\boldsymbol{\theta} = \mathbf{P}, \quad \mathbf{I} = \frac{\mathbf{W}^{T*}\mathbf{W}}{N}$$

While $\mathbf{Q}$ cannot be determined from this equation, the vector $\mathbf{P}$ can be determined by: $\mathbf{P} = \mathbf{W}^{T*}\boldsymbol{\theta}/N = \mathbf{IQ}$. This expression is not invertible because the matrix $\mathbf{I}$ has no inverse. We will, however, seek the theta-function parameters directly from $\mathbf{P}$.

Now let us address the properties of the matrix $\mathbf{I}$, which I now write explicitly:

$$\mathbf{I} = \frac{\mathbf{W}^{T*}\mathbf{W}}{N} \quad \text{or} \quad \{\mathbf{I}\}_{ij} = \frac{\{\mathbf{W}^{T*}\mathbf{W}\}_{ij}}{N} \tag{21.29}$$

Note that the matrix $\mathbf{W}$ has the matrix elements:

$$\{\mathbf{W}\}_{ij} = w^{iI_j}, \quad \mathbf{W}^{T} \Rightarrow \{\mathbf{W}\}_{ji} = w^{I_j i} \quad \mathbf{W}^{T*} \Rightarrow \{\mathbf{W}^{*}\}_{ji} = w^{-I_j i}$$

for $w = e^{2\pi i/N}$. Hence

$$\{\mathbf{I}\}_{ij} = \frac{\{\mathbf{W}^{T*}\mathbf{W}\}_{ij}}{N} = \frac{\sum_{k=1}^{N} w^{-I_i k} w^{kI_j}}{N} = \frac{\sum_{k=1}^{N} w^{k(I_j - I_i)}}{N} = \frac{\sum_{k=1}^{N} e^{2\pi i k(I_j - I_i)/N}}{N}$$

Now this last expression tells us that if $I_j - I_i = nN$ for $n = 0, \pm1, \pm2, \ldots$ then

$$e^{2\pi i k(I_j - I_i)/N} = 1$$

This then leads to the explicit results:

$$\{\mathbf{I}\}_{ij} = I_{ij} = \frac{\sum_{k=1}^{N} e^{2\pi i k(I_j - I_i)/N}}{N} = \begin{cases} 1 & \text{when } I_j - I_i = nN, \quad \text{for } n = 0, \pm 1, \pm 2, \ldots \\ 0 & \text{otherwise} \end{cases}$$

$$\tag{21.30}$$

The matrix I_{ij} has diagonal elements which are 1, but there are also off-diagonal elements of the matrix which are 1 for particular values of i, j. The fact that I_{ij} is *not* an identity matrix should be clear and indeed it should also be clear from the above definition for I_{ij} *is not invertible*.

This is a result we have anticipated because we know that we cannot determine the vector $\mathbf{Q}$ from the vector $\boldsymbol{\theta}$ because these vectors are not of equal length, for $\mathbf{Q}$ is a very long vector of an astronomical number of terms, say billions or trillions of typical applications, while $\boldsymbol{\theta}$ will typically have only a few thousand points. Nevertheless, the vector $\mathbf{Q}$ is determined from a small number of theta-function parameters for the period matrix, $\mathbf{B}$ ($N(N-1)/2$ terms) and phases $\boldsymbol{\phi}$ (N terms). In spite of these properties of the matrix I_{ij} we now find this matrix is indispensable for determining the period matrix and phases from particular elements of the $\mathbf{Q}$ matrix as we now discuss.

A Numerical Procedure for Computing the Period Matrix, B, and Phases, ϕ

Given the vector $\boldsymbol{\theta} = \{\theta_k,\ k = 0, 1, \ldots, N - 1\}$, assume that we have reasonable numerical values for the number of degrees of freedom, $N = N_{\text{dof}}$, and the theta-function limits, $\mathbf{M} = [M, M, \ldots, M]$, then:

(1) Compute $\mathbf{W}$ using the theta summation. First compute

$$J = \frac{1}{2}[(2M + 1)^N - 1]$$

Then, sum over $-\mathbf{M}$, $\mathbf{M}$ in the theta function for the full number of degrees of freedom. Use the dummy summation index j that runs from $-J$ to J, and find the values of the integer vectors $\mathbf{m}^j$ for which we determine the *integer wavenumbers* for the theta function:

$$I_j = \sum_{n=1}^{N} m_n^j n$$

This then gives the elements of the matrix $\mathbf{W}$:

$$W_{kj} = e^{2\pi i k I_j / N}$$

(2) From $\mathbf{W}$ one then computes the vectors

$$\mathbf{P} = \frac{\mathbf{W}^{T*}}{N}\boldsymbol{\theta}, \quad \mathbf{I} = \frac{\mathbf{W}^{T*}\mathbf{W}}{N}$$

(3) We now seek the period matrix $\mathbf{B}$ and the phases $\boldsymbol{\phi}$ from the vector $\mathbf{Q} = \mathbf{Q}(\mathbf{B}, \boldsymbol{\phi})$, where $\mathbf{Q}$ is a solution of the equation

$$\mathbf{IQ} = \mathbf{P}$$

This latter equation *can be solved for only a limited number of the elements of* $\mathbf{Q}$. This is consistent with our understanding that $\mathbf{Q}$ cannot be computed strictly from the vector $\boldsymbol{\theta}$, except in a least squares sense. But this limited number of elements is enough to finally compute the period matrix and phases.

The brute-force computation of the DTF has a power law $(2M + 1)^N$. To see how many terms this can amount to see Table 21.2 with two power laws: powers of two, 2^n, and three, 3^n:

A true fast DTF would consume computer time like $N \ln_2 M$ instead of the NM number of operations as suggested by the matrix formulation given above. This clearly offers a great savings in computer time. Suppose that $N = 1024$ (number of points in the space/time series) and $M = 2.5 \times 10^{10}$. Then $NM = 2.56 \times 10^{13}$.

Table 21.2 Explosive Nature of Two Power Laws to Illustrate the Rapid Increase in the Number of Complex Exponentials in a Partial Theta-Function Summation

n	2^n	3^n
1	2	3
2	4	9
3	8	27
4	16	81
5	32	243
6	64	279
7	128	2187
8	256	6561
9	512	19,683
10	1024	59,049
11	2048	177,147
12	4096	531,441
13	8192	1,594,323
14	16,384	4,782,969
15	32,768	14,348,907
16	65,536	43,046,721
17	131,072	129,140,163
18	262,144	387,420,489
19	524,288	1,162,261,467
20	1,048,576	3,486,784,401
21	2,097,152	10,460,353,203
22	4,194,304	31,381,059,609
23	8,388,608	94,143,178,827
24	16,777,216	2.82430×10^{11}
25	33,554,432	8.47289×10^{11}
26	67,108,864	2.54187×10^{12}
27	134,217,728	7.62560×10^{12}
28	268,435,456	2.28768×10^{13}
29	536,870,912	6.86304×10^{13}
30	1,073,741,824	2.05891×10^{14}
31	2,147,483,648	6.17673×10^{14}
32	4,294,967,296	1.85302×10^{15}
33	8,589,934,592	5.55906×10^{15}
34	17,179,869,184	1.66772×10^{16}
35	34,359,738,368	5.00315×10^{16}

However, $N \ln_2 M \approx 1024 \times 24.5 = 25{,}088$, about a billion times faster than the NM estimate. Note that for a typical FFT for 1024 points, $N \ln_2 M \approx 1024 \times 10 = 10{,}240$. So doing the theta function for 25 billion theta-function components would be only about 2.5 times slower than a typical FFT for 1024 points! Can this rapid calculation be realized in practice? The problem still seems to be open, but it does not hurt to look into what difficulties there are.

We need to study the properties of the matrix $W_{kj} = w^{kI_j}$. To do so let us consider some examples (Table 21.3):

Example 1: $N = 8$, $M = 1$, $N = 2$ (two degrees of freedom summed over -1, 1) so that $M = (2M + 1)^N = 9$. Use

$$Q_j \Rightarrow e^{-(m_1^2 B_{11} + 2m_1 m_2 B_{12} + m_2^2 B_{22})/2 + i(m_1 \phi_1 + m_2 \phi_2)}$$

With regard to the inverse problem it is worthwhile noting that the vector of Q's has nine elements, while there are eight equations. In reality there are five numbers we seek, the three elements of the period matrix and the two phases. Furthermore, note that $Q_0 = 1$, so we really have eight elements and eight equations. Let us follow this direction for a moment. Note that

$$\ln Q_{-1} = -B_{22}/2 - i\phi_2, \quad \ln Q_1 = -B_{22}/2 + i\phi_2$$

Sum these to get

$$B_{22} = -(\ln Q_{-1} + \ln Q_1) = -\ln(Q_{-1}Q_1) \tag{21.31}$$

Likewise

$$\ln Q_{-2} = -(B_{11} - 2B_{12} + B_{22})/2 + i(-\phi_1 + \phi_2)$$

$$\ln Q_2 = -(B_{11} - 2B_{12} + B_{22})/2 + i(\phi_1 - \phi_2)$$

Table 21.3 A Simple DTF for $M = 1$, $N = 2$

l	j	(m_1, m_2)	Coefficients Q_j	I_j	Theta Summation Element, w^{I_j}
1	-4	$(-1, -1)$	$Q_{-4} = e^{-(B_{11} + 2B_{12} + B_{22})/2 - i(\phi_1 + \phi_2)}$	-3	$e^{-3 \cdot 2\pi i/N}$
2	-3	$(-1, 0)$	$Q_{-3} = e^{-B_{11}/2 - i\phi_1}$	-1	$e^{-1 \cdot 2\pi i/N}$
3	-2	$(-1, 1)$	$Q_{-2} = e^{-(B_{11} - 2B_{12} + B_{22})/2 + i(-\phi_1 + \phi_2)}$	1	$e^{1 \cdot 2\pi i/N}$
4	-1	$(0, -1)$	$Q_{-1} = e^{-B_{22}/2 - i\phi_2}$	-2	$e^{-2 \cdot 2\pi i/N}$
5	0	$(0, 0)$	$Q_0 = 1$	0	1
6	1	$(0, 1)$	$Q_1 = e^{-B_{22}/2 + i\phi_2}$	2	$e^{2 \cdot 2\pi i/N}$
7	2	$(1, -1)$	$Q_2 = e^{-(B_{11} - 2B_{12} + B_{22})/2 + i(\phi_1 - \phi_2)}$	-1	$e^{-1 \cdot 2\pi i/N}$
8	3	$(1, 0)$	$Q_3 = e^{-B_{11}/2 + i\phi_1}$	1	$e^{1 \cdot 2\pi i/N}$
9	4	$(1, 1)$	$Q_4 = e^{-(B_{11} + 2B_{12} + B_{22})/2 + i(\phi_1 + \phi_2)}$	3	$e^{3 \cdot 2\pi i/N}$

Add these to get

$$\ln Q_{-2} + \ln Q_2 = -(B_{11} - 2B_{12} + B_{22})$$

Do the same for the next pair and find

$$\ln Q_{-3} = -B_{11}/2 - i\phi_1, \quad \ln Q_3 = -B_{11}/2 + i\phi_1$$

so that

$$B_{11} = -(\ln Q_{-3} + \ln Q_3) = -\ln(Q_{-3}Q_3) \tag{21.32}$$

Now solve for the off-diagonal term of the period matrix:

$$\ln Q_{-2} + \ln Q_2 = -[-(\ln Q_{-3} + \ln Q_3) - 2B_{12} - (\ln Q_{-1} + \ln Q_1)]$$

and find

$$B_{12} = \frac{1}{2}\left(\frac{\ln Q_{-2} + \ln Q_2}{\ln Q_{-3} + \ln Q_3 + \ln Q_{-1} + \ln Q_1}\right) = \frac{1}{2}\left(\frac{\ln(Q_{-2}Q_2)}{\ln(Q_{-3}Q_3Q_{-1}Q_1)}\right) \tag{21.33}$$

Likewise the two equations:

$$\ln Q_{-1} = -B_{22}/2 - i\phi_2, \quad \ln Q_1 = -B_{22}/2 + i\phi_2$$

can be subtracted to give the phase

$$\phi_2 = -\frac{i}{2}(\ln Q_1 - \ln Q_{-1}) = -\frac{i}{2}\ln\frac{Q_1}{Q_{-1}} \tag{21.34}$$

Finally, subtract the two equations:

$$\ln Q_{-3} = -B_{11}/2 - i\phi_1, \quad \ln Q_3 = -B_{11}/2 + i\phi_1$$

and get the other phase

$$\phi_1 = -\frac{i}{2}(\ln Q_3 - \ln Q_{-3}) = -\frac{i}{2}\ln\frac{Q_3}{Q_{-3}} \tag{21.35}$$

So, this illustrates how to get the theta-function parameters from the Q's. Notice that we did not use all the Q's, that is, we did not use Q_0, Q_{-4}, and Q_4. The particular chosen Q's give the exact number of equations to get the parameters of the theta function. Can we generally reduce the number of linear equations necessary for inverting the DTF? I show these results in Table 21.5. Six Q's, Q_{-3}, Q_{-2}, Q_{-1}, Q_1, Q_2, Q_3, give the five theta-function parameters, B_{11}, B_{22}, B_{12}, ϕ_1, and ϕ_2. Notice that we did not need to compute the other diagonal term B_{21}, because $B_{12} = B_{21}$ since the period matrix is symmetric. This will of course always be the case. For example, suppose in general we have

N degrees of freedom. Thus, we need to compute N^2 parameters for the elements of the period matrix, plus the N phases. So the number of Q's we need is

$$N_Q = N^2 + N = N(N+1)$$

To compute the diagonal and upper triangle of the period matrix we need $N(N+1)/2$ parameters plus the N phases. The total number of parameters is then

$$N_{\text{parm}} = \frac{1}{2}N(N+1) + N = \frac{1}{2}(N^2 + 3N)$$

In the present case we have for $N = 2$, $N_Q = 5$ as we have discovered above by the simple example. The message is that, in general, we need only a small fraction of the Q's since the number of useful Q's, $N_Q = N^2 + N = N(N+1)$, is much less than the total number of Q's, $M = (2M+1)^N$: $N_Q \ll M$. A summary is given in Table 21.5.

Now construct the actual matrix form of the DTF:

$$
\begin{bmatrix} \theta_0 \\ \theta_1 \\ \theta_2 \\ \theta_3 \\ \theta_4 \\ \theta_5 \\ \theta_6 \\ \theta_7 \end{bmatrix}
=
\begin{bmatrix}
W_{00} & W_{01} & W_{02} & W_{03} & W_{04} & W_{05} & W_{06} & W_{07} & W_{08} \\
W_{10} & W_{11} & W_{12} & W_{13} & W_{14} & W_{15} & W_{16} & W_{17} & W_{18} \\
W_{20} & W_{21} & W_{22} & W_{23} & W_{24} & W_{25} & W_{26} & W_{27} & W_{28} \\
W_{30} & W_{31} & W_{32} & W_{33} & W_{34} & W_{35} & W_{36} & W_{37} & W_{38} \\
W_{40} & W_{41} & W_{42} & W_{43} & W_{44} & W_{45} & W_{46} & W_{47} & W_{48} \\
W_{50} & W_{51} & W_{52} & W_{53} & W_{54} & W_{55} & W_{56} & W_{57} & W_{58} \\
W_{60} & W_{61} & W_{62} & W_{63} & W_{64} & W_{65} & W_{66} & W_{67} & W_{68} \\
W_{70} & W_{71} & W_{72} & W_{73} & W_{74} & W_{75} & W_{76} & W_{77} & W_{78}
\end{bmatrix}
\begin{bmatrix} Q_{-4} \\ Q_{-3} \\ Q_{-2} \\ Q_{-1} \\ Q_0 \\ Q_1 \\ Q_2 \\ Q_3 \\ Q_4 \end{bmatrix}
$$

$$
=
\begin{bmatrix}
1 & 1 & 1 & 1 & 1 & 1 & 1 & 1 & 1 \\
w^{-3} & w^{-1} & w^1 & w^{-2} & 1 & w^2 & w^{-1} & w^1 & w^3 \\
w^{-6} & w^{-2} & w^2 & w^{-4} & 1 & w^4 & w^{-2} & w^2 & w^6 \\
w^{-1} & w^{-3} & w^3 & w^{-6} & 1 & w^6 & w^{-3} & w^3 & w^1 \\
w^{-4} & w^{-4} & w^4 & 1 & 1 & 1 & w^{-4} & w^4 & w^4 \\
w^{-7} & w^{-5} & w^5 & w^{-3} & 1 & w^3 & w^{-5} & w^5 & w^7 \\
w^{-2} & w^{-6} & w^6 & w^{-4} & 1 & w^4 & w^{-6} & w^6 & w^2 \\
w^{-5} & w^{-7} & w^7 & w^{-6} & 1 & w^6 & w^{-7} & w^7 & w^5
\end{bmatrix}
\begin{bmatrix} Q_{-4} \\ Q_{-3} \\ Q_{-2} \\ Q_{-1} \\ Q_0 \\ Q_1 \\ Q_2 \\ Q_3 \\ Q_4 \end{bmatrix}
$$

21.7 General Procedure for Computing the Period Matrix and Phases from the Q's

Let us now see if there exists a general procedure for extracting the Riemann spectrum from the Q's. We have

$$Q_j = e^{-(m_1^2 B_{11} + 2m_1 m_2 B_{12} + m_2^2 B_{22})/2 + i(m_1\phi_1 + m_2\phi_2)}$$

Table 21.4 A Simple DTF Summation. The logarithms of the Q's are explicitly given.

l	j	(m_1, m_2)	Coefficients $\ln Q_j$	I_j	Theta Summation Element, w^{I_j}
1	-4	$(-1, -1)$	$\ln Q_{-4} = -(B_{11} + 2B_{12} + B_{22})/2 - i(\phi_1 + \phi_2)$	-3	$e^{-3 \cdot 2\pi i/N}$
2	-3	$(-1, 0)$	$\ln Q_{-3} = -B_{11}/2 - i\phi_1$	-1	$e^{-1 \cdot 2\pi i/N}$
3	-2	$(-1, 1)$	$\ln Q_{-2} = -(B_{11} - 2B_{12} + B_{22})/2 + i(-\phi_1 + \phi_2)$	1	$e^{1 \cdot 2\pi i/N}$
4	-1	$(0, -1)$	$\ln Q_{-1} = -B_{22}/2 - i\phi_2$	-2	$e^{-2 \cdot 2\pi i/N}$
5	0	$(0, 0)$	$\ln Q_0 = 0$	0	1
6	1	$(0, 1)$	$\ln Q_1 = -B_{22}/2 + i\phi_2$	2	$e^{2 \cdot 2\pi i/N}$
7	2	$(1, -1)$	$\ln Q_2 = -(B_{11} - 2B_{12} + B_{22})/2 + i(\phi_1 - \phi_2)$	-1	$e^{-1 \cdot 2\pi i/N}$
8	3	$(1, 0)$	$\ln Q_3 = -B_{11}/2 + i\phi_1$	1	$e^{1 \cdot 2\pi i/N}$
9	4	$(1, 1)$	$\ln Q_4 = -(B_{11} + 2B_{12} + B_{22})/2 + i(\phi_1 + \phi_2)$	3	$e^{3 \cdot 2\pi i/N}$

We have from Table 21.3 the example for two degrees of freedom ($N = 2$) and summation from $-1, 1$ ($M = 1$). Now use the logarithms of the Q's (Table 21.4):

We have

$$\ln Q_j = -(m_{j1}^2 B_{11} + 2m_{j1}m_{j2}B_{12} + m_{j2}^2 B_{22})/2 + i(m_{j1}\phi_1 + m_{j2}\phi_2)$$

from which the following steps are clear:

$$\ln Q_j = \left(-m_{j1}^2 \quad -2m_{j1}m_{j2} \quad -m_{j2}^2 \quad m_{j1} \quad m_{j2}\right) \begin{pmatrix} \frac{1}{2}B_{11} \\ B_{12} \\ \frac{1}{2}B_{22} \\ i\phi_1 \\ i\phi_2 \end{pmatrix}$$

or alternatively

$$\begin{pmatrix} \ln Q_{-3} \\ \ln Q_{-2} \\ \ln Q_{-1} \\ \ln Q_1 \\ \ln Q_2 \\ \ln Q_3 \end{pmatrix} = \begin{pmatrix} m_{11}^2 & m_{11}m_{12} & m_{12}^2 & m_{11} & m_{12} \\ m_{21}^2 & m_{21}m_{22} & m_{22}^2 & m_{21} & m_{22} \\ m_{31}^2 & m_{31}m_{32} & m_{32}^2 & m_{31} & m_{32} \\ m_{51}^2 & m_{51}m_{52} & m_{52}^2 & m_{51} & m_{52} \\ m_{61}^2 & m_{61}m_{62} & m_{62}^2 & m_{61} & m_{62} \\ m_{71}^2 & m_{71}m_{72} & m_{72}^2 & m_{71} & m_{72} \end{pmatrix} \begin{pmatrix} -\frac{1}{2}B_{11} \\ -B_{12} \\ -\frac{1}{2}B_{22} \\ i\phi_1 \\ i\phi_2 \end{pmatrix}$$

We see that there is a judicial choice of the integer vector **m**:

$$
\begin{pmatrix} \ln Q_{-3} \\ \ln Q_{-2} \\ \ln Q_{-1} \\ \ln Q_1 \\ \ln Q_2 \\ \ln Q_3 \end{pmatrix} = \begin{pmatrix} 1 & 0 & 0 & -1 & 0 \\ 1 & -1 & 1 & -1 & 1 \\ 0 & 0 & 1 & 0 & -1 \\ 0 & 0 & 1 & 0 & 1 \\ 1 & -1 & 1 & 1 & -1 \\ 1 & 0 & 0 & 1 & 0 \end{pmatrix} \begin{pmatrix} -\frac{1}{2}B_{11} \\ -B_{12} \\ -\frac{1}{2}B_{22} \\ i\phi_1 \\ i\phi_2 \end{pmatrix}
$$

$$
\begin{pmatrix} \ln Q_{-3} \\ \ln Q_{-2} \\ \ln Q_{-1} \\ \ln Q_1 \\ \ln Q_2 \\ \ln Q_3 \end{pmatrix} = \begin{pmatrix} -\dfrac{1}{2}B_{11} - i\phi_1 \\[2mm] -\dfrac{1}{2}B_{11} + B_{12} - \dfrac{1}{2}B_{22} - i\phi_1 + i\phi_2 \\[2mm] -\dfrac{1}{2}B_{22} - i\phi_2 \\[2mm] -\dfrac{1}{2}B_{22} + i\phi_2 \\[2mm] -\dfrac{1}{2}B_{11} + B_{12} - \dfrac{1}{2}B_{22} + i\phi_1 - i\phi_2 \\[2mm] -\dfrac{1}{2}B_{11} + i\phi_1 \end{pmatrix}
$$

$$
\begin{pmatrix} \ln Q_{-3} + \ln Q_3 \\ \ln Q_{-2} + \ln Q_2 \\ \ln Q_{-1} + \ln Q_1 \\ \ln Q_1 - \ln Q_{-1} \\ \ln Q_2 - \ln Q_{-2} \\ \ln Q_3 - \ln Q_{-3} \end{pmatrix} = \begin{pmatrix} -B_{11} \\ -B_{11} + 2B_{12} - B_{22} \\ -B_{22} \\ 2i\phi_2 \\ 2i(\phi_1 + \phi_2) \\ 2i\phi_1 \end{pmatrix}
$$

So what I have done is to:

- Take all combinations of the **m** vector components which have 1 or −1 as elements and write down the associated value for the ln Q's.
- Take all the Q's associated with *only one component of* **m** *equal to* 1, that is, [0, 0,± 1, 0, ..., 0]. By *adding* all the Q's symmetric about the central Q, that is, *that* for the zero vector "[0, 0, ..., 0]," in the above table we can *compute all the diagonal elements of the period matrix.*
- By *differencing* all of the same elements about the central Q we get the *phase vector.*

- By taking other combinations of the elements we get the *off-diagonal elements of the period matrix. These correspond to only two elements of the vector* m *equal to 1 or −1.* The example $[0, 0, \pm 1, 0, \pm 1, 0, 0, 0]$ gives the B_{12} off-diagonal term in the period matrix.
- Some of the elements are not included after the sums and differences, but these do not contribute because they just give sums of the phases (the term $\ln Q_2 - \ln Q_{-2} = 2i(\phi_1 + \phi_2)$ in the above last equation).

The final results are given in Table 21.5.

Summary: By considering the m vectors with only 1-wise interactions $[0, 0, \pm 1, 0, \ldots, 0]$ we are able to compute the *diagonal elements of the period matrix and the phases.* Likewise, by considering the m vectors with 2-wise interactions $[0, \pm 1, 0, 0, \pm 1, 0, \ldots, 0]$ we are able to compute the *off-diagonal elements.* Formally speaking *this set of Q's is unique,* that is, for those with 1-wise and 2-wise interactions we can always find the theta-function parameters (period matrix and phases). Therefore, *this set of Q's is identical to the theta parameters.* There is no fundamental difference between the two. All other Q's can be derived from these.

Some other interesting results are the following. Add the logs of the first and third Q's to get:

$$\ln Q_1 + \ln Q_3 = -\frac{1}{2}(B_{11} + B_{22}) + i(\phi_1 + \phi_2)$$

and

$$\ln Q_{-3} + \ln Q_{-1} = -\frac{1}{2}(B_{11} + B_{22}) - i(\phi_1 + \phi_2)$$

Add these to get the *trace of the period matrix*:

$$B_{11} + B_{22} = -\ln(Q_{-3}Q_3 Q_{-1} Q_1)$$

Table 21.5 Explicit Differences and Sums from Table 21.3 which Give the Parameters in the Riemann Spectrum

$(-3+3)$	$B_{11} = -(\ln Q_{-3} + \ln Q_3) = -\ln(Q_{-3}Q_3)$
$(-1+1)$	$B_{22} = -(\ln Q_{-1} + \ln Q_1) = -\ln(Q_{-1}Q_1)$
$(-2+2)$	$B_{12} = \frac{1}{2}\left(\dfrac{\ln Q_{-2} + \ln Q_2}{\ln Q_{-3} + \ln Q_3 + \ln Q_{-1} + \ln Q_1}\right) = \frac{1}{2}\left(\dfrac{\ln(Q_{-2}Q_2)}{\ln(Q_{-3}Q_3 Q_{-1}Q_1)}\right)$
$(3-(-3))$	$\phi_1 = -\frac{i}{2}(\ln Q_3 - \ln Q_{-3}) = -\frac{i}{2}\ln\dfrac{Q_3}{Q_{-3}}$
$(-1-1)$	$\phi_2 = -\frac{i}{2}(\ln Q_1 - \ln Q_{-1}) = -\frac{i}{2}\ln\dfrac{Q_1}{Q_{-1}}$

This result is to be compared to the *sum of the phases*:

$$\phi_1 + \phi_2 = \frac{i}{2} \ln \frac{Q_2}{Q_{-2}}$$

Nota Bene: We can always evaluate the off-diagonal terms with Schottky uniformization (Chapter 15), algebraic-geometric loop integrals (Chapter 14), or the method of Nakamura-Boyd (Chapter 16). Thus, in the procedure outlined above one does not even need to compute the off-diagonal elements of the Riemann matrix from the Q's, because the off-diagonal elements can be computed from the methods of Chapters 14–16.

21.8   Embedding the Discrete Theta Function

We have the DTF

$$\theta(x) = \sum_{j=0}^{(2M+1)^N - 1} Q_j \exp\left(i\frac{2\pi}{L}J_j x\right) = \sum_{j=0}^{M-1} Q_j \exp\left(i\frac{2\pi}{L}J_j x\right), \quad M = (2M+1)^N$$

The goal of this section is to use the *embedding procedure* (Ruelle and Takens, 1971) to construct an extended space series of the form:

$$\theta(x_k), \theta(x_k + L), \theta(x_k + 2L), \dots, \theta(x_k + nL)$$

Thus, the original vector for the DTF is appended with another which is phase shifted from the first (assuming wrap around to obey the periodic boundary conditions), subsequently followed by a doubly phase shifted version of the theta function, etc. The embeddings in terms of Fourier analysis take the form:

$$\theta(x + nL) = \sum_{j=0}^{M-1} Q_j \exp\left(i\frac{2\pi}{L}I_j(x + nL)\right), \quad M = (2M+1)^N$$

This is interesting because we can write this as

$$\theta(x + nL) = \sum_{j=0}^{M-1} P_j \exp\left(i\frac{2\pi}{L}I_j x\right), \quad P_j = Q_j \exp\left(i\frac{2\pi}{L}I_j nL\right)$$

Therefore, the embedding procedure just phase shifts the Q_j's. Thus, we can use the embedding procedure to provide an invertible transform, that is

$$\theta = WQ, \quad Q = W^{-1}\theta$$

where now $\theta = [\theta(x_k), \theta(x_k + L), \theta(x_k + 2L), \ldots, \theta(x_k + nL)]$ and the $\mathbf{W}$ matrix is padded with additional terms so that we now have $M \times M$ system of equations, $M = (2M + 1)^N$. The advantage of the approach is that the linear equations are now invertible. The disadvantage is that there are a lot of them! Computation is now proportional to M^2 for the brute-force case, but this can be reduced to a *fast computation* with $M \ln M$ rather than M^2 terms by using the standard FFT algorithm.

21.9 A Numerical Example for Extracting the Riemann Spectrum from the Q's

For those brave enough to attempt to program the approach of Section 21.7 for computing the Riemann spectrum from the Q's I give some details about a more complex case for which the number of discrete points in the theta function is $N = 8$, but for which the number of Q's is 25. This case is rich enough that one can see all of the structure of the DTFs. Indeed it becomes clear how to find the particular terms in Equations (9.1) and (9.2) to demonstrate the fact that the theta function is nothing more than a linear Fourier transform with particular coefficients. Likewise I give a table of the I_{ij} for aid in selecting how to extract the particular Q's which are useful for computing the Riemann spectrum.

Example: $N = 8$, $M = 2$, $N = 2$ (two degrees of freedom summed from -2 to 2) so that $M = (2M + 1)^N = 25$. Use Table 21.6 for aid in program checkout.

Table 21.7 is a table of the matrix I_{ij}. This should provide all the aid one needs for program development and checkout. Note that the matrix I_{ij} has ones on the diagonal elements and a pattern of ones in the off-diagonal elements.

Or one can use the *truncated version* of the matrix I_{ij} (Table 21.8):

Table 21.6 Table of Theta-Summation Parameters for the Special Case of $N = 8$ and $M = 25$

l	j	j Centered	(m_1, m_2)	Coefficients $\ln Q_j$	I_j	Theta Summation Element, w^{l_j}
1	0	-12	$(-2, -2)$	$\ln Q_0 = -2(B_{11} + 2B_{12} + B_{22})/2$ $- i(2\phi_1 + 2\phi_2)$	-6	$e^{-6 \cdot 2\pi i/N}$
2	1	-11	$(-2, -1)$	$\ln Q_1 = -(4 B_{11} + 4B_{12} + B_{22})/2$ $- i(2\phi_1 + \phi_2)$	-4	$e^{-4 \cdot 2\pi i/N}$
3	2	-10	$(-2, 0)$	$\ln Q_2 = -2B_{11} - 2i\phi_1$	-2	$e^{-2 \cdot 2\pi i/N}$
4	3	-9	$(-2, 1)$	$\ln Q_3 = -(4B_{11} - 4B_{12} + B_{22})/2$ $- i(2\phi_1 - \phi_2)$	0	1

Continued

Table 21.6 Table of Theta-Summation Parameters for the Special Case of $N = 8$ and $M = 25$

l	j	j Centered	(m_1, m_2)	Coefficients ln Q_j	I_j	Theta Summation Element, w^{I_j}
5	4	−8	(−2, 2)	$\ln Q_4 = -2(B_{11} - 2B_{12} + B_{22})$ $- i(2\phi_1 - 2\phi_2)$	2	$e^{2 \cdot 2\pi i/N}$
6	5	−7	(−1, −2)	$\ln Q_5 = -(B_{11} + 4B_{12} + 4B_{22})/2$ $- i(\phi_1 + 2\phi_2)$	−5	$e^{-5 \cdot 2\pi i/N}$
7	6	−6	(−1, −1)	$\ln Q_6 = -(B_{11} + 2B_{12} + B_{22})/2$ $- i(\phi_1 + \phi_2)$	−3	$e^{-3 \cdot 2\pi i/N}$
8	7	−5	(−1, 0)	$\ln Q_7 = -B_{11}/2 - i\phi_1$	−1	$e^{-1 \cdot 2\pi i/N}$
9	8	−4	(−1, 1)	$\ln Q_8 = -(B_{11} - 2B_{12} + B_{22})/2$ $- i(\phi_1 - \phi_2)$	1	$e^{1 \cdot 2\pi i/N}$
10	9	−3	(−1, 2)	$\ln Q_9 = -(B_{11} - 4B_{12} + 4B_{22})/2$ $- i(\phi_1 - 2\phi_2)$	3	$e^{3 \cdot 2\pi i/N}$
11	10	−2	(0, −2)	$\ln Q_{10} = -2B_{22} - 2i\phi_2$	−4	$e^{-4 \cdot 2\pi i/N}$
12	11	−1	(0, −1)	$\ln Q_{11} = -B_{22}/2 - i\phi_2$	−2	$e^{-2 \cdot 2\pi i/N}$
13	12	0	(0, 0)	$\ln Q_{12} = 0$	0	1
14	13	1	(0, 1)	$\ln Q_{13} = -B_{22}/2 + i\phi_2$	2	$e^{2 \cdot 2\pi i/N}$
15	14	2	(0, 2)	$\ln Q_{14} = -2B_{22} + 2i\phi_2$	4	$e^{4 \cdot 2\pi i/N}$
16	15	3	(1, −2)	$\ln Q_{15} = -(B_{11} - 4B_{12} + 4B_{22})/2$ $+ i(\phi_1 - 2\phi_2)$	−3	$e^{-3 \cdot 2\pi i/N}$
17	16	4	(1, −1)	$\ln Q_{16} = -(B_{11} - 2B_{12} + B_{22})/2$ $+ i(\phi_1 - \phi_2)$	−1	$e^{-1 \cdot 2\pi i/N}$
18	17	5	(1, 0)	$\ln Q_{17} = -B_{11}/2 + i\phi_1$	1	$e^{1 \cdot 2\pi i/N}$
19	18	6	(1, 1)	$\ln Q_{18} = -(B_{11} + 2B_{12} + B_{22})/2$ $+ i(\phi_1 + \phi_2)$	3	$e^{3 \cdot 2\pi i/N}$
20	19	7	(1, 2)	$\ln Q_{19} = -(B_{11} + 4B_{12} + 4B_{22})/2$ $+ i(\phi_1 + 2\phi_2)$	5	$e^{5 \cdot 2\pi i/N}$
21	20	8	(2, −2)	$\ln Q_{20} = -2(B_{11} - 2B_{12} + B_{22})$ $+ i(2\phi_1 - 2\phi_2)$	−2	$e^{-2 \cdot 2\pi i/N}$
22	21	9	(2, −1)	$\ln Q_{21} = -(4B_{11} - 4B_{12} + B_{22})/2$ $+ i(2\phi_1 - \phi_2)$	0	1
23	22	10	(2, 0)	$\ln Q_{22} = -2B_{11} + 2i\phi_1$	2	$e^{2 \cdot 2\pi i/N}$
24	23	11	(2, 1)	$\ln Q_{23} = -(4B_{11} + 4B_{12} + B_{22})/2$ $+ i(2\phi_1 + \phi_2)$	4	$e^{4 \cdot 2\pi i/N}$
25	24	12	(2, 2)	$\ln Q_{24} = -2(B_{11} + 2B_{12} + B_{22})/2$ $+ i(2\phi_1 + 2\phi_2)$	6	$e^{6 \cdot 2\pi i/N}$

Table 21.7 The Matrix I_{ij} for the Special Case of $N = 8$ and $M = 25$

q_j		25	24	23	22	21	20	19	18	17	16	15	14	13	12	11	10	9	8	7	6	5	4	3	2	1	
q_{-12}	4	0	0	1	0	0	0	0	0	0	0	0	1	0	0	0	0	0	0	0	0	1	0	0	0	1	1
q_{-11}	4	0	1	0	0	0	0	0	0	0	0	1	0	0	0	1	0	0	0	0	0	0	0	0	1	0	2
q_{-10}	4	1	0	0	0	1	0	0	0	0	0	0	0	0	1	0	0	0	0	0	0	0	0	1	0	0	3
q_{-9}	3	0	0	0	1	0	0	0	0	0	0	0	0	1	0	0	0	0	0	0	0	0	1	0	0	0	4
q_{-8}	4	0	0	1	0	0	0	0	0	0	0	0	1	0	0	0	0	0	0	0	0	1	0	0	0	1	5
q_{-7}	3	0	0	0	0	0	0	1	0	0	0	0	0	0	0	0	1	0	0	0	1	0	0	0	0	0	6
q_{-6}	3	0	0	0	0	0	1	0	0	0	1	0	0	0	0	0	0	0	0	1	0	0	0	0	0	0	7
q_{-5}	2	0	0	0	0	0	0	0	0	1	0	0	0	0	0	0	0	0	1	0	0	0	0	0	0	0	8
q_{-4}	2	0	0	0	0	0	0	0	1	0	0	0	0	0	0	0	0	1	0	0	0	0	0	0	0	0	9
q_{-3}	3	0	0	0	0	0	0	1	0	0	0	0	0	0	0	0	1	0	0	0	1	0	0	0	0	0	10
q_{-2}	4	0	1	0	0	0	0	0	0	0	0	1	0	0	0	1	0	0	0	0	0	0	0	0	1	0	11
q_{-1}	4	1	0	0	0	1	0	0	0	0	0	0	0	0	1	0	0	0	0	0	0	0	0	1	0	0	12
q_0	3	0	0	0	1	0	0	0	0	0	0	0	0	1	0	0	0	0	0	0	0	0	1	0	0	0	13
q_1	4	0	0	1	0	0	0	0	0	0	0	0	1	0	0	0	0	0	0	0	0	1	0	0	0	1	14
q_2	4	0	1	0	0	0	0	0	0	0	0	1	0	0	0	1	0	0	0	0	0	0	0	0	1	0	15
q_3	3	0	0	0	0	0	1	0	0	0	1	0	0	0	0	0	0	0	0	1	0	0	0	0	0	0	16
q_4	2	0	0	0	0	0	0	0	0	1	0	0	0	0	0	0	0	0	1	0	0	0	0	0	0	0	17

Continued

Table 21.7 The Matrix I_{ij} for the Special Case of $N = 8$ and $M = 25$

	1	2	3	4	5	6	7	8	9	10	11	12	13	14	15	16	17	18	19	20	21	22	23	24	25		q_j
18	0	0	0	0	0	0	0	0	1	0	0	0	0	0	0	0	0	1	0	0	0	0	0	0	0	2	q_5
19	0	0	0	0	0	1	0	0	0	1	0	0	0	0	0	0	0	0	1	0	0	0	0	0	0	3	q_6
20	0	0	0	0	0	0	1	0	0	0	0	0	0	0	0	1	0	0	0	1	0	0	0	0	0	3	q_7
21	0	0	1	0	0	0	0	0	0	0	0	1	0	0	0	0	0	0	0	0	1	0	0	0	1	4	q_8
22	0	0	0	1	0	0	0	0	0	0	0	0	1	0	0	0	0	0	0	0	0	1	0	0	0	3	q_9
23	1	0	0	0	1	0	0	0	0	0	0	0	0	1	0	0	0	0	0	0	0	0	1	0	0	4	q_{10}
24	0	1	0	0	0	0	0	0	0	0	1	0	0	0	1	0	0	0	0	0	0	0	0	1	0	4	q_{11}
25	0	0	1	0	0	0	0	0	0	0	0	1	0	0	0	0	0	0	0	0	1	0	0	0	1	4	q_{12}

Table 21.8 The Truncated Matrix I_{ij} for the Special Case of $N = 8$ and $M = 25$														
	1	2	3	4	5	6	7	8	9	10	11	12	No.	q_j	$\overline{I}Q$
1	1	0	0	0	1	0	0	0	0	0	0	0	2	q_{-12}	$q_{-12} + q_{-8}$
2	0	1	0	0	0	0	0	0	0	0	1	0	2	q_{-11}	$q_{-11} + q_{-2}$
3	0	0	1	0	0	0	0	0	0	0	0	1	2	q_{-10}	$q_{-10} + q_{-1}$
4	0	0	0	1	0	0	0	0	0	0	0	0	1	q_{-9}	q_{-9}
5	1	0	0	0	1	0	0	0	0	0	0	0	2	q_{-8}	$q_{-12} + q_{-8}$
6	0	0	0	0	0	1	0	0	0	1	0	0	2	q_{-7}	$q_{-7} + q_{-3}$
7	0	0	0	0	0	0	1	0	0	0	0	0	1	q_{-6}	q_{-6}
8	0	0	0	0	0	0	0	1	0	0	0	0	1	q_{-5}	q_{-5}
9	0	0	0	0	0	0	0	0	1	0	0	0	1	q_{-4}	q_{-4}
10	0	0	0	0	0	1	0	0	0	1	0	0	2	q_{-3}	$q_{-7} + q_{-3}$
11	0	1	0	0	0	0	0	0	0	0	1	0	2	q_{-2}	$q_{-11} + q_{-2}$
12	0	0	1	0	0	0	0	0	0	0	0	1	2	q_{-1}	$q_{-10} + q_{-1}$

22 Summing Riemann Theta Functions over the *N*-Ellipsoid

22.1 Introduction

To numerically sum the theta functions we clearly cannot sum over an infinite range. And even summing over constant limits (rather than infinity) in a "brute-force" algorithm (Chapter 20) leads to very large computer times. Therefore, an important question is: Over what range can we sum the theta function in order to guarantee convergence, but at the same time to save computer time, sum over as small a number of terms as possible? This chapter addresses one approach to this question. Clearly some consideration of errors in the computation of the theta function must also be made. Perhaps the simplest, although quite naïve approach, would be to first consider the case for the one-dimensional partial theta summation given by

$$\theta(x) = \sum_{n=-M}^{M} q^{n^2} e^{inkx} = 1 + 2\sum_{n=1}^{M} q^{n^2} \cos(nkx), \quad q = e^{-(1/2)B_{11}}, \qquad (22.1)$$

where M is a finite integer. We can treat this by simply assuming that $q^{M^2} < c$ for some value of M, where c is a small number, perhaps $\sim 10^{-10}$. Here B_{11} is the associated diagonal element of the (1×1) period matrix. The limit M then ensures that we get most of the theta-function summation, out to an error c that we *a priori* choose for a given value of B_{11}.

If we carry out a similar process for each of the diagonal elements of the period matrix then it seems reasonable to write (see Chapter 20)

$$\theta(x, t | \mathbf{B}, \boldsymbol{\phi}) = \sum_{m_1=-M_1}^{M_1} \sum_{m_2=-M_2}^{M_2} \cdots \sum_{m_N=-M_N}^{M_N} \exp\left\{-\frac{1}{2}\sum_{m=1}^{N}\sum_{n=1}^{N} m_m m_n B_{mn}\right\}$$

$$\times \exp\left\{i\sum_{n=1}^{N} m_n \kappa_n x - i\sum_{n=1}^{N} m_n \omega_n t + i\sum_{n=1}^{N} m_n \phi_n\right\}. \qquad (22.2)$$

Here each of the summation limits guarantees the convergence of the associated one degree-of-freedom theta function related to each of the diagonal elements of the period matrix, B_{ii}. In this way, we are led immediately to an *N-rectangular parallelepiped* for the partial theta-function summation (each

Doi: 10.1016/S0074-6142(10)97022-8

side of the N-dimensional figure has length $2M_n$). The number of terms in the above partial summation is then given by Osborne (1995a,b):

$$N_{\text{rect}} = \prod_{n=1}^{N} (2M_n + 1).$$
(22.3)

For the special case when $M=M_1=M_2=L=M_N$ we have a summation over the *hypercube* and the latter equation reduces to

$$N_{\text{cube}} = (2M + 1)^N.$$
(22.4)

Thus in this case the number of terms in the summation of the theta function occurs over a hypercube in lattice space, a number that increases exponentially with the number of degrees of freedom N! This is of course bad news for those who would seek to compute theta functions numerically. The focus of this chapter is to show how to make these computations *polynomial* rather than *exponential* in character, thus rendering the approach practical from a computer-time perspective.

Clearly, the hypercube *circumscribes* an N-sphere or *hypersphere*. For the purposes of numerical computation we can compare the number of terms in the partial theta-function summation in the hypercube $N_{\text{cube}} = (2M + 1)^N$ to the number of terms inside the enclosed hypersphere, N_{sphere}. As seen in the next section $N_{\text{sphere}} \ll N_{\text{cube}}$, thus offering large savings in computer time for those who might be erroneously tempted to use the N-cube for computing theta functions. Subsequently, in later sections, I discuss how an associated N-*ellipsoid* (circumscribed by an N-*rectangular parallelepiped*) is even more efficient at selecting the appropriate terms in the theta function, a method that results in even greater savings in computer time.

22.2   Summing over the N-Sphere or Hypersphere

Here is the partial theta summation over the N-cube:

$$\theta(x, t | \mathbf{B}, \boldsymbol{\phi}) = \sum_{m_1=-M}^{M} \sum_{m_2=-M}^{M} \cdots \sum_{m_N=-M}^{M} \exp\left\{ -\frac{1}{2} \sum_{m=1}^{N} \sum_{n=1}^{N} m_m m_n B_{mn} \right\}$$

$$\exp\left\{ i \sum_{n=1}^{N} m_n \kappa_n x - i \sum_{n=1}^{N} m_n \omega_n t + i \sum_{n=1}^{N} m_n \phi_n \right\}.$$
(22.5)

When we say that this summation is over an N-cube we mean that we are summing over the components of the $\mathbf{m}$ vector in an N-dimensional lattice space where the dimension of one side of the N-cube is $2M$. In writing the above expression we assume that the integer M has been selected based upon the smallest element on the diagonal of the period matrix, that is, the element that is most nonlinear. Convergence in this case is also assured for all the other elements in the period matrix.

Now I would like to sum the theta function over the associated N-sphere that is circumscribed by an N-cube. The N-circles and spheres are given by the successive simple formulas:

$$
\begin{aligned}
m_1^2 &= R^2 & \text{line(1D)} & \Rightarrow m_1 = R = M, \\
m_1^2 + m_2^2 &= R^2 & \text{circle(2D)} & \Rightarrow m_2 = \sqrt{R^2 - m_1^2}, \\
m_1^2 + m_2^2 + m_3^2 &= R^2 & \text{sphere(3D)} & \Rightarrow m_3 = \sqrt{R^2 - m_1^2 - m_2^2}, \\
m_1^2 + m_2^2 + m_3^2 + \ldots + m_N^2 &= R^2 & N - \text{sphere(ND)} & \Rightarrow m_N = \sqrt{R^2 - m_1^2 - \ldots - m_{N-1}^2}.
\end{aligned}
$$

$$(22.6)$$

Then the theta function has the following partial summation over the N-sphere:

$$
\theta(x, t | \mathbf{B}, \boldsymbol{\phi}) = \sum_{m_1=-R}^{R} \sum_{m_2=-R_1}^{R_1} \cdots \sum_{m_N=-R_{N-1}}^{R_{N-1}} \exp\left\{ -\frac{1}{2} \sum_{m=1}^{N} \sum_{n=1}^{N} m_m m_n B_{mn} \right\}
$$

$$(22.7)$$

$$
\times \exp\left\{ i \sum_{n=1}^{N} m_n \kappa_n x - i \sum_{n=1}^{N} m_n \omega_n t + i \sum_{n=1}^{N} m_n \phi_n \right\},
$$

where

$$
\begin{aligned}
R_1 &= I\sqrt{M^2 - m_1^2}, \\
R_2 &= I\sqrt{M^2 - m_1^2 - m_2^2}, \\
&\vdots \\
R_N &= I\sqrt{M^2 - m_1^2 - m_2^2 - \ldots - m_N^2},
\end{aligned}
$$

$$(22.8)$$

so that the partial theta-function summation (22.7) becomes

$$
\theta(x, t) = \sum_{m_1=-M}^{M} \sum_{m_2=-I\sqrt{M^2-m_1^2}}^{I\sqrt{M^2-m_1^2}} \sum_{m_3=-I\sqrt{M^2-m_1^2-m_2^2}}^{I\sqrt{M^2-m_1^2-m_2^2}} \cdots \sum_{m_N=-I\sqrt{M^2-m_1^2-m_2^2-\ldots-m_{N-1}^2}}^{I\sqrt{M^2-m_1^2-m_2^2-\ldots-m_{N-1}^2}} .
$$

$$
\times \exp\left(-\frac{1}{2} \sum_{m=1}^{N} \sum_{n=1}^{N} m_m m_n B_{mn} \right) \exp\left(i \sum_{n=1}^{N} m_n \kappa_n x - i \sum_{n=1}^{N} m_n \omega_n t + i \sum_{n=1}^{N} m_n \phi_n \right).
$$

$$(22.9)$$

Here $I(\ )$ means "integer part of." By summing over the N-sphere we are "cutting off the corners of the N-cube" to save summing over this extra volume which is not needed because the actual volume to be summed over, and the associated error, depends only on the radial coordinate, a subject discussed in more detail below. The two-dimensional case is illustrated in Figure 22.1 and helps provide motivation for the method. Note that the number of lattice points inside the circle is less than the number of lattice points inside the circumscribing square. This does not result in great savings in computing time for two dimensions, but for a large number of dimensions the savings are considerable as we now see.

Now consider a specific case such that $M = 2$, $N = 50$, that is, we have a 50-cube of side $2M = 4$ so that the enclosed 50-sphere has a diameter 4 in lattice space. Figure 22.2 is a graph of the number of terms to be included in the partial theta summation (the number of complex exponential terms in Equation (22.9)) in the N-dimensional lattice space versus the number of degrees of freedom N (number of summations in the theta function). Note that the number of terms is $N_{sphere}=1{,}923{,}350$ for 50 degrees of freedom. On the other hand, a simple calculation reveals that the number of terms for the circumscribing cube is $N_{cube} = 8.88 \times 10^{34}$! The ratio of the n-sphere volume to the n-cube volume is $N_{sphere}/N_{cube} = 2.166 \times 10^{-29}$, a huge savings in computer time.

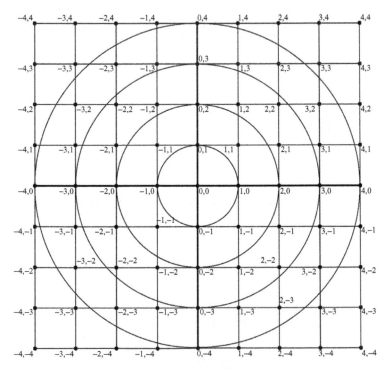

Figure 22.1 Two-dimensional squares and circumscribed circles.

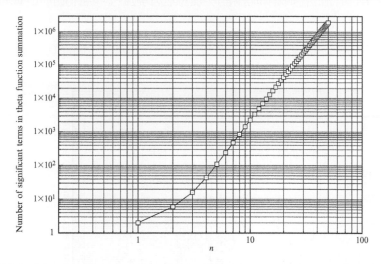

Figure 22.2 Number of terms in partial theta-function sum on an N-sphere in N dimensions in lattice space.

To aid in getting a quick look at the computer-time saving I graph, in Figure 22.3, the *ratio* of the volume of the N-sphere to the N-cube as a function of the number of degrees of freedom. This surprising graph reflects the relative computer time between the computation of the theta function for the 50-sphere and the computation for the 50-cube. This is of course a relative measure of the computer time, but it is suggestive that by computing over the N-sphere rather than the N-cube we can save large amounts of computer time. It is not hard to show that this graph is independent of the parameter M. Thus the relative computer time is always the same, but the actual computer time will of course depend on M.

It is worthwhile considering additional simple examples. The first has $M = 2$, $n = 10$ for which we see from the graph in Figure 22.3 that the ratio of the n-sphere volume to the n-cube is 10^{-3}, implying that to sum the theta function over the cube takes 1000 times more than the number of operations for summing over the hypersphere, that is, we have, for 10 degrees of freedom, a factor of 1000 savings in computer time. The cube summation requires $(2M + 1)^N = 5^{10} = 9.76 \times 10^6$ operations, while summing over the sphere is about 1000 times fewer operations. Another example is that for $n = 13$ and we see that summing over the N-sphere requires about 10,000 times less computer time.

Consider the example for $M = 4$, $n = 20$, this means $(2M + 1)^N = 9^{20} = 1.21 \times 10^{19}$ operations. The ratio of the volume of the sphere to the cube is 10^{-9}. Once again this ratio is independent of M.

We now show that additional computer time can be saved by summing over the N-ellipsoid rather than the N-sphere.

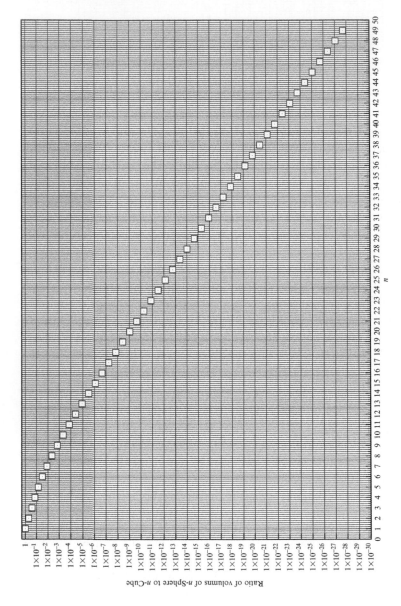

Figure 22.3   Ratio of N-sphere to N-cube for a 50-dimensional lattice space.

22.3 The Ellipse in Two Dimensions

The above results suggest that summing over the N-sphere rather than the N-cube can save huge amounts of computer time. Considerably greater amounts of computer time can be saved because that the N-rectangular parallelepiped circumscribes a smaller volume N-ellipsoid. It is this latter volume in lattice space that is most appropriate for summing theta functions. To see this consider the generalized nome defined by

$$q = \exp\left\{-\frac{1}{2}\sum_{i=1}^{N}\sum_{j=1}^{N} m_i m_j B_{ij}\right\}, \tag{22.10}$$

where $B_{ij} = B_{ji}$ because the period matrix is symmetric. In two dimensions ($N = 2$) this is

$$q = \exp\left\{-\frac{1}{2}\sum_{i=1}^{N}\sum_{j=1}^{N} m_i m_j B_{ij}\right\} = \exp\left\{-\frac{1}{2}m_1^2 B_{11} - m_1 m_2 B_{12} - \frac{1}{2}m_2^2 B_{22}\right\}. \tag{22.11}$$

Taking the logarithm gives the *two-dimensional ellipse* in the lattice (integer) space (m_1, m_2) we have:

$$m_1^2 B_{11} + 2m_1 m_2 B_{12} + m_2^2 B_{22} = -2\ln q_0 \tag{22.12}$$

We are in *lattice space* where m_1, m_2 are *orthogonal, integer coordinates.* Notice that q_0 is here taken to be the numerical "zero" in the computations, typically on the order of, say, 10^{-10}. We are of course interested in summing the two-dimensional theta function over all lattice points *inside* this ellipse.

More generally, we have the N-dimensional ellipse given by

$$\sum_{i=1}^{N}\sum_{j=1}^{N} m_i m_j B_{ij} = 2|\ln q_0|. \tag{22.13}$$

The absolute value expression on the right has been used because $q_0 < 1$ always. Equation (22.13) is the basis for the computation of N-dimensional theta functions and is discussed in detail below. First, to provide some intuition, I look at some of the details of the two-dimensional case in the following section.

22.4 Principal Axis Coordinates in Two Dimensions

It is easy to find a set of principal axis coordinates in which the ellipse can be written in "standard form." In what follows I set

$$x = m_1, \quad y = m_2$$

and temporarily forget that we have an integer space. Then

$$ax^2 + bxy + cy^2 = d, \quad \text{where} \quad a = B_{11}, b = 2B_{12}, c = B_{22}, d = -2 \ln q_0$$

$$(22.14)$$

This is just an ellipse in two dimensions. Rotating to a new system of coordinates gives

$$\begin{bmatrix} x \\ y \end{bmatrix} = \begin{bmatrix} \cos \theta & -\sin \theta \\ \sin \theta & \cos \theta \end{bmatrix} \begin{bmatrix} x' \\ y' \end{bmatrix}.$$

The above equation for the ellipse can be written in the new coordinate system:

$$(a \cos^2 \theta + b \sin \theta \cos \theta + c \sin^2 \theta)x'^2$$
$$+ [-2a \sin \theta \cos \theta + b(\cos^2 \theta - \sin^2 \theta) + 2c \sin \theta \cos \theta]x'y' \qquad (22.15)$$
$$+ (a \sin^2 \theta - b \sin \theta \cos \theta + c \cos^2 \theta)y'^2 + d = 0.$$

To find the *principal axes frame* we need to determine the angle for which we have the ellipse in *standard form*, that is, for which the cross term $x'y'$ has zero coefficient:

$$-2a \sin \theta \cos \theta + b(\cos^2 \theta - \sin^2 \theta) + 2c \sin \theta \cos \theta = 0, \qquad (22.16)$$

so that the angle of the major axis of the ellipse with respect to the x-axis is:

$$\cot 2\theta = \frac{a - c}{b}. \qquad (22.17)$$

Using the triangle of Figure 22.4 we find

$$\sin 2\theta = \frac{b}{\sqrt{(a - c)^2 + b^2}}, \quad \cos 2\theta = \frac{a - c}{\sqrt{(a - c)^2 + b^2}}.$$

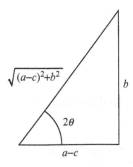

Figure 22.4   Triangle for rotating an ellipse to principal axis coordinates.

Translating back to the elements of the period matrix gives

$$\sin 2\theta = \frac{2B_{12}}{\sqrt{(B_{11} - B_{22})^2 + 4B_{12}^2}}, \quad \cos 2\theta = \frac{B_{11} - B_{22}}{\sqrt{(B_{11} - B_{22})^2 + 4B_{12}^2}} \quad (22.18)$$

or

$$\tan 2\theta = \frac{2B_{12}}{B_{11} - B_{22}} \quad (22.19)$$

See the triangle of Figure 22.5.

An interesting limit occurs when $B_{11}=B_{22}$ (all period matrix elements have the same amplitude) and for "no interactions" we have off-diagonal terms $B_{12} = 0$. The physics of nonlinear interactions actually precludes this, but we assume in calculations of this type that the off-diagonal elements of the period matrix are small with respect to the diagonal elements, i.e., we are near the linear limit). To this end, the ellipse in the principal-axis frame is given by

$$Ax'^2 + By'^2 = d \quad (22.20)$$

where

$$A = a \, \cos^2 \theta + b \, \sin \theta \cos \theta + c \, \sin^2 \theta,$$
$$B = a \, \sin^2 \theta - b \, \sin \theta \cos \theta + c \, \cos^2 \theta,$$

or in terms of the period matrix:

$$A = B_{11} \, \cos^2 \theta + 2B_{12} \, \sin \theta \cos \theta + B_{22} \, \sin^2 \theta$$
$$B = B_{11} \, \sin^2 \theta - 2B_{12} \, \sin \theta \, \cos \theta + B_{22} \, \cos^2 \theta \quad (22.21)$$

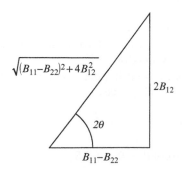

Figure 22.5 Triangle for rotating an ellipse to principal axis coordinates in lattice space of the theta function.

In the principal axis frame we have the ellipse equation:

$$\frac{Ax'^2}{d} + \frac{By'^2}{d} = 1 \tag{22.22}$$

or returning to integer lattice coordinates:

$$\frac{m_1^2}{\alpha^2} + \frac{m_2^2}{\beta^2} = 1, \tag{22.23}$$

where

$$\alpha = \sqrt{\frac{d}{A}} = \sqrt{\frac{2|\ln q_0|}{B_{11}\cos^2\theta + 2B_{12}\sin\theta\,\cos\theta + B_{22}\sin^2\theta}}$$

$$\beta = \sqrt{\frac{d}{B}} = \sqrt{\frac{2|\ln q_0|}{B_{11}\cos^2\theta - 2B_{12}\sin\theta\,\cos\theta + B_{22}\sin^2\theta}} \tag{22.24}$$

I have used absolute values because the logarithm of q_0 is always negative since $q_0 < 1$, which means $|\ln q_0| = -\ln q_0$. Of course the simplest case occurs for no interactions, that is, $B_{12} = 0$ and when $B_{11} = B_{22}$ we have:

$$\alpha = \sqrt{\frac{d}{A}} = \sqrt{\frac{2|\ln q_0|}{B_{11}}}$$

$$\beta = \sqrt{\frac{d}{B}} = \sqrt{\frac{2|\ln q_0|}{B_{22}}} = \alpha \tag{22.25}$$

which is the case for a *circle* in 2D with radius

$$R = \alpha = \beta = \sqrt{\frac{2|\ln q_0|}{B_{11}}}$$

and we have the N-sphere in two dimensions. Again the value q_0 is just the precision one wants, that is, say 10^{-10}. Furthermore to compute R we take the integer part of the above expression.

Now let us take a look at what happens when we have $B_{11} = B_{22}$, but retain interactions, that is, $B_{12} \neq 0$. This gives

$$2\theta = 90° \Rightarrow \theta = 45°$$

$$A = B_{11} + 2B_{12}\sin\theta\cos\theta = B_{11} + 2B_{12}\frac{\sqrt{2}}{2}\frac{\sqrt{2}}{2} = B_{11} + B_{12}$$

$$B = B_{22} - 2B_{12}\sin\theta\cos\theta = B_{22} - 2B_{12}\frac{\sqrt{2}}{2}\frac{\sqrt{2}}{2} = B_{22} - B_{12}$$

so that

$$\alpha = \sqrt{\frac{d}{A}} = \sqrt{\frac{2|\ln q_0|}{B_{11} + B_{12}}}, \quad \beta = \sqrt{\frac{d}{B}} = \sqrt{\frac{2|\ln q_0|}{B_{22} - B_{12}}}.$$

We thus have an ellipsoid at an angle of 45° with major and minor axes which differ due the interaction term. This case corresponds to the ellipsoid circumscribed by a sphere, circumscribed in turn by a cube. Therefore, the ellipsoid requires even less computer time than a sphere or cube.

22.5 Solving for the Coordinate m_2 in Terms of m_1

To set the stage for further calculations, consider the following example where I set $B_{11} = 6.5879$, $B_{22} = 6.8551$, and $B_{12} = 2.1972$ where the value $q_0 = 10^{-12}$ is also taken. The ellipse has been computed using the formulas above and is graphed below. For this example, the left most point has coordinates $m_1 = -4.05893$, $m_2 = 1.9147$, which is the leftmost point just inside of the ellipse. Clearly m_1 should be summed from $m_1 \Rightarrow -4, 4$. The other limits for m_2 are computed as a function of m_1 by the method described below. In any event, it is easy to see from the graph in Figure 22.6 which lattice points to sum over, that is, they are just those inside the ellipse corresponding to the desired error in the theta-function summation; the diagonals are the principal axes. The dots correspond to the relevant points in lattice space. Note that only half of them are shown, as the others obey the simple transformation $\mathbf{m} \to -\mathbf{m}$; this observation saves a factor of 2 in computer time. This is because all of the terms denoted by dots can be computed and the remainder can be computed by letting $\mathbf{m} \to -\mathbf{m}$. Furthermore, the point at $m_1 = 0$, $m_2 = 0$ just gives a contribution "1" to the theta function, and this can be added in later. Therefore, the lattice points to be summed over in the theta function are those given in Table 22.1 and Figure 22.6.

Given a value of the integer index m_1 (the first summation in the theta function) we would like to find the appropriate analytical values to be summed over for the second integer index m_2 as shown in the above figure and in the table. The equation of the ellipse is

$$B_{22}m_2^2 + (2B_{12}m_1)m_2 + B_{11}m_1^2 + 2\ln q_0 = 0,$$

which we would like to solve for m_2 in terms of m_1 as a quadratic equation. Hence we write

$$ay^2 + by + c = 0, \quad y = \frac{-b \pm \sqrt{b^2 - 4ac}}{2a},$$

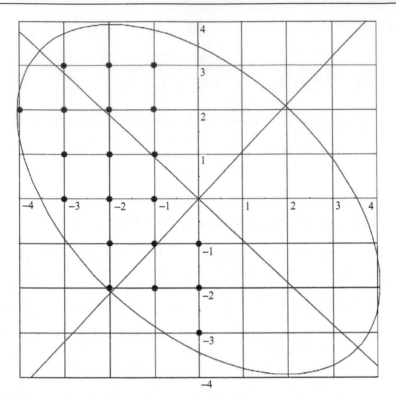

Figure 22.6 Rotation of an ellipse to principal axis coordinates in lattice space for a two-dimensional Riemann matrix, artificial zero 10^{-12}.

where

$$y = m_2, \quad x = m_1$$

and

$$a = B_{22}, \quad b = 2B_{12}m_1, \quad c = B_{11}m_1^2 + 2\ln q_0.$$

The solution to the above quadratic equation is, using the fact that $\ln q_0 < 0$:

$$m_2 = \frac{-B_{12}m_1}{B_{22}} \pm \sqrt{\frac{2|\ln q_0|}{B_{22}} - \frac{(B_{22}B_{11} - B_{12}^2)}{B_{22}^2} m_1^2}$$

This result is programmed "on the fly," that is, as m_1 changes (is iterated over) then m_2 also changes by the above formula. This formula can be seen to easily predict the correct lattice points for the numerical example given in Table 22.1.

Table 22.1 Lattice Points to be
Summed Over in the Theta Function
for the Case of Figure 22.6

m_1	m_2
−4	2
−3	0
−3	1
−3	2
−3	3
−2	−2
−2	−1
−2	0
−2	1
−2	2
−2	3
−1	−2
−1	−1
−1	0
−1	1
−1	2
−1	3
0	−3
0	−2
0	−1

Consider the previous example for the special case $B_{12}=0$ and $B_{11}=B_{22}$. This gives

$$m_2 = \pm\sqrt{R^2 - m_1^2}, \quad R = \sqrt{\frac{2|\ln q_0|}{B_{22}}},$$

which are just the results for the 2-sphere. Note that

$$R = I\left(\sqrt{\frac{2|\ln q_0|}{B_{22}}}\right),$$

where I means "integer part of" and that m_1 and m_2 are integers in the lattice space.

22.6 The Case for Three and N Degrees of Freedom

Now let us go to the next step, for *three degrees of freedom*. The equation of the ellipsoid in three dimensions takes the convenient form:

$$B_{11}m_1^2 + 2B_{12}m_1m_2 + B_{22}m_2^2 + 2B_{13}m_1m_3 + B_{33}m_3^2 + 2B_{23}m_2m_3 + 2\ln q_0 = 0$$

which we would like to solve for m_3 in terms of m_1, m_2 as a quadratic equation. Let us rewrite the above in the following form:

$$B_{33}m_3^2 + 2(B_{13}m_1 + B_{23}m_2)m_3 + B_{11}m_1^2 + 2B_{12}m_1m_2 + B_{22}m_2^2 + 2\ln q_0 = 0.$$

Hence we have

$$ay^2 + 2by + c = 0,$$

$$y = \frac{-2b \pm \sqrt{4b^2 - 4ac}}{2a} = \frac{-b \pm \sqrt{b^2 - ac}}{a} = -\frac{b}{a} \pm \sqrt{\frac{b^2}{a^2} - \frac{c}{a}},$$

where

$$y = m_3$$

and

$$a = B_{33}, \quad b = B_{13}m_1 + B_{23}m_2, \quad c = B_{11}m_1^2 + 2B_{12}m_1m_2 + B_{22}m_2^2 + 2\ln q_0.$$

Use the solution to the above quadratic equation and get

$$m_3 = -\frac{(B_{13}m_1 + B_{23}m_2)}{B_{33}}$$

$$\pm \sqrt{\frac{2|\ln q_0|}{B_{33}} + \frac{(B_{13}m_1 + B_{23}m_2)^2}{B_{33}^2} - \frac{(B_{11}m_1^2 + 2B_{12}m_1m_2 + B_{22}m_2^2)}{B_{33}}}$$

Let us consider the special case for which $B_{12} = B_{13} = B_{23} = 0$ and $B_{11} = B_{22} = B_{33}$. This gives

$$m_3 = \pm\sqrt{\frac{2|\ln q_0|}{B_{33}} - \frac{(B_{11}m_1^2 + B_{22}m_2^2)}{B_{33}}} = \pm\sqrt{R^2 - m_1^2 - m_2^2}, \quad R = \sqrt{\frac{2|\ln q_0|}{B_{33}}},$$

which is just the case for the 3-sphere.

From the above cases for two and three dimensions it is easy to write by inspection the case for computing all the points *inside the N-ellipsoid*:

$$B_{nn}m_n^2 + 2\left(\sum_{j=1}^{n-1} m_j B_{jn}\right)m_n + \sum_{j=1}^{n-1}\sum_{k=1}^{n-1} m_j m_k B_{jk} + 2\ln q_0 = 0$$

Dividing by $2 \ln q_0$ gives the *equation for the N-ellipsoid*:

$$C_{nn}m_n^2 + 2\left(\sum_{j=1}^{n-1} m_j C_{jn}\right)m_n + \sum_{j=1}^{n-1}\sum_{k=1}^{n-1} m_j m_k C_{jk} = 1, \quad C_{jk} = \frac{B_{jk}}{2|\ln q_0|}$$

The *summation limits in the theta function* M_n (which bound the N-ellipsoid) then solve

$$C_{nn}M_n^2 + 2\left(\sum_{j=1}^{n-1} m_j C_{jn}\right)M_n + \sum_{j=1}^{n-1}\sum_{k=1}^{n-1} m_j m_k C_{jk} = 1, \quad C_{jk} = \frac{B_{jk}}{2|\ln q_0|}$$

This can also be written in terms of the M_n limits of the theta function as a function of the period matrix, so we have the alternative form:

$$B_{nn}M_n^2 + 2\left(\sum_{j=1}^{n-1} m_j B_{jn}\right)M_n + \sum_{j=1}^{n-1}\sum_{k=1}^{n-1} m_j m_k B_{jk} + 2\ln q_0 = 0$$

This latter equation is the most important one in this chapter. It describes the upper (+) and lower (−) limits of the theta-function summations $M_n^{\pm}$ given M_1. The signs are taken from the two signs in front of the square root term in the solution of the quadratic equation. Here are the results for the first three dimensions:

One dimension:

$$B_{11}M_1^2 + 2\ln q_0 = 0 \Rightarrow M_1 = \sqrt{\frac{2|\ln q_0|}{B_{11}}}$$

Two dimensions:

$$B_{22}M_2^2 + 2m_1 B_{12}M_2 + m_1^2 B_{11} + 2\ln q_0 = 0$$

Three dimensions:

$$B_{33}M_3^2 + 2(m_1 B_{13} + m_2 B_{23})M_3 + m_1^2 B_{11} + 2m_1 m_2 B_{23} + m_2^2 B_{22} = 2|\ln q_0|$$

Rescale by $2|\ln q_0|$ to get

$$C_{33}M_3^2 + 2(m_1 C_{13} + m_2 C_{23})M_3 + m_1^2 C_{11} + 2m_1 m_2 C_{23} + m_2^2 C_{22} - 1 = 0.$$

These results demonstrate that the previously studied cases for two and three dimensions are recovered from the general formula given above. The general equation for N-dimensions is

$$C_{nn}M_n^2 + 2\left(\sum_{j=1}^{n-1} m_j C_{jn}\right)M_n + \sum_{j=1}^{n-1}\sum_{k=1}^{n-1} m_j m_k C_{jk} = 1, \quad C_{jk} = \frac{B_{jk}}{2|\ln q_0|}$$

We can rewrite it as

$$a_n M_n^2 + b_n M_n + c_n = 0$$

where

$$a_n = C_{nn}$$

$$b_n = 2\left(\sum_{j=1}^{n-1} m_j C_{jn}\right)$$

$$c_n = \sum_{j=1}^{n-1}\sum_{k=1}^{n-1} m_j m_k C_{jk} - 1$$

Finally,

$$M_n^{\pm} = -\left(\frac{\sum_{j=1}^{n-1} m_j C_{jn}}{C_{nn}}\right) \pm \sqrt{R_{nn}^2 + \left(\frac{\sum_{j=1}^{n-1} m_j C_{jn}}{C_{nn}}\right)^2 - \frac{\sum_{j=1}^{n-1}\sum_{k=1}^{n-1} m_j m_k C_{jk}}{C_{nn}}},$$

$$R_{nn}^2 = \frac{1}{C_{nn}} = \frac{2|\ln q_0|}{B_{nn}}$$

This expression gives the upper and lower limits of the theta-function summation in terms of the normalized period matrix $\mathbf{C}$.

22.7   Summation Values for m_1

To get the *starting value* of m_1 we need to look at the leftmost point of the ellipse, where

$$\frac{dm_1}{dm_2} = 0.$$

To this end, we would like to have the solution of the ellipse equation for m_1 in terms of m_2. We have in two dimensions:

$$B_{11}m_1^2 + (2B_{12}m_2)m_1 + B_{22}m_2^2 + 2\ln q_0 = 0$$

We solve for m_1 in terms of m_2 as a quadratic equation, to give starting values for m_1 for the first summation in the theta function. The solution to the above quadratic equation is

$$M_1 = -\frac{B_{12}m_2}{B_{11}} \pm \sqrt{R_{11}^2 - \left(\frac{B_{22}B_{11} - B_{12}^2}{B_{11}^2}\right)m_2^2}, \quad R_{11}^2 = \frac{2|\ln q_0|}{B_{11}}$$

Now take the derivative with respect to m_2 and set this to zero, $dm_1/dm_2 = 0$, to obtain the leftmost and rightmost points of the ellipse in two dimensions. Here is the final result of the coordinate at the leftmost *or* rightmost point of the ellipse, m_1, m_2 at the upper (plus sign) or lower (minus sign) point of the ellipse:

$$M_2 = \pm\sqrt{\frac{B_{11}}{B_{22}}\left(\frac{B_{12}^2}{B_{22}B_{11} - B_{12}^2}\right)}R_{11}$$

Note that as the interactions go to zero ($B_{12} = 0$) we get $M_2 = 0$, the correct value for a circle (22.6).

The coordinate M_1 at the leftmost/rightmost point of the ellipse is found by inserting this value into the above equation:

$$M_1 = -\frac{B_{12}m_2}{B_{11}} \pm \sqrt{R_{11}^2 - \left(\frac{B_{22}B_{11} - B_{12}^2}{B_{11}^2}\right)m_2^2}, \quad R_{11}^2 = \frac{2|\ln q_0|}{B_{11}}$$

For $m_2 = 0$ we get for M_1 the results for the 2-sphere:

$$M_1 = \pm R_{11}.$$

We can simplify the value of M_1:

$$M_1 = \pm\left[\frac{B_{12}^2}{B_{11}B_{22}}\sqrt{\left(\frac{B_{11}B_{22}}{B_{22}B_{11} - B_{12}^2}\right)} + \sqrt{\frac{B_{11}B_{22} - B_{12}^2}{B_{11}B_{22}}}\right]R_{11}$$

This is the coordinate M_1 at the left ($-$) or rightmost point ($+$) of the ellipse.

The point on the left side of the ellipse is given by the pair:

$$M_1 = -\left[\frac{B_{12}^2}{B_{11}B_{22}}\sqrt{\left(\frac{B_{11}B_{22}}{B_{22}B_{11} - B_{12}^2}\right)} + \sqrt{\frac{B_{11}B_{22} - B_{12}^2}{B_{11}B_{22}}}\right]R_{11},$$

$$M_2 = \sqrt{\frac{B_{11}}{B_{22}}\left(\frac{B_{12}^2}{B_{22}B_{11} - B_{12}^2}\right)}R_{11}$$

and the pair on the right side is

$$M_1 = \left[\frac{B_{12}^2}{B_{11}B_{22}} \sqrt{\left(\frac{B_{11}B_{22}}{B_{22}B_{11} - B_{12}^2} \right)} + \sqrt{\frac{B_{11}B_{22} - B_{12}^2}{B_{11}B_{22}}} \right] R_{11},$$

$$M_2 = -\sqrt{\frac{B_{11}}{B_{22}} \left(\frac{B_{12}^2}{B_{22}B_{11} - B_{12}^2} \right)} R_{11}.$$

This is an important result as it gives the starting values of the limits of the two sums in the theta function in two dimensions.

Another case for the ellipse, when $q_0 = 10^{-16}$, for a two-dimensional lattice is shown in Figure 22.7.

Extending these results to N dimensions to determine the starting value of M_1 is straightforward and will be left as an exercise to the reader.

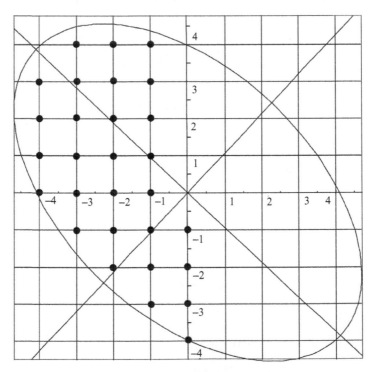

Figure 22.7 Rotation of an ellipse to principal axis coordinates in lattice space for a two-dimensional Riemann matrix, artificial zero 10^{-16}.

22.8 Summary of Theta-Function Summation over Hyperellipsoid

What follows is a summary of the results obtained in this chapter:

Summary of Limits to the Theta-Function Summation

The quadratic equation for the theta-function limits M_n is given by

$$B_{nn}M_n^2 + 2\left(\sum_{j=1}^{n-1} m_j B_{jn}\right)M_n + \sum_{j=1}^{n-1}\sum_{k=1}^{n-1} m_j m_k B_{jk} + 2\ln q_0 = 0$$

Here are the limits $M_n^{\pm}$ in explicit form as a solution of the above quadratic equation:

$$M_n^{\pm} = -\frac{\sum_{j=1}^{n-1} m_j B_{jn}}{B_{nn}} \pm \sqrt{R_{nn}^2 - \frac{\sum_{j=1}^{n-1}\sum_{k=1}^{n-1} m_j m_k B_{jk}}{B_{nn}} + \left(\frac{\sum_{j=1}^{n-1} m_j B_{jn}}{B_{nn}}\right)^2},$$

$$R_{nn}^2 = \frac{2|\ln q_0|}{B_{nn}}$$

Here the values of the limits are assumed to be the integer part of the above formulas. The *minus sign is the lower limit* of the theta-function summation and the *plus sign is the upper limit*. The theta function then has the form:

$$\theta(x,t) = \sum_{m_1=-M_1^-}^{M_1^+}\sum_{m_2=-M_2^-}^{M_2^+}\cdots\sum_{m_N=-M_N^-}^{M_N^+} q_{m_1 m_2,\ldots,m_N}$$

$$\times \exp\left\{i\sum_{j=1}^{N} m_j k_j x - i\sum_{j=1}^{N} m_j\omega_j t + i\sum_{j=1}^{N} m_j\phi_j\right\}$$

Here, the values of $M_1^{\pm}$ for the leftmost point of the ellipsoid, which obeys $\nabla m_1 = 0$, as discussed in the last section. The resultant summation is over the appropriate *N-ellipsoid* that depends on the desired "zero" q_0 (typically, say, 10^{-10}) and on the period matrix B_{ij}. The summations are not symmetric because the N-ellipsoid is not symmetric in lattice space.

Continued

Furthermore, the limits are not constants, but must be computed "on the fly." The result is a theta function that instead of requiring exponential computer time requires *polynomial time provided the spectrum decreases rapidly enough at large wavenumber*! This does not mean that computing theta functions in the soliton limit would not be time consuming, but at least it would be polynomial for a rapidly decreasing spectrum at high wavenumber or frequency. Modular transformations of the Riemann spectrum help relieve many of the difficulties of computing theta functions in the soliton limit (Chapter 8). Likewise Poisson summation can also be very helpful (also in Chapter 8).

Another way to write the above limits is the following:

$$M_n^{\pm} = -\left(\frac{b_{n-1}}{B_{nn}}\right) \pm \sqrt{R_{nn}^2 - \frac{d_{n-1}}{B_{nn}} + \left(\frac{b_{n-1}}{B_{nn}}\right)^2}, \quad R_{nn}^2 = \frac{2|\ln q_0|}{B_{nn}}$$

$$b_{n-1} = \sum_{j=1}^{n-1} m_j B_{jn}$$

$$d_{n-1} = \sum_{j=1}^{n-1} \sum_{k=1}^{n-1} m_j m_k B_{jk}$$

One dimension:

$$M_1 = -\left(\frac{b_0}{B_{11}}\right) \pm \sqrt{R_{11}^2 - \frac{d_0}{B_{11}} + \left(\frac{b_0}{B_{11}}\right)^2},$$

$$b_0 = 0,$$

$$d_0 = 0,$$

$$M_1^{\pm} = \pm\sqrt{\frac{2|\ln q_0|}{B_{11}}} = \pm R_{11}$$

In the numerical computations, I often use a different scaling (a factor of 2π) for the period matrix and so have the formula:

$$\ln q_0 = -\pi B_{11} R_{11}^2, \quad q_0 = e^{-\pi B_{11} R_{11}^2} \Rightarrow R_{11} = \sqrt{\frac{2|\ln q_0|}{B_{11}}} \Rightarrow R_{11} = \sqrt{\frac{|\ln q_0|}{\pi B_{11}}}.$$

Suppose we take $q_0 = 10^{-17}$, $B_{11} = 0.7$ then I get: $R_{11} = 4.093 \Rightarrow 4$, that is, one must sum over four terms in the one degree-of-freedom theta function to retain an accuracy of $q_0 = 10^{-17}$.

Two dimensions:

$$M_2 = -\left(\frac{b_1}{B_{22}}\right) \pm \sqrt{R_{22}^2 - \frac{d_1}{B_{22}} + \left(\frac{b_1}{B_{22}}\right)^2},$$

$$b_1 = m_1 B_{12},$$

$$d_1 = m_1^2 B_{11},$$

$$M_2^\pm = -\left(\frac{m_1 B_{12}}{B_{22}}\right) \pm \sqrt{R_{22}^2 - \frac{B_{11} m_1^2}{B_{22}} + \left(\frac{m_1 B_{12}}{B_{22}}\right)^2}$$

$$= -\left(\frac{m_1 C_{12}}{C_{22}}\right) \pm \sqrt{R_{22}^2 - \frac{C_{11} m_1^2}{C_{22}} + \left(\frac{m_1 C_{12}}{C_{22}}\right)^2}.$$

For the 2-sphere case ($B_{11}=B_{22}$, $B_{12}=0$):

$$M_2 = \pm\sqrt{R_{22}^2 - m_1^2}, \quad R_{22} = \sqrt{\frac{2|\ln q_0|}{B_{22}}}.$$

Three dimensions:

$$M_3 = -\left(\frac{b_2}{B_{33}}\right) \pm \sqrt{\frac{2|\ln q_0|}{B_{33}} - \frac{d_2}{B_{33}} + \left(\frac{b_2}{B_{33}}\right)^2},$$

$$b_2 = m_1 B_{13} + m_2 B_{23},$$

$$d_2 = \sum_{j=1}^{2}\sum_{k=1}^{2} m_j m_k B_{jk} = m_1^2 B_{11} + 2 m_1 m_2 B_{12} + m_2^2 B_{22},$$

$$M_3^\pm = -\left(\frac{m_1 B_{13} + m_2 B_{23}}{B_{33}}\right)$$

$$\pm \sqrt{\frac{2|\ln q_0|}{B_{33}} - \left(\frac{m_1^2 B_{11} + 2 m_1 m_2 B_{12} + m_2^2 B_{22}}{B_{33}}\right) + \left(\frac{m_1 B_{13} + m_2 B_{23}}{B_{33}}\right)^2}.$$

For the 3-sphere case ($B_{11} = B_{22} = B_{33}, B_{12} = B_{23} = B_{13} = 0$):

$$M_3 = \pm\sqrt{\frac{2|\ln q_0|}{B_{33}} - m_1^2 - m_2^2} = \pm\sqrt{R^2 - m_1^2 - m_2^2}, \quad R = \sqrt{\frac{2|\ln q_0|}{B_{33}}}.$$

22.9   Discussion of Convergence of Summation Method

Let us consider "white noise" (for concreteness consider the KdV or KP equations) which have spectral amplitudes given approximately by (for a rough estimate the logarithm in the Hirota transformation has been ignored):

$$a_n \simeq k_n^2 q_n = k_n^2 e^{-(1/2)B_{nn}} = C.$$

This can be solved for the diagonal elements of the period matrix:

$$q_n = e^{-(1/2)B_{nn}} = \frac{C}{k_n^2} \Rightarrow -\frac{1}{2}B_{nn} = \ln\left(\frac{C}{k_n^2}\right).$$

Therefore, the values of the period matrix are given by

$$B_{nn} = -2\ln\left(\frac{k_n^2}{C}\right)$$

for white noise. In this case, the B_{nn} increase monotonically with increasing k_n; this implies that even for a spectrum which is flat, that is, for "white noise" (the spectral amplitudes are all constant, $a_n = C$) we have decreasing nonlinearity with increasing wavenumber. This is a very important result, since it guarantees that there is a kind of "cutoff" in nonlinearity for a physically realizable spectrum, which means that we have a natural truncation of the number of contributing terms in the theta function at high wavenumber. To illustrate this idea, Figure 22.8 is a graph of B_{nn} versus the wavenumber k_n.

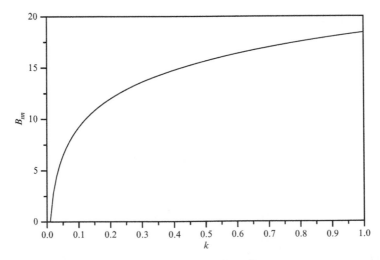

Figure 22.8   Amplitudes of the diagonal elements of the period matrix B_{nn} versus the wavenumbers k_n.

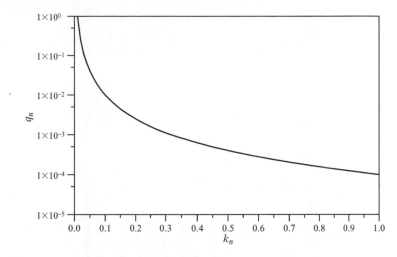

Figure 22.9   Decay of the nomes q_n versus the wavenumbers k_n.

Of course ocean surface waves have spectra which decrease with a power law, $\sim k^{-3}$, and this gives an even faster rate of convergence of the Riemann spectrum with high wavenumber, thus reducing the number of points in lattice space for the theta-function summation. For realistic ocean wave spectra this is an important consideration.

This result bodes well for computation on the N-ellipsoid because the nomes decrease with increasing wavenumber. The nomes have the form:

$$q_n = e^{-(1/2)B_{nn}} = \frac{C}{k_n^2}$$

as a function of wavenumber as shown in Figure 22.9.

But the most important message is that for white noise we have a quadratic reduction in the amplitudes of the q_n as wavenumber increases so that the N-ellipsoid is correspondingly reduced in size at increasing wavenumber; indeed we should have the linear spectral problem at some relatively large wavenumber (for which only one summation term contributes to the theta function for each wavenumber, just as in linear Fourier analysis). A white noise spectrum is therefore not as hard to compute as one might have anticipated.

22.10   Example Problem

I have computed a numerical run using a 20×20 period matrix and the results are given in Table 22.2.

Table 22.2 Results of a Numerical Run for the Computation of a Theta Function

q_0	N_{dof}	Number of m-Vectors	Number in N-Cube	Volume N-Ellipsoid/N-Cube	CPU Time
1.0D-02	20	79	$0.17433922 \times 10^{10}$	$0.45313957 \times 10^{-7}$	<0.01 s
1.0D-03	20	900	$0.47683716 \times 10^{14}$	$0.18874368 \times 10^{-10}$	0.01 s
1.0D-04	20	6597	$0.47683716 \times 10^{14}$	$0.13834912 \times 10^{-9}$	0.1 s
1.0D-05	20	37,462	$0.47683716 \times 10^{14}$	$0.78563508 \times 10^{-9}$	0.4 s
1.0D-06	20	171,249	$0.47683716 \times 10^{14}$	$0.35913518 \times 10^{-8}$	1.8 s
1.0D-07	20	680,186	$0.39896133 \times 10^{17}$	$0.17048920 \times 10^{-10}$	6.9 s
1.0D-08	20	2,352,275	$0.39896133 \times 10^{17}$	$0.58959975 \times 10^{-10}$	22.3 s
1.0D-09	20	7,270,680	$0.39896133 \times 10^{17}$	$0.18224022 \times 10^{-9}$	64.8 s
1.0D-10	20	20,429,358	$0.39896133 \times 10^{17}$	$0.51206361 \times 10^{-9}$	2 min 51.2 s
1.0D-11	20	52,431,062	$0.39896133 \times 10^{17}$	$0.13141891 \times 10^{-8}$	6 min 58.5 s
1.0D-12	20	124,916,152	$0.60788327 \times 10^{19}$	$0.20549365 \times 10^{-10}$	15 min 55.5 s
1.0D-13	20	278,073,914	$0.60788327 \times 10^{19}$	$0.45744623 \times 10^{-10}$	34 min 10.8 s

The number of terms summed in the theta function (number of m-vectors in the N-ellipsoid, column 3) is computed as a function of the artificial zeroes (column 1). The number of m-vectors in the N-ellipsoid is seen to be quite small compared to the corresponding number of m-vectors in the N-cube (column 4). In the last column is the computer time for an Intel 3.3 GHz processor.

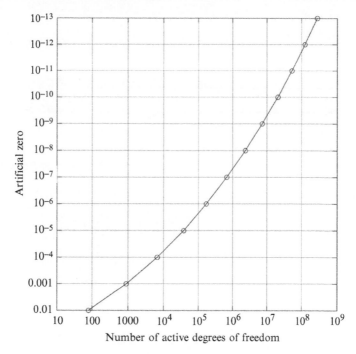

Figure 22.10 Number of active degrees of freedom (number of significant terms in theta function as computed inside the N-ellipsoid) as a function of the size of the artificial zero for a case in 20 dimensions.

Figure 22.10 is a graph of the number of degrees of freedom (number of terms in the theta function summation inside the N-ellipsoid) as a function of the accuracy desired. These results can give some idea of the relative computational effort involved for this highly nonlinear case.

23 Determining the Riemann Spectrum from Data and Simulations

23.1 Introduction

Measuring and analyzing wave data and *developing numerical models* of ocean waves are among the most familiar activities in the field of physical oceanography. *Data assimilation* is also an important aspect of the real-time assessment of the oceanic environment. And of course taking the ordinary linear Fourier transform of data or simulations is a common procedure. In this chapter, I discuss one way to obtain the *nonlinear Fourier transform (Riemann spectrum) of data and simulations* and how to *assimilate data* in real time. One begins with a space or time series $(\eta(x, 0), \eta(0, t))$ in one or two spatial dimensions $(\eta(x, y, 0), \eta(0, y, t))$ and then computes the *Riemann spectrum* (the Riemann matrix and phases). The Riemann spectrum is the nonlinear Fourier spectrum that constitutes a matrix of spectral "amplitudes" and a vector of phases. In this remarkable set of coordinates one can analyze the nonlinear wave dynamics in many ways to understand the underlying physics. At this level one knows what types of nonlinear spectral components there are (including positive and negative solitons (holes), "table top" solitons, Stokes waves, unstable "rogue" modes, vortices, etc.) in the data or simulations provided one a priori assumes a particular nonlinear wave equation (Chapters 33 and 34).

The procedure just referred to for obtaining the Riemann spectrum from data I call nonlinear adiabatic annealing (NLAA) or nonlinear adiabatic annealing on a Riemann surface (NLAARS). The method is designed primarily for the nonlinear space and/or time series analysis of data, or the establishment of the Riemann spectrum for some Cauchy initial value problem for hyperfast numerical modeling purposes. It is an iterative procedure that allows us to slowly vary a "temperature" or "nonlinearity parameter" λ while iterating to improve on an estimate of the Riemann spectrum of the input data. One begins the procedure with knowledge of the actual value of the nonlinear parameter, λ (in terms of the water depth, say) but the annealing process is initiated with an actual value of the parameter which is much smaller, $\lambda_0 \ll \lambda$. It is assumed that the starting value λ_0 is so small that the linear Fourier approximation works quite well for the input data. This means that initially the Riemann

Doi: 10.1016/S0074-6142(10)97023-X

spectrum can be assumed to be the linear Fourier spectrum provided that λ_0 is small enough. Then by using standard nonlinear parameter fitting techniques (see, e.g., the book *Numerical Recipes*; Press et al., 1992), one is able to slowly increase λ_0, stopping to estimate the Riemann spectrum at every stage. Thus, by slowly (adiabatically) increasing the "temperature" (a nonlinearity parameter in the present case), one is able to obtain convergence to the Riemann spectrum of the input data.

In the annealing process it is important to remember the *Novikov conjecture*, that is, that *physical wave motion (corresponding to the solution of a nonlinear partial differential equation) must lie on an underlying Riemann surface*. For many practical purposes this implies that *the off-diagonal terms of the Riemann matrix must be computable from the diagonal terms* using information about the underlying Riemann surface on which the integrable equation lies. Thus, to insure that the Riemann matrix lies on a Riemann surface at every stage of the annealing process, I compute the off-diagonal terms of the matrix from the most recent estimation of the diagonal terms, using, for example, the Schottky uniformization (Chapter 15) or the Nakamura-Boyd (Chapter 16) procedures. This is an important feature of the approach discussed in this chapter.

The NLAARS approach discussed here is explicitly given for the Korteweg-deVries (KdV) equation but is easily applicable to any nonlinear integrable equation. The NLAARS method itself gives numerical results which are identical to the usual *theoretical approach* used in the study of nonlinear, integrable PDEs with periodic boundary conditions: One solves the Floquet problem for the eigenvalue problem (essentially the spatial part of the Lax pair) and the resultant eigenvalue spectrum is then used to compute loop integrals for the Riemann matrix, wavenumbers, frequencies, and phases. For the KdV equation these results are discussed in Chapters 17 and 19; for the nonlinear Schrödinger equation see Chapters 18 and 24. Extension of NLAA to *nonintegrable*, nonlinear wave equations awaits documentation elsewhere.

23.2   Space Series Analysis

We consider the case of shallow-water dynamics described to leading order by the *space* Korteweg-deVries (sKdV) equation:

$$\eta_t + c_0\eta_x + \alpha\eta\eta_x + \beta\eta_{xxx} = 0 \tag{23.1}$$

with coefficients $c_0 = \sqrt{gh}, \alpha = 3c_0/2h, \beta = c_0h^2/6$. The linear dispersion relation has the well-known form $\omega = c_0k - \beta k^3$. The Cauchy or initial value problem implies that given $\eta(x, 0)$ one can compute the solution for all space and time $\eta(x,t)$. For periodic boundary conditions the solution of Equation (23.1) in terms of the inverse scattering transform is discussed in Chapters 10, 14, 17, and 19. The numerical modeling of (23.1) is discussed in Chapter 32.

This procedure assumes that one measures a space series $\eta(x, 0)$ and then seeks to determine the Riemann matrix and phases of the data, in complete analogy with the determination of the Fourier spectrum using ordinary linear Fourier analysis. To be concrete, assume that we measure a single space series, $\eta(x, 0)$. Let us use (Chapter 10)

$$\eta(x, 0) = \frac{2}{\lambda} \partial_{xx} \ln \theta(x, 0 | \mathbf{B}, \boldsymbol{\phi}) \tag{23.2}$$

where $\lambda = 3/2h^3$ and h is the water depth. $\theta(x, 0 | \mathbf{B}, \boldsymbol{\phi})$ is the Riemann theta function. The above expression (23.2) can be inverted to give

$$\theta(x, 0 | \mathbf{B}, \boldsymbol{\phi}) = \exp\left(\frac{\lambda}{2} \iint_x \eta(x', 0) \mathrm{d}x' \mathrm{d}x''\right) \tag{23.3}$$

The double integration is accomplished numerically by the fast Fourier transform with a simple integration filter $-1/k^2$. At this point the emphasis changes from the measured space series in a shallow water environment, $\eta(x,0)$, to the space series $\theta(x,0|\mathbf{B},\boldsymbol{\phi})$ that is "theta-function like." We now seek to analyze $\theta(x,0| \mathbf{B},\boldsymbol{\phi})$ for its spectral content, that is, we seek the Riemann matrix $\mathbf{B}$ and phases $\boldsymbol{\phi}$ from the "measured space series" $\theta(x,0|\mathbf{B},\boldsymbol{\phi})$.

One first sets $\lambda = \lambda_0 \ll 3/2h^3$, essentially linearizing $\theta(x, 0|\mathbf{B},\boldsymbol{\phi})$ in the exponent in Equation (23.3). One then takes the linear Fourier transform of $\theta(x,0|\mathbf{B},\boldsymbol{\phi})$ to determine an initial estimate of the Riemann matrix diagonal elements, $B_{ii} = -2 \ln a_i$, and phases, ϕ_i. The a_i are (in this linear approximation) just the linear Fourier amplitudes and the ϕ_i are the linear Fourier phases.

The next step is to increase λ by a small amount and then to recompute $\theta(x, 0|\mathbf{B},\boldsymbol{\phi})$ by Equation (23.3). The Riemann matrix and phases are then estimated again by making a small correction (this is just a least squares problem, which can benefit also from the *embedding procedure* as discussed in Chapter 21). One continues to iterate until the final value of λ is reached ($\lambda = 3/2h^3$) and the Riemann matrix and phases are computed as final values.

Conceptually, one might rewrite Equation (23.3) as a Taylor series:

$$\theta(x, 0 | \mathbf{B}, \boldsymbol{\phi}) = \exp\left(\frac{\lambda}{2} \iint_x \eta(x', 0) \mathrm{d}x' \mathrm{d}x''\right)$$

$$\cong 1 + \frac{\lambda}{2} \iint_x \eta(x', 0) \mathrm{d}x' \mathrm{d}x'' + \frac{1}{2}\left(\frac{\lambda}{2} \iint_x \eta(x', 0) \mathrm{d}x' \mathrm{d}x''\right)^2 + \ldots$$

Suppose the smallest value λ_0 is such that $\lambda_0^2 \ll 1$, so that only the first two terms are important in the expansion: this provides the first estimate of the Riemann spectrum. Then, we increase the estimate so that $\lambda_0^3 \ll 1$, such that the first three terms in the expansion are important: this provides the second estimate of the Riemann spectrum. It is in this sense that we iterate on the spectrum until the correct, physical value of λ ($= 3/2h^3$) has been reached for which we have the final and correct estimate of the Riemann spectrum. We are literally "pulling" the Riemann spectrum "up" from the linear Fourier spectrum computed for λ_0. The procedure ensures that we stay in the vicinity of the "correct" region of the space of Riemann parameters that we seek. If one does not carry out the adiabatic process correctly there is the possibility of not finding the appropriate Riemann spectrum, that is, the one which grows from the linear Fourier transform and which assumes the *oscillatory basis* on the *Riemann surface*.

In the NLAARS procedure the oscillatory basis is assured as we adiabatically anneal the data starting with the estimates from the linear Fourier transform and because we use the commensurable wavenumbers $k_n = 2\pi n/L$. We maintain the Riemann spectrum on a Riemann surface during the annealing process because we use Schottky uniformization (Chapter 15) to determine the off-diagonal elements of the Riemann matrix from the diagonal elements (see Section 23.6).

It is further important to remember that in linear Fourier analysis we take N points for a space or time series and we compute N points for the Fourier transform. In the nonlinear Fourier procedure given here we instead determine N points for the Riemann spectrum, that is, we determine the diagonal elements and phases. Thus, the number of actual fitted parameters are the same in the linear and nonlinear Fourier analysis procedures at N wavenumbers. The off-diagonal elements of the Riemann spectrum are then computed by the methods of Chapters 14–16. Formally speaking, therefore, the *genus* of the numerical solutions we are computing is given by N. However, as a practical consideration we can apply an appropriate filtering algorithm to reduce the genus to a much smaller number for practical computations, as discussed in Section 23.5.

23.3   Time Series Analysis

Experimentally, we record wave amplitudes as a function of time at a single spatial location, thus implying the need to determine the scattering transform (Riemann spectrum) of a time series, $\eta(0, t)$. The goal is to seek the Riemann matrix and phases of the data, in complete analogy with the determination of the Fourier spectrum using ordinary linear Fourier analysis. The time-like KdV equation used in this context (tKdV) (Osborne and Petti, 1993) is given by

$$\eta_x + c_0'\eta_t + \alpha'\eta\eta_t + \beta'\eta_{ttt} = 0 \qquad (23.4)$$

where $c_0' = c_0^{-1} = 1/\sqrt{gh}, \alpha' = -\alpha/c_0^2 = -3/2hc_0, \beta' = -\beta/c_0^4 = -h^2/6c_0^3$. The linear dispersion relation is $k = \omega/c_0 + (\beta/c_0^4)\omega^3$. The boundary value problem is well defined for Equation (23.4) and this implies that given $\eta(0, t)$ one can compute the solution for all space and time $\eta(x, t)$. For periodic boundary conditions ($\eta(x, t) = \eta(x, t + T)$), the solution of Equation (23.4) in terms of the inverse scattering transform is discussed in Chapters 10, 14, 17, and 19; this solution is the same as that for the space KdV equation (23.1) except for the obvious changes of variable and constants apparent between sKdV and tKdV. To be concrete with regard to the problem of time series analysis, assume that we measure a single space series, $\eta(0, t)$. Let us now use the tKdV Hirota transformation ($x \rightarrow t, t \rightarrow x, \omega \rightarrow k, k \rightarrow \omega, \lambda \rightarrow c_0^2 \lambda$):

$$\eta(0, t) = \frac{2}{c_0^2 \lambda} \partial_{tt} \ln \theta(0, t | \mathbf{B}, \boldsymbol{\phi}) \tag{23.5}$$

Inverting Equation (23.5) gives

$$\theta(0, t | \mathbf{B}, \boldsymbol{\phi}) = \exp\left(\frac{c_0^2 \lambda}{2} \int \int_t \eta(0, t') \mathrm{d}t' \mathrm{d}t''\right) \tag{23.6}$$

As above one first sets $\lambda = \lambda_0 \ll 3 / 2h^3$, essentially "linearizing" the time series $\theta(0,t|\mathbf{B},\boldsymbol{\phi})$. The first step is to take the linear Fourier transform of $\theta(0,t|\mathbf{B},\boldsymbol{\phi})$, to determine an initial estimate of the Riemann matrix diagonal elements, $B_{ii} = -2 \ln a_i$, and phases, ϕ_i. The a_i are (in this linear approximation) just the linear Fourier amplitudes and the ϕ_i are the linear Fourier phases.

The next step is to increase λ by a small amount and then to recompute $\theta(0,t \mid \mathbf{B},\boldsymbol{\phi})$. The Riemann matrix and phases are then estimated again by making a small correction. One continues to iterate until the final value of λ is reached and the Riemann matrix and phases converge to their final values. As before the underlying Riemann surface is maintained by computing the off-diagonal terms of the Riemann matrix using Schottky uniformization (Chapter 15).

23.4 Nonlinear Adiabatic Annealing

Since the space and time series approaches given above are the same except for a simple change of variables, I now apply the method directly only to space series, leaving the time series as a simple exercise for the reader. The inverse problem for the theta function (determine the "Riemann" spectrum) requires that we compute the period matrix, $\mathbf{B}$, and phases, $\boldsymbol{\phi}$, from a space (or time) series, $\eta(x)$. The procedure I have developed requires, as a first step, that we compute from a space series, $\eta(x, t = 0)$, an estimate of the associated theta function, $\theta(x, t = 0 | \mathbf{B}, \boldsymbol{\phi})$. This is done by inverting the relation

$$\eta(x, 0) = \frac{2}{\lambda} \partial_{xx} \ln \theta(x, 0) \tag{23.7}$$

to get

$$\Theta(x,0) = \exp\left(\frac{\lambda}{2}\int\int_x \eta(x',0)dx'dx''\right) \tag{23.8}$$

The problem with the inversion is that it differs from the actual theta function by a multiplicative constant, a:

$$\Theta(x,0) = a\theta(x,0) = \exp\left(\alpha + \frac{1}{2}\lambda\int\int_x \eta(x',0)dx'dx''\right), \quad a = e^{\alpha} \tag{23.9}$$

Here α is just the arbitrariness in computing the double antiderivative. Another integration constant is set to zero to ensure that Equation (23.9) be periodic. The as yet unknown constant, α, has no influence on the solution to the KdV equation since:

$$\frac{\lambda}{2}\eta(x,0) = \partial_{xx}\ln\Theta(x,0) = \partial_{xx}\ln a\theta(x,0) = \partial_{xx}(\ln a + \ln\theta(x,0)) = \partial_{xx}\ln\theta(x,0).$$

In terms of the *discrete theta function* (Chapter 21), $\boldsymbol{\theta} = \mathbf{WQ}$, we then have:

$$\boldsymbol{\Theta} = a\boldsymbol{\theta} = a\mathbf{WQ}$$

so that

$$\boldsymbol{\Theta} = a\boldsymbol{\theta} = a\mathbf{WQ} = a\sum_{m=-M}^{M} Q_m\, e^{im\cdot kx}, \quad Q_m = e^{-\frac{1}{2}m\cdot Bm + im\cdot\phi}$$

where the matrix $\mathbf{W}$ has the elements

$$W_{kj} = w^{I_j k} = w^{kI_j}, \quad I_j = \sum_{n=1}^{N} m_n^j n$$

Finally, after multiplying by $\mathbf{W}^{T*}$ (T means transpose and * means complex conjugate):

$$\mathbf{W}^{T*}\boldsymbol{\Theta} = a\mathbf{W}^{T*}\boldsymbol{\theta} = a\mathbf{IQ} \equiv \mathbf{P}$$

where $\mathbf{I} = \mathbf{W}^{T*}\mathbf{W}$.

    We now have to solve the following *nonlinear equations*:

$$a\mathbf{IQ}(\mathbf{B}, \boldsymbol{\phi}) = \mathbf{P} \tag{23.10}$$

where

$$\mathbf{I} = \mathbf{W}^{T*}\mathbf{W}, \quad \mathbf{P} = \mathbf{W}^{T*}\boldsymbol{\Theta}, \quad Q_m = e^{-\frac{1}{2}m \cdot \mathbf{B}m + im \cdot \boldsymbol{\phi}} \tag{23.11}$$

Here $\mathbf{Q}$ is a vector of length $\boldsymbol{M} = (2M + 1)^N + 1$ (Chapter 21) and the particular terms summed over are found by summation over an N-ellipsoid as discussed in Chapter 22. Equation (23.10) must be solved for the parameters: a, $\mathbf{B}$, $\boldsymbol{\phi}$. This means that we must numerically compute a even though it does not play a role in the solution of the KdV equation. A similar integration constant appears in the Nakamura-Boyd approach (Chapter 16), that is, the constant must be solved for along with the Riemann spectrum, but afterwards the constant can be discarded.

Is there a simple way to estimate the parameter a on each iteration? We know that

$$\Theta(x, t) = a\theta(x, t)$$

and assume that we are working with $\Theta(x, t) = a\theta(x, t)$ at zero time: $\Theta(x, 0) = a\theta(x, 0)$. Since we are working with a space series we can also consider the first point, $x = 0$, and evaluate the expression thusly: $\Theta(0, 0) = a\theta(0, 0)$. This means the parameter a has the unique value

$$a = \frac{\Theta(0,0)}{\theta(0,0)}$$

that is, the ratio of the first point in the data, $\Theta(0,0)$, divided by the first point in the theta function, $\theta(0,0)$, which is based upon the most recent estimate of the Riemann spectrum. Notice that

$$\theta(x, t) = \sum_{m=-M}^{M} Q_m e^{im \cdot kx - im \cdot \omega t}, \quad Q_m = Q_m(\mathbf{B}, \boldsymbol{\phi}) = e^{-\frac{1}{2}m \cdot \mathbf{B}m + im \cdot \boldsymbol{\phi}}$$

Hence

$$\theta(0, 0) = \sum_{m=-M}^{M} Q_m(\mathbf{B}, \boldsymbol{\phi})$$

This latter expression is one of the most important of the so-called *theta constants*. Therefore, given estimates of $\mathbf{B}$, $\boldsymbol{\phi}$ we can compute a as

$$a \equiv a(\mathbf{B}, \boldsymbol{\phi}) = \frac{\Theta(0,0)}{\theta(0,0)} = \frac{\Theta(0,0)}{\sum_{m=-\infty}^{\infty} Q_m(\mathbf{B}, \boldsymbol{\phi})} \tag{23.12}$$

Thus, a is the ratio of the first point in the doubly integrated, exponentiated measured space series, $\Theta(0, 0)$, and the sum of the phased-nomes: $Q_m(\mathbf{B}, \boldsymbol{\phi})$.

The nonlinear equations then become:

$$\mathbf{I}\frac{\mathbf{Q}(\mathbf{B}, \boldsymbol{\phi})}{\theta(0, 0)} = \mathbf{P} = \frac{\mathbf{W}^{T*}\boldsymbol{\Theta}}{\Theta(0, 0)}$$

or

$$\mathbf{I}\frac{\mathbf{Q}(\mathbf{B}, \boldsymbol{\phi})}{\sum_{m=-M}^{M} Q_m(\mathbf{B}, \boldsymbol{\phi})} = \mathbf{P} = \frac{\mathbf{W}^{T*}\boldsymbol{\Theta}}{\Theta(0, 0)}$$

Finally,

$$\mathbf{I}\mathbf{Q}(\mathbf{B}, \boldsymbol{\phi}) - A(\mathbf{B}, \boldsymbol{\phi})\frac{\mathbf{W}^{T*}\boldsymbol{\Theta}}{\Theta(0, 0)} = 0, \quad A(\mathbf{B}, \boldsymbol{\phi}) = \theta(0, 0) = \sum_{m=-M}^{M} Q_m(\mathbf{B}, \boldsymbol{\phi})$$

where

$$A(\mathbf{B}, \boldsymbol{\phi}) = \theta(0, 0) = \sum_{m=-M}^{M} Q_m(\mathbf{B}, \boldsymbol{\phi}) = 1 + \sum_{m=0}^{M} Q_m(\mathbf{B}, \boldsymbol{\phi}) + Q_m^*(\mathbf{B}, \boldsymbol{\phi})$$

$$= 1 + 2\sum_{m=0}^{M} q_m \cos(\phi_m)$$

where the ordinary nome is

$$q_m = e^{-\frac{1}{2}m \cdot Bm}$$

As a consequence of the above relations we obtain the NLAA procedure shown in the flowchart in Figure 23.1; we compute a from the most recent estimates of the period matrix and phases at each iteration.

23.5   Outline of Nonlinear Adiabatic Annealing on a Riemann Surface

Let us now discuss the flowchart in Figure 23.1 that gives a rough overview of the NLAARS algorithm. As mentioned above, we are implementing an annealing algorithm that "lifts" the Riemann spectrum (matrix and phases) up from an initial estimate using linear Fourier analysis to the actual values that we seek. In the process, of course, we could use the method of Chapter 14, for

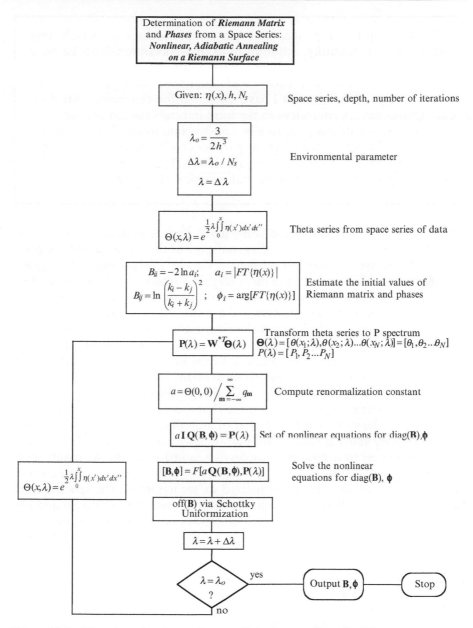

Figure 23.1 Flowchart for the nonlinear adiabatic annealing algorithm.

which the algebraic-geometric loop integrals guarantee that we are on a Riemann surface. Likewise we could use Schottky uniformization (Chapter 15) or the Nakamura-Boyd approach (Chapter 16) at each iteration to ensure that we are on a Riemann surface. In Figure 23.1 I use Schottky uniformization to insure that the estimate of the spectrum always remains on a Riemann surface.

> ***Nota Bene***: Up to the present time I know of no theorems which have established that Schottky uniformization and Nakamura-Boyd lie on a Riemann surface even though they most likely do. Nevertheless, after extensive numerical simulations and analytical computations using FORTRAN and Mathematica I have arrived at the conclusion that these two approaches are equivalent to the loop integrals (which of course are known to have a Riemann surface by construction to satisfy the Novikov conjecture). Therefore, for practical purposes I will use these approaches interchangeably in all the work discussed here. The actual proof of these assertions presents an outstanding mathematical challenge.

Another practical aspect of the work given here is that we are using the NLAARS approach to give us the diagonal part of the Riemann matrix (diag (**B**)) and the phases to match a particular Cauchy initial condition or given space/time series. Then one of the methods of Chapters 14–16 can be applied to give the off-diagonal elements. We are thus estimating two vectors (diag (**B**), $\boldsymbol{\phi}$), equivalent to a complex vector, for a space/time series. This is equivalent to linear Fourier analysis in which the space/time series is used to get the Fourier spectrum (a complex vector). The off-diagonal elements are "constrained" to assume particular values in terms of diag(**B**) by the theories of Chapters 14–16, thus giving us reasonable assurance that we are on a Riemann surface, and hence have a physically realizable problem.

In the flowchart of Figure 23.1, we see that the inputs are the space series, $\eta(x)$, the water depth, h, and the number of annealing iterations, N_s. This provides us with the starting value of the nonlinearity parameter $\lambda_0 = (3 / 2h^3)/N_s$ that we iterate on to determinate successive estimates of the Riemann spectrum. Equation (23.3) is then used to find $\theta(x,0)$ with $\lambda = \Delta\lambda$ (a small parameter). $\theta(x,0)$ is subsequently renormalized as we iterate on the spectrum via the parameter a. We first take the linear Fourier transform of $\theta(x,0)$ and use the results to estimate diag(**B**) and $\boldsymbol{\phi}$. The off-diagonal elements are estimated on the first iteration from the *leading-order Schottky formula* (see Chapters 15 and 16):

$$B_{ij} \cong \ln\left(\frac{k_i - k_j}{k_i + k_j}\right)^2$$

The so-called **P** spectrum (Chapter 21) is then computed and, from the initial estimate of the Riemann spectrum, the normalization constant a is computed as a theta constant. Then the set of nonlinear equations

$$a\mathbf{IQ}(\mathbf{B}, \boldsymbol{\phi}) = \mathbf{P}$$

is used to estimate the Riemann spectrum. The "slowness" of the adiabatic annealing process ensures that we stay near the most recent estimate of the function $\eta(x, 0)$ by making only small corrections to the Riemann spectrum on each iteration.

An alternative approach to adiabatic annealing can be developed using the Riemann theta function itself to build an algorithm for NLAARS (as opposed to the nonlinear equations introduced above). This might be approached using the fast algorithms of Chapters 8, 9, and 22 to fit to the input data $\eta(x, 0)$ and to directly apply the nonlinear annealing procedure given in *Numerical Recipes* (1994) or some other preferred approach.

Nota Bene: Estimation of the Genus of a Space/Time Series

Let us first consider the case for a *one degree-of-freedom theta function*. Thus, the genus of the solution is *one* by definition. For the KdV equation this is a Stokes wave solution and let us assume that N terms in the theta series need to be summed to give us the precision we desire. This means there are N linear Fourier modes. What genus do we choose for iterating the NLAA process? The answer is easy: We chose the genus to have the value of N. This is because when we use the linear Fourier transform to analyze the space series $\eta(x, 0)$ for our initial guess of λ_0 small, we will find N linear Fourier components. To be consistent with the NLAA iteration we need to keep a Riemann matrix that is $N \times N$ to retain all of the information in the input space series. Of course, as we iterate we will find that the values of the diagonal elements of the Riemann matrix will converge to only *one* value that has physical significance, that is, the other diagonal elements of the Riemann matrix will become so large that they do not contribute significantly to the wave train. Thus, by starting with N as an estimate of the genus, we conclude after the NLAARS iterations that the genus is 1.

An important issue in the *analysis of data* is to estimate the genus of an input wave train. I do not have a *general* estimate of the genus of an arbitrary wave train, but I have a procedure for estimating the genus of a given measured space/time series. First, I assume that the genus is just the number of linear Fourier components in a given series. However, a space series with $N = 1000$ points will have 1000 Fourier components. More than likely choosing the genus to be 1000 will be wasteful of computer time and resources and in all probability many of the diagonal elements will be void of important physical content. As I discuss in detail in Chapter 28 there is a simple procedure to reduce the genus to a reasonable, physically significant value. One first determines the value of the wavenumber or frequency beyond which the KdV equation is no longer applicable; this is done using the linear dispersion relation such that one keeps the second-order term small. Once one has this estimate, one can easily notice that all Fourier components beyond this KdV "cutoff" *are small and therefore linear.* Suppose that there are N_{KdV} linear Fourier components to the left of the KdV cutoff. Maintaining a Riemann matrix for linear modes is not necessary and thus one can simply retain a Riemann matrix of $N_{KdV} \times N_{KdV}$. Providing that $N_{KdV} \ll N$ one therefore realizes significant improvement in algorithm performance. Before beginning the NLAA process it is prudent to first *low-pass filter the input time series* to remove small linear Fourier modes beyond the KdV cutoff. In Chapter 28, I analyze a 500 point time series and show how to reduce the physically significant Riemann matrix from 500×500 to 50×50 by a simple linear filtering technique.

23.6 Establishing the Riemann Spectrum for the Cauchy Problem

Given the Cauchy initial condition $\eta(x, 0)$, we seek to discover the wave evolution in all space and time, $\eta(x, t)$. To use the IST we need to apply

$$\eta(x, t) = \frac{2}{\lambda} \partial_{xx} \ln \theta(x, t | \mathbf{B}, \boldsymbol{\delta})$$

which means that we also need the Cauchy initial condition for the theta function, $\theta(x, 0)$. We have seen that there are several methods for computing the Riemann spectrum, $\mathbf{B}$, $\boldsymbol{\phi}$. These include the method discussed in this chapter (NLAARS), and the three approaches in Chapters 14–16. Clearly, for the Cauchy problem the procedure is the same as that for data analysis as shown in Figure 23.1.

An additional approach to the Cauchy problem would be to choose the Riemann spectrum that you desire and then to evaluate the nonlinear wave motion numerically. Here is how this might be implemented numerically:

(1) Specify the wavenumbers to be commensurable as in ordinary linear Fourier analysis.
(2) Pick the diagonal elements of the Riemann matrix to give us the cnoidal wave amplitudes desired. Also select the phases to take on any values, perhaps random, that are desired.
(3) Then use Schottky uniformization (or loop integrals or Nakamura-Boyd) to compute the off-diagonal elements. Recall that to leading order these depend only on the wavenumbers. At higher order they also include the diagonal elements of the Riemann matrix.
(4) Then use Schottky uniformization (or loop integrals or Nakamura-Boyd) to compute the frequencies. These approaches will generally include a nonlinear correction to the linear dispersion relation.

The use of (1) loop integrals, (2) Schottky uniformization, and (3) Naka-mura-Boyd all should guarantee that we stay on a particular Riemann surface and therefore that we have a physically realizable solution to a particular wave equation. For a numerical check of the approach one can use, for example, the Fornberg-Whitham approach (1978) to numerically integrate the wave equation using the fast Fourier transform to compare to the periodic IST prediction.

23.7 Data Assimilation

The data assimilation process is an interesting and important one because in a particular sea state we may wish to continually update our knowledge of the Riemann spectrum, from say radar measurements or other types of local instruments such as buoys, etc. During a storm, say, one might have already established a good estimate of the Riemann spectrum for the particular moment in time. Then as more measurements come in during further development of the storm one could continue the iteration process without heavy computing requirements simply by employment of the iteration loop of Figure 23.1. One could then continually monitor the Riemann matrix for the presence of "rogue" elements in the sea state, be they rogue waves, holes in the sea, etc. Data assimilation could therefore be a way to provide up-to-date conditions for assessing possible threats to ship integrity in real time.

Part Eight

Theoretical and Experimental Problems in Nonlinear Wave Physics

To study nonlinear waves using the inverse scattering transform (IST) in the context of physical oceanography, I address a number of problems of theoretical and experimental interest.

In Chapter 24 I discuss the *nonlinear instability analysis* of the nonlinear Schroedinger (NLS) equation using IST. In contrast to *linear* instability analysis, the approach provides a full understanding of all smooth, nonsingular solutions of the equation. Perhaps the most important result is the explicit analytical form of the Riemann spectrum that allows for physical interpretation of the unstable mode "rogue wave" solutions. The main conclusion of this chapter is that there are *two kinds of NLS spectrum*: (1) small amplitude *nearly linear Stokes wave trains* and (2) *large amplitude unstable modes* that are often interpreted as "rogue wave" solutions of the equation. Only above a *particular amplitude threshold* do the rogue wave solutions occur. Of course the nonlinear instability analysis provides us with the unique perspective that initial, small-amplitude modulations that grow exponentially do not grow forever. Instead they reach a maximum height and then return, via FPU recurrence, to almost nearly the initial state. These are of course the "rogue wave" solutions of the NLS equation.

Chapter 25 discusses application of the IST to internal wave data with solitons. This chapter analyzes data from the Andaman Sea using the IST. An overview of some of the more interesting nonlinear physical problems for internal waves is also given. Chapter 26 gives a theoretical and numerical exposition of acoustic wave propagation in the oceanic waveguide. A number of novel applications are discussed. Vortex dynamics are considered from the point of view of soliton theory in Chapter 27. What relationship do vortices have to

Doi: 10.1016/S0074-6142(10)97047-2

solitons? The answer to this question is given an explicit theoretical foundation and the Riemann spectrum is described.

The analysis of shallow-water ocean surface waves is given in Chapter 28. Indeed, the Riemann spectrum of the data is computed and nonlinear filtering techniques are used to enhance the physical interpretation of the data in terms of the cnoidal wave components in the spectrum. Chapter 29 gives an overview of laboratory experiments on rogue waves conducted at Marintek in Trondheim, Norway. Explicit techniques are given for generating in a wave tank several nonlinear Fourier components in the IST spectrum for the NLS equation. These are particular kinds of rogue waves that occur in surface water waves. The IST is used to analyze the data to aid physical understanding and to provide specific predictions about the behavior of individual rogue waves. Techniques are discussed for generating random wave trains in a wave tank which are dominated by rogue waves. The data are analyzed using the IST. Chapter 30 describes the IST analysis of data from Duck Pier in 8 m depth. This is an extreme situation in which solitons dominate the spectrum. Finally, in Chapter 31 I discuss harmonic generation in shallow-water waves. This is a classical problem in the study of nonlinear waves and data are presented which, when analyzed with the IST, provide for new perspective on shallow-water wave motion.

24 Nonlinear Instability Analysis of Deep-Water Wave Trains

24.1 Introduction

I have previously discussed how *unstable modes* of the nonlinear Schrödinger (NLS) equation may be viewed as simple prototypical candidate solutions for certain types of "extreme," "rogue," or "freak" ocean surface waves (Osborne et al., 2000; Osborne, 2001, 2002) (see also Chapters 12 (where special solutions of NLS are addressed) and 18 (where the structure of the spectral eigenvalue problem is addressed)). This approach suggests that extreme waves can arise physically from the *modulational (Benjamin-Feir) instability* mechanism (Benjamin and Feir, 1967; Zakharov, 1968; Lighthill, 1965, 1967) and are "nonlinear Fourier components" in the inverse scattering transform (IST) solution of the NLS equation, whose unstable modes are extremely rich in their dynamical behavior, vastly more so than had previously been suspected. This chapter discusses how to compute the stable and unstable spectral solutions for the NLS equation for periodic boundary conditions. While many researchers have focused primarily upon a limited subset of solutions that physically correspond to relatively small-amplitude, initial modulations, there is also a class of solutions of *large-amplitude modulations* that have much larger rogue wave maximum amplitudes.

In the present chapter, I discuss IST for the periodic NLS equation (Kotljarov and Its, 1976; Tracy, 1984; Tracy and Chen, 1988; Belokolos et al., 1994; Calini et al., 1996; Islas and Schober, 2005; Calini and Schober, 2002; Schober, 2006) that describes the dynamics of deep-water gravity wave trains to leading order in cubic nonlinearity:

$$i\left(\psi_t + C_g\psi_x\right) + \mu\psi_{xx} + \nu|\psi|^2\psi = 0, \quad \psi(x,t) = \psi(x+L,t) \tag{24.1}$$

with coefficients for infinite water depth given by

$$\mu = -\frac{\omega_0}{8k_0^2}, \quad \nu = -\frac{\omega_0 k_0^2}{2}, \quad C_g = \frac{1}{2}\frac{\omega_0}{k_0} \tag{24.2}$$

where ω_0, k_0 are the frequency and wavenumber of the carrier wave; the dispersion relation is $\omega_0^2 = gk_0$ for g the acceleration of gravity. By considering the following transformation:

$$u = \lambda\psi, \quad X = x - C_g t, \quad T = \beta t \tag{24.3a}$$

where

$$\lambda = \sqrt{\frac{v}{2\mu}} = \sqrt{2}k_0^2, \quad \beta = \mu = -\frac{\omega_0}{8k_0^2}$$

one finds the rescaled NLS equation:

$$iu_T + u_{XX} + 2|u|^2 u = 0 \tag{24.3b}$$

This is a convenient form of the equation for making calculations with the IST. IST for NLS with periodic boundary conditions was first derived by Kotljarov and Its (1976). Essentially, the theory integrates NLS using the Zakharov-Shabat eigenvalue problem (Zakharov and Shabat, 1972) and prescribes *two inverse problems* in terms of the *hyperelliptic* and *Riemann theta-function representations*. I use the theta-function representation herein to compute properties of extreme waves.

The legacy of recent work on deep-water wave trains rests with the discovery of the modulational instability (Lighthill, 1965; Benjamin and Feir, 1967; Zakharov, 1968) and the subsequent discovery of the high-order dynamical equations of motion (Zakharov, 1968), followed by experimental and theoretical analyses conducted by Yuen and Lake (1982). An interesting historical perspective was provided by Yuen (1991) who is a pioneer in the study of instabilities in deep-water wave trains.

In this chapter I discuss how unstable solutions of the NLS equation in $1 + 1$ dimensions can be found using the periodic IST formulation. I then discuss how IST can be used to make specific predictions about the behavior of NLS unstable modes in both space and time. In particular, I numerically compute a number of solutions to the equation and give graphical examples. I also give a general equation for determining the maximum unstable-mode wave height; a simple formula predicts the ratio of maximum amplitude, A_{max}, to carrier wave height, a_0, in terms of the modulation wavenumber, K, and the carrier wavenumber, k_0. *Linear instability analysis* is of course able to address only the *initial exponential growth* of a small-amplitude modulation; the work described in this chapter may be viewed as a kind of *nonlinear instability analysis* that allows all dynamical properties of the space-time dynamics of the NLS equation to be computed for all time. A typical small-amplitude modulation of a carrier wave train is shown in Figure 24.1.

24.2 Unstable Modes and Their IST Spectra

The periodic IST (see an overview in the Appendix and Chapters 12 and 18) describes a very rich and complex structure for the nonlinear Fourier solutions of the NLS equation with periodic boundary conditions. In linear Fourier

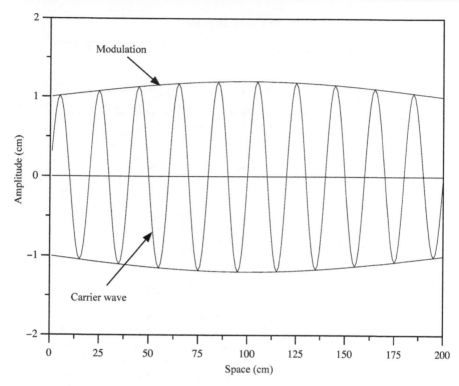

Figure 24.1 The simplest way to think about modulations of wave train solutions of the NLS equation is to begin with a small-amplitude modulation.

analysis one computes Fourier amplitudes and phases as a function of the wavenumber or frequency. In nonlinear Fourier analysis for the NLS equation the spectral parameter (wavenumber or frequency) is *a complex number*, leading to unstable modes or "breathers" in the time evolution of the wave train. One can think of the spectrum in the complex plane which contains the real and imaginary parts of the complex wavenumber, $\lambda = \lambda_R + i\lambda_I$, where the real and imaginary parts have been specified by subscripts. This spectral plane is referred to as the λ-plane and is shown in Figure 24.2.

The simplest kinds of spectra occur only on the real and imaginary axes. The more complicated generic solutions occur anywhere in the complex plane. An *unmodulated carrier wave* has eigenvalues at $\pm iA$, where A is the amplitude of the *plane carrier wave* solution to the NLS equation (24.3):

$$u(x,t) = A\,e^{2iA^2T} \tag{24.4}$$

Note that in Figure 24.2 the carrier eigenvalues $\pm iA$ are connected by a vertical straight line, referred to as a spine. A *simple modulation* of this carrier is manifested by *two points of simple spectrum* in the complex plane connected by a

The IST lambda plane

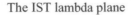
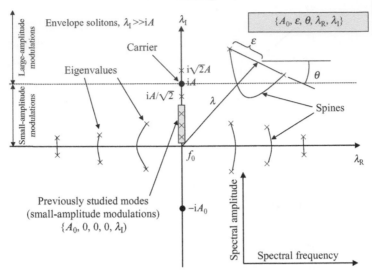

Figure 24.2 Properties of the IST spectrum for solutions of the NLS equation with periodic boundary conditions. The IST spectrum lies in the complex plane referred to as the λ-plane. The stable mode solutions, denoted graphically as points of simple spectrum, $\times$, one above and the other below the real axis, are connected by a curved line known as a spine which crosses the real axis. Simple unstable modes are two points of simple eigenvalues connected by a spine that does not cross the real axis. The eigenvalues $\pm iA$ correspond to the carrier wave. The eigenvalues $\pm iA/\sqrt{2}$ correspond to an unstable solution of the NLS equation (see Chapter 12, Figure 12.5, Equation (12.25)); the eigenvalues $\pm iA\sqrt{2}$ correspond to a different unstable mode solution (see Chapter 12, Figure 12.7, Equation (12.27)).

spine, the curved line shown in Figure 24.2 (see Chapter 12 for a detailed discussion of spines). In Figure 24.2 the *centroid* of the two points of simple spectrum (shown in the first quadrant) is given by the complex number λ.

The IST *eigenvalues* for the two points of simple spectrum are given by $\lambda \pm \varepsilon$, where ε is a complex number defining the "amplitude" of the stable or unstable mode, $\varepsilon = |\varepsilon|\exp(i\theta)$. When $i0 < \lambda < iA$, on the imaginary axis, $|\varepsilon|$ is the actual (small-amplitude) modulation amplitude; when $\lambda_I > A$ no small initial modulation generates the motion of the wave; only *large-amplitude modulations* occur (with highly nontrivial unstable mode wave behavior, see below and Chapter 12, Figure 12.7, Equation (12.27)).

Some nomenclature and physical perspective is in order: When $\theta = 0$ the unstable mode is denoted a *cross state.* When $\theta = \pi/2$ the unstable mode is a *slot state.* These states appear either parallel along the t-axis in the x-t plane or are right-left phase shifted by 90° as they propagate along the time axis. Other values of θ refer to a *slant state* (either "slant right" or "slant left"). These states have diagonal trajectories in the x-t plane.

The stability of the NLS modes is determined by the *spine*. A *single mode* of the NLS equation typically consists of *two points of simple spectrum connected by a spine*. When the spine crosses the real axis the mode is said to be *stable* and the solution to the NLS equation is a Stokes wave. When the spine does not cross the real axis the mode is *unstable* and the solution to NLS is an *unstable mode*. When λ lies out in the complex plane such that $\lambda = \lambda_R + i\lambda_I$ then we interpret λ_I as having information about the *amplitude* of the unstable mode and λ_R refers to the *group velocity relative to the linear group velocity*, $C_g = \omega_0/2k_0$; modes to the right of the imaginary axis move faster than C_g, while modes to the left of the imaginary axis move slower than C_g. Note that the IST spectrum in the upper half plane has its *specular image* in the lower half plane. Horizontal "bands" of spectrum effectively characterize the physics of nonlinear deep-water wave trains (see Figure 24.2):

(1) A small-amplitude band around the real axis defines the *nearly linear, stable modes* of a particular solution of NLS. Modes on the real axis (for which the two points of simple spectrum are connected by a spine across the real axis) are always stable. These modes are ordinary Stokes waves.

(2) Further from the origin is a large-amplitude horizontal band effectively containing the *unstable modes*, often referred to as possible candidates for "rogue" waves in deep-water wave trains. A spine that does not cross the real axis connects the two points of simple spectrum.

(3) Still further from the real axis is the *soliton regime* in which no instabilities occur, only *wave group solitons*. This regime occurs far from the influence of the carrier iA ($\lambda \gg iA$).

Nota Bene: The physical interpretation of the periodic IST spectrum in the λ-plane is governed by some simple, useful rules:

(1) A single nonlinear Fourier mode consists of *two points of spectrum* (simple eigenvalues) connected by a *spine*.

(2) If the spine connects two points of spectrum in the upper and lower half planes (i.e., the *spine crosses the real axis*), the nonlinear Fourier mode is *stable*, that is, it corresponds to a *simple Stokes wave*.

(3) If the spine connects two points of spectrum in the upper (and lower) half planes (i.e., *the spine does not cross the real axis*), the nonlinear Fourier mode is *unstable* and may be a candidate for a "rogue" wave.

(4) Only if the energy of the unstable modes is substantially larger than the energy of the stable Stokes waves will one observe the "rogue" waves in an experiment. This is because in a sea state where the unstable modes are small with respect to the stable modes one finds that the stable modes will swamp the unstable modes and one effectively sees only a "normal" near-Gaussian sea state for the wave amplitudes. On the other hand, if the energy of the unstable modes is much larger than the stable modes then one sees large extreme wave amplitudes exceeding the Gaussian probability density function in the right hand, large-amplitude tail.

All of the rogue waves discussed herein are unstable mode solutions that are natural nonlinear Fourier components of the periodic IST. There are an infinite number of these components each characterized by the five-parameter family $(A, \lambda_R, \lambda_I, |\varepsilon|, \theta)$. The modes that are unstable ("rogue waves") occur when $\lambda_I \neq 0$ and when the spine does not cross the real axis. Referring back to Figure 24.2 it is interesting to note that the spectral choice which we often find of interest herein *consists of a nonlinear interaction between the carrier wave, iA, and the spectral component corresponding to the modulation of the carrier wave, $\lambda \pm \varepsilon$*. This point cannot be overemphasized, for it is the major source of *nonlinear instability* in the NLS equation. Only when iA and $\lambda \pm \varepsilon$ are of the same order of magnitude will the nonlinear instability mechanism dominate the motion. In the limit $\lambda \pm \varepsilon \gg iA$, the presence of the carrier is no longer significant and one has the *soliton limit*, that is, the instability disappears because the carrier wave has an amplitude which is small with respect to the component, $\lambda \pm \varepsilon$, which is now just a *cnoidal wave solution* to NLS (Yuen, 1991). It goes without saying that if $A \to 0$ the cnoidal wave is an exact solution of the NLS equation; the cnoidal wave can range from a linear sine wave up through the Jacobian *dn* function and finally up to the envelope soliton solutions.

24.3   Properties of Unstable Modes

In this section, a simple formula is given for the *maximum amplitude of an unstable mode* of the NLS equation. The Appendix provides several details of the calculations in this Section. For typical dimensional numerical examples and laboratory experiments, one normally considers a small-amplitude modulation, ε, of a carrier wave of amplitude, a_0, with wavenumber, $k_0 = 2\pi/L_0$, and frequency $\omega_0 = \sqrt{gk_0}$. The modulation wavelength is L (the modulation wavenumber is $K = 2\pi/L$). One then computes the IST of a complex waveform (single degree of freedom) solution of the NLS equation in terms of an eigenvalue that is purely imaginary, $\lambda = i\lambda_I$. If one assumes the restriction $0 < \lambda_I < A$, then the modulations are initially small provided ε is small. The main results are that the *wavenumber*, for an infinitesimal-amplitude, unstable mode is given by

$$K = \frac{2\pi}{L} = 2\sqrt{A^2 - \lambda_I^2} + O(\varepsilon) \qquad (24.5)$$

This equation has been computed from the appropriate "algebraic-geometric loop integral" to leading order in perturbation amplitude, ε. The maximum amplitude of the unstable mode during its evolution is given by

$$u_{\max} = A + 2\lambda_I + O(\varepsilon^2) \qquad (24.6)$$

I have computed this result from the theta-function solution of the NLS equation, where λ_I is the imaginary part of the eigenvalue.

Thus, the maximum amplitude has the formula (eliminate λ_I from Equations (24.5) and (24.6)):

$$\frac{A_{\max}}{A} = 1 + 2\sqrt{1 - \left(\frac{K}{2A}\right)^2} = 1 + 2\sqrt{1 - \left(\frac{\pi}{AL}\right)^2} \tag{24.7}$$

Dimensional forms can be derived for Equations (24.5)–(24.7) using the transformation equation (24.3a). One is lead to the *dimensional form for Equations (24.5) and (24.6)*:

$$K = 2\sqrt{2}k_0^2\sqrt{a_0^2 - \lambda_0^2} + O(\varepsilon) \tag{24.8}$$

$$u_{\max} = \lambda A_{\max} = A + 2\lambda_0$$

where $A = \sqrt{2}k_0^2 a_0$ and $\lambda_I = \sqrt{2}k_0^2\lambda_0$, so that

$$A_{\max} = a_0 + 2\lambda_0 \tag{24.9}$$

and from the first of Equation (24.8) find

$$\lambda_0 = \sqrt{a_0^2 - \left(\frac{K}{2\sqrt{2}k_0^2}\right)^2} \tag{24.10}$$

The dimensional ratio of maximum amplitude to carrier amplitude is then (insert Equation (24.10) into the second of Equation (24.8)):

$$\frac{A_{\max}}{a_0} = 1 + 2\frac{\lambda_0}{a_0} = 1 + 2\sqrt{1 - \left(\frac{K}{2\sqrt{2}k_0^2 a_0}\right)^2} \tag{24.11}$$

where $0 < K < 2\sqrt{2}k_0^2 a_0$. Equation (24.11) is one of the most important results of this chapter and predicts the ratio of the maximum amplitude, $A_{\max}$, of the rogue wave to the carrier wave amplitude, a_0, as a function of dimensionless perturbation wavenumber. This expression has been derived from the *nonlinear instability analysis* described herein, which provides knowledge of the space/time evolution of the wave train. Linear instability analysis provides information about the *initial, small-amplitude growth* of the wave and is not valid for large values of time. The expression (24.11) is graphed in Figure 24.3. This graph says that a wave train which is a small-amplitude modulation (say, $\varepsilon \sim 10^{-5}$) will grow to a maximum amplitude which depends only on the perturbation wavenumber, K, itself. When $K/2k_0^2 a_0 \geq \sqrt{2}$ there is no enhancement in

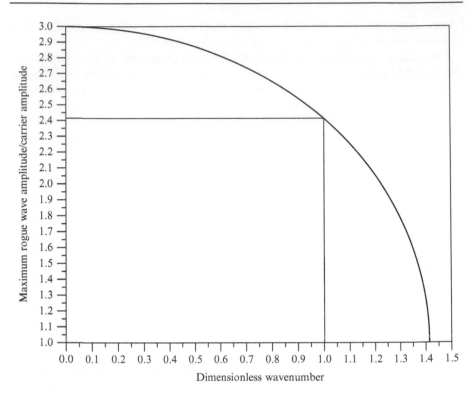

Figure 24.3   The maximum height of a rogue mode relative to the carrier amplitude, A_{max}/a_0, as a function of normalized modulational wavenumber, $K/2k_0^2 a_0$ (Equation (24.11)).

the wave height during its dynamical evolution (the maximum amplitude is no longer real). As the perturbation wavenumber decreases toward zero (the perturbation wavelength tends to infinity), the amplitude enhancement increases to *three times* the carrier amplitude, a_0. This corresponds to letting the eigenvalue vary on the interval, $0 < \lambda_I < A$, in the spectral plane. For $K/2k_0^2 a_0 = 1$ we have the *maximum growth rate* that occurs at $\lambda = i/\sqrt{2} \sim 2.41i$; this particular value was discussed by Yuen (1991) and the solution to the NLS equation is given by Equation (12.25) and is shown in Figure 12.6 of Chapter 12.

The perturbation dispersion relation from IST has been derived by computing the appropriate algebraic-geometric loop integral to leading order in ε (discussed in the Appendix):

$$\omega = 2\lambda K + O(\varepsilon) \qquad\qquad (24.12)$$

To make this expression dimensional, note that $\omega T = \Omega t$ where $T = \beta t (\beta = \mu)$, so that

$$\Omega = \omega \beta = 2\beta \lambda K \qquad\qquad (24.13)$$

Now use the first of Equation (24.8) and rescale the eigenvalue for convenience

$$\lambda = i\sqrt{2}k_0^2\lambda_0$$

Then, the dimensional dispersion relation becomes:

$$\Omega = i\omega_0 k_0^2 \lambda_0 \sqrt{a_0^2 - \lambda_0^2}$$

Use Equation (24.10) and find

$$\Omega = i\omega_0 k_0^2 a_0^2 \left(\frac{K}{2\sqrt{2}k_0^2 a_0}\right)\sqrt{1 - \left(\frac{K}{2\sqrt{2}k_0^2 a_0}\right)^2} \tag{24.14}$$

Squaring this gives the form of the perturbation dispersion relation used by Yuen (1991):

$$\Omega^2 = \frac{\omega_0^2}{8k_0^2}\left[\frac{K^2}{8k_0^2} - k_0^2 a_0^2\right]K^2 \tag{24.15}$$

Note that for $K/2k_0^2 a_0 = 1$, Equation (24.15) has the maximum growth rate, $\text{Im}\,\Omega_{max} = \omega_0 k_0^2 a_0^2/2$; this verifies the claim above that the eigenvalue $\lambda = iA/\sqrt{2}$ (with amplitude enhancement of $1 + \sqrt{2}$) corresponds to the maximum growth rate. Therefore, the maximum growth rate does *not* generate the largest unstable mode ($A_{max}/a_0 \approx 2.41$, see Figure 24.4); instead the largest unstable mode arises for $K \approx 0$ ($A_{max}/a_0 = 3$) (Figure 24.3).

The derivation of Equations (24.14) and (24.15) emphasizes an important aspect of the *nonlinear instability analysis of the NLS equation*. The result just derived, to $O(\varepsilon)$, agrees with the linear instability analysis conducted by Yuen (1991). The advantage of the IST is that it allows one to make a fully nonlinear instability analysis for *all* smooth, nonsingular solutions of the NLS equation to $O(\varepsilon)$ in the Riemann spectrum. This implies that the higher order terms in the expansion of the wavenumber (Equation (24.8)) and frequency (Equation (24.12)) can be computed to arbitrary order in the parameter ε, including interactions among all the components in the spectrum. This aspect of the theory is illustrated in numerical examples below.

The results just derived work in the *spectral band* $0 < \lambda < iA$, that is, they work for spectral components *below the carrier wave* in the complex plane. What happens when we have a component *above the carrier*? Note that Equation (24.6), $u_{max} = A + 2\lambda_I$ and Equation (24.13), $\omega = 2\lambda K + O(\varepsilon)$ are completely general and work anywhere in the complex plane. However, the expression (24.5) for the wavenumber, in the band $iA < \lambda < \infty$, becomes in dimensional form:

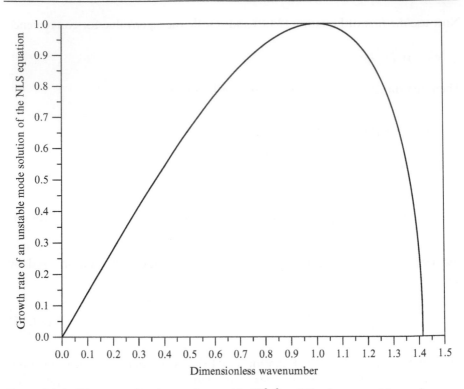

Figure 24.4 The normalized growth rate $(\Omega_I/(k_0^2 a_0^2 \omega_0/2))$ of an unstable mode solution of the NLS equation arising from a small-amplitude modulation as a function of normalized initial perturbation wavenumber, K (the imaginary part of Equation (24.14) as a function of dimensionless wavenumber, $K/2k_0^2 a_0$).

$$K = 2i\sqrt{2}k_0^2\sqrt{\lambda_0^2 - a_0^2} + O(\varepsilon) \qquad (24.16)$$

This leads to the modulation dispersion relation:

$$\Omega = \omega_0 \lambda_0 k_0^2 \sqrt{\lambda_0^2 - a_0^2} \qquad (24.17)$$

Thus, moving from the band below the carrier, $0 < \lambda < iA$, to the band above the carrier, $iA < \lambda < \infty$, means the wavenumber changes from being real to imaginary and the frequency goes from imaginary to real! The waves are *still unstable above the carrier but it is a spatial instability, whereas below the carrier the instability is temporal in character.* The final *modulation dispersion relation* is then given by

$$\left(\frac{K}{2\sqrt{2}k_0^2 a_0}\right)^2 = \frac{1}{2}\sqrt{1 + \left(\frac{2\Omega}{a_0^2 k_0^2 \omega_0}\right)^2} - \frac{1}{2} \qquad (24.18)$$

In consequence of this analysis it is now easy to see why there is *no small-amplitude spatial modulation* that can be used to induce instabilities in the IST spectrum *above the carrier* in the complex plane. Thus, experimentally, the tried and true method for generating unstable modes in the laboratory (e.g., small-amplitude modulations which grow into unstable modes) can never be used to access the really large modes that occur in the complex plane above the carrier. Here is one of the most important conclusions of this chapter: *If one seeks to make really large rogue waves in the laboratory he needs the full complexity of the IST to access these modes experimentally.* This point is illustrated graphically in Figure 12.8 of Chapter 12 where at $t = 0$ we see that the *Cauchy initial state* is a large-amplitude hump; this contrasts to the $t = 0$ state shown in Figure 12.6 of Chapter 12 that is a small-amplitude modulation (too small to see by eye at the scale of the figure). Both initial conditions grow into a large-amplitude rogue wave, but the large hump initial condition of Figure 12.8 of Chapter 12 grows into a wave of height about $3.8a_0$ compared to about $2.4a_0$ for the small-amplitude modulation. Bigger rogue waves grow out of the nontrivial, large-amplitude initial conditions characterized by Equation (24.18) and solutions of the NLS equation discussed below in terms of Riemann theta functions for modes above the carrier in the λ-plane.

How can we also predict the maximum amplitude of rogues above the carrier in the complex plane? This is easily done, for the formula $u_{max} = A + 2\lambda_I$ still works there. In the case of Figure 12.8 of Chapter 12, $\lambda_I = iA\sqrt{2}$, $A = 1$, so that $u_{max} = 1 + 2\sqrt{2} \approx 3.8284$. One can of course substitute Equation (24.17) into this formula to obtain u_{max} in terms of the temporal modulation. However, please note that the results presented in this chapter are based upon the Cauchy problem for which $u(x, 0)$ specifies the initial condition for the (space-like) NLS equation (24.2). To apply these results to laboratory work one instead uses the boundary value condition $u(0, t)$ that solves the time-like NLS equation (see Chapters 12 and 29).

We cannot say at the present time whether the large-amplitude rogue modes of the type in Figure 12.8 of Chapter 12 can exist in the ocean; this is as yet an open problem and is discussed further in Chapter 20.

24.4 Formulas for Unstable Modes and Breathers

In this section, we focus on the spectral solutions of the NLS equation (assumed to have periodic boundary conditions in the spatial variable, X, as given by Equation (24.3b)) that have the general formulation:

$$\psi(X, T) = A \frac{\theta(X, T | \tau, \delta^-)}{\theta(X, T | \tau, \delta^+)} e^{2iA^2T} \tag{24.19}$$

The $\theta(X,T|\tau, \delta)$ are generalized Fourier series known as N-dimensional Riemann theta functions:

$$\theta(X, T|\tau, \delta^{\pm}) = \sum_{m_1=-\infty}^{\infty} \sum_{m_2=-\infty}^{\infty} \cdots \sum_{m_N=-\infty}^{\infty} \exp i \left[\sum_{n=1}^{N} m_n K_n X + \sum_{n=1}^{N} m_n \Omega_n T \right.$$
$$\left. + \sum_{n=1}^{N} m_n \delta_n^{\pm} + \sum_{j=1}^{N} \sum_{k=1}^{N} m_j m_k \tau_{jk} \right]$$

$$(24.20)$$

where K_n are the wavenumbers, Ω_n are frequencies, $\delta_n^{\pm}$ are phases, and the $N \times N$ matrix τ_{mn} is the "period matrix" or "Riemann matrix." The diagonal elements of the period matrix refer to the "modes" of the NLS equation and the off-diagonal elements refer to nonlinear interactions amongst the modes. Details are given in the Appendix for the determination of the "theta parameters," K_n, Ω_n, δ_n, and τ_{mn}.

To provide some perspective on the above theta-function solution (24.19) of the NLS equation (24.3b), note that a simple "carrier wave" solution is given by

$$\psi(X, T) = A \, e^{2iA^2 T} \qquad\qquad\qquad (24.21)$$

Thus, the theta ratio, $\theta(X,T|\tau, \delta^-)/\theta(X,T|\tau, \delta^+)$, can be viewed as a long-wave modulation of the carrier wave; indeed the case $\theta(X,T|\tau, \delta^-)/\theta(X,T|\tau, \delta^+) = 1$ corresponds to no modulation at all. It should be emphasized that the solution (24.19) is a very general one for the NLS equation with quasi-periodic or periodic boundary conditions, $\psi(x, t) = \psi(x+L, t)$, for L the spatial period.

The unstable modes and breathers discussed in this chapter are characterized by a 2×2 "period matrix," τ, which implies that $N = 2$. This means the theta function reduces to

$$\theta(X, T|\tau, \delta^{\pm}) = \sum_{m_1=-\infty}^{\infty} \sum_{m_2=-\infty}^{\infty} \exp i \left[\sum_{n=1}^{2} m_n K_n X + \sum_{n=1}^{2} m_n \Omega_n T + \sum_{n=1}^{2} m_n \delta_n^{\pm} \right.$$
$$\left. + \sum_{j=1}^{2} \sum_{k=1}^{2} m_j m_k \tau_{jk} \right]$$

$$(24.22)$$

There are essentially five spectral parameters in the formulation which describe a single mode: A, λ_R, λ_I, ε_0, 0, where A is real, $\lambda = \lambda_R + i\lambda_I$, and $\varepsilon = \varepsilon_0 \exp(i\theta)$ (see Figure 24.2). The spectrum is symmetric about the real axis, so that iA is paired by $-iA$, λ by λ^*, and ε by ε^* (here "i" is the imaginary unit, "*" refers to complex conjugate, and the subscripts "R, I" refer to real and imaginary parts. The first parameter in the spectrum, A, describes the carrier wave. All other parameters describe the long-wave modulation of the carrier. Physically, the NLS equation describes the space-time dynamics of the nonlinear interaction between the carrier wave and the modulation. It is this dynamics that leads to the unstable modes and breather wave trains discussed herein. In the present analysis, to obtain concrete formulas for these modes, we have sought a singular asymptotic expansion in terms of the parameter ε where we assume $\varepsilon \ll 1$; the limit as ε tends to zero is singular. The wavenumbers, frequencies, phases, and period matrix are given to order $O(\varepsilon)$ are found to be (see Appendix):

Expansion parameter and sign of Riemann sheet index:

$$\varepsilon_1 = \varepsilon_0 \, e^{i\theta}, \quad \varepsilon_2 = \varepsilon_1^*$$
$$\sigma_1 = 1, \quad \sigma_2 = -1 \tag{24.23}$$

Spectral eigenvalue:

$$\lambda_1 = \lambda_R + i\lambda_I, \quad \lambda_2 = \lambda_1^* \tag{24.24}$$

Spectral wavenumber:

$$K_1 = -2\sqrt{A^2 + \lambda_1^2}, \quad K_2 = -2\sqrt{A^2 + \lambda_2^2} \tag{24.25}$$

Spectral frequency:

$$\Omega_1 = 2\lambda_1 K_1, \quad \Omega_2 = 2\lambda_2 K_2 \tag{24.26}$$

Period matrix:

$$\tau_{11} = \frac{1}{2} + \frac{i}{\pi} \ln\left(\frac{K_1^2}{\varepsilon_1}\right), \quad \tau_{12} = \frac{i}{2\pi} \ln\left(\frac{1 + \lambda_1\lambda_2 + \frac{1}{4}K_1K_2}{1 + \lambda_1\lambda_2 - \frac{1}{4}K_1K_2}\right) \tag{24.27}$$

$$\tau_{21} = \tau_{12}, \qquad \tau_{22} = \frac{1}{2} + \frac{i}{\pi} \ln\left(\frac{K_2^2}{\varepsilon_2}\right) \tag{24.28}$$

Phases:

$$\delta_1^+ = \pi + i \ln(\lambda_1 - \tfrac{1}{2}K_1) + i \ln(\sigma_1\lambda_1 + \tfrac{1}{2}K_1)$$
$$\delta_1^- = \pi + i \ln(\lambda_1 + \tfrac{1}{2}K_1) + i \ln(\sigma_1\lambda_1 + \tfrac{1}{2}K_1)$$
$$\delta_2^+ = \pi + i \ln(\lambda_2 - \tfrac{1}{2}K_2) + i \ln(\sigma_2\lambda_2 - \tfrac{1}{2}K_2) \tag{24.29}$$
$$\delta_2^- = \pi + i \ln(\lambda_2 + \tfrac{1}{2}K_2) + i \ln(\sigma_2\lambda_2 - \tfrac{1}{2}K_2)$$

Application of the above formulas in the theta function provides a simple way to compute the unstable modes and breathers for the particular case of a modulated plane wave carrier.

Nota Bene: The above formulation leads to the conclusion that there must be an infinite number of periodic IST nonlinear modes (Fourier components) in the (Riemann) spectrum of a solution to the NLS equation! How can we possibly use a spectral theory with an infinite number of component solutions? This does not at all seem practical. The answer is quite simple. In linear Fourier analysis the sine wave components are characterized by parameters consisting of amplitude, wavenumber, frequency, and phase and therefore, with all the varieties of these parameters possible there are effectively an infinite number of kinds of linear Fourier components, all of them being sine waves. In the nonlinear theory for the periodic NLS equation the same is true, except that the *shape* of the spectral components is different depending upon the parameters (see some characteristic formulas in Chapter 12). Not only is the shape of the nonlinear Fourier components different, but there are also two types of spectrum: (1) *stable components* that are *classical Stokes waves* and (2) *unstable components* that are the *unstable, rogue component solutions of NLS*. What about the infinite number of these components? In any application there are a finite number of modes N, simply because of the discrete nature of the space and/or time series of data or in simulations. These are the major physical differences between linear and nonlinear Fourier analysis in deep-water wave trains.

Of course, the way to compose solutions for the nonlinear problem is not as a superposition of sine waves, but is instead the *nonlinear superposition formula for theta functions* given above (Equation (24.19)). The way to *numerically compute theta functions* is discussed in Chapters 20–22 and the way to *compute the nonlinear (Riemann) spectrum* is discussed in Chapters 14–16, 23, and 24. Applications to the *analysis of particular data sets* are discussed in Chapters 25 and 28–31. Applications to *hyperfast modeling* are discussed in Chapters 32–34.

24.5 Examples of Unstable Mode (Rogue Wave) Solutions of NLS

The first example is effectively a demonstration of the existence of particular *nonhomoclinic solution* of the NLS equation (those for which $|\varepsilon| \neq 0$). To this end consider a *slot state* for which $L = \sqrt{2}\pi \cong 4.44$. An example spectrum is given in Figure 24.5 which has eight degenerate stable modes on the real axis

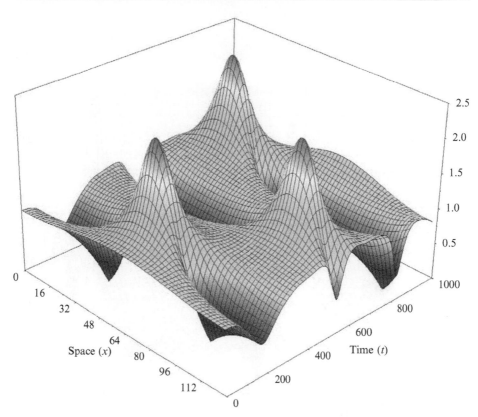

Figure 24.5 Space/time evolution of a slot-state rogue wave. Here $\lambda = iA/\sqrt{2}$ and $|\varepsilon| = 0.05$. Note that the wave is periodic in space, but alternates its phase along the time axis during the evolution. (See color plate).

(they have zero amplitudes and do not contribute to the dynamics), the carrier is $A = 1$ and two other eigenvalues occur at 0.76 and 0.66 ($\lambda_I \cong 0.707$) on the imaginary axis. Therefore, $|\varepsilon| = (0.76 - 0.66)/2 = 0.05$, verifying a finite value for $|\varepsilon|$. The space/time evolution of this rogue wave state is shown in Figure 24.5. Note that the wave rises up to a maximum height of $1 + \sqrt{2}$, but one of the interesting things about this solution, in contrast to the homoclinic state shown in Figure 12.6 of Chapter 12, is the fact that the present state is not purely periodic in time, but subsequent temporal appearances of the state are out of phase from the first appearance. This example also demonstrates that the splitting of a degenerate double point in the spectrum does not influence the amplitude of the resultant rogue state at leading order. However, as we see in the next example, if there are *two* degrees of freedom with finite values of $|\varepsilon|$, then these two states can interact to create a wave in configuration space which is higher than the amplitude law discussed in the last section.

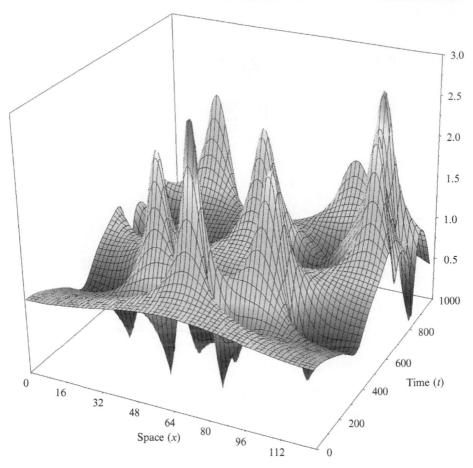

Figure 24.6   Space/time evolution of two slot-state rogue waves. Note that the maximum amplitude is 3.06. The space/time evolution is quite complex in this case. (See color plate).

In the second example, we have two slot states. IST, to $O(\varepsilon^2)$, predicts the maximum to be

$$A_{\max} \cong A_1 + A_1|\varepsilon_1| + A_2|\varepsilon_2| + O(\varepsilon^2)$$

Applying the IST one finds: $A = 1.019$, $A_1 = 2.9012$, $A_2 = 2.4058$, $|\varepsilon_1| = 0.04955$, and $|\varepsilon_2| = 0.0095$. These combine to give a maximum amplitude $A_{\max} = 3.068$ (see Figure 24.6). This compares to the value $A_{\max} = 3.062$ determined from the numerical simulation of the NLS equation. Extending the above formula beyond $O(\varepsilon^2)$ is hardly necessary in the present case.

Now let us show a couple of cases for NLS modes out in the complex plane. The first example has the spectral eigenvalue $\lambda = 0.1+i$. The maximum rogue wave amplitude is three (just use the formula $u_{\max} = A + 2 \lambda_I$). Because of the

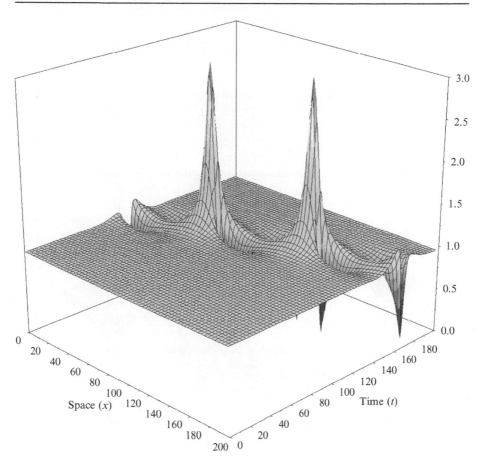

Figure 24.7 Space/time evolution of a rogue wave solution of the NLS equation for the spectral eigenvalue $\lambda = 0.1 + i$. (See color plate).

inclusion of a finite real part for the eigenvalue the waves have a group velocity *greater* than the velocity of the carrier. This is easily seen in Figure 24.7.

A second example is shown in Figure 24.8, where the spectral eigenvalue has been taken to be $\lambda = 1.6 + 1.6i$ for which the amplitude enhancement gives $u_{max} = 4.2$. In this example, the motion lies above the carrier wave in the complex plane, where $A = 1$. These large, broad crested waves are quite manifestly different from those that occur below the carrier. One can of course make even higher rogue waves just by venturing further out into the complex plane and by including more phased locked modes in the IST spectrum. It appears that the true limits on rogue wave heights are physical and likely dominated by wave breaking.

Where does the *envelope soliton* enter in all of this very general formulation for the NLS equation for periodic boundary conditions? An example is shown in Figure 24.9 where $\lambda = 10 + 20i$. In this case the carrier wave, $A = 1$, is quite small with respect to the maximum amplitude $u_{max} = 1 + 2(20) = 41$.

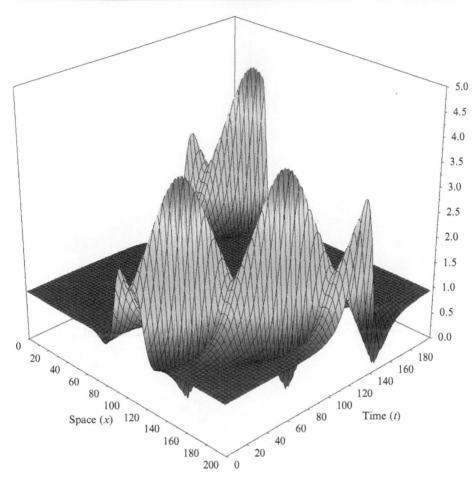

Figure 24.8   Space/time evolution of a rogue wave solution of the NLS equation for the spectral eigenvalue $\lambda = 1.6 + 1.6i$. (See color plate).

Therefore, the nonlinear interaction between the carrier and spectral component is rather small, evidence of which is indicated by the small oscillation in the soliton amplitude shown in Figure 24.9. Whether such large envelope soliton packets could exist in the real ocean is uncertain. For the physical parameters chosen in the present example the wave would clearly exceed the wave-breaking limit.

24.6   Summary and Discussion

The IST in the theta-function formulation is seen to provide new insight and perspective about the nonlinear dynamics of deep-water wave trains. While

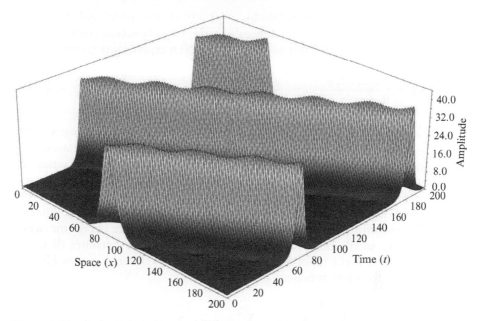

Figure 24.9   Space/time evolution of a solution of the NLS equation that is near the soliton limit. (See color plate).

space is limited for this monograph, more results will soon appear on the vast richness and complexity of rogue wave solutions of the NLS equation (Osborne, 2010).

The present results allow some suggestions to be made about future measurement programs both in the laboratory and in the ocean. With regard to laboratory experiments the approaches developed herein will certainly help study the many kinds of possible rogue wave modes and their approximate validity for describing real water wave dynamics. However, another aspect of this work is the continued development of time series analysis methods for the study of measured time series of data. The goal would be to obtain the nonlinear Fourier (IST) spectra directly from the data to assess the dynamical behavior of a particular wave train and to also nonlinearly filter the data to investigate particular rogue wave components.

Appendix   Overview of Periodic Theory for the NLS Equation with Theta Functions

The IST in the theta-function formulation is seen to provide new insight and perspective about the NLS equation. A *brief* and *brisk* review is given here of some of the principles (Kotljarov and Its, 1976; Tracy, 1984; Tracy and Chen,

1988; see the monograph by Belokolos et al. (1994) for additional discussion and as a guide to additional references). In the theta-function representation the spectral (nonlinear Fourier) solution of the NLS equation is given by

$$
u(x,t) = \frac{\theta_N(x,t|\tau,\delta^-)}{\theta_N(x,t|\tau,\delta^+)}
\tag{24.A.1}
$$

where $\theta_N(x,t|\tau,\delta)$ are N-degree-of-freedom Riemann theta functions: τ is the interaction matrix and $\delta^\pm = [\delta_1^\pm, \delta_2^\pm, \ldots, \delta_N^\pm]$ are two vectors of phases. The Riemann theta functions are a generalization of ordinary Fourier series. In the *linear case* one computes Fourier amplitudes, wavenumber, frequencies, and phases to solve a *linear wave equation* for particular initial conditions. For the NLS equation the task is similar, except that the nonlinear Fourier amplitudes constitute a matrix rather than a vector. In fact, when the wave amplitudes are small the interaction matrix, τ, effectively becomes diagonal and the Riemann theta function becomes an ordinary Fourier series. Here is the expression for theta functions used herein:

$$
\theta(x,t|\tau,\delta^\pm) = \sum_{m_1=-\infty}^{\infty} \sum_{m_2=-\infty}^{\infty} \cdots \sum_{m_N=-\infty}^{\infty} \exp i \left[\sum_{n=1}^{N} m_n K_n x + \sum_{n=1}^{N} m_n \Omega_n t \right.
$$
$$
\left. + \sum_{n=1}^{N} m_n \delta_n^\pm + \sum_{j=1}^{N} \sum_{k=1}^{N} m_j m_k \tau_{jk} \right]
\tag{24.A.2}
$$

Here $\mathbf{K} = [K_1, K_2, \ldots, K_N]$ is a vector of wavenumbers and $\Omega = [\Omega_1, \Omega_2, \ldots, \Omega_N]$ is a vector of frequencies. This function was central to the mathematical solution of the *Jacobian inverse problem* that lay in the domain of the field of pure mathematics (algebraic geometry) in the last half of the nineteenth century. The work of Jacobi, Poincaré, Baker, and many others was crucial in the solution of this fundamental mathematical problem.

To compute a solution of the NLS equation one has the following recipe. We provide also the special definitions needed to construct the unstable solutions described in the body of this chapter. We note that the IST spectrum for the types of wave we are dealing with consists of a conjugate pair of band edges at $\pm iA$ due to the carrier, and nearly degenerate pairs of band edges at $\lambda_{2j-1}^0 \pm \varepsilon_{2j-1}$ and their conjugates at $\lambda_{2j}^0 \pm \varepsilon_{2j} = \lambda_{2j-1}^{0*} \pm \varepsilon_{2j-1}^*$. We adopt the convention that the odd-numbered points lie in the upper complex plane, while the even-numbered points lie in the lower complex plane. We refer reader to Tracy and Chen (1988) for more complete details.

(a) Choose an appropriate set of holomorphic differentials and loop cycles on the two-sheeted Riemann surface (denoted R) given by the polynomial expression:

$$R(\lambda) = \prod_{k=1}^{4N+2} (\lambda - \lambda_k) = (\lambda^2 + A^2) \prod_{k=1}^{2N} (\lambda - \lambda_k^0 + \varepsilon)(\lambda - \lambda_k^0 - \varepsilon) \qquad \text{(24.A.3)}$$

where λ_k are the spectral eigenvalues in the complex plane. The two-sheeted nature of the Riemann surface arises because of the square-root in this expression. On R one can construct N linearly independent holomorphic (regular or analytic) differentials. The most common choice is the following "oscillatory basis":

$$dU_j \equiv \frac{\prod_{i \neq k}^{N} \left(\lambda - \lambda_j^0\right)}{\sqrt{R(\lambda)}} \, d\lambda \qquad \text{(24.A.4)}$$

One can show that R is topologically equivalent to a sphere with N handles attached to it. On this surface there exist $2N$ distinct *closed curves* which are split into two classes which are called a- and b-cycles (the actual notation is a_j, b_j, $j = 1, \ldots, N$). The a- and b-cycles are said to be orthogonal to each other for they obey the following orthogonality conditions:

$$a \circ a = 0, \quad b \circ b = 0, \quad a_j \circ b_k = \delta_{jk} \qquad \text{(24.A.5)}$$

The meaning of this shorthand notation is supplied by the following rules:

(1) The a_j do not cross any other a_j cycles. The b_j do not cross any other b_j cycles.
(2) The cycle a_j intersects b_j only once and intersects no other b-cycle. The intersection point of these two cycles, after deforming them to cross at a right angle, has the property that the tangent to a_j is coincident with the tangent to b_j after a counterclockwise rotation by $\pi/2$.

An appropriate choice of a- and b-cycles for the present calculation is given by choosing the cycle a_k to encircle the nearly degenerate pair of band edges $\lambda_k^0 \pm \varepsilon_k$ in a counterclockwise sense. This choice of a-cycles automatically induces a related set of b-cycles through the orthogonality relations given above.

(b) Compute the invertible matrix of "a-periods":

$$A_{kj} \equiv \int_{a_j} dU_k = \delta_{kj} + O(\varepsilon) \qquad \text{(24.A.6)}$$

(c) Compute the invertible matrix of "b-periods":

$$B_{kj} \equiv \int_{b_j} dU_k \qquad \text{(24.A.7)}$$

(d) Compute the Riemann matrix τ by the following procedure. First compute the $\mathbf{C}$ ($\mathbf{C} = \mathbf{A}^{-1}$) and τ matrices:

$$\boldsymbol{\tau} = \mathbf{A}^{-1} \mathbf{B} \qquad \text{(24.A.8)}$$

But, since the matrix of a-periods is equal to the identity matrix to order ε, we have that the τ matrix is equal to the matrix of b-periods to that same order. The τ matrix is most easily expressed in terms of the wavenumbers, with diagonal elements:

$$\tau_{jj} = \frac{1}{2} + \frac{i}{\pi} \ln\left(\frac{K_j^2}{\varepsilon_j}\right) \tag{24.A.9}$$

and off-diagonal elements:

$$\tau_{jk} = \frac{i}{2\pi} \ln\left(\frac{1 + \lambda_j\lambda_k + \frac{1}{4}K_jK_k}{1 + \lambda_j\lambda_k - \frac{1}{4}K_jK_k}\right) \tag{24.A.10}$$

The wavenumbers K_j are given below.

(e) Compute the wavenumbers and frequencies.

One first computes the argument of the θ-function:

$$W_j(x, t) = \sum_{k=1}^{N} \int_{p_0}^{\mu_k(x, t)} d\psi_j \tag{24.A.11}$$

which can be written ($\mu_k(x, t)$ are the hyperelliptic functions):

$$W_j(x, t) = \sum_{k=1}^{N} \int_{p_0}^{\mu_k(x, t)} d\psi_j = \sum_{k=1}^{N} \sum_{m=1}^{N} C_{jm} \int_{p_0}^{\mu_k} \frac{\lambda^{m-1} d\lambda}{R(\lambda)}$$

for which we find

$$W_j(x, t) = \frac{1}{2\pi}(K_j x + \Omega_j t + \delta_j^{\pm}) \tag{24.A.12}$$

where the wavenumbers have the form:

$$K_j = -4\pi i C_{jN} \cong -2\sqrt{A^2 + \lambda_j^{02}} \tag{24.A.13}$$

and the frequencies are

$$\Omega_j = -8\pi i \left[C_{j, N-1} + \left(\frac{1}{2}\sum_{k=1}^{2N+2} \lambda_k\right) C_{j, N} \right]$$

and we find the modulation dispersion relation

$$\Omega_j = -2\lambda_j^0 K_j \tag{24.A.14}$$

and the phases are computed from the hyperelliptic function initial conditions. One finds the following phase:

$$\delta_j^\pm = 2\pi \left[\int_{p_0}^{\infty\pm} d\psi_j - \sum_{k=1}^{N} \left(\int_{p_0}^{\mu_j(0,\,0)} d\psi_j - \int_{a_k} \psi_j d\psi_k \right) \right]$$

which can be approximated by

$$\delta_j^\pm = \pi + i \, \ln\left[\sigma_j (\lambda_j^0 \pm \tfrac{1}{2}K_j)(\lambda_j^0 + \tfrac{1}{2}K_j) \right] \tag{24.A.15}$$

Here the constant $\sigma_i = \pm 1$ is a *Riemann sheet index*. This completes specification of solutions to the NLS equation. The simple formulas developed throughout this chapter have been derived from the formulation in this Appendix.

25 Internal Waves and Solitons

25.1 Introduction

The scope of this chapter is to illustrate some of the possibilities for analyzing data containing long internal wave motions using the inverse scattering transform (IST). I use the Andaman Sea data (Osborne et al., 1978; Osborne and Burch, 1980) as an example of internal soliton propagation. I do not attempt to review the field in this or other chapters, but instead hope to aid the researcher in the analysis of data and in modeling of long internal waves.

The dynamics of internal solitons were encountered many years ago by the oil industry in the Andaman Sea, offshore Thailand. A preliminary measurement program, conducted by Exxon Production Research Company, revealed currents as high as 1.6 m/s (see Figure 25.1 for the locations of the measurements obtained in the southern Andaman Sea, indicated by "+" signs on the map). Subsequent drilling by Esso Exploration Company, the exploration arm of Exxon Corporation, verified the intensity of these currents during drilling operations in 1100 m depths beginning in January 1976 for over a year and a half.

It was conjectured by the oceanography group at Exxon Production Research Company that the source of these currents was due to internal wave activity in the area. This interpretation was in part motivated by the papers by Perry and Schimke (1965) and Hunkins and Fliegel (1973) and by the actual measurements themselves. A subsequent detailed measurement program and drilling operations in the area by the drill ship *Discoverer 534* lead to the conclusion that energetic solitons with amplitudes as high as 150 m were responsible for currents up to 2.3 m/s (4.47 knots) (Osborne et al., 1978; Osborne and Burch, 1980).

A large body of experiments has been conducted since that time in a wide variety of offshore locations and it has become clear that *internal solitons or solitary waves* are ubiquitous features in most continental shelf/slope regions around the world (see the online *Internal Wave Atlas*; Jackson, 2004 and references cited therein). In recent years there have been many advances in the field, including theoretical, numerical and experimental work (Lee and Beardsley, 1974; Ostrovsky, 1978; Helfrich et al., 1984; Holloway, 1987; Apel, 1988; Farmer and Army, 1988; Army and Farmer, 1988; Grimshaw and Melville, 1989; Helfrich and Melville, 1989; Helfrich, 1992; Lamb and Yan, 1996; Holloway et al., 1997; Grimshaw and Ostrovsky, 1997a,b; Grue et al., 1997; Grimshaw et al., 1998; Stanton and Ostrovsky, 1998; Liu et al., 1998; Holloway et al., 1999; Choi and Camassa, 1999; Farmer and Armi, 1999; Grue et al., 1999;

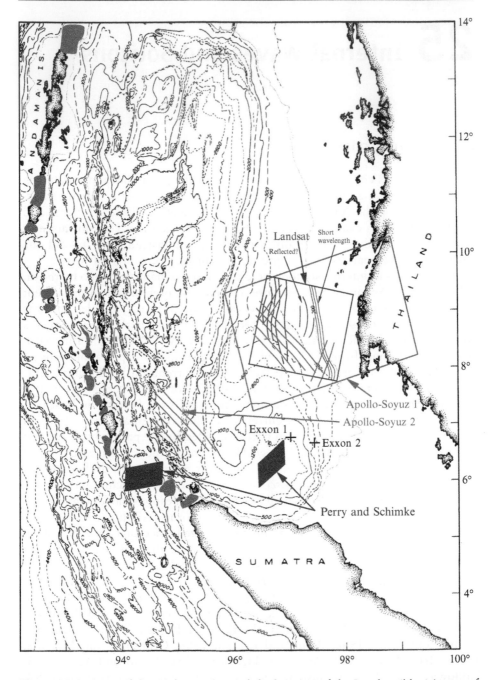

Figure 25.1 Map of the Andaman Sea and the location of the Landsat (blue) image of Figure 25.3 and the Apollo-Soyuz photographs of Figures 25.4 (Apollo-Soyuz 1, red), 25.5 (Apollo-Soyuz 2, red). The "rip zones" of Perry and Schimke (1965) are shown as blue rectangles. (See color plate).

Helfrich et al., 1999; Grue et al., 2000; Holloway et al., 2001; Mitrool'sky, 2001; Lamb, 2002; Grimshaw et al., 2002; Staquet and Sommeria, 2002; Ostrovsky and Grue, 2003; Lamb, 2003; Moum et al., 2003; Grimshaw et al., 2004; Duda et al., 2004; Lamb, 2004, Lamb and Wilkie, 2004; Klymak et al., 2006; Apel et al., 2007; Scotti et al., 2007; El, 2007; Hawkins et al., 2008; Scotti et al., 2008; Christov, 2009). The recent excellent review by Helfrich and Melville (2007) provides an overview of experiments and theories plus a list of important references.

The influence of internal solitons and solitary waves on various kinds of operations on and below the ocean's surface has become increasingly important in the last few decades as man has explored increasingly greater depths in search of ocean resources and for defense purposes. Particle velocities in long internal waves can generate forces on ships, on fixed and compliant surface piercing and/or underwater structures or floating vehicles. Hence, there is the potential for exceeding design loads in structures and for disturbing the ballast of floating underwater vessels of all types. Figure 25.2 shows an incident in the Strait of Gibraltar in which one possible explanation for the damage to the Victor class Russian submarine was that it might have been carried to the surface by an *upwardly mobile internal soliton*, followed by subsequent impact with a surface ship. While this possibility is unconfirmed it does illustrate the potential for energetic internal waves to influence submarine ballasting.

Figure 25.2   A Victor class Russian submarine on the surface in the Straights of Gibraltar. One possible scenario for the damage is that the submarine encountered an upward moving internal wave forcing it to make brisk contact with a surface ship (Office of Naval Research). (See color plate).

A full physical understanding of internal solitons is still lacking. Considerable progress in the modeling of large-amplitude internal waves (Camassa and Choi, 1999; Scotti et al., 2007, 2008), in the theory, in the development of nonlinear time series analysis techniques and in experimental methods will be required in the future. There are a number of fundamental open questions: For example, given the current state of theoretical knowledge, how does one know, in an experimental context, what a soliton is? Is a particular feature in a data set really a soliton? Given the number of types of solitons that have been found theoretically (positive and negative pulses, kinks, "table-top" solitons, unstable modes, vortices, etc.) (Ostrovsky, 1978; Grimshaw, 1985, 1997, 2001, 2005, 2007; Holloway, 1987; Holloway et al., 1997, 1999, 2001; Grimshaw et al., 1998; Ostrovsky and Stepanyants, 2005; Helfrich and Melville, 2007), how do we know that a time series has any or all of these kinds of coherent structures in them? Is a particular feature in a time series a soliton? Are solitons hidden by other nonlinear Fourier components in the data and hence not visible to the eye? Might these hidden solitons appear later in the evolution of the system dynamics? These are very challenging problems and work of this type is addressed at length in this book using the IST. In this chapter some aspects of the analysis of data are discussed, together with particular aspects of the theoretical development and modeling of long internal waves.

To set the stage for the nonlinear spectral approach in this chapter, recall that we are familiar with the idea that Fourier analysis may be used to analyze a time series, that is, that a time series can be decomposed into a large number of sine waves with particular amplitudes, frequencies, and phases. On the other hand, we do not actually observe sine waves when we look at a time series, yet, based upon Fourier theory we understand that sine waves are the fundamental building blocks for describing linear wave motion. These ideas permeate research activities in many fields of science including the field of physical oceanography.

On the other hand, *nonlinear time series* can be decomposed into *nonlinear basis functions*. These include Stokes waves, cnoidal waves, and various kinds of positive and negative solitons, kinks, table-top solitons, unstable modes, vortices, etc. There exists the possibility that these kinds of coherent structures can physically overlap in a time series. Hence, we may not be able to tell by eye which features in a time series are solitons, what kinds of solitons they are, etc. Of course, we are all familiar with the fact that on the shelf slope the solitons tend to be rank ordered (see, e.g., the work of Apel, 2003), for physical reasons related to their *adiabatic evolution* up the continental slope). This fact often helps us feel that we have rather complete understanding, but in reality up on the continental shelf in relatively shallow water the situation may be quite different. In this region one expects that there might be solitons of different types that are not at all rank ordered but are instead mixed together in a quite complex way. It is therefore important to have a kind of nonlinear Fourier analysis that can tell us what the actual physical "solitonic" structure of a nonlinear wave train is. We are far from a full understanding of this problem, but it is hoped that the present effort could aid the researcher to move in this direction.

25.2 The Andaman Sea Measurements

Large-amplitude internal solitons were observed in the Andaman Sea north of Sumatra in 1093 m of water (Osborne et al., 1978; Osborne and Burch, 1980). It was demonstrated that these data contained most of the qualitative features of the theory of the Korteweg-deVries (KdV) equation and some quantitative comparisons were also carried out. The area in which the measurements were made was in the southern Andaman Sea ("+" signs in Figure 25.1). Surface striations referred to as "surface rips" consisted of 100-200 km long bands of surface waves 300-400 m wide and were easily visible in both Landsat images (Figure 25.3) and in Apollo-Soyuz photographs (Figures 25.4 and 25.5) (see also the online *Internal Wave Atlas*; Jackson, 2004). Perry and Schimke (1965) had observed surface rips in this area a few years before and the location of their observations is shown in the blue parallelograms in Figure 25.1; they also made bathythermograph measurements of the underlying internal waves.

Figure 25.3 Landsat image of the Andaman Sea surface showing long striations thought to be surface rips associated with internal soliton activity. The white regions are clouds. North is upward (Courtesy of Al Strong, NOAA).

Figure 25.4 Apollo-Soyuz photograph of the surface of the Andaman Sea showing surface striations associated with internal wave activity. North is to the left (Courtesy of NASA, Johnson Spacecraft Center). (See color plate).

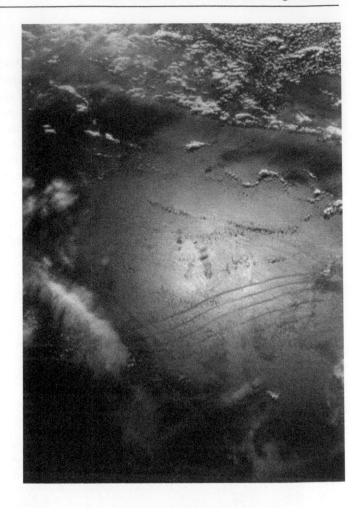

Landsat images turned out to be a big help in understanding the spatial scales of the Andaman Sea waves. Landsat was designed to take images of land areas, but when the instrument focused on areas of the ocean the resultant negatives were totally black. Delicate work in the dark room allowed these images to be over exposed, leading to results such as that in Figure 25.3 which shows a number of striations thought to be related to internal wave activity; a large packet of waves at the center left center is quite typical of Landsat images in the region. A large number of short wavelength waves are clustered about the 200 m contour in Figure 25.3; this is best seen when comparing this image with Figure 25.1 (See color plate) in which the image boundary is shown as a blue square and the largest packet is also shown in blue. In addition to the short internal waves near the 200 m contour there are also other striations that appear to be westward moving waves that may have been reflected from the bathymetric contours near the coast of Thailand.

Figure 25.5 Apollo-Soyuz photograph of the surface of the Andaman Sea showing surface striations associated with internal wave activity. This photograph was taken a few minutes before that in Figure 25.4. North is to the left (Courtesy of NASA, Johnson Spacecraft Center). (See color plate).

Figure 25.4 (See color plate) is an Apollo-Soyuz photograph (a Hasselblad image taken by astronaut Vance Brand) of an area in the Andaman Sea. A large packet is seen propagating toward Thailand; the Phuket peninsula is easily seen on the upper right. The position of this photograph is shown as a parallelogram outlined in red in Figure 25.1; the red striations inside the parallelogram are estimates of the wave positions in Figure 25.4. Figure 25.5 (See color plate) is another Apollo-Soyuz photograph taken a few minutes earlier than the one in Figure 25.4. The upper packet is that shown in Figure 25.4; the lower packet trails the first by about 12 h. Additional red striations correspond to this trailing packet, below and to the left of the first Apollo-Soyuz photograph, are also shown on the map in Figure 25.1 (See color plate). All three of these images show similar long, parallel striations evidently related to "surface rips" on the ocean surface, suggesting the presence of internal solitons below the surface.

The series of photographs in Figure 25.6 were taken onboard the survey vessel *Oil Creek* by the author; these photographs are of "surface rips" associated with long internal wave activity and are believed to result from a near resonant interaction between long internal waves and short surface waves above it (Phillips, 1974); the resonant interaction is most efficient when the phase speed of the long internal waves is equal to the group speed of the surface waves. This connection between the Andaman Sea internal waves and the surface rips was pointed out by Perry and Schimke (1965) and Osborne and Burch (1980) and is illustrated in Figure 25.10 where the rip is found to be on the leading edge of the soliton above the convergence zone in the particle velocity field.

The temperature measurements, obtained at location "Exxon 1" in Figure 25.1, are given in Figure 25.7. The time series spanned about 4 days and were made at five different depths: 53, 87, 116, 164, and 254 m. The measurements are punctuated by temperature enhancements that occur in groups separated by 12.4 h. Increases in temperature occur because, for the density structure in the Andaman Sea, the waves are displaced downward, pulling warm near-surface waters down to the depths of the instruments. The tendency for the waves to be rank ordered (the largest wave in a packet is faster and outruns the others, thus appearing to the left, while smaller more slowly moving waves trail out to the right) (Osborne and Burch, 1980) was considered to be evidence of the behavior of the Cauchy problem for waves obeying near-KdV behavior as predicted by the IST (Gardner et al., 1967). The temperature structure found at the measurement location was generally typical of that shown in Figure 25.8, constructed from an expendable bathythermograph (XBT) cast. An expanded view of the first 12.4 h segment of temperature time series taken at 164 m from the surface is shown in Figure 25.9. This will be analyzed for nonlinear spectral content below.

25.3   The Theory of the KdV Equation as a Simple Nonlinear Model for Long Internal Wave Motions

The history of soliton systems is considered by many to begin with the work of Russell (1844), who was the first to document his observations of the "solitary wave" in a canal near Edinburgh, Scotland, in about 1825. Russell was inspired by these observations and designed and subsequently carried out a series of exacting laboratory measurements (Russell, 1844) in which he was able to determine a surprising number of soliton properties. He called his discovery, in the hyperbole of Victorian times, the *great solitary wave of translation*. While the solitary wave was not immediately accepted as a viable waveform and in fact

Figure 25.6 Surface rips associated with long internal wave activity beneath the Andaman Sea surface. This sequence of photographs was taken aboard the survey vessel Oil Creek by the author. The time between each photograph is about 1 min. (See color plate).

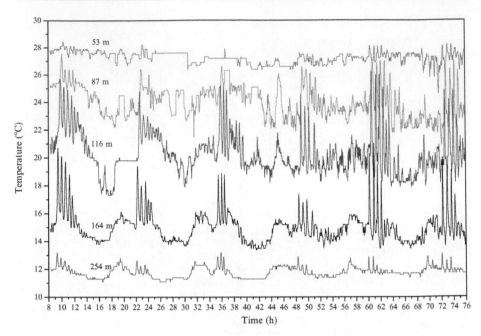

Figure 25.7   Andaman Sea time series of temperature taken at various depths in the period October 24-28, 1976. (See color plate).

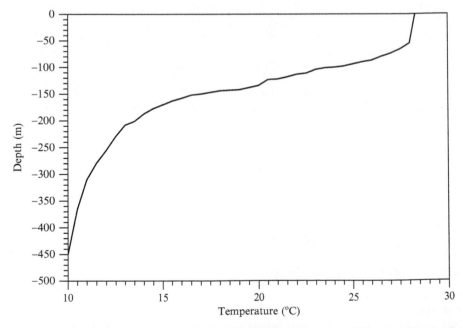

Figure 25.8   Andaman Sea XBT temperature data as a function of depth, taken in the period October 24-28, 1976.

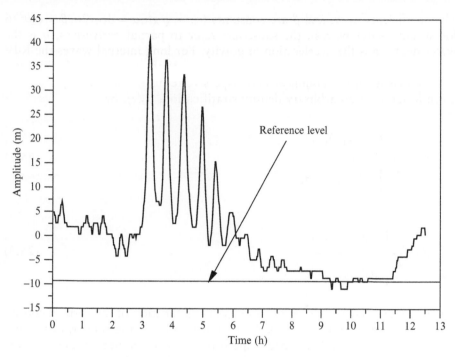

Figure 25.9 Andaman Sea thermistor temperature time series at 164 m depth, beginning at 10:20 h on October 24, 1976.

the idea met with some resistance (Emmerson, 1977), it was eventually found theoretically by Rayleigh (1876) and Boussinesq (1872) in the mid-1870s (see Whitham, 1974; Miles, 1981) for modern accounts of this work and a guide to the references). The work of Korteweg and deVries (1895) resulted in the prototypical partial differential equation for soliton systems, now known as the KdV equation.

I do not delve deeply into the theory of the KdV equation here, see instead Chapters 10, 14, 17, and 19–23. However, I do discuss how the KdV equation can be extended to arbitrarily high order using Lie-Kodama transforms (Kodama, 1985a,b) (see, e.g., Osborne, 1997). In this way, we extinguish any idea that the *theory of the KdV equation* might not be able to handle highly nonlinear waves in the ocean. The natural *hierarchical* extension of the KdV equation and its associated theory to 2 + 1 dimensions and to higher order is discussed in Chapters 32 and 33. See also the discussion in Section 25.6.

Consider the model equation for long *internal wave* propagation given by Korteweg and deVries (Benny, 1974) (see Osborne and Burch, 1980 for a guide to the earlier references):

$$\eta_t + c_0\eta_x + \alpha\eta\eta_x + \beta\eta_{xxx} = 0 \tag{25.1}$$

The coefficients c_0, α, and β are constants ($c_0 = \sqrt{gh}, \alpha = 3c_0/2h, \beta = c_0h^2/6$ for surface water waves), the subscripts refer to partial derivatives, h is the water depth, g is the acceleration of gravity. For long internal waves the KdV equation describes the vertical displacement $\eta(x, t)$ of an isopycnal. In this case, the coefficients of the nonlinear and dispersive terms of the KdV equation are given in terms of an arbitrary density stratification, $\rho_0(z)$, by

$$\alpha = \frac{3c_0}{2} \frac{\displaystyle\int_0^h \rho_0(z)\phi_z^3(z)dz}{\displaystyle\int_0^h \rho_0(z)\phi_z^2(z)dz} \tag{25.2}$$

$$\beta = \frac{c_0}{2} \frac{\displaystyle\int_0^h \rho_0(z)\phi^2(z)dz}{\displaystyle\int_0^h \rho_0(z)\phi_z^2(z)dz} \tag{25.3}$$

The undisturbed density stratification $\rho_0(z)$ is a function of the vertical coordinate z (measured upwards from the bottom) and h is the total water depth. The constant c_0 is the linear phase speed of the wave motion; c_0 and $\phi(z)$ are the eigenvalues and eigenfunctions of the eigenvalue problem:

$$(\rho_0\phi_z)_z - (g/c_0^2)\rho_{0,z}\phi = 0 \tag{25.4}$$

where the surface and bottom boundary conditions are

$$\phi(0) = \phi(h) = 0 \tag{25.5}$$

The stream function associated with the flow is given by

$$\psi(x, z, t) = c_0\phi(z)\eta(x, t) \tag{25.6}$$

Horizontal and vertical particle velocities are computed from

$$u = \psi_z, \quad w = -\psi_x \tag{25.7}$$

When the density stratification can be approximated by a *two-layer fluid*, the expressions for c_0, α, and β simplify; if the upper layer has density ρ_1 and thickness h_1 and the lower layer has density ρ_2 and thickness h_2 for $h_1 < h_2$ and $\rho_1 \sim \rho_2 \sim \rho$, then the coefficients of the KdV equation are given approximately by

$$c_0 = \left[g\left(\frac{\Delta\rho}{\rho}\right)\frac{h_1}{1+\rho}\right]^{1/2} \tag{25.8}$$

$$\alpha = -3c_0 \frac{1-r}{2h_1} \tag{25.9}$$

$$\beta = c_0 \frac{h_1 h_2}{6} \tag{25.10}$$

where $r = h_1/h_2$ and $\Delta\rho = \rho_1 - \rho_2 \ll 1$.

It is clear from the above considerations that *the KdV equation is valid for arbitrary stratification*. The above results can be extended to arbitrary order in nonlinearity via the Whitham hierarchy (Whitham, 1974; Lee and Beardsley, 1974), as discussed in more detail below.

The general single-soliton solution to the KdV equation is

$$\eta(x,t) = -\eta_1 \, \text{sech}^2[(x - c_1 t)/L_1] = -\eta_1 \, \text{sech}^2[(K_1 x - \Omega_1 t)] \tag{25.11}$$

The depth variation in this solution is given by

$$\eta(x,z,t) = \phi(z)\eta(x,t) \tag{25.12}$$

This latter equation says that the amplitude of an isopycnal, $\eta(x, z, t)$, can be obtained by multiplying the amplitude of the KdV solution, $\eta(x, t)$, by the eigenfunction solution of Equation (25.4), $\phi(z)$. The maximum amplitude for the first mode eigenfunction is normalized to one, hence the internal wave amplitudes $\eta(x, z, t)$ are a maximum at some depth z_0 ($\phi(z_0) = \phi_{max} = 1$) and decrease above and below this depth to zero at the free surface and the bottom (since $\phi(0) = \phi(h) = 0$).

The wavelength, phase speed, and frequency of the soliton are

$$L_1 = \left(\frac{12\beta}{h_1|\alpha|}\right)^{1/2} = K_1^{-1}$$

$$c_1 = c_0\left(1 + \frac{|\alpha|}{3c_0}\eta_1\right) \tag{25.13}$$

$$\Omega_1 = \frac{c_1}{L_1}$$

In Equations (25.11)–(25.13) the eigenfunctions, eigenvalues, and coefficients of the KdV equation may be determined for general stratification (by Equations (25.2)–(25.5)) or for a two-layer fluid (by Equations (25.8)–(25.10)). Note that the negative sign before the amplitude η_1 in Equation (25.11) means that the soliton is a negatively (downwardly) displaced wave, a result of the assumption that $h_1 < h_2$; the sign is reversed when the upper layer is thickest. The parameter α is negative for internal waves when the upper layer is thinner than the lower. The associated velocity field (the contours are referred to as isovels) is computed from Equations (25.6) and (25.7) (Figure 25.10).

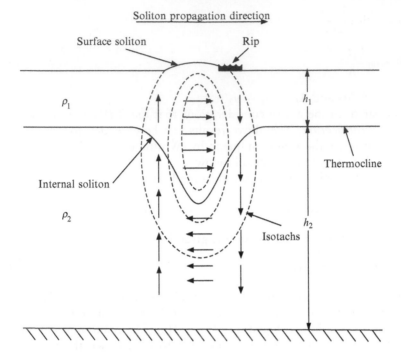

Figure 25.10   An internal soliton in a two-layer ocean.

25.4   Background on KdV Theory and Solitons

A wide range of applications for the KdV equation and the solitary wave occurred in the first half of the twentieth century as is well documented in the important work of Munk (1949). Up until this time, however, there were only a few hints as to the truly fundamental nature of the solitary wave (Miura, 1974). It is safe to say that the issues of the nonlinear stability of these waves remained an open problem and it was generally conceded that solitary waves would normally self-destruct under mutual collisions with each other.

The work of Fermi et al. (1955) was seminal to further progress. Numerical experiments for the thermalization of neutrons in an atomic reactor, essentially a discretization of the KdV equation, revealed an amazing result: A sine wave initial condition did not thermalize (i.e., render the Fourier spectrum "white noise"), as initially predicted by them, but instead returned almost nearly to the initial conditions after a certain "recurrence time." This was surely a surprising result at the time and FPU recurrence stimulated the amazing work of Zabusky and Kruskal (1965) who observed, from numerical solutions of KdV with periodic boundary conditions, that the solitary waves (there were nine of them, with only small radiation residuals) behaved *elastically* during the collision processes: This observation prompted them to rename the solitary wave solutions of the KdV equation *solitons*. One of the important properties of the solitons is that they survive collisions with each other, maintaining their amplitudes and speeds afterward, but at the same time they undergo a *phase shift* with respect to each other: the larger soliton is shifted forward and the smaller soliton is shifted backward relative to what their positions would be in the absence of nonlinear interactions.

Subsequent to the numerical discovery of the soliton a pioneering break-through was made with regard to the mathematical physics: Gardner et al. (1967) discovered the exact solution to the KdV equation for the *Cauchy problem* on the *infinite line*: An initial waveform, suitably localized in space on the interval $-\infty < x < \infty$ at time $t = 0$, will subsequently evolve into a *sequence of rank-ordered solitons* followed by a *radiation tail*. This result is shown schematically in Figure 25.11.

For the study of internal solitons in the ocean the above picture of a rank-ordered sequence of solitons does not always occur. When rank ordering does occur it tends to happen on the shelf slope where adiabatic evolution occurs toward decreasing water depth, an effect that tends to preserve the rank ordering. However, on the continental shelf, the more appropriate boundary conditions may be *periodic* with a typical period of 12.4 h. This fact, together for the case with a relatively flat bottom, means that rank ordering does not often occur. Zabusky and Kruskal (1965) studied periodic boundary conditions in their numerical simulations in which they discovered the soliton. They published a film, issued by Zabusky et al. (1963) that is instructive in this regard. We show four panels from our numerical simulations of their results in Figure 25.12. The initial condition is a simple sine wave (Figure 25.12A). At

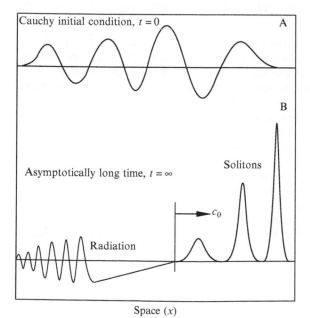

Cauchy initial condition, $t = 0$ A

B

Asymptotically long time, $t = \infty$

Solitons

c_0

Radiation

Space (x)

Figure 25.11 An arbitrary waveform at time $t = 0$ (A), evolves into a sequence of rank-ordered solitons plus a radiation tail as t tends to ∞ (B).

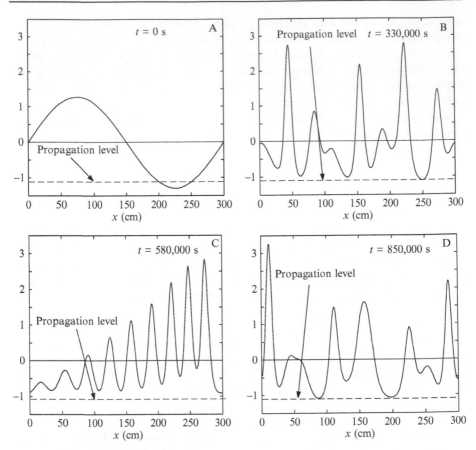

Figure 25.12   Illustration of the discovery of the soliton by Zabusky and Kruskal (1965). The KdV equation is integrated numerically with periodic boundary conditions, which are appropriate for some simple kinds of internal soliton dynamics on the continental shelf. Four time values are shown in the numerical solutions for a sine wave (internal tide) initial condition (A). At a later time one observes that solitons emerge from the sine wave and undergo quite complex interactions (B). At a particular instant of time the nine solitons are well delineated, indeed rank ordered as in the infinite-line case (C). The soliton complex interaction structure is again seen at later times (D). Note the propagation level shown in each panel. The area in the initial wave train (A) under the sine wave and above this level provides the water mass that leads to solitons. This contrasts to the case for infinite-line boundary conditions where the soliton mass is bounded by the sine wave and the zero level. Thus, the case for periodic boundary conditions (often appropriate for internal wave motions) always leads to much larger solitons than the case for infinite-line boundary conditions because the reference level is always below the zero level.

a later time the soliton interactions have the form shown in Figure 25.12B; note that at this moment there is no rank ordering of the solitons. In Figure 25.12C a *particular moment* is shown when the solitons are all nicely rank ordered. Later, however (Figure 25.12D) another complex situation is found. Indeed, we suggest that many complex oceanic observations may have this complicated structure. Identification and understanding of this complex structure by eye is not so easy. One focus of this chapter is that *nonlinear Fourier analysis* (based upon the IST) can serve as a *time series analysis* tool for understanding the nonlinear structure shown in Figure 25.12 and in oceanic internal wave data.

An interesting problem is the propagation of internal tidal waves up the continental shelf toward shore. Numerically, this problem can be attacked by assuming that a periodic oscillating paddle motion exists offshore and that these "tidally generated" waves then propagate toward shore into decreasing water depth. The simplest simulation of this problem would be to use the "time" KdV equation as discussed elsewhere in this monograph (e.g., see Chapter 28.3). Then one would simulate the motion by slowly changing the coefficients as the waves propagate toward shore (see Chapters 32–34 for applications to surface waves with directional spreading and at higher order).

25.5   Nonlinear Fourier Analysis of Soliton Wave Trains

The spectral decomposition in terms of cnoidal waves can of course be used for trains of solitons. In this case, the spectral decomposition consists of the solitons themselves plus nonlinear interactions (interaction phase shifts). An example is shown in Figure 25.13, a three-soliton decomposition shown in the *soliton basis* (each component is seen to be a soliton, not an oscillatory wave train, see Chapter 8). The solitons are shown in top-to-bottom vertical order from largest to smallest. Below the solitons are the nonlinear interactions and finally the three-soliton wave train solution of the KdV equation is shown. Note how important the nonlinear interactions are. Excluding the nonlinear interactions gives the wave train denoted by the dotted line. Including the nonlinear interactions phase shifts the largest soliton *forward* and the two smaller solitons *backward*. The characteristic *nonlinear interaction function* is that shown in the figure, that is, to phase shift a soliton a "phase shift function" consisting of a negative pulse followed by a positive pulse (a forward phase shift) occurs in the periodic IST. The phase shift function consisting of a positive pulse followed by a negative pulse generates a backward phase shift. The results in Figure 25.13 are from a simple nonlinear spectral analysis of a solitonic wave train by the IST for periodic boundary conditions in the soliton basis (Chapters 10, 14, 17, and 20–22). The results are consistent with the predictions of phase shifting on the infinite line by the infinite-line IST. Clearly, nonlinear superposition is the only way to have a correct spectral decomposition of the wave train.

How was the graph in Figure 25.13 generated? First a simulation of a three-soliton case was conducted using the Hirota method for the KdV equation

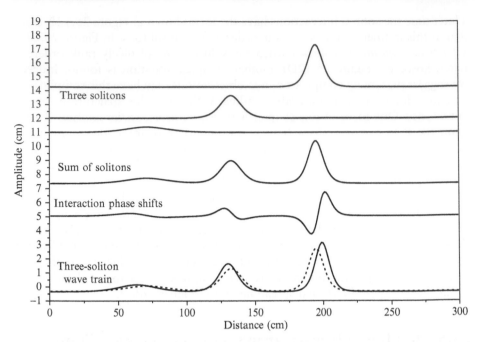

Figure 25.13 Three-soliton decomposition. Shown is the IST spectral decomposition. From top to bottom are the three solitons, the nonlinear interactions (phase shifts) and in the bottom two graphs are the constructed soliton train (solid line, via the IST, which is the linear summation of the solitons and interactions) and the linear superposition of the solitons (dotted line). The largest soliton is phase shifted forward, the two smaller solitons are phase shifted backward. This result emphasizes the important role that nonlinear interaction phase shifts play in the nonlinear dynamics.

(Chapter 4); while this procedure assumes infinite-line boundary conditions the solitons are placed on a finite interval, but far from the boundaries and themselves to simulate periodic boundary conditions. This gave the bottom solid-line curve for the three-soliton wave train. Then a spectral analysis was conducted using the spectral eigenvalue problem for the KdV equation (Chapter 17). The Riemann theta function loop integrals were computed to get the Riemann spectrum (Chapter 19). Finally, the simulation and spectral decomposition of Figure 25.13 was generated with the Riemann theta functions (Chapter 20) in the soliton basis (Chapter 8).

25.6 Nonlinear Spectral Analysis of Andaman Sea Data

Observations of large-amplitude internal solitons were made in the Andaman Sea north of Sumatra in 1093 m of water (Osborne et al., 1978; Osborne and Burch, 1980). In these papers it was demonstrated that the Andaman Sea data set contained most of the qualitative features of the IST theory of Gardner et al. (1967) for the KdV equation. Additional quantitative comparisons were also made. In this section, I use methods developed elsewhere in this book (Chapters 10, 14, 17, 20, and 22) to analyze 12.4 h segments of the data and to interpret the results physically in terms of internal solitons with amplitudes up to about 51 m.

The basis of the spectral analysis is the Schroedinger eigenvalue problem, the spatial part of the Lax pair, that is used to integrate the KdV equation. The role of the eigenvalue problem is discussed in Chapter 3 (on the infinite line) and in Chapter 17 (periodic boundary conditions). Here, the periodic problem is of course appropriate because the wave trains are tidally generated with a period of 12.4 h. We will also use the Riemann spectrum: See Chapters 14–16 for computing the Riemann spectrum; Chapters 10, 20–22, and 32 discuss how to compute the Riemann theta-function solution of the KdV equation.

In the data analysis of the Andaman Sea time series it was found that a two-layer model was insufficient, that is, two layers would not adequately describe the physics and so the continuous model was used. This means that we must rely on the eigenvalue problem (Equations (25.4), (25.5)) and the integrals (25.2), (25.3) to furnish the linear phase speed, c_0, the coefficient of the nonlinear term in KdV, α, and the coefficient of the dispersive term, β.

To show how these coefficients were computed and how the data analysis was conducted it is worthwhile revisiting the measurement program. Temperature sensors were placed on a taut subsurface mooring at approximate depths below the ocean surface of 53, 87, 116, 164, and 254 m. These were designed to measure time series of water temperature $T(t)$ at the respective vertical locations during the 4-day deployment period. The temperatures were recorded every 75 s during this time. The thermal time constant of the instruments was of the order of 2 min; thus temperature fluctuations occurring faster that this period were filtered out and aliasing is not a problem given the 75 s sampling internal. Time series, each roughly 3800 points long, were recorded at the five depths.

Prior to the arrival of each packet of waves (we obtained advanced warning from another survey vessel several kilometers to the west of our own), we launched an XBT. These devices consist of a weighted thermistor that falls vertically into the ocean and transmits water temperature to the surface through a small diameter wire. This allowed measurement of water temperature versus depth during the "quiet time" when large internal waves were not present. We had also established the salinity–temperature (S-T) curve for the region by making conductivity, temperature, and depth (CTD) casts. These results allow computation of the spatial variation of the water density $\rho_0(z)$ directly from the XBT measurements.

The procedure for the nonlinear spectral analysis of a single 12.4 h record (595 points) at a depth below the surface of 164 m is the following:

(1) Establish the S-T curve from the CTD casts.
(2) Use the "quiet time" XBT to get temperature as a function of depth $T_0(z)$ in the absence of major internal wave activity.

(3) Combine the S-T curve and the $T_0(z)$ to get the undisturbed density profile $\rho_0(z)$.
(4) Solve the eigenvalue problem, Equations (25.4), (25.5), for the linear phase speed c_0 and eigenfunction $\phi(z)$. Here only the first mode was necessary since it energetically dominated the dynamics. Furthermore the position of the temperature sensor at 164 m is near a zero in the second mode eigenfunction.
(5) Numerically compute the integrals (25.2) and (25.3) to get the KdV coefficients α and β.
(6) Compute the physical amplitude of the recorded waves $\eta(x = 0, z_0, t)$ (the mooring location is assumed to be at $x = 0, z_0$) by the relation:

$$\eta(x = 0, z_0, t) = -\left[\frac{T(t) - T_0}{(dT_0/dz)_0 \phi(z_0)}\right]$$

where the temperature gradient at $z_0 = 929$ m (i.e., 164 m below the surface) is $(dT_0/dz)_0 \approx 0.12\,°\text{C/m}$.
(7) Compute the spectral parameter used to multiply $\eta(x = 0, z_0, t)$ as it appears in the spectral eigenvalue problem: $\lambda' = c_0^2\lambda = c_0^2\alpha/6\beta$.

The Andaman Sea time series, whose nonlinear spectral analysis we seek, is shown in Figure 25.9. The series, recorded on October 24, 1976 beginning at 10:20 h is 12.4 h long. The quiescent temperature was estimated to be $14.8 \pm 0.2\,°\text{C}$ and this number was used to establish the background level of the signal. Variation of the measured temperature signal $T_0(z_0)$ by the temperature uncertainty of $\pm 0.2\,°\text{C}$ did not appreciably affect our spectral analysis results. The general appearance of the recorded signal is that of a relatively calm background punctuated by rank-ordered depressions in the thermocline; we have previously interpreted these depressions as internal solitons (Osborne et al., 1978; Osborne and Burch, 1980). This section presents results that support that conclusion using the IST.

Before proceeding with the nonlinear spectral analysis it is worthwhile looking at the data for scale lengths consistent with the KdV equation. First note that the dispersion relation for a soliton solution is given by

$$\Omega_n = c_0 K_n + 4\beta K_n^3$$

Introducing the notation $\tau_n = \Omega_n^{-1}, L_n = K_n^{-1}$ and noting the constraint relation between the amplitude and wavenumber, $K_n^2 = \lambda\eta_n/2$, we have

$$\frac{\tau_n}{\tau_0} = \frac{1}{1 + \alpha\eta_n/3c_0}\left(\frac{L_n}{L_0}\right)$$

where $\tau_0 = L_0/c_0$ and

$$\frac{L_n}{L_0} = \left(\frac{\eta_n}{\bar{h}}\right)^{-1/2}, \quad L_0 = \sqrt{\frac{12\beta}{|\alpha|\bar{h}}}$$

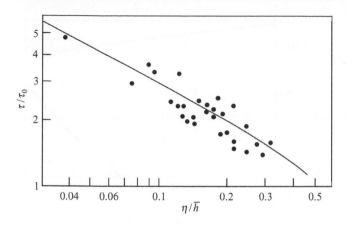

Figure 25.14 Soliton amplitude and temporal width for 29 pulses in the Andaman Sea at 164 m below the surface. The estimates were made directly from the time series in Figure 25.9.

Here $\bar{h}$ is an arbitrary constant that is chosen to be the depth of the maximum in the first mode eigenfunction.

The ideal result at this point would be to graph L_n/L_0 (dimensionless wavelength) versus $\eta_n/\bar{h}$ (dimensionless soliton amplitude). Experimentally, however, we can estimate η_n and τ_n (the half-width at half-maximum of each "solitonic feature" in a recorded time series). A graph of τ_n/τ_0 versus $\eta_n/\bar{h}$ (which theoretically deviates slightly from a straight line on a log-log plot) is thus straightforward in an experimental context. We see in Figure 25.14 results from the major "wavelike" or "solitonic" features observed in the temperature time series recorded at a depth of 164 m for the entire 4-day period. Twenty-nine soliton-like features (six groups of waves) were identified and although there was some scatter, the data are well characterized by the amplitude/wavelength relation and hence have scale lengths consistent with the assumptions implicit in the KdV equation. One concludes that these features may be solitons. Given only the information contained in the single record, we may use spectral methods to ensure that these waves are solitons. As seen below, however, the soliton properties determined directly from the time series are only approximate. The spectral analysis ensures better estimates of the soliton amplitudes relative to the reference level.

The spectral analysis of data was conducted by computing the elements of the Riemann matrix and the associated phases using the methods of Chapters 17 and 19. The stratification eigenfunction analysis of the density structure (Figure 25.15) gave a linear phase speed of $c_0 = 2.17$ m/s for the first mode eigenfunction. The maximum of the first mode eigenfunction occurred at $\bar{h} = 235$ m in water of total depth of 1093 m.

From the Riemann matrix I graph in Figure 25.16 the diagonal element amplitudes as a function of their associated frequencies. The diagonal element amplitudes are of course the amplitudes of the individual cnoidal waves in the spectrum. I also graph the modulus of the cnoidal waves, see scale on

Figure 25.15 First mode eigenfunction in the Andaman Sea.

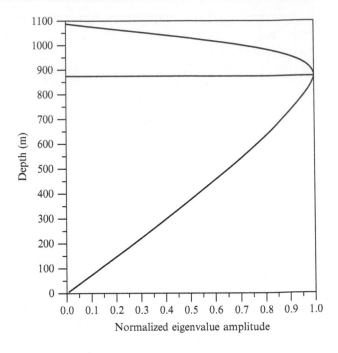

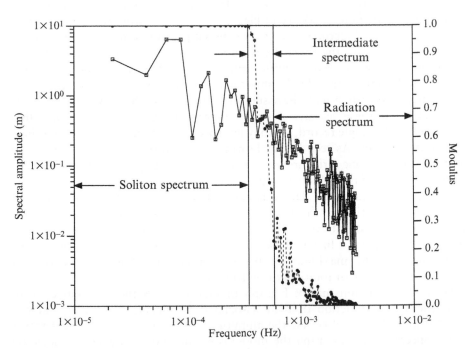

Figure 25.16 IST spectrum of a temperature time series at 164 m depth in the Andaman Sea.

the right-hand side of Figure 25.16. Recall that the modulus lies between 0 and 1: for sine waves it is small, for Stokes waves it is ~0.5, and for solitons it is ~1. We see in Figure 25.16 that the spectrum of the cnoidal waves rapidly decreases with increasing frequency. The modulus is very near 1 for the first 16 components and this implies that these components are solitons. A drop off occurs in the modulus with increasing frequency. The implication is that there are several components (~8) that are Stokes waves and the remaining components are very nearly sine waves (117). Thus, we have 16 soliton components plus other components that are much less nonlinear that we might call radiation modes. Of course, the distinction between solitons and radiation is not as well defined for the periodic problem (with respect to the infinite line problem) because of the "intermediate cnoidal wave spectrum" of Stokes waves, that is, those components having intermediate values of the spectral modulus m.

There are only six solitons visible by eye in Figure 25.16. Why does the spectral analysis of Figure 25.16 tell us that there are 16? This is because we are in mid evolution of the initial waveform (whose shape we do not know) and the solitons have not all grown up to their maximum amplitudes. While the solitons appear nicely rank ordered, they are not fully developed. This is because the density of solitons in time is so great at the moment of observation that they are not well separated. Instead, the solitons are "bunched together" and partially overlapping in the measured wave train.

Can we get a better idea about how the energy is partitioned between the solitons and radiation in the measured time series of Figure 25.9? This can be done by nonlinearly filtering the waves into soliton and radiation components using the IST. In Figure 25.16, we see that there are 16 soliton components. For the present analysis the Riemann matrix is 141×141 and the upper-left submatrix of 16×16 components corresponds to the solitons in the spectrum; the remaining 125×125 submatrix corresponds to the combined intermediate and radiation components. In Figure 25.17, I show how the soliton 16×16 submatrix (the black curve labeled "soliton components" in the figure) looks when computed as a time series using the Riemann theta function. The components 17-30 are also shown and labeled "radiation components." Most of the energy of course lies in the soliton part of the spectrum, while only a small part lies in the radiation part, for which the latter is in reality only two closely spaced spectral components of "intermediate spectrum" as seen in Figure 25.16 for which the moduli are >0.9 but less than 0.99. The packet of radiation components shown in Figure 25.17 is in reality dominated by these two components. Thus, the spectrum of the Andaman Sea time series is almost totally dominated by 16 solitons bound up together, with only six solitons showing in the measurements. To see the rest of the 16 solitons in the measured data it would have been necessary to take additional measurements further from the source region, which of course was not done. The area below the curve for the soliton part of the spectrum in Figure 25.17 and the reference level below provides the water mass for the additional solitons.

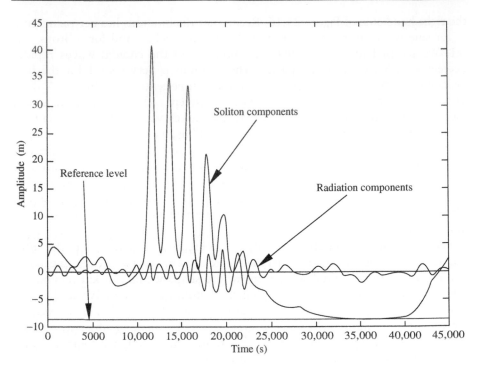

Figure 25.17 The time series of Andaman Sea data in Figure 25.9 has been filtered into soliton and radiation parts.

25.7   Extending the KdV Model to Higher Order

It is not hard to derive a wave equation that describes internal wave behavior to one order of nonlinear approximation higher than the KdV equation (see Lee and Beardsley, 1974; Whitham, 1974):

$$\eta_t + c_0\eta_x + \alpha\eta\eta_x + \beta\eta_{xxx} = \lambda_1\eta_{xxxxx} + \lambda_2\eta\eta_{xxx} + \lambda_3\eta_x\eta_{xx} + \lambda_4\eta^2\eta_x \quad (25.14)$$

Note that this is just the KdV equation with higher order terms on the right-hand side; the coefficients are functions of the density structure, bathymetry, and wave amplitude in appropriate situations. *This extension can be carried to arbitrarily high order in nonlinearity,* something we do not discuss in detail here (see also Osborne, 1997, 2001 and cited references).

One can write the extensions of the KdV equation in rather a general way using the Whitham (1974) hierarchy:

The KdV equation (Whitham 1):

$$u_t + \varepsilon(6uu_x + u_{xxx}) = O(\varepsilon^2)$$

Extended KdV equation (Whitham 2):

$$u_t + \varepsilon(6uu_x + u_{xxx}) + \varepsilon^2(\alpha_1 u_{5x} + \alpha_2 uu_{xxx} + \alpha_3 u_x u_{xx} + \alpha_4 u^2 u_x) = O(\varepsilon^3)$$

Higher extended KdV equation (Whitham 3):

$$u_t + \varepsilon(6uu_x + u_{xxx}) + \varepsilon^2(\alpha_1 u_{5x} + \alpha_2 uu_{xxx} + \alpha_3 u_x u_{xx} + \alpha_4 u^2 u_x)$$
$$+ \varepsilon^3(\beta_1 u^3 u_x + \beta_2 u_x^3 + \beta_3 uu_x u_{xx} + \beta_4 u^2 u_{xxx} + \beta_5 u_{xx} u_{xxx}$$
$$+ \beta_6 u_x u_{xxxx} + \beta_7 u_x u_{xxxxx} + \beta_8 u_{7x}) = O(\varepsilon^4)$$

Note that equation "Whitham 2" is just Equation (25.23) rescaled. The equation "Whitham 3" is one higher order than Equation (25.23). One can continue to as high an order as one wants using this hierarchy. These equations constitute the leading order contributions to the higher KdV hierarchy for the Euler equations, a formal expansion about zero wavenumber, $k \sim 0$. Each equation has been computed to some order in a power of ε. The fact that these equations can be related to the so-called *Lax hierarchy* of integrable equations (by the IST) using Lie-Kodama transforms is a beautiful result and leads to the generalized nonlinear Fourier analysis of wave trains to arbitrary order (Osborne, 1995a,b, 1997, 2001). As far as the author knows at the present time, the coefficients for a generalized density structure have been derived only for the first two equations, although additional effort might be applied to the third equation. Here is a table that summarizes some of the advances:

Theory of the KdV equation and its asymptotic integrability to arbitrary order:

(1) *Coefficients are computed from the actual continuous density stratification.*
(2) *Whitham hierarchy is parallel to the Lax hierarchy* and hence *integrability* and *asymptotic integrability* can be studied. This leads to higher order nonlinear Fourier analysis of time series.
(3) Easily related to *Lie-Kodama transform* and *nonlinear signal processing.*
(4) Related to *integrability* via the *IST.*
(5) Additional higher order equations in $1 + 1$ dimensions include the modified KdV equation and the Gardner equations (Chapter 2). In $2 + 1$ one finds the KP equation and the $2 + 1$ Gardner equation (Chapters 2, 11, 32 and 33).

In Figure 25.18 I show the evolution of a *negative soliton state* that evolves from a sine wave initial condition for the Whitham 2 equation. I refer to the negative state as a "hole." For the case considered one could image a condition on the continental shelf where nonlinear effects are particularly large, which can increase the amplitude of the higher order terms in the Whitham 2 equation. Note that the hole is seen to exist simultaneously with a positive pulse soliton. In particular, the two emerge from the initial condition at the same time forming *a positive pulse/negative hole pair*! Of course in the internal wave field, when the upper layer is thinner than the lower layer, the hole would be a *positively buoyant, upwardly mobile soliton*. This kind of hydrodynamic behavior could present possible ballasting problems for submarine or underwater untethered vehicle activity in the upper ocean.

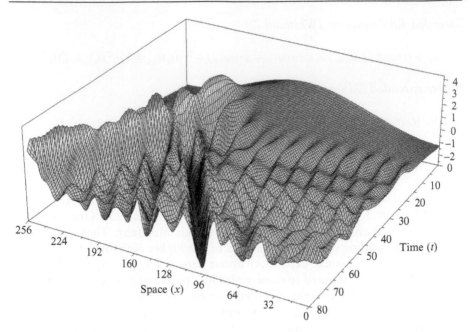

Figure 25.18   Evolution of a hole state from the W2 equation. The hole is seen as a channel beginning near space coordinate 96 and time 80. A hole state in the internal wave field, when the upper layer is thinner than the lower layer, is a positively buoyant, positive soliton pulse. (See color plate).

26 Underwater Acoustic Wave Propagation

26.1 Introduction

This chapter discusses acoustic wave propagation in the oceanic sound field where the sound speed and hence also the index of refraction vary as a function of depth and range. In this brief and brisk application of multidimensional Fourier methods I address the exact solutions of the parabolic wave equation (PE) and how these solutions can be used to improve our understanding of the physics of acoustic wave propagation in the oceanic wave guide, how to model acoustic wave propagation two and three orders of magnitude faster than conventional fast Fourier transform (FFT) approaches (Tappert, 1977, Collins, 1989) and how to analyze acoustic wave data.

Three methods for solving the acoustic Schrödinger or parabolic wave equation are discussed. The particular boundary conditions used are those for acoustic propagation in the oceanic sound channel, although any other set of boundary conditions could be used just as well. The solution of the PE is a complex function $\psi(r,z)$ of depth z and range r that propagates in a potential, $U(r,z)$. *The first method* reduces the acoustic propagation to a set of ordinary differential equations (ODEs) that evolve the Fourier coefficients as a function of range. *The second of the two methods* are based upon a kind of nonlinear separation of variables technique motivated by the Padé approximation, $\psi(r, z) = G(r,z)/F(r,z)$. Thus, the fields $\psi(r,z)$, $U(r,z)$ are replaced by the auxiliary fields $G(r,z)$, $F(r,z)$, where we find the relation $U(r,z) = 2\partial_{zz}\ln F(r,z)$. Since $U(r, z)$ and $F(r,z)$ are uniquely related to one another we are left with solving the resultant nonlinear equation for $G(r,z)$ which has coefficients in terms of $F(r, z)$ and its r and z derivatives. The equation for $G(r,z)$ is a homogeneous bilinear form and *two methods of exact solutions* are discussed, one based upon the reduction to a set of linear equations the other based upon the use of Riemann theta functions. Indeed the novelty of this work is the application of the theta functions to the solution of the PE, although the other two methods have both theoretical and numerical utility.

The new methods that have been developed here have enabled a number of advances in the understanding of the physics of nonlinear acoustic propagation in media with variable index of refraction. There are several goals that are related to this work:

Doi: 10.1016/S0074-6142(10)97026-5

(1) *Developing the physics of acoustic wave propagation* as a kind of *nonlinear Fourier analysis with nonlinear basis functions.*

(2) *Developing nonlinear Fourier data analysis capability* to the same physical order as the parabolic equation and to its higher order, large-angle extensions.

(3) Applying nonlinear Fourier analysis to develop *hyperfast codes for the numerical modeling of acoustic propagation.*

The following areas have been investigated theoretically and numerically in some detail and as a consequence they appear feasible in real-time applications, either in fixed or floating coordinate frames, for *in situ* operations.

(1) *Physics of nonlinear mode solutions of the PE.* Many of the nonlinear modes may at first seem "exotic" in their shape, form, and dynamics and have not been dealt with previously in the literature. They may be used to analytically construct all smooth, nonsingular solutions of the PE using *Riemann theta functions.*

(2) *Numerical methods for exactly solving the PE.* These are methods developed to efficiently deal with *numerical multidimensional Fourier analysis.*

(3) *Data and signal processing analysis.* These procedures exploit the multidimensional Fourier methods introduced here.

(4) *Hyperfast modeling of the PE.* Primarily consists of the application of the multidimensional Fourier transforms using the methods given in Chapters 9 and 20–22.

(5) *Time reversal mirroring technology.* Often used to communicate to a specific point in the ocean (consisting of a local source in combination with a transducer array down range). The proposed approaches allow communication to *any point in the environment, rather than only to the source point, in the vertical measurement plane. This capability comes about because it is possible to adjust the nonlinear phase information in the formulation to treat the entire transducer array as a lens with a focus at any desired point in the plane of measurements.*

(6) *Acoustic communications.* Further advancements using *amplitude modulations* allow for complex, large bandwidth, *encrypted communications.* Whereas current encryption methods are one-dimensional, the proposed methods are ∼1000-dimensional and *cannot be clandestinely decrypted* by modern day conventional technology. The encryption methods described here are applicable to all areas and types of communications.

(7) *Noise reduction. Nonlinear filtering methods* allow for the removal of certain bothersome, physically undesirable modes from the nonlinear Fourier spectrum, that is, from the *Riemann matrix.*

(8) *Imaging inside shadow zones.* Application of the method to the *PE for forward scattering or to the Helmholtz equation for back scattering from a solid body,* combined with *Riemann matrix filtering,* allows for *underwater detection and imaging of targets* even in *highly variable environments.*

(9) *Lens design and construction by nonlinear filtering.* Since particular nonlinear modes of the PE have lens-like properties, it is natural to think of the oceanic environment as made up of the nonlinear superposition of complex lenses with variable index of refraction (they are nonlinear modes and are *not* sine waves). These lenses are physically located by their phase and can be used in real time for the assessment of environmental properties and carrying out actions such as information retrieval, communication, and detection.

(10) *Kalman filtering of locally measured environmental variables,* including acoustic properties, density, and salinity structure: The approach provides for onboard capability for *real-time data assimilation to determine the nonlinear set of basis functions* (or Riemann matrix and phases) for understanding and exploiting the acoustic environment in terms of items (1)-(9) above and other issues not mentioned.

Many of the above topics are important for particular operations in the oceanic environment. The multidimensional Fourier method is superior to the conventional Fourier transform in many ways. The Riemann spectrum contains more information and is for most practical purposes invariant in depth and range, a capability that the linear Fourier transform is unable to provide. Therefore, the Riemann spectrum is the natural one for describing the solutions to acoustic wave equations such at the Helmholtz and parabolic equations.

Nota Bene: The work presented in this chapter is based upon the so-called *parabolic equation* (PE) first introduced by Tappert (1977) to the oceanic acoustic community. Tappert's approach is to use the FFT to numerically integrate the PE. There are several theoretical approaches to solving this equation, formally equivalent to the Schrödinger equation of quantum mechanics. The best-known method is that of *eigenfunction analysis* in which the solution of the PE is given as a linear superposition of the eigenstates. Another well-known method is that of *Feynman functional integration,* most often applied in the field of quantum mechanics. The third method, and that which is used herein, is the method based upon the use of *Riemann theta functions* (multidimensional Fourier series) to solve the problem. This chapter is devoted to a discussion of this method to the solution of the PE. Use of the method to solve extensions of the PE to larger angles will be discussed elsewhere.

26.2 The Parabolic Equation

A common starting place for the study of acoustic wave propagation is the *parabolic (Schrödinger) equation* (Leontovich and Fock, 1946; Tappert, 1977), whose solution $\psi(r,z)$ is a function of range r and depth z

$$2ik_0\psi_r + \psi_{zz} + U(r,z)\psi = 0, \quad U(r,z) = U(r,-z), \quad \psi(r,z) = -\psi(r,-z) \quad (26.1)$$

where the range-dependent *potential function* is given by

$$U(r,z) = k_0^2[n^2(r,z) - 1] \tag{26.2}$$

and the *index of refraction* $n(r,z)$ is written in terms of the *sound speed field* $c(r,z)$ by

$$n(r,z) = \frac{c_0}{c(r,z)}$$

c_0 is a reference sound speed. Note that in Equation (26.1) the boundary conditions are constructed by extending the acoustic propagation problem into a domain that is the spectral image of the ocean's surface. This requires that the potential be an even function in z, $U(r,z) = U(r,-z)$, and that the solution of the acoustic problem be an odd function of z, $\psi(r,z) = -\psi(r,-z)$.

We assume all fields are range r and depth z dependent. Equation (26.1) is a *Cauchy problem* in the sense that given $\psi(0,z)$ (the source field with, say, a Gaussian depth profile and oscillation frequency, ω_0; $k_0 = \omega_0/c_0$), one seeks the complex sound field $\psi(r,z)$ over all range, r, and depth, z, given the environment defined by $n^2(r,z)$. Note that for *small variations in the sound speed relative to the reference sound speed* the potential has the approximate form:

$$U(r,z) = k_0^2[n^2(r,z) - 1] = k_0^2\left[\frac{c_0^2}{c^2(r,z)} - 1\right] \cong -2\frac{k_0^2}{c_0}[c(r,z) - c_0] \quad (26.3)$$

The associated *pressure field* is given by

$$p(r,z) = \psi(r,z)\sqrt{\frac{2}{\pi k_0 r}}\, e^{i(k_0 r - \pi/4)} \quad (26.4)$$

and therefore $\psi(r,z)$ may be interpreted as a *complex pressure amplitude or modulation* riding on a (cylindrically) range-decaying carrier wave. The dispersion relation relating the reference sound speed, c_0, wavenumber, k_0, and frequency, ω_0, is given by $\omega_0 = c_0 k_0 = 2\pi f_0$.

One normally thinks of the PE (26.1) as a linear equation with variable coefficient, $U(r,z)$. Alternatively, Equation (26.1) is a nonlinear equation (with a range/depth dependent coefficient) that can be addressed with nonlinear methods as suggested below. The emphasis herein is on the Cauchy problem for Equation (26.1). An important observation is that the potential of the PE (26.1) is $U(r,z) = 4\pi^2 f_0^2[n^2(r,z) - 1]/c_0^2$, so that nonlinear effects in the PE, for a given index of refraction $n(r,z)$, are proportional to the squared frequency, f_0^2. Therefore, for a particular sound speed field we expect the solution to the PE to be more nonlinear the larger is the frequency, ω_0 or wavenumber, k_0. This is the same kind of nonlinear dependence that occurs in the nonlinear Schrödinger equation for deep-water surface waves (Chapter 12, Equation (12.3)).

One important standard for obtaining the numerical solutions of Equation (26.1) is the split-step, range-marching algorithm developed by Tappert (Hardin and Tappert, 1973; Tappert, 1977), where the FFT is applied to numerically integrate the sound field envelope, $\psi(r,z)$, over range and depth given a set of measured or theoretical sound speed depth profiles specified over a given range distance. It is of course well known that the physical applicability of the PE can be improved by modifying the equation for higher angle propagation and by modeling attenuation in the fluid and bottom, but these are not discussed in this brief introduction. Suitable adjustment of the input environment, $n^2(r,z)$, to include attenuation and a bottom model are natural extensions. The more modern work of Collins (1989) is of fundamental importance in this regard and the book by Jensen et al. (2000) provides a marvelous and thorough overview of the field and contains a long list of important references.

It is natural to focus on two important problems in ocean acoustics:

(1) *Shallow-water acoustics*: Generally speaking, shallow water is a low-frequency environment (10 Hz to several kHz). At low frequency the potential is less nonlinear than at high frequency. This does not simplify the problem however, as the full formulation is still necessary, also because the environment $n(r,z)$ can be quite complex in shallow-water acoustics.

(2) *High-frequency acoustics*: The range of high-frequency acoustics lies from a few kHz to a thousand kHz. Because the potential is proportional to frequency squared, f_0^2, we expect very high nonlinear effects in this regime. The approach using Riemann theta functions provides a number of advantages over conventional numerical integrations using the FFT, including the fact that there is no degradation of the computed sound field due to numerical noise problems at large range.

26.3 Solving the Parabolic Equation with Fourier Series

To provide perspective on what follows, I now discuss how to solve the PE using ordinary Fourier series. It is natural to think of the potential as a *range-dependent Fourier series*:

$$U(r,z) = \sum_{n=-\infty}^{\infty} U_n(r)e^{ik_n z} = a_0(r) + \sum_{n=1}^{\infty} a_n(r)\cos k_n z + b_n(r)\sin k_n z \quad (26.5)$$

Here $a_0(r)$ is the average value of $U(r,z)$ with respect to z:

$$a_0(r) = \langle U(r,z) \rangle = \frac{1}{L_z}\int_0^{L_z} U(r,z)\mathrm{d}z$$

Note that $a_0(r)$ is *range dependent in general*. Smith and Tappert (1994) suggest that $a_0(r)$ be set to zero to determine the reference sound speed as a function of range $c(r,z)$, although this is not the choice made by many investigators (Jensen et al., 2000).

The motivation for the selection of range-dependent coefficients in Equation (26.5) is because the *ocean is generally range dependent*, leading to the notion that the range variation can be put into the Fourier coefficients themselves. One can think of making sound speed measurements as a function of depth over many kilometers of range. This provides a sound speed field, $c(r,z)$, which can be used to compute the potential by Equation (26.2). Thus, the range-dependent coefficients $U_n(r)$ can, for many applications, often be viewed as previously determined from experiment. The Fourier coefficients in Equation (26.5) are interrelated by the expressions:

$$a_n(r) = U_n(r) + U_{-n}(r), \quad b_n(r) = i[U_n(r) - U_{-n}(r)], \quad U_n(r) = \frac{1}{2}[a_n(r) - ib_n(r)]$$

Likewise we can think of a *range-dependent Fourier series for the complex sound field*:

$$\psi(r,z) = \sum_{n=-\infty}^{\infty} \psi_n(r)e^{ik_nz} = c_0(r) + \sum_{n=1}^{\infty} c_n(r)\cos k_n z + d_n(r)\sin k_n z \quad (26.6)$$

Here again $c_0(r)$ is the mean of $\psi(r,z)$ over z. The Fourier coefficients in Equation (26.6) are interrelated by the following expressions:

$$c_n(r) = \psi_n(r) + \psi_{-n}(r), \quad d_n(r) = i[\psi_n(r) - \psi_{-n}(r)], \quad \psi_n(r) = \frac{1}{2}[c_n(r) - id_n(r)]$$

The use of Fourier series for the two fundamental fields in the PE is of course quite natural in view of application of the FFT for solving range-dependent PE problems, namely the split-step FFT algorithm (Harden and Tappert, 1973; Tappert, 1977; Smith and Tappert, 1994). In algorithms of this type the range coefficients $\psi_n(r)$ are computed by numerical integration using the FFT. Here, they are *treated analytically* as the exact solution of the PE in terms of *inverse scattering variables* the main subject of this chapter.

Applying the boundary conditions in Equation (26.1) for the complex sound envelope and potential we find:

$$U_{\text{sym}}(r,z) = \sum_{n=1}^{\infty} a_n(r)\cos(k_n z) = \sum_{n=1}^{\infty}[U_n(r) + U_{-n}(r)]\cos(k_n z) \quad (26.7)$$

$$\psi_{\text{asym}}(r,z) = \sum_{n=1}^{\infty} d_n(r)\sin(k_n z) = i\sum_{n=1}^{\infty}[\psi_n(r) - \psi_{-n}(r)]\sin(k_n z) \quad (26.8)$$

By inserting Equations (26.5) and (26.6) into the PE (26.1) we obtain the following set of ODEs for the range evolution of the coefficients of the complex sound envelope:

$$\frac{d\psi_n(r)}{dr} + i\frac{k_n^2}{2k_0}\psi_n(r) - \frac{i}{2k_0}\sum_{m=-\infty}^{\infty} U_{n-m}(r)\psi_m(r) = 0 \quad (26.9)$$

Thus, one possible approach for numerically integrating the PE is, given the range-dependent Fourier coefficients of the potential $U_n(r)$ and the starter field

$\psi(0,z)$ (and the starter field Fourier coefficients $\psi_n(0)$), integrate Equation (26.9) to obtain the range evolution of the Fourier coefficients $\psi_n(r)$. Then, the solution to the PE is given by the Fourier series (Equation (26.8)). Of course, the convolution operation in Equation (26.9) is computed using the FFT (Press et al., 1992).

It is worth pointing out that while the numerical model Equations (26.7)–(26.9) has been developed here primarily for instructive purposes, the actual computer code executes about 20% slower than the FFT solution of the PE by the split-step method. The reason that Equation (26.9) is competitive with the FFT approach is that the convolution summation can be executed using the FFT, thus rendering Equation (26.9) much faster than might at first seem possible. I am not recommending that Equation (26.9) be an alternative to other methods. The advantage of Equation (26.9) is that it leads to two other *exact numerical approaches* discussed in the sections that follow, which do have decided advantages over a number of other methods used today in the field of ocean acoustics.

There are two special cases of interest for Equation (26.9). The first occurs if the potential is zero, that is, when the index of refraction is a constant. In this case, the ODEs become

$$\frac{\mathrm{d}\psi_n(r)}{\mathrm{d}r} + \mathrm{i}\frac{k_n^2}{2k_0}\psi_n(r) = 0 \tag{26.10}$$

These have solution

$$\psi_n(r) = \psi_n(0)\mathrm{e}^{-\mathrm{i}k_n^2 r/2k_0} \tag{26.11}$$

When substituted into Equation (26.6) we have the well-known solution to the PE for zero potential:

$$\psi(r,z) = \sum_{n=-\infty}^{\infty} \psi_n(0)\mathrm{e}^{\mathrm{i}k_n z - \mathrm{i}k_n^2 r/2k_0} \tag{26.12}$$

Thus, if the initial condition is a Gaussian, then so too is the solution to the PE a Gaussian (the Fourier transform of a Gaussian is a Gaussian). This is just the evolution of a Gaussian packet as presented in quantum mechanical textbooks. For the ocean acoustics problem this case has been discussed by Jensen et al. (2000) and is well worth reading.

The second case occurs when the potential is not range dependent ($U_n(r) = U_n$) so that

$$U(r,z) = \sum_{n=-\infty}^{\infty} U_n\mathrm{e}^{\mathrm{i}k_n z} \tag{26.13}$$

Then the ODEs (26.9) become

$$\frac{d\psi_n(r)}{dr} + i\frac{k_n^2}{2k_0}\psi_n(r) - \frac{i}{2k_0}\sum_{m=-\infty}^{\infty} U_{n-m}\psi_m(r) = 0 \tag{26.14}$$

This is of course another well-known case for which ordinary separation of variables works for the PE. Jensen et al. (2000) is a valuable reference also for this case.

26.4 Solving the Parabolic Equation Analytically

I now use a kind of *nonlinear separation of variables* to solve the PE (26.1). I assume the solution to be the ratio of two functions of range and depth $G(r,z)$, $F(r,z)$:

$$\psi(r,z) = a\frac{H(r,z)}{F(r,z)}\exp\left(i\frac{K_r^2}{2k_0}r\right) = \frac{G(r,z)}{F(r,z)} \tag{26.15}$$

where

$$G(r,z) = aH(r,z)\exp\left(i\frac{K_r^2}{2k_0}r\right)$$

Here $F(r,z)$, $G(r,z)$, and $H(r,z)$ are functions to be found. This means that Equation (26.4) takes the form

$$p(r,z) = a\frac{H(r,z)}{F(r,z)}\sqrt{\frac{2}{\pi k_0 r}}e^{i(\kappa r - \pi/4)}, \quad \kappa = k_0 + \Delta k, \quad \Delta k = \frac{K_r^2}{2k_0} \tag{26.16}$$

and we are able to interpret the wavenumber as having been corrected for the *nonlinear Stokes effect*. Thus, $\Delta k = K_r^2/2k_0$ is the *small (Stokes) correction* to the *range-decaying carrier wave* (see Chapter 2, Equation (2.43) for the correction for ocean surface waves):

$$a\sqrt{\frac{2}{\pi k_0 r}}e^{i[(k_0 + \Delta k)r - \pi/4]} \tag{26.17}$$

It is natural to say that the wavenumber k_0 has been "shifted" by the nonlinear correction Δk. Thus, the range-decaying carrier (Equation (26.17)) is modulated by the ratio $G(r,z)/F(r,z)$ (Equation (26.16)).

We call this approach a kind of *nonlinear separation of variables*. As will be seen below we can treat $F(r,z)$ as a real function and $G(r,z)$ as a complex function in the case when the potential $U(r,z)$ is real; otherwise, in particular when attenuation in present, $U(r,z)$ must be treated as complex. Now insert the *ansatz* Equation (26.15), $\psi(r,z) = G(r,z)/F(r,z)$, into Equation (26.1) to get:

$$\frac{G}{F^3}(2F_z^2 + F^2 U) + \frac{2ik_0(FG_r - GF_r) - 2F_z G_z - GF_{zz} + FG_{zz}}{F^2} = 0 \quad (26.18)$$

It is convenient to add $2GF_{zz}/F^2$ to the second term and to subtract $2GF_{zz}/F^2$ from the first term:

$$\frac{G}{F^3}(2F_z^2 + F^2 U - 2FF_{zz}) + \frac{2ik_0(FG_r - GF_r) - 2F_z G_z + GF_{zz} + FG_{zz}}{F^2} = 0$$

$$(26.19)$$

This implies the following nonlinear separation of variables:

$$2F_z^2 + F^2 U - 2FF_{zz} = -\lambda F^2 \tag{26.20a}$$

$$2ik_0(FG_r - GF_r) - 2F_z G_z + GF_{zz} + FG_{zz} = \lambda FG \tag{26.20b}$$

Here λ is a constant to be determined. The Equations (26.20a) and (26.20b) solve Equation (26.19), for indeed by inserting Equations (26.20) into Equation (26.19) we have:

$$-\frac{G}{F^3}\lambda F^2 + \frac{\lambda FG}{F^2} = 0$$

Equations (26.20) can be put in Hirota operator form (see Chapters 4, 6) (Hirota, 2004):

$$D_z^2 F \cdot F - UF \cdot F = \lambda F \cdot F$$

$$(2ik_0 D_r + D_z^2)G \cdot F = \lambda G \cdot F$$

where the explicit operator notation is

$$D_r G \cdot F = FG_r - GF_r$$

$$D_z^2 G \cdot F = FG_{zz} - 2G_z F_z + GF_{zz}$$

Nota Bene: When the potential is assumed to be real, then so too is the function F; the constant λ is also real for real $U(r, z)$. In cases where *sound absorption* occurs this simple assumption cannot be made and the functions U and F must be viewed as being complex. For example, suppose that F has real and imaginary parts, $F(r, z) = f(r, z) + ig(r, z)$, and then we have

$$\ln F(r, z) = \ln [f(r, z) + ig(r, z)]$$
$$= \frac{1}{2} \ln [f^2(r, z) + g^2(r, z)] + i \, \text{Arg}[f(r, z) + ig(r, z)]$$

Generally speaking, the imaginary part (corresponding to absorption) can be considered to be small in acoustic wave propagation (Jensen et al., 2000).

Note that Equation (26.20a) has derivatives with respect to depth z only and Equation (26.20b) also contains derivatives with respect to the depth z and range r. From Equation (26.20a) we can solve for the potential U:

$$U(r, z) = \frac{D_z^2 F \cdot F - \lambda F \cdot F}{F \cdot F} = 2 \left(\frac{F F_{zz} - F_z^2}{F^2} \right) - \lambda \tag{26.21}$$

which can also be written

$$U(r, z) = 2 \partial_{zz} \ln F - \lambda \tag{26.22}$$

nonlinear separation of variables (Equations (26.20)) has allowed us to rewrite the PE (26.1) as two Equations (26.20b) and (26.22). Equation (26.22) provides a simple transformation from the known potential $U(r,z)$ to the function $F(r,z)$. Therefore, $F(r,z)$ can be viewed as just being a "transformed potential." Once $F(r,z)$ is found from $U(r,z)$, Equation (26.20b) can then be used to find $G(r,z)$ and Equation (26.15) gives the solution to the PE (26.1).

Summing Up:
The function $G(r, z)$ is governed by

$$2ik_0(FG_r - GF_r) - 2F_zG_z + GF_{zz} + FG_{zz} = \lambda FG \tag{26.23}$$

Note that the functions F and G can be rescaled arbitrarily because any rescaling can be canceled from Equation (26.23). Any rescaling of F will disappear from Equation (26.22) because of the double derivative. Notice that Equation (26.15) says that any rescaling must be the same for F and G, otherwise an arbitrary phase will appear in the solution of the PE, but this is of course not crucial because it can be easily accounted for. Note further that λ is a constant to be determined for each Cauchy problem.

Thus, from Equation (26.22) we see that the function $F(r,z)$ depends explicitly on the potential $U(r,z)$, and therefore the two functions are equivalent. Either of the functions $F(r,z)$, $U(r,z)$ can be used to describe the environment (with its variable index of refraction) via Equations (26.2), (26.3), and (26.22). To express $F(r,z)$ in terms of $U(r,z)$ we invert Equation (26.22) to find:

$$F(r, z) = \exp\left(\frac{1}{2} \int \int_z [U(r, z') + \lambda] \mathrm{d}z' \mathrm{d}z''\right) \tag{26.24}$$

In the general case the double integral is most easily done numerically using the FFT with a simple integration filter, $1/k^2$. Here, $U(r,z)$ is taken to be periodic in z, $U(r,z) = U(r,-z)$. This means that because of the double integration in Equation (26.24) we have *two* arbitrary constants α, β as a result of the double integrable which leads to the terms $\alpha + \beta z$. The constant β is set to zero to ensure periodic boundary conditions. The second integration constant is arbitrary because the $F(r,z)$ can be arbitrarily rescaled as discussed above.

Equation (26.23) can be used to find the unknown function $G(r,z)$. Stated differently the function $F(r,z)$ and its derivatives with respect to r and z are the variable coefficients of a partial differential equation (26.23) for the function $G(r,z)$. One might argue that Equation (26.23) is more complex than the PE (26.1). Amazingly, however, Equation (26.1) is solvable for $G(r,z)$ assuming the environment $F(r,z)$ is given, as seen below. Equation (26.23) is a linear equation in $G(r,z)$, with spatially *variable coefficients* given by $F(r,z)$ and its derivatives. Furthermore, Equation (26.23) is *homogeneous in the fields*, that is, each term is a product of two fields $(G(r,z),F(r,z))$ and/or their derivatives. One refers to equations of this type as a *bilinear form* (Hirota, 2004).

Plane Wave Solution: The particular case for $F(r, z) \sim 1$ means $\lambda \sim 0$, $U(r, z) \sim 0$, that is, no variation of the index of refraction as a function of r, z. In such a case we have for Equation (26.1):

$$2ik_0\psi_r + \psi_{zz} = 0$$

which has the *plane wave solution*:

$$\psi(r,z) = \psi_0 \exp\left(ikz - i\frac{k^2}{2k_0}r\right)$$

Thus, without a modulation (which implies that $G(r,z)/F(r,z) \sim 1$) we have the simple plane wave, thus motivating the form of Equation (26.15).

Continued

For $F(r,z) \sim 1$ and small-amplitude acoustics the equation for ψ (Equation (26.1)) is the same as the equation for G (Equation (26.23)), namely:

$$2ik_0 G_r + G_{zz} = 0$$

This can be seen by noting that all derivatives of $F(r,z)$ are zero in Equation (26.23). Thus, for constant index of refraction ($F(r,z) \sim 1$, $U(r,z) \sim 0$), there is no advantage of the method given here over conventional methods. *When there is a spatially variable index of refraction the nonlinear methods discussed herein are worth pursuing.*

26.5 The Functions $F(r,z)$ and $G(r,z)$ as Ordinary Fourier Series; Solution of the PE in Terms of Matrix Equations

We now return to the exact homogenous bilinear form for the PE given by Equation (26.23):

$$2ik_0(FG_r - F_rG) - 2F_zG_z + F_{zz}G + FG_{zz} - \lambda FG = 0 \qquad (26.23)$$

We will attempt to solve Equation (26.23) by using Fourier series for the functions F and G. This exercise will require that we be able to take derivatives and products of Fourier series and to apply a kind of "algebra of bilinear forms." It is important to recognize at this point that Equation (26.23) is more complex than the original PE. However, Equation (26.23) had the important property that it is *homogeneous*, that is, that each term in the equation is a *quadratic form*, thus allowing us to make computations not easily made directly with the PE, which is inhomogeneous, leading to the ODEs (26.14) with the convolution term.

The goal is to solve Equation (26.23) using ordinary Fourier series for the functions F and G. To this end we introduce Fourier series that have *range-dependent coefficients*:

$$F(r,z) = \sum_{n=-\infty}^{\infty} F_n(r)e^{k_n z}, \quad G(r,z) = \sum_{n=-\infty}^{\infty} G_n(r)e^{k_n z}$$

The range dependence is necessary because Equation (26.23) is a nonlinear partial differential equation in both depth z and range r. Since the function $F(r,z)$ is computed from the potential $U(r,z)$ (see Equation (26.22) above), we are assuming the range-dependent coefficients $F_n(r)$ to be known. The goal is then, given $F_n(r)$, to determine the range-dependent coefficients $G_n(r)$ from the bilinear form (26.23). Note that in the inverse operation (26.24) we have two arbitrary integration constants. One of these constants is set to zero to avoid a linearly growing term in the exponential of Equation (26.24) for we are assuming periodic boundary conditions in the range r. The other constant is arbitrary for it can be canceled from the bilinear form (26.23); here it is set to one. In the multidimensional approach used elsewhere in this monograph, the constant becomes an important ingredient for determining the Riemann spectrum, see also Chapter 23.

It is important to distinguish between the above two Fourier transforms for the functions $F(r,z)$ and $G(r,z)$. The function $F(r,z)$ is computed from the potential and the range dependence in the coefficients reflects what is in the environment, which we assume comes from a measurement. The potential then has the Fourier series:

$$U(r,z) = \sum_{m=-\infty}^{\infty} \sum_{n=-\infty}^{\infty} U_{mn}(r)e^{ik_m z + il_n r} = \sum_{m=-\infty}^{\infty} U_m(r)e^{k_m z}$$

where we assume an even grid for the range and depth, so that the wavenumbers are commensurable. Equation (26.24) leads to the Fourier transform for the function $F(r,z)$:

$$F(r,z) = \sum_{m=-\infty}^{\infty} \sum_{n=-\infty}^{\infty} F_{mn}(r)e^{ik_m z + il_n r} = \sum_{m=-\infty}^{\infty} F_m(r)e^{k_m z}$$

where

$$U_m(r) = \sum_{n=-\infty}^{\infty} U_{mn}(r)e^{il_n r}, \quad F_m(r) = \sum_{n=-\infty}^{\infty} F_{mn}(r)e^{il_n r}$$

No dynamical relationship between the range r and the depth z is assumed. We have simply made an experimental measurement of the potential and then computed its Fourier transform. $F(r,z)$ is related to the potential $U(r,z)$ via Equation (26.22).

The range-dependent coefficients $G_n(r)$ is however dynamic. This is because the solution to the PE comes from the evolution of the sound field in the potential and therefore must encounter range nonlinearity and dispersion.

The solution to the PE is then given by Equation (26.15). To this end we must solve Equation (26.23) for G given F. We thus require products of the type

$$F(r,z)G(r,z) = \sum_{m=-\infty}^{\infty} \sum_{n=-\infty}^{\infty} F_m(r)G_n(r)e^{i(k_m + k_n)z} = \sum_{n=-\infty}^{\infty} \left[\sum_{m=-\infty}^{\infty} F_{n-m}(r)G_m(r) \right] e^{ik_n z}$$

$$(26.25)$$

The right-hand term illustrates how the *product of a Fourier series* can be written as a *Fourier series* (Zygmund, 1959). The *Fourier coefficients of a product FG* are computed from the *convolution of the Fourier coefficients* for F and G themselves even when there are range-dependent coefficients. The coefficients of each of the bilinear products in Equation (26.23) are also easily found (see Appendix):

$$FG_r = \sum_{n=-\infty}^{\infty} \left[\sum_{m=-\infty}^{\infty} F_{n-m}(r) \frac{dG_m(r)}{dr} \right] e^{ik_n z}$$

$$F_r G = \sum_{n=-\infty}^{\infty} \left[\sum_{m=-\infty}^{\infty} \frac{dF_{n-m}(r)}{dr} G_m(r) \right] e^{ik_n z}$$

$$F_z G_z = -\sum_{n=-\infty}^{\infty} \left[\sum_{m=-\infty}^{\infty} k_m k_{n-m} F_{n-m}(r) G_m(r) \right] e^{ik_n z}$$

$$F_{zz} G = -\sum_{n=-\infty}^{\infty} \left[\sum_{m=-\infty}^{\infty} k_{n-m}^2 F_{n-m}(r) G_m(r) \right] e^{ik_n z}$$

$$FG_{zz} = -\sum_{n=-\infty}^{\infty} \left[\sum_{m=-\infty}^{\infty} k_m^2 F_{n-m}(r) G_m(r) \right] e^{ik_n z}$$

Insert these into Equation (26.23) and get

$$\sum_{n=-\infty}^{\infty} \sum_{m=-\infty}^{\infty} \{ 2ik_0 [F_{n-m}(r) G'_m(r) - F'_{n-m}(r) G_m(r)] + 2k_m k_{n-m} F_{n-m}(r) G_m(r)$$
$$-k_{n-m}^2 F_{n-m}(r) G_m(r) - k_m^2 F_{n-m}(r) G_m(r) - \lambda F_{n-m}(r) G_m(r) \} e^{ik_n z} = 0$$

where the prime notation means: $' = d/dr$. Take the inverse Fourier transform

$$\sum_{m=-\infty}^{\infty} \{ 2ik_0 F_{n-m}(r) G'_m(r) + [(2k_m k_{n-m} - k_{n-m}^2 - k_m^2 - \lambda) F_{n-m}(r)$$
$$-2ik_0 F'_{n-m}(r)] G_m(r) \} = 0$$

which we then write in the form of (infinite-dimensional) matrices times vectors:

$$\tilde{\alpha} G' + \tilde{\beta} G = 0$$

We seek to solve this set of ODEs for the range-dependent Fourier coefficients for $G(r,z)$ which have the notation

$$\{ G(r) \}_n = \{ G_n(r), n = -\infty, \ldots, -1, 0, 1, \ldots, \infty \}$$
$$= \{ G_{-\infty}(r), \ldots, G_{-1}(r), G_0(r), G_1(r), G_2(r), \ldots, G_\infty(r) \}$$

The matrices are given by

$$\{\tilde{\alpha}\}_{mn} = \alpha_{mn} = 2ik_0 F_{n-m}(r)$$

$$\{\tilde{\beta}\}_{mn} = \beta_{mn} = -\left[\left(\frac{2\pi}{L_z}\right)^2 (n-2m)^2 + \lambda\right] F_{n-m}(r) - 2ik_0 F'_{n-m}(r)$$

where the indices have the range $m, n = -\infty, \ldots, -1, 0, 1, \ldots, \infty$. I have used $k_n = 2\pi n/L_z$. Thus, all matrices and vectors in the present formulation are infinite, although in practical applications they are taken to be finite. To solve $\tilde{\alpha}G' + \tilde{\beta}G = 0$ for $G(r)$ we write

$$G' + i\tilde{\gamma}G = 0, \quad \tilde{\gamma} = -i\tilde{\alpha}^{-1}\tilde{\beta} \tag{26.26a}$$

which has the formal solution

$$G(r) = e^{i\tilde{\alpha}^{-1}\tilde{\beta}r}G(0) = e^{-i\tilde{\gamma}r}G(0)$$

$$= [\cos\tilde{\gamma}r - i\sin\tilde{\gamma}r]G(0) = \left[1 - \frac{i\tilde{\gamma}r}{1!} - \frac{\tilde{\gamma}^2 r^2}{2!} + \frac{i\tilde{\gamma}^3 r^3}{3!} + \cdots\right]G(0) \tag{26.26b}$$

Here the $G(0)$ are the values of the Fourier coefficients of $G(r,z)$ at zero range: $G_n(0)$. To carry out these computations we have assumed that the matrix $\tilde{\alpha}$ has an inverse, which of course must be true from Equation (26.22). This, then, is the exact range evolution of the Fourier coefficients of the function $G(r,z)$. All that preceded above was for *infinite matrices*.

The function $G(r,z)$ has the Fourier series:

$$G(r,z) = \sum_{n=-\infty}^{\infty} G_n(r)e^{k_n z}$$

Here

$$\{G_n(r), \quad n = -\infty, \ldots, -1, 0, 1, \ldots, \infty\}$$

where the $G_n(r)$ are the elements of the infinite-length vector G. Then the solution to the PE is given by

$$\psi(r,z) = \frac{G(r,z)}{F(r,z)}$$

We have thus exactly solved the PE to all orders using ordinary Fourier analysis (with range-dependent coefficients) and the algebra of bilinear forms.

Of course a further step is to rewrite all of the above calculations in terms of the discrete Fourier transform, which has a finite number of degrees of freedom where the indices have a finite range that we might think of writing as $m,n = -M/2,\ldots,-1,0,1,\ldots,M/2$ or, more commonly, $m,n = 0,1,\ldots,M-1$. The numerical computations can now be made in terms of finite-dimensional matrices. The computer computations are now standard and the solution of Equation (26.26a) is easily found by the usual methods (Press et al., 1992). Note that the matrix $\tilde{\gamma} = -i\tilde{\alpha}^{-1}\tilde{\beta}$ is in this regard complex. Numerical solution follows by diagonalizing the matrix $\tilde{\gamma}$ via a transformation of the form $\mathbf{P}^{-1}\tilde{\gamma}\mathbf{P}$. Thus, one has no accuracy degradation due to a numerical integrator. One constructs the exact solutions for each value of the range r (Equation (26.26b)).

I now show how to solve the PE using multidimensional Fourier series.

26.6   Solving the Parabolic Equation in Terms of Multidimensional Fourier Series

It is useful to test the hypothesis that the functions $G(r,z), F(r,z)$ can be written as *Riemann theta functions* with *independent Riemann matrices*, $\hat{F}_{mn}$, $\hat{G}_{mn}$. The vector form of the equations is

$$F(r,z) = \theta(r,z|\hat{\mathbf{F}}, \boldsymbol{\phi}) = \sum_{\mathbf{m}\in\mathbb{Z}} e^{-\frac{1}{2}\mathbf{m}\cdot\hat{\mathbf{F}}\mathbf{m}} e^{i\mathbf{m}\cdot\boldsymbol{\kappa}z + i\mathbf{m}\cdot\boldsymbol{\lambda}r + i\mathbf{m}\cdot\boldsymbol{\phi}} \tag{26.27a}$$

$$G(r,z) = \theta(r,z|\hat{\mathbf{G}}, \boldsymbol{\gamma}) = \sum_{\mathbf{m}\in\mathbb{Z}} e^{-\frac{1}{2}\mathbf{m}\cdot\hat{\mathbf{G}}\mathbf{m}} e^{i\mathbf{m}\cdot\boldsymbol{\kappa}z + i\mathbf{m}\cdot\boldsymbol{\lambda}r + i\mathbf{m}\cdot\boldsymbol{\gamma}} \tag{26.28a}$$

and the scalar form is

$$F(r,z) = \theta(r,z|\hat{\mathbf{F}}, \boldsymbol{\phi}) = \sum_{m_1=-\infty}^{\infty} \sum_{m_2=-\infty}^{\infty} \cdots \sum_{m_N=-\infty}^{\infty} \exp\left(-\frac{1}{2}\sum_{m=1}^{N}\sum_{n=1}^{N} m_m m_n \hat{F}_{mn}\right)$$

$$\times \exp\left[i\left(\sum_{n=1}^{N} m_n\kappa_n\right)z + i\left(\sum_{n=1}^{N} m_n\lambda_n\right)r + i\left(\sum_{n=1}^{N} m_n\phi_n\right)\right] \tag{26.27b}$$

$$G(r,z) = \theta(r,z|\hat{\mathbf{G}}, \boldsymbol{\gamma}) = \sum_{m_1=-\infty}^{\infty} \sum_{m_2=-\infty}^{\infty} \cdots \sum_{m_N=-\infty}^{\infty} \exp\left(-\frac{1}{2}\sum_{m=1}^{N}\sum_{n=1}^{N} m_m m_n \hat{G}_{mn}\right)$$

$$\times \exp\left[i\left(\sum_{n=1}^{N} m_n\kappa_n\right)z + i\left(\sum_{n=1}^{N} m_n\lambda_n\right)r + i\left(\sum_{n=1}^{N} m_n\gamma_n\right)\right] \tag{26.28b}$$

See Baker (1897), Belokolos et al. (1994), Osborne (1995a,b, 2002), and Osborne et al. (2000) as useful references together with Chapters 7–9 and

20–23. And therefore the solution to the PE consists of the ratio of two (multidimensional) Fourier series (i.e., Riemann theta functions) given in Equations (26.27) and (26.28). From Equations (26.1), (26.15), and (26.22) we see that the boundary conditions for these two functions are

$$F(r,z) = F(r, -z)$$

$$G(r,z) = -G(r, -z)$$

For periodic or antiperiodic boundary conditions the components of the wave vector $\mathbf{k}$ are commensurable. In view of the numerous applications of the PE using numerical methods and in view of data analyses based upon application of the FFT (which is periodic), we find the commensurability assumption to be reasonable. Note that the γ used in the last subsection in the matrix solution of the PE, by abuse of notation, is not the γ used in this subsection.

Equations (26.27) and (26.28) are *multidimensional Fourier series* (Baker, 1897): The multidimensional spectra consist of (Riemann or symmetric) matrices, F_{mn}, G_{mn}, and phases ϕ_n, γ_n, where $m,n = 1,2,\ldots,N$ for N the number of degrees of freedom for a particular application. While these expressions are quite intimidating at first glance, it is worthwhile noting that they are just convenient book keeping formulas for explicitly keeping *nonlinearity in the problem* by including *all possible sums and differences of wavenumbers, frequencies, and phases*. Use of the Riemann matrix is actually a simplification, such that the Fourier coefficients in Equations (26.27) and (26.28) are related to $N \times N$ matrices rather than an otherwise infinite number of multidimensional Fourier coefficients encountered in the more general theory of Chapter 7. These arguments are discussed in more detail in Chapters 7 and 8.

Theorem: Equations (26.27) and (26.28) together with Equation (26.15) solve the parabolic Equation (26.1) with the boundary conditions given in Equation (26.1) provided that $\hat{G} = \hat{F}$ and $\phi \neq \gamma$. The matrix $\hat{F}$ and phases ϕ are the multidimensional Fourier spectrum for the potential $U(r,z)$ via Equations (26.22) and (26.24). The matrix $\hat{G} = \hat{F}$ and phases γ are the spectrum of $G(r,z)$. Furthermore, the phases γ can be computed so that the solution to Equation (26.1) satisfies the Cauchy initial condition, $\psi(0, z)$. The proof of this theorem follows Krichever (1989, 1992) (see also Belokolos et al., 1994 for a review, extensions and a full list of references).

Of course, the *Novikov conjecture* requires that the system be associated with a Riemann surface and one must use loop integrals (Chapter 14), Schottky uniformization (Chapter 15), Nakamura-Boyd (Chapter 16) or, for many numerical and experimental applications, adiabatic annealing (Chapter 23) to compute the associated Riemann spectrum *to guarantee one has solutions of the PE*. Arbitrary choices for $\hat{F}$, ϕ, and γ will *not* in general give one a solution of the PE.

Therefore, the form of the functions $F(r,z)$ and $G(r,z)$ appropriate for solving the PE are then taken to be

$$F(r, z) = \theta(r, z | \hat{\mathbf{F}}, \boldsymbol{\phi}) = \sum_{m \in \mathbb{Z}} e^{-\frac{1}{2} \mathbf{m} \cdot \hat{\mathbf{F}} \mathbf{m}} e^{i \mathbf{m} \cdot \boldsymbol{\kappa} z + i \mathbf{m} \cdot \boldsymbol{\lambda} r + i \mathbf{m} \cdot \boldsymbol{\phi}} \tag{26.27c}$$

$$G(r, z) = \theta(r, z | \hat{\mathbf{G}}, \boldsymbol{\gamma}) = \sum_{m \in \mathbb{Z}} e^{-\frac{1}{2} \mathbf{m} \cdot \hat{\mathbf{F}} \mathbf{m}} e^{i \mathbf{m} \cdot \boldsymbol{\kappa} z + i \mathbf{m} \cdot \boldsymbol{\lambda} r + i \mathbf{m} \cdot \boldsymbol{\gamma}} \tag{26.28c}$$

We must be sure that these equations satisfy the boundary conditions: $F(r,z) = F(r,-z)$ and $G(r,z) = -G(r,-z)$. This will be done in Section 26.8. Before taking this step we derive essential results necessary for satisfying the boundary conditions.

26.7   Rewriting the Theta Functions in Alternative Forms

Equations (26.27) and (26.28) can be written in terms of *ordinary linear Fourier series* with *range-dependent coefficients*:

$$F(r, z) = \sum_{n=-\infty}^{\infty} F_n(r) e^{i k_n z} \tag{26.29}$$

$$G(r, z) = \sum_{n=-\infty}^{\infty} G_n(r) e^{i k_n z} \tag{26.30}$$

To see how Equations (26.29) and (26.30) arise from the theta function formulas (26.27) and (26.28), we rewrite Equation (26.27) as a single series rather than a multidimensional one (see Chapter 8):

$$F(r, z) = \sum_{j=-\infty}^{\infty} Q_j(r) e^{i K_j z} \tag{26.31}$$

$$Q_j(r) = q_j e^{i \Lambda_j r}, \quad q_j = \exp\left(-\frac{1}{2} \sum_{m=1}^{N} \sum_{n=1}^{N} m_m^j m_n^j \hat{F}_{mn} + i \Phi_j \right)$$

where

$$K_j = \sum_{n=1}^{N} m_n^j \kappa_n, \quad \Lambda_j = \sum_{n=1}^{N} m_n^j \lambda_n, \quad \Phi_j = \sum_{n=1}^{N} m_n^j \phi_n \tag{26.32}$$

Of course one must retain the appropriate book keeping in the multidimensional Fourier series of Equations (26.27) and (26.28) to coincide with the terms in the single summation over index j (Osborne, 1995a,b). Likewise, Equation (26.28) can be rewritten:

$$G(r,z) = \sum_{j=-\infty}^{\infty} P_j(r)e^{iK_jz} \tag{26.33}$$

$$P_j(r) = p_je^{i\Lambda_j r}, \quad p_j = \exp\left(-\frac{1}{2}\sum_{m=1}^{N}\sum_{n=1}^{N} m_m^j m_n^j \hat{G}_{mn} + i\Gamma_j\right), \quad \Gamma_j = \sum_{n=1}^{N} m_n^j \gamma_n$$

To carry out the operations given above we compactify the nested summations in the Riemann theta functions to give a single summation over j. In this process we associate the integer $j(-\infty<j<\infty)$ with each vector $\mathbf{m}^j = [m_1^j, m_2^j, \ldots, m_N^j]$, that is, there is a one-to-one correspondence with j and $\mathbf{m}^j$. Now if we assume that the wavenumbers are commensurable then $\kappa_n = 2\pi n/L$ and we have for the z coordinate wavenumbers

$$K_j = \sum_{n=1}^{N} m_n^j \kappa_n = \frac{2\pi}{L_z}I_j, \quad \text{where } I_j = \sum_{n=1}^{N} m_n^j n \tag{26.34}$$

Therefore, the wavenumbers K_j in the above single series (Equations (26.31), (26.33)) are $2\pi/L_z$ times the (positive and negative) integers, but the K_j are often repeated (Osborne, 1995a,b). If we sum over the repeated wavenumbers corresponding to each integer n in the above ordinary linear Fourier series (Equations (26.29), (26.30)), we can write the Fourier coefficients in terms of the parameters of the theta functions. Then, the Fourier coefficient functions $F_n(r)$, $G_n(r)$ in Equations (26.29) and (26.30) are given by the exact, analytic forms:

$$F_n(r) = \sum_{\{j\in\mathbb{Z}:I_j=n\}} q_je^{i\Lambda_j r}, \quad q_j = \exp\left(-\frac{1}{2}\sum_{m=1}^{N}\sum_{n=1}^{N} m_m^j m_n^j \hat{F}_{mn} + i\Phi_j\right) \tag{26.35a}$$

$$G_n(r) = \sum_{\{j\in\mathbb{Z}:I_j=n\}} p_je^{i\Lambda_j r}, \quad p_j = \exp\left(-\frac{1}{2}\sum_{m=1}^{N}\sum_{n=1}^{N} m_m^j m_n^j \hat{G}_{mn} + i\Gamma_j\right)$$

The operation $\{j\in\mathbb{Z}:I_j=n\}$ under the summation sign means "sum j over all integers $(-\infty,\ldots,-2,-1,0,1,2,\ldots,\infty)$ for which $I_j=n$." Summations of this type give us the ordinary Fourier coefficients associated with the wavenumbers, $\kappa_n = 2\pi I_j/L_z$ for particular n and j.

Nota Bene: Formally speaking, the Riemann spectra in the two expressions of Equation (26.35a) are different: The function $F(r,z)$ has the spectrum $\hat{F}_{mn}, \phi_n$ (Riemann matrix and phases) and the function $G(r,z)$ has the spectrum $\hat{G}_{mn}, \gamma_n$. But the theorem of Section 26.6 tells us that $\hat{G}_{mn} = \hat{F}_{mn}$ and that the phases ϕ_n and γ_n are different. The phases ϕ_n are the phases of the field $F(r,z)$ (describing the environment or sound speed structure of the domain over which we solve the PE) and the phases γ_n describe the initial Cauchy field $\psi(0,z)$ (the initial Gaussian, say, of the sound speed source). The arbitrariness and specificity of the phases is allowed by Krichever's solution scheme (Krichever, 1989, 1992).

We then have the form of $F_n(r)$, $G_n(r)$ in Equations (26.29) and (26.30) that is used in this chapter for solving the PE:

$$F_n(r) = \sum_{\{j\in\mathbb{Z}:I_j=n\}} q_j e^{i\Lambda_j r + i\phi_j}, \quad q_j = \exp\left(-\frac{1}{2}\sum_{m=1}^{N}\sum_{n=1}^{N} m_m^j m_n^j \hat{F}_{mn}\right) \quad (26.35b)$$

$$G_n(r) = \sum_{\{j\in\mathbb{Z}:I_j=n\}} q_j e^{i\Lambda_j r + i\gamma_j}$$

For numerical computations of theta functions the relations (26.35b) are quite useful, for they reduce *multidimensional theta functions* to *ordinary Fourier series* (Equations (26.29) and (26.30)), thus reducing the computer time by many orders of magnitude for typical applications. The procedures for computing theta functions as ordinary Fourier series are discussed in Chapter 9; see also Chapter 32.

In the *linear limit* for small-amplitude acoustics we have $U(r,z)\simeq0$ and $F(r,z)\simeq1$. We then have the simplified ordinary Fourier coefficients:

$$G_n(r) = a_n e^{il_n r + i\phi_n}, \quad a_n = e^{-\frac{1}{2}\hat{F}_{nn}}$$

where $l_n = -k_n^2/2k_0$ is the "dispersion relation." In the usual linear Fourier notation:

$$G(r,z) = \sum_{n=-N_z/2}^{N_z/2} G_n(0) e^{ik_n z - ik_n^2 r/2k_0}, \quad G_n(0) = a_n e^{i\phi_n} \quad (26.36)$$

which is the solution $\psi(r,z) = G(r,z)$ of the PE when the sound speed is everywhere constant.

Let us now write some of the alternative useful forms for the theta functions as *linear Fourier series*. Consider the *partial theta-function summation* in the following form:

$$\theta(r,z|\mathbf{B}) = \sum_{m_1=-M}^{M} \sum_{m_2=-M}^{M} \cdots \sum_{m_N=-M}^{M} \exp\left(-\frac{1}{2}\sum_{m=1}^{N}\sum_{n=1}^{N}m_m m_n B_{mn}\right)$$

$$\times \exp\left[i\left(\sum_{n=1}^{N}m_n\kappa_n\right)z + i\left(\sum_{n=1}^{N}m_n\lambda_n\right)r + i\sum_{n=1}^{N}m_n\phi_n\right]$$

where M is an integer limit which we take to be finite since ∞ is not possible in numerical computations. The partial theta summation can be written over a single sum:

$$\theta(r,z|\mathbf{B},\phi) = \sum_{j=-J}^{J} q_j e^{iK_j z + i\Lambda_j r + i\Phi_j}, \quad J = [(2M+1)^N - 1]/2$$

$$q_j = e^{-\frac{1}{2}\mathbf{m}^j \cdot \mathbf{B}\mathbf{m}^j} = \exp\left(-\frac{1}{2}\sum_{m=1}^{N}\sum_{n=1}^{N}m_m^j m_n^j B_{mn}\right)$$

$$K_j = \mathbf{m}^j \cdot \boldsymbol{\kappa} = \sum_{n=1}^{N}m_n^j \kappa_n \qquad (26.37)$$

$$\Lambda_j = \mathbf{m}^j \cdot \boldsymbol{\lambda} = \sum_{n=1}^{N}m_n^j \lambda_n$$

$$\Phi_j = \mathbf{m}^j \cdot \boldsymbol{\phi} = \sum_{n=1}^{N}m_n^j \phi_n$$

A slight modification in this notation gives the alternative form:

$$\theta(r,z|\mathbf{B},\boldsymbol{\phi}) = \sum_{j=-J}^{J} Q_j e^{iK_j z + i\Lambda_j r}, \quad J = [(2M+1)^N - 1]/2$$

$$Q_j = \exp\left(-\frac{1}{2}\sum_{m=1}^{N}\sum_{n=1}^{N}m_m^j m_n^j B_{mn} + i\Phi_j\right) \qquad (26.38)$$

Here the range r is a parameter that needs to be iterated across the range resolution of choice for the depth resolution required. The depth wavenumbers κ_n are taken to be commensurable and therefore so too are the K_j commensurable. The range wavenumbers λ_n are incommensurable while the Λ_j are also incommensurable.

At the sound source, $r = 0$, the above expression becomes:

$$\theta(0, z|\mathbf{B}, \boldsymbol{\phi}) = \sum_{j=-J}^{J} Q_j e^{iK_j z}, \quad J = [(2M + 1)^N - 1]/2 \tag{26.39}$$

We can also write the above theta function by introducing Fourier coefficients which vary as a function of range, r:

$$\theta(r, z|\mathbf{B}) = \sum_{j=-J}^{J} Q_j(r) e^{iK_j z}, \quad Q_j(r) = q_j e^{i\Lambda_j r + i\Phi_j} \tag{26.40}$$

We can also write the above sum as an ordinary Fourier series:

$$\theta(r, z|\mathbf{B}, \boldsymbol{\phi}) = \sum_{n=-N_z}^{N_z} \theta_n(r) e^{ik_n z}$$

$$\theta_n(r) = \sum_{\{j \in \mathbb{Z} : I_j = n\}} q_j e^{i\Lambda_j r + i\Phi_j}, \quad q_j = \exp\left(-\frac{1}{2}\sum_{m=1}^{N}\sum_{n=1}^{N} m_m^j m_n^j B_{mn}\right) \tag{26.41}$$

This is just an ordinary Fourier series where each of the coefficients is itself a Fourier series in the range variable r. Thus, this is just an *ordinary Fourier series summed over commensurable wavenumbers* $k_n = 2\pi n/L_z$. The leap from Equations (26.40) to (26.41) is fundamental (see Figures 8.5 and 8.6 in Chapter 8). To be concrete look at Figure 8.6 which is a histogram of the number of occurrences at each wavenumber of the $Q_j(r = 0)$ in Equation (26.40) for $(2M+1)^N = 16,807$ points. We see that the histogram has been taken over the number of occurrences for each of the integer wavenumbers $I_j = L_z K_j/2\pi$ (note that $l = J + j$ in the notation of Chapter 8). To obtain Equation (26.41), all the amplitudes terms in a column of the histogram must be summed to give the coefficient $\theta_n(0)$. Consider the column for I_{10} in Figure 8.6 of Chapter 8: All the amplitude terms in the column must be summed to give the coefficient $\theta_{10}(0)$. This is of course to be done for all columns and all values of the range r. Thus, we see the meaning of the summation notation in the second equation of Equation (26.41).

Equation (26.41) is an interesting result because it says that the range dependence for the Fourier coefficients is analytic *in terms of inverse scattering variables*, that is, in terms of q_j, Λ_j, Φ_j or in terms of the Riemann spectrum $\mathbf{B}, \boldsymbol{\phi}$, and wavenumbers $\boldsymbol{\lambda}$. Chapter 9 discusses the physical behavior of the range-dependent coefficients $\theta_n(r)$ for a simulation of the Korteweg-deVries(KdV) equation.

Finally, let us notice that $q_j = q_{-j}$, $K_{-j} = -K_j$, $\Lambda_{-j} = -\Lambda_j$ (see Figure 8.4, Chapter 8) and therefore we get the form for the theta function that resembles the one degree-of-freedom theta function (Whittaker and Watson, 1902):

$$\theta(r, z | \mathbf{B}, \boldsymbol{\phi}) = 1 + 2 \sum_{j=1}^{J} q_j \cos(K_j z + \Lambda_j r + \Phi_j) \tag{26.42}$$

It should be noted that the limit $J = [(2M+1)^N - 1]/2$ is typically a large number. For example, if $M = 4$ (the number of Stokes harmonics) and $N = 10$ (the number of nonlinear modes or "nonlinear" sine waves or modes in the spectrum), then $J = (9^{10} - 1)/2 \simeq 1.7 \times 10^9$.

We can reduce the above to an ordinary Fourier series with range-dependent coefficients by noticing that

$$\cos(K_j z + \Lambda_j r + \Phi_j) = \cos(\Lambda_j r) \cos(K_j z + \Phi_j) - \sin(\Lambda_j r) \sin(K_j z + \Phi_j)$$

Fourier Series for the Theta Function

$$\theta(r, z | \mathbf{B}, \boldsymbol{\phi}) = 1 + \sum_{j=1}^{J} a_j(r) \cos(K_j z + \Phi_j) + b_j(r) \sin(K_j z + \Phi_j) \tag{26.43}$$

where

$$a_j(r) = 2q_j \cos(\Lambda_j r), \quad b_j(r) = -2q_j \sin(\Lambda_j r)$$

These expressions say that the range dependence for the Fourier coefficients is analytic in terms of inverse scattering variables, that is, in terms of q_j, Λ_j.

An alternative form puts the phases into the Fourier coefficients is

$$\cos(K_j z + \Lambda_j r + \Phi_j) = \cos(\Lambda_j r + \Phi_j) \cos(K_j z) - \sin(\Lambda_j r + \Phi_j) \sin(K_j z)$$

Another Fourier Series for the Theta Function is

$$\theta(r, z | \mathbf{B}, \boldsymbol{\phi}) = 1 + \sum_{j=1}^{J} a_j(r) \cos(K_j z) + b_j(r) \sin(\Lambda_j z) \tag{26.44}$$

where

$$a_j(r) = 2q_j \cos(K_j r + \Phi_j), \quad b_j(r) = -2q_j \sin(\Lambda_j r + \Phi_j)$$

This is the form I prefer for numerical computations and have used for some time. Note that the coefficients look like *random phaser sums* (Beckmann, 1967).

Now write the theta function over the commensurable, ordered wavenumbers k_n for sine/cosine series arises from Equation (26.41) where $A_n(r) = \theta_n(r) + \theta_{-n}(r)$ and $B_n(r) = i[\theta_n(r) - \theta_{-n}(r)]$ together with the inverse relations $\theta_n(r) = [A_n(r) - iB_n(r)]/2$ and $\theta_{-n}(r) = [A_n(r) + iB_n(r)]/2$ ordinary Fourier series for the theta function:

$$\theta(r, z|\mathbf{B}, \boldsymbol{\phi}) = A_0(r) + \sum_{n=1}^{N_z} A_n(r)\cos(k_n z) + B_n(r)\sin(k_n z), \quad k_n = \frac{2\pi n}{L_z}$$

$$A_0(r) = \sum_{\{j\in\mathbb{Z}:I_j=0\}} a_j(r) = \sum_{\{j\in\mathbb{Z}:I_j=0\}} q_j\cos(\Lambda_j r + \Phi_j)$$

$$A_n(r) = \sum_{\{j\in\mathbb{Z}:I_j=n\}} a_j(r) = 2\sum_{\{j\in\mathbb{Z}:I_j=n\}} q_j\cos(\Lambda_j r + \Phi_j)$$

$$B_n(r) = \sum_{\{j\in\mathbb{Z}:I_j=n\}} b_j(r) = -2\sum_{\{j\in\mathbb{Z}:I_j=n\}} q_j\sin(\Lambda_j r + \Phi_j)$$

$$q_j = \exp\left(-\frac{1}{2}\sum_{m=1}^{N}\sum_{n=1}^{N} m_m^j m_n^j B_{mn}\right), \quad \Lambda_j = \sum_{n=1}^{N} m_n^j \lambda_n, \quad \Phi_j = \sum_{n=1}^{N} m_n^j \phi_n$$

$$(26.45)$$

The above equations form the basis of the theta-function algorithm used in Osborne (1995a,b). An important feature of this program is that the j are summed over $-J \le j \le -1$ (corresponding to a set of lattice vectors $\mathbf{m}$). The remaining terms are those for $j = 0$ (corresponding to $\mathbf{m} = 0$) and for $1 \le j \le J$ (corresponding to integer lattice vectors $-\mathbf{m}$). The combination of terms with lattice vectors $\mathbf{m}$ and $-\mathbf{m}$ gives the factor of 2 on the right-hand side of the coefficients $A_n(r)$ and $B_n(r)$ of Equation (26.45).

It is worthwhile noting that all summations given herein over the integers j are such that $-J \le j \le J = [(2M+1)^N - 1]/2$ or over $1 \le j \le J$; this summation notation is used herein as a single-sum representation of the multidimensional theta function. Likewise summation over "n" such that $0 \le n \le N_x$ is used to sum over terms in the ordinary Fourier transform. Note that N_x is a relatively small number in numerical calculations ($\sim 10^3$ or 10^4) while the integer J is exponentially large ($\sim 10^{10}$). Also N is used to represent the number of nonlinear modes in the Riemann theta function (number of nested summations).

Note that the integer wavenumbers are defined by: $I_j = K_j/\Delta k$; $\Delta k = 2\pi/L_z$ and it follows that the wavenumbers I_j are of course commensurable because the K_j and the κ_n are commensurable. Note further that the wavenumbers Λ_j are *not* commensurable because the λ_n ($\sim -k_n^2/2k_0$) are not commensurable. Here the sums are over the usual ordinary Fourier transforms for periodic boundary conditions:

$$k_n = \frac{2\pi n}{L_z}, \quad \Delta k_z = \frac{2\pi}{L_z}, \quad L_z = N\Delta k_z$$

Let us now renormalize the theta functions in the following way:

$$\theta(r, z|\mathbf{B}, \boldsymbol{\phi}) = A_0(r)\left[1 + \sum_{n=1}^{N} \alpha_n(r)\cos(k_z, nz) + \beta_n(r)\sin(k_z, nz)\right]$$

where

$$\alpha_n(r) = \frac{A_n(r)}{A_0(r)}, \quad \beta_n(r) = \frac{B_n(r)}{A_0(r)}$$

Take the log of the theta function:

$$\ln \theta(r, z | \mathbf{B}, \boldsymbol{\phi}) = \ln \left\{ A_0(r) \left[1 + \sum_{n=1}^{N} \alpha_n(r) \cos(k_{z,n} z) + \beta_n(r) \sin(k_{z,n} z) \right] \right\}$$

$$= \ln A_0(r) + \ln \left[1 + \sum_{n=1}^{N} \alpha_n(r) \cos(k_{z,n} z) + \beta_n(r) \sin(k_{z,n} z) \right]$$

The potential is then

$$U(r, z) = 2\partial_{zz} \ln \theta(r, z | \mathbf{B}) = 2\partial_{zz} \ln \left[1 + \sum_{n=1}^{N_n} \alpha_n(r) \cos(k_{z,n} z) + \beta_n(r) \sin(k_{z,n} z) \right]$$

$$\alpha_n(r) = \frac{A_n(r)}{A_0(r)}, \quad \beta_n(r) = \frac{B_n(r)}{A_0(r)} \tag{26.46}$$

We have thus found the Fourier series of Equation (26.46), using recursion for the logarithm (additional discussion of this point is given in Chapter 32 for the KdV and KP equations).

26.8 Applying Boundary Conditions to the Theta Functions

We now consider the boundary conditions that need to be applied to Equations (26.27) and (26.28) as given in Equation (26.1). The domain of integration is conveniently reflected about the free surface at $z = 0$. Then the potential is symmetric about the sea surface, $U(r,z) = U(r,-z)$. By Equation (26.22) this means that the function F is also symmetric: $F(r,z) = F(r,-z)$. The complex sound field is antisymmetric: $\psi(r,z) = -\psi(r,-z)$. Therefore, the boundary condition for the function G is also antisymmetric: $G(r,z) = -G(r,-z)$.

Let us apply the symmetric boundary condition, $F(r,z) = F(r,-z)$, to Equation (26.27). First we write Equation (26.27) in vector form:

$$F(r, z) = \sum_{m=-\infty}^{\infty} e^{-\frac{1}{2}m \cdot \hat{F}m} e^{im \cdot \kappa z + im \cdot \lambda r + im \cdot \phi} \tag{26.47a}$$

This summation can be broken into three parts:

$$F(r,z) = \sum_{m<0} e^{-\frac{1}{2}m \cdot \hat{F}m} e^{im \cdot \kappa z + im \cdot \lambda r + im \cdot \phi} + \sum_{m=0} e^{-\frac{1}{2}m \cdot \hat{F}m} e^{im \cdot \kappa z + im \cdot \lambda r + im \cdot \phi}$$
$$+ \sum_{-m} e^{-\frac{1}{2}m \cdot \hat{F}m} e^{im \cdot \kappa z + im \cdot \lambda r + im \cdot \phi} \qquad (26.47b)$$

The term for $m = 0$ is just 1. The first term is summed over a subset m, while the last term is summed over the corresponding negative m, namely $-m$. This gives

$$F(r,z) = 1 + \sum_{m \in \mathbb{Z}} \left[e^{-\frac{1}{2}m \cdot \hat{F}m} e^{im \cdot \kappa z + im \cdot \lambda r + im \cdot \phi} + e^{-\frac{1}{2}m \cdot \hat{F}m} e^{-im \cdot \kappa z - im \cdot \lambda r - im \cdot \phi} \right]$$
$$\qquad (26.48)$$
$$F(r,z) = 1 + 2 \sum_{m \in \mathbb{Z}}' e^{-\frac{1}{2}m \cdot \hat{F}m} \cos(m \cdot \kappa z + m \cdot \lambda r + m \cdot \phi)$$

Here $\mathbb{Z} = \{-\infty, \ldots, -2, -1, 0, 1, 2, \ldots, \infty\}$. To understand the meaning of the prime on the summation see Figure 8.4 of Chapter 8. To this end return to Equation (26.47b); the first term corresponds to summing over all terms in Figure 8.4, Chapter 8 up to but not including $m = 0$; the next term is for $m = 0$ (which is just "1") and the last term is for all points to the right of $m = 0$. The *prime in the summation* of the second equation in Equation (26.48) means *sum over all points to the left* of $m = 0$. Note that $m = 0 = [0, 0, \ldots, 0]$ is the zero lattice vector. The second of Equation (26.48) can be written

$$F(r,z) = 1 + 2 \sum_{m \in \mathbb{Z}}' e^{-\frac{1}{2}m \cdot \hat{F}m} [\cos(m \cdot \lambda r + m \cdot \phi) \cos(m \cdot \kappa z)$$
$$- \sin(m \cdot \lambda r + m \cdot \phi) \sin(m \cdot \kappa z)]$$

To obtain the part of this function that is symmetric about z (i.e., to select the terms in the above Fourier series so that $F(r, -z) = F(r, z)$), we write

$$F_{\text{sym}}(r, z) = \frac{1}{2}[F(r, z) + F(r, -z)] \qquad (26.49)$$

and find

$$F_{\text{sym}}(r, z) = 1 + 2 \sum_{m \in \mathbb{Z}}' e^{-\frac{1}{2}m \cdot \hat{F}m} \cos(m \cdot \lambda r + m \cdot \phi) \cos(m \cdot \kappa z) \qquad (26.50)$$

Another form for this latter expression is

$$F_{\text{sym}}(r, z) = 1 + 2 \sum_{m \in \mathbb{Z}}' F_m(r) \cos(m \cdot \kappa z), \quad F_m(r) = e^{-\frac{1}{2}m \cdot \hat{F}m} \cos(m \cdot \lambda r + m \cdot \phi)$$

Recall the vector form of $G(r,z)$ (Equation (26.28a)):

$$G(r, z) = \sum_{m=-\infty}^{\infty} e^{-\frac{1}{2}\mathbf{m} \cdot \hat{\mathbf{F}}\mathbf{m}} e^{i\mathbf{m} \cdot \boldsymbol{\kappa} z + i\mathbf{m} \cdot \boldsymbol{\lambda} r + i\mathbf{m} \cdot \boldsymbol{\gamma}}$$

Next we apply the antisymmetric boundary condition, $G(r,z) = -G(r,-z)$, to get

$$G_{\text{asym}}(r, z) = \frac{1}{2}[G(r, z) - G(r, -z)] \tag{26.51}$$

One finds easily:

$$G_{\text{asym}}(r, z) = -2 \sum_{\mathbf{m}\in\mathbb{Z}}' e^{-\frac{1}{2}\mathbf{m} \cdot \hat{\mathbf{F}}\mathbf{m}} \sin(\mathbf{m} \cdot \boldsymbol{\lambda} r + \mathbf{m} \cdot \boldsymbol{\gamma}) \sin(\mathbf{m} \cdot \boldsymbol{\kappa} z) \tag{26.52a}$$

Another form for this latter expression is

$$G_{\text{asym}}(r, z) = 2 \sum_{\mathbf{m}\in\mathbb{Z}}' G_{\mathbf{m}}(r) \sin(\mathbf{m} \cdot \boldsymbol{\kappa} z), \quad G_{\mathbf{m}}(r) = -e^{-\frac{1}{2}\mathbf{m} \cdot \hat{\mathbf{F}}\mathbf{m}} \sin(\mathbf{m} \cdot \boldsymbol{\lambda} r + \mathbf{m} \cdot \boldsymbol{\gamma}) \tag{26.52b}$$

Therefore, the analytical solution to the PE that satisfies the boundary conditions in Equation (26.1) is given by

$$\psi(r, z) = \frac{G_{\text{asym}}(r, z)}{F_{\text{sym}}(r, z)}$$

so that the exact multidimensional solution to the PE (26.1) is given by

$$\psi(r, z) = \frac{-2 \sum_{\mathbf{m}\in\mathbb{Z}}' e^{-\frac{1}{2}\mathbf{m} \cdot \hat{\mathbf{F}}\mathbf{m}} \sin(\mathbf{m} \cdot \boldsymbol{\lambda} r + \mathbf{m} \cdot \boldsymbol{\gamma}) \sin(\mathbf{m} \cdot \boldsymbol{\kappa} z)}{1 + 2 \sum_{\mathbf{m}\in\mathbb{Z}}' e^{-\frac{1}{2}\mathbf{m} \cdot \hat{\mathbf{F}}\mathbf{m}} \cos(\mathbf{m} \cdot \boldsymbol{\lambda} r + \mathbf{m} \cdot \boldsymbol{\phi}) \cos(\mathbf{m} \cdot \boldsymbol{\kappa} z)} \tag{26.53a}$$

Nota Bene: How to use the exact solution (26.53a) to the PE. Equation (26.53a) provides *all exact, smooth, nonsingular solutions of the PE!* It is the main result of this chapter. Equation (26.53a) satisfies the boundary conditions of the PE (26.1): $U(r, z) = U(r, -z)$ and $\psi(r, z) = -\psi(r, -z)$. The Riemann spectrum $\hat{\mathbf{F}}, \boldsymbol{\phi}$ is that for the *sound speed field* (or the *potential* through Equation (26.2)). The computation of $\hat{\mathbf{F}}, \boldsymbol{\phi}$ would best be made from the measured sound speed field using the methods of Chapter 23 with the aid of the Nakamura-Boyd approach in Chapter 16. The phases $\boldsymbol{\gamma}$ are determined from the Cauchy initial values of the source field $\psi(0, z)$

Continued

using the approach of Chapter 23. Therefore, when applying Equation (26.53a) one must first determine $\hat{\mathbf{F}}$, $\boldsymbol{\phi}$ from the *sound speed field* and then determine the phases $\boldsymbol{\gamma}$ from the shape of the *sound speed field at the source* $\psi(0, z)$.

Nota Bene: Time reversal symmetry for the exact solution (26.53) to the PE. It is interesting at this juncture to note how *time reversal symmetry* can be used to communicate to a single point in space r_0, z_0 that we will associate with a localized source function $\psi(r_0, z_0)$. Each source function is assumed to be localized in space, that is, to be perhaps a small-width Gaussian. The expanding source "explosion" is measured down field by a vertical transducer array that records the sound pressure field $p(r_R, z_n)$ (here r_R is the range distance of the transducer array and $z_n (n = 1, 2, \ldots, N)$ are the vertical positions of the N transducers). By "time reversing" the measured signals we find that the signal converges to the source point once again, thus forming the basis for "time reversal symmetry" and its use for communicating to a point in space r_0, z_0. Now suppose the point r_0, z_0 is slowly moving in space: $r_0(t), z_0(t)$. This means that we can still continue to focus on this moving point (perhaps a "whale" or UUV, Section 26.18) and continue to communicate with it as it moves, by time evolving the set of phases $\boldsymbol{\gamma}$! From the point of view of the multidimensional approach we are treating the transducer array as a "lens" which can be focused to any desired point in space by adjusting the phases $\boldsymbol{\gamma}$. Thus, after an initial explosion by the moving point to establish the "time reversal" phases $\boldsymbol{\gamma}$, only sonar information of the trajectory of the moving point would be necessary for maintaining communications.

It is worthwhile giving the *ordinary Fourier transform representation of the solution of the PE*. This is done by writing Equation (26.53) as ordinary Fourier series. Begin with Equation (26.45) and we have the function F:

$$F(r, z) = A_0(r) + \sum_{n=1}^{N_z} A_n(r) \cos(k_n z) + B_n(r) \sin(k_n z)$$

$$G(r, z) = C_0(r) + \sum_{n=1}^{N_z} C_n(r) \cos(k_n z) + D_n(r) \sin(k_n z)$$

Apply the boundary conditions $F(r, z) = F(r, -z)$ to get

$$F_{\text{sym}}(r, z) = A_0(r) + \sum_{n=1}^{N_z} A_n(r) \cos(k_n z)$$

Apply the boundary conditions $G(r,z) = -G(r,-z)$ to get

$$G_{\text{asym}}(r,z) = \sum_{n=1}^{N_z} D_n(r) \sin(k_n z)$$

Finally, the *solution to the PE* ($\psi(r,z) = G_{\text{asym}}(r,z)/F_{\text{sym}}(r,z)$) *becomes* in terms of linear Fourier transforms:

$$\psi(r,z) = \frac{\sum_{n=1}^{N_z} D_n(r) \sin(k_n z)}{A_0(r) + \sum_{n=1}^{N_z} A_n(r) \cos(k_n z)}$$

$$A_0(r) = \sum_{\{j\in\mathbb{Z}:I_j=0\}} q_j \cos(\Lambda_j r + \Phi_j)$$

$$A_n(r) = 2 \sum_{\{j\in\mathbb{Z}:I_j=n\}} q_j \cos(\Lambda_j r + \Phi_j) \qquad (26.53b)$$

$$D_n(r) = -2 \sum_{\{j\in\mathbb{Z}:I_j=n\}} q_j \sin(\Lambda_j r + \Gamma_j)$$

This is a fundamental result. It is the solution of the PE as a ratio of two Fourier series and is equivalent to Equation (26.53a). This means that the ratio of Fourier series in the solution (26.53b) must exist. But of course the solution exists because the two Fourier series in Equation (26.53b) above are *special*, that is, they are theta functions and we know that the ratio of two theta functions is meromorphic (Mumford, 1983, 1984 and 1991). Zygmund (1959) also discusses in a general way when two Fourier series may be divided. Why write the solution of the PE in terms of ordinary Fourier series rather than as Riemann theta functions? Because most of the numerical operations can be computed using the FFT algorithm, results we use many times in this book.

Nota Bene: Commensurability and Periodicity of Solutions of the PE
The inverse scattering transform (IST) z wavenumbers κ_n have the form

$$\kappa_n = \frac{2\pi n}{L_z}$$

or in vector form

$$\boldsymbol{\kappa} = \frac{2\pi[1, 2, \ldots, N]}{L_z}$$

Therefore, the κ_n are commensurable, that is, they are *one-to-one with the integers*. The z wavenumbers in the *theta function* are

---

Continued

$$K_{m^j} = K_j = \sum_{n=1}^{N} m_n^j \kappa_n = \frac{2\pi}{L_z} \sum_{n=1}^{N} m_n^j n = \frac{2\pi}{L_z} I_j$$

Then

$$I_j = \sum_{n=1}^{N} m_n^j n$$

Therefore, the wavenumbers K_j fall on the integers, but are *not* one-to-one with the integers. Instead the K_j are integer multiples of $2\pi/L_z$ and are *often repeated* (see Figures 8.4 and 8.5 of Chapter 8). Finally, note that the $k_n = 2\pi n/L_z$ are the linear Fourier wavenumbers and they too are commensurable and one-to-one with the integers. Clearly, for this case $k_n = \kappa_n$. Why have I given them separate names if they are the same? Because in general one can treat the quasi-periodic problem for which the κ_n are not commensurable and the k_n do not formally exist! (See also discussion with regard to Table 32.1). This is because in the quasi-periodic case the actual Fourier series is *not reducible to an ordinary periodic Fourier transform*, but the theta function is itself the *only* generic Fourier transform. The number of Fourier components is formally speaking $2J+1 \sim 10^N$, a huge number and therefore the wavenumbers K_j become *dense in wavenumber space*, that is, much denser than the κ_n. One could therefore study, by assuming that the κ_n are incommensurable, subgrid scale phenomena many orders of magnitude smaller than the scale $\kappa_{n+1} - \kappa_n$.

What about the wavenumbers λ_n? They are *computed from the recipe of the inverse scattering transform* and therefore have the form $\lambda_n = -k_n^2/2k_0 + \cdots$, where the dots indicate nonlinear interactions among the degrees of freedom or modes of the Schrödinger equation. Since the λ_n are proportional to the squared wavenumber they are of course not commensurable and neither are the

$$\Lambda_j = \sum_{n=1}^{N} m_n^j \lambda_n$$

All this of course means that the solution to the PE is periodic in the depth z but quasi-periodic in the range r.

Note Bene: Linear Limit of the Solution of the PE
Equation (26.12) gives the solution of the PE when the potential is zero. Can we recover the linear solution directly from the general solution of the PE (26.53)? Let us look in particular at Equation (26.53b) in this limit. Clearly, when the potential tends to zero the modes of the PE become infinitesimal because we have removed the nonlinearity from the problem. Thus, we need to look at the case where $q_j \to 0$. This is a well-understood limit of the theta-function formulation. As $q_j \to 0$ the diagonal elements of

the Riemann matrix grow to dominate the off-diagonal elements and the size of the modes q_j decrease until the nonlinear interactions in the theta function become insignificant. Thus, the form of Equation (26.53b) becomes:

$$A_0(r) = 1 \qquad\qquad (26.53b)$$

$$A_n(r) = 2q_n \cos(l_n r + \phi_n), \quad q_n = e^{-\frac{1}{2}B_{nn}}, \quad l_n = \frac{-k_n^2}{2k_0}$$

$$D_n(r) = -2q_n \sin(l_n r + \gamma_n)$$

so that

$$\psi(r,z) \simeq \frac{-2\sum_{n=1}^{N_z} q_n \sin(l_n r + \gamma_n)\sin(k_n z)}{1 + 2\sum_{n=1}^{N_z} q_n \cos(l_n r + \phi_n)\cos(k_n z)}$$

where we have used the fact that the only contributing **m** vectors are those of the form $\mathbf{m} = [0,0,\ldots,\cdot,\ldots,0,0]$ where the only nonzero n component is indicated with the symbol "$\cdot$". Because of the smallness of the summation in the denominator in the limit $q_j \to 0$, this becomes:

$$\psi(r,z) \simeq -2\sum_{n=1}^{N_z} q_n \sin(l_n r + \gamma_n)\sin(k_n z)$$

How does this compare with Equation (26.12) after we apply the asymmetric boundary condition to the latter equation? Let us write Equation (26.12) in the following form, making the phases explicit:

$$\psi(r,z) = \sum_{n=-\infty}^{\infty} \psi_n(r)e^{ik_n z}, \quad \psi_n(r) = \psi_n(0)e^{il_n r + i\gamma_n}, \quad l_n = \frac{-ik_n^2 r}{2k_0}$$

This can be written in the form

$$\psi(r,z) = \sum_{n=0}^{\infty} a_n(r)\cos k_n z + b_n(r)\sin k_n z$$

where $\psi_n(r) = [a_n(r) - ib_n(r)]/2$, that $a_n(r) = 2\psi_n(0)\cos(l_n r + \gamma_n)$ and $b_n(r) = -2\psi_n(0)\sin(l_n r + \gamma_n)$. Apply the boundary condition $\psi(r,z) = -\psi(r,-z)$ and we get

$$\psi(r,z) = \sum_{n=1}^{\infty} b_n(r)\sin k_n z = -2\sum_{n=0}^{\infty} \psi_n(0)\sin(l_n r + \gamma_n)\sin k_n z$$

Continued

This is identical to the limit of the IST given above if we recognize $\psi_n(0) = q_n$. Thus, we have the important result that the exact solution of the PE reduces to the linear Fourier transform solution in the small-amplitude limit when the potential is zero, that is, for a homogeneous halfspace.

Nota Bene: Determining the Spectrum for a Homogeneous Halfspace

When we do not have a potential or when we have only a very small potential it would best be prudent to modify the procedure given previously in the chapter for the determination of the Riemann spectrum. In this case it is the source function $\psi(0,z)$ that should dominate the determination of the spectrum, which of course should be used to determine $\hat{F}, \gamma$. This exercise gives us some interesting perspective. It says that there is a fundamental Riemann matrix $\hat{F}$ from which we can simultaneously fit the environment (either the potential $U(r,z)$ or its auxiliary function $F(r,z)$) by adjusting the phases ϕ; the source function $\psi(0,z)$ can be satisfied by adjusting the phases γ. There are of course a number of methods for doing this, including loop integrals, Schottky uniformization and the Nakamura-Boyd approach (see Chapters 14–16) or nonlinear adiabatic annealing (Chapter 23).

Nota Bene: Time Reversal Symmetry

The perspective just given is that one has a fundamental Riemann matrix $\hat{F}$ and particular phases for the environment ϕ and other phases for the source function γ. In the time reversal symmetry problem, provided we remain in the plane of the measurements that established the sound speed field over depth and range, a moving target could be followed and communicated with simply by adjusting the phases γ. This idea extends the notion of time reversal symmetry beyond the original concept that uses a source at a particular location and a transducer array downrange. Without a measurement of the environment one records the signals occurring at many vertical locations in the array and time reverses them to send information. Because of time reversal symmetry the signals all converge at the source once again.

The modification of the time reversal symmetry approach using the new methods given herein is now discussed. One uses knowledge of the source and the measurements at each of the vertically placed instruments in the array to establish the Riemann matrix and the phases for the environment. One then computes the appropriate set of phases to converge the return signal to the source. One then follows the source with sonar and keeps its location known. One then continues communications by updating the phases γ to keep the time-reversed signal focused at the moving source. A section below gives more details.

Now we address the *source function*. We have for the functions F, G:

$$F(0, z) = 1 + 2 \sum_{m \in \mathbb{Z}}{}' e^{-\frac{1}{2} \mathbf{m} \cdot \hat{\mathbf{F}} \mathbf{m}} \cos(\mathbf{m} \cdot \boldsymbol{\phi}) \cos(\mathbf{m} \cdot \boldsymbol{\kappa} z)$$

$$G(0, z) = -2 \sum_{m \in \mathbb{Z}}{}' e^{-\frac{1}{2} \mathbf{m} \cdot \hat{\mathbf{F}} \mathbf{m}} \sin(\mathbf{m} \cdot \boldsymbol{\gamma}) \sin(\mathbf{m} \cdot \boldsymbol{\kappa} z)$$

(26.54)

so that

$$\psi(0, z) = \frac{-2 \sum_{m \in \mathbb{Z}}{}' e^{-\frac{1}{2} \mathbf{m} \cdot \hat{\mathbf{F}} \mathbf{m}} \sin(\mathbf{m} \cdot \boldsymbol{\gamma}) \sin(\mathbf{m} \cdot \boldsymbol{\kappa} z)}{1 + 2 \sum_{m \in \mathbb{Z}}{}' e^{-\frac{1}{2} \mathbf{m} \cdot \hat{\mathbf{F}} \mathbf{m}} \cos(\mathbf{m} \cdot \boldsymbol{\phi}) \cos(\mathbf{m} \cdot \boldsymbol{\kappa} z)}$$

Likewise the Cauchy condition for Equation (26.53b) is given by

$$\psi(0, z) = \frac{\sum_{n=1}^{N_z} D_n(0) \sin(k_n z)}{A_0(0) + \sum_{n=1}^{N_z} A_n(0) \cos(k_n z)}$$

Here is a procedure for determining the Riemann matrix, $\mathbf{F}$, and phases $\boldsymbol{\phi}$, $\boldsymbol{\gamma}$.

(1) Fit the environment to $F(0,z)$ to obtain $\mathbf{F}$ and $\boldsymbol{\phi}$. The procedure is referred to as "nonlinear adiabatic annealing" and is discussed in Chapter 23.
(2) Decide the form of the initial condition, for example, a Gaussian starter, for $\psi(0,z)$. Then form $G(0,z) = \psi(0,z)F(0,z)$ and determine the phases $\boldsymbol{\gamma}$ to this latter function by adiabatic annealing to match the initial conditions.

We thus have the exact solution of the PE for a particular environment and source field!

Two special cases are of interest. The *first occurs if the potential is zero*, that is, when the index of refraction is a constant. In this case we have $U(r,z) \simeq 0$, $F(r,z) \simeq 1$. This happens in the linear limit that is discussed in detail below.

The *second case occurs when the potential is not range dependent* so that the function $F(r,z)$ becomes:

$$F(r, z) = 1 + 2 \sum_{m \in \mathbb{Z}}{}' e^{-\frac{1}{2} \mathbf{m} \cdot \hat{\mathbf{F}} \mathbf{m}} \cos(\mathbf{m} \cdot \boldsymbol{\phi}) \cos(\mathbf{m} \cdot \boldsymbol{\kappa} z)$$

(26.55)

The function $G(r,z)$ is then

$$G(r, z) = -2 \sum_{m \in \mathbb{Z}}{}' e^{-\frac{1}{2} \mathbf{m} \cdot \hat{\mathbf{F}} \mathbf{m}} \sin(\mathbf{m} \cdot \boldsymbol{\lambda} r + \mathbf{m} \cdot \boldsymbol{\gamma}) \sin(\mathbf{m} \cdot \boldsymbol{\kappa} z)$$

(26.56)

Finally, the complex sound field has the form

$$\psi(r,z) = \frac{-2\sum_{m\in\mathbb{Z}}'e^{-\frac{1}{2}m\cdot\hat{F}m}\sin(m\cdot\boldsymbol{\lambda}r + m\cdot\boldsymbol{\gamma})\sin(m\cdot kz)}{1 + 2\sum_{m\in\mathbb{Z}}'e^{-\frac{1}{2}m\hat{F}m}\cos(m\cdot\boldsymbol{\phi})\cos(m\cdot kz)} \tag{26.57}$$

Note that the range dependence occurs only in the numerator, not the denominator. This means the environment (the denominator) does not have range dependence, but the solution of the PE does of course have range dependence as seen in the numerator.

26.9   One Degree-of-Freedom Case

The one degree-of-freedom case reduces identically to the following exact forms:

$$F(r,z) = \sum_{n=-\infty}^{\infty} q^{n^2} e^{in(\kappa z+\lambda r+\phi)}, \quad q = e^{-\frac{1}{2}B_{11}}$$

$$G(r,z) = \sum_{n=-\infty}^{\infty} q^{n^2} e^{in(\kappa z+\lambda r+\gamma)}$$

Of course it is clear that $\lambda = -\kappa^2/2k_0 + \lambda'$, where the additional term λ' is q dependent (this is the nonlinear correction to the Stokes wave wavenumber, see, e.g., the single degree-of-freedom case for the KdV equation, Equation (16.29)). Apply the boundary conditions $F(r,z) = F(r,-z)$:

$$F(r,z) \rightarrow \frac{1}{2}[F(r,z) + F(r,-z)]$$

to get

$$F(r,z) = 1 + 2\sum_{n=1}^{\infty} q^{n^2} \cos[n(\lambda r + \phi)]\cos(n\kappa z)$$

and apply the boundary conditions $|\psi(x,t)|$:

$$G(r,z) \rightarrow \frac{1}{2}[G(r,z) - G(r,-z)]$$

for which

$$G(r,z) = -2\sum_{n=1}^{\infty} q^{n^2} \sin[n(\lambda r + \gamma)]\sin(n\kappa z)$$

The one degree-of-freedom solution to the PE can then be written:

$$\psi(r,z) = \frac{-2\sum_{n=1}^{\infty} q^{n^2} \sin\left[n(\lambda r + \gamma)\right] \sin(n\kappa z)}{1 + 2\sum_{n=1}^{\infty} q^{n^2} \cos\left[n(\lambda r + \phi)\right] \cos(n\kappa z)}$$

Here I have not included the Stokes correction (Equation (26.17)), which is easily done, and includes the wavenumber correction λ' to the "dispersion relation" $\lambda = -\kappa^2/2k_0 + \lambda'$. Suppose that the initial condition is given by (I have arbitrarily chosen this):

$$\psi(0,z) = \frac{-2\sum_{n=1}^{\infty} q^{n^2} \sin(n\kappa z)}{1 + 2\sum_{n=1}^{\infty} q^{n^2} \cos(n\kappa z)}$$

This expression therefore fixes the phase shifts to be $\phi = 0$, $\gamma = -\pi/2$. Thus, with the possible exception of the Stokes correction, we have a one degree-of-freedom solution. This solution might be checked by substitution into the bilinear form (see Chapter 16 on the Nakamura-Boyd procedure in which the Stokes correction would automatically appear). A larger number of degrees of freedom requires that we must fix a larger number of phases ϕ, γ. In this simple one degree-of-freedom case there are no off-diagonal elements in the Riemann matrix; the only diagonal element you can select yourself.

For the case of small nonlinearity (small variations in the index of refraction) we have $q \ll 1$ and we have the following approximate relations:

$$F(r,z) \simeq 1 + 2q\cos(\lambda r + \phi)\cos(\kappa z)$$

$$G(r,z) \simeq -2q\sin(\lambda r + \gamma)\sin(\kappa z)$$

$$\psi(r,z) \simeq \frac{-2q\sin(\lambda r + \gamma)\sin(\kappa z)}{1 + 2q\cos(\lambda r + \phi)\cos(\kappa z)}$$

$$U(r,z) = 2\partial_{zz}\ln F(r,z) = -4\kappa^2 q\cos(\lambda r + \phi)\left\{\frac{\cos\kappa z + 2q\cos(\lambda r + \phi)}{[1 + 2q\cos(\lambda r + \phi)\cos\kappa z]^2}\right\}$$

These expressions can be quite useful in many cases for computing back-of-the-envelope results for the small-amplitude, single degree-of-freedom case. These expressions are now quite explicit. The parameters are q (the single Riemann matrix element) and the phases ϕ, γ. Of course the single degree-of-freedom case is not very flexible for describing acoustic waves in an arbitrary range-dependent ocean with arbitrary sound speed structure. But you will find that the above equations are fun to plot and one sees the nonlinearity already in this simple case.

26.10   Linear Limit of the Theta-Function Formulation

I have visited the linear limit of the exact solution of the PE above, but I would now like to revisit the problem once more with a somewhat different perspective.

The linear limit follows from the general solution of the PE after applying the boundary conditions. This limit corresponds to the case where the diagonal elements of the Riemann matrix dominate the spectrum, leaving it essentially equivalent to the linear Fourier transform with small range dependence in the Fourier coefficients other than the sinusoidal dependence of the linear problem. In this case all of the "cross terms" in the theta function make no significant contribution and *only the diagonal elements are important*. The linear limit occurs when the potential satisfies:

$$U(r, z) = k_0^2[n^2(r, z) - 1] \cong -2\frac{k_0^2}{c_0}[c(r, z) - c_0] \ll 1$$

Since in the ocean the index of refraction for sound waves does not differ very much from unity (i.e., the range-depth dependent sound speed does not differ much from the reference sound speed), the potential will be small for relatively low-frequency sound waves, $\omega_0 = c_0 k_0$, thus $k_0^2 = (\omega_0/c_0)^2$. It is rare that $n^2(r,z) - 1$ exceeds about 10% in the ocean, so we are left with the requirement that

$$2\left(\frac{\omega_0}{c_0}\right)^2 \frac{c(r, z) - c_0}{c_0} \ll 1$$

which gives a rough estimate of the upper frequency for which nonlinearity is small:

$$f_0 = \frac{c_0}{2\sqrt{2\pi}} < 300 \text{ Hz}$$

For frequencies only marginally larger there can possibly form "coherent structures" in the spectrum. An example of such a structure is given in Section 26.15.

We have, by inspection, the following expressions in the linear limit (see Chapter 8 in which I argue that in this limit the off-diagonal elements can be neglected):

$$F(r, z) = 1 + 2\sum_{n=1}^{\infty} e^{-\frac{1}{2}n^2 \hat{F}_{nn}} \cos(\lambda_n r + \phi_n) \cos(\kappa_n z)$$

$$G(r, z) = -2\sum_{n=1}^{\infty} e^{-\frac{1}{2}n^2 \hat{F}_{nn}} \sin(\lambda_n r + \gamma_n) \sin(\kappa_n z)$$

(26.58)

In the linear limit (where the diagonal terms dominate the off-diagonal terms of the matrix **F**) we invoke the following useful expressions:

$$\mathbf{m} \cdot \mathbf{1} \Rightarrow l_n, \quad \mathbf{m} \cdot \boldsymbol{\phi} \Rightarrow \phi_n, \quad \mathbf{m} \cdot \mathbf{k} \Rightarrow k_n \qquad (26.59)$$

The solution to the PE in this limit has the form of the ratio of two Fourier series:

$$\psi(r,z) = \frac{-2\sum_{n=1}^{\infty} e^{-\frac{1}{2}n^2 \hat{F}_{nn}} \sin(\lambda_n r + \gamma_n) \sin(\kappa_n z)}{1 + 2\sum_{n=1}^{\infty} e^{-\frac{1}{2}n^2 \hat{F}_{nn}} \cos(\lambda_n r + \phi_n) \cos(\kappa_n z)} \tag{26.60}$$

Now if we continue to make the environmental variations smaller and smaller we get

$$\psi(r,z) = -2\sum_{n=1}^{\infty} e^{-\frac{1}{2}n^2 \hat{F}_{nn}} \sin(\lambda_n r + \gamma_n) \sin(\kappa_n z) \tag{26.61}$$

The linear limit for the potential, using $\ln[1+\delta] \simeq \delta$ (for $\delta \ll 1$), is

$$U(r,z) \simeq -4\sum_{n=1}^{\infty} k_n^2 e^{-\frac{1}{2}n^2 \hat{F}_{nn}} \cos(\lambda_n r + \phi_n) \cos(\kappa_n z) \tag{26.62}$$

We therefore see that the solution to the PE in the linear Fourier limit is given by ordinary Fourier series. This is the *Born approximation* of the Schrödinger equation, that is, the solution for which the scattering occurs at small angles (Morse and Feshbach, 1953).

26.11 Implementation of Multidimensional Fourier Methods in Acoustics

Let me briefly summarize what has been accomplished up to this point in this chapter. I have developed a method to solve the PE from a set of ODEs (26.9) that has numerical computational speeds that are competitive with Tappert's scheme. While this approach is *not* recommended to replace the Tappert approach, nevertheless the discovery of this scheme lays the foundation for the development of *two new analytical approaches* for solving the PE based upon *Hirota bilinear forms* (Chapters 4, 6). The *first approach* (Section 26.5) develops matrix equations from the bilinear forms for analytically solving the PE based upon ordinary linear Fourier transforms. The *second approach* (Section 26.6) uses multidimensional Fourier transforms (Riemann theta functions) to solve the bilinear form (and hence the PE) analytically. It is easy to show that the two approaches (one is based upon ordinary linear Fourier series, the other on multidimensional Fourier series) are equivalent because the theta functions can be rewritten as ordinary Fourier series, and thus it becomes clear how to relate the two methods. I now give details on why and how these two approaches are equivalent and how to use these ideas to numerically implement the methods for work in ocean acoustics. Below is a brief discussion of the advantages of using theta functions to do acoustics over other methods.

When we build the potential $U(r,z)$ (from temperature, density, or sound speed measurements), we normally think (as an oceanographical exercise) of commensurable wavenumbers for both depth and range. This means we deal with a regular grid in r, z space. This of course has a two-dimensional Fourier transform with which we normally take with an FFT algorithm. *However, from the perspective of solving the PE, the range variable does not have commensurable wavenumbers, but the range variable is often taken to lie on an equally space grid.* This is because, even in the linear case, the range wavenumber has the dispersion relation $l_n = -k_n^2/2k_0$, that is proportional to the square of the z wavenumber.

We thus have a subtle mathematical perspective between theoretical and experimental procedures that I would now like to explore. To implement the procedures discussed in this chapter we must first consider the mapping (Equation (26.24)) from the potential $U(r,z)$ to the function $F(r,z)$. If $U(r,z)$ is assumed to be periodic or quasi-periodic then so too is $F(r,z)$. Recall the definition of the function $F(r,z)$ as a multidimensional Fourier series:

$$F(r, z) = \theta(r, z | \hat{\mathbf{F}}, \boldsymbol{\phi}) = \sum_{\mathbf{m} \in \mathbb{Z}} e^{-\frac{1}{2}\mathbf{m} \cdot \hat{\mathbf{F}}\mathbf{m}} e^{i\mathbf{m} \cdot \boldsymbol{\kappa}z + i\mathbf{m} \cdot \boldsymbol{\lambda}r + i\mathbf{m} \cdot \boldsymbol{\phi}} \tag{26.63}$$

The linear Fourier counterpart of this function, especially for experimental purposes, is given by the standard formulas:

$$F(r, z) = \sum_{n=-\infty}^{\infty} \sum_{m=-\infty}^{\infty} F_{nm}(r) e^{ik_n z + il_m r} = \sum_{n=-\infty}^{\infty} F_n(r) e^{k_n z} \tag{26.64}$$

where

$$F_n(r) = \sum_{m=-\infty}^{\infty} F_{nm}(r) e^{il_m r} \tag{26.65}$$

We often assume that $k_n = 2\pi n/L_z$ and $l_m = 2\pi m/L_r$, where L_z, L_r are the sides of the depth-range box in which we have represented the potential $U(r,z)$ (recall that the function $F(r,z)$ is derived from the potential by Equation (26.22)). Now it is easy to see that we have explicitly made k_n, l_m commensurable, so we have a rectangular grid for defining $U(r,z)$ and $F(r,z)$, standard procedures in physical oceanography. This is because the above expressions are the usual ones we deal with when we apply ordinary linear Fourier analysis in two dimensions. All operations arise from use of the discrete Fourier transform (the FFT) for numerical computations.

However, and in addition to the above considerations, the function $F(r,z)$ has linear Fourier coefficients derivable from the multidimensional Fourier series written above:

$$F_n(r) = \sum_{\{j \in \mathbb{Z}: I_j = n\}} q_j e^{i\Lambda_j r}, \quad q_j = \exp\left(-\frac{1}{2}\sum_{m=1}^{N}\sum_{n=1}^{N} m_m^j m_n^j \hat{F}_{mn} + i\Phi_j\right) \tag{26.66}$$

for

$$\Lambda_j = \sum_{n=1}^{N} m_n^j \lambda_n, \quad \Phi_j = \sum_{n=1}^{N} m_n^j \phi_n \tag{26.67}$$

We have two formulas for the range varying coefficients of the environmental function $F(r,z)$:

$$F_n(r) = \sum_{m=-\infty}^{\infty} F_{nm}(r) e^{i l_m r} \tag{26.68a}$$

$$F_n(r) = \sum_{\{j \in \mathbb{Z} : I_j = n\}} q_j e^{i \Lambda_j r} \tag{26.68b}$$

The first is from *ordinary two-dimensional Fourier series*, the second from the *multidimensional Fourier formulation* as given above and elsewhere (in particular see Chapter 9). Let us see in what sense the two expressions can be made to be effectively *equivalent* for applications. The first says we are dealing with commensurable wavenumbers and therefore live on a regular grid for the depth/range variables z, r. The second says two things: (1) the Fourier coefficients themselves are Fourier series over the range variable (there are *many* terms in these series, $\sim 10^N$, for N the number of nonlinear Fourier modes) and (2) the range wavenumbers λ_n (and hence the Λ_j) are *not commensurable*. In typical applications there may be *billions of components* $q_j \exp[i\Lambda_j r]$ and the wavenumbers Λ_j are *dense* on the wavenumber axis, *not separated* by large and equally spaced Λ_j (as is the case for the depth wavenumbers κ_n), and this means formally that the $F_n(r)$ are *not periodic* in r, but instead are *quasi-periodic* in r. The first expression for $F_n(r)$ above (Equation (26.28)) is periodic in r, the second is quasi-periodic, an important difference.

What is the advantage in using the quasi-periodic formulation in terms of the spectrum of the multidimensional Fourier series instead of the usual linear Fourier spectrum on a rectangular grid? Because we get the exact solution of the PE! This means we get a whole host of advantages over conventional linear Fourier analysis: (1) nonlinear filtering, (2) hyperfast modeling, (3) nonlinear data assimilation, (4) the direct acquisition of coherent structures in the oceanic acoustic field, etc.

Thus, the multidimensional Fourier methods given herein mean that we must take an additional step, that is, to compute $F(r,z)$ on a grid which is commensurable in the depth wavenumbers κ_n but which is incommensurable in the range wavenumbers λ_n.

How can we use multidimensional Fourier series to work on acoustic problems? We have to go to the measured data for the acoustic field and compute the FFT:

$$F_n(r) = \sum_{m=-\infty}^{\infty} F_{nm}(r) e^{i l_m r} \quad \text{(periodic in } r) \tag{26.69a}$$

using standard procedures. Then we have to compute the Riemann spectrum (on the right-hand side) of Equation (26.68b) above:

$$F_n(r) = \sum_{\{j \in \mathbb{Z} : I_j = n\}} q_j e^{i\Lambda_j r} \quad \text{(quasi-periodic in } r) \tag{26.69b}$$

The procedure for doing this combines the *nonlinear adiabatic annealing* (Chapter 23) and the *Nakamura-Boyd equations* for the determination of the Riemann spectrum (Chapter 16). One effectively computes the $F_n(r)$ from sound speed data and then computes the Riemann matrix $\hat{F}_{mn}$, wavenumbers λ_n, and phases ϕ_n via the methods of Chapter 23 from $F_n(r)$. Then one uses the same approach to compute the phases γ_n in the function $G(r,z)$ for, say, a Gaussian starter and finally one has the exact solution to the PE, Equation (26.15). There you have it, the power of Riemann theta functions to make important new contributions to ocean acoustics.

Nota Bene: In the acoustic problem we assume that the potential $U(r, z)$ is known. The potential would come from physical measurements of sound speed profiles made over a regular depth/range grid r, z from which we would compute the potential $U(r, z)$. To solve the PE with this information one would have to extract the period matrix **B** and the phases $\boldsymbol{\phi}$ (phase locking is important for the "unstable acoustic modes") from the input potential $U(r, z)$. One would thus find the multidimensional Fourier series on a grid that is regular in depth/range but which is *commensurable* in the depth z wavenumbers κ_n and which is *incommensurable* in the range r wavenumbers λ_n. One could use the *nonlinear adiabatic annealing approach* of Chapter 23 to determine the Riemann spectrum **B**, $\boldsymbol{\phi}$ with the aid of one of the methods of Chapters 14–16 to estimate the *off-diagonal elements of the Riemann matrix*, the *relative phases* (again phase locking is important for unstable modes) and the *frequencies*. An *alternative procedure* would be to first determine the *linear two-dimensional Fourier transform* of $U(r, z)$ on a regular grid and to then "fit" the Riemann spectrum, with incommensurable range wavenumbers, to this rectangular grid.

    Finally, we would find the phases in the function $G(x, t)$ to satisfy the form of the source function, $\psi(0, z)$. The solution to the PE is then given by: $\psi(r, z) = G(r, z)/F(r, z)$, where $U(r, z) = 2\partial_{xx} \ln F(r, z)$.

Nota Bene: The *acoustic problem for the PE* is like the *ocean surface wave problem* for the NLS equation, but with a fundamental and important difference. In the acoustic problem we assume that the potential $U(r, z)$ is known. For *the NLS problem* this would be like *prespecifying* the space-time dependence of the squared modulus $U(r, t) = \nu|\psi(x, t)|^2$! Of course this specification would come from physical, perhaps remotely sensed measurements, made over a regular space/time grid: The waves themselves would be measured and then the Hilbert transform (Chapter 13) would be used to determine the modulus of the complex envelope $|\psi(x, t)|$ from the surface elevation $\eta(x, t)$. Thus, to solve the NLS equation with this information one would have to extract the period matrix $\mathbf{B}$, the phases $\boldsymbol{\phi}$ (phase locking is important for the "unstable modes"), and frequencies $\boldsymbol{\omega}$ from the input "data" $|\psi(x, t)|^2$. One would thus find the multidimensional Fourier series on a grid that is regular in space/time but which is *commensurable* in the x wavenumbers κ_n and which is *incommensurable* for the t frequencies ω_n. One could use the *nonlinear adiabatic annealing approach* of Chapter 23 with the aid of one of the methods of Chapters 14–16 to determine the *off-diagonal elements of the Riemann matrix*, the *relative phases* (again phase locking is important for unstable modes) and the *frequencies*. An *alternative procedure* would be to first determine the *linear two-dimensional Fourier transform* of $|\psi(x, t)|^2$ on a regular grid and to then "fit" the Riemann spectrum, with incommensurable frequencies, to this rectangular grid.

The solution to the NLS equation is given by: $\psi(x, t) = G(x, t)/F(x, t)$; $|\psi(x, t)|^2 = 2\partial_{xx} \ln F(x, t)/\lambda$, $\lambda = \sqrt{v/2\mu}$ (see Equations (24.1)–(24.3)). Finally, to simulate the actual motion we would find the phases in the function $G(x, t)$. It is in this way that we see the connection between the NLS equation for ocean surface waves and the PE acoustic problem.

26.12 Physical Interpretation of the Exact Solution of the PE

Let us revisit some of the physical considerations for ocean acoustics and the solutions of the PE. The exact general solution of the PE is given by the expression

$$\psi(r, z) = \frac{-2 {\sum_{m \in \mathbb{Z}}}' e^{-\frac{1}{2}\mathbf{m} \cdot \hat{\mathbf{F}}\mathbf{m}} \sin(\mathbf{m} \cdot \boldsymbol{\lambda} r + \mathbf{m} \cdot \boldsymbol{\gamma}) \sin(\mathbf{m} \cdot \mathbf{k}z)}{1 + 2 {\sum_{m \in \mathbb{Z}}}' e^{-\frac{1}{2}\mathbf{m} \cdot \hat{\mathbf{F}}\mathbf{m}} \cos(\mathbf{m} \cdot \boldsymbol{\phi}) \cos(\mathbf{m} \cdot \mathbf{k}z)} \tag{26.70}$$

This is of course the most important result of this chapter. Clearly, this expression satisfies the boundary conditions for the potential and the complex sound field at the free surface. To satisfy the boundary conditions for the sound field at the bottom one needs to add a suitable bottom model. However, a very interesting *simple case* occurs when the bottom is a perfectly reflecting boundary, so that there is a pressure release condition also at the bottom, $\psi(r, z = Z) = 0$. Introducing a bottom model (see below) modifies these boundary conditions because of the presence of the potential in the region of the bottom. The bottom model modifies the shape of the potential and hence the function $F(r, z)$; this occurs simply by modifying the Riemann spectrum to "fit" the shape of the potential. Determination of the Riemann spectrum requires the use of the Schottky procedure (Chapter 15), the Nakamura-Boyd method (Chapter 16), in combination nonlinear adiabatic annealing (Chapter 23). Dissipation is also an ingredient that is simple to add by requiring a small imaginary contribution to the potential.

Now the expression (26.69) above for the solution to the complex sound field may have both the *stable* and *unstable modes,* in analogy with the nonlinear Schrödinger equation (see Chapters 12, 18, 24 and 26). Not only is the period matrix complex, but also the frequencies and phases may be complex. This case can be studied naturally using any of the approaches in Chapters 14–16.

Now the expression (26.70) provides a natural sequence of steps to follow to *solve the PE* for a simple problem:

(1) *Given the potential* (or the temperature or density fields) over an equally spaced grid in z and r. Use *nonlinear adiabatic annealing to get the Riemann matrix and phases* for the function $F(r,z)$ ($\hat{F}$, φ). To this end one might pick a set of wavenumber pairs (κ_m, λ_n) in the wavenumber plane, where the λ_n are of course incommensurable and would also have to be a part of the fitting process. The Riemann matrix diagonal elements are one-to-one with the number of wavenumber pairs. For 10 pairs of wavenumbers, say, the Riemann matrix is 10 × 10. The chosen Riemann matrix, wavenumber pairs, and phases define the *potential function* of the PE. Submatrices which are 2 × 2 will be phase locked when *unstable modes* occur, leading to a threshold effect for the creation of "rogue modes" or "lenses" in the sound field (see Chapters 12, 18, 24, and 30).

(2) Insert the *Riemann matrix* $\hat{F}$ into the function $G(r,z)$ with *unknown phases* γ. Then, using the above expression for the complex sound field, fit the phases via the adiabatic annealing procedure to compute the γ for the *Cauchy initial conditions.*

(3) Evaluate the sound field on the r, z grid of choice! Of course one graphs the *transmission loss* as a function of r, z.

26.13   Solving the PE for a Given Source Function

We have seen that the process of obtaining the Riemann matrix and phases F, φ (that describe the environment or sound speed field) is a straightforward application of Chapter 23. However, to solve the PE we still need the phase vector γ that assures that we satisfy a particular *source function.* These parameters are easily seen in the solution of the PE given in terms of Riemann theta functions (Equation (26.70)). This section discusses how to compute the phases γ for a source function $\psi(0,r)$. To this end we use the form of the solution of the PE that in terms of *ordinary linear Fourier series* is given by (Equation (26.53b)):

$$\psi(r,z) = \frac{\sum_{n=1}^{N_z} D_n(r)\sin(k_n z)}{A_0(r) + \sum_{n=1}^{N_z} A_n(r)\cos(k_n z)} \tag{26.71}$$

The denominator is of course assumed known because it describes the sound speed field. Our goal is to find the appropriate coefficients $D_n(0)$ to satisfy the chosen initial conditions $\psi(0,z)$. This means that, given the form for the $D_n(0)$, we must find the appropriate phases Γ_j (and from these the parameters γ_j) in the following relation (the last equation in Equation (26.53b)):

$$D_n(r) = -2 \sum_{\{j \in \mathbb{Z}: I_j = n\}} q_j \sin(\Lambda_j r + \Gamma_j), \quad \Gamma_j = \sum_{n=1}^{N} m_n^j \gamma_n \qquad (26.72)$$

To find the $D_n(0)$ I follow the procedure given below. To obtain the γ_j from the $D_n(0)$ I use adiabatic annealing as described in Chapter 23 and also discussed below.

Here is the procedure for determining the $D_n(0)$ from knowledge of the sound speed field and the source function. Suppose that the solution of the PE is assumed to have a Fourier series:

$$\psi(r, z) = \sum_{n=1}^{N_z} d_n(r) \sin(k_n z) \qquad (26.73)$$

This of course obeys the asymmetric boundary condition at the free surface. The Cauchy source condition can be written from Equation (26.73)

$$\psi(0, z) = \sum_{n=1}^{\infty} d_n(0) \sin(k_n z) \qquad (26.74)$$

where for a Gaussian "explosion" at $r = 0$ near some depth $z = z_d$ we know that the $d_n(0)$ are also Gaussian and known (Jensen et al., 2000).

Now insert Equation (26.74) into Equation (26.71) to get

$$\sum_{n=1}^{\infty} d_n(0) \sin(k_n z) = \frac{\sum_{n=1}^{N_z} D_n(0) \sin(k_n z)}{A_0(0) + \sum_{n=1}^{N_z} A_n(0) \cos(k_n z)} \qquad (26.75)$$

We can change the limits slightly because the term for $n = 0$ of the sine series is zero $(d_0(0) = D_0(0) = 0)$ and the term for $n = 0$ for the cosine series $(A_0(r))$ is known from Equation (26.53b):

$$\sum_{n=0}^{N_z} d_n(0) \sin(k_n z) = \frac{\sum_{n=0}^{N_z} D_n(0) \sin(k_n z)}{\sum_{n=0}^{N_z} A_n(0) \cos(k_n z)} \qquad (26.76)$$

For which

$$\sum_{n=0}^{N_z} D_n(0) \sin(k_n z) = \sum_{n=0}^{N_z} d_n(0) \sin(k_n z) \sum_{m=0}^{N_z} A_m(0) \cos(k_m z) \qquad (26.77)$$

Thus, we encounter a situation in which the product of two Fourier series is a Fourier series (see discussion with regard to Equation (26.53b) about dividing Fourier series). Using Section 6 of Tolstov (1962) (see also Zygmund, 1959) we find the coefficients $D_n(0)$ in terms of the coefficients of the source function $(d_n(0))$ and the environment $(A_n(r))$:

$$D_n(0) = \frac{1}{2} \sum_{m=1}^{N_z} d_n(0)[A_{m-n}(0) - A_{m+n}(0)] \tag{26.78}$$

This expression is of course just a *convolution for a sine series*; the *general form* for determining the coefficients of a Fourier series resulting from the product of two Fourier series is given in Equations (26.9) and (26.25) and the Appendix. Here the $d_n(0)$ are the Fourier coefficients for the Gaussian initial condition. The $A_n(0)$ correspond to the environment at $r = 0$, that is, they contain the Riemann matrix and phases $\hat{\mathbf{F}}, \boldsymbol{\phi}$ of the sound speed field at $r = 0$ as computed in Equation (26.53b).

We still need to determine the phases γ_j. To do this combine Equation (26.72) with Equation (26.78) to get

$$\sum_{m=1}^{N_z} d_n(0)[A_{m+n}(0) - A_{m-n}(0)] = 4 \sum_{\{j \in \mathbb{Z} : I_j = n\}} q_j \sin \Gamma_j \tag{26.79}$$

This expression is to be solved for the phases γ_j. To this end recall that $\Gamma_j = \Gamma_j(\gamma_n)$, so we have the relation (26.79) which can, with adiabatic annealing (Chapter 26), be solved for the γ_j. Note that the Riemann matrix and phases $\hat{\mathbf{F}}, \boldsymbol{\phi}$ are used to compute the input $A_n(0)$ by

$$A_n(0) = 2 \sum_{\{j \in \mathbb{Z} : I_j = n\}} q_j \cos \Phi_j, \quad q_j = \exp(-\frac{1}{2} \sum_{m=1}^{N} \sum_{n=1}^{N} m_m^j m_n^j \hat{F}_{mn})$$

$$\Phi_j = \mathbf{m}^j \cdot \boldsymbol{\phi} = \sum_{n=1}^{N} m_n^j \phi_n, \quad \Gamma_j = \mathbf{m}^j \cdot \boldsymbol{\gamma} = \sum_{n=1}^{N} m_n^j \gamma_n$$

The $d_n(0)$ are the Fourier transform of the Gaussian source $\psi(0,z)$ and are of course assumed known. While the adiabatic annealing process can be used to determine the γ_j, one could also obtain aid from the approaches of Chapter 14–16, that is, loop integrals, Schottky uniformization, and the Nakamura-Boyd method.

> *Nota Bene*: We now have procedures for modeling acoustic wave propagation in a given sound speed field $c(r, z)$. One first computes the potential Equation (26.2) and then computes the Riemann spectrum $\hat{\mathbf{F}}, \boldsymbol{\phi}$ (the spectrum of the environment) using adiabatic annealing (Chapter 26). Then one assumes a form for the source function $\psi(0, z)$. The procedure given

above allows us to compute the source function phases γ. Finally, Equation (26.53b) can be used to generate solutions of the PE for each value of the range parameter r. Since the solution is *exact* and requires only the FFT, the method is fast, lacking any form of numerical integration. This means that range degradation of the general solution to the PE given herein is not a problem.

26.14 Range-Independent Problem

For a *range-independent* ocean we have the potential:

$$V(z) = n^2(z) - 1 = \frac{c_0^2}{c^2(z)} - 1 \tag{26.80}$$

The acoustic wave equation therefore becomes:

$$2ik_0 \frac{\partial \psi}{\partial r} + \frac{\partial^2 \psi}{\partial z^2} + k_0^2 V(z)\psi = 0 \tag{26.81}$$

which we can solve by separation of variables:

$$\psi(r, z) = R(r)Z(z)$$

Thus, we have

$$R(r) = R_0 \exp\left(-i\frac{\kappa^2}{2k_0} r\right)$$

so the solution is oscillatory in r, the range. This leads to the *Schrödinger eigenvalue equation*

$$\frac{d^2 Z}{dz^2} + [k_0^2 V(z) + \kappa^2]Z = 0, \quad Z(0) = Z(h) = 0 \tag{26.82}$$

Now let us change the notation a bit in the eigenvalue problem. We write

$$k_0^2 V(z) + \kappa^2 = \frac{\omega_0^2}{c^2(z)} - k_r^2 \tag{26.83}$$

Hence the eigenvalue problem becomes:

$$\frac{d^2 Z}{dz^2} + \left[\frac{\omega_0^2}{c^2(z)} - k_r^2\right] Z = 0 \tag{26.84}$$

This is an alternative form that may be preferred in some cases. The dispersion relation is simple:

$$k_z^2 + k_r^2 = \frac{\omega_0^2}{c_0^2}$$

(26.85)

The algebraic-geometric spectral decompositions for the range-independent case include the sound speed profiles:

$$U(z) = k_0^2 V(z) = 2\frac{\partial^2}{\partial z^2} \ln \theta(z|\mathbf{B}, \boldsymbol{\phi})$$

$$\theta(z|\mathbf{B}, \boldsymbol{\phi}) = \sum_{m_1=-\infty}^{\infty} \sum_{m_2=-\infty}^{\infty} \cdots \sum_{m_N=-\infty}^{\infty} \exp\left(i\sum_{n=1}^{N} m_n(\kappa_n z + \phi_n) - \frac{1}{2}\sum_{m=1}^{N}\sum_{n=1}^{N} m_m m_n B_{mn}\right)$$

(26.86)

The *single acoustic soliton* for the range-independent, single degree-of-freedom case has the potential function (Landau and Lifshitz, 1958):

$$U(z) = k_0^2 V(z) = 2K^2 \text{sech}^2(Kz)$$

Thus, the potential in Equation (26.82) is

$$V(z) = 2\left(\frac{K}{k_0}\right)^2 \text{sech}^2(Kz)$$

(26.87)

Since

$$U(z) = k_0^2 V(z) = k_0^2 \left[\frac{c_0^2}{c^2(z)} - 1\right] \simeq -2\frac{k_0^2}{c_0}[c(z) - c_0]$$

then the associated sound speed field has the form

$$c(z) \simeq c_0 \left[1 - \left(\frac{K}{k_0}\right)^2 \text{sech}^2(Kz)\right]$$

(26.88)

The potential for the acoustic soliton and its sound speed field are shown in Figures 26.1 and 26.2.

26.15   Determination of the Environment from Measurements

In the special case when we can take the environment $U(r,z)$ to be a known numerical function then we can invert Equation (26.22) to find the alternative numerical function that describes the environment, $F(r,z)$ (Equation (26.24)).

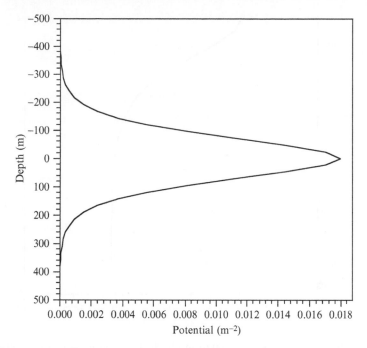

Figure 26.1 Potential function for a single range-independent acoustic soliton. Soliton wavenumber is $K = 1/100\text{m}^{-1}$, sound speed $c = 1500\text{m/s}$, source frequency $f_0 = 25\text{Hz}$. See Equation (26.87).

Once we know $F(r,z)$ we could take its derivatives with respect to r and z and then insert these into Equation (26.23) and then numerically solve for $G(r,z)$. The solution to the PE is then given by Equation (26.15). I have explained how to numerically solve the PE in this way not because I think it would be a good approach in practice, but because it gives some of the spirit of the mathematics and its potential to solve acoustics problems. In reality one would instead begin with the measured potential $U(r,z)$, numerically compute $F(r,z)$ from Equation (26.24) and then (see Chapter 23 for the nonlinear adiabatic annealing procedure) determine the environmental parameters $(\hat{F}_{mn}, \phi_n)$. Finally, the parameters for sound propagation in Equation (26.21), γ_n, would be found from the initial condition at a source, $G(0,z) \sim \psi(0,z)F(0,z)$ (Equation (26.15)), where for example $\psi(0,z)$ might be a Gaussian source or alternatively determined from measurement.

This means that to use this formulation one path to follow would be to determine the function $F(r,z)$ (or equivalently the potential $U(r,z)$) by estimating the Riemann matrix, $\hat{F}$, and phases, $\boldsymbol{\phi}$, from data, perhaps from *in situ* measurements at selected points on a grid in r, z. Alternatively, one could send UUV's around a loose grid to make density and sound speed measurements and then to construct the potential, $U(r,z)$, and the associated Riemann matrix, $\hat{F}$,

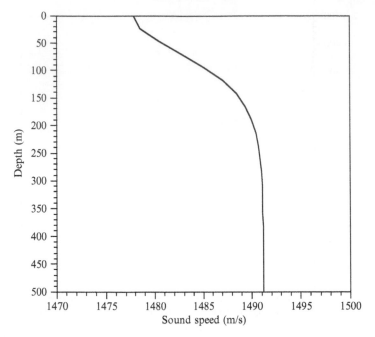

Figure 26.2 Sound speed profile (Equation (26.88)) associated with the acoustic soliton of Figure 26.1. Notice the surface duct associated with this exact solution of the PE.

and phases, ϕ. Alternatively, one might use airborne expendable bathythermo-graphs (AXBTs) to get the necessary density structure and sound speed profiles. Likewise, one could make a large number of XBTs (expendable bathythermo-graphs) or CTDs (conductivity-temperature-depth) measurements from a ship, or any alternative method that provides temperature and/or density structure as a function of depth and range. Once the environment is known (i.e., once $\hat{F}$ and ϕ are computed from the data) then continued *sparse measurements* by under-water autonomous vehicles would allow for the application of *data assimila-tion* techniques to be used to *continuously update* $\hat{F}$ and ϕ over time. This means that we would always have available in real time the environmental properties ready for application of the new methods given herein.

An alternative is to set up an experimental situation in which there are two spatially separated vertical transducer arrays. Each array provides measure-ments of both the environment and the sound field, thus providing estimates of $\hat{F}$, ϕ, and γ. These estimates can be refined by combining the information from the two arrays with numerical procedures based upon the model given above. This appears to be a very practical solution for estimating the environment.

26.16 Coherent Modes in the Acoustic Field

One of the important aspects of the present work is the surprising nature of the exact solutions of the PE (26.1) and the spectral decomposition for both the envelope of the pressure field and the potential $V(r,z)$. One of the most interesting results relates to the discovery of an *algebraic soliton solution* to the PE. This structure may be viewed as a single nonlinear Fourier *component of the potential of the PE* (26.1). I graph this structure in Figure 26.3. We see the potential $V(r,z)$ graphed vertically as a function of range, r, and depth, z. Note that the potential is related to the sound speed profiles by $k_0[c(r,z)-c_0]$ $\cong -V(r,z)$. Thus, a reversal in sign, a translation by the reference speed, c_0, and a rescaling relates the sound speed profile to the potential (we are assuming that the potential is generally small, a result which is true of the world's oceans). The contours of the potential are shown in Figure 26.4, further illustrating the

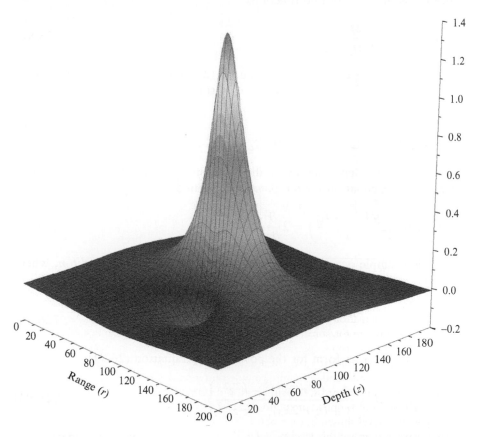

Figure 26.3 Potential function of the PE (see Equation (26.93)) for a single nonlinear, range-dependent Fourier mode. This case corresponds to an algebraic soliton for the sound wave field (not shown, see discussion with regard Equations (26.89)–(26.95)). (See color plate).

complex nature of the algebraic soliton. Finally, in Figure 26.5 I show some characteristic sound speed profiles associated with the solitonic structure. Note that the sound speed profiles near the peak of the soliton (labeled 102, 96, 93, 90) have the appearance of arctic-like sound speed profiles. Further from the peak (labels 85, 80, 70, 60) we see the conversion to profiles that are more like those observed during the summer months. Finally, in the tail region of the soliton the profiles are nearly isovelocity. Additional work will entail a detailed investigation into the full set of nonlinear Fourier modes of the PE and their influence on sound propagation.

We have already shown in Section 26.4 how nonlinearity enters in the solutions of the PE. We further saw in Section 26.5 that it is natural to write the exact solutions of the PE in terms of multidimensional Fourier series. Now let's see if we can discover, in a general fashion, how this nonlinear spectral structure arises. As already discussed earlier in this chapter let us write the solution to the PE as the ratio of two functions:

$$\psi(r,z) = \frac{G(r,z)}{F(r,z)} \tag{26.89}$$

If we substitute this expression into the PE (26.1) we find that it is natural to separate the result into two decoupled equations:

$$2FF_{zz} - 2F_z^2 - UF^2 = 0$$
$$2ik_0(FG_r - F_rG) + FG_{zz} - 2F_zG_z + F_{zz}G = 0 \tag{26.90}$$

The first equation depends only on the function $F(r,z)$ and the potential $U(r,z)$. If we solve this equation for the potential, we find

$$U(r,z) = 2\left(\frac{FF_{zz} - F_z^2}{F^2}\right) = 2\frac{\partial^2}{\partial z^2}\ln F(r,z) \tag{26.91}$$

A specific example occurs for the case of an *algebraic acoustic soliton*, where we find:

$$F(r,z) = 1 + \left(\frac{k^2r}{2k_0}\right)^2 + k^2z^2 \tag{26.92}$$

which gives the exact form for the potential via Equation (26.91):

$$U(r,z) = 4k^2\left\{\frac{1 + (k^2r/2k_0)^2 - k^2z^2}{[1 + (k^2r/2k_0)^2 + k^2z^2]^2}\right\} \tag{26.93}$$

The solution for $G(r,z)$ is then found from the second of Equation (26.90):

$$G(r,z) = 1 + \frac{k^4r^2}{4k_0^2} + k^2z^2 - 4\left[1 + i\left(\frac{k^2r}{2k_0}\right)\right] \tag{26.94}$$

And finally the symmetric part of the solution of the PE is

$$\psi(r,z) = \frac{G(r,z)}{F(r,z)} = a_0 \left[1 - \frac{4(1 + i(k^2 r/2k_0))}{1 + (k^2 r/2k_0)^2 + k^2 z^2} \right] \qquad (26.95)$$

where a_0 is an arbitrary amplitude of the modulation.

Note that by inserting Equations (26.92) and (26.93) in the first of Equation (26.90) verifies we have a solution to the latter equation. Likewise inserting Equations (26.92) and (26.94) in the second of Equation (26.90) indicates we have an exact solution of the latter equation. It goes without saying that the potential $U(r,z)$ (Equation (26.93)) and its associated eigenfunction solution (Equation (26.95)) solve the PE (26.1). We graph relevant properties of the algebraic soliton in Figure 26.3–26.5.

This example of an algebraic acoustic soliton has been given because it corresponds to a *single nonlinear Fourier mode*, together with the associated eigenvalue solution of the PE (antisymmetric about the sea surface at $z = 0$). It suffices to say that the derivation has been accomplished on the infinite r, z plane, rather than with use of theta functions (for which the Riemann matrix is 2×2). This has been done to provide the simple formulas given above.

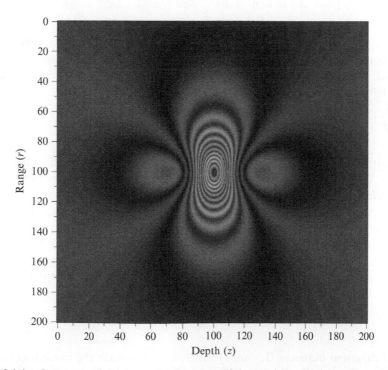

Figure 26.4 Contours of the potential function of the PE for a single nonlinear Fourier mode as shown in the figure. (See color plate).

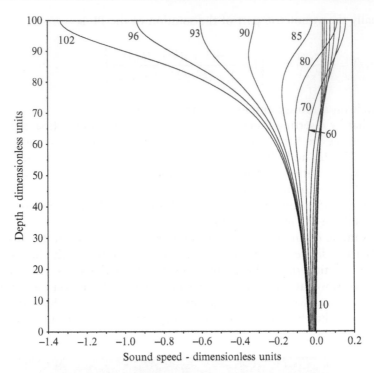

Figure 26.5   Sound speed profiles for the algebraic soliton solution of the PE as shown in Figure 26.3. Dimensionless, shifted units are used for sound speed. Graphed is $k_0[c(r,z)-c_0]$. The curves are labeled by dimensionless down range distance, 0 is the reference range, 102 is the largest.

26.17   Shadow Zone Analysis

It is worthwhile discussing briefly the nonlinear filtering method in a few particulars. When the filtering algorithm is applied it is important to note that we are *simultaneously removing* certain information (1) in the sound speed profile and (2) in the solution to the PE, while leaving other information untouched. This means that the *Riemann matrix* must be separated into two parts, one that we want to keep and the other which should be removed in the filtering process. We can think of dividing the Riemann matrices into two submatrices, B_1, B_2, containing the sum total of diagonal elements of the original:

$$B = \begin{bmatrix} B_1 & B_{12} \\ B_{12} & B_2 \end{bmatrix}$$

The off-diagonal matrices B_{12} are symmetric and contain the nonlinear interaction terms coupling B_1, B_2. Formally speaking, this filtering process means that we have two sound speed profiles, say $c_1(r,z)$ and $c_2(r,z)$ and two solutions to the PE, $\psi_1(r,z)$ and $\psi_2(r,z)$, that, thanks to the nonlinear nature of the filtering process, can be written:

$$c(r,z) = c_1(r,z) + c_2(r,z) + c_{int}(r,z)$$
$$\psi(r,z) = \psi_1(r,z) + \psi_2(r,z) + \psi_{int}(r,z)$$

Each of $c_1(r,z)$ and $\psi_1(r,z)$ have the Riemann matrix $\mathbf{B}_1$, while $c_2(r,z)$, $\psi_2(r,z)$ have $\mathbf{B}_2$. The interaction terms, $U_{int}(r,z)$, $\psi_{int}(r,z)$ are related to $\mathbf{B}_{12}$. We now address application of these ideas to some very simple aspects of the study of shadow zones.

We have considered a range-independent case where the sound speed profile has a surface duct as seen in Figure 26.6. We take the source to lie at 40 m below the surface assuming a Gaussian form with frequency $f_0 = 800$Hz. We see the transmission loss as a function of depth and range in Figure 26.7. Note that the rays either remain in the surface duct (within about 150 m of the surface), or they bend toward bottom reflecting paths. The region bounded roughly by the depth of the surface duct, 150 m, and the region to the right of a range of about 5 km is referred to as a shadow zone because the level of sound there is very much lower than in the other regions. Of course there is some level of sound in the shadow zone, as can be seen in Figure 26.7. To give some perspective about the physics of Figure 26.8, let us look at the isovelocity

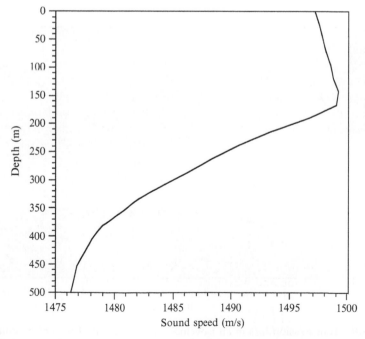

Figure 26.6 Sound speed profile with a surface duct in the upper 150 m of the water column.

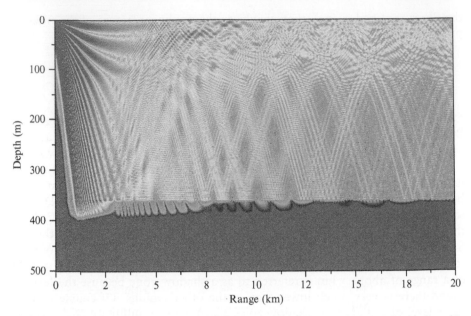

Figure 26.7 Transmission loss in an environment described by the sound speed profile given in the figure. The central sound frequency is 800 Hz. A simple constant profile bottom and bottom sponge are used. (See color plate).

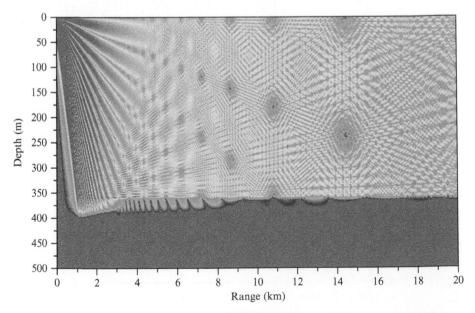

Figure 26.8 Transmission loss in an isovelocity environment. The central sound frequency is 800 Hz. A simple constant profile bottom and bottom sponge are used. (See color plate).

case, shown in Figure 26.8. Here the effect of a surface duct no longer plays a role and the ray paths have no preferential directions and can be seen to reflect off the bottom causing an interference pattern. While some zones are low in sound, the overall picture is that the entire region is nicely insonified.

Let us now use the notion of nonlinear filtering to investigate how one might go about "seeing" into the shadow zone. We filtered components from the Riemann matrix to render the sound speed profile approximately isovelocity in the upper 150 m, thus removing the surface duct. We then computed the linear Fourier coefficients of the starting field $\psi(r = 0, z)$ and then we integrated the field according to the exact solution of the PE. The resultant transmission loss is shown in Figure 26.9. We find, by comparison to Figure 26.8, that the new sound speed profile insonifies the shadow zone slightly better. Nevertheless, a "duct" still appears, even though it is somewhat deeper in this case.

We considered another case where the original sound speed profile is filtered to provide a new profile that *increases* from 150 m up to the surface. This "forward leaning" profile gives the transmission loss shown in Figure 26.10. Again, some improvement is noted in the insonization of the shadow zone.

These results suggest that at the present time it is an art to choose the best Riemannian filter to give the best insonization of a shadow zone. Nevertheless, it is important to note that filtering the Riemann matrix allows one to tailor the sound speed environment, by extracting a particular sound speed profile which best insonifies a particular region.

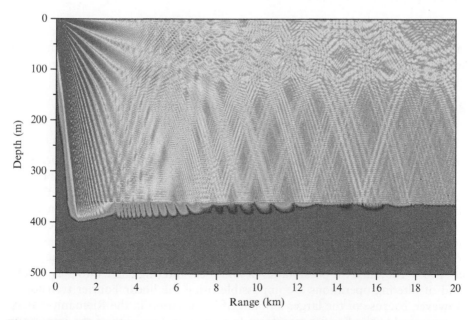

Figure 26.9 Transmission loss in environment similar to Figure 26.8 except that the profile is filtered to be approximately isovelocity within 150 m of the surface. The central sound frequency is 800 Hz. A simple constant profile bottom and bottom sponge are used. (See color plate).

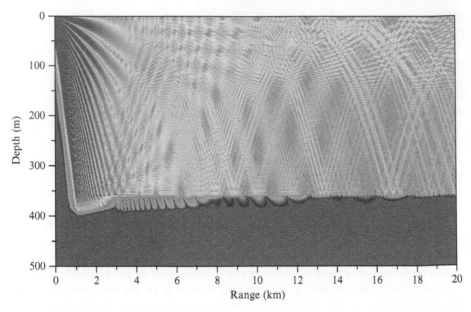

Figure 26.10 Transmission loss in environment similar to Figure 26.8 except that the sound speed profile is filtered to give a linear sound intensity from 150 m up to the surface. The central sound frequency is 800 Hz. A simple constant profile bottom and bottom sponge are used. (See color plate).

Filtering the actual nonlinear Fourier components in the Riemann spectrum, rather than the sound speed profiles themselves, is discussed briefly in the next Section.

26.18 Application to Unmanned, Untethered, Submersible Vehicles

Of the many possible applications there are those that can arise from processing of partial information from the environment (data assimilation) that is obtained by local measurements made during the movements of an unmanned, untethered, submersible vehicle (UUSV). This information could prove very useful in assessing the local environment and its residents. How can we arrive at this conclusion? The reason is quite simple. Once the average local information has been condensed into the form of a Riemann matrix of some reasonable dimension, say 100×100, then the possibilities of nonlinear filtering come into play. These kind of filtering operations are impossible using the linear Fourier transform. However, because of the larger amount of information in the Riemann matrix as compared to the Fourier transform (a vector) and because of the important physical complexity in the Riemann theta functions, one can expect to be able to address filtering applications that cannot be addressed by simpler methods.

The idea, for example, of removing acoustic modes which block the possibility of seeing into certain zones is a laudable goal.

How are such filtering operations to be carried out? Since the Riemann matrix contains the acoustic modes on the diagonal, we can focus on these elements to pick out the particular modes we seek to filter. Nonlinear interactions are contained in the off-diagonal elements and are addressed separately. There are two kinds of modes, the first is viewed as a kind of "unstable mode" in the theory and these latter contain the solitonic modes discussed earlier. Modes of this type have a local Riemann matrix that is 2×2, where the off-diagonal part provides the nonlinear coupling necessary to keep the mode "coherent" (localized in r, z space). The second kind of acoustic mode is more familiar, that is, it is nearly linear, being a quasi-sine mode. These modes require only one element on the diagonal of the Riemann matrix.

Use of the nonlinear filtering approach can be carried out on successive passes and, as the vehicle obtains additional information over time, this information can be incorporated into an update of the environmental information that is contained in the Riemann matrix. It is conceivable that a UUSV could thus carry out a greatly improved estimation of its environment in this way (Figure 26.11).

As seen in the previous sections there are now available, thanks to the methods discussed herein, a whole new body of approaches for the study of ocean acoustics. From the previous section it is now clear that the process of nonlinear filtering using the Riemann matrix is still in the beginning and the "selection" of filters and appropriate sound speed profiles to aid the process of "seeing" into shadow zones is still in its infancy. Nevertheless, we can foresee the day when software and methods will have been refined to the point where applications will become possible in a number of areas, including onboard processing of data in UUSVs.

26.19 Application to Communications, Imaging, and Encryption

Time reversal mirror technology has many potential applications. One of the main advantages of this approach is that it does, at first order, not require detailed knowledge of the oceanic environment. Nevertheless, our own perspective is that the processing of the available information at both a vertical-receive array (VRA) and a source-receive array (SRA) could be desirable. Formally speaking, the time series obtained at these two spatial locations is enough to construct all of the information necessary to compute the Riemann theta functions, and hence the solution of the PE. One computes specifically the Riemann matrix, the wavenumbers and the phases. Therefore, one can construct all the information necessary for a full understanding of the acoustic dynamics both during the forward and reversed phases of the motion.

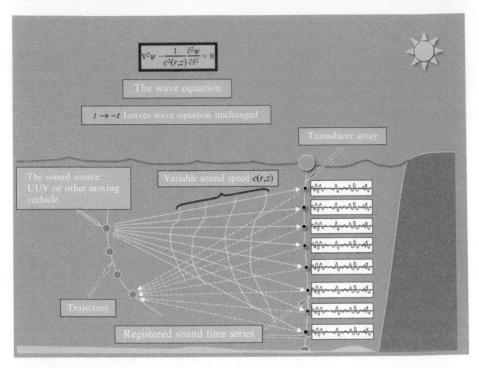

Figure 26.11 A UUSV moves along a trajectory providing a sound source for time reversal symmetry. After an initial burst of sound the UUV moves silently, but its trajectory is followed as a reflected sound source. Distant communications are maintained by sonar updating of the UUSV trajectory and modification of the IST phases of the exact solution of the PE. (See color plate).

At this stage we can address two aspects of the problem that may be useful. The first is to refine the return signal by nonlinear filtering techniques and the second is to "attach" additional, perhaps encrypted, information to the signal via amplitude modulation of the "carrier" of the arriving and retransmitted signals. Such approaches would also benefit by applying the ideas of nonlinear filtering to the Riemann theta functions to bring out cleaner signals in the presence of noise.

It seems clear that the present approach, particularly in situations where the oceanic environment is extremely complicated such as up on the continental shelf, can also be applied to underwater imaging. This is because of the possibility of using the complex spectral structure of Riemann theta function to construct acoustic lenses that can be a consequence of the nonlinear filtering procedure. This idea might be useful for removing background noise and focusing on those parts of the Riemann matrix that have the information required to construct and enhance a particular image.

Nota Bene: The material in this section is particularly suited for *acoustic communications* using the PE. In this scenario we have a range-decaying acoustic carrier wave whose amplitude is modulated by the complex solution of the PE (26.4). We thus have a *nonlinear modulation theory for AM communications using acoustic waves*. The broadcast radio antenna is a loudspeaker and the receiver antenna is a microphone. The broadcast acoustic carrier corresponds to the frequency of the "radio station." The message or "voice communication" is put into the complex envelope (solution of the PE), a narrow banded oscillation about the carrier. To communicate with the *linear modulation theory* of AM radio you need the ordinary linear Fourier transform (Beckmann, 1967). To communicate with *nonlinear* modulation theory (the kind that governs acoustic waves described herein), you need the IST. Many of the methods of this monograph show how ordinary linear Fourier series can be used to compute Riemann theta functions and hence to do the necessary operations for understanding physics, analyzing data, and hyperfast modeling of nonlinear systems. The same tools can be used for communications with oceanic acoustic waves. Note that linear communications theory often uses the concept of random phasor sums. Equations (26.53b) for $A_n(r), D_n(r)$ generalize the concept of random phasor sums for nonlinear communications.

Nota Bene: It is worthwhile noting that *encryption* could also be made an integral part of *nonlinear communications*. This is because ordinary encryption is based upon the algebra of prime numbers. This algebra is based upon the Riemann zeta function that is the Mellin transform of the one-dimensional theta function. There exists an analogous N-dimensional zeta function that is the N-dimensional Mellin transform of the N-dimensional Riemann theta function. Thus, "almost impossible to break" encryption would be feasible and would provide possible added value to the method. Why "almost impossible"? Because without a quantum computer (which do not yet exist) I do not foresee modern computers that could break an N-dimensional encryption scheme: Estimated computer times would require $\sim 10^{100}$ operations, more than say $\sim 10^{70}$ Universal lifetimes on the fastest existing computers and for the even faster computation capability which might be developed over the next few decades.

Appendix: Products of Fourier Series

We look momentarily at how to form the product of two Fourier series, say
$W(x) = U(x)V(x)$, where

$$U(x) = \sum_{n=-\infty}^{\infty} U_n e^{ik_n x}, \quad V(x) = \sum_{n=-\infty}^{\infty} V_n e^{ik_n x}$$

A useful reference is Zygmund (1959). We then have

$$W(x) = U(x)V(x) = \sum_{m=-\infty}^{\infty} \sum_{j=-\infty}^{\infty} U_m V_j e^{i(k_m + k_j)x}$$

Note that $k_n = 2\pi n/L$, where L is the length of the periodic box for the Fourier
transform. Then set $k_m + k_j = k_n$ and note that this is equivalent to $m+j=n$.
Substitute $j = n-m$ into the above equation and get

$$W(x) = U(x)V(x) = \sum_{m=-\infty}^{\infty} \sum_{n-m=-\infty}^{\infty} U_m V_{n-m} e^{ik_n x} = \sum_{m=-\infty}^{\infty} \sum_{n=-\infty}^{\infty} U_m V_{n-m} e^{ik_n x}$$

$$= \sum_{n=-\infty}^{\infty} \left[\sum_{m=-\infty}^{\infty} U_m V_{n-m} \right] e^{ik_n x} = \sum_{n=-\infty}^{\infty} W_n e^{ik_n x}$$

Thus, *the product $U(x)V(x)$ of two Fourier series is also a Fourier series* whose
coefficients

$$W_n = \sum_{m=-\infty}^{\infty} U_m V_{n-m}$$

are the *convolution of the Fourier coefficients* of the two functions $U(x), V(x)$.
   Now let us look at derivatives of products and products of derivatives of
Fourier series. To do this we consider the second derivative the product
$W(x) = U(x)V(x)$:

$$W'' = (UV)'' = UV'' + 2U'V' + U''V$$

The primes refer to derivatives with respect to the spatial variable, x. We would
like to substitute the Fourier series for $U(x), V(x)$ (and their derivatives) into the
expression above and verify that we have solution. Note that

$$U' = i \sum_{n=-\infty}^{\infty} k_n U_n e^{ik_n x}, \quad U'' = -\sum_{n=-\infty}^{\infty} k_n^2 U_n e^{ik_n x}$$

$$V' = i \sum_{n=-\infty}^{\infty} k_n V_n e^{ik_n x}, \quad V'' = -\sum_{n=-\infty}^{\infty} k_n^2 V_n e^{ik_n x}$$

$$W' = i \sum_{n=-\infty}^{\infty} k_n W_n e^{ik_n x}, \quad W'' = -\sum_{n=-\infty}^{\infty} k_n^2 W_n e^{ik_n x}$$

Now let us substitute these series into $W'' = (UV)'' = UV'' + 2U'V' + U''V$, using the convolution theorem above. We have

$$(UV)'' = -\sum_{n=-\infty}^{\infty}\left[\sum_{m=-\infty}^{\infty} k_n^2 U_m V_{n-m}\right] e^{ik_n x}$$

$$UV'' = -\sum_{n=-\infty}^{\infty}\left[\sum_{m=-\infty}^{\infty} k_{n-m}^2 U_m V_{n-m}\right] e^{ik_n x}$$

$$U'V' = -\sum_{n=-\infty}^{\infty}\left[\sum_{m=-\infty}^{\infty} k_m k_{n-m} U_m V_{n-m}\right] e^{ik_n x}$$

$$U''V = -\sum_{n=-\infty}^{\infty}\left[\sum_{m=-\infty}^{\infty} k_m^2 U_m V_{n-m}\right] e^{ik_n x}$$

Substitution into $(UV)'' = UV'' + 2U'V' + U''V$ gives

$$\sum_{n=-\infty}^{\infty}\left[\sum_{m=-\infty}^{\infty} k_n^2 U_m V_{n-m}\right] e^{ik_n x}$$

$$= \sum_{n=-\infty}^{\infty}\left[\sum_{m=-\infty}^{\infty} k_{n-m}^2 U_m V_{n-m} + 2\sum_{m=-\infty}^{\infty} k_m k_{n-m} U_m V_{n-m} + \sum_{m=-\infty}^{\infty} k_m^2 U_m V_{n-m}\right] e^{ik_n x}$$

Take the inverse Fourier transform:

$$\sum_{m=-\infty}^{\infty} (k_n^2 - k_{n-m}^2 - 2k_m k_{n-m} - k_m^2) U_m V_{n-m} = 0$$

Finally, we can write this as

$$\sum_{m=-\infty}^{\infty} A_{mn} U_m V_{n-m} = 0 \quad \text{where} \quad A_{mn} = k_n^2 - k_{n-m}^2 - 2k_m k_{n-m} - k_m^2$$

The above equation must be true provided $A_{mn} = 0$ for all m, n. Thus,

$$k_n^2 = k_{n-m}^2 + 2k_m k_{n-m} + k_m^2 \quad \text{or} \quad n^2 = (n-m)^2 + 2m(n-m) + m^2$$

(where we have used $k_n = 2\pi n/L$) are the conditions which guarantee $A_{mn} = 0$. We can verify this identically with simple algebra:

$$A_{mn} = n^2 - (n-m)^2 - 2m(n-m) - m^2 = 0$$

Of course this result assumes that the formal products in the above formulation exist. This is true provided that the coefficients of their Fourier series decay as $n \to \infty$.

Clearly, the following relation hold for solving the bilinear form for the function G:

$$FG_r = \sum_{m=-\infty}^{\infty} \sum_{n=-\infty}^{\infty} F_m(r) \frac{dG_n(r)}{dr} e^{i(k_m+k_n)x} = \sum_{n=-\infty}^{\infty} \left[\sum_{m=-\infty}^{\infty} F_{n-m}(r) \frac{dG_m(r)}{dr} \right] e^{ik_n x}$$

$$F_r G = \sum_{m=-\infty}^{\infty} \sum_{n=-\infty}^{\infty} \frac{dF_m(r)}{dr} G_n(r) e^{i(k_m+k_n)x} = \sum_{n=-\infty}^{\infty} \left[\sum_{m=-\infty}^{\infty} \frac{dF_{n-m}(r)}{dr} G_m(r) \right] e^{ik_n x}$$

$$F_x G_x = - \sum_{m=-\infty}^{\infty} \sum_{n=-\infty}^{\infty} k_m k_n F_m(r) G_n(r) e^{i(k_m+k_n)x} = - \sum_{n=-\infty}^{\infty} \left[\sum_{m=-\infty}^{\infty} k_m k_{n-m} F_{n-m}(r) G_m(r) \right] e^{ik_n x}$$

$$F_{xx} G = - \sum_{m=-\infty}^{\infty} \sum_{n=-\infty}^{\infty} k_m^2 F_m(r) G_n(r) e^{i(k_m+k_n)x} = - \sum_{n=-\infty}^{\infty} \left[\sum_{m=-\infty}^{\infty} k_{n-m}^2 F_{n-m}(r) G_m(r) \right] e^{ik_n x}$$

$$FG_{xx} = - \sum_{m=-\infty}^{\infty} \sum_{n=-\infty}^{\infty} k_n^2 F_m(r) G_n(r) e^{i(k_m+k_n)x} = - \sum_{n=-\infty}^{\infty} \left[\sum_{m=-\infty}^{\infty} k_m^2 F_{n-m}(r) G_m(r) \right] e^{ik_n x}$$

27 Planar Vortex Dynamics

27.1 Introduction

Vortex dynamics play an important role in the study of turbulent fluid motions (Batchelor, 1967; Lesieur, 1990; Saffman, 1992). Indeed the study of *coherent vortices* in two dimensions is undergoing a resurgence because knowledge of particular vortex solutions in various special cases is often important for understanding the coherent component of turbulent flows. Applications of coherent vortices include those occurring in rotating and stratified fluids (Hopfinger and Van Heijst, 1993), in geophysical flows (Friedlander, 1980; Flierl, 1987; Cushman-Roisin, 1994; Pedlosky, 1998, 2003; Solomon, 1998; McWilliams, 2006; Vallis, 2006), and fluid layers with electromagnetic external forces (Marteau et al., 1995).

The importance of atmospheric and oceanic eddy formation in all aspects of their fluid dynamical motions and in climate studies is irrefutable. Figure 27.1 shows an example of ocean eddy formation near the Gulf Stream off the United States east coast. It goes without saying that the eddy dynamics clearly play a significant role in the geophysical fluid dynamical turbulence of oceanic processes. Additional examples are shown in Figure 27.2 (turbulence in the Jovian atmosphere) and in Figure 27.3 (hurricane Bill off the coast of Florida).

This chapter provides an overview of some of the work on *vortex dynamics* from the point of view of *soliton theory*. Among the important questions to ask are: What is the relationship, if any, of a vortex to a soliton? In particular, soliton interactions are *elastic*, that is, their properties, such as the amplitude and phase speed, are conserved under interactions. However, vortex interactions are typically *inelastic*, that is, vortex properties are often not preserved during collisions with each other, for example, they can merge. Early perspective on this problem was provided by Zabusky in his classic paper on *vortons* or *V-states* (Deem and Zabusky, 1978). *Vortex patches* consist of a connected region of finite area containing uniform vorticity surrounded by an irrotational fluid (Saffman, 1992). Indeed the stability of these patches is an important and nontrivial area of research (Dritschel, 1990; Dhanak and Marshall, 1993).

A relevant modern perspective on this problem relates to the fact that solitons are known to be solutions of *integrable wave equations*, while vortices are typically solutions of *nonintegrable equations*. Integrability in this case

Doi: 10.1016/S0074-6142(10)97027-7

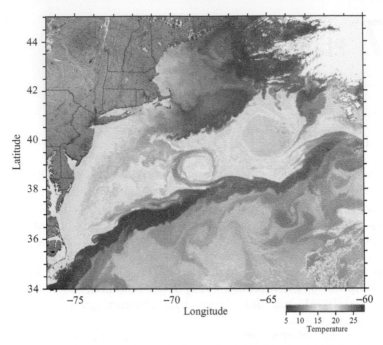

Figure 27.1 AVHRR (advanced very high resolution radiometer) image taken aboard a NOAA satellite on June 11, 1997. The color-coding indicates temperature (see color bar) where the yellow and orange colors indicate warmer water and blue colors indicate cooler water. Note the two large warm water eddies to the north of the Gulf Stream, seen as a warm front (dark red) on the image. Below and within the front are shown the complex geophysical fluid dynamical turbulence typical of oceanic processes. (See color plate).

relates to the use of the inverse scattering transform to solve the associated wave equation. However, there are a number of integrable cases describing vortices and in these cases the most typical scenario is that vortices correspond to *two solitons in a bound state, that is, two solitons that are phase locked with each other.*

This chapter investigates vortex dynamics in terms of the inverse scattering transform and Riemann theta functions. The dynamics are reduced to planar, inviscid motion and two cases are investigated: (1) the Navier-Stokes equations and (2) the two-dimensional Schrödinger equation. Both cases reduce to the classical *Poisson equation for particular functionals* (Batchelor, 1967; Saffman, 1992). In the present chapter the vortex is first found to be amenable to the Hirota method for particular *integrable functionals*, for which we discuss how a single vortex can be described as two solitons that are phase-locked with each other. Extension to the periodic problem generalizes this to vortex dynamics described by two phase-locked nonlinear modes characterized by

Figure 27.2   Image of the Great Jupiter Red spot taken February 25, 1979 by Voyager 1. The image was taken at a distance of 9.2 million km from Jupiter. Note the complex wave dynamics to the left of the red spot, which is more that 300 years old and is larger than the Earth. Geophysical fluid dynamics on Jupiter is more energetic than on the Earth due to the large size of the Jovian planet and to its fast rotation (courtesy NASA). (See color plate).

Figure 27.3   Hurricane Bill off the coast of Florida in August 2009. The hurricane was category 4 at the time of this image taken from a satellite of the National Oceanic and Atmospheric Administration (NOAA). (See color plate).

a 2×2 Riemann matrix. The vortex dynamics in the plane is analogous to unstable nonlinear Schrödinger equation (NLS) mode dynamics on the line. Indeed "rogue waves" are in this sense analogous to "vortices" as they both have similar 2×2 Riemann matrices.

27.2   Derivation of the Poisson Equation for Vortex Dynamics in the Plane

We begin with the *Navier-Stokes (the momentum) equation*:

$$\frac{d\mathbf{u}}{dt} = \frac{\partial \mathbf{u}}{\partial t} + \mathbf{u}\cdot\nabla\mathbf{u} = -\frac{\nabla p}{\rho} + \nabla\phi + \frac{v}{\rho}\nabla^2\mathbf{u} + \frac{\mathbf{F}_{ext}}{\rho} \tag{27.1}$$

and the *continuity equation*:

$$\frac{\partial \rho}{\partial t} + \nabla\cdot(\rho\mathbf{u}) = 0 \tag{27.2}$$

Here $p = p(x,y,z,t)$ is the pressure, $\rho = \rho(x,y,z,t)$ is the density, v is the molecular viscosity, $\mathbf{u} = \mathbf{u}(x,y,z,t)$ is the three-dimensional velocity field, $\phi = \phi(x,y,z,t)$ is an external conservative potential (e.g., the gravitational potential), and $\mathbf{F}_{ext} = \mathbf{F}_{ext}(x,y,z,t)$ is an external force.

Now let us find the *vorticity equation* using the following vector identity:

$$\boldsymbol{\omega} \times \mathbf{u} = (\mathbf{u}\cdot\nabla)\mathbf{u} - \frac{1}{2}\nabla|\mathbf{u}|^2 \tag{27.3}$$

Then the momentum equation becomes:

$$\frac{\partial \mathbf{u}}{\partial t} + \boldsymbol{\omega} \times \mathbf{u} = -\frac{\nabla p}{\rho} + \nabla\left(\phi - \frac{1}{2}|\mathbf{u}|^2\right) + \frac{v}{\rho}\nabla^2\mathbf{u} + \frac{\mathbf{F}_{ext}}{\rho}$$

Take the curl of the above equation to get:

$$\frac{\partial \nabla \times \mathbf{u}}{\partial t} + \nabla \times (\boldsymbol{\omega} \times \mathbf{u}) = \frac{\nabla\rho \times \nabla p}{\rho^2} + \mu\nabla^2(\nabla \times \mathbf{u}) + \nabla \times \frac{\mathbf{F}_{ext}}{\rho} \tag{27.4}$$

Here we have set $\mu = v/\rho$, the *kinematic viscosity*. Now use the *definition of vorticity*

$$\boldsymbol{\omega} = \nabla \times \mathbf{u}$$

to find

$$\frac{\partial \boldsymbol{\omega}}{\partial t} + \nabla \times (\boldsymbol{\omega} \times \mathbf{u}) = \frac{\nabla\rho \times \nabla p}{\rho^2} + \mu\nabla^2\boldsymbol{\omega} + \nabla \times \frac{\mathbf{F}_{ext}}{\rho}$$

To complete the derivation of the vorticity equation, use the vector identity:

$$\nabla \times (\mathbf{A} \times \mathbf{B}) = \mathbf{A}(\nabla\cdot\mathbf{B}) + (\mathbf{B}\cdot\nabla)\mathbf{A} - \mathbf{B}(\nabla\cdot\mathbf{A}) - (\mathbf{A}\cdot\nabla)\mathbf{B}$$

Then

$$\nabla \times (\boldsymbol{\omega} \times \mathbf{u}) = \boldsymbol{\omega}(\nabla \cdot \mathbf{u}) + (\mathbf{u} \cdot \nabla)\boldsymbol{\omega} - (\boldsymbol{\omega} \cdot \nabla)\mathbf{u}$$

since $\nabla \cdot \boldsymbol{\omega} = \nabla \cdot \nabla \times \mathbf{u} = 0$ (the divergence of any curl is zero). We then have

$$\underbrace{\frac{\partial \boldsymbol{\omega}}{\partial t} + \mathbf{u} \cdot \nabla \boldsymbol{\omega}}_{\dfrac{d\boldsymbol{\omega}}{dt}} + \boldsymbol{\omega}(\nabla \cdot \mathbf{u}) - (\boldsymbol{\omega} \cdot \nabla)\mathbf{u} = \frac{\nabla \rho \times \nabla p}{\rho^2} + \mu \nabla^2 \boldsymbol{\omega} + \nabla \times \frac{\mathbf{F}_{\text{ext}}}{\rho}$$

or

$$\frac{d\boldsymbol{\omega}}{dt} = \boldsymbol{\omega} \cdot \nabla \mathbf{u} - \boldsymbol{\omega}(\nabla \cdot \mathbf{u}) + \frac{\nabla \rho \times \nabla p}{\rho^2} + \mu \nabla^2 \boldsymbol{\omega} + \nabla \times \frac{\mathbf{F}_{\text{ext}}}{\rho} \quad \text{(3D Vorticity Equation)}$$

$$(27.5)$$

This is the *vorticity equation* that we have sought. It is fully functional in three dimensions for variable (dynamic) $\rho = \rho(\mathbf{x}, t)$, finite kinematic viscosity μ, external forcing $\mathbf{F}_{\text{ext}}$, and $\mathbf{x} = [x, y, z]$.

Now let us go to *planar flow*, such that the velocity vector has no z component, $\mathbf{u} = [u_x, u_y, 0]$, and the vorticity vector has only a z component, $\boldsymbol{\omega} = [0, 0, \omega]$. We further assume incompressible flow so that $\rho = 1$, hence $\nabla \cdot \mathbf{u} = 0$. Then the vorticity equation in the plane becomes:

$$\frac{d\boldsymbol{\omega}}{dt} = \mu \nabla^2 \boldsymbol{\omega} + \nabla \times \frac{\mathbf{F}_{\text{ext}}}{\rho}$$

Finally

$$\frac{d\omega}{dt} = \frac{\partial \omega}{\partial t} + \mathbf{u} \cdot \nabla \omega = \mu \nabla^2 \omega + G_{\text{ext}} \quad \text{(Vorticity Equation in Planar Flow)} \quad (27.6)$$

where

$$G_{\text{ext}} \hat{\mathbf{k}} = \nabla \times \mathbf{F}_{\text{ext}}$$

where $\hat{\mathbf{k}} = [0, 0, 1]$. Now for two-dimensional, incompressible fluid motion one can introduce the *stream function* for the velocity field (two dimensional in the x, y plane) and vorticity (one dimensional in the z direction):

$$\mathbf{u} = \mathbf{k} \times \nabla \psi = [\psi_y, -\psi_x, 0]$$

$$\boldsymbol{\omega} = [0, 0, -\nabla^2 \psi]$$

Using these in the planar vorticity equation (27.6) we have

$$\frac{d\nabla^2\psi}{dt} = \frac{\partial\nabla^2\psi}{\partial t} + [\psi_y, -\psi_x]\cdot\left(\frac{\partial}{\partial x}\nabla^2\psi, \frac{\partial}{\partial y}\nabla^2\psi\right) = \nu\nabla^4\psi + G_{ext}$$

so that

$$\frac{d\nabla^2\psi}{dt} = \frac{\partial\nabla^2\psi}{\partial t} + J(\nabla^2\psi, \psi) = \nu\nabla^4\psi + G_{ext} \tag{27.7}$$

For two-dimensional, incompressible fluid motion one has the equation for the *stream function* $\psi(x,y,t)$:

$$\frac{\partial\nabla^2\psi}{\partial t} + J(\nabla^2\psi, \psi) = \nu\nabla^4\psi + F_{ext} \quad \text{(Planar Vortex Dynamics)} \tag{27.8}$$

This expression is obtained directly from the Navier-Stokes equations by eliminating the pressure term, as seen above. Note also that $J(f,g) = f_x g_y - f_y g_x$ is the *Jacobian* or *Poisson bracket*.

When viscosity and external forcing are absent this expression reduces to:

$$\frac{\partial\nabla^2\psi}{\partial t} + J(\nabla^2\psi, \psi) = 0 \tag{27.9a}$$

For a stationary or steady flow the time derivative disappears and we have

$$J(\nabla^2\psi, \psi) = 0$$

This is simply a statement that the vorticity $\omega = -\nabla^2\psi$ is constant along the contours of the *stream function* $\psi(x,y)$. If we write

$$\nabla^2\psi = F(\psi(x,y))$$

where $F(\psi(x,y,t))$ is a functional of the stream function, this then implies that

$$\begin{aligned}J(\nabla^2\psi, \psi) &= J(F(\psi), \psi) = F_x(\psi)\psi_y - F_y(\psi)\psi_x \\ &= F'\psi_x\psi_y - F'\psi_y\psi_x = F'(\psi_x\psi_y - \psi_y\psi_x) \\ &= 0\end{aligned}$$

where $F_x(\psi) = F'\psi_x$ and $F_y(\psi) = F'\psi_y$ for $F' = \partial F / \partial\psi$.

Therefore, we have the Poisson equation

$$\nabla^2\psi = F(\psi) \quad \text{(Poisson Equation)} \tag{27.10}$$

well known to provide the means to study *planar, steady-state vortex flows* of an *ideal incompressible liquid*. The analytic behavior (of Equation (27.10)) can be studied by the Hirota N-soliton solutions and for periodic boundary conditions using the Riemann theta function, as discussed below.

27.3 Poisson Equation for Schrödinger Dynamics in the Plane

Is there a regime in which we can study the Schrödinger equation using soliton methods? Here is the well-known form of the Schrödinger equation in the plane:

$$i\psi_t + \nabla^2\psi + U(x, y, t)\psi = 0 \quad \text{(Schrödinger Equation)} \tag{27.11}$$

where for many purposes it is appropriate to take

$$U(x, y, t) = -\frac{F(\psi)}{\psi} \quad \text{(Potential as Functional)} \tag{27.12}$$

This expression connects the potential with the functional. Note that the potential can include problems with external shear, bathymetry, and nonuniform boundary conditions (e.g., see Chapters 2, 26, and 33). The advantage of the Schrödinger equation is that it puts time evolution into the vortex interactions for problems described by Equation (27.11). The stationary case is just the time-independent Poisson equation:

$$\nabla^2\psi + U(x, y)\psi = 0 \tag{27.10}$$

For the particular case for $U(x,y,t) = 2|\psi|^2 + U_{ext}(x,y,t)$ one has for example Bose-Einstein condensation; $U_{ext}(x,y,t)$ is an external confining potential. For oceanic surface wave applications the external potential can include external forcing such as the wind, dissipation, and bathymetric effects. In this case, however, the Laplacian operator becomes *hyperbolic*: $\nabla^2\psi \Rightarrow \psi_{xx} - 2\psi_{yy}$ (Yuen and Lake, 1982).

Cases for further study are given by Stuart (1967), Pasmanter (1994), Dauxois et al. (1995).

27.4 Specific Cases of the Poisson Equation for Vortex Dynamics in the Plane

The stationary Schrödinger equation for two-dimensional flows is given by Equation (27.12). Here are some interesting special cases:
The Helmholtz Equation: $U(x,y) = 1$

$$\nabla^2\psi = -\psi \tag{27.13}$$

Kolmogorov flows, cellular structures with square and hexagonal cells and quasicrystal patterns have been found as solutions (Zaslavsky et al., 1991).

The Sinh-Poisson Equation: $U(x,y) = [(1 - \varepsilon^2)/2]\sinh(2\psi)/\psi$

$$\nabla^2\psi = -\frac{1 - \varepsilon^2}{2}\sinh(2\psi) \qquad (27.14)$$

The Kelvin-Stuart Cat's Eyes: $U(x,y) = -(1 - \varepsilon^2)e^{-2\psi}/\psi$

$$\nabla^2\psi = (1 - \varepsilon^2)e^{-2\psi} \qquad (27.15)$$

The Fourth Case: $U(x,y) = -(A\,e^{2\psi} + B\,e^{-2\psi})/\psi$

$$\nabla^2\psi = A\,e^{2\psi} + B\,e^{-2\psi} \qquad (27.16)$$

The solutions of all of the above cases are given in Dauxois et al. (1995) (see also Stuart, 1967). Of course the integral sine-Gordon (or sine-Poisson) equation, $u_{tt} - u_{xx} + \sin u = 0$, is also related to the sinh-Poisson equation by a simple transformation (see below and Forest and McLaughlin, 1983).

27.5 Geophysical Fluid Dynamics

The potential vorticity equation (PVE) (see Pedlosky, 1979 and cited references) offers a vast, rich source of nonlinear phenomena to be studied in the context of large and meso-scale motions in the ocean and the atmosphere:

$$\frac{\partial}{\partial t}(\nabla^2\psi - F\psi) + J(\psi, \nabla^2\psi) + \beta\psi_x = 0 \qquad (27.17)$$

The equation as written here is in the β-plane, which is tangent to the earth's surface at some latitude θ; ψ is the stream function, $F = (L / R)^2$, $R = \sqrt{gD}/f$ (Rossby deformation radius), $f = 2\Omega\sin\theta$ (Coriolis parameter), $\beta = \beta_0 L^2/U$ (β-plane parameter), $\beta_0 = (2\Omega/r_E)\cos\theta$. Here r_E is the radius of the earth, g is the acceleration of gravity, L is a typical length scale, U is a typical velocity scale, D is the water depth, and Ω is the rotation rate of the earth. The Jacobian of the stream function and its Laplacian are given by

$$J(A, B) = \frac{\partial A}{\partial x}\frac{\partial B}{\partial y} - \frac{\partial A}{\partial y}\frac{\partial B}{\partial x} \qquad (27.18)$$

The stream function $\psi(x,y,t)$ and the sea surface elevation $\eta(x,y,t)$ are related by the relation:

$$\eta(x, y, t) = \psi(x, y, t) \qquad (27.19)$$

The horizontal velocity field due to the presence of the dynamics of $\psi(x,y,t)$ (or that of the surface elevation $\eta(x,y,t)$) is given by

$$
u = -\frac{\partial \psi}{\partial y}
$$

$$
v = \frac{\partial \psi}{\partial x} \tag{27.20}
$$

Lagrangian fluid particle motions (i.e., the motion of passive tracer particles placed in the flow) have the requisite (Hamiltonian) equations of motion:

$$
\dot{x} = -\frac{\partial \psi}{\partial y}
$$

$$
\dot{y} = \frac{\partial \psi}{\partial x} \tag{27.21}
$$

The above formulation means that we have at our disposal the mathematical machinery to compute the stream function $\psi(x,y,t)$ by Equation (27.17), the sea surface elevation by Equation (27.19), the horizontal velocity field $\mathbf{u} = [u,v]$ by Equation (27.20), and the trajectory $\mathbf{x}(t)$ of passive tracer particles by Equation (27.21).

27.5.1 Linearization of the Potential Vorticity Equation

The potential vorticity equation is easily linearized by removing the Jacobian term in Equation (27.17):

$$
\frac{\partial}{\partial t}(\nabla^2 \psi - F\psi) + \beta \psi_x = 0 \tag{27.22}
$$

If we consider solutions of the type

$$
\psi(x,y,t) = \psi_o \, e^{i\mathbf{k}\cdot\mathbf{x} - \omega t} = \psi_o \, e^{i[k_x x + k_y y - \omega t]} \tag{27.23}
$$

and insert Equation (27.23) into Equation (27.22) we obtain the dispersion relation

$$
\omega = -\frac{\beta k_x}{k^2 + F} \tag{27.24}
$$

The general Fourier sum solution to Equation (27.22) for periodic boundary conditions ($0 \le x \le L_x$, $0 \le y \le L_y$) is given by

$$
\psi(\mathbf{x},t) = \sum_{m=-\infty}^{\infty} \sum_{n=-\infty}^{\infty} \psi_{mn} \cos(\mathbf{k}_{mn} \cdot \mathbf{x} - \omega_{mn} t + \phi_{mn}) \tag{27.25}
$$

where $\psi_{mn} = [2P(k_m, l_{yn})\Delta k_m \, \Delta l_n]$, $\mathbf{k}_{mn} = [k_m, l_n]$, and $\omega_{mn} = \omega_{mn}(\mathbf{k}_{mn})$ is the dispersion relation (27.24). The power spectrum of ψ_{mn} is often assumed to have the form $P(k, l) \sim k^{-\gamma}$ $(k = |\mathbf{k}|)$ where the phases ϕ_{mn} are uniform random numbers on $(0, 2\pi)$. The spectrum ranges between (k_o, k_N), where $k_o = 2\pi \, / \, L$ corresponds to the largest spatial scale L and k_N to the smallest.

27.5.2   The KdV Equation as Derived from the Potential Vorticity Equation

First note that the linearized potential vorticity equation (27.22) can be rewritten:

$$\nabla^2 \psi_t - F\psi_t + \beta\psi_x = 0 \qquad\qquad (27.26)$$

If we reduce the influence of the y coordinate by a suitable scaling then

$$\nabla^2 \psi_t \simeq \psi_{xxt} \qquad\qquad (27.27)$$

then Equation (27.26) can then be put into the form

$$\psi_t + c_o\psi_x + \beta'\psi_{xxt} = 0 \qquad\qquad (27.28)$$

which is the *linearized Benjamin-Bona-Mohoney (1972, BBM) equation* $(c_o = -\beta/F$ and $\beta' = -1/F)$, with the dispersion relation:

$$\omega = -\frac{\beta k_x}{k_x^2 + F} \qquad\qquad (27.29)$$

If we set to leading order

$$\psi_t + c_o\psi_x \simeq 0 \qquad\qquad (27.30)$$

in the dispersive term of Equation (27.28) we find the linearized Korteweg-deVries (KdV) equation:

$$\psi_t + c_o\psi_x + \beta''\psi_{xxx} = 0 \qquad\qquad (27.31)$$

where $\beta'' = -c_o\beta'$, and the dispersion relation is given by

$$\omega = c_o k_x - \beta'' k_x^3 \qquad\qquad (27.32)$$

Thus, the linearized PVE reduces to the linearized BBM equation and then to the linearized KdV with a simple rescaling of the y axis so that the x motion dominates.

In the same way one can begin with the PVE equation (27.17) and, via a suitable multiscale expansion (Benny, 1967; Malanotte Rizzoli, 1982), show

$$J(\psi, \nabla^2 \psi) \rightarrow \alpha \psi \psi_x \tag{27.33}$$

where α is a constant, so that we arrive at the BBM equation directly:

$$\psi_t + c_0 \psi_x + \alpha \psi \psi_x + \beta' \psi_{xxt} = 0 \tag{27.34}$$

and with (27.30) we get the KdV equation:

$$\psi_t + c_0 \psi_x + \alpha \psi \psi_x + \beta'' \psi_{xxx} = 0 \tag{27.35}$$

The rigorous derivation of KdV from the PVE was first done by Benny (1967) (see also Redekopp and Weidmann, 1978). In this derivation one retains only linear Fourier modes in the y coordinate, such that solutions to the potential vorticity equation are given by

$$\psi(x, y, t) \Rightarrow Y(y)\psi(x, t)$$

where $Y(y)$ is the solution to a simple time-independent eigenvalue problem. The resultant motion occurs in an east-west channel for $\psi(x,t)$ and the north-south direction is governed by $Y(y)$. Malanotte Rizzoli (1982) gives numerous numerical examples of these kinds of vortices.

The discussion above shows how vortices arise from the potential vorticity equation as KdV-type solitons that are "closed" by an eigenvalue problem in the north-south direction. This is perhaps the simplest vortex that can arise from these complex dynamics. However, the potential vorticity equation itself can for stationary flows be reduced to a variation of the Poisson equation, whose vortex structure is very rich. Indeed for a reasonable choice for the functional, the Riemann spectrum consists of both vortices (which have a 2×2 Riemann matrix) and nonlinear waves (which have a 1×1 Riemann matrix). Thus, the full range of nonlinear interactions includes wave-vortex interactions. Time evolution of the Riemann spectrum is of course governed by the potential vorticity equation; this step in the mathematics would therefore be useful for *inelastic* vortex-vortex and vortex-wave interactions. Inclusion of currents, bathymetry, and other external forcing such as wind can be handled in the usual way (Pedlosky, 1979), although to my knowledge application of the methods given herein, for say a power-law spectrum, including the nonlinear dynamics of vortex-wave interactions has never been done. Applying the methods given herein would allow the complex dynamics of many vortices and nonlinear waves (characterized by a Riemann spectrum) to be included as an exercise in nonlinear dynamics. Applications can be made to problems in the analysis of data (determination of the nonlinear spectrum from remotely sensed data, see, e.g., Figure 27.1) or in the hyperfast modeling of the vortex dynamics.

27.6   The Poisson Equation for the Davey-Stewartson Equations

The Davey-Stewartson equations, discussed in detail in Chapters 2 and 33, are given by

$$i\psi_t - \psi_{xx} + \psi_{yy} - \delta|\psi|^2\psi - \Phi_x\psi = 0$$

$$\Phi_{xx} + \Phi_{yy} = -2(|\psi|^2)_x$$

(27.36)

These equations describe the two-dimensional nonlinear dynamics of water waves as a narrow-banded envelope equation in the absence of surface tension. The shallow water case is known to be integrable (Ablowitz and Segur, 1981). Here $\psi(x,y,t)$ is the complex envelope function for which the surface elevation is given by

$$\eta(x, y, t) = Re\{\psi(x, y, t)\, e^{ik_0 x - i\omega_0 t}\}$$

(27.37)

and $\Phi(x,y,t)$ is the low frequency part of the velocity potential. The second of Equations (27.36) for $\Phi(x,y,t)$ is the Poisson equation with a slowly varying "potential" driven by the complex envelope function. Therefore, it should be no surprise to find that *coastal zone dynamics* should be characterized by *time varying vortices in the low wavenumber part of the velocity potential*. Chapter 33 shows how to access this dynamics using Riemann theta functions.

If in Equation (27.36) we assume there is no directional spreading so that the terms including y derivatives disappear and we can integrate the second equation to get $\Phi_x = -2|\psi|^2$ (radiation stress). Then we get for the first of Equation (27.36) the *shallow-water nonlinear Schrödinger equation*:

$$i\psi_t - \psi_{xx} + |\psi|^2\psi = 0$$

This is of course the *defocusing case* that is *stable to small perturbations* in the envelope.

27.7   Nonlinear Separation of Variables for the Schrödinger Equation

I now explore how to solve the Schrödinger equation using the Hirota approach (see Chapters 4, 6, and 24). Of course the specific goal is to gain physical understanding and hyperfast numerical capability. Given the *Schrödinger equation*, with complex solution $\psi(x,t)$

$$i\psi_t + \psi_{xx} + \sigma\psi_{yy} + U(x, y, t)\psi = 0$$

(27.38)

seek a solution as the ratio of two functions:

$$\psi(x, y, t) = \frac{G(x, y, t)}{F(x, y, t)} \tag{27.39}$$

We can think of this approach as a kind of nonlinear separation of variables, as will now be demonstrated. We can treat $F(x,y,t)$ as a real function and $G(x,y,t)$ as a complex function in the case when $U(x,y,t)$ is real; otherwise, in particular when dissipation is present, $U(x,y,t)$ must be treated as complex. Assume

$$U(x, y, t) = 2\partial_{xx} \ln F(x, y, t) + 2\sigma\partial_{yy} \ln F(x, y, t) \tag{27.40}$$

We see that the operator in (27.40) is elliptic (the Laplacian) for $\sigma = 1$ (Bose-Einstein condensation) and hyperbolic for $\sigma = -1$ (ocean surface waves).

Use Equations (27.39) and (27.40) in (27.38) to get:

$$i(FG_t - GF_t) + (FG_{xx} + GF_{xx} - 2F_x G_x) + \sigma(FG_{yy} + GF_{yy} - 2F_y G_y) = \lambda FG \tag{27.41}$$

Use the Hirota operator

$$\begin{aligned}
D_t G{\cdot}F &= FG_t - GF_t \\
D_x^2 G{\cdot}F &= FG_{xx} + GF_{xx} - 2G_x F_x \\
D_y^2 G{\cdot}F &= FG_{yy} + GF_{yy} - 2G_y F_y
\end{aligned} \tag{27.42}$$

to write Equation (27.41) in Hirota operator form:

$$\left(iD_t + D_x^2 + \sigma D_y^2\right) G \cdot F = \lambda G \cdot F \tag{27.43}$$

The bilinear forms of Equations (27.41) and (27.43) are *not* generally integrable. The *stationary case* (for stationary vortex dynamics), Equation (27.45) below, is integrable and has been studied in particular detail.

The goal is to pick a reasonable potential $U(x,y,t)$ (which may include dynamics, bathymetry, arbitrary boundary (coastline) shape, etc.) and to compute $F(x,y,t)$ by inverting Equation (27.40) (see Chapters 14–16 and 23). Then we solve Equations (27.41) or (27.43) to get $G(x,y,t)$. Then we have solutions to (27.38) given by Equation (27.39).

In the *sinh-Poisson or sine-Gordon equations* we would then have for stationary solutions of Equation (27.38) (see Equation (27.10))

$$(FG_{xx} + GF_{xx} - 2F_x G_x) + \sigma(FG_{yy} + GF_{yy} - 2F_y G_y) = \lambda FG \tag{27.44}$$

or

$$\left(D_x^2 + \sigma D_y^2\right) G \cdot F = \lambda G \cdot F \tag{27.45}$$

These latter equations give analytical expressions for the *stationary vortex solutions* in terms of Riemann theta functions for *specific potentials U(x,y)*. One of these forms for the potential is particularly important, that for the sinh-Poisson equation (Ting et al., 1984) (see Sections 27.8 and 27.9) below. The approach can always be used to compute numerical solutions for the stationary case. Analytically, there are about a half dozen forms of the potential $U(x,y)$ that can be exactly integrated (Section 27.4). The special case for *soliton-like vortex solutions*, which occurs for $\lambda = 0$, is now discussed in Section 27.8.

27.8 Vortex Solutions of the sinh-Poisson Equation Using Soliton Methods

A key idea in the implementation of the soliton methods in this Chapter is that the sinh-Poisson equation has soliton solutions, but the vortex solution consists of bound solitons with phase locking, that is, two phase-locked solitons create a particular vortex solution. Four solitons, phase-locked in pairs, lead to two interacting vortices, etc.

The sinh-Poisson equation used in this section has the form:

$$\psi_{xx} + \psi_{yy} + \sinh\psi = 0 \tag{27.46}$$

The two-soliton solution is given by (Chow et al., 1997):

$$\psi = 4\tanh^{-1}\left(\frac{g_2}{f_2}\right), \quad g_2 = e^{X_1} + e^{X_2}, \quad f_2 = 1 + A\,e^{X_1+X_2}$$

$$X_n = k_n x + l_n y, \quad k_n^2 + l_n^2 = -1, \quad n = 1,2 \tag{27.47}$$

$$A = -\frac{(k_1 - k_2)^2 + (l_1 - l_2)^2}{(k_1 + k_2)^2 + (l_1 + l_2)^2}$$

Note that we could have chosen the dynamical phases to be $X_n = k_n x + l_n y + \phi_n$, for an arbitrary choice of the phases ϕ_n, where to have phase locking we require $\phi_1 = \phi_2$. In Equation (27.47) I have set $\phi_1 = \phi_2 = 0$. Let us now choose the "reality constraint"

$$k_1 = k_2^* = i\sqrt{1 + l^2}, \quad l_1 = l_2 = l \tag{27.48}$$

for l real, and the solution to the sinh-Poisson equation becomes:

$$\psi = 4\tanh^{-1}\left[\frac{l\,\cos\left(\sqrt{1 + l^2}\,x\right)}{\sqrt{1 + l^2}\,\cosh(ly)}\right] \tag{27.49}$$

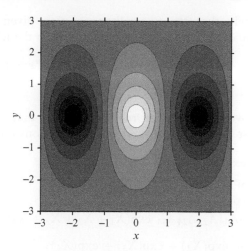

Figure 27.4 Streamlines for the Mallier-Maslowe vortex for $\sqrt{1+l^2} = \pi/2$.

This is the stream function of a vortex solution of the sinh-Poisson equation known as the Mallier-Maslowe vortex. A graph of the contours of the stream function is given in Figure 27.4. Note that this solution is periodic in x, but decays exponentially in y.

Note further that due to the dispersion relation $k_n^2 + l_n^2 = -1$ either k_n or l_n must be complex; furthermore, to guarantee that the *stream function be real* the wavenumbers k_n, l_n must be taken in complex conjugate pairs (thus the use of the phrase "reality constraint" with regard to Equation (27.48) above). This means that only an even number of solitons can give real vortex solutions of the sinh-Poisson equation.

Let us now go on to the four-soliton solution of Equation (27.46). The Hirota transformation is (Chow et al., 1997, 1998) (see Chapters 4 and 6):

$$\psi = 4\tanh^{-1}\left(\frac{f}{g}\right) = 2\ln\left(\frac{f+g}{f-g}\right) \tag{27.50}$$

Note that the functions $f + g$ and $f - g$ are the soliton limits of the two theta functions in the solution to the periodic problem in Section 27.9. Equation (27.50) leads to the two bilinear forms which I write in Hirota operator notation:

$$\left(D_x^2 + D_y^2\right)(g{\cdot}g + f{\cdot}f) = 0, \quad \left(D_x^2 + D_y^2\right)(g{\cdot}f) = -g{\cdot}f \tag{27.51}$$

Here the following identity is found useful:

$$2(\ln F)_{xx} = \frac{D_x^2 F{\cdot}F}{F^2}$$

We see immediately that the two-soliton solution given above, Equation (27.47), solves Equation (27.51). This happens due to the identity

$$D_x^m D_y^n \exp(ax + ry) \, \exp(bx + sy) = (a - b)^m (r - s)^n \exp[(a + b)x + (r + s)y]$$

$$(27.52)$$

As mentioned above the dispersion relation requires that complex conjugate wavenumbers must be used. Since there are four solitons one takes

$$
\begin{aligned}
f = {}& 1 + A_{12} \, \exp(X_1 + X_2) + A_{13} \, \exp(X_1 + X_3) \\
& + A_{14} \, \exp(X_1 + X_4) + A_{23} \, \exp(X_2 + X_3) \\
& + A_{24} \, \exp(X_2 + X_4) + A_{34} \, \exp(X_3 + X_4) \\
& + A_{12}A_{13}A_{14}A_{23}A_{24}A_{34} \, \exp(X_1 + X_2 + X_3 + X_4)
\end{aligned}
\tag{27.53}
$$

$$
\begin{aligned}
g = {}& \exp(X_1) + \exp(X_2) + \exp(X_3) + \exp(X_4) \\
& + \chi_1 \, \exp(X_2 + X_3 + X_4) + \chi_2 \, \exp(X_1 + X_3 + X_4) \\
& + \chi_3 \, \exp(X_1 + X_2 + X_4) + \chi_4 \, \exp(X_1 + X_2 + X_3)
\end{aligned}
\tag{27.54}
$$

The phases occur in complex conjugate pairs:

$$
\begin{aligned}
X_1 &= X = p_1 x + q_1 y, & X_3 &= X^* \\
X_2 &= Z = p_2 x + q_2 y, & X_4 &= Z^*
\end{aligned}
\tag{27.55}
$$

With the choices $q_1 = \alpha$ (real) and $q_2 = \beta$ (real), the remaining parameters are

$$
\begin{aligned}
& p_n^2 + q_n^2 = -1, \quad n = 1, 2 \\
& p_1 = i\sqrt{1 + \alpha^2}, \quad q_1 = \alpha, \quad p_2 = i\sqrt{1 + \beta^2}, \quad q_2 = \beta \\
& A_{ij} = \frac{S_{ij} + 1}{S_{ij} - 1}, \quad S_{ij} = p_i p_j + q_i q_j \\
& \chi_1 = \chi_3 = n_1 = A_{12}A_{14}A_{24}, \quad \chi_2 = \chi_4 = n_2 = A_{12}A_{13}A_{23}
\end{aligned}
\tag{27.56}
$$

The algebra is somewhat involved and symbolic computation has been used to arrive at the final answer:

$$
\begin{aligned}
g = 2 \Bigg\{ & \exp(\alpha y) \cos\left(\sqrt{1 + \alpha^2}\, x\right) + \exp(\beta y) \cos\left(\sqrt{1 + \beta^2}\, x\right) \Bigg\} \\
& + \left(\frac{\alpha - \beta}{\alpha + \beta}\right)^2 \exp[(\alpha + \beta)y] \left[\left(1 + \frac{1}{\beta^2}\right) \exp(\beta y) \cos\left(\sqrt{1 + \alpha^2}\, x\right) \right] \\
& + \left[\left(1 + \frac{1}{\alpha^2}\right) \exp(\alpha y) \cos\left(\sqrt{1 + \beta^2}\, x\right) \right]
\end{aligned}
$$

$$(27.57)$$

$$f = 1 + \left(1 + \frac{1}{\alpha^2}\right) \exp(2\alpha y) + \left(1 + \frac{1}{\beta^2}\right) \exp(2\beta y)$$

$$+ 2 \left[\frac{\alpha\beta - \sqrt{1+\alpha^2}\sqrt{1+\beta^2} + 1}{\alpha\beta - \sqrt{1+\alpha^2}\sqrt{1+\beta^2} - 1}\right] \exp[(\alpha+\beta)y] \cos\left[\left(\sqrt{1+\alpha^2} + \sqrt{1+\beta^2}\right)x\right]$$

$$+ 2 \left[\frac{\alpha\beta + \sqrt{1+\alpha^2}\sqrt{1+\beta^2} + 1}{\alpha\beta + \sqrt{1+\alpha^2}\sqrt{1+\beta^2} - 1}\right] \exp[(\alpha+\beta)y] \cos\left[\left(\sqrt{1+\alpha^2} - \sqrt{1+\beta^2}\right)x\right]$$

$$+ \left(1 + \frac{1}{\alpha^2}\right)\left(1 + \frac{1}{\beta^2}\right)\left(\frac{\alpha-\beta}{\alpha+\beta}\right)^4 \exp[2(\alpha+\beta)x]$$

$$(27.58)$$

A graph of the two-vortex solutions to the sinh-Poisson equation using Equations (27.57) and (27.58) is given in Figure 27.5.

27.9 Vortex and Wave Solutions of the sinh-Poisson Equation Using Algebraic Geometry

It is of course a real triumph to connect vortex dynamics with soliton solutions of the sinh-Poisson equation using the Hirota method. The long debate as to whether vortices are solitons has now been resolved for a large class of problems.

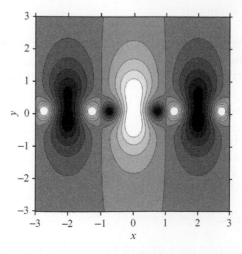

Figure 27.5 Streamlines for the two-vortex solution of the sinh-Poisson equation for $\sqrt{1+\alpha^2} = \pi/2$ and $\sqrt{1+\beta^2} - 3\pi/2$.

We have seen that a *vortex solution of the sinh-Poisson equation (27.14)
corresponds to a two-soliton, phase-locked solution* of the equation. Thus,
two solitons are found to arise in a bound pair and appear in physical space
as a vortex. Clearly these bound solitons are not the only types of vortices.
In addition to the soliton limit, we can have other types of Riemann spectra
that can obey phase locking. Furthermore, in the study of *turbulence* we need
not only paired vortices and paired "special functions" of the theory but
also nonlinear waves, because a full spectrum consists of both *vortices* and
nonlinear waves in a complete nonlinear *Riemann spectrum* of the problem.

A general solution of the vortex dynamics consists of a *time-varying Riemann spectrum*, rooted in the evolution of the *Schrödinger equation* (27.11),
the *planar vorticity equation* (27.8), or the *potential vorticity equation*
(27.17), depending upon the application. Furthermore, we need approaches
which allow us to nonlinearly Fourier analyze turbulent data for its nonlinear
spectral content (vortices and nonlinear waves) and to perform numerical simulations. Once again we would also like to have hyperfast modeling as a tool for
the numerical computations. As for many other problems in this monograph I
consider how to accomplish these goals using Riemann theta functions and the
methods of algebraic geometry.

The sinh-Poisson equation has been solved for all real solutions for periodic/
quasiperiodic boundary conditions and for zero boundary conditions (Ting
et al., 1984). The later boundary condition, $u = 0$ on the sides of a rectangle,
means that we need to consider the real, periodic, odd-parity solutions of the
sinh-Poisson equation (see Chapter 26 on acoustic wave propagation in the
ocean which has similar boundary conditions). The form of the equation that
is most general is given by:

$$u_{xx} + u_{yy} + \lambda^2 \sinh u = 0 \tag{27.59}$$

The constant λ, by abuse of notation, is not related to its assumed values used
elsewhere in this monograph.

The starting point is the sine-Gordon (or sine-Poisson) equation

$$u_{tt} - u_{xx} + \lambda^2 \sin u = 0$$

which is completely integrable for periodic/quasiperiodic boundary conditions (Kozel and Kotljarov, 1976; McKean, 1980, 1981; Date, 1982; Forest and
McLaughlin, 1982, 1983; Ting et al., 1984, 1987; Ercolani and Forest, 1985).

Note that under the transformation

$$t \to y, \quad x \to \pm ix, \quad u \to \pm iu$$

the sine-Gordon equation goes over to the sinh-Poisson equation. Thus, Ting
et al. (1984) carried out this transformation on the Lax pair for the sine-
Gordon equation to obtain the Lax pair for the sinh-Poisson equation:

$$\left\{ \sigma_y \partial_x - \frac{1}{4}(u_y + iu_x)\sigma_x + \frac{\lambda^2}{32p}e^{-u}(1 + \sigma_z) + \frac{\lambda^2}{32p}e^{u}(1 - \sigma_z) - p \right\}\phi = 0$$

$$\left\{ -i\sigma_y \partial_x - \frac{1}{4}(u_y + iu_x)\sigma_x - \frac{\lambda^2}{32p}e^{-u}(1 + \sigma_z) - \frac{\lambda^2}{32p}e^{u}(1 - \sigma_z) - p \right\}\phi = 0$$

$$(27.60)$$

where the Pauli matrices are given by

$$\sigma_x = \begin{pmatrix} 0 & 1 \\ 1 & 0 \end{pmatrix}, \quad \sigma_y = \begin{pmatrix} 0 & -i \\ i & 0 \end{pmatrix}, \quad \sigma_z = \begin{pmatrix} 1 & 0 \\ 0 & -1 \end{pmatrix} \tag{27.61}$$

The compatibility condition for the Lax pair (27.60) gives the sinh-Poisson equation. Note that if we let $x \to x / \lambda$ and $y \to y / \lambda$, we find that the sinh-Poisson equation becomes $u_{xx} + u_{yy} + \sinh u = 0$, so that we can rescale λ^2 to 1. We write this formally as

$$u(\lambda, x, y) = u(1, \lambda x, \lambda y) \tag{27.62}$$

Thus, in the calculations below we can take $\lambda^2 = 1$ and use Equation (27.62) to compute solutions for other values of λ^2.

The full analysis in Ting et al. (1984) follows Forest and McLaughlin (1983) closely and will not be reproduced here. Essentially, one carries out the usual Floquet analysis of the eigenvalue problem in Equation (27.60) and determines the main and auxiliary spectrum with the aid of squared eigenfunctions. The auxiliary spectrum of the "gammas" $\gamma_i(x,y)$ is found to evolve in terms of both x and y and the equations of motion are found to be

$$\frac{d\gamma_i}{dx} = \frac{\left[\left(\prod_{k \neq i}^{N} \gamma_k / 8Q^{1/2} \right) - 2 \right] \left[\prod_{l=0}^{2N}(\gamma_i - E_l) \right]^{1/2}}{\prod_{j \neq i}^{N}(\gamma_i - \gamma_j)}$$

$$\frac{d\gamma_i}{dy} = i\frac{\left[\left(\prod_{k \neq i}^{N} \gamma_k / 8Q^{1/2} \right) + 2 \right] \left[\prod_{l=0}^{2N}(\gamma_i - E_l) \right]^{1/2}}{\prod_{j \neq i}^{N}(\gamma_i - \gamma_j)}$$

$$(27.63)$$

$$Q^{1/2} = + \left(\prod_{l=1}^{2N} E_l \right)^{1/2}, \quad E_0 = 0 \tag{27.64}$$

This is a set of simple nonlinear first-order ordinary differential equations that evolve on a two-sheeted Riemann surface. For particular initial conditions associated with the main spectrum $\{E_i\}$, Equation (27.63) can be formally integrated to give $\gamma_i(x,y)$. Then the solution to the sinh-Poisson equation is given by

$$u = u(x, y) = \ln\left[\frac{\prod_{i=1}^{N}\gamma_i(x, y)}{\left(\prod_{i=1}^{2N}E_i\right)^{1/2}}\right] \qquad (27.65)$$

Therefore, it is clear that $\gamma_i(x,y)$ are the *nonlinear (hyperelliptic) modes* of the sinh-Poisson equation. However, as with the KdV equation and other nonlinear evolution equations, these modes, for a general N degree-of-freedom case, are not physical, that is, each of the modes is not, in and of itself, a solution of the sinh-Poisson equation. However, a *linearization* of these modes leads to the *Riemann theta function solution of the sinh-Poisson equation*, which does have modes that are, individually, solutions of the equation. We have seen this happen in many nonlinear wave equations in this monograph and the main conclusion is that for the physical understanding of the solutions of nonlinear equations the theta functions are crucial. For example, the KdV and NLS equations are hyperelliptic (as is the sinh-Poisson equation) and the passage to the theta functions through a dependent variable transformation is fundamental. *Most problems in higher dimensions are not hyperelliptic*, but the theta functions solve the problem nevertheless (see Krichever (1988) and Chapter 32 on the KP equation). Indeed for applications in the physical sciences I often express the opinion that, *if you are not in the theta function business, you are not in business*. Hyperelliptic functions can take you only so far in the analysis of data. No matter how much you may love the hyperelliptic functions, eventually theta functions do much more in terms of the physics, higher dimensional problems, data analysis, and hyperfast modeling.

Let us now proceed with showing how the theta function solution of the sinh-Poisson equation arises. This happens by an "algebro-geometric" linearization of the $\gamma_i(x,y)$ equations (27.63). First note in Figure 27.6 the main spectrum. It consists of a branch line along the positive real axis and the main spectrum of the $E_i = p_i^2$ consists of the quartet

$$\left\{E_i, E_i^*, \frac{1}{16^2 E_i}, \frac{1}{16^2 E_i^*}\right\} \qquad (27.66)$$

as shown in Figure 27.6. Note the branch lines between pairs of eigenvalues in the upper and lower half planes. These lie on particular lines radiating from the origin. The above main spectrum defines *all* the physical properties and shapes of the vortices, as discussed below.

In order to carry out the linearization of the equations of motion (27.63), it is important to recognize that the auxiliary spectrum $\gamma_i(x,y)$ evolves on a two-sheeted Riemann surface defined by the function:

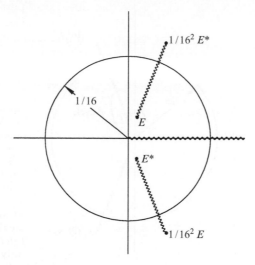

Figure 27.6 Main spectrum in the E plane.

$$R(E) = \left[\prod_{i=0}^{2N}(E - E_i)\right]^{1/2}, \quad E_0 = 0 \qquad (27.67)$$

There are several results that can be defined on the Riemann surface which will prove useful later. Particular path or "cycle" integrals can be defined on the Riemann surface. These include the so-called α and β cycles as shown respectively in Figures 27.7 and 27.8. The α-cycle is a closed curve that always stays

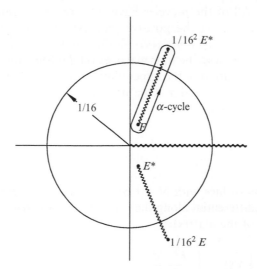

Figure 27.7 α-cycle in the E plane.

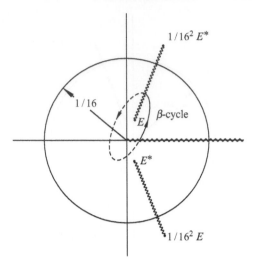

Figure 27.8 β-cycle in the E plane.

on one of the two sheets and encircles one of the branch cuts (denoted in the figures by a squiggly line). The number of the cycle has the same number as the branch cut. The β-cycle is likewise a closed path that goes through two branch cuts so that part of the path is on one sheet of the Riemann surface while the remaining segment of the path lies on the other. This is made clear in Figure 27.8 where part of the β path is a solid line (it lies on one Riemann surface) and the other part of the path is a dotted line (it lies on the other Riemann surface). All of the β-cycles have one common branch cut which is the one that extends along the positive real axis; the β-cycles do not cross one another.

We are now ready to describe four matrices on the Riemann surface that will be useful in the linearization of the hyperelliptic flow $\gamma_i(x,y)$. These are matrices which are called **A, B, C,** and τ. One at this point introduces certain holomorphic (analytic) differentials:

$$d\Omega_i = \frac{E^{N-i}dE}{R(E)} \tag{27.68}$$

valid for a Riemann surface with $N + 1$ branch cuts. The **A** matrix results from integrating the ith differential along the jth α-cycle (the integral is defined to be the (i,j)th element of the matrix):

$$A_{ij} = I_i(\alpha_j) = \int_{\alpha_j} d\Omega_i = \int_{\alpha_j} \frac{E^{N-i}dE}{R(E)} \tag{27.69}$$

Likewise the **B** matrix arises if we integrate the same differentials along the β-cycles:

$$B_{ij} = I_i(\beta_j) = \int_{\beta_j} \mathrm{d}\Omega_i = \int_{\beta_j} \frac{E^{N-i}\mathrm{d}E}{R(E)} \tag{27.70}$$

Now it is convenient to normalize the **A** matrix by forming its inverse $\mathbf{C} = \mathbf{A}^{-1}$ so that

$$\mathbf{CA} = \mathbf{I} \tag{27.71}$$

where **I** is the identity matrix. Finally, when the **C** matrix is multiplied by the **B** matrix, we get a fundamental matrix in the theory of theta functions, the *Riemann matrix*:

$$\boldsymbol{\tau} = \mathbf{CB} \tag{27.72}$$

All of the above matrices are constants for a particular *main spectrum* ($E_i, i = 1,\ldots,2N$). The matrices characterize the Riemann surface on which the auxiliary spectrum evolves. In particular, the **A** and $\boldsymbol{\tau}$ matrices are related to the periods of the auxiliary spectral components, $\gamma_i(x,y)$. The elements of the Riemann matrix $\boldsymbol{\tau}$ are "periods" on the Riemann surface, and the fact that $\boldsymbol{\tau}$ appears explicitly in the theta functions, has resulted in it being also called the "period matrix" (Baker, 1897).

At this point it is convenient to introduce the "phases" of the auxiliary dynamics of the $\gamma_i(x,y)$. Thus, the linearization of the equation of motion of the $\gamma_i(x,y)$ (27.63) requires a transformation of the $\{\gamma_i(x,y)\}$ to a set of N phases $\{X_i(x,y)\}$. The latter are integrals of linear combinations of the Abelian differentials with the elements of the **C** matrix serving as coefficients. The integrals are not on closed contours, because otherwise they would reduce to the $\boldsymbol{\tau}$ matrix elements or to Kronecker delta functions. The phase integrals begin with an arbitrary fixed point γ_0 on the Riemann surface and end at the auxiliary spectral points for the $\gamma_i(x,y)$:

$$\begin{aligned}
X_j(x,y) &= -\sum_{k=1}^{N} \int_{\gamma_0}^{\gamma_k(x,\,t)} \sum_{m=1}^{N} C_{jm}\,\mathrm{d}\Omega_m \\
&= -\sum_{k=1}^{N} \int_{\gamma_0}^{\gamma_k(x,\,t)} \sum_{m=1}^{N} C_{jm} \frac{E^{N-i}\mathrm{d}E}{R(E)}
\end{aligned} \tag{27.73}$$

The equations of motion that the $\{X_i(x,y)\}$ obey are obtained by taking the derivatives with respect to x and y. In particular, the X_{jx}, X_{jy} are related to the γ_{jx}, γ_{jy} by applying Leibniz's rule for differentiating an integral:

$$X_{jx} = -\sum_{m=1}^{N} C_{jm} \sum_{k=1}^{N} \frac{\gamma_k^{N-m}}{R(\gamma_k)} \gamma_{kx}$$

$$X_{jy} = -\sum_{m=1}^{N} C_{jm} \sum_{k=1}^{N} \frac{\gamma_k^{N-m}}{R(\gamma_k)} \gamma_{ky} \tag{27.74}$$

The final equations can be obtained by inserting the auxiliary equations of motion (27.63) into these expressions. These resultant equations can be simplified (Ting et al., 1984) to give:

$$X_{jx} = (-1)^N \frac{C_{jN}}{8Q^{1/2}} + 2C_{j1} \equiv k_j$$

$$X_{jy} = i(-1)^N \frac{C_{jN}}{8Q^{1/2}} - 2iC_{j1} \equiv \omega_j \tag{27.75}$$

These are constants (wavenumbers) and they are given the usual physical meanings that should be obvious from this "straight line" flow. The resulting equations can be immediately integrated to give

$$X_j = k_j x + \omega_j y + X_{0j} \tag{27.76}$$

where the X_{0j} are integration constants determined by the boundary conditions. Finally we have

$$X_j(x, y) = -\sum_{k=1}^{N} \int_{\gamma_0}^{\gamma_k(x, t)} \sum_{m=1}^{N} C_{jm} \frac{E^{N-i} dE}{R(E)} = k_j x + \omega_j y + X_{0j} \tag{27.77}$$

Thus, this approach has succeeded in linearizing the equations of motion of the auxiliary spectrum (27.63) by the transformation (27.73). Since the dependence of the phases on x and y in Equation (27.77) is trivial, the only remaining step is to invert this equation to determine the dependence of the $\gamma_i(x,y)$ on x and y. Once this explicit result is obtained the solution to sinh-Poisson is given by Equation (27.65).

This problem is known as the classical "Jacobian inversion problem" of algebraic geometry. The goal is to find the end points of the integrations of Equation (27.77) (the $\gamma_i(x,y)$) when the values of the Abelian integrals are given (right-hand side of Equation (27.77), $k_j x + \omega_j y + X_{0j}$). This problem can be solved by the Riemann theta functions (Siegel, 1969b; Baker, 1897). In the present notation the theta functions have the form:

$$\theta(\mathbf{X}, \tau) = \sum_{m_1, \dots, m_N = -\infty}^{\infty} \exp\left[2\pi i \sum_{i=1}^{N} m_i X_i + \pi i \sum_{i=1}^{N} \sum_{j=1}^{N} m_i m_j \tau_{ij}\right] \tag{27.78}$$

Using the theta function and applying well-known mathematics from the nineteenth century (Baker, 1897), we see that Equation (27.65) reduces to

$$u(x, y) = 2 \ln \left[\frac{\theta(\mathbf{X} + 1/2)}{\theta(\mathbf{X})} \right], \quad \mathbf{1} = [1, \ldots, 1]^{\mathrm{T}} \tag{27.79}$$

This is the solution of the sinh-Poisson equation for periodic/quasiperiodic and zero boundary conditions on a rectangle. This is the miracle of algebraic geometry and the application of the "Jacobian inverse problem" to the linearization of the hyperelliptic flow for the $\gamma_i(x,y)$.

To compute numerically Equation (27.79), we note that vortex solutions come in pairs with the phases

$$X_1 = k_1 x + \omega_1 y + \phi_1$$

$$X_1 = k_1 x - \omega_1 y + \phi_1$$

so that the Riemann matrix is 2×2 for a real vortex solution. For real solutions of the sinh-Poisson equation $u(x,y)$, the elements of the period matrix must be purely imaginary.

Now let us look at some numerical examples of vortex solutions of the sinh-Poisson equation (Figures 27.9–27.14). Figure 27.9 shows a one-band solution (a 2×2 Riemann matrix with phase-locked components). Here the x and y wavenumbers are equal. The eigenvalues and branch cuts lie on the imaginary axis. In Figure 27.10 we have a one-band solution in which the eigenvalues and branch cut lie on a radius from the origin in the first and forth quadrants.

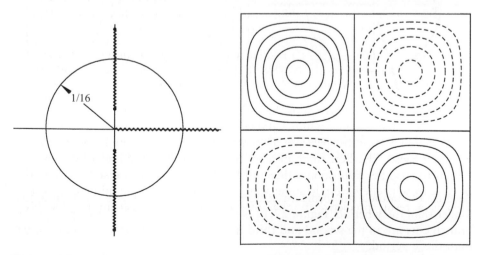

Figure 27.9 Case 1—E-plane spectrum and vortices for a one-band solution for which the eigenvalues and branch cuts are on the imaginary axis.

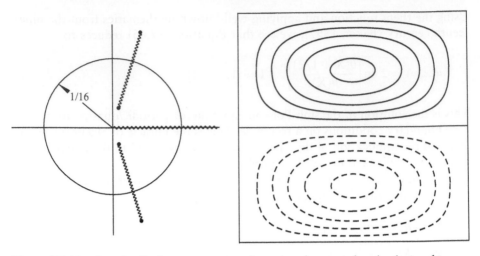

Figure 27.10 Case 2—E-plane spectrum and vortices for a one-band solution for which the eigenvalues and branch cuts are in the first and forth quadrants.

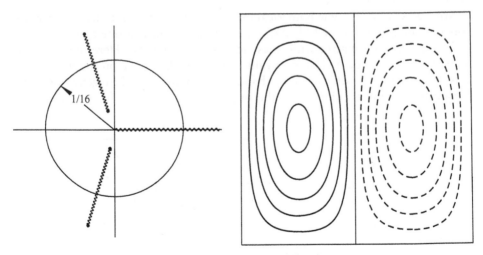

Figure 27.11 Case 3—E-plane spectrum and vortices for a one-band solution for which the eigenvalues and branch cuts are in the second and third quadrants.

Because the branch cut forms an acute angle with the real axis the x period is longer than the y period. Figure 27.11 is similar to Figure 27.10 except that the branch cuts are in the second and third quadrants so that the y period is larger than the x period.

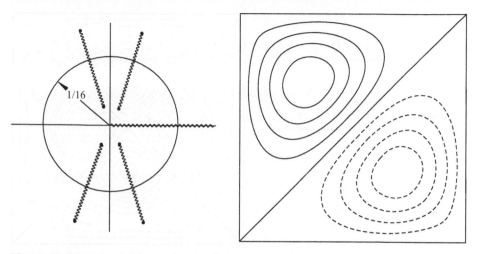

Figure 27.12 Case 4—A two-band solution that is the nonlinear superposition of the one-band solutions in Figures 27.10 and 27.11.

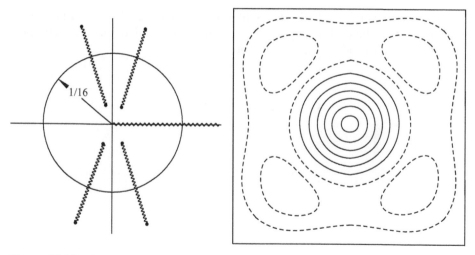

Figure 27.13 Case 5—Two-band solution for which the y period is three times the x period.

Figure 27.12 shows a vortex pattern that is a two-band solution and is effectively a nonlinear superposition of the two cases of Figures 27.10 and 27.11. In Figure 27.13 is a two-band solution for which the y period is three times the x period. Finally, in Figure 27.14 we have a two-band solution that has a y period that is five times the x period.

Figure 27.14   Case 6—Two-band solution for which the y period is five times the x period.

Now the reader has all the tools to do the nonlinear Fourier analysis of vortices in the plane. The physics, nonlinear analysis of data, and hyperfast modeling are all products of the method.

28 Nonlinear Fourier Analysis and Filtering of Ocean Waves

28.1 Introduction

One of the most common tools for the analysis of ocean surface waves is the linear Fourier transform. This important procedure not only provides a way to characterize natural wave trains in terms of a linear superposition of sine waves, but also plays a theoretically important role in the solution of the Cauchy problem for linear wave equations. In spite of the great success of the linear Fourier approach, it is well known that ocean surface waves can instead be highly nonlinear. This chapter addresses a technique for the *nonlinear Fourier analysis* of shallow-water surface wave trains. The method is based upon the inverse scattering transform (IST) for the Korteweg-deVries (KdV) equation with periodic boundary conditions. We employ the following *nonlinear Fourier decomposition: A measured signal consisting of dynamically evolving unidirectional, shallow-water waves can be decomposed into a linear superposition of cnoidal waves* (the traveling wave solution of the KdV equation) *plus mutual nonlinear interactions amongst the cnoidal waves.* Thus, nonlinear Fourier analysis as used here extends and generalizes the usual linear sum of sine waves. In the small-amplitude limit the cnoidal waves become sine waves and the interactions disappear; in this way the formalism also naturally contains and includes ordinary linear Fourier analysis. Numerical implementation of the IST provides a technique for nonlinear Fourier analysis in which measured time series are effectively *projected onto the cnoidal wave basis functions.* We show how to use the IST to nonlinearly low-pass, band-bass, and high-pass filter measured time series. We apply the method to data obtained in the Adriatic Sea in a depth of 16.5 m about 20 km from Venice, Italy. This chapter is a description and extension of several papers in the literature (Osborne et al., 1988, 1991, 1996, 1998). These papers are the result of a long and fruitful collaboration with L. Cavaleri. Much of the earlier work on applications of the IST to physical problems and data analysis are given in Osborne, 1982; Osborne et al., 1982a,b,c, 1983; Osborne and Boffetta, 1989, Osborne, 1989), Osborne et al., 1990, 1998; Osborne, 1990; Osborne and Segre, 1993; Osborne and Petti, 1994. The spirit of analyzing random wave trains using IST has been stimulated immensely by the important papers by Long and Resio (2004, 2007).

Doi: 10.1016/S0074-6142(10)97028-9

28.2   Preliminary Considerations

Linear Fourier methods are easily applied to solve the full water wave problem for the *linearized Euler equations* (Whitham, 1974). In this formulation the sine wave is the basis function onto which the wave motion may be projected for both the surface elevation and the velocity potential. Theoretically, therefore, the sine wave basis provides a powerful tool for the study of linear partial differential equations (PDEs) (with well-defined dispersion relations) and their associated Cauchy (initial value) problem. All solutions are constructed as a linear superposition of sine waves. In this way the Fourier amplitudes and phases are *constants of the motion* for linear wave dynamics. Thus, the Fourier spectrum forms a unique representation of the linear wave dynamics for all space and time.

The Fourier approach and its many applications, with the invention of modern computing machinery, has also become one of the most powerful and useful of all data analysis tools. The method has been applied to a wide range of problems in oceanography and ocean engineering, including extensions of the approach to power spectral analysis, linear filtering, and the computation of transfer functions in floating, tethered, and compliant surface and subsurface vessels of all types (Sarpkaya and Isaacson, 1981; Bendat and Piersol, 1986). Not only are data analyses fundamentally dependent on the approach but also wind-wave modeling, hindcasting, and prediction depend crucially on Fourier methods (Komen et al., 1994). Numerically, the Fourier transform provides the basis of the "spectral methods" so useful in the numerical integration of nonlinear PDEs (Fornberg and Whitham, 1978). Much of our understanding of higher order effects also comes from applying the Fourier approach to nonlinear problems. For example, observation of higher harmonics provides useful information about nonlinear effects and it is often convenient to assume time-varying Fourier coefficients and phases in order to study the nonlinear dynamics (e.g., see Chapter 32).

However, appealing the linear Fourier method may be, it has long been known that ocean waves are fundamentally nonlinear. In fact, John Scott Russell made the first observations of solitary waves in the mid-1800s (Russell, 1838, 1844) and the Stokes wave and solitary wave were found theoretically a few years later (Boussinesq, 1871; Rayleigh, 1876). Then an important nonlinear shallow-water wave equation was found by Korteweg and deVries (1895). The traveling wave solution of the KdV equation embraces both Stokes waves and solitary waves in terms of the simple Jacobian elliptic *cn* function or "cnoidal wave" (Whitham, 1974; Miles, 1980), which has also been used extensively in practical aspects of oceanography and ocean engineering (Munk, 1949; Weigel, 1964; Sarpkaya and Isaacson, 1981; Elgar and Guza, 1986; Dean and Dalrymple, 1991). For small-amplitude waves the cnoidal wave tends to a sine wave, for intermediate amplitude the cnoidal wave approximates the finite order Stokes wave, and for large amplitude one obtains the solitary wave (Whitham, 1974). Thus, the cnoidal wave can be viewed as a nonlinear generalization of the ordinary sine wave for unidirectional, small-but-finite amplitude waves in shallow water for the KdV equation. Directionally spread wave trains in shallow water are described by the KP equation (Chapter 32).

The present chapter focuses on an approach that *generalizes linear Fourier analysis* to the nonlinear domain: Instead of sine wave basis functions we use cnoidal waves as the fundamental basis onto which we *project measured time series*. We focus on *improving linear Fourier analysis procedures* by including a major source of nonlinearity in the method, that is, the quadratic nonlinearity in the KdV equation for shallow-water wave dynamics. We apply the method to the analysis of data obtained on the offshore tower of the Italian National Research Council (Consiglio Nazionale delle Ricerche) in 16.5 m depths about 20 km from Venice (Cavaleri and Zecchetto, 1987).

How does a new data analysis approach for the *nonlinear Fourier transform* arise in the context of shallow-water wave dynamics? During the last few decades a number of important discoveries have been made with regard to the physical and mathematical structure of so-called "integrable" nonlinear wave equations. These results include the identification of large classes of nonlinear PDEs that have been found to be integrable by a relatively new method of mathematical physics known as the *inverse scattering transform* (Newell, 1983, 1985; Newell and Moloney, 1992; Lonngren and Scott, 1978; Lamb, 1980; Novikov et al., 1984; Ablowitz and Segur, 1981; Eilenberger, 1981; Calogero and Degasperis, 1982; Dodd et al., 1982; Matsuno, 1984; Novikov et al., 1984; Tracy, 1984; Faddeev and Takhtajan, 1987; Drazin and Johnson, 1989; Fordy, 1990; Infeld and Rowlands, 1990; Makhankov, 1990; Ablowitz and Clarkson, 1991; Dickey, 1991; Gaponov-Grekhov and Rabinovich, 1992; Belokolos et al., 1994; Deconinck, 1998; Remoissenet, 1999; Ablowitz et al., 2004; Dauxois and Peyrard, 2004; Hirota, 2004; Gesztesy et al., 2008). Of particular interest in this regard is the prototypical integrable PDE, the KdV equation (Korteweg-deVries, 1895), which describes the propagation of unidirectional, nonlinear, small-but-finite amplitude waves in shallow water. Herein, the importance of this equation rests with the fact that it derives from the Euler equations in one spatial dimension using a singular perturbation expansion about *zero wavenumber*, $k \sim 0$ (see, e.g., Ablowitz and Segur, 1981). The expansion is carried out to third order in the linear dispersion relation, for example, $\omega = c_o k - \beta k^3$. Here $c_o = \sqrt{gh}$ is the linear phase speed and $\beta = c_o h^2/6$; g is the acceleration of gravity and h is the water depth. The KdV equation is valid as long as the motion is (sufficiently) unidirectional, the wave amplitude, a, is small with respect to the depth, $h (a \ll h)$, the wavelength, l, is large with respect to the depth, $h (l \gg h)$, and the second term in the dispersion relation is small with respect the first, $h^2 k^2/6 \ll 1$ (Whitham, 1974; Miles, 1980).

The dimensional form of the (space-like) KdV equation (also referred to as sKdV) is given by

$$\eta_t + c_o \eta_x + \alpha \eta \eta_x + \beta \eta_{xxx} = 0, \quad \eta(x,t) = \eta(x+L,t) \tag{28.1}$$

$\eta(x,t)$ is the space/time evolution of the free surface elevation, where $\eta(x,0)$ is assumed given, $\alpha = 3c_o/2h$ and the constants c_o and β are given above. Note that Equation (28.1) can be linearized by removal of the nonlinear term $\alpha \eta \eta$; the resultant linear PDE has the linear dispersion relation discussed above. Associated with sKdV is the *space-like Ursell number*:

$$U_s = \frac{3\eta_n}{8k_n^2 h^3} \tag{28.2}$$

Equation (28.1) solves the *Cauchy problem*: given the form of the wave train at time $t = 0$, namely $\eta(x,0)$, Equation (28.1) determines the solution at later time: $\eta(x,t)$. We call $\eta(x,0)$ a *space series*; as we see below with regard to the IST, Equation (28.1) provides the mathematical machinery to spectrally analyze any function $\eta(x,0)$ or an associated space series (presumably measured by remote sensing technology).

Normally, we write nonlinear wave equations in their space-like form. But for many data analysis purposes we record data as time series rather than as space series. For the analysis of *time series data*, it is more appropriate to consider the *time-like* KdV equation (tKdV) (Karpman, 1975; Ablowitz and Segur, 1981):

$$\eta_x + c_o' \eta_t + \alpha' \eta \eta_x + \beta' \eta_{xxx} = 0, \quad \eta(x,t) = \eta(x, t+T) \tag{28.3}$$

where $\eta(0,t)$ is assumed given and $c_o' = 1/c_o$, $\alpha' = -\alpha/c_o^2$, and $\beta' = -\beta/c_o^4$; Equation (28.3) has the linearized dispersion relation $k = \omega/c_o + (\beta/c_o^4)\omega^3$. The IST of Equation (28.3) is easily obtained from Equation (28.1) by a simple change of variables (Osborne, 1993a,b,c,d,e): $x \to t$, $t \to x$, $k \to \omega$, and $\omega \to k$. Equation (28.3) solves the *boundary value problem*: given the solution of the equation at some specific spatial location, $x = 0$, namely $\eta(0,t)$, Equation (28.3) determines the solutions at all other spatial points $x \neq 0$: $\eta(x,t)$. In an experimental context we call $\eta(0,t)$ a *time series*. We discuss in detail below how IST allows us to use Equation (28.3) to spectrally analyze any function $\eta(0,t)$ or its associated time series. Associated with tKdV is the *time-like Ursell number*:

$$U_t = \frac{3c_o^2}{32\pi^2} \frac{\eta_n}{f_n^2 b^3} \tag{28.4}$$

Here $f_j = j/T_j$, where T_j is the period of the time series. This expression will have utility in the data analysis below.

While the KdV equation was first found to be integrable for *infinite-line boundary conditions* (Gardner et al., 1967), herein we are primarily interested in the associated IST for *periodic boundary conditions* (Dubrovin and Novikov, 1975a,b; Dubrovin et al., 1976). Periodic IST for the KdV equation is one of the important topics in this book and is covered in Chapters 10, 14–17, and 19–23. The nonlinear spectral method has been applied to the analysis of data in Chapters 25 and 28–31.

The remainder of the chapter is organized as follows. In the following sections we discuss linear Fourier analysis (Section 28.3), nonlinear Fourier analysis using the IST with periodic boundary conditions (Section 28.4), relevant theoretical results that are useful for the time series analysis of data (Section 28.5), and important physical considerations regarding the applicability of the nonlinear Fourier approach (Section 28.6). Finally, we discuss the analysis of measured time series from the Adriatic Sea (Section 28.7). A summary and discussion is given in Section 28.8.

28.3   Sine Waves and Linear Fourier Analysis

Fourier analysis allows the construction of linear wave trains, formally functions of space and time, $\eta(x,t)$, by a linear superposition of sine waves:

$$\eta(x,t) = \sum_{n=1}^{N} \eta_n \cos\left(k_n x - \omega_n t + \phi_n\right) \tag{28.5}$$

In the present case there are N sine waves that are interpreted as "degrees of freedom" or "Fourier components" in the wave train. In Equation (28.5) the η_n are the Fourier amplitudes, the k_n are the wavenumbers, the ω_n are the frequencies, and the ϕ_n are the phases. The commensurable spectral frequencies are given by $f_n = \omega_n/2\pi$ for $1 \le n \le N$ (f_N is the *Nyquist frequency*). The relationship between the frequencies ω_n and the wavenumbers k_n is given by the *dispersion relation*, $\omega_n = \omega_n(k_n)$. The dispersion relation defines the physics via the correspondences:

$$\frac{\partial}{\partial t} \leftrightarrow -i\omega, \quad \frac{\partial}{\partial x} \leftrightarrow ik$$

For example, the simple dispersion relation for *unidirectional, long waves in shallow water*, $\omega = c_o k - \beta k^3$, has the associated PDE (the linearized KdV equation):

$$\eta_t + c_o \eta_x + \beta \eta_{xxx} = 0 \tag{28.6}$$

The simplest periodic solution to Equation (28.6) is a traveling sine wave

$$\eta(x,t) = \eta_o \cos\left(k_o x - \omega_o t + \phi_o\right) \tag{28.7}$$

from which the general Fourier solution for N components may be constructed by Equation (28.5) for the particular case when there are N degrees of freedom in the wave train. The important point is that the *amplitudes of the sine waves and their phases are constants of the wave motion, provided that the motion is linear*. Thus, the Fourier spectrum forms a unique representation of the wave dynamics for all space and time.

   In oceanic applications one is often interested in the analysis of time series, that is, measurements of the wave amplitude, $\eta(0,t)$, taken at a fixed spatial location over some convenient time interval T; this implies setting $x = 0$ in Equations (28.5) and (28.7). An important issue is that Equation (28.5) and its numerical implementation in terms of the fast Fourier transform (FFT) (Bendat and Piersol, 1986) obey *periodic boundary conditions*.

28.4   Cnoidal Waves and Nonlinear Fourier Analysis

For nonlinear problems one is faced with a number of limitations of the linear Fourier approach. First, the linear Fourier amplitudes and phases are no longer constants of the motion. Formally speaking, this means that the Fourier spectrum is itself a function of space and time for nonlinear problems and, consequently *averaged* Fourier spectra are often found to be more useful (Bendat and Piersol, 1986). Second, the sine wave is *not* the appropriate basis function for nonlinear wave propagation. For example, if one takes the Fourier spectrum of a *Stokes wave* or *solitary wave* one finds that there are, effectively, a very large number of components in the spectrum; on the other hand, if one takes the IST of either of these latter waveforms then one finds only *one* component in the spectrum. This is because the IST essentially *glues together* linear Fourier components that are *phase locked with each other.* Since a pure Stokes wave consists of a large number of linear Fourier components, all phase locked with one another, IST views the Stokes waves as a single component. On the other hand, IST views different Stokes waves or different cnoidal waves, each with its own unique amplitude and phase, as being *free waves* (they generally propagate with relative to one another because they can have different phase speeds), and each of the free waves corresponds to a different component in the IST spectrum. All the components in a single Stokes or solitary wave are called *bound waves* or modes. Since the cnoidal waves in general require a unique amplitude, phase, and modulus, a graph of each of these quantities as a function of frequency constitutes the full IST spectrum of cnoidal waves. We now discuss some of the details of the IST and how the method can be used as a data analysis tool.

Korteweg and deVries found the simple periodic solution of Equation (28.1) which is known as the *cnoidal wave*:

$$\eta(x,t) = \frac{4k^2}{\lambda} \sum_{n=1}^{\infty} \frac{n(-1)^n q^n}{1-q^{2n}} \cos(nkx) = 2\eta_0 cn^2\{(K(m)/\pi)[k(x-C_0 t)]; m\}$$

$$(28.8)$$

where $\lambda = \alpha/6\beta$. The Jacobian elliptic function cn has modulus m given by

$$mK^2(m) = \frac{3\pi^2 \eta_0}{2k^2 h^3} = 4\pi^2 U, \quad U = \frac{3\eta_0}{8k^2 h^3}$$

$$(28.9)$$

η_0 is the amplitude of the cnoidal wave, U is the Ursell number, k is the wavenumber, and h is the water depth. $K(m)$ is an elliptic integral (Abramowitz and Stegun, 1964). The nonlinear phase speed C_0 of the cnoidal wave has the formula:

$$C_0 = c_0\{1 + 2A_0/h - 2k^2 h^2 K^2(m)/3\pi^2\}$$

$$(28.10)$$

The series representation given in Equation (28.8) for the cnoidal wave is just the infinite order *Stokes series* solution to the KdV equation (Whitham, 1974). The modulus m of the Jacobian elliptic function cn and the nonlinear phase speed C_o are well-known relations that are a function of the amplitude η_o (see, e.g., Osborne, 1993a,b,c,d,e). When the modulus $m \to 0$ the cnoidal wave reduces to a sine wave; intermediate values of the modulus correspond to the Stokes wave; when $m \to 1$ the cnoidal wave approaches a solitary wave or soliton.

We view the cnoidal wave as a kind of *basis function* for the KdV equation (just as the sine wave is a basis function for the *linearized* KdV equation) and a long sought goal has been to develop a simple approach to *nonlinear Fourier analysis* in terms of these fundamental waveforms (Osborne, 1991). Formally speaking, the discovery of the theoretical formulation for the periodic IST resolved this problem about three decades ago (Dubrovin and Novikov, 1975a,b; Its and Matveev, 1975; Date and Tanaka, 1976; Dubrovin et al., 1976; Flaschka and McLaughlin, 1976; McKean and Trubowitz, 1976; Belokolos et al., 1994). To see how this formulation comes about we address the general solution to the tKdV equation (28.3) in terms of the so-called θ-*function representation*

$$\eta(x,t) = \frac{2}{c_o^2 \lambda} \partial_{tt} \ln \theta_N(x,t), \tag{28.11}$$

where the function is given by

$$\theta_N(x,t) = \sum_{m_1,\,\ldots,\,m_N=-\infty}^{\infty} \exp\left[i \sum_{n=1}^{N} m_n X_n(x,t) + \frac{1}{2} \sum_{m=1}^{N} \sum_{n=1}^{N} m_m m_n B_{mn} \right] \tag{28.12}$$

The integer N is the number of degrees of freedom (i.e., the number of cnoidal waves) in a particular solution to the KdV equation. The θ-function argument resembles that of linear Fourier analysis:

$$X_n(x,t) = k_n x - \omega_n t + \phi_n \tag{28.13}$$

The *period or interaction matrix*, $\mathbf{B} = \{B_{mn}\}$, the *wavenumbers*, k_n, the *frequencies*, ω_n, and the *phases*, ϕ_n are constants which depend upon particular methods whose determination is discussed in Chapters 14–16, 19, and 32. The period matrix $\mathbf{B}$ has elements that are *negative definite* to insure mathematical convergence of the θ-function (Equation (28.12)). The *cnoidal wave amplitudes* are determined by the diagonal elements of $\mathbf{B}$ and the *nonlinear interactions* are determined by the off-diagonal terms. From the point of view of mathematical physics and applied mathematics, the discovery of the periodic IST for the KdV equation has revealed remarkable relationships among the Riemann theory of Abelian functions, the spectral theory of the Schrödinger operator and algebraic geometry (Belokolos et al., 1994). Practical implementation of the approach is described here. An important result is that the nonlinear Fourier technique is formulated in a physical and mathematical form simple enough that practical data analysis applications can now be made.

With linear Fourier analysis one projects a wave train onto the sine wave modes of Equation (28.5). With nonlinear Fourier analysis one projects a wave train onto the cnoidal wave basis functions of Equations (28.11) and (28.12). We now discuss some of the basic results for numerical methods and data analysis procedures which allow us to accomplish the projection of time series onto nonlinear basis functions.

28.5   Theoretical Background for Data Analysis Procedures

This section is devoted to a statement of particular theoretical results that will be used in the analysis of data in Section 28.7. The mathematical structure for the Riemann theta-function provides the basis of data analysis procedures.

28.5.1   Cnoidal Wave Decomposition Theorem for θ-Functions

The θ-function solution (28.11, 28.12) of the tKdV equation (28.3) is readily written in the following form (Osborne, 1995):

$$\eta(x,t) = \frac{2}{c_0^2 \lambda} \frac{\partial^2}{\partial t^2} \ln \theta_N(x,t) = \underbrace{\eta_{cn}(x,t)}_{\substack{\text{Linear superposition} \\ \text{of cnoidal waves}}} + \underbrace{\eta_{\text{int}}(x,t)}_{\substack{\text{Nonlinear interactions} \\ \text{among the cnoidal waves}}}$$

$$(28.14)$$

Therefore: *Unidirectional, long shallow-water wave trains can be represented by a linear superposition of cnoidal waves plus their mutual nonlinear interactions.* Fluid dynamicists, oceanographers, ocean engineers, and others are all familiar with the cnoidal wave as a fundamental manifestation of nonlinearity in shallow-water wave motion (Weigel, 1964; Dean and Dalrymple, 1991). Now it is possible to conduct nonlinear Fourier analysis with these fundamental traveling waves as basis functions. Note that the cnoidal wave basis functions in Equation (28.14) provide a unique and general way to write the spectral decomposition for solutions of the KdV equation.

In Equation (28.14) one has for the linear superposition of the cnoidal waves the following expression:

$$\eta_{cn}(x,t) = 2\sum_{n=1}^{N} \eta_n cn^2\{(K(m_n)/\pi)[k_n(x - C_n t) + \phi_n]; m_n\} \qquad (28.15)$$

where the moduli, m_n, the wavenumbers, k_n, frequencies, ω_n, and phases, ϕ_n are given by algebro-geometric loop integrals (Dubrovin and Novikov, 1975a,b) (see Chapters 10, 14, and 19). The nonlinear interaction term in Equation (28.14) is given by (Osborne, 1995):

$$\eta_{\text{int}}(x,t) = \frac{2}{c_0^2 \lambda} \frac{\partial^2}{\partial t^2} \ln\left\{1 + \frac{F(x,t,G)}{F(x,t,1)}\right\} \tag{28.16}$$

where

$$F(x,t,G) = \sum_{m_1,\dots,m_N=-\infty}^{\infty} G \exp\left[i \sum_{n=1}^{N} m_n X_n(x,t) + \frac{1}{2}\sum_{m=1}^{N}\sum_{n=1}^{N} m_m m_n D_{mn}\right] \tag{28.17}$$

$$G = \exp\left[\frac{1}{2}\sum_{m=1}^{N}\sum_{n=1}^{N} m_m m_n O_{mn}\right] - 1 \tag{28.18}$$

where the interaction matrix **B** is written as the sum of diagonal ($\mathbf{D} = \{D_{mm}\}$) and off-diagonal ($\mathbf{O} = \{O_{mn}\}$) parts: $\mathbf{B} = \mathbf{D} + \mathbf{O}$. The phase variable in Equation (28.17) is given by $X_n(x,t) = k_n x - \omega_n t + \phi_n$. An important limit occurs when the *wave amplitudes of the KdV equation are small with respect to the depth*: the cnoidal waves tend to sine waves and the interactions tend to zero. In this way one recovers from (Equation (28.14)) ordinary, linear Fourier analysis in the linear limit (Equation (28.5)). The latter result provides one of the reasons why periodic IST can be viewed as a *nonlinear generalization of linear Fourier series*.

It is interesting to note that for the interactions to be identically zero in Equation (28.14) one requires that the off-diagonal terms $\{O_{mn}\}$ be zero in Equation (28.18) so that $G = F = O$ and therefore $\eta_{\text{int}}(x,t) = 0$. This of course cannot happen except in the small amplitude, linear limit. It is nevertheless tempting to interpret $\eta_{\text{int}}(x,t)$ as being perturbative in character, that is, small with respect to the summed cnoidal waves. Instead, in particular for large amplitude waves, the interactions can be of the *same order as the summed cnoidal waves* ($O(1)$), particularly in the soliton limit.

Equation (28.14) is remarkably simple from a physical point of view. It is in some sense the ultimate synthesis of a complex series of experimental, physical, and mathematical steps begun over 200 years ago and includes the fundamental work of Cauchy, Poisson, Airy, Russell, Stokes, Boussinesq, Riemann, Baker, and Korteweg and deVries in the nineteenth century, just to name a few. More recent theoretical progress includes the discovery of the IST (Gardner et al., 1967), the understanding of the Floquet analysis of the periodic Schrödinger equation, the development of a general theory for the reduction of integrable PDEs to hyperelliptic functions, the exploitation of the N-dimensional Jacobian transformation on a two-sheeted Riemann surface as a linearization of the hyperelliptic flow, and the explicit representation of general periodic solutions of the KdV equation in terms of N-dimensional Riemann θ-functions, all of which occurred in the last three decades (see Belokolos et al., 1994 and cited references). Crucial to this chapter is that such a complicated recipe, the IST for periodic boundary conditions, has been fully harnessed to address the analysis of experimental data from the Adriatic Sea. To this end the result (Equation (28.14)) is of prime importance not only from a theoretical point of view but also for applications in the nonlinear Fourier analysis of broad-spectrum oceanic surface wave motions in shallow water and for numerical modeling (Chapter 32).

It is worthwhile discussing the physical interpretation of the nonlinear interactions. It is well known that interactions in the KdV equation take on the form of *phase shifts*. For example, when two solitons collide they emerge from the collision completely intact with their phase speeds the same as before the collision: However, the larger soliton is phase shifted *forward* and the small soliton is phase shifted *backward* in the spatial coordinate with respect to where they would have been in the absence of collisions (see, e.g., Ablowitz and Segur, 1981 and cited references). This phase-shifting property of the KdV equation is found in the *nonlinear interaction contribution* of Figure 25.13 in Chapter 25. In terms of this simple example, a *negative pulse* in the nonlinear interactions, followed by a *positive pulse*, will *phase shift a soliton forward*. A *positive pulse followed by a negative pulse* generates a backward phase shift of a soliton.

28.5.2   Nonlinear Filtering with θ-Functions

A theoretical result which is quite powerful for the analysis of time series data is the following: By *not summing over a particular degree of freedom* in the θ-function (Equation (28.12)), one removes to a high order of approximation the associated degree of freedom (i.e., the cnoidal wave) from the filtered wave train. This happens because when the nonlinearity is not too great, as is the case with the Adriatic Sea data, the off-diagonal elements are nearly independent of the spectral amplitudes (see Chapters 15, 32). One can always be assured that the filtering operation is complete for large nonlinearity, simply by iterating the filtering operation several times. Stated differently, the associated *open band in the Floquet spectrum* is thereby rendered *degenerate* and the wave train is reconstructed *without* the associated cnoidal wave degree of freedom (Osborne, 1995). Effectively, this operation is equivalent to reducing the interaction matrix $\mathbf{B}$ from an $N \times N$ matrix to an $(N-1) \times (N-1)$ matrix. In this way a specific cnoidal wave is removed from the IST spectrum. Of course any combination of cnoidal waves can be removed in the same way. We refer to the removal of selected cnoidal wave components from the IST spectrum as *nonlinear filtering*.

There are three filtering applications of the θ-functions that we exploit in the data analysis below. The *first filtering application* is used for the extraction of *soliton components from a measured wave train*. One uses knowledge of the *theta function modulus* to identify the number of solitons in the spectrum, N_{sol} (Osborne and Bergamasco, 1986). One establishes the soliton components by a very simple rule first introduced by Ferguson et al. (1982). A soliton on the periodic interval $(0, T)$ is just a cnoidal wave with modulus very nearly 1 such that the wave is graphically indistinguishable from a soliton on the infinite line. Thus, the $N_{sol} \times N_{sol}$ submatrix (in the upper left hand corner of the interaction matrix **B**) contains all the information about the soliton components in the spectrum. In fact, by numerically executing Equations (28.11) and (28.12) for this $N_{sol} \times N_{sol}$ submatrix one easily extracts the N-soliton wave train from a measured time series.

The *second filter application* using the θ-function representation is as a *low-pass filter*. Here one effectively computes the period matrix out to the physically relevant characteristic value $N = N_{char}$ for KdV evolution (see Section 28.7.1); this reduces the numerical problem to an $N_{char} \times N_{char}$ interaction matrix. This procedure is useful for saving computer time, that is, consider a 2000 point time series. In this case the interaction matrix is 1000×1000. However, by recognizing that long waves in shallow water provide a characteristic frequency considerably lower than the Nyquist frequency, one can, say, reduce the interaction matrix to 50×50. This results in considerable savings in computer time as discussed below with regard to the Adriatic Sea analysis.

Finally, the *third filter application* is one in which one seeks to *band-pass filter* the measured wave train. In this case a lower cutoff, N_{low}, and an upper cutoff, N_{high} are identified on the basis of *physical considerations*. This means that one seeks to extract a band of spectral components (cnoidal waves) from the time series with width $(N_{high} - N_{low})$. The interaction matrix is now $(N_{low} - N_{high}) \times (N_{low} - N_{high})$ and Equations (28.11) and (18.12) are evaluated as before to get the band-pass-filtered wave train. *High-pass* filtering is a simple and obvious extension of band-pass filtering. Examples of these approaches are given below.

On the basis of the topics just discussed, we suggest another way to formally decompose a measured wave train that, in its simplest form, we write:

$$\eta(x, t) = \eta_{sol}(x, t) + \eta_{rad}(x, t) + \eta_{int}(x, t) \tag{28.19}$$

Here we have placed the soliton, $\eta_{sol}(x,t)$, and radiation parts, $\eta_{rad}(x,t)$, as *separate contributions* and we have included the *interaction terms* between these fundamental components to complete the formulation. To make this expression concrete suppose that the number of the solitons is found to be N_{sol}, while the remaining components are identified with the radiation modes for which there are N_{rad} terms in the spectrum. Then we make the following observation: the wave train $\eta_{sol}(x,t)$ is computed from its $N_{sol} \times N_{sol}$ period matrix $\mathbf{B}_{sol}$, a square submatrix in the upper left hand corner of **B**. Likewise the radiation components are computed from its $N_{rad} \times N_{rad}$ period matrix $\mathbf{B}_{rad}$, which is a square submatrix in the lower right-hand corner of $\mathbf{B}$. Both $\mathbf{B}_{sol}$ and $\mathbf{B}_{rad}$ are submatrices centered on the diagonal of $\mathbf{B}$. All of the off-diagonal terms which have not been included in $\mathbf{B}_{sol}$ and $\mathbf{B}_{rad}$ then occur as off-diagonal terms in $\mathbf{B}_{int}$ which may be used to compute the interaction contribution, $\eta_{int}(x,t)$. The interaction matrix has the form:

$$\mathbf{B} = \mathbf{B}_{sol} + \mathbf{B}_{rad} + \mathbf{B}_{int} \tag{28.20}$$

We now extend and generalize this approach and in the following concrete example we discuss an interaction matrix $\mathbf{B}$ that is 9×9. We have, for illustrative purposes, elected to *low-pass filter* the first three components, *band-pass filter* the second two components, and *high-pass filter* the final four components in the cnoidal wave spectrum. To this end we have identified the three interaction matrices corresponding to these three filter bands. The period matrix then has the following form:

$$
\begin{bmatrix}
\begin{bmatrix} B_{11} & B_{12} & B_{13} \\ B_{21} & B_{22} & B_{23} \\ B_{31} & B_{32} & B_{33} \end{bmatrix} & \begin{matrix} B_{14} & B_{15} \\ B_{24} & B_{25} \\ B_{34} & B_{35} \end{matrix} & \begin{matrix} B_{16} & B_{17} & B_{18} & B_{19} \\ B_{26} & B_{27} & B_{28} & B_{29} \\ B_{36} & B_{37} & B_{38} & B_{39} \end{matrix} \\
\begin{matrix} B_{41} & B_{42} & B_{43} \\ B_{51} & B_{52} & B_{53} \end{matrix} & \begin{bmatrix} B_{44} & B_{45} \\ B_{54} & B_{55} \end{bmatrix} & \begin{matrix} B_{46} & B_{47} & B_{48} & B_{49} \\ B_{56} & B_{57} & B_{58} & B_{59} \end{matrix} \\
\begin{matrix} B_{61} & B_{62} & B_{63} \\ B_{71} & B_{72} & B_{73} \\ B_{81} & B_{82} & B_{83} \\ B_{91} & B_{92} & B_{93} \end{matrix} & \begin{matrix} B_{64} & B_{65} \\ B_{74} & B_{75} \\ B_{84} & B_{85} \\ B_{94} & B_{95} \end{matrix} & \begin{bmatrix} B_{66} & B_{67} & B_{68} & B_{69} \\ B_{76} & B_{77} & B_{78} & B_{79} \\ B_{86} & B_{87} & B_{88} & B_{89} \\ B_{96} & B_{97} & B_{98} & B_{99} \end{bmatrix}
\end{bmatrix}
$$

and the corresponding frequencies and phases are vectors

$$[f_1, f_2, f_3, f_4, f_5, f_6, f_7, f_8, f_9]$$

$$[\phi_1, \phi_2, \phi_3, \phi_4, \phi_5, \phi_6, \phi_7, \phi_8, \phi_9]$$

In this wave we have identified the appropriate filter matrices along the diagonal elements of the period matrix. The filter submatrix, the frequencies, and the phases for the first three *low-pass components* (1, 2, 3) in the cnoidal wave spectrum (soliton spectrum) are

$$
\begin{bmatrix} B_{11} & B_{12} & B_{13} \\ B_{21} & B_{22} & B_{23} \\ B_{31} & B_{32} & B_{33} \end{bmatrix}, \quad [f_1, f_2, f_3], \quad [\phi_1, \phi_2, \phi_3]
$$

The filter submatrix, the frequencies, and the phases for the *band-pass components* (4, 5) (intermediate spectrum) are

$$\begin{bmatrix} B_{44} & B_{45} \\ B_{54} & B_{55} \end{bmatrix}, \quad [f_4, f_5], \quad [\phi_4, \phi_5]$$

and the filter submatrix, the frequencies, and the phases for the *high-pass components* (6, 7, 8, 9) (radiation spectrum) are

$$\begin{bmatrix} B_{66} & B_{67} & B_{68} & B_{69} \\ B_{76} & B_{77} & B_{78} & B_{79} \\ B_{86} & B_{87} & B_{88} & B_{89} \\ B_{96} & B_{97} & B_{98} & B_{99} \end{bmatrix}, \quad [f_6, f_7, f_8, f_9], \quad [\phi_6, \phi_7, \phi_8, \phi_9]$$

Each of these submatrices, together with their associated frequencies and phases, corresponds to exact solutions of the KdV equation at $x = 0$. The solution to the KdV equation, corresponding to the full 9×9 period matrix can then be written in the following notation:

$$\eta(0, t) = \eta_{1-3}(0, t | 3 \times 3) + \eta_{4-5}(0, t | 2 \times 2) + \eta_{6-9}(0, t | 4 \times 4) + \eta_{\text{int}}(0, t)$$

$$(28.21)$$

Note that $x = 0$ is the location of a presumed wave staff, pressure recorder, or other measuring instrument. In the above equation the subscripts refer to the range of frequency components; in the parenthesis, to the left of the vertical bar we note the space, $x = 0$, and time, t, dependencies. To the right of the vertical bar is the size of the associated interaction matrix. This contrasts to the sum of cnoidal waves plus interactions in Equation (28.14), for now we have a *sum of sequential submatrix solutions* plus *nonlinear interactions among the submatrix solutions*. The *interaction terms* now come from those terms in the original period matrix that are *not* included in the three submatrices.

28.6 Physical Considerations and Applicability of the Nonlinear Fourier Approach

In this section we discuss some of the physical considerations that one should address before analyzing time series data with cnoidal wave basis functions. First let us reflect on a comparison between linear Fourier analysis and the IST to see how their usage contrasts in the analysis of measured time series.

28.6.1 Properties of the Nonlinear Fourier Approach

(1) Since ocean waves are nonlinear, one knows that the linear Fourier amplitudes will vary in time (see Chapters 31 and 32 for discussions). The time-varying coefficients occur because the wave dynamics occurs at higher nonlinear order than the associated linear system (Whitham, 1974).

(2) Fourier series can be used to solve linear PDEs, such as in the linearized water wave equations (Whitham, 1974). Here the perspective is that one is *projecting time series onto the linear Fourier modes* in order to gain insight about the physical behavior of the measured wave trains.

(3) In a similar way, by going to higher nonlinear order with cnoidal wave basis functions, we are simply trying to improve the ability of spectral analysis procedures to access nonlinear effects in data. We hope to gain more information from measured wave trains by projecting them onto cnoidal waves rather than onto sine waves. We are trying to learn new things about nonlinear ocean wave dynamics that we have not been able to learn using the linear superposition of sine waves.

(4) One often uses linear Fourier analysis to look for higher order nonlinear behavior, in particular for higher harmonics due to the Stokes wave nonlinearity (this is done, e.g., in the bi- and tri-spectral analysis of waves). In nonlinear Fourier analysis the *Stokes wave itself is a single component* in the IST spectrum. This occurs because IST pastes together *phase locked Fourier components* into a single waveform, that is, the Stokes wave, which is then treated as a single nonlinear Fourier spectral component.

(5) Some of the *higher order nonlinear effects that one is able to address with IST* in the analysis of time series data are readily characterized: (a) The modulus of each cnoidal wave tells us whether the component is a sine wave, Stokes wave, solitary wave, etc. (b) The *nonlinear interactions* among the cnoidal waves tell us about the effect of *global phase shifting of cnoidal waves* in the spectrum. (c) The pair-wise, triple-wise, etc. interactions among selected cnoidal wave components are related to 3-wave and 4-wave interactions among cnoidal waves. (d) The nonlinear interactions *among different block submatrix components* in the spectrum tell us about interactions between the solitons and radiation, or between solitons and intermediate spectrum, etc. (e) The spectrum is *coherent-structure dominated* when the soliton dynamics comprise most of the energy in the spectrum. (f) The nonlinear effects *above the order of the KdV equation* may be important and may be characterized in terms of *space-/time-varying cnoidal wave amplitudes* (Osborne et al., 1998).

28.6.2 Preliminary Tests of the Time Series

In order to best understand the use of IST as a time series analysis tool one needs to consider a number of preliminary tests that should be made on the data in order to insure that the physical behavior of the data analysis is consistent with cnoidal wave basis functions. Furthermore, when applying IST to the analysis of data there are a number of reasons why the analysis should be done with care. The purpose the following paragraphs is to provide a physical basis for the data analysis and to address a number of important issues before the IST analysis begins. In effect, the measured wave train must approximately satisfy the following criteria (Osborne et al., 1998):

(a) The data should be tested to make sure that they are essentially unidirectional. Extension of the approach to the case with directional spreading is underway and will be reported on in the future. For the present work we adhere to the unidirectionality assumption.

(b) The wave amplitudes should be small, but finite in amplitude.

(c) The waves should be long with respect to the depth.

(d) The waves should approximately obey the dispersion relation of the linearized KdV equation. This means that there exists a region in the spectral (frequency) domain where KdV dynamics should approximately hold.

A perfect selection of appropriate data is probably not possible because Mother Nature does not often constrain herself to *a priori* specifications for perfect or near perfect data sets (this normally occurs only in carefully controlled laboratory experiments). Nevertheless, we feel that the above rules provide reasonableness criteria for projecting data onto cnoidal wave basis functions. In future work the above limitations will be slowly removed and the applicability of the method extended to increasingly more general cases including finite amplitude waves and directional spreading. At this stage of development, however, we prefer to remain rather conservative and to use the methods where they are most appropriate.

28.6.3 The Use of Periodic Boundary Conditions

There are several reasons why periodic boundary conditions for IST is important in the analysis of unidirectional, shallow-water wave data:

(1) Linear Fourier analysis, in the guise of the *discrete Fourier transform* (DFT, a linear, *periodic algorithm*, numerically implemented as the FFT), has played an important and ubiquitous role in the processing of a large variety of signals (Bendat and Piersol, 1986). Furthermore, the small-amplitude, *linear limit of periodic IST* is given by the DFT (Osborne and Bergamasco, 1985, 1986; see also Chapter 10). Thus, the nonlinear Fourier approach used herein reduces identically to ordinary linear (discrete) Fourier analysis when the wave amplitudes are small.

(2) Many types of wave motion, such as oceanic surface and internal waves, are often viewed as approximations of *stationary* and *ergodic random processes*, that is, as infinitely long stochastic processes. Motions of this type are normally approximated with periodic (Osborne, 1982; Bendat and Piersol, 1986) boundary conditions using the linear Fourier transform with phases which are uniformly distributed random numbers on the interval $(0, 2\pi)$; one uses the linear *periodic* FFT algorithm to address problems of this type. In this case the temporal period, T, of the wave train is *formally* taken in the limit $T \rightarrow \infty$; however, for practical considerations, T is often chosen to be from a few minutes to several hours and averaging procedures (over subintervals either in time or in frequency) are employed to improve spectral estimates of the $T \rightarrow \infty$ limit. Recently, a nonlinear generalization of such processes was considered by Osborne (1993a,b,c,d,e). We therefore feel that it is appropriate to use periodic boundary conditions for many *nonlinear* data analysis applications.

(3) Many kinds of natural, nonlinear wave phenomena have a dominant period, such as the 12.4-h tidal period often found for internal wave motions in the ocean. Periodic IST can play a natural role in the understanding of such processes (Osborne and Burch, 1980; Osborne, 1990).

On the basis of the above and previous results and on work presented herein, a simple guiding criterion for the *development of numerical data analysis algorithms for IST* is the following: *a numerical, nonlinear (IST) Fourier algorithm (numerical IST) should be discrete and periodic and reduce to the DFT in the small-amplitude limit.* Practical implementation of numerical IST parallels in many respects the linear Fourier analysis of time series with the DFT.

28.7   Nonlinear Fourier Analysis of the Data

The analysis of a data set from the oceanic environment using the full mathematical and numerical structure of the periodic IST is the main goal of this chapter. To this end we consider a surface wave train measured in the Northern Adriatic Sea in 16.5 m depth. A time series of $N = 1000$ points has been selected for the present study and is shown in Figure 28.1A. The data were recorded and digitized at temporal intervals of $\Delta t = 0.25$ s for a total period of $T = 250$ s. The linear Fourier transform of the series is given in Figure 28.1B. Note that the dominant spectral energy lies roughly in an interval from about 8 to 10 s (corresponding to a spectral peak ~ 0.10-0.12 Hz); the significant wave height is $H_s = 2.5$ m and the zero crossing period is $T_z = 10.2$ s.

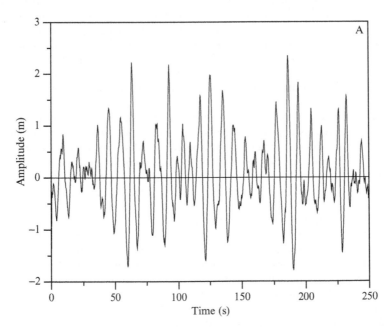

Figure 28.1   (A) Time series measured in the Adriatic Sea. (B) Linear Fourier transform of the measured time series in (A).

(Continued)

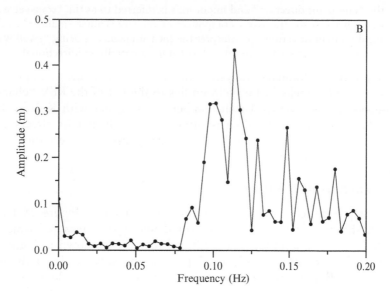

Figure 28.1 Cont'd

28.7.1 Applicability of the Nonlinear Fourier Approach

A number of considerations must be addressed before proceeding with the IST procedure. An important part of the data analysis is to first establish whether the measured time series can be characterized *approximately* in terms of the physics of the KdV equation. To this end a number of tests are made on the data in order to ensure that the physical scales of interest are consistent with unidirectional, small-but-finite amplitude, long-wave motion in shallow water. There are at least three tests that the data should satisfy:

(1) A rough measure of the applicability of the KdV equation is given by the frequency $f_{char} \sim 1.14 c_o / 2\pi h \sim 0.14$ Hz that covers the cnoidal wave spectrum under consideration here. The most important range in the spectrum of the data is then $0 \leq f \lesssim f_{char}$. Inverse scattering theory assumes that the wave components to the right of this interval are essentially sine waves, as seen in the analysis below.

(2) The time-like Ursell number, $U = 3 c_o^2 H_s T_z^2 / 32\pi^2 h^3 \sim 0.1$, should lie in the range of KdV behavior (Osborne et al., 1999). Here H_s is the significant wave height and T_z is the zero crossing period. This estimate of the Ursell number suggests that the waves are only moderately nonlinear. Furthermore, this analysis verifies that the waves are small-but-finite in amplitude and are long with respect to the depth.

(3) The amount of directional spreading is rather small in the measured wave train. Only 3% of the wave energy is in the direction transverse to the dominant wave motion. This estimate was made using orthogonal measurements of particle velocity in the horizontal plane beneath the free water surface (Cavaleri and Zecchetto, 1987). The measured particle velocity components are then rotated to the "principal axis coordinate frame" for which the major axis is termed

the "dominant direction" and minor axis is referred to as the "transverse direction." Because the transverse component of velocity is much smaller than that in the dominant direction, we interpret the measurements as being "swell waves," which for our purposes are treated as being essentially unidirectional.

As a result of the preliminary analysis of the data, we conclude that about 94% of the energy in the measured wave train lies to the left of the KdV "characteristic frequency," the average Ursell number is consistent with KdV evolution and the degree of wave spreading is sufficiently small that its effects can be neglected. We now proceed with the data analysis using the periodic IST with cnoidal wave basis functions.

28.7.2 Analysis of the Data

The next step is to low-pass filter the input wave train of Figure 28.1A. The results are shown in Figure 28.2: the dotted line is the measured time series of Figure 28.1A, the solid line is the filtered data. Also shown in Figure 28.2 is the long, low-frequency "solitonic" part of the wave train; a discussion and further analysis is given below. The low-pass filtering is done using both

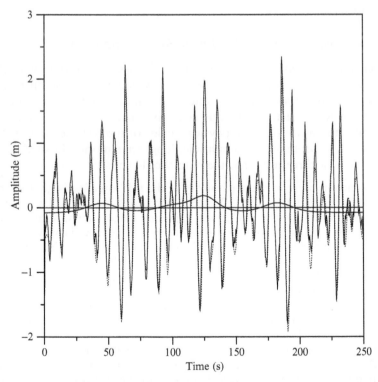

Figure 28.2   Input time from the Adriatic Sea as shown in Figure 28.1A (dotted line) has been low-pass filtered on the interval 0-0.2 Hz (solid line). The long, low-amplitude wave train is the solitonic component in the cnoidal wave spectrum.

the linear Fourier transform and the hyperelliptic functions (for the latter see Chapter 17). The fact that both filtering methods provide the same results supports the conclusion that the high frequency tail of the spectrum is essentially linear and can therefore be truncated. This fact is consistent with our estimated value of f_{char} for the KdV equation; frequencies higher than this number will of course be linear as required by the IST for KdV. This filtering step is essential because it allows us to reduce the size of the Riemann matrix from 500×500 to 50×50, resulting in a consistent physical analysis and at the same time saving considerable amounts of computer time.

We are now prepared to make a full investigation of the nonlinear physics in the measured wave train using the θ-*function representation*. This is accomplished by first computing the *interaction matrix* **B** for the *filtered wave train* of Figure 28.2, which has exactly 50 degrees of freedom (e.g., 50 hyperelliptic functions and 50 cnoidal waves). The interaction matrix **B** is then exactly 50×50. The matrix elements B_{ij} are graphed as a function of i and j in Figure 28.3. Note that all the elements in the **B** matrix are negative definite (see vertical axis; negative definiteness is a theoretical and numerical requirement

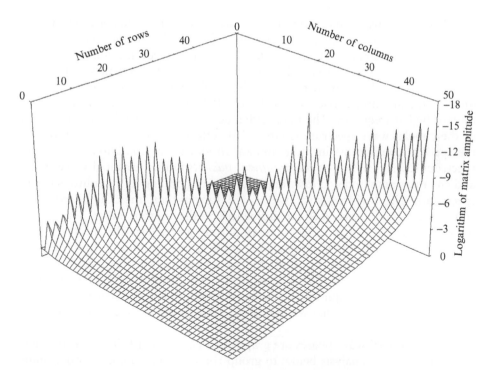

Figure 28.3   The amplitudes of the elements of the period matrix for the Adriatic Sea data, which is 50×50. The diagonal elements, directly related to the cnoidal wave amplitudes, form a jagged "backbone" on the surface. The higher the diagonal elements, the smaller are the associated cnoidal waves.

for the θ-series (Equation (28.12)) to be convergent). The dominant feature of Figure 28.3 is the *ridge* or *backbone* along the diagonal of the matrix; each diagonal matrix element determines the mathematical quantity referred to as the *nome* of a cnoidal wave (Abramowitz and Stegun, 1964) and hence, through a simple computation, provides the amplitude and modulus of the cnoidal wave (Osborne, 1993a,b,c,d,e, 1995) (see Chapter 10).

Before considering in detail the cnoidal wave spectrum, however, it is worthwhile noting that one can easily "read" the interaction-matrix graph in Figure 28.3. When the diagonal matrix elements are large the cnoidal wave amplitudes are small; likewise when the diagonal matrix elements are small the cnoidal wave amplitudes are large. This is because each cnoidal wave has amplitude $\sim \exp[-B_{ii}/2]$. The solitons correspond to the first four small-amplitude elements in the upper left hand corner of the interaction matrix as shown on the left of Figure 28.3. The low region near the center of the diagonal corresponds to the *radiation modes* that lie at the peak of the IST spectrum (Figure 28.4A). The high region to the left of the interaction matrix in Figure 28.3 consists of the small modes separating the solitons and the radiation. The high region to the right is the low amplitude, high frequency tail of the spectrum.

The cnoidal wave spectrum and its associated modulus are shown in Figure 28.4A. The spectrum itself is somewhat similar to the linear Fourier spectrum in appearance (Figure 28.1B). There are, however, a number of fundamental differences between the two types of spectra. Qualitatively, both spectra have similar shapes, for example, small amplitudes occur at low frequency; the amplitudes rise to a peak near 0.12 Hz and finally, slowly decay toward higher frequency. The main differences come about when the modulus of the cnoidal wave spectrum is large. This effect increases the amplitudes of the spectral components in the dominant modes of the *radiation* part of the spectrum (corresponding to the *dominant packets* in the measured wave train) and enhances amplitudes at low frequency (corresponding to the solitons in the spectrum). The low-frequency waves are longer than their high-frequency counterparts and hence feel the influence of the bottom more, leading to increased nonlinear effects. The high-amplitude waves near the spectral peak, being larger than the other components, are also more nonlinear. Both of these results can quantitatively be attributed to an increase in the *spectral Ursell number*, $U_j = 3g\eta_j / 32\pi^2 f_j^2 h^2$. Consequently, one should anticipate that low frequency and large amplitude waves have large moduli and hence large nonlinearity; this is exactly what we find in the analysis of the measured wave data (Figure 28.4).

The IST cnoidal wave phases are graphed in Figure 28.4B. It has been found convenient for the analysis below to group the frequency domain into a number of physically interesting regions called (1) solitons, (2) intermediate spectrum, (3) low-level radiation, (4) dominant packet radiation, (5) spectral tail and (6) linear tail. These regions are indicated in the figure and will be explained further below.

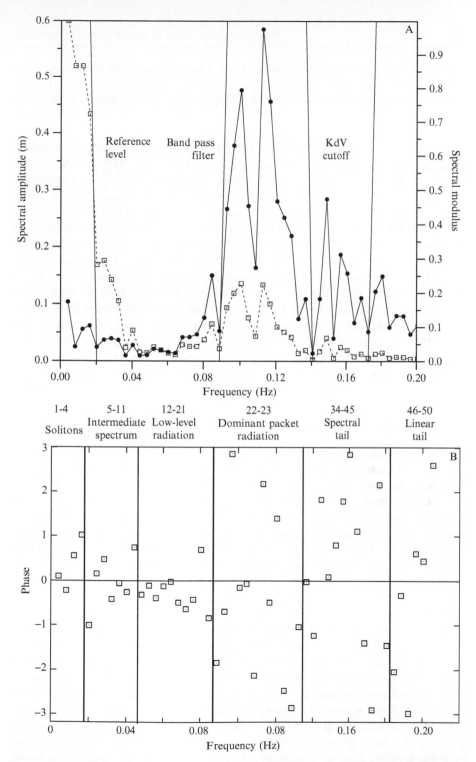

Figure 28.4 The cnoidal wave spectrum of the Adriatic Sea time series (A) and the phase spectrum (B). The frequency domain has been divided into six somewhat arbitrary regions selected primarily upon certain aspects of the physics of the measured wave train.

In Figure 28.5 we show the results of the θ-function analysis of the *low-pass filtered* Adriatic Sea wave train of Figure 28.2. We have used the cnoidal wave decomposition discussed in Section 28.5 (Equation (28.14)), that is, a shallow-water wave train can be decomposed into its constituent cnoidal waves plus their mutual nonlinear interactions. The upper three-fourths of Figure 28.5 shows the 50 cnoidal waves extracted from the filtered wave train of Figure 28.2. Below the cnoidal waves are shown the *linear superposition of the cnoidal waves*, their *mutual nonlinear interactions* and, finally, at the bottom the *reconstructed Adriatic Sea (low-pass filtered) wave train* of Figure 28.2. Note that the wave train corresponding to the linear superposition of the cnoidal waves is somewhat *larger* than the low-pass filtered wave train. This leads to the following physical conclusion: *the effect of the nonlinear interactions is to reduce the amplitudes of the summed cnoidal waves in the reconstruction process of* Equations (28.11) and (28.12). Therefore, the nonlinear interactions are *out of phase* (by $\sim 180°$) with the summed cnoidal waves. This effect can be easily seen by comparing the summed cnoidal waves with the nonlinear interactions in Figure 28.5. While this may at first seem surprising, it is enough to recall the simple two-soliton interaction in which the soliton amplitudes *decrease* during a collision (see, e.g., the discussion in Osborne and Bergamasco (1985, 1986) and Figure 25.13). One may attribute the nonlinear interactions to an effective *global spatial/temporal phase shifting* of the summed cnoidal waves relative to one another.

28.7.3 Nonlinear Filtering

This section is devoted to the study of the nonlinear behavior of the Adriatic Sea wave train in frequency bands of physical interest in the spectral domain. This is a concept quite similar to linear band-pass filtering using the linear Fourier transform: One selects a relatively narrow band of frequencies in the spectrum and excludes the remaining components in the Fourier inversion process. The resultant wave train contains only those frequencies included in the filter band. In the simplest case one takes a "perfect filter" which means only the sine waves associated with the filter band are summed in the Fourier transform, that is, no accounting is made of the Gibbs effect which occurs due to the sharp shoulder at the filter edges.

A similar procedure can be constructed for the IST as discussed previously in Section 28.5. It is this kind of spectral decomposition that we now consider for the Adriatic Sea data. We begin with 50 frequencies and a 50×50 period matrix (frequency interval (0, 0.2 Hz). Figure 28.4A and B gives both the cnoidal wave spectral amplitudes and moduli, together with the cnoidal wave *phases* which are shown at the bottom of the figure (we momentarily postpone discussion of the phases to a paragraph below). Shown in the figure are the submatrix partitions that we intend to use in the nonlinear frequency domain analysis. We have given a name to each frequency partition that is intended to indicate what kind of nonlinear physics that might be expected to occur there. Table 28.1 is used for organizational purposes, giving the nonlinear physically significant regions in the spectrum.

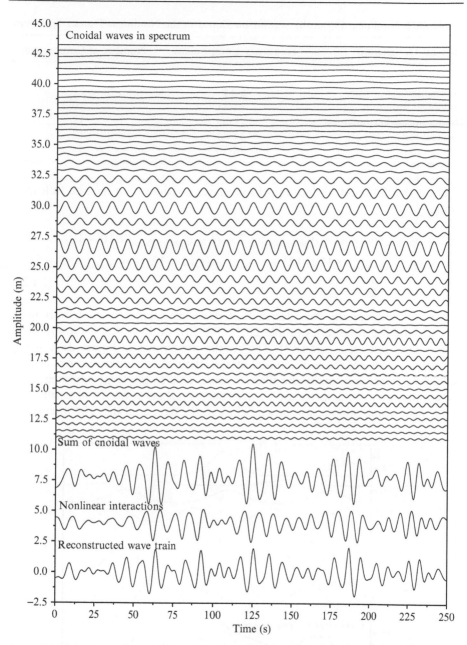

Figure 28.5 The cnoidal waves in the inverse scattering transform spectrum of the filtered Adriatic Sea wave train of Figure 28.2. The 50 cnoidal waves are shown in the upper three-fourths of the figure. The lower three curves are the sum of the cnoidal waves, the nonlinear (phase shifting) interactions, and the reconstructed, filtered Adriatic Sea wave train.

Table 28.1 Nonlinear Physically Significant Zones in the Nonlinear Fourier Spectrum

Physical Designation	Frequency Components	Number of Components
Soliton spectrum	1-4	4
Intermediate spectrum	5-11	7
Low-level radiation	12-21	10
Dominant packet radiation	22-33	12
Spectral tail	34-45	12
Linear tail	46-50	5

Thus, the IST spectral decomposition under consideration has the following form:

$$\eta(t) = \eta_{\text{Solitons}}(t|4 \times 4) + \eta_{\text{Intermediate}}(t|7 \times 7) + \eta_{\text{Low radiation}}(t|10 \times 10)$$
$$+ \eta_{\text{Dominant packet}}(t|12 \times 12) + \eta_{\text{Spectral tail}}(t|12 \times 12) + \eta_{\text{Linear tail}}(t|5 \times 5) + \eta_{\text{int}}$$
$$(28.22)$$

We now sequentially consider each of these six submatrix contributions to the Adriatic Sea time series (Table 28.1). We first address the *"solitonic" part of the spectrum*, $\eta_{\text{Solitons}}(t|4 \times 4)$. The results are shown in Figure 28.6 where in vertical order are the four cnoidal waves, their linear superposition, nonlinear interactions, and finally the filtered wave. Due to the large nonlinearity of the cnoidal waves, the interaction wave train is *not perturbative in nature* as it is for most of the other cases shown below in the data analysis. The interaction

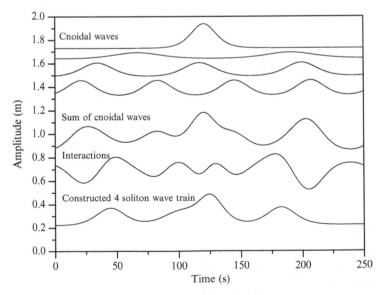

Figure 28.6 The cnoidal wave decomposition of the "solitonic" part of the spectrum in Figure 28.4.

train is instead quite large, even though the moduli of some of the components are somewhat smaller than 1, about the same size as the summed cnoidal waves themselves.

The results for the *intermediate spectrum*, $\eta_{\text{Intermediate}}(t|7\times7)$, are shown in Figure 28.7. The term "intermediate spectrum" was first used by Osborne and Bergamasco (1985, 1986) to describe the periodic IST spectrum lying immediately between the solitonic and radiation components. In this case there are seven cnoidal waves that are arranged at the top of Figure 28.7 in vertical order, from lower to higher frequency. The linear superposition of the cnoidal waves, their interactions and the constructed wave train corresponding to the intermediate spectrum are also shown. Note that the nonlinear interactions are rather small in this case; most of the nonlinear effects are due to the first four cnoidal waves which have moduli ~ 0.3 (see Figure 28.4). To see how nonlinear are the cnoidal waves turn the page upside down to see if the waves are up-down symmetric. If the waves are indeed up-down symmetric they are sine waves and hence linear.

The analysis for the *low radiation components*, $\eta_{\text{Low radiation}}(t|10\times10)$, is given in Figure 28.8; here there are 10 cnoidal waves, their sum, their interactions, and the final constructed wave train. Note that the interactions are quite small here. In fact, the only significant contribution to the nonlinearity occurs from the last cnoidal wave which has a moderate value for its modulus ~ 0.1.

The *dominant packet radiation*, $\eta_{\text{Dominant packet}}(t|12\times12)$, occurs in the frequency range 0.088-0.132 Hz (frequency numbers 22-33). These components taken together are the most nonlinear of the radiation components in the spectrum, with moduli up to about 0.23. The cnoidal waves, sum of cnoidal waves,

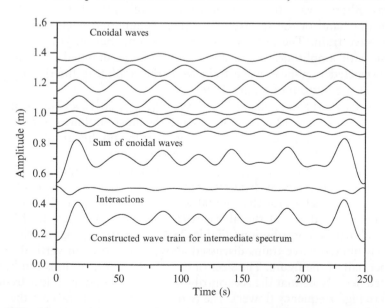

Figure 28.7 The cnoidal wave decomposition of the intermediate spectrum for which there are seven cnoidal waves in the spectrum of Figure 28.4.

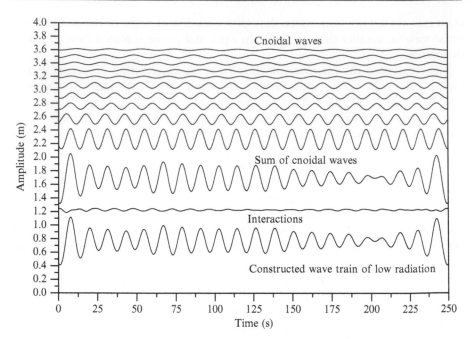

Figure 28.8 The cnoidal wave decomposition of the low level radiation components in the spectrum of Figure 28.4.

and nonlinear interactions, are vertically ordered, together with the associated band-pass filtered wave train are shown in Figure 28.9. There are 12 components in this band of the spectrum and their summation is clearly a narrow-banded wave train. The nonlinear interaction contribution is rather large in this case with amplitudes up to about 40% of the summed cnoidal waves. The constructed nonlinear wave train from the nonlinear filtering operation is shown at the bottom of Figure 28.9.

The results of the *spectral tail* analysis, $\eta_{\text{Spectral tail}}(t|12 \times 12)$, are given in Figure 28.10; there are 12 cnoidal waves and the most nonlinear of these waves (the fourth) has a modulus of only 0.1. Note that the nonlinear interactions are rather small in this case, a result that is not surprising.

The *linear tail* analysis, $\eta_{\text{Linear tail}}(t|5 \times 5)$, is given in Figure 28.11. These waves are virtually linear as demonstrated by the small interaction contribution and by the small values for the cnoidal wave moduli shown in Figure 28.4.

A complete synthesis of the above results is given in Figure 28.12. All of the six submatrix contributions (Table 28.1), corresponding to the selected band-pass filtered wave trains discussed above, are shown in vertical order in the upper half of the figure. The sizes of the relative contributions can be easily compared here. Note that the six contributions are ordered from low frequency (upper) to high frequency (lower). The two most energetic bands are the "dominant packet radiation" and the "spectral tail." From our previous analysis we

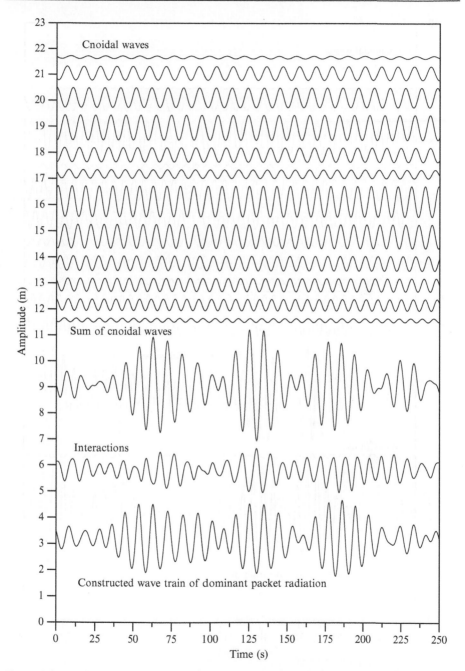

Figure 28.9 The cnoidal wave decomposition of the dominant packet radiation in the spectrum of Figure 28.4.

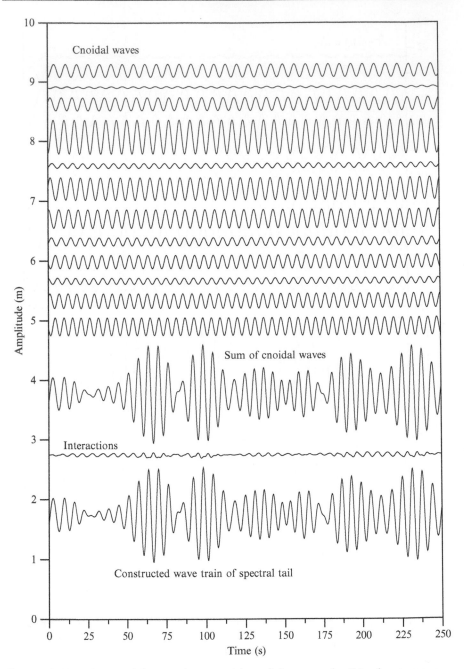

Figure 28.10 The cnoidal wave decomposition of the spectral tail in the spectrum of Figure 28.4.

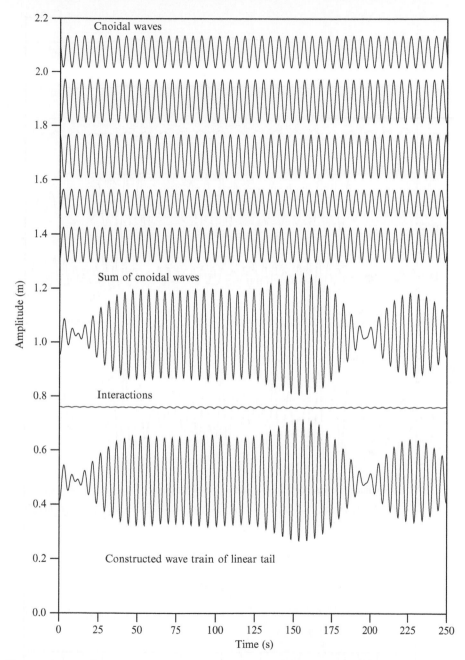

Figure 28.11 The cnoidal wave decomposition of the linear tail in the spectrum of Figure 28.4.

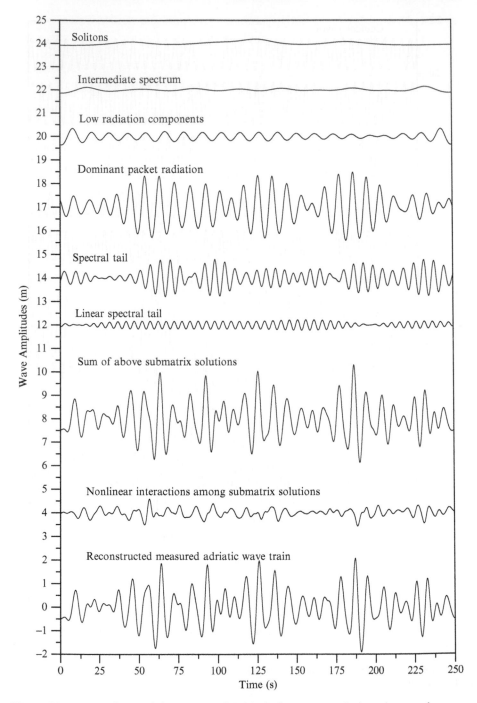

Figure 28.12 Synthesis of the measured Adriatic Sea wave train based upon the submatrix solutions associated with the physical regions of the cnoidal wave spectrum of Figure 28.4.

have seen that, among the six submatrix contributions, the dominant packet radiation has the greatest nonlinear interaction contribution (Figure 28.9).

The sum of the submatrix contributions is also given in Figure 28.12 (see Equation (28.14)), together with the nonlinear interaction phase shifts and, finally, the reconstructed, filtered input wave train is shown at the bottom. While it would be tempting to conclude that the nonlinear interactions shown in Figure 28.12 are the sum of those in the previous six figures, this is not the case as we now discuss. A comparison of Figure 28.5 (sum of cnoidal waves) with Figure 28.12 (the submatrix analysis) indicates that most of the nonlinear interactions are included in the six submatrix contributions. As a result, when we calculate the interactions *among* the six submatrix contributions, we find that the nonlinear interactions are substantially reduced from the contributions seen in the cnoidal wave decomposition of Figure 28.5. We find that the interactions are smaller in the submatrix filtering analysis (Figure 28.12) because a large part of the nonlinear effects reside in the *off-diagonal terms of the submatrices*. In the cnoidal wave analysis of Figure 28.5, there are $(50^2 - 50 = 2450)$ *off-diagonal terms*; in the submatrix analysis of Figure 28.12, 428 of these off-diagonal terms reside *within* the block matrices themselves and hence do not contribute to the interactions outside the block submatrices. Thus, the nonlinear interaction contribution in Figure 28.12 arises simply because fewer off-diagonal terms have been included in the interactions in the *six band-pass filtering operations*.

We now discuss briefly the *IST phases* shown in Figure 28.4B. First recall that a typical result of analyzing ocean waves with the linear Fourier transform is that the Fourier phases are often found to be uniformly distributed random numbers on the interval $(-\pi, \pi)$; this result thus provides support for the *random phase approximation* for ocean waves. We had therefore expected the cnoidal wave phases to also be random in our IST data analysis. Instead the results that we found are definitely not random as seen in Figure 28.4. There is some evidence for clustering of the phases lying to the left of the dominant peak in the spectrum. We have found this effect in quite a number of analyzed time series. The phases under the spectral peak and tail instead are typically found to be relatively random in appearance. We have not yet made a systematic study of this nonrandomness at low frequency. We suspect that it is a nonlinear phenomenon, but at the present time a full explanation of the effect is unknown.

28.8 Summary and Discussion

We have discussed the application of periodic inverse scattering theory to the analysis of unidirectional, nonlinear, shallow-water ocean surface waves. Recent theoretical advances allow shallow-water wave trains to be represented in terms of a kind of nonlinear Fourier analysis that uses cnoidal waves as the fundamental basis functions (Osborne, 1995b). The IST technique gives new perspective to the problem because it provides the capability for physically representing a measured wave train in terms of a linear superposition of cnoidal wave plus their mutual nonlinear interactions. The physical simplicity of these results is of course rather surprising in view of the fact that periodic inverse scattering theory is quite rich in its mathematical complexity, encompassing the fields of algebraic geometry, the Floquet analysis of the Schrödinger operator, and the Riemann theory of Abelian functions (Dubrovin, 1981; Dubrovin and Novikov, 1975a, 1975b; Dubrovin et al., 1976).

We feel that the results of this analysis are rather satisfying because we have learned a number of new important features about shallow-water, nonlinear wave dynamics: (a) The wave trains may be *synthesized from cnoidal waves* (rather than from sine waves). (b) The spectrum of the waves consists of the cnoidal wave (1) amplitudes, (2) moduli, and (3) phases as a function of frequency. (c) For the moderate-amplitude measured wave trains analyzed herein most of the spectrum is describable in terms of moderate amplitude Stokes waves. One may interpret the nonlinear interaction contribution as a kind of "glue" that binds together the cnoidal waves and provides the role of *global, temporally distributed phase shifts*. The sum of the cnoidal waves is actually larger than the measured wave train itself. Hence the nonlinear interactions reduce the overall amplitude of the superposed cnoidal waves in the process of temporal phase shifting which occurs during the theoretical/numerical reconstruction of the measured wave train. The reduction in the summed cnoidal wave amplitudes occurs primarily because the *interaction contribution is roughly 180° out of phase with the summed cnoidal waves*.

For the Adriatic Sea time series analyzed herein the nonlinearity is not very large. Aside from four small-amplitude cnoidal (solitonic or Stokes) waves (see Figure 28.4), the remaining components have moduli $\lesssim 0.23$. Therefore, most of the energy in the wave train can be characterized as being moderately nonlinear: a majority of the cnoidal wave components can be viewed as moderate amplitude Stokes waves which interact with one another. This result is in stark contrast to what one might expect for wave trains in much shallower water. A preliminary result (Osborne et al., 1996) indicates that wave trains typical of the one analyzed herein become, in fact, *soliton dominated* as they propagate into even shallower water. One of the primary goals of this research is to apply new methods to even shallower water wave trains; in effect a systematic study of wave trains with higher Ursell numbers needs to be made. Indeed, Chapter 30 provides the analysis of Duck Pier data that is soliton dominated.

While the nonlinear interactions in the present analysis are significantly large, one would naturally, from a physical standpoint, expect to have even larger, longer, and significantly more nonlinear waves in even shallower water. This is because of the dependence of the Ursell number on the parameters H_s (significant wave height), L (wave length, $\sim c_o T$), and h (depth): $U = 3H_s L^2 / 32\pi^2 h^3$. Therefore, other experiments with higher Ursell number will exhibit more nonlinear effects, with a much more energetic low frequency solitonic part. In particular, the nonlinear evolution of shoaling waves as they propagate into shallow water is an important area of investigation.

The results in this chapter might be well compared to the seminal work of Elgar and Guza, 1986.

29 Laboratory Experiments of Rogue Waves

29.1 Introduction

The study of instabilities of deep-water wave trains has its foundations in the paper by Lighthill (1965) and the subsequent papers by Benjamin and Feir (1967) and Zakharov (1968). A great leap forward was made by Yuen and Lake (1982) from theoretical, numerical, and experimental points of view. Indeed, many of the contributions in this area of research in the past decade have included the rediscovery of the *modulational instability*, its application to the study of *random wave trains*, and the renaming of the "Yuen parameter" the "Benjamin-Feir index." Tulin and Waseda (1999) have also made fundamental wave tank experiments. The application of the periodic inverse scattering transform to this area of research was made by Osborne et al. (2000) and Osborne (2001, 2002) and in Chapters 12, 18, 24 and 29 in this book.

The goal of this chapter is to apply periodic IST to the analysis of laboratory data that has previously been studied for its statistical content (Onorato et al., 2001, 2002, 2004, 2005, 2006; Mori, 2007). Theoretically, the *spectral theory* of the *nonlinear Schrödinger equation* (NLS) is addressed herein (Zakharov, 1968; Hasimoto and Ono, 1972; Zakharov and Shabat, 1972; Whitham, 1974; Kotljarov and Its, 1976; Tracy, 1984; Tracy et al., 1984; Tracy and Chen, 1988; Belokolos et al., 1994). This theory is known as the *periodic inverse scattering transform*, a kind of generalized, *nonlinear Fourier analysis*. At the level of the NLS equation there are essentially four kinds of nonlinear physical effects that can contribute to the formation of extreme, unidirectional surface waves:

(1) The superposition of weakly nonlinear Fourier components.
(2) The Stokes wave nonlinearity.
(3) The modulational instability.
(4) The wave/current interaction.

In the context of measured nonlinear wave trains, item (3) is of prime interest in this chapter because the theory predicts the existence of *unstable wave packets* that constitute a *second population* of nonlinear waves, the *first population* being the weakly nonlinear superposition of sine waves (item (1) above) together with the Stokes-wave correction (item (2)). The second population unstable wave packets, which can rise up to more than twice the significant wave height, are a *new kind of nonlinear spectral component* predicted by the NLS equation (Kotljarov and Its, 1976; Tracy, 1984; Akhmediev et al., 1985, 1987, 1990, 2000; Akhmediev, 1986, 2001; Tracy and Chen, 1988; Ablowitz and Herbst, 1990; Akmediev and Mitskevich, 1991; Belokolos et al., 1994; Akhmediev and Ankiewicz, 1997; Henderson et al., 1999; Akhmediev and Soto-Crespo, 2008).

Do second population waves actually exist in water waves? To answer this question we conducted a number of experiments in the wave flume at Marintek in Trondheim, Norway, and assessed the results in terms of NLS spectral theory. It is found that the second population waves not only exist, but they can also, under particular circumstances discussed below, dominate the energetics of the wave train leading to a condition which might be called a *rogue sea* which is defined more concretely below and is exemplified by Figure 1.8B and Figures 1.13–1.15 of Chapter 1 and the discussions of Chapters 12, 18 and 24.

This chapter addresses a new approach for the Fourier analysis of deep-water wave trains. The method is intrinsically nonlinear and describes a *nonlinear Fourier spectrum* for a *nonlinear random wave train* which includes the four types of nonlinearity discussed above:

(1) The *superposition of weakly nonlinear quasi-sine wave components*. The small amplitude limit of this part of the theory is just ordinary linear Fourier analysis.
(2) The *Stokes wave* nonlinearity is a correction of the solution of the Schrödinger equation on the order of 10-15% (see Equation (2.38) of Chapter 2) (Stokes, 1845).
(3) The *nonlinear superposition of nonlinear, unstable wave packets*. Unstable packets are a type of nonlinear Fourier component in NLS spectral theory. This spectral contribution arises from the *Benjamin-Feir instability* and obeys a *threshold effect*: Only large enough wave fields have unstable modes in the spectrum.
(4) The influence of the *wave/current interaction* on the nonlinear spectrum. This effect appears in corrections to the coefficients of the NLS equation or as an external forcing term.

Items (1), (2), and (4) are well known in linear and nonlinear wave trains and are not discussed here in detail. While it is natural to refer to the traditional "linear superposition of sine waves" as *first population* waves, this category also includes the small perturbations that give rise to the Stokes wave nonlinearity. Small Stokes-like nonlinearities in random wave trains, leading to small deviations in the Rayleigh distribution of wave heights, belong to this class of wave motions.

The focus in this chapter is on the *dynamical* behavior of the *second population* waves, that is, unstable wave packets governed by the Benjamin-Feir/modulational instability. In the NLS equation spectral theory, the first and second population waves are different kinds of spectral components, uniquely distinguishable from each other. In the historical context of the study of ocean waves, most time series contain only first population waves, but for certain particular kinds of sea states one can also have a contribution from second population waves. Indeed, for sea states that I refer to as *rogue seas* the *second population waves energetically dominate the first population waves* in the nonlinear spectrum. In this case second population waves lead to an enhanced tail in the distribution of wave and crest heights (Onorato et al., 2001, 2002, 2004, 2005, 2006; Mori, 2007). Rogue seas are thus characterized by extreme waves that are much larger than those found in a weakly linear sea state with Gaussian crests or Rayleigh wave heights.

The NLS theoretical formulation *covers all of the physical effects* in the list above (1)-(4). The *nonlinear spectral analysis* of measured wave trains reduces to an *ordinary linear Fourier series for sufficiently small waves*. On the other hand, for sufficiently large amplitude waves, when a *nonlinear threshold* is reached, then the second population waves begin to appear in the nonlinear Fourier spectrum. Once this threshold is sufficiently surpassed, a rogue sea is seen to naturally develop. This *modulational threshold* (first identified by Tracy, 1984) occurs when the modulational parameter exceeds 1. Thus, a natural measure, in $1 + 1$ dimensions, of the existence of rogue wave packets is the *modulational parameter*, discussed in detail in the following sections.

29.2 Linear Fourier Analysis and the Nonlinear Schrödinger Equation

It is common to consider *envelope equations* for describing wave trains with a narrow-banded spectrum. Let us consider the linear case first, which constitutes a linear superposition for the surface wave elevation in deep water:

$$\eta(x,t) = \sum_{n=-\infty}^{\infty} \eta(k_n) \, e^{ik_n x - i\omega_n(k_n)t + i\phi_n} \tag{29.1}$$

$\omega_n = \omega_n(k_n)$ is the linear dispersion relation. Now suppose that the spectrum $\eta_n = \eta(k_n)$ is narrow-banded, that is, most of the energy is centered near the (carrier) wavenumber k_o with frequency ω_o, where in deep water $\omega_o^2 = gk_o$. Thus, the range of frequencies in the spectrum are distributed about ω_o such that a simple series might be substituted for ω:

$$\omega = \omega_o + (k - k_o)\omega_o' + \frac{1}{2}(k - k_o)^2 \omega_o'' + \cdots$$

If we also introduce a *complex modulation function*, $\psi(x,t)$, then the surface elevation is given by

$$\eta(x,t) = \text{Re}\Big[\psi(x,t) \, e^{ik_o x - i\omega_o t}\Big] \tag{29.2}$$

We set $\Delta k_n = k_n - k_o$ (a small wavenumber difference about the peak wavenumber k_o), and we get

$$\psi(x,t) = \sum_{n=-\infty}^{\infty} \eta(k_o + \Delta k_n) \, e^{i\Delta k_n x - i\Delta \omega_n(k_n)t + i\phi_n}$$

which has the free field *Schrödinger equation of motion*

$$i(\psi_t + C_g\psi_x) + \mu\psi_{xx} = 0 \tag{29.3}$$

where $C_g = \omega'_o$ and $\mu = \omega''_o/2$ (primes mean d/dk_o) so that the linear dispersion relation is

$$\Delta\omega_n = \omega(k_n) - \omega_o = C_g\Delta k_n + \frac{1}{2}\omega''_o(\Delta k_n)^2$$

The surface elevation has the equation of motion:

$$i\eta_t - (\omega_o - k_o\omega'_o + \frac{1}{2}k_o^2\omega''_o)\eta + i(\omega'_o - k_o\omega''_o)\eta_x - \frac{1}{2}\omega''_o\eta_{xx} = 0 \tag{29.4}$$

with dispersion relation $\omega = \omega_o + (k - k_o)\omega'_o + (k - k_o)^2\omega''_o/2 + \cdots$.

In summary, Equation (29.3) is a linear wave equation describing the complex envelope function $\psi(x, t)$ of the surface elevation $\eta(x, t) = \text{Re}[\psi(x, t)\,e^{ik_o x - i\omega_o t}]$ that is assumed to be a narrow-banded, linear wave train. Thus, $\psi(x, t)$ describes wave groups with group speed $C_g = \omega'_o$ that linearly disperse as they propagate and the carrier wave $e^{ik_o x - i\omega_o t}$ has dispersion relation $\omega_o^2 = gk_o$ in deep water. Equation (29.3) is equivalent to the Schrödinger equation of quantum mechanics in free space. Because Equations (29.3) and (29.4) are linear equations, their Fourier structure is simple: The usual linear Fourier transform solves Equation (29.3) for all (Cauchy) initial conditions. So, assuming that Equation (29.3) is true, then (1) modeling of wave trains is straightforward (the fast Fourier transform algorithm suffices for computing the space/time evolution), and (2) the Fourier analysis of oceanic data is also straightforward. It is hard to imagine having a better theory than Equations (29.3) and (29.4). We have all the tools for understanding how linear waves behave in the oceanic environment. Of course, Equation (29.3) is narrow banded, but we can always improve this feature to arbitrary order by adding additional linear dispersive terms to the equation. However, ocean waves are *nonlinear*, so it is appropriate to move in that direction.

Recently, a number of authors have presented the case for understanding a number aspects of oceanic rogue wave dynamics through application of the modulational instability (Trulsen and Dysthe 1996, 1997a,b; Osborne et al., 2000; Trulsen et al., 2000; Onorato et al., 2001, 2002, 2004, 2005, 2006; Osborne, 2001, 2002; Janssen, 2003; Mori, 2007)). The basic idea is that one can simply modify Equation (29.3) to obtain leading order nonlinear effects not present in the linear approximation. To modify Equation (29.3) for nonlinearity, one adds a simple cubic term (Zakharov, 1968; Whitham, 1973):

$$i(\psi_t + C_g\psi_x) + \mu\psi_{xx} + v|\psi|^2\psi = 0 \tag{29.5}$$

where $v = -\omega_o k_o^2/2$. This is the so-called *NLS equation*, the simplest possible *nonlinear* wave equation for deep-water wave dynamics. This is just a small step toward full understanding of nonlinear water waves, but it is an important step, because as there is much to learn from Equation (29.5) and many of the properties of Equation (29.5) are important in the study of ocean waves.

Equation (29.5) has the list of surprising properties (1)-(4) given above (inclusion of currents requires either modification of the coefficients or the addition of an external forcing term (Johnson, 1997), or adding an external potential (Stocker and Peregrine, 1999)). These properties arise as a consequence of the exact solution of Equation (29.5) for periodic boundary conditions using the *inverse scattering transform*, the main topic of this book. The NLS equation and its spectral structure are discussed in Chapters 12, 18, and 24.

29.3 Nonlinear Fourier Analysis for the Nonlinear Schrödinger Equation

It is convenient to write the NLS equation in a form that is simpler for theoretical calculations:

$$iu_t + u_{xx} + 2|u|^2u = 0 \tag{29.6}$$

This equation arises from Equation (29.5) by a simple rescaling and Galilean transformation:

$$u = \lambda\psi, \quad x \to x - C_g t, \quad t \to \mu t \tag{29.7}$$

where $\lambda = \sqrt{v/2\mu}$.

The Fourier solution of the NLS equation (29.6) for periodic boundary conditions is given by Kotljarov and Its (1976), Tracy (1984), and Tracy and Chen (1988) (see also Belokolos et al., 1994):

$$u(x,t) = A\frac{\theta(x,t|\mathbf{B},\delta^-)}{\theta(x,t|\mathbf{B},\delta^+)}\, e^{2iA^2t} \tag{29.8}$$

where the Riemann theta functions, $\theta(x,t|\mathbf{B},\delta^\pm)$, are given by

$$\theta(x,t) = \sum_{m_1=-\infty}^{\infty}\sum_{m_2=-\infty}^{\infty}\cdots\sum_{m_N=-\infty}^{\infty}\exp\left[2\pi i\sum_{n=1}^{N}m_nX_n + \frac{1}{2}\sum_{m=1}^{N}\sum_{n=1}^{N}m_mm_n\tau_{mn}\right] \tag{29.9}$$

where

$$X_n = K_nx - \Omega_nt - \delta_n^\pm$$

The wavenumbers, K_n, frequencies, Ω_n, and phases, δ_n are computed by the methods of algebraic geometry (see Chapter 24). It should be noted that the theta functions (Equation (29.9)) are just generalized Fourier series, where the *spectral amplitudes* correspond to a (Riemann) *matrix*, τ, rather than to a *vector*, $\eta(k_n)$, $n = -\infty, \ldots, -2, -1, 0, 1, 2, \ldots, \infty$, as in linear Fourier analysis (Equation (29.1)).

To better understand the solutions of Equation (29.6) using the nonlinear Fourier decomposition (Equation (29.9)), note that the ratio of theta functions, $\theta(x, t | \mathbf{B}, \delta^-)/\theta(x, t | \mathbf{B}, \delta^+)$, is the *complex modulation envelope function*. When there is no modulation, $\theta(x, t | \mathbf{B}, \delta^-)/\theta(x, t | \mathbf{B}, \delta^+) = 1$, we have

$$u(x, t) = A \, e^{2iA^2 t} \tag{29.10}$$

This is the so-called *plane wave solution of the NLS equation*. It corresponds to an unmodulated carrier wave and physically provides the nonlinear Stokes wave correction to the frequency in the surface elevation.

Details of the spectral theory for the NLS equation are given in Chapters 12, 18, and 24. It is prudent to read these chapters before reading the rest of this chapter.

29.4 Marintek Wave Tank

This section describes results from experiments at Marintek (Norwegian Marine Technology Research Institute), a company of the SINTEF Group, located in Trondheim, Norway. Various statistical analyses of the data have been conducted and published elsewhere (Onorato et al., 2001, 2002, 2004, 2005, 2006; Mori, 2007). In particular, it has been shown not only theoretically, but also experimentally, that knowledge of the behavior of the solutions of the NLS equation is important for the development of laboratory procedures for generating rogue waves which arise as a consequence of the nonlinear behavior in surface water waves. Many of these procedures are described elsewhere in this monograph and are reviewed in this chapter and in the references.

The technical details of the facility can be found on the Web site http://www.sintef.no/Home/Marine/Marintek. The tank is under the directorship of Dr Carl Trygve Stansberg who in this capacity was crucial to the success of the experiments; he was also a participating scientist on the project. The tank is 260 m long and consists of two sections. The first section after the wave maker is 85 m long and has a width of 10.5 m and a depth of 10 m. The second section is 175 m long, a width of 10.5 m, and a depth of 5.6 m. Figure 29.1 shows the facility from the wave maker end of the tank. One can see the platform on which much of the instrumentation and computer facilities are housed. The wave maker (10.5 m wide and 10 m deep) is behind the viewer in Figure 29.1 and consists of two separately driven sections that are computer controlled. The probe locations used in the present experiments are given in Table 29.1.

Figure 29.1   The wave flume at Marintek, which is 10.5 m × 10 m × 260 m (the last 175 m are 5.6 m deep), as seen from the end of the tank near the wave maker. Many instruments and computers are located on the green carriage suspended above the tank in which a sinusoidal wave can be seen which reflects the ceiling lights. The carriage moves on rails for the full 260 m of the tank. (See color plate).

29.5   Deterministic Wave Trains as Time Series

The deterministic experiments analyzed here had periods in the range 1.3-1.5 s and wave amplitudes of about 20-25 cm, although some large waves had amplitudes up to 45 cm. The group speed and dispersion relation are given by

$$C_g = \frac{1}{2}\frac{\omega_0}{k_0}, \quad \omega_0^2 = gk_0, \quad \omega_0 = 2\pi f_0, \quad f_0 = \frac{1}{T_0}$$

so that for periods of $T_0 = 1.3$ s the group speed was about $C_g \cong 1.015$ m/s, while for $T_0 = 1.5$ s the group speed was $C_g \cong 1.171$ m/s. Phase speeds were twice these values, near 2 m/s.

We now compute the "time-like" Benjamin-Feir (BF) parameter, useful for the generation of extreme waves. Recall that the "space-like" parameter has the value

$$I_{BF} = 2\sqrt{2}\,\frac{k_0 a_0}{\Delta k/k_0} > n \sim \frac{\text{Carrier wave steepness } k_0 a_0}{\text{Spectral bandwidth } \Delta k/k_0}$$

Table 29.1 Table of the probe positions used in the Marintek experiment. Estimates of the time that a wave group propagates from one probe to another are also given in the right had column

Probe #	Distance - m	Time Between Probes
1	10	
2	29	17.08 sec
3	35	4.27 sec
4	40	4.27 sec
5	45	4.27 sec
6	60	12.81 sec
7	65	4.27 sec
8	70	4.27 sec
9	75	4.27 sec
10	75	0
11	75	0
12	80	4.27 sec
13	85	4.27 sec
14	115	25.62 sec
15	120	4.27 sec
16	160	34.16 sec
17	160	0
18	160	0
19	200	34.16 sec

Consider a simple deterministic wave train (a time series) with carrier period T_o and modulation period T_{mod}. Use the fact that $\Delta k/k_o = 2\Delta f/f_o$ (where $\Delta f = 1/T_{\text{mod}}$, $\Delta k \equiv K$ is the modulation wavenumber) and the *time-like* BF parameter becomes:

$$I_{\text{BF}} = \sqrt{2}\,\frac{k_o a_o}{\Delta f/f_o} > n \sim \frac{\text{Carrier wave steepness } k_o a_o}{\text{Spectral bandwidth } \Delta f/f_o} \tag{29.11}$$

or in terms of the number of carrier oscillations under the modulation, N_t:

$$I_{\text{BF}} = \sqrt{2}N_t k_o a_o > n \sim (\text{Number carrier oscillations } N_t)\,(\text{steepness } k_o a_o) \tag{29.12}$$

where the number of carrier oscillation in a time series is half that in a space series:

$$N_t = \frac{f_o}{\Delta f} = \frac{T_{\text{mod}}}{T_o} = \frac{1}{2}N_x \tag{29.13}$$

It is common to take the *characteristic wave steepness* in a *random wave train* in the form (we use these results below):

$$k_o a_o = \frac{\omega_o^2}{g} a_o = \frac{4\pi^2}{g} a_o f_o^2 = \frac{4\pi^2}{g} \frac{a_o}{T_o^2} \tag{29.14}$$

How high can deterministic wave packets become with respect to the carrier amplitude? From Equation (12.14):

$$\frac{A_{\max}}{a_o} = \left(1 + 2\frac{\lambda_I}{a_o}\right) = 1 + 2\sqrt{1 - \left(\frac{K}{2\sqrt{2}k_o^2 a_o}\right)^2} \tag{29.15}$$

The experiment I would now like to discuss is shown in Figure 29.2. The carrier period is 1.3 s and the modulation period is $T_{\mathrm{mod}} = 19.5$ s. The ratio of the modulation period T_{mod} to the carrier period T_o has the value $N_t = N_{\mathrm{mod}}/T_o = 15$. We used a carrier amplitude of $a_o = 4$ cm. The wavenumber is $k_o = \omega_o^2/g = 2.3813$ m^{-1} and the steepness is $k_o a_o = 4\pi^2 a_o/gT_o^2$ $= 0.10414$ m^{-1}. This gives the BF parameter $I_{\mathrm{BF}} = \sqrt{2}N_t k_o a_o = 2.2$, which means that NLS theory predicts that there are *two unstable modes in the measured wave train*. This suggests that two unstable wave packets should appear as the wave train propagates down the tank from the probe at 10 m.

The experiment shown in Figure 29.2 consists of a small amplitude modulation as generated by the wave maker. Already by the first probe, 10 m from the wave maker, we see that the modulation has grown considerably. Further down the tank each of the maxima in the modulation envelope has developed into quite concentrated, localized packets with relatively large amplitude waves arising as a consequence. Such behavior, the local focusing of wave packets in this fashion, is viewed as a result of the modulational instability. In this section I discuss a new method for the study of such wave trains, a method that can be referred to as *nonlinear Fourier analysis for the NLS equation*. In this context we see elsewhere in this monograph that a wave train with nonlinear Fourier modes that are "BF unstable" and "breathe in time" are indicative of NLS behavior. Of course, water waves in principle are described in $1 + 1$ dimensions by NLS only at leading nonlinear order. Higher order effects will clearly have an influence on the behavior of deep-water wave trains, as well as the influence of wave breaking and the presence of wind.

It is also useful to look at the first two "constants of the motion" for the NLS equation to see if indeed they are constant in the present experiments, a possible measure of the correctness of NLS as a physical model. These constants (for the time NLS equation, tNLS, see Chapter 12) are computed by

$$\int_0^T \psi\psi^* \, dt = \text{constant}, \qquad \int_0^T (\psi^*\psi_t - \psi\psi_t^*) \, dt = \text{constant}$$

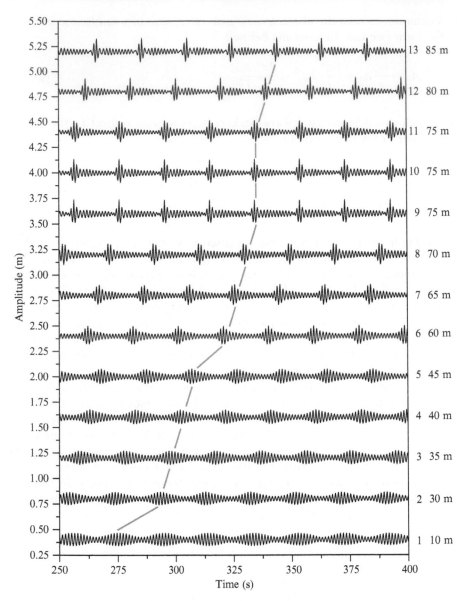

Figure 29.2   Experiment conducted at Marintek as measured by 19 probes distributed along the length of the tank, Table 29.1. This is a simple small-amplitude modulation. The line segments have been added to aid the eye in following the evolution of a single packet.

In the experiments of Figure 29.2 the carrier wave period was 1.3 s and the modulation period was 19.5 s. By looking at the time series at the individual probes and computing the standard deviation, I found the results given in the upper curve of Figure 29.3. The standard deviation of the waves (equivalent to the first constant of the motion, easily proved by the properties of the

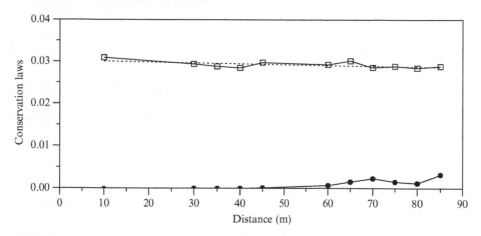

Figure 29.3 Decay of wave trains due to dissipation as measured by the standard deviation of the time series registered at each probe.

Hilbert transform, see Chapter 13) was about 3.0 cm with an error of 2.2%. Shown is an exponential curve that has been fit to the data for which the $1/e$ decay constant was 1500.8 m. This decay is due to all sources of dissipation in the tank and hence the standard deviation of the waves would be expected to be reduced to $1/e = 0.3679$ of its value at the wave maker after about 1.5 km, well beyond the 260 m length of the tank. The standard deviation decreases about 7% as the wave train propagates from the probe at 10 m to the probe at 85 m. The second constant of the motion is shown as the lower curve (marked by solid dots). It is seen that there is a slow *increase* in the value of this integral over the first 85 m of the tank. Strictly speaking, neither of the two integrals is a true constant of the motion and therefore the waves are *not* governed by the NLS equation. Segur et al. (2008) have studied the particular case where the BF rise time/distance is on the order of the dissipation decay time/distance, which occurs for small amplitude waves, $\sim$2 mm; in this case the dissipation dominates the dynamics and the BF instability is stabilized.

However, as seen below our perspective should be modified by noting that the actual wave dynamics correspond very nearly to an NLS equation whose dynamics are *slowly varying* around the actual NLS dynamics. The inverse scattering transform provides a tool to give us this perspective. In Figure 29.4 are the wave trains that I am analyzing by the IST of NLS. I have translated the wave trains into a system of coordinates that moves with the group speed of the measured packets (I actually use the measured packet speed, not the linear group speed, although they do not significantly differ). Also I show only one period of the packet trains. In the lower panel (at 10 m from the wave maker), we see a modulated wave packet that has been generated by the wave maker as a small amplitude modulation. As the packet propagates down the tank it focuses, becomes narrower and higher as a consequence of the modulational instability. There are 11 probes under investigation in the present experiment. Is the nonlinear

Figure 29.4 Sequence of time series of length equal to one modulation wavelength as they propagate down the tank. The data have been temporally translated so that the peaks of the largest waves are centered in the time series.

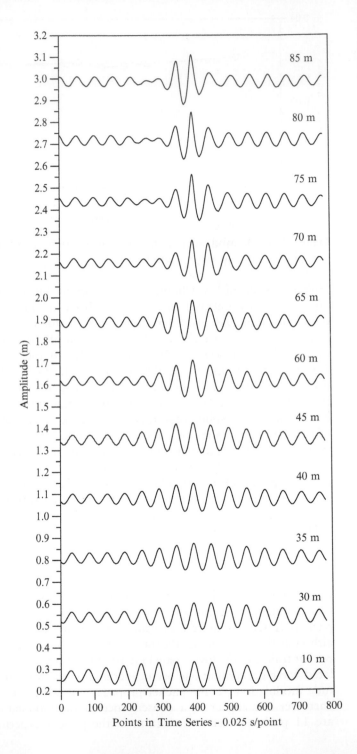

evolution of Figures 29.2 and 29.4 described exactly by the NLS equation? Probably not, but is there instead a sense in which the predictions of the NLS equation can be useful to us?

We will see that the answer to the last question is "yes" and the tool we use to study the data of Figure 29.4 is the Zakharov-Shabat eigenvalue problem for periodic boundary conditions (Floquet analysis, see Chapters 12, 18, and 24) in the complex lambda plane (see discussion in particular with regard to the spines) for each of the wave trains in Figure 29.4. These results are shown in Figure 29.5A-K. Let us first look at Figure 29.5A that shows the IST spectrum of the first wave train at the bottom of Figure 29.4 at 10 m from the wave maker. It is useful to look at Figure 18.5 of Chapter 18 to interpret the results of Figure 29.5 of this chapter with regard to "slot states." The spectrum of Figure 29.5A consists of a carrier wave (the uppermost eigenvalue " $\times$ " in the figure) and two unstable wave packets that are slot states below the carrier. Thus, the modulated packet at the bottom of Figure 29.4 will presumably "fission" into two unstable packets during its evolution down the wave tank. Over what distance will the unstable packets appear? Use of the "time-like" BF parameter (Equation (29.11)) together with Equations (12.16) and (12.17) and Figure 12.4 of Chapter 12 tells us that the "rise distance" for the large slot state in Figure 29.5A is about 70 m. We also expect that the two packets in the spectrum will be separated from each other by mutual repulsion and appear clearly to the eye as the wave train propagates down the tank (see Figure 29.6 for an enlarged view of the wave train as it appears at the probe 85 m from the wave maker).

What happens as the wave train propagates down the tank? Does the spectrum remain time invariant, that is, does NLS evolution govern the motion of the waves? The answer is probably "no," but we anticipate and hope that the evolution can be viewed approximately as NLS dynamics with slowly varying spectral components, at least slower than their linear Fourier modes which clearly evolve during the evolution of unstable wave trains. In the present case we watch the *nonlinear Fourier components of the NLS spectrum* as the wave train evolves down the tank. If the eigenvalues and spines are fixed during the evolution then we can definitely say that the wave trains obey the NLS equation. In Figure 29.5B we see that the wave train has evolved over 20 m to a condition somewhat different: the carrier amplitude has diminished slightly, while the average distance between the simple eigenvalues in the smaller unstable mode has decreased.

Moving to Figure 29.5C we see that the smallest unstable (slot) mode has disappeared. Some energy has been transferred to two smaller sidebands, but the interesting thing here is that we have lost an unstable mode, at least momentarily. This same spectral picture is shown in Figure 29.5D where one unstable mode and a large carrier dominate the spectrum. However, to continue with the surprises, the next frame in Figure 29.5E (45 m from the wave maker) indicates that the smaller unstable mode reappears (at the expense of the two sidebands), returning almost to the situation in the initial wave train at 10 m. As the wave train continues to propagate down the tank it maintains

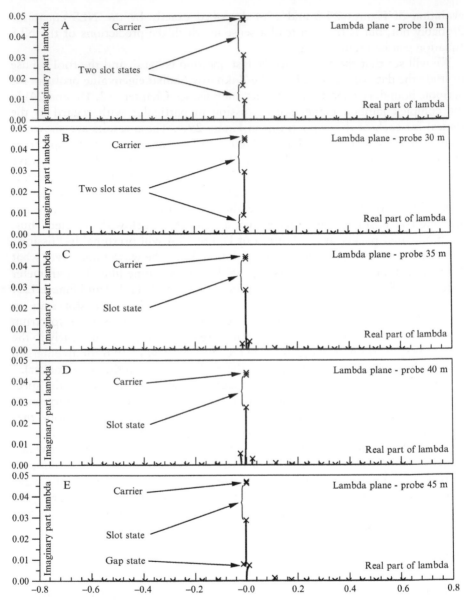

Figure 29.5   (A)-(K) IST of data from probes in Figure 29.4 shown in the lambda plane.
(*Continued*)

the same three modes, that is, a carrier and two unstable modes. There is, however, some growth in the small amplitude side bands (nearly linear, stable Fourier modes) during the evolution. At the probe at 85 m we have an interesting situation, that is, the two unstable modes become visible, showing that the IST for NLS has predicted correctly the two unstable mode amplitudes, which

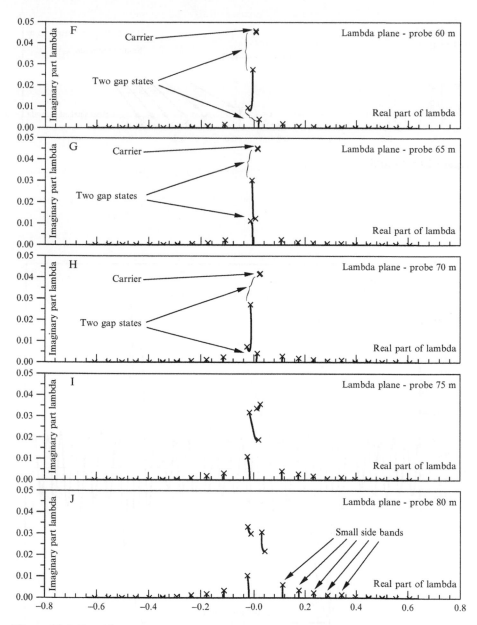

Figure 29.5 Cont'd

have emerged into view (Figure 29.6). One can conclude that the dynamics of
the sequence of time series in Figure 29.4 was governed by an evolution which
was consistent with the focusing of two unstable modes in the presence of a
carrier wave, but that the evolution of the spectral properties of these three
components (an effect beyond the order of the NLS equation) is also important

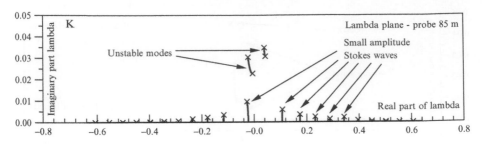

Figure 29.5 Cont'd

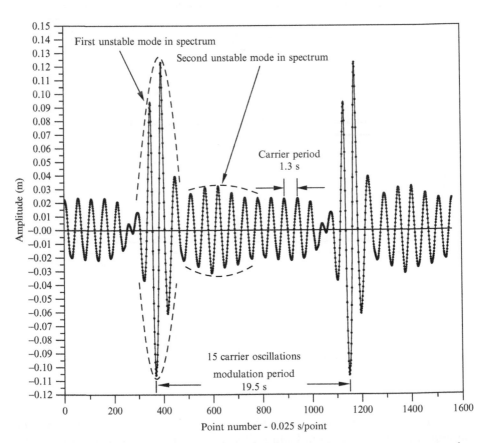

Figure 29.6 Two periods of the deterministic wave train at 85 m from the wave maker (from Figure 29.4). The dotted envelopes are added to aid the eye in locating the two unstable packets seen in the NLS spectrum in Figure 29.5.

for describing the dynamics. This evolution beyond NLS appears to be slow, in some sense adiabatic, in the present case, although discontinuous evolution (note the disappearance of the smaller unstable mode at 35 m from the wave maker) is also evident. It is important to note that the five eigenvalues dominant the motion for the entire set of experimental observations: there is one eigenvalue for the carrier and two each for each of the unstable modes. These results suggest that NLS is a good place to start to understand nonlinearities in deep-water wave trains, but that higher order evolution is also important. Indeed, one could be motivated to develop a set of ordinary differential equations for the *time evolution of the NLS spectrum* in order to include higher order effects. We have seen that while dissipation is important in the present experiments, it does not preclude the natural evolution of the modulational instability. This is of course normal in the present case because the BF growth distance is in the order of 85 m, while the decay distance is in the order of 1.5 km. The modulational instability occurs on spatial and temporal scales that are much faster than the spatial/temporal scales of dissipation, thus allowing time for the instability to carry out its evolution before the waves decay.

In order to aid us in physically viewing the appearance of the modulational instability in the present experiments, I show in Figure 29.7 the evolution of the *surface elevation* of the wave group as it propagates down the tank. Each

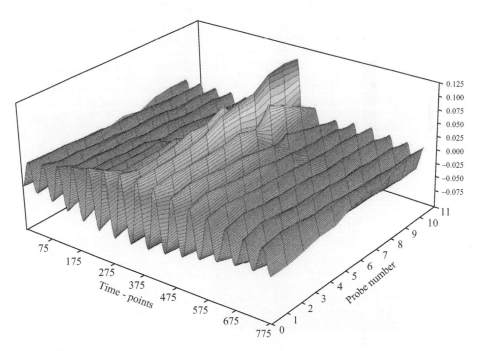

Figure 29.7   Surface wave evolution of deterministic wave train out to 85 m from the wave maker. Surface constructed from individual time series in Figure 29.4. (See color plate).

of the time series in Figure 29.4 is graphed as a function of time and of the probe positions. One can easily see the growth of the packet maximum as well as the decrease in the packet width during the evolution of the wave train. This provides a qualitative view of the modulational instability in the presence of small dissipation. The contours of Figure 29.7 are shown in Figure 29.8. These two figures illustrate the focusing action of the BF instability, that is, the initial broad, low packet becomes narrower, and higher during the evolution.

While the evolution of the wave train in Figure 29.4 maintains a consistent picture from the initial paddle-generated wave train out to about 70 m, the subsequent evolution of the NLS spectrum undergoes drastic change as can be seen between Figure 29.5H and I. During this point in the evolution the carrier drops rapidly and the open gap in the largest unstable mode decreases its value of ε and the value of the angle θ in the λ plane rotates about 80°. Indeed, it is unclear whether the idea of a slot state is still appropriate; one might reinterpret the dynamics in terms of another path for the loop integrals and hence to

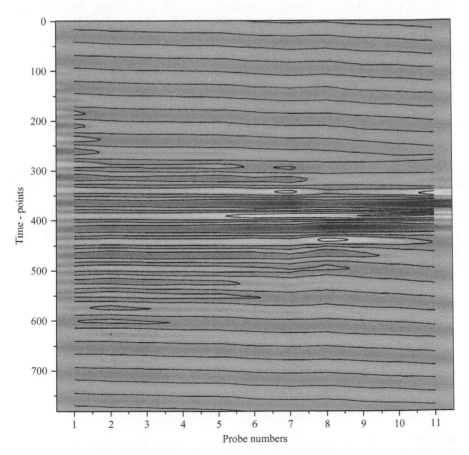

Figure 29.8 Contours of wave evolution of Figure 29.7 of a deterministic wave train out to 85 m from the wave maker. Surface and contours constructed from individual time series in Figure 29.4. (See color plate).

describe the spectrum as a small amplitude carrier plus two unstable modes with points of simple spectrum connected by spines. Physically, this complex, non-NLS behavior seems to be occurring during the "fissioning" of the two packets between the probe measurement at 70 and 75 m. Evidently, nonlinear effects beyond the order of the NLS equation will be necessary to describe the physics in detail. These results indicate an interesting direction for future research.

29.6 Random Wave Trains

This section is dedicated to the study of nonlinear random wave trains in the Marintek tank. As seen for the case of deterministic wave trains we can take the nonlinear Fourier transform of time series data, just as we can take the linear Fourier transform. However, the information is quite different because of the possible presence of packets that are BF unstable even though the wave trains are random. We begin with a discussion of the application of the theory of the NLS equation to random wave trains and then proceed to the analysis of the experimental data.

29.6.1 Characteristics of Random Wave Trains Using IST for NLS

Random wave trains from the point of view of the inverse scattering transform arise when the phases $\delta_n^{\pm}$ in Equation (29.9) are taken to be uniformly distributed random numbers. Of course one realizes that there must also be a constraint relation between the two sets of phases as discussed in detail in Chapter 24 and therefore the two sets of phases are not independent of one another. Each set of dual phases $\delta_n^{\pm}$ leads to a single realization for a surface wave train governed by the nonlinear Fourier representation (Equation (29.8)). What are some of the properties of wave trains governed by this formulation? We now discuss a few.

We focus on a time series of length T, significant wave height $H_s = 4\sigma$ (σ is the *standard deviation* of the *random time series*) for which f_o is the peak spectral frequency. Use the fact that $\Delta k / k_o = 2\Delta f / f_o$ (where $\Delta f = 1/T$ is the spectral frequency resolution and $\Delta k \equiv K$ is simultaneously the wavenumber resolution and the minimum modulation wavenumber) and the BF parameter becomes:

$$I_{BF} = \sqrt{2} \frac{k_o a_o}{\Delta f / f_o} > n \sim \frac{\text{Carrier wave steepness } k_o a_o}{\text{Spectral bandwidth } \Delta f / f_o} \tag{29.16}$$

or in terms of the number of carrier oscillations under the modulation, N_t:

$$I_{BF} = \sqrt{2} N_t k_o a_o > n \sim (\text{Number carrier oscillations } N_t)(\text{steepness } k_o a_o) \tag{29.17}$$

where the number of carrier oscillations in a time series is half that in a space series:

$$N_t = \frac{f_o}{\Delta f} = \frac{T}{T_o} = \frac{1}{2} N_x = \frac{1}{2} \frac{k_o}{K} \tag{29.18}$$

It is common to take the *characteristic wave steepness* in a random wave train in the form:

$$k_o a_o = \frac{\sqrt{2}\pi^2}{g} H_s T_o^{-2} = \frac{\sqrt{2}\pi^2}{g} H_s f_o^2 \tag{29.19}$$

where we have used $a_o = \sqrt{2}\sigma$ and $k_o = \omega_o^2/g$. This definition is convenient because for a sine wave of amplitude a we have $\sigma = \sqrt{2}a/2$ and hence an estimate of the carrier amplitude gives the required result $a_o = \sqrt{2}\sigma = a$. We are left with *an estimate of the BF parameter for a random wave train*:

$$I_{BF} = \frac{2\pi^2}{g} \frac{H_s f_o^3}{\Delta f} = \frac{2\pi^2}{g} H_s f_o^3 T > n \tag{29.20}$$

This expression provides a convenient way to estimate the number of unstable wave packets $n = [I_{BF}]$ in a random time series (the square brackets mean "integer part of"). Note that the number of unstable modes I_{BF} is proportional to the length T of the time series and to the significant wave height H_s. The cubic dependence on the frequency means that in growing, nonequilibrium sea states, where the dominant frequency is larger, one can have some very interesting rogue waves! To this end, Figure 29.9 shows some of the important aspects of a time series (or space series) and its spectrum. The figure shows a

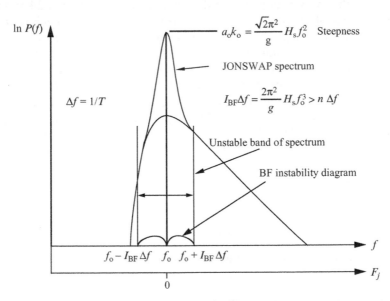

Figure 29.9 A JONSWAP frequency power spectrum with enhancement parameter $\gamma = 6$. The small double lobes to each side of the dominant frequency represent the instability diagram from the NLS equation. Shown are the necessary parameters for computing the Benjamin-Feir nonlinearity parameter: $I_{BF} = (2\pi^2/g)H_s f_o^3/\Delta f$. The actual form of the BF instability diagram shown above is given in Fig. 12.4.

JONSWAP power spectrum with $\gamma = 6$. It is easy to see why enhancing γ increases the BF parameter and therefore increases the number of unstable packets in a wave train. This occurs because enhancing γ increases the steepness and decreases the bandwidth of the spectrum.

How high can unstable wave packets become with respect to significant wave height? Use $a_0 = \sqrt{2}\sigma$, $H_s = 4\sigma$ in Equation (24.9) we find:

$$H_{max} = \frac{\sqrt{2}}{2}\left(2\frac{\lambda_I}{a_0} + 1\right)H_s \tag{29.21}$$

where λ_I is the imaginary part of the λ plane eigenvalue. For example, with $\lambda_I = a_0/\sqrt{2}$ we have $H_{max} = 1.704H_s$, for $\lambda_I = a_0$ we get $H_{max} = 2.121H_s$, and for $\lambda_I = \sqrt{2}a_0$ then $H_{max} = 2.707H_s$. These may be compared to an often-assumed "definition" of a rogue wave: $H_{max} > 2H_s$.

What does the estimate of the BF parameter (Equation (29.20)) have to do with the JONSWAP spectral domain of Figure 29.9, the instability diagram (Figure 12.4 of Chapter 12) and the maximum height of an unstable mode (Figure 12.5 of Chapter 12)? I now discuss Figure 29.9 and how to get all of this information out of the spectrum. In Figure 29.9 there is a *band of unstable spectrum about the peak* f_0. The width of the band of unstable spectrum is proportional to the BF parameter:

$$I_{BF}\,\Delta f = \frac{2\pi^2}{g}H_sf_o^3 > n\,\Delta f$$

In Figure 29.9 we see the double-lobed structure of the BF instability diagram below the spectrum (compare to Figure 12.4 where the left hand lobe is the mirror image of the right hand lobe). Thus, we see that the unstable band lies in the frequency range $f_0 \pm I_{BF}\Delta f$. This says that nonlinearity in the NLS equation lies in a *band of spectrum* (width $2I_{BF}\Delta f = 2n\Delta f$) about the peak; there are n Fourier components to the right of the peak that are unstable and n to the left. The width of the band is proportional to the BF parameter, which is after all a *global* estimate of how nonlinear a sea state is. If $I_{BF}\Delta f > n$ is large then we have a *rogue sea*, that is, there are many rogue wave components in the spectrum. The larger is the BF parameter the more nonlinear modes there are. The smaller is the BF parameter the fewer unstable modes there are. If I_{BF} is small enough then no spectral components are unstable and the spectrum can be computed with linear Fourier analysis. We have mentioned that there are two kinds of spectrum: (1) linear or almost linear stable sine or Stokes waves and (2) unstable "rogue" modes centered in a band about the center of the spectrum. Thus, the linear components are outside the band of unstable modes. Basically, *nonlinearity in the NLS equation occurs in the unstable band* about the peak of the spectrum. Spectral components in this spectral band can "breathe," that is, oscillate up to their maximum height (and be interpreted as a "rogue wave" or "rogue packet") and back down again (FPU recurrence). Here are some properties of the rogue waves in the spectrum of Figure 29.9:

The number of rogue waves in the wave train (Equation (29.20)):

$$[I_{\mathrm{BF}}] = \left[\frac{2\pi^2}{g}\frac{H_s f_o^3}{\Delta f}\right] = \left[\frac{2\pi^2}{g}H_s f_o^3 T\right] = n$$

where $[\cdot]$ means "integer part of." We see that the integer part of I_{BF} is proportional to the significant wave height and length of a time series, but is inversely proportional to the cubed period of the carrier. If one wants more rogue waves in an experiment, one needs to increase the significant wave height and increase the number of carrier oscillations under the modulation (that is make the spectrum narrower).

The instability diagram for jth unstable mode growth rate for distance:

$$K = i\omega_o k_o^2 a_o^2 \left(\frac{2\pi F_j}{\sqrt{2}f_o k_o a_o}\right)\sqrt{1 - \left(\frac{2\pi F_j}{\sqrt{2}f_o k_o a_o}\right)^2}$$

Here j is an integer over the range $1 \le j \le n$.

The maximum wave height of jth unstable mode:

$$H_{\mathrm{max},j} = \frac{\sqrt{2}}{2}\left(2\frac{\lambda_j}{a_o} + 1\right)H_s, \quad \frac{\lambda_j}{a_o} = \sqrt{1 - \left(\frac{2\pi F_j}{2\sqrt{2}f_o k_o a_o}\right)^2}$$

where λ_j is the amplitude of the centroid of the two points of simple spectrum connected by a spine (an unstable mode). Here $F_j = f_j - f_o$.

Distance to maximum amplitude of unstable mode:

$$\varepsilon\, e^{\gamma_j X} \sim O(1)$$

$$\gamma_j = \omega_o k_o^2 a_o^2\left(\frac{2\pi F_j}{\sqrt{2}f_o k_o a_o}\right)\sqrt{1 - \left(\frac{2\pi F_j}{\sqrt{2}f_o k_o a_o}\right)^2}$$

$$T_{\mathrm{BF},j} = |\varepsilon_j|/\gamma_j$$

Again j is the jth component from the dominant frequency of the spectrum. Here γ_j is the inverse growth rate from the instability diagram; γ_j is not related to the JONSWAP parameter γ.

In summary, there is a band of unstable Fourier modes about the dominant frequency in the spectrum. The width of this band depends on the size of the BF parameter, from which one can make estimates of the growth rate, maximum rogue wave amplitude ($\sim H_{\mathrm{max},j}$), and distance required for the rogue wave to rise up from the initial conditions ($\sim 1/K = 1/\gamma_j$). An *ordinary Fourier spectrum* then provides *all the information one could want about any rogue waves in the spectrum*. Of course, one should instead use the inverse scattering transform to give more refined estimates, results we see in Section 29.6.3.

29.6.2 Measured Random Wave Trains

We have conducted a number of deep-water, random wave experiments in the facility at Marintek in Trondheim, Norway. We conducted the experiments discussed herein using standard software for wave generation with random Fourier phases and the JONSWAP power spectrum:

$$
P(f) = \frac{g^2 \alpha^*}{(2\pi)^4 f^5} \exp\left[-\frac{5}{4}\left(\frac{f_d^*}{f}\right)^4\right] \gamma^{\exp\left[-\frac{1}{2}\left(\frac{f-f_d^*}{\sigma_0 f_d^*}\right)^2\right]}
\tag{29.22}
$$

For present purposes we varied only the parameters γ, α^*, the others remained their standard values. Nineteen probes were placed along the tank and time series of one-half hour were recorded at a rate of 40 Hz. A typical experiment is shown in Figure 29.10–29.13, where we used $\gamma = 6$.

29.6.3 Nonlinear Spectral Analysis of the Random Wave Trains

First let us look at the evolution of a single packet in a random wave train, shown in Figure 29.10. As before the data are placed in a coordinate frame for which the packet maximum wave height is centered on the measurement interval, here windowed to a 25 s interval. We see the initial packet at the bottom of the figure propagate down the tank becoming more narrow and enhancing its height up to a maximum value. The wave then decreases its amplitude and broadens as it concludes an FPU cycle. The surface associated with this evolution is given in Figure 29.11 and the contours are shown in Figure 29.12. Even in this random case we see the quite classical action of the BF instability. Now let us look at full, nonlinear random time series obtained in the Marintek tank.

Figure 29.13 shows the results of one experiment with random wave trains, generated by Marintek software, for which the JONSWAP power spectrum with random Fourier phases was used to drive the wave maker. Line segments are used to help the eye follow the evolution of wave packets in the measured data. At the first probe at 10 m from the wave maker (the lowest time series in Figure 29.13 and shown enlarged in Figure 29.14), we see considerable packet structure, but the packets are relatively low and broad, although some nonlinear evolution has already occurred. In contrast is the signal measured at probe 8 (Figure 29.15) that is 70 m from the wave maker. In this latter time series we see packets that are quite narrow and considerably higher than those at probe 1. Indeed, one packet has a central wave height of 32.53 cm that is over 5 standard deviations. Another packet has a height of 31.07 cm or 4.8 standard deviations. A third packet has a wave height of 23.75 cm or about at 3 standard deviations. The behavior at probe 1, only 10 m from the wave maker is relatively broad and low, and is nearly Gaussian in its statistical behavior; the behavior at probe 8 is highly non-Gaussian in the amplitudes and non-Rayleigh in the wave heights. The important point is that by ensuring that the modulational parameter was large for these random wave experiments, we were able to elicit the BF

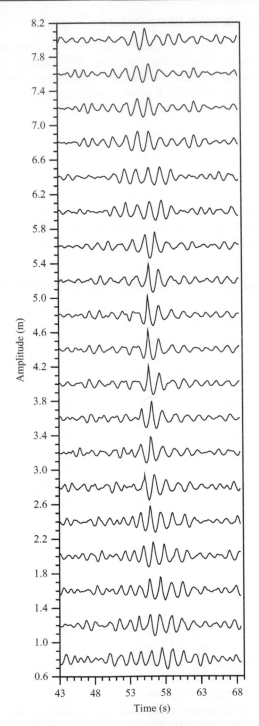

Figure 29.10 Evolution of an unstable wave packet in a random wave train. See Figure 29.13 for the probe distances for each time series.

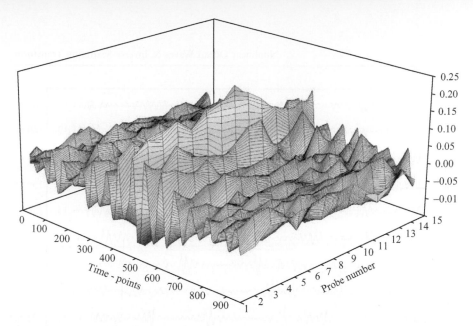

Figure 29.11 Surface of the random wave packet of Figure 29.10 shown as a function of probe number and time. Note that the redundant probes have been removed. (See color plate).

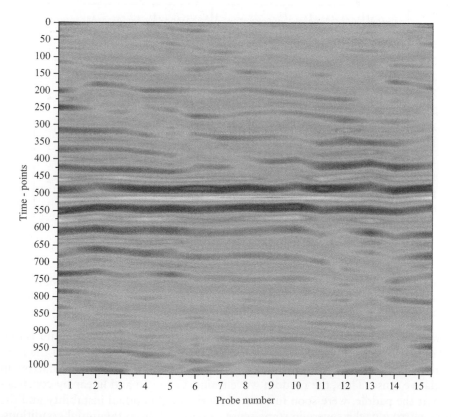

Figure 29.12 Contours of the random wave packet of Figure 29.10 shown as a function of probe number and time. Note that the redundant probes have been removed. (See color plate).

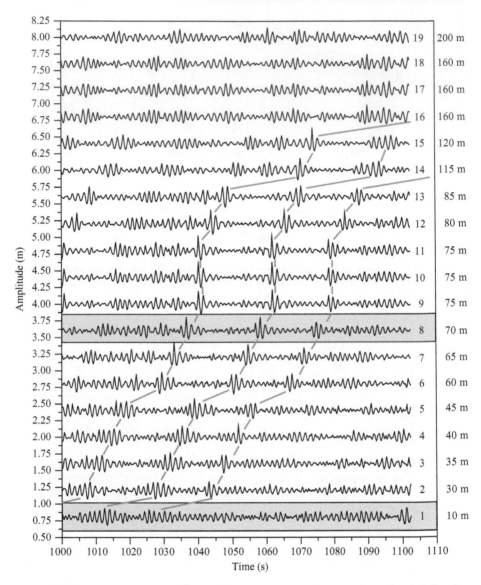

Figure 29.13   A 100 s section of a random wave experiment conducted at Marintek. We used the value $\gamma = 6$ for the JONSWAP power spectrum. Note that all the probes, even those which are redundant at 75 and 160 m are shown.

instability in random wave trains. Indeed, the single most important point is that we enhanced the constant γ in the JONSWAP spectrum in order to add the non-equilibrium peak on the spectrum, thus making the spectrum high and narrow in a relative sense. Thus, the random wave trains, Gaussian and linear by construction at the paddle, were soon influenced by the modulational instability and the large narrow packets were in some sense "fissioned" from the initial conditions. This is a result predicted by the inverse scattering transform for the NLS

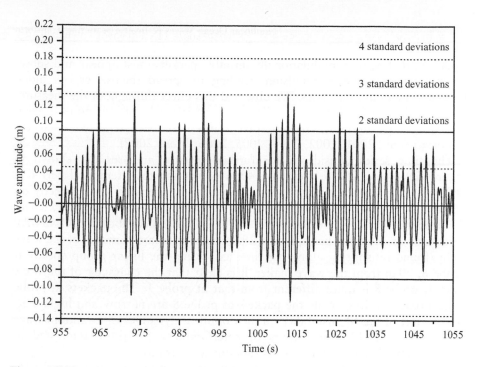

Figure 29.14 A 4096 point time series from probe 1 at 10 m from the wave maker.

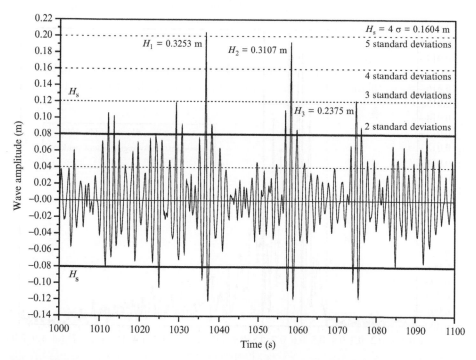

Figure 29.15 A 4096 point time series from probe 8 at 70 m from the wave maker. Three extreme waves have amplitudes that are greater than three standard deviations. One of the waves is greater than twice the significant wave height.

equation, that is, that modulationally unstable packets must appear and subsequently undergo mutual repulsion, tending to spread themselves out along measured random time series. We observed this effect throughout the measurements at Marintek. Statistical analyses emphasizing the strong non-Gaussian behavior of BF unstable random wave trains have been reported elsewhere (Onorato et al., 2001, 2002, 2004, 2005, 2006; Mori, 2007).

The time series considered for further spectral analyses are shown in Figures 29.14 and 29.15; they have 4096 points and their temporal period is 102.4 s. For reference we put several properties of the wave train directly on Figures 29.14 and 29.15. The time series in Figure 29.14 is at probe 1, where the properties of the waves are still quite like those expected of the JONSWAP spectrum. On the other hand, Figure 29.15 shows the same interval of the wave train (found by shifting along the time axis using the linear group speed) at probe 8, 70 m from the wave maker. It is clear that the character of the wave train at probe 8 is quite different from that at probe 1. The packets at probe 1 are broad and low, while the packets at probe 8 are narrow and high. This is the effect of the BF instability on the nonlinear dynamics of a random wave train. Figure 29.13 offers hours of entertainment for those interested in learning how this instability affects random wave trains.

I show the linear Fourier transform of the probe 8 time series in Figure 29.16. Also shown are the bounds of the band-pass filter used to remove the Stokes

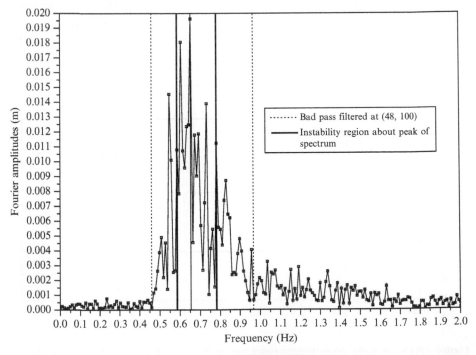

Figure 29.16 Fourier transform of time series at probe 8 in Figure 29.15. The location of the band-pass filter that removes the Stokes contribution is also shown.

contribution to the wave dynamics. This is a necessary step, because the NLS equation does not directly contain the Stokes effect, which is instead included only in Equation (29.2) and its higher order contributions (Equation (2.38), Chapter 2).

The filtered wave train is shown in Figure 29.17, along with the modulus of the envelope of the wave train, which has been computed using the Hilbert transform. The standard deviation σ of the wave train and the "amplitude of the carrier wave" $a_o = \sqrt{2}\sigma$ are shown in the figure.

In Figure 29.18 I show the results of the inverse scattering transform computation on the time series of Figure 29.17. I now discuss briefly how to interpret this interesting nonlinear spectrum. Note that the horizontal frequency axis is centered at the peak of the spectrum where the frequency is taken to be zero; this choice of the horizontal axis is also shown in Figure 29.9. There are two kinds of IST spectrum. The first kind of spectrum has simple sine waves (or at most low-amplitude Stokes wave components) that are shown connecting to the frequency axis by a line, a "spine." These are the low-lying components to the right and left of the spectrum (they are denoted by crosses) and one can think of them as being like ordinary linear Fourier components. The other kind

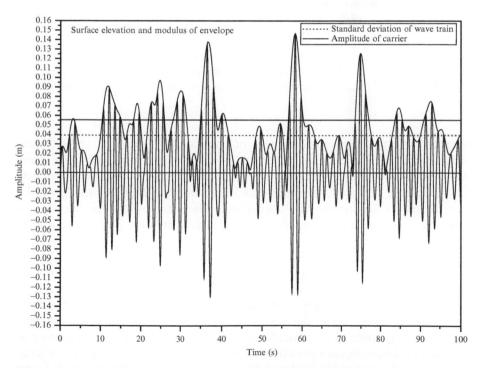

Figure 29.17 Application of the Hilbert transform to the band-pass filtered time series at probe 8 in Figure 29.15 to determine the modulus of the envelope of the wave train. This step also includes the filtering operation discussed in Figure 29.16. Note that the time series is now up-down symmetric thanks to the filtering operation which removed the Stokes wave nonlinearity.

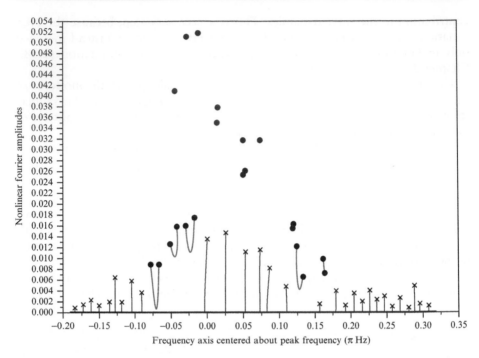

Figure 29.18 Inverse scattering transform spectrum of the band-pass filtered time series at probe 8 in Figure 29.17. The red eigenvalue pairs connected by spines correspond to unstable wave packets. Note that the frequency axis is centered at the peak frequency.

of spectrum is totally new and consists of unstable wave packets. These consist of two points of simple spectrum (denoted by large dots) connected by a spine. When the two points of spectrum are degenerate, no spine can be seen because the two points lie almost on top of each other. In other cases the spines can be seen clearly connecting the two eigenvalues. There are 14 unstable modes, the larger of which are candidates for extreme waves at some point during their nonlinear evolution.

To properly interpret the IST spectrum of Figure 29.18, we compare to the linear Fourier spectrum on the same scale, see Figure 29.19. All of the modes are stable and consist of sine waves. How can the nonlinear spectrum in Figure 29.18 have so many unstable wave packets? Because, simply put, they have robbed energy from the surrounding linear Fourier modes. Since the maximum amplitude of the unstable modes is proportional to the centroid of the pairs of simple eigenvalues connected by a spine, we can compare the large amplitudes of the unstable modes of Figure 29.18 to the small amplitude linear Fourier modes of Figure 29.19. There is a considerable difference and we are led to the conclusion that the unstable modes can be very large with respect to their linear Fourier counterparts.

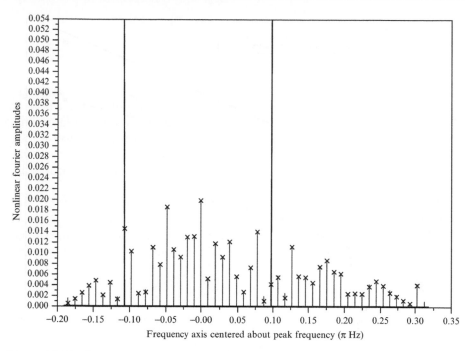

Figure 29.19 Linear Fourier spectrum of the time series at probe 8 in Figure 29.17. The scale is the same as the IST spectrum in Figure 29.18, so that comparison of the two can be made.

We finally compare the heights of the largest observed packets in the time series of Figure 29.17 with inverse scattering theory using Figure 29.18 and Equation (29.21). The results are shown in Figure 29.20. The theory of Equation (29.21) is shown as a solid line. The wave heights measured from Figure 29.17 (which has been filtered for the Stokes effect) are shown as solid squares. The wave heights measured from Figure 29.15 (no filtering for the Stokes effect) are shown as open squares. One does not expect perfect agreement between theory and experiment because the measurements at probe 8 give the packet heights only *at one spatial location*. Since the packets are unstable their amplitudes are undergoing considerable space/time dynamics and we cannot expect that they will all be at their maximum heights at *any particular spatial location*. The results of Figure 29.20 show the observed wave heights to fall below their maximum theoretical heights, an expected result.

It is interesting to note that the BF parameter, as computed by Equation (29.20), is $I_{BF} = 9.79$ for the measured time series. This result is based upon linearized modulation theory and should be compared to the actual number of unstable packets in Figure 29.18, namely, 14. Complete inverse scattering theory contains the full NLS spectrum, including *large amplitude modulations*. In the present case the number of fully nonlinear modes is 14, larger than the nine modes estimated by the BF parameter.

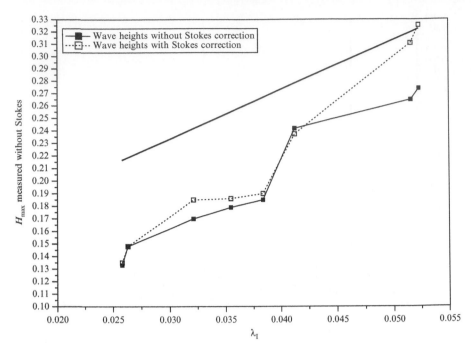

Figure 29.20 The largest packet heights predicted by theory (solid line) to the actual packet heights measured from Figure 29.15 (with Stokes contribution) and Figure 29.17 (Stokes contribution filtered out).

29.7   Summary and Discussion

We have used the periodic inverse scattering transform to study the nonlinear dynamics of deep-water wave trains, both theoretically and experimentally. Experimentally, we have used IST as a time series analysis tool to enhance our understanding of measured wave trains in the wave tank facility at Marintek in Trondheim, Norway. We have discussed how deep-water wave trains have two kinds of spectrum, namely, a near linear component and a separate component of unstable wave packets. These packets are discrete nonlinear Fourier components of the IST spectrum; they have their own nonlinear space/time dynamics and also nonlinearly interact with one another and the near-linear background sea state.

When does the NLS equation fail in the analysis of the Marintek data using the nonlinear Fourier transform of NLS? What kinds of higher order non-linearity do we see beyond the NLS equation? Generally speaking we might expect NLS to fail when the nonlinearity is large, but more specifically:

(1) When the waves have large steepness, $k_o a_o \sim 0.3$.
(2) When the waves become asymmetrical starting from symmetrical initial conditions (Lo and Mei, 1985).

(3) When there are nonlinear interactions, fissioning, and collisions.

(4) When there is wave breaking (Tulin and Waseda, 1999).

For the kinds of dynamics seen in the present experiments and analysis, the wave trains seem to show signs of evolving by higher order dynamics and not NLS dynamics at certain times during their evolution:

(1) The waves become asymmetrical for symmetrical initial conditions, see Figure 29.2.

(2) A mode disappears during evolution and manifests itself as side bands, see Figure 29.5B and C, and then the unstable mode reappears, Figure 29.5E.

(3) The unstable modes move rapidly around the lambda plane during interactions or fissioning, see Figure 29.5H-K. The explanation of the "brisk movements" in the NLS spectrum, at higher order than the NLS equation, is unknown.

(4) Considerable energy is injected into the radiation modes at higher frequency with respect to the unstable modes, a tendency for dynamical evolution toward "broad bandedness." The stable radiation or Stokes modes grow during evolution as seen in Figure 29.5A-K.

Clearly, the results shown here are indicative of a further need to address higher order nonlinearities in deep-water wave trains in future studies.

30 Nonlinearity in Duck Pier Data

30.1 Introduction

This chapter discusses analysis of ocean surface wave data taken at Duck Pier at the U.S. Army Corp of Engineers Field Research Facility, Kitty Hawk, North Carolina. The scope of the present work is to analyze data that was previously analyzed for very precise spectral content (Long and Resio, 2007). In this chapter I analyze a time series using the *nonlinear spectral approach* given in this monograph (Chapters 17, 19 and 23). While it is natural to use the *elliptic modulus* as an indication of the nonlinearity of the nonlinear Fourier components, I also include the Ursell number in the present analysis. While the Ursell number is mathematically equivalent to the modulus, we find that the Ursell number often provides a more physically reasonable indicator of nonlinearity. The main conclusions of this chapter are:

(1) *Quadratic nonlinearity* in the surface wave field is included by spectrally analyzing the data using the nonlinear spectral theory of the Korteweg-deVries (KdV) equation.
(2) We find that the *Ursell number* provides a valuable indicator of nonlinearity in the data analysis.
(3) The two major physical conclusions of the study are: A Duck Pier time series of only moderate nonlinearity has been analyzed ($H_s \sim 2.4$m, $T_d \sim 15.0$s, Ursell number ~ 0.8) and it is found that (1) *nonlinear spectral effects occur in a broad frequency range extending from low frequency to a frequency well to the right of the peak of the (cnoidal wave) spectrum.* Furthermore (2) *the low frequency region of the nonlinear spectrum can be characterized as soliton dominated (a "power law" or "white noise" soliton contribution which might be referred to as a "nonlinear soliton gas" over the low frequency range).* The Ursell number has typical power law, f^{-2}, behavior at low frequency.

In what follows we use the nonlinear spectral analysis approach based upon the KdV equation, that is, that analysis which is appropriate for long, unidirectional nonlinear waves in shallow water for which the theoretical method is valid as an expansion of the Euler equations about wavenumber $k \sim 0$. The theory and numerical methods (Chapters 10, 14, 17, 19–23, and 32) tell us that the wave field can be spectrally decomposed into a *sum of cnoidal wave basis functions plus nonlinear interactions among the basis functions.* This approach contrasts to the more common spectral decomposition in terms of a linear superposition of sine waves. Therefore, the motivation for this study of wave measurements at Duck Pier is that we can learn more about the nonlinear

Doi: 10.1016/S0074-6142(10)97030-7

physics and behavior of shallow water waves using the cnoidal wave decomposition. The spectral decomposition in terms of cnoidal waves uses these familiar functions discussed in detail in Bob Wiegel's book *Oceanographical Engineering* (1964) many decades ago. The review in Dean and Dalrymple (1991) gives an excellent engineering perspective. Likewise those familiar with the use of the Ursell number for characterizing nonlinearity in coastal zone wave fields will also feel at home with the approach given herein.

One can ask the following question: Is the cnoidal wave decomposition of high enough order to address legitimate problems in coastal engineering? Certainly, the method is better than the sine wave decomposition, that is, it gives us a nonlinear spectral perspective not available with linear Fourier analysis. However, this is not the end of the possibilities. In Chapters 32 and 33, I give approaches which go to higher order and also give spectral wave spreading using the KP and 2 + 1 Gardner equations. Thus, the nonlinear spectral approach seems limitless for future work. In the present Chapter, I limit the analysis to simple cnoidal wave decompositions that may, for practical purposes, be sufficient for a number of coastal engineering problems.

30.2   The Ursell Number

The "space-like" Ursell number is given by

$$U_{\text{space}} = \frac{3a}{4k^2 h^3} = \frac{3}{16\pi^2} \left(\frac{a}{h}\right)\left(\frac{L}{h}\right)^2 = \frac{\lambda a}{2k^2}, \quad \lambda = \frac{3}{2h^3} \tag{30.1}$$

where a is the amplitude of a sine wave, k is the wavenumber, and h is the water depth. The wavenumber is related to the wavelength, L, by $k = 2\pi/L$. The formula given above is for the analysis of a wave whose spatial extent, L, is known and is therefore appropriate for *space series analysis*. In the analysis below a is to be interpreted as the amplitude of a particular nonlinear wave component in the spectrum (a *cnoidal wave*), k is the wavenumber of the component, and L its associated wavelength. Here λ is the parameter that the nonlinear spectral decomposition "sees" (in the spectral analysis using the Schroedinger equation, Chapter 17). One can see that as the water depth decreases, the nonlinearity parameter λ increases as $\sim 1/h^3$, that is, halving the depth increases the nonlinearity by a factor of 8, almost a factor of ten!

For the analysis of a *time series*, a major goal of this chapter, the Ursell number needs to be modified to include the frequency rather than the wavenumber. This gives (using the approximate shallow water dispersion relation, $\omega \cong c_o k; c_o = \sqrt{gh}$):

$$U_{\text{time}} = \frac{3c_o^2 a}{4\omega^2 h^3} = \frac{3}{16\pi^2} \left(\frac{a}{h}\right)\left(\frac{c_o T}{h}\right)^2 = \frac{\lambda' a}{2\omega^2} = \frac{\lambda' a}{8\pi^2 f^2}, \quad \lambda' = c_o^2 \lambda = \frac{3c_o^2}{2h^3} \tag{30.2}$$

In this case, for the analysis of time series, the amplitude a is taken to be the amplitude of a cnoidal wave component in the nonlinear spectrum and f is the frequency of the component which has period $T = 1/f$. In this case the parameter λ' governs the nonlinearity of the spectral decomposition of the time series, that is, one computes the solution to the spectral eigenvalue problem using the time series $\lambda' \eta(0, t)$. At high frequency the dispersion relation is better approximated by $\omega^2 = gk \tanh kh$, although the resultant corrections to the Ursell number are at most about 7% or 8% (upward) in the present analysis (see Figure 2.1 in Chapter 2). The approximate (KdV-like) dispersion relation arises from a Taylor series expansion about small wavenumber: $\omega = \sqrt{gk \tanh kh} \approx c_0 k - \beta k^3$, $\beta = c_0 h^2 / 6$.

30.2.1 Cnoidal Waves and the Spectral Ursell Number

It is useful to read Sections (10.6.11, 12) of Chapter 10 to gain familiarity with the cnoidal wave solution of the KdV equation and to learn how to compute properties of these waves. See also Weigel (1964) and Dean and Dalrymple (1991).

In classical Jacobian elliptic function notation, the cnoidal wave is given by (Whitham, 1974):

$$\eta(x, t) = 2\eta_c cn^2 \{ [K(m)/\pi](kx - \omega_c t)|m \} - \bar{\eta} \tag{30.3}$$

where $K(m)$ is the elliptic integral of the first kind and ω_c is the dispersion relation corrected for nonlinearity. Here $\bar{\eta}$ is the *cnoidal function mean*, which is removed in order to give a zero mean wave field for $\eta(x, t)$. The *cnoidal wave amplitude*, η_c, is related to the *modulus*, m, by

$$mK^2(m) = \left(\frac{3\pi^2}{2k^2 h^3} \right) \eta_c, \quad 0 \leq m < 1 \tag{30.4}$$

If we define the *nonlinear Ursell number* to be

$$U_c = \frac{3\eta_c}{4k^2 h^3} \tag{30.5}$$

then *the relationship between the Ursell number, U_c, and the modulus, m, is*

$$mK^2(m) = 2\pi^2 \left(\frac{3\eta_c}{4k^2 h^3} \right) = 2\pi^2 U_c \tag{30.6}$$

or

$$U_c = \frac{mK^2(m)}{2\pi^2} \tag{30.7}$$

One of the major conclusions is that *both the modulus, m, and the Ursell number, U_c are equally good indicators of the nonlinearity in a measured wave train*, that is, given one of them you can always compute the other by the above formulas. Use of both is recommended in order to understand all of the physics. One is accustomed to using the modulus, m, to give an indication of how nonlinear a cnoidal wave is, that is, for small m ($\sim$0.1) we have a *sine wave*, for moderate m ($\sim$0.5) we have a *Stokes wave*, and for $m \sim 1$ we have a *soliton*. A continuous deformation of the waveform from sine wave to soliton occurs as m is slowly changed from 0 to 1. Of course the cnoidal wave also depends on the wavenumber or frequency and oscillates the appropriate number of times inside a single period, L or T, just as a sine wave component does in linear Fourier analysis.

A main focus of this chapter is to discuss how the *equivalence between Ursell number and modulus* can be used to improve understanding of nonlinear wave data. Another focus is to show how new numerical methods for nonlinear Fourier analysis allow us to compute the *nonlinear cnoidal wave spectrum* of a measured wave train, even when the physics is extremely nonlinear (i.e., soliton dominated). It should be noted that this capability for high nonlinearity is a new result, as in the past it has not been possible to spectrally analyze data with high nonlinearity.

Now let us discuss the behavior of the Ursell number for small m and for large $m \sim 1$. When m is small we have

$$K \approx \frac{\pi}{2} \tag{30.8}$$

so that

$$m \approx 8U_c, \quad U_c \approx m/8, \quad m \ll 1 \tag{30.9}$$

Likewise when $m \approx 1$ we have the well-known relation:

$$K \approx \frac{1}{2}\ln\left(\frac{16}{1-m}\right) \quad \text{or} \quad K = \ln\left(\frac{16}{1-m}\right)^{1/2} \tag{30.10}$$

Therefore,

$$m\ln\left(\frac{16}{1-m}\right) \cong 2\pi^2 U_c$$

so that

$$U_c = \frac{mK^2(m)}{2\pi^2} \cong \frac{1}{2\pi^2}m\left[\ln\sqrt{\frac{16}{1-m}}\,\right]^2, \quad m \cong 1 \tag{30.11}$$

Table 30.1 Table of Elliptic Function Moduli Near 1 and the Associated Ursell Number

m	U_c
0.9	0.294
0.99	0.682
0.999	1.186
0.999 9	1.818
0.999 99	2.585
0.999 999	3.485
0.999 999 9	4.520
0.999 999 99	5.689
0.999 999 999	6.992
0.999 999 999 9	8.429
0.999 999 999 99	10.001
0.999 999 999 999	11.707
0.999 999 999 999 999 9	19.770
0.999 999 999 999 999 999 999 999 999	53.767

Table 30.1 gives the modulus for the case $m \approx 1$ for a single cnoidal wave component and its Ursell number. Note that $U_c \approx 1$ for $m \approx 0.9978$. So, the conclusion is that for more than about *three nines* in the modulus we have Ursell numbers greater than 1. From Table 30.1, if there are *16 nines*, we have about the maximum nonlinearity possible in a double precision computation in modern computers, which is an Ursell number ~ 20. The largest value of nonlinearity rapidly computable in a modern computer (in quadruple precision, ~ 34 decimals) is for m with about *27 nines*, where $U_c \sim 54$. Figure 30.1 is a graph of the Ursell number versus the number of nines in the modulus, that is, a graph of Equation (30.11) and Table 30.1.

The way to interpret Figure 30.1 is to assume that we have a single cnoidal wave component in the cnoidal wave spectrum with modulus m. Then we have associated with it an Ursell number that we can pick off of the graph. This approach is quite nice, since we can easily estimate the Ursell number, whereas we would need to use the amplitude of the component, its wavenumber (or frequency), and the depth, to compute the parameter.

It is worthwhile looking at the theoretical possible Ursell numbers in the nonlinear spectrum. We are analyzing time series so the formula from (30.2) above is

$$U_{\text{time}} = \frac{\lambda' a}{8\pi^2 f^2}, \quad \lambda' = c_o^2 \lambda = \frac{3c_o^2}{2h^3} \tag{30.12}$$

Note that the Ursell number has the frequency dependence: $U_{\text{time}} \sim f^{-2}$ if the spectral amplitudes are kept constant, $a \sim$ constant (white noise). Below we analyze a Duck Pier time series of length T that is discretized into N points. This means we have discrete frequencies:

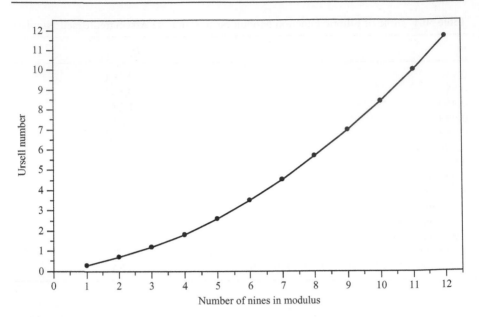

Figure 30.1   Graph of the Ursell number as a function of the number of nines in the modulus, m.

$$f_n = n\Delta f = \frac{n}{T}$$

Therefore, the spectral Ursell number as a function of the frequency, f_n, is

$$U_{\text{time},\,n} = \frac{\lambda' a_n}{8\pi^2 f_n^2} = \frac{\lambda' a_n T^2}{8\pi^2 n^2} \tag{30.13}$$

where a_n is the cnoidal wave spectral amplitude at frequency f_n. Therefore, the spectral Ursell number for a particular component in the cnoidal spectrum will be large provided that: (1) the cnoidal wave spectral amplitude is large, (2) the frequency is small, and (3) the water depth is shallow.

The *Ursell number at the smallest frequency component*

$$f_1 = \Delta f = \frac{1}{T}$$

is

$$U_{\text{time},\,1} = \frac{\lambda' a_1}{8\pi^2 f_1^2} = \frac{\lambda' a_1 T^2}{8\pi^2} \tag{30.14}$$

30.3 Estimates of the Ursell Number from Duck Pier Data

It is worthwhile first computing a *characteristic spectral Ursell number*, U_{char}, based upon *global parameters of a measured wave train* such as significant wave height, H_s and dominant period, T_d:

$$a = H_s/2, \quad f = f_d, \quad T = T_d = 1/f_d \tag{30.15}$$

so that:

$$U_{char} = \frac{3c_o^2 H_s}{8\omega_d^2 h^3} = \frac{3}{32\pi^2}\left(\frac{H_s}{h}\right)\left(\frac{c_o T_d}{h}\right)^2 = \frac{\lambda' H_s}{4\omega_d^2} = \frac{\lambda' H_s}{16\pi^2 f_d^2}, \quad \lambda' = c_o^2 \lambda = \frac{3c_o^2}{2h^3} \tag{30.16}$$

In the present case, for a particular Duck Pier wave train (see Figure 30.2), we have

$$H_s = 2.403\,\text{m}, \quad h = 7.87\,\text{m}, \quad c_o = 8.787\,\text{m/s}$$

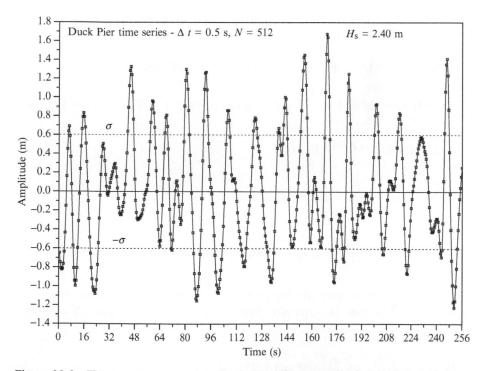

Figure 30.2 Time series measured at Duck Pier. The water depth is 7.87 m and the significant wave height is 2.403 m. The time series is 256 s long and is digitized at 0.5 s.

$$a = H_s/2 = 1.2017, \quad f = f_d = 0.0664\,\text{Hz}, \quad T = T_d = 1/f_d = 15.15\,\text{s}$$

$$L_d = c_o T_d = 132.33\,\text{m}$$

We find

$$\frac{H_s}{h} = 0.3053, \quad \frac{c_o T_d}{h} = 16.815, \quad \lambda' = c_o^2 \lambda = \frac{3 c_o^2}{2 h^3} = 0.2376$$

Finally,

$$U_{char} = \frac{3}{32\pi^2}\left(\frac{H_s}{h}\right)\left(\frac{c_o T_d}{h}\right)^2 = \frac{3}{32\pi^2}(0.3053)(16.815)^2 = 0.8201$$

Likewise with the other formulas:

$$U_{char} = \frac{\lambda' H_s}{16\pi^2 f_d^2} = 0.8201$$

Let us look at a *hypothetical maximum Ursell number case for Duck Pier,* $U = U_{\text{maximum}}$, for a large wave train:

$$H_s = h/2 = 3.935\,\text{m}, \quad h = 7.87\,\text{m}, \quad c_o = 8.787\,\text{m/s}$$

$$a = H_s/2 = 1.9675\,\text{m}, \quad f = f_d = 0.0526\,\text{Hz}, \quad T = T_d = 1/f_d = 19.00\,\text{s}$$

$$L_d = c_o T_d = 166.953\,\text{m}$$

$$\frac{H_s}{h} = 0.5, \quad \frac{c_o T_d}{h} = 21.214, \quad \lambda' = c_o^2 \lambda = \frac{3 c_o^2}{2 h^3} = 0.2376$$

$$U_{\text{maximum}} = \frac{\lambda' H_s}{16\pi^2 f_d^2} \approx 2.14 \quad \text{(Rough Maximum Ursell Number)}$$

Now the *Ursell number for the first component in the nonlinear spectrum* (where the modulus is typically very near 1) is (for, say, $N = 4096$, $T_{\text{series}} = 2048\,s$):

$$U_{\text{time}, 1} = \frac{\lambda' a_1}{8\pi^2 f_1^2} = \frac{\lambda' a_1 T_{\text{series}}^2}{8\pi^2} \approx 13,220.24 a_1$$

When this component is of even moderate size we can expect to get large Ursell numbers; this is because the modulus can be very near 1 at low frequency. In double precision we might expect to get, say, 12 *nines* in the modulus, so that the Ursell number would be ~12 (see Fig. 30.1 and Table 30.1).

For the time series analyzed below we have $N = 512$, $T_{series} = 256$ s and find:

$$U_{time, 1} = \frac{\lambda' a_1}{8\pi^2 f_1^2} = \frac{\lambda' a_1 T_{series}^2}{8\pi^2} \approx 197.213a$$
$$= 45(\sim 24 \text{ nines, quadruple precision required!})$$

Thus, if we want to get the modulus from the time series we would need to compute the cnoidal wave spectrum using quadruple precision! Of course we try not to use the spectral algorithm (Chapter 17) in this case, but instead use the adiabatic annealing algorithm of Chapter 23 which does not require a computation of the modulus, but instead only of the Riemann spectrum and therefore can be computed in double precision. This avoids the difficulties of computing in quadruple precision and leads us to the Ursell number as a characteristic parameter for describing the waves. Indeed, after the fact one can estimate the modulus from the diagonal elements of the Riemann matrix in quadruple precision using the formulas (10.157, 10.158) of Section 10.6.11 for a nome of $q_n = \exp[-B_{nn}/2]$ for the diagonal element B_{nn} (nth wavenumber of frequency component) of the Riemann matrix.

30.4   Analysis of Duck Pier Data

We now analyze a time series from Duck Pier in 7.87 m of water. The wave train has a significant wave height of 2.403 m, a dominant period of 15.15 s, and is 256 s long (see Figure 30.2). The characteristic Ursell number for this wave train is ~0.82.

The linear Fourier spectrum of the Duck data in Figure 30.2 is shown in Figure 30.3. A relatively narrow-banded spectrum with peak near 0.066 Hz is seen. Note also the Stokes harmonic at double this frequency. A substantial energetic contribution is also found at low frequency, an additional indication of shallow-water nonlinearity in the data.

In order to fully check out the algorithms used in the analysis, I first analyzed the Duck time series in *quadruple precision* (solving the KdV eigenvalue problem of Chapter 17), suffering through the need to see all the details of the nonlinear cnoidal wave spectral analysis (indeed the quadruple precision computation is quite simple using Intel Fortran 77 and the code runs only about three times slower than the associated double precision code). The first check on the analysis of the above time series is shown in Figure 30.4, where the so-called *Floquet spectrum* is shown. This oscillation (known as the half-trace of the mondromy matrix) is graphed as a function of squared

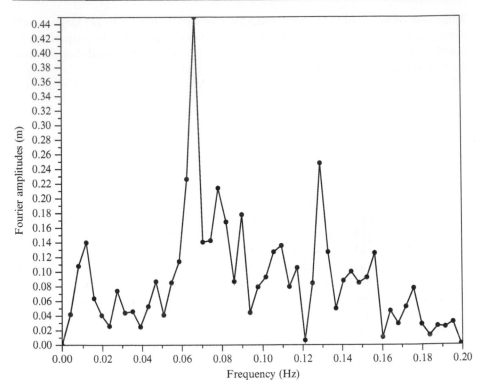

Figure 30.3   Linear Fourier transform of time series of measured Duck Pier data (Figure 30.2).

frequency, f^2. The vertical axis is logarithmic so that the variations are shown to vary over a factor of $\sim 10^{50}$. Intersections of this function with ± 1 (the horizontal dotted lines) constitute the main spectrum eigenvalues of the Floquet analysis. The extreme variations of the half-trace (which swings widely over many orders of magnitude) are a simple diagnostic of soliton behavior in the wave train. One can estimate the number of solitons simply by counting these brisk oscillations that generally fall to the left of the diagram. The much smaller oscillations to the right correspond to small amplitude Stokes or sine waves (for a complete interpretation of Floquet analysis of the KdV equation, see Chapter 17, in particular Figure 17.1 of that chapter is very informative). Of course the main reason for running this diagnosis is to make sure that none of the oscillations in the Floquet diagram get missed; indeed in the present case the resolution is quite high and one sees no possibility of missed eigenvalues and indeed the *algorithm is automatic*, finding all the eigenvalues is a practical result of the computation (Chapter 17).

An initial estimate of the nonlinear spectrum can be computed directly from the main eigenvalues of the Floquet spectrum (these are known as the so-called hyperelliptic function amplitudes). Figure 30.5 gives the amplitudes of the

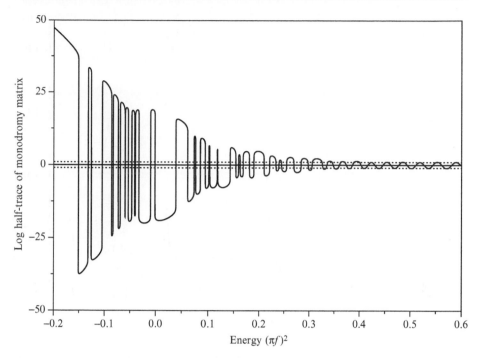

Figure 30.4 The nonlinear time series analysis algorithm has been used to compute the Floquet diagram of the time series of Figure 30.2. The eigenvalues of the Floquet problem occur when the half-trace of the monodromy matrix (graphed logarithmically on the vertical axis) crosses ±1 (horizontal dotted lines). Soliton behavior occurs when there are excursions (across zero) of many orders of magnitude in the trace of the monodromy matrix, here seen to range up to 47 orders of magnitude. The current case, as emphasized by these extreme excursions, is therefore dominated by soliton activity.

hyperelliptic functions (solid line) and of their modulus (dotted line). As previously suspected the modulus is near one for over half of the frequency range in the spectrum due to the large nonlinearity in the measured time series.

We now look at the *cnoidal wave (fully nonlinear) spectrum* (computed using the adiabatic algorithm of Chapter 23) of the time series in Figure 30.2. This is shown in Figure 30.6, where both the cnoidal wave amplitudes and the moduli are graphed as a function of frequency. The spectrum is seen to have large energetic components at low frequency. The modulus is near 1 in this low frequency range and so one is forced to conclude that these cnoidal waves are solitons. The tail of the spectrum is seen to have moduli that are relatively small and hence one concludes that the components at high frequency are essentially sine waves. The intermediate components, which have moduli ~0.3-0.7, are interpreted as Stokes waves.

In Figure 30.7 we give the cnoidal wave spectrum together with the Ursell number, both as a function of frequency. Surprisingly, large Ursell numbers

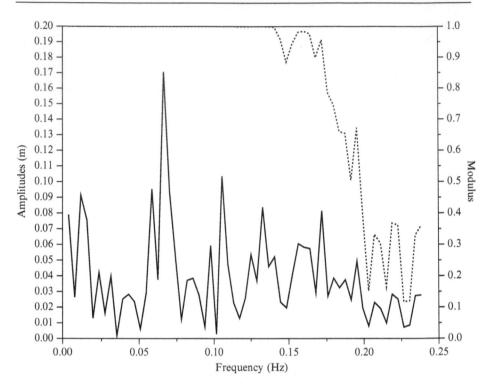

Figure 30.5   An initial estimate of the spectrum (solid line) is shown here, where the bandwidths in the Floquet spectrum are graphed as spectral amplitudes (these are the amplitudes of the hyperelliptic function spectral decomposition). The modulus (dotted line) is also graphed. The fact that the moduli are near 1 for more than half of the frequency range is an indication of soliton behavior.

are found from low frequency out to the peak of the spectrum, indicating (as the moduli did above) that these frequency components are quite nonlinear. Note that the scale of the Ursell numbers is logarithmic, due to the large variation in their values as a function of frequency.

In Figure 30.8 we show the same graph as Figure 30.7, but all axes are now given as logarithmic. Of specific note is that at low frequencies in the cnoidal wave spectrum one finds a broad plateau of roughly equal amplitudes; these are the solitons in this spectrum which may be characterized as "low frequency white noise," or as a "nonlinear soliton gas." In fact the gas extends to the right of the peak of the spectrum. Further note that the Ursell numbers are described by a power law with amplitudes $\sim f^{-2}$.

Figure 30.9 shows the actual *cnoidal wave components* in the spectrum graphed vertically from low frequency (top) to high frequency (bottom), all graphed as a function of time. A total of 50 total spectral components are shown.

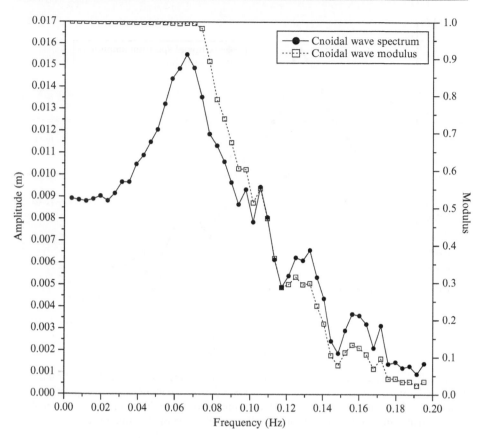

Figure 30.6   The cnoidal wave spectrum of the time series in Figure 30.2. Shown are the cnoidal wave amplitudes and the modulus of each cnoidal wave. Note that the modulus is above 0.9 even out near the peak of the spectrum, indicating extreme nonlinear behavior in the dynamics of the time series.

The first component is clearly a single soliton, the second is a soliton train that oscillates twice, the third a soliton train that oscillates three times, etc. The first 35 or so components from the top can be seen to be nonlinear as their shape is significantly different from that of a sine wave. The first 18 can be interpreted as solitons (see the moduli ~ 1 in Figure 30.6 and the large Ursell numbers in Figures 30.7 and 30.8). The remaining 17 components to the right of the solitonic part of the spectrum (immediately to the right of the components for which $m \sim 1$ in the spectrum in Figures 30.6–30.8) are Stokes waves. The remaining 15 or so components are effectively sine waves.

Nonlinear wave trains such as the one analyzed herein have nonlinear components that are quite different in their behavior from linear sine waves. Note that the components are quite nonlinear in the upper half of the figure where

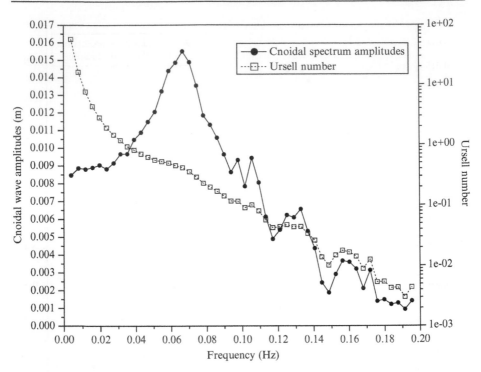

Figure 30.7   The cnoidal wave spectrum of the time series in Figure 30.2. Shown are the cnoidal wave amplitudes and the Ursell number of each cnoidal wave, graphed as a function of frequency. Note that the Ursell number is above 0.3 even out to the peak of the spectrum, indicating the extreme nonlinear behavior of the time series.

the spectral component moduli are near 1 and the Ursell numbers are large. Below the spectral components is the sum of the cnoidal waves. Below this latter curve are the nonlinear interactions and the last and lowest curve is the actual measured wave train of Figure 30.2. The large amplitude of the nonlinear interactions is indicative of the large degree of nonlinearity in the measured wave train from Duck Pier (Figure 30.2). To test by eye whether a single component is nonlinear turn the page upside down: If the wave is nonlinear it will appear up-down asymmetric. Thus, Figure 30.9 has the simple interpretation that the sum of the cnoidal waves, plus their mutual nonlinear interactions gives the measured Duck Pier wave train.

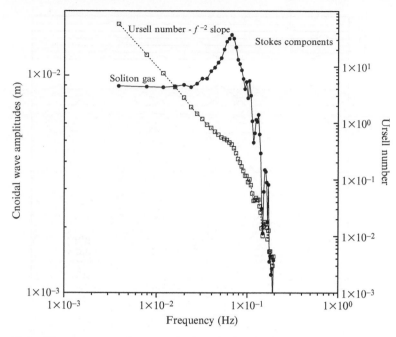

Figure 30.8 The spectrum of Figure 30.7 graphed in log-log coordinates. Note that the cnoidal wave spectrum is characterized by a broad low frequency plateau of solitonic activity, that is, a region for which one has solitonic near "white noise." In this same region the Ursell number is seen to have a power law or scaling behavior, $\sim f^{-2}$. The tail of the cnoidal wave spectrum is roughly f^{-4}, consistent with known results for the linear Fourier spectrum.

One of the surprising results of the present analysis is the dominance of solitons to the left of the peak of the spectrum. This highly energetic, nonlinear part of the spectrum might be referred to as a *soliton gas* in a statistical mechanical interpretation of the Riemann spectrum in which the phases are effectively random. Such an interpretation of the dynamics of shallow water waves will have a future impact on our description of their dynamics and will clearly impact our approach to many coastal dynamics problems.

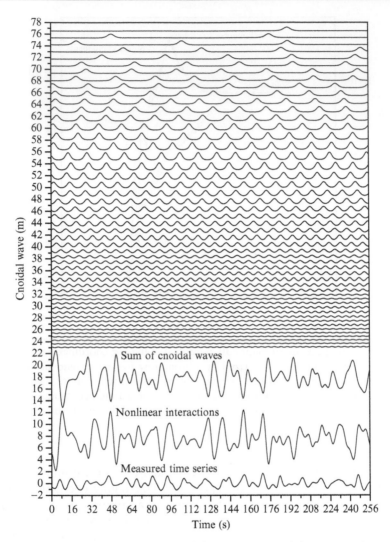

Figure 30.9   Cnoidal waves in the spectral decomposition of the measured wave train of Figure 30.2 as a function of time.

31 Harmonic Generation in Shallow-Water Waves

31.1 Introduction

This chapter addresses harmonic generation in shallow-water wave trains. The idea is quite simple (Mei, 1983). A Cauchy initial condition, assumed to be a sine wave, generates harmonic components in the Fourier spectrum due to shallow-water nonlinearities. It is convenient to address the problem from the point of view of the Korteweg-deVries (KdV) equation and to write its solutions in terms of the linear Fourier transform with time-varying coefficients. Then harmonic generation occurs in the time evolution of the Fourier modes. This chapter emphasizes that the harmonic generation is a consequence of observing the system in a sinusoidal basis set, that is, energy is transferred from one harmonic to another due to the nonlinear interactions.

However, a different perspective is also appropriate, one that uses the *cnoidal wave basis functions* of the KdV equation. To this end I apply the spectral theory of the KdV equation (incorporating Riemann θ-functions) and analyze numerical solutions of the equation and experimental laboratory data. In this way I show that the higher order basis of cnoidal waves allows us to view shallow-water wave motions as being *nonresonant*. Indeed, shallow-water waves, at the order of the KdV equation, are *resonant only in the sinusoidal wave basis*. The cnoidal waves have constant amplitudes and phases during the time evolution.

The periodic KdV equation is (Korteweg and deVries, 1895):

$$\eta_t + c_o\eta_x + \alpha\eta\eta_x + \beta\eta_{xxx} = 0, \quad \eta(x,t) = \eta(x+L,t) \tag{31.1}$$

where the constant coefficients are given by: $c_o = \sqrt{gh}$, $\alpha = 3c_o/2h$, and $\beta = c_oh^2/6$; h is the water depth and g is the acceleration of gravity. A normalized version of this equation, commonly used in inverse scattering theory, is given by ($u = \lambda\eta$, $x \to x - c_ot$, $t \to \beta t$ for $\lambda = \alpha/6\beta$):

$$u_t + 6uu_x + u_{xxx} = 0 \tag{31.2}$$

I exploit the fact that the KdV equation is completely integrable by the inverse scattering transform and address the particular results for periodic boundary conditions.

31.2   Nonlinear Fourier Analysis

The approach is based upon the general periodic solution to the KdV equation (31.1) in terms of the Riemann θ-*function*:

$$\eta(x,t) = \frac{2}{\lambda}\frac{\partial^2}{\partial x^2}\ln\theta_N(x,t) \qquad (31.3)$$

where

$$\theta_N(x,t) = \sum_{m_1=-\infty}^{\infty}\sum_{m_2=-\infty}^{\infty}\cdots\sum_{m_N=-\infty}^{\infty}\exp\left[i\sum_{n=1}^{N}m_nX_n + \frac{1}{2}\sum_{m=1}^{N}\sum_{n=1}^{N}m_mm_nB_{mn}\right] \quad (31.4)$$

Here N is the number of *cnoidal waves in a broad-spectrum solution* to the KdV equation. The *summation indices* m_n $(1 \leq n \leq N)$ are integers summed from $-\infty$ to ∞. The θ-function *phases* have the same familiar form as that used in linear Fourier analysis: $X_n = k_nx - \omega_nt + \phi_n$. Explicit computation of the *period (interaction) matrix*, $\mathbf{B} = \{B_{mn}\}$, the wavenumbers, k_n, the frequencies, ω_n, and the phases, ϕ_n are discussed in Chapters 10, 14–16, and 23. The numerical algorithms for θ-functions are discussed in Chapters 19–23.

31.3   Nonlinear Spectral Decomposition

The θ-function solutions (31.3), (31.4) to the KdV equation (31.1) can be written in the following form:

$$\eta(x,t) = \frac{2}{\lambda}\frac{\partial^2}{\partial x^2}\ln\theta_N(x,t) = \underbrace{\eta_{cn}(x,t)}_{\substack{\text{Linear superpostion} \\ \text{of N cnoidal waves}}} + \underbrace{\eta_{int}(x,t)}_{\substack{\text{Nonlinear interactions} \\ \text{among the N cnoidal waves}}} \qquad (31.5)$$

$$= 2\sum_{n=1}^{N}\eta_ncn^2\{(K(m_n)/\pi)[k_nx - \omega_nt + \phi_n]; m_n\} + \eta_{int}(x,t)$$

where the particular expression for the interaction terms $\eta_{int}(x,t)$ is given in Chapter 10. This result essentially states that *Shallow-water wave trains governed by the KdV equation can be represented by a linear superposition of N cnoidal waves plus their mutual nonlinear interactions*. Essentially, the nth cnoidal wave derives from the particular diagonal element B_{nn} of the Riemann matrix. Likewise each off-diagonal term B_{mn} contributes to the nonlinear interactions between the mth and nth cnoidal waves. How is this nonlinear Fourier formulation in terms of Riemann θ-functions related to ordinary linear Fourier analysis? This is seen by letting the wave amplitudes become so small that the cnoidal wave components become sine waves and the nonlinear interactions tend to zero (Chapter 10). In this way linear Fourier analysis is recovered from the nonlinear theory. An important aspect of Equation (31.5) is that the *amplitudes of the cnoidal waves, η_n, and their phases, ϕ_n, are constants of the motion for KdV evolution.*

31.4   Harmonic Generation in Shallow Water

A well-known procedure for numerically solving the KdV equation is to assume a Fourier series for the solution:

$$\eta(x,t) = \sum_{n=-N_x/2}^{N_x/2} \eta_n(t) e^{ik_n x - i\omega_n t + i\phi_n} \tag{31.6}$$

Here the Fourier coefficients are taken to be a function of time due to the nonlinear term in KdV. The integer $N_x = L/\Delta x$, Δx is the spatial discretization of x, where L is the period of the wave train and $\eta(x,t) = \eta(x+L,t)$ is assumed. By inserting the above Fourier series into the KdV equation, one naturally finds ordinary differential equations (ODEs) for the evolution of the Fourier coefficients:

$$\dot{\eta}_n + i\omega_n \eta_n + i\alpha \sum_{m=-\infty}^{\infty} k_n \eta_n \eta_{n-m} = 0 \tag{31.7}$$

where $\omega_n = c_o k_n - \beta k_n^3$ and the over dot is a time derivative. These latter equations may be solved numerically to obtain $\eta_n(t)$ as a function of time; the solution of KdV $\eta(x,t)$ then follows from the Fourier transform (Equation (32.6)). Assuming time evolution for the Fourier spectrum is the perspective of many modern approaches to the modeling of nonlinear wave equations, including the so-called higher order spectral methods (Dommermuth and Yue, 1987; West et al., 1987). In the special case for linear motion, one sets $\alpha = 0$ and finds that the time evolution of the Fourier coefficients is given by the elementary result $\eta_n(t) = \eta_n(t) e^{-i\omega_n t}$.

If the nonlinearity is not too great the importance of higher order harmonics should diminish rapidly with n and therefore allow truncation of the Fourier series after a small number of terms N. The resulting set of N ODEs gives the solution for the N time-dependent Fourier coefficients, $\eta_n(t)$, in Equation (31.6) which are clearly nonlinearly coupled via Equation (31.7). This approach, that is, to introduce a Fourier series with time-dependent coefficients into a nonlinear partial differential equation (PDE) in order to "reduce" the system to a set of ODEs (formally $N \to \infty$) is quite standard in mathematical physics.

From the point of view of the above set of ODEs (31.5), one finds that it is quite simple to introduce the concept of *resonances* and *harmonic generation*. For short-time solutions, one finds that certain Fourier series coefficients grow at the expense of others. For long-time behavior the above ODEs must describe, for periodic boundary conditions, Fourier coefficients that are oscillatory, alternatively growing, and decaying in time (Fermi-Pasta-Ulam (FPU) recurrence). In what follows we present a new prospective that describes the nonlinear interactions in terms of a *cnoidal wave basis*.

31.5   Periodic Inverse Scattering Theory

The cnoidal waves do not exchange energy amongst one another *and therefore IST for the KdV equation does not give rise to either* resonances or harmonic generation. The cnoidal waves interact nonlinearly with one another via Equation (31.5), but their amplitudes are constants of the motion. This is one of the most important conclusions of this chapter: By studying KdV evolution using a sine-wave basis (with time-varying Fourier components (Equation (31.6))), one has resonances and harmonic generation since the spectral amplitudes change as a function of time; if instead one uses a cnoidal wave basis (via the periodic inverse scattering transform; Equations (31.3) and (31.4)), one does not have resonances or harmonic generation since the spectral amplitudes are constants. This illustrates the convenience of using higher order basis functions and inverse scattering transform theory in the present case.

31.6   Classical Harmonic Generation and FPU Recurrence in a Simple Model Simulation

Now consider a numerical example to illustrate how Fourier amplitudes vary in time in the evolution of a sine-wave initial condition. Figure 31.1 shows a series of wave trains (A)-(E) which were computed by numerically integrating the KdV equation (using the algorithm of Fornberg and Whitham, 1978) from a sine-wave initial condition in 15 cm depth (A); the initial wave has amplitude 1 cm and wave length 256 cm. The wave train has been numerically integrated and halted at $t = 3.2$ s (B), $t = 6.3$ s (C), $t = 9.5$ s (D), and $t = 12.9$ s (E). The initial sine wave is seen to evolve into a Stokes wave plus something else (B). The wave motion in panel (C) evidently is a Stokes-like waveform plus some another small component(s). The smaller wave migrates somewhat to the left in (D). Finally, in panel (E) the wave train has almost returned to the sine-wave initial conditions, that is, we have an example of FPU recurrence (Fermi et al., 1955).

In Figure 31.2A-E we show the Fourier spectra for each of the wave trains in Figure 31.1A-E. In the first panel (A) we find of course only a single sine wave that matches the initial conditions. As the waveform evolves from Figure 31.1A to

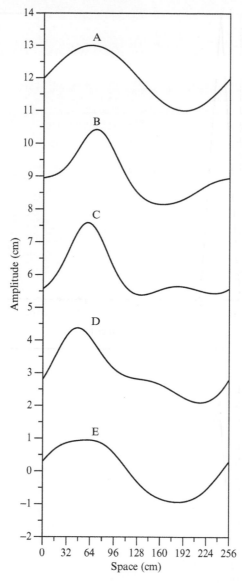

Figure 31.1 Numerical solution of the KdV equation in 15 cm depth. The sine-wave initial condition has amplitude 1 cm and wavelength 256 cm (A). The integration was stopped at 3.2 s (B), 6.3 s (C), 9.5 s (D), and 12.9 s (E).

B, the Fourier spectrum reduces the amplitude of the first component and broadens as shown in Figure 31.2B. The spectrum of the waveform in Figure 31.1C is shown in Figure 31.2C and is seen to further broaden, but only slightly. However, by the time the evolution arrives at Figure 31.1D, the Fourier spectrum has begun to narrow again and its peak amplitude increases (Figure 31.2D). Finally, for the

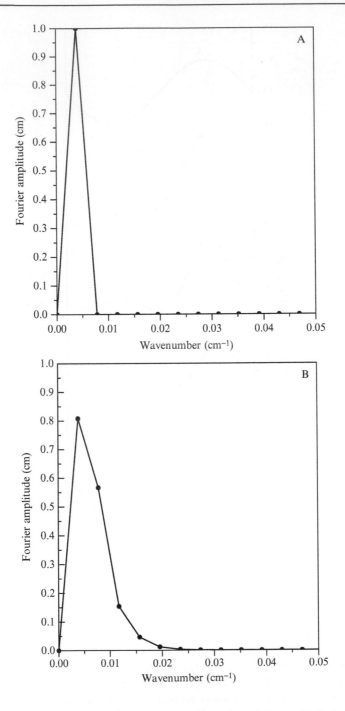

Figure 31.2 Fourier transforms for the five space series of Figure 31.1. A sine-wave initial condition (A) is seen to undergo harmonic generation in (B)-(D). Then the waveform almost returns to the initial conditions via FPU recurrence (E).

(Continued)

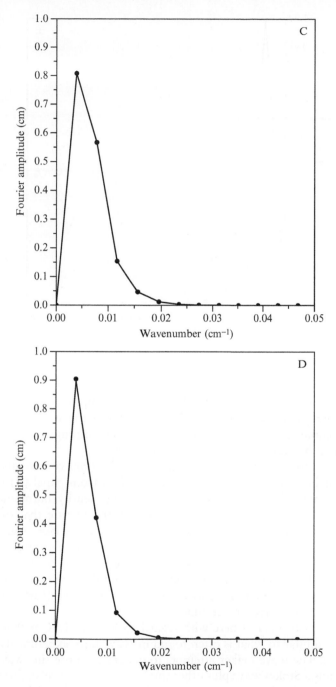

Figure 31.2 Cont'd

(Continued)

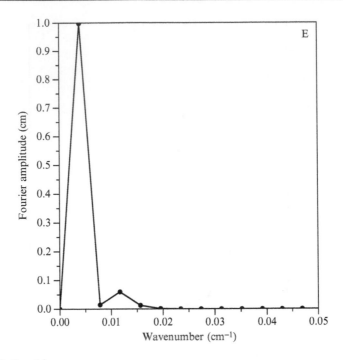

Figure 31.2 Cont'd

final wave train in Figure 31.1E, the Fourier spectrum, Figure 31.2E, is seen to almost return to that in Figure 31.2A. This is the Fourier view of FPU recurrence, that is, an initial, delta-function spectrum broadens during the evolution, but eventually (almost) returns to the initial conditions.

How do we interpret the numerical simulation in terms of resonances? Resonances occur because, as the wave motion evolves from panel (A) to (B), the initial sine wave is seen to loose energy and the nearby harmonics instead increase their energy with time. This is exactly the picture that one obtains by doing a multiscale expansion of the KdV equation (see, e.g., Mei and Ünlüata, 1972; Bryant, 1973; Mei, 1983). However, in the present example we have numerically continued the evolution to long times, beyond the applicability of a truncated multiscale expansion, and find that the evolution, via FPU recurrence, reverses itself and the wave motion returns very nearly to the initial conditions after some recurrence time.

Another point of interest here with regard to Figure 31.1C is that the space series is very similar to what is often found in wave tank experiments (see, e.g., Section 31.7). The wave maker attempts to generate a pure sine wave and one instead obtains a Stokes wave plus other small amplitude wave(s). From the multiscale analysis this phenomenon is easily described in terms of the resonance effect, that is, higher order harmonics grow at the expense of the initial sine wave.

Let us now give an alternative interpretation of this simple numerical experiment using the inverse scattering transform. Instead of taking the Fourier transform of each of the space series in Figure 31.1, we now take their inverse scattering transforms. The results are shown in Figures 31.3–31.7. The first result given in Figure 31.3 is the cnoidal wave decomposition for the space

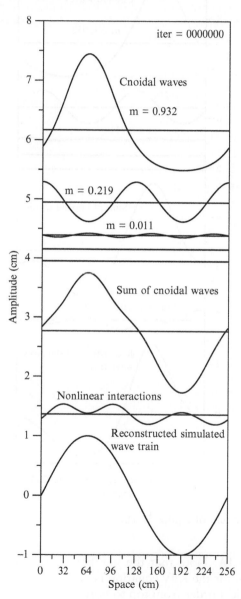

Figure 31.3 IST spectrum of Figure 31.1A.

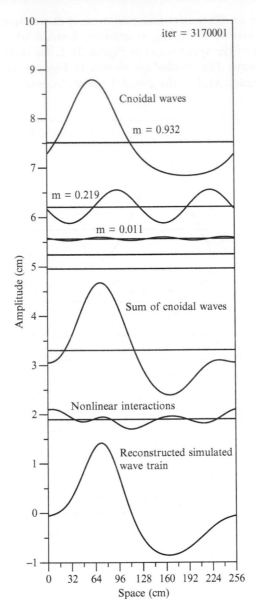

Figure 31.4   IST spectrum of Figure 31.1B.

series in Figure 31.1A. Note that the number of iterations in the numerical (spectral) algorithm for the KdV equation is given in the upper right hand corner. Shown, in vertical order from top to bottom, are the five cnoidal waves in the spectrum, the sum of these cnoidal waves, the nonlinear interactions, and the reconstructed wave train. Note that only three of the cnoidal waves make

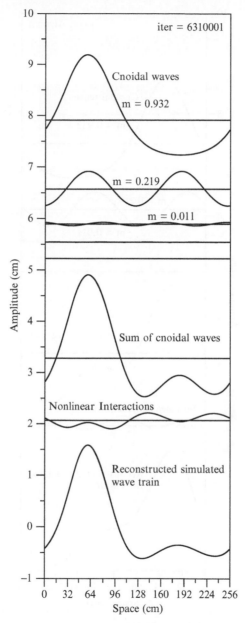

Figure 31.5   IST spectrum of Figure 31.1C.

a substantial contribution to the dynamics: the highest wave is a large amplitude Stokes wave with modulus 0.931; the next wave is a Stokes wave with
modulus 0.219 and finally there is a small amplitude sine wave with modulus
0.011. Two other sine-wave components are so small that they are not visible

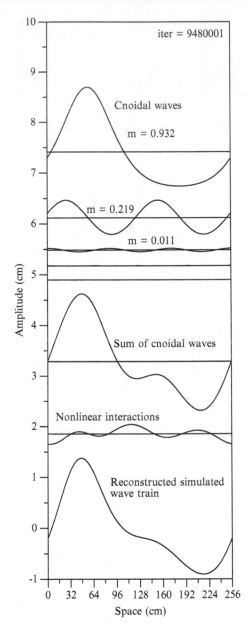

Figure 31.6 IST spectrum of Figure 31.1D.

at the graphical resolution given here. Observe that the sum of the cnoidal waves does not do a very good job of recovering the initial sine wave. Only by adding the nonlinear interactions is the sine wave fully reconstructed. This is an important example problem: *the nonlinear spectral decomposition of a*

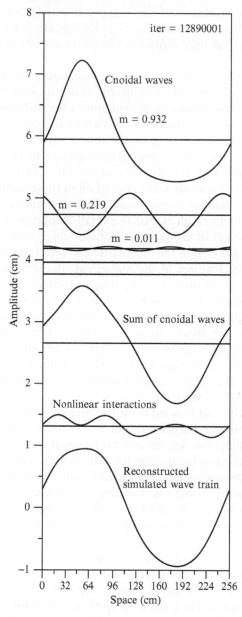

Figure 31.7 IST spectrum of Figure 31.1E.

sine wave! The nonlinear dynamics are seen to be dominated by a rather larger Stokes wave and a smaller Stokes wave. This result confirms our physical intuition with regard to the space series in Figure 31.1C: An initial sine wave evolves into a large Stokes wave plus a smaller Stokes wave.

The spectral decomposition of the space series of Figure 31.1B is given in Figure 31.4. Note that the cnoidal waves in the spectrum are the same as those found in Figure 31.3 (as they should be) but they are phase shifted (they have propagated at their phase speeds) relative to their former positions. However, the nonlinear interactions have changed substantially. As a consequence, the wave motion is seen to be governed by the time-phase-shifted cnoidal waves together with their appropriate nonlinear interactions. Note, further, that the nonlinear interactions are different for different values of time, even though the cnoidal waves have the same amplitudes and moduli, that is, the nonlinear interactions depend crucially on the relative phases of the cnoidal waves in the nonlinear dynamics of the KdV equation. Of course, it is the Riemann θ-function which automatically takes care of all of these nonlinear interactions.

In Figure 31.5 we give the cnoidal wave decomposition of the wave train in Figure 31.1C. The main result is that the small amplitude Stokes wave has its maximum in the trough of the large amplitude Stokes wave, thus creating the small peak on the right side of the reconstructed wave train, thereby recovering one of the important features in the numerical simulations. Note once again that the nonlinear interactions are required to recover all the subtle features in the simulated wave train. In Figure 31.6 we see the inverse scattering results for panel (D) of Figure 31.1. Basically, this is very similar to the previous case except that the smaller Stokes wave has shifted slightly to the left. In Figure 31.7 we give the decomposition of Figure 31.1E; these results are seen to be very similar to those of Figure 31.3. This should not be surprising since we have almost recovered the initial conditions at this point in time for the nonlinear evolution (FPU recurrence).

It is now worthwhile addressing what the inverse scattering transform interpretation of this example has done to our perception of KdV resonances. First note that the basis functions are the cnoidal waves and that they all have amplitudes that are *constants of the motion* for evolution governed by the KdV equation. The physics has been completely and fully captured by the nonlinear spectral decomposition. This is as it should be because the cnoidal waves are the *exact* physical basis functions of the KdV equation.

This perspective contrasts with our previous interpretation of resonances in terms of sine-wave basis functions in which some waves gain or lose energy at the expense of others. However, the sine waves are not generally solutions of the KdV equation and, as pointed out with regard to Equations (31.6) and (31.7), *this requires that the linear Fourier decomposition of KdV has temporally varying amplitudes*. Thus, the traditional interpretation (multiscale analysis) is that resonances happen quite naturally. However, the inverse scattering transform with cnoidal wave basis functions does not predict the existence of resonances.

We conclude that, given the nonlinear cnoidal basis functions, one does not have resonances because there is no transfer of energy from one mode to the other, that is, the mode amplitudes are simply constants of the motion that interact with one another via the off-diagonal terms in the Riemann matrix.

Both the multiscale approach and the inverse scattering transform approach are equally good methods to apply to the problem of nonlinear, shallow-water wave evolution. However, in the present case, inverse scattering modes are *the natural modes* of the physical problem and therefore no resonances occur from the point of view of this completely integrable Hamiltonian system.

31.7 Search for Harmonic Generation in Laboratory Data

We now consider the results of a simple wave tank experiment (for more details on the facility, see Osborne and Petti, 1994). The water depth was set to 40 cm and the wave maker was programmed to generate a sinusoidal initial wave. The period was selected to be 4 s and we show in Figure 31.8 the results of four probe measurements at 4.25, 7.01, 11.02, and 15.02 m from the wave maker. This experiment is somewhat more nonlinear that the computer experiments reported above. Figure 31.9 shows the IST spectrum at the first probe. Shown are the cnoidal wave amplitudes and their moduli as a function of frequency. Note that we have taken *two periods of the generated wave trains*, that is, the temporal length of the measured time series is 8 s rather than 4 s. It is easy to see that wave trains of this type are *quasiperiodic* rather than perfectly periodic, since perfect periodicity is an improbable occurrence in a laboratory environment. For this reason the IST spectrum in Figure 31.9 has the characteristic that the odd frequencies (1, 3, 5, 7, etc.) are much smaller than the even frequencies (2, 4, 6, 8, etc.). Perfect periodicity would give zero amplitudes for the odd frequencies. Further note that the first two components at low frequency have moduli that are very nearly 1, and hence these components are very nearly solitons. Thus, a low frequency soliton ($\Delta\omega = 2\pi/T$ for the period, $T = 8$ s) is found to contribute to the lack of perfect periodicity in the measured wave trains.

In these experiments even the first probe is quite far from the sine-wave initial conditions and the other probes are even further away. Furthermore, the presence of a ramp at the end of the flume kept the test section to less than 20 m of the total 50 m length of the facility. For this reason we were unable to investigate the possible presence of FPU recurrence. Nevertheless, it is clear that the linear Fourier spectra for the present experiment were quite broad banded and increasing in bandwidth as the waves propagated down the tank.

Figures 31.10–31.13 provide the details of the cnoidal wave, inverse scattering transform decomposition of the four time series of Figure 31.8. Figure 31.10 gives the analysis for probe 1. Shown in the figure (in vertical order starting at the top) are the 12 cnoidal waves that comprise the spectrum, the sum of these cnoidal waves, the nonlinear interactions among the cnoidal waves, and finally the reconstructed measured wave train. As mentioned above the first two components are essentially solitons and this is easily seen at the top of Figure 31.10.

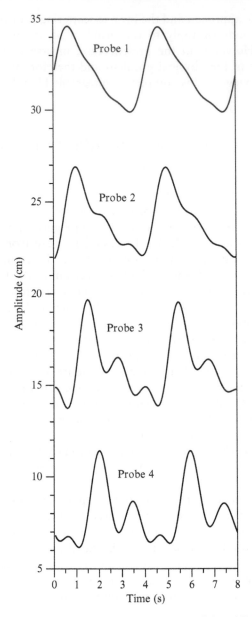

Figure 31.8   Measurements of wave trains from a sinusoidal paddle motion in the wave tank in Florence, Italy.

Note that the top curve is a single soliton whereas the second curve is a double peaked soliton train. Furthermore, the single soliton coincides with one of the maxima in the measured wave train; this can be verified to also be the case in Figures 31.11–31.13. The third component from the top of Figure 31.10 is a

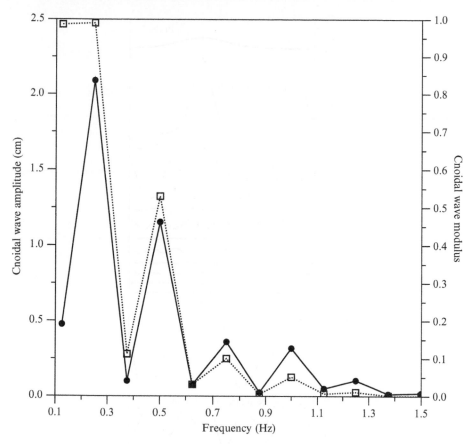

Figure 31.9 Cnoidal wave spectrum and modulus at probe 1 of the measurements of Figure 31.8.

small amplitude sine wave, but the fourth component, with a modulus of about 0.55, is rather a robust Stokes wave (invert the figure to better see the Stokes shape of this component). The remaining components have much smaller moduli, that is, less than about 0.15, and are either Stokes waves or weak sine waves. The summation of the cnoidal wave components is shown near the middle of Figure 31.10; this result differs quite a lot from the (reconstructed) measured wave train at the bottom of the figure. Only by adding the influence of the nonlinear interactions do we get good agreement between the reconstructed wave train and the measured wave train itself.

Figure 31.11 gives the IST spectral decomposition at probe 2. These results are quite similar to those in Figure 31.10, except that there is more structure in the measured wave train and this is accounted for by the combined action of the temporally phased cnoidal wave components and the nonlinear interactions. The observable structure in the measured wave train is seen to be

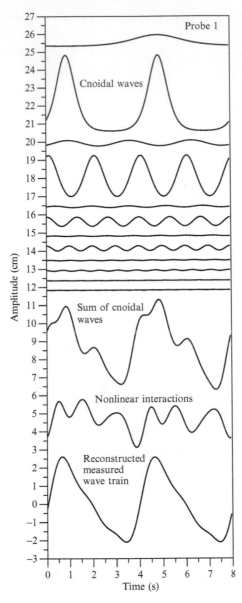

Figure 31.10   IST spectrum of probe 1 (Figure 31.8).

considerably enhanced in Figures 31.12 and 31.13. Additional peaks are found to emerge from the initial wave train as the temporal phases of the cnoidal wave components shift relative to one another. These dynamics are seen to be quite complex and, to us, quite lovely.

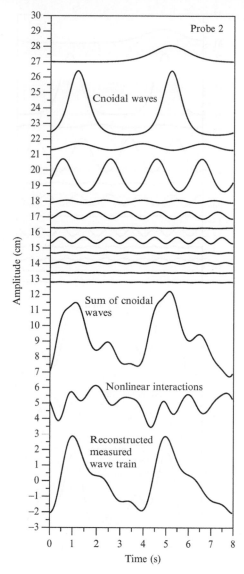

Figure 31.11 IST spectrum of probe 2 (Figure 31.8).

In spite of the complex nonlinear interaction dynamics in the measured wave trains, the dynamics of the inverse scattering transform cnoidal wave amplitudes are seen to be quite simple as shown in Figure 31.14. Here the main cnoidal wave amplitudes are found to change very little as the wave train propagates down the canal, that is, they are essentially constants of the motion for KdV evolution and are nearly constant for the present wave tank

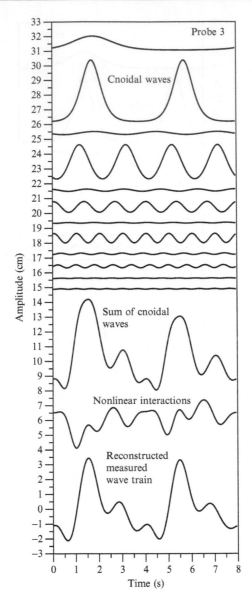

Figure 31.12   IST spectrum of probe 3 (Figure 31.8).

experiments. This is the miracle of using IST as a tool for signal processing of shallow-water waves: the spectral/cnoidal wave decomposition demonstrates nearly constant amplitudes during the complex evolution of an experimentally measured shallow-water wave train. This is in contrast to the situation with sine-wave basis functions which have considerable time evolution in their amplitudes as Figure 31.15 demonstrates.

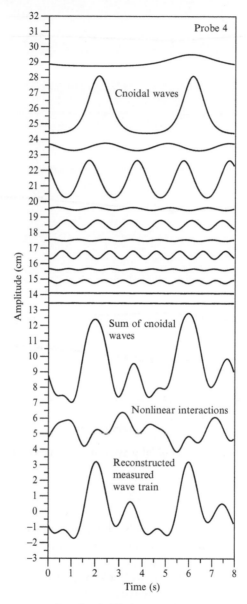

Figure 31.13 IST spectrum of probe 4 (Figure 31.8).

31.8 Summary and Discussion

The inverse scattering transform in the θ-function formulation is seen to provide new insight and perspective about the nonlinear dynamics of shallow-water wave trains. The propagation of shallow-water waves is seen to be

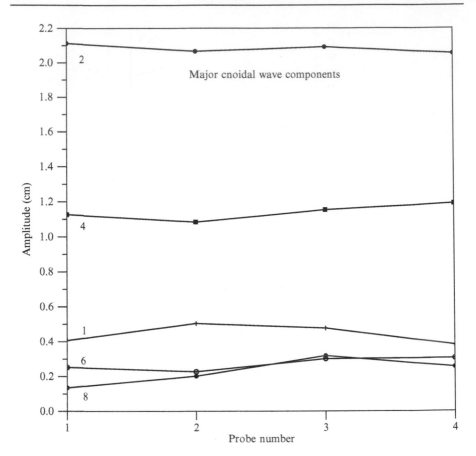

Figure 31.14 Time variation of cnoidal wave components as a function of probe number.

governed by two considerations, namely, the space/time evolution of individual cnoidal waves with particular constant amplitudes and moduli, plus their mutual nonlinear interactions. The actual spectrum of the cnoidal wave components is seen to depend on the period or interaction matrix, **B**, whose *diagonal elements* define the cnoidal wave amplitudes and moduli and whose *off-diagonal elements* govern the nonlinear interactions. Both the synthesis of computer generated, shallow-water wave trains and the analysis of data are found to be feasible using methods developed in this monograph.

Analytical, numerical, and experimental results demonstrate that the nonlinear dynamics of the KdV equation have no resonances in the θ-function formulation. However, for example, cnoidal wave resonances might themselves occur during adiabatic shoaling of nonlinear wave trains in shallow water regions.

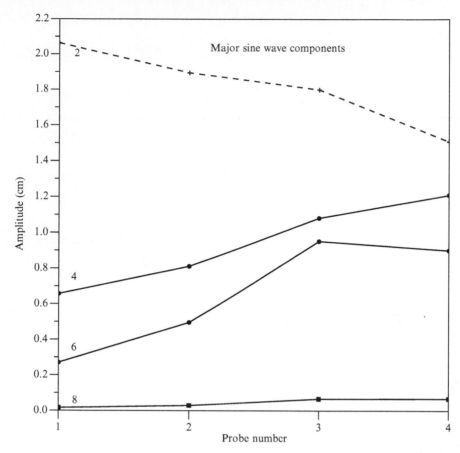

Figure 31.15   Time variation of sine-wave components as a function of probe number.

This is because KdV evolution is only approximately correct over variable bathymetry and the space/time evolution of the wave train would necessarily require variations in the cnoidal wave amplitudes during the dynamical motion.

Part Nine

Nonlinear Hyperfast Numerical Modeling

This part of the book addresses some aspects of the *hyperfast modeling* of surface and internal waves. Chapter 32 addresses, in great detail, the *KdV and KP equations*. On this basis Chapter 33 discusses a numerical model for the *2 + 1 Gardner equation*. By a simple selection of the constant coefficients of the equation one can separately or simultaneously simulate the KdV, KP, modified KdV, and Gardner equations. The 2 + 1 Gardner equation is suggested as a possible important model equation also for internal waves. The *Davey-Stewartson equations* are studied in Chapter 34 and a theoretical model for the simultaneous simulation of the *surface wave elevation* and the *velocity potential* for wave trains with narrow-banded spectra is given and a hyperfast numerical model is described.

32 Hyperfast Modeling of Shallow-Water Waves: The KdV and KP Equations

32.1 Introduction

A new numerical model has been developed for shallow-water waves based upon the Kadomtsev-Petviashvili (KP) equation (see Chapter 11). While the model is "conventional" in the sense of the KP approximation, the actual algorithm is quite different than that typically developed using the fast Fourier transform (FFT): The new algorithm is based upon the mathematics of the periodic inverse scattering transform (IST) and employs the *Riemann theta function*, which has a *spectral matrix* (Riemann matrix) rather than the *spectral vector* of the linear Fourier transform.

I develop several algorithms for computing theta functions in this monograph, two of which are useful for modeling purposes, the first of which is used in this chapter and is based upon three steps:

(1) Reduce the theta function to a *Fourier series with time varying coefficients*. The advantage of the method is that the theta function becomes just an ordinary Fourier series (Chapters 7–9) for which the FFT can be used to make many of the numerical computations.
(2) Take a particular *modular transformation* of the Riemann matrix (Chapter 8).
(3) *Sum the theta functions over the n-ellipsoid* (Chapter 22).

These three steps constitute a fast algorithm for the theta function. As described in this chapter, the above three steps are combined in a "preprocessor" algorithm; the remaining computations are completed with the FFT. I have also developed a *discrete Riemann theta function* that employs the results of Chapters 8 and 21 and provides similar functionality and speed.

The above two algorithms for *accelerated* or *fast theta functions* (FTF) are described as being "hyperfast" in this monograph. These approaches are substantially faster than the typical "brute force" theta function algorithms discussed in Chapter 20.

The KP model is *phase resolving* in the sense that it is not statistical but instead deterministic in nature, actually solving the KP equation for a chosen initial condition. The model is coded in its most natural form, meaning that

Doi: 10.1016/S0074-6142(10)97032-0

if one chooses he can calculate only the *two-dimensional wave spectrum* as a function of time. Otherwise it is a simple matter to compute the *surface elevation* as an additional output. For those interested in output that is *not* phase resolving, it is a simple matter to achieve this feature by internal manipulation of the variables with no sacrifice in computer time: To this end one analytically computes the correlation function of the theta functions, from which the correlation function of the KP equation is then determined. The new model is found to be about two to three orders of magnitude faster than models that employ the FFT (see e.g. Fornberg and Whitham, 1978).

The methods of this chapter allow us to study the nonlinear behavior of shallow-water surface waves using the IST for the KP equation with periodic boundary conditions. The KP equation is a completely integrable Hamiltonian system whose solutions describe the dynamics of shallow-water wave trains in $2 + 1$ dimensions in terms of Riemann theta functions for which the associated Riemann matrix characterizes the nonlinear directional spectrum. The Riemann spectrum is computed by the method of Nakamura and Boyd in which the spectral parameters are determined via the Hirota approach (Chapter 16); and by the Schottky uniformization procedure in which the spectral parameters are computed as Poincaré series (Chapter 15). Both approaches are found to be numerically equivalent and hence both methods ensure that the wave dynamics are those of the KP equation and are therefore associated with a Riemann surface, in agreement with the Novikov conjecture.

I show in Figure 32.1 a photograph of two crossed cnoidal waves in shallow water in which there is clear evidence of a *Mach stem* (Miles, 1977) or *dromion* (Fokas and Santini, 1989, 1990) at the intersection point. I thank Dr Paul Palo for this extraordinary photograph; his observational powers at an opportune moment have allowed the capture of an amazing natural phenomenon. This appears to be a classical genus-two solution of the KP equation as shown in Figure 32.3. It should be pointed out that, as a result of the research conducted in this chapter, typical shallow-water waves have *many nonlinear directional wave components*, not just two as shown in the special case of this photograph. Patterns such as those which appear in the photograph are therefore not considered generic, although they may often occur, but are instead considered to be exceptional, especially in water depths greater than a few meters. This is no surprise to oceanographers who have long known that ocean waves have complex fully directional spectra.

32.2   Overview of the Literature

I discuss, both theoretically and numerically, the modeling of directionally spread, shallow-water wave trains using the KP equation, a generalization of the Korteweg-deVries (KdV) equation from $1 + 1$ dimensions (x, t) to $2 + 1$ dimensions (x, y, t). I first address shallow-water wave dynamics from the point of view of the KdV equation. I then discuss the KP equation, which includes

Figure 32.1   Photograph of a simple shallow water case (San Nicholas, California) that appears to be two crossed cnoidal waves with a Mach stem or dromion at the intersection point. Photograph courtesy of Dr Paul Palo, Naval Facilities Engineering Service Center, Port Hueneme, California. (See color plate).

directional spreading, albeit for spreading angles less than about ±20°; larger spreading angles require nonlinear corrections not addressed in this chapter (see instead Chapter 33 for the 2 + 1 Gardner equation). Both the KdV and KP equations are known to be integrable Hamiltonian systems (Ablowitz and Segur, 1981; Novikov et al, 1984; Ablowitz and Clarkson, 1991). These equations have been solved for periodic boundary conditions by the IST (Dubrovin and Novikov, 1975a,b; Dubrovin et al., 1976; Its and Matveev, 1975; Krichever, 1981; Belokolos et al., 1994).

Previous research provides a rich foundation for the computation of Riemann theta functions. Several authors, including Nakamura (1980), Nakamura and Matsuno (1980), Hirota and Ito (1983), Boyd (1984a,b,c), Bobenko (1987), Bobenko and Bordag (1987, 1989a,b), Osborne (1995a,b), Deconinck et al. (2004) established procedures for numerically computing Riemann theta functions and for computing the spectral parameters of the theta function (Riemann matrix, wavenumbers, frequencies, and phases) to ensure that one has solutions for KdV and KP. The Riemann spectrum can be computed using any of several methods. These include the method of loop integrals (Dubrovin and Novikov, 1975a,b), Schottky uniformization (Bobenko and Bordag, 1989), and the method of Nakamura (1980) and Boyd (1984a,b,c, 1990, 1998). In this monograph I use all three approaches to determine the spectrum of the simulations. After some experience in determining spectra using these approaches, I have found that Nakamura-Boyd provides a quick and relatively easy method to extract the physics of the spectrum of a particular nonlinear, integrable wave equation. Stated another way: Given a new equation that I have not addressed in the past I find the NB method gives me a rapid way to develop an estimate of the Riemann spectrum and to construct a hyperfast numerical model, that is, typically only a few days are necessary for model construction and checkout after the mathematical analysis has been done.

An outline of the chapter is as follows. I give a brief discussion of the IST, with periodic boundary conditions, for the KdV and KP equations in Section ~ 32.3. This section describes the nonlinear spectral representation for shallow-water wave trains in terms of the Riemann (spectral) matrix and a set of arbitrary (possibly random) phases. In Section 32.4, I discuss important useful properties of Riemann theta functions that lead to the new numerical algorithm. In particular, I show how to write the theta function as a two-dimensional, ordinary linear Fourier series with time varying coefficients that have the form of temporal Fourier series with amplitudes that are a function of the Riemann matrix and with incommensurable frequencies. In Section 32.5, I discuss the Schottky uniformization procedure that allows one to easily characterize the Riemann spectrum in terms of the physically and numerically convenient uniformization parameters. In Section 32.6, I write the leading order contribution of the Schottky Poincaré series, valid when the waves are not too large. In Section 32.7, I discuss implementation of the method of Nakamura (1980) and Boyd (1984a,b,c), a kind of "physical effectivization" for computing the Riemann spectrum. In Section 32.8 a procedure is given for computing the linear Fourier components for the wave field solution of the KP equation in terms of IST parameters. I then address in Section 32.9 numerical procedures based on Schottky uniformization for computing the Riemann spectrum. Section 32.10 discusses the procedure for fast computation of Riemann theta functions; details for the computation of nonlinear, directional KP wave trains from directional spectra are given in Section 32.11. Finally, in Section 32.12, I give a numerical example of the approach for a 30×30 Riemann matrix. The new algorithm is compared to the well-known Fornberg and Whitham (1978) approach for computation of solutions of the KP equation.

32.3 The Inverse Scattering Transform for Periodic Boundary Conditions

In shallow water it is often convenient to address the physics of nonlinear wave propagation in terms of the KdV equation (unidirectional propagation) and the KP equation (which contains directional spreading). These equations result from the application of the method of multiple scales to the Euler equations for small but finite amplitude waves in shallow water (Ablowitz and Segur, 1981; Johnson, 1997). From the point of view of the mathematical physics, one views this exercise as an expansion of the equations near zero wavenumber, $k \sim 0$. The cases are distinct enough to consider separately, although the KdV equation is contained within the KP formulation when there is no directional spreading, that is, when the y component wavenumber $l_n = 0$. I first discuss the case for KdV.

32.3.1 The KdV Equation

The KdV equation (Korteweg and deVries, 1895) describes unidirectional wave propagation in shallow water. It is given by

$$\eta_t + c_o\eta_x + \alpha\eta\eta_x + \beta\eta_{xxx} = 0 \tag{32.1}$$

where $c_o = \sqrt{gh}$, $\alpha = 3c_o/2h$, $\beta = c_oh^2/6$, h the water depth and g the acceleration of gravity. Periodic boundary conditions are assumed, $\eta(x, t) = \eta(x + L, t)$. By neglecting the nonlinear term (set $\alpha = 0$), the solution of the associated linear KdV equation gives the linear dispersion relation: $\omega = c_ok - \beta k^3$.

The nonsingular (physical) solutions of KdV are written in terms of multidimensional Fourier series (Riemann theta functions) given by Dubrovin and Novikov (1975a,b) and Its and Matveev (1975):

$$\eta(x,t) = \frac{2}{\lambda}\partial_{xx}\ln\theta(x,t) = \frac{2}{\lambda}\left(\frac{\theta\theta_{xx} - \theta_x^2}{\theta^2}\right) \tag{32.2}$$

where $\lambda = \alpha/6\beta = 3/2h^3$. Here the Riemann theta function has the form of a multidimensional Fourier series:

$$\theta(x,t) = \sum_{m_1=-\infty}^{\infty}\sum_{m_2=-\infty}^{\infty}\cdots\sum_{m_N=-\infty}^{\infty}\exp\left(-\frac{1}{2}\sum_{m=1}^{N}\sum_{n=1}^{N}m_mm_nB_{mn}\right)$$
$$\times\exp\left(i\sum_{n=1}^{N}m_n\kappa_nx - i\sum_{n=1}^{N}m_n\omega_nt + i\sum_{n=1}^{N}m_n\phi_n\right) \tag{32.3a}$$

or in vector form:

$$\theta(x,t|\mathbf{B},\boldsymbol{\phi}) = \sum_{\mathbf{m}\in\mathbb{Z}}q_{\mathbf{m}}\exp(i\mathbf{m}\cdot\boldsymbol{\kappa}x - i\mathbf{m}\cdot\boldsymbol{\omega}t + i\mathbf{m}\cdot\boldsymbol{\phi}), \quad q_{\mathbf{m}} = e^{-\frac{1}{2}\mathbf{m}\cdot\mathbf{Bm}} \tag{32.3b}$$

The B_{mn} are the elements of the Riemann matrix, k_n are the IST wavenumbers, ω_n are the frequencies, and ϕ_n are the phases in the IST spectrum. Rules for the determination of these parameters are discussed in Sections 32.5–32.10. Note that the Riemann matrix plays the role of a spectrum for nonlinear, integrable partial differential equations; the theta function solution of the KdV equation ((32.2), (32.3)) reduces to the linear Fourier transform in the small-amplitude limit, that is, when the off-diagonal terms, B_{mn}, are small relative to the diagonal terms B_{nn}. The case for the KdV equation with a 2×2 Riemann matrix has been studied extensively by Boyd (1984a,b,c). The cases for 2×2 and 3×3 Riemann matrices are discussed by Segur and Finkel (1982) and Dubrovin et al. (1997). Deconinck et al. (2004) also studied an approach for the precise computation of the set of $\mathbf{m} = [m_1, m_2, \ldots, m_N]$ vectors over which to sum (Equation (32.3)).

Due to the periodic boundary conditions one has the following rule for computing the wavenumbers: $\kappa_j = 2\pi_j/L$, where L is the period of the wave train; this coincides with linear Fourier analysis. The number of nested sums in the theta function, N, provides the number of degrees of freedom (cnoidal waves) or components in the IST spectrum. Because of the structure of the theta function, the multiple (nested) summations occur over *all possible sums and differences of the wavenumbers and frequencies*. It is for this reason that the theta functions physically describe nonlinear interactions so well.

As noted above, the spectrum for the KdV equation is a matrix (the Riemann matrix) rather than a vector (as for the linear FFT). The elements on the diagonal of the Riemann matrix correspond to cnoidal waves (see Weigel, 1964; Dean and Dalrymple, 1991 for a discussion of cnoidal waves in oceanographic and ocean engineering contexts), which are the N nonlinear modes of the KdV equation; these contrast with the sine wave modes of linear Fourier analysis. The nonlinear interactions are contained in the off-diagonal terms of the Riemann matrix. Since the cnoidal waves depend on their modulus, m, which varies between 0 and 1, the nonlinear modes of KdV can be sine waves ($m \sim 0$), Stokes waves ($m \sim 0.5$), and solitons ($m \sim 1$). Consequently, as the modulus is continuously increased from 0 to 1 the cnoidal wave varies smoothly from a sine wave to a Stokes wave to a soliton. It is often convenient to write the solutions to the KdV equation ((32.2), (32.3)) as a linear sum of cnoidal waves plus nonlinear interactions (Osborne, 1995a,b). The more nonlinear modes the KdV equation has in a particular application, the more terms one needs to sum in the theta function (see Osborne, 1995a,b, 2002) and Chapter 20).

A well-known procedure for numerically solving the KdV equation is to assume a Fourier series for the solution:

$$\eta(x,t) = \sum_{n=-N_x/2}^{N_x/2} \eta_n(t) e^{ik_n x - i\omega_n t + i\phi_n} \tag{32.4}$$

Here the Fourier coefficients are taken to be a function of time due to the nonlinear term in KdV. By inserting the above Fourier series into the KdV equation, one naturally finds ordinary differential equations (ODEs) for the evolution of the Fourier coefficients:

$$\dot{\eta}_n + i\omega_n\eta_n + i\alpha \sum_{m=-\infty}^{\infty} k_n\eta_n\eta_{n-m} = 0 \tag{32.5}$$

where $\omega_n = c_0 k_n - \beta k_n^3$ and the "over dot" is a time derivative. These latter equations may be solved numerically to obtain $\eta_n(t)$ as a function of time; the solution of KdV $\eta(x, t)$ then follows from the Fourier transform (Equation

(32.4)). In the special case for linear motion, one sets $\alpha = 0$ and finds that the time evolution of the Fourier coefficients is given by the elementary result $\eta_n(t) = \eta_n(0)\,e^{-i\omega_n t}$. General time evolution of the Fourier spectrum is also the goal of many modern approaches to the modeling of nonlinear wave equations, including the so-called higher order spectral methods (West et al.; Yue et al.).

In Section 32.8 I discuss how the ODEs (32.5) can be solved in a natural way using IST.

32.3.2 The Kadomtsev-Petviashvili Equation

The KP equation (Kadomtsev-Petviashvili, 1970), which describes the nonlinear dynamics of directionally spread surface waves in shallow water, is written as:

$$(\eta_t + c_o\eta_x + \alpha\eta\eta_x + \beta\eta_{xxx})_x + \gamma\eta_{yy} = 0 \tag{32.6}$$

where the coefficients c_o, α, β are the same as in the KdV equation above and $\gamma = c_o/2$. The wave field also depends on the y coordinate. Periodic boundary conditions, $\eta(x, y, t) = \eta(x + L_x, y, t) = \eta(x, y + L_y, t)$, are assumed for x, y. Wave evolution in the time coordinate is quasiperiodic.

Note that the KP equation (32.6) is often referred to as KP II in the literature. KP II is valid for water depths substantially greater than about half a centimeter, that is, when surface tension can be neglected. When surface tension dominates, for depths less than about half a centimeter, the sign of γ is negative and the resultant mathematical/physical behavior of KP I takes on a different character than that of KP II.

It is natural to write the solution to the KP equation in terms of the linear Fourier transform with time varying coefficients:

$$\eta(x, y, t) = \sum_{m=-M/2}^{M/2} \sum_{n=-N/2}^{N/2} \eta_{mn}(t)e^{i(k_m x + l_n y - \omega_{mn}t + \varphi_{mn})} \tag{32.7}$$

The periodic boundary conditions mean that the wavenumber pairs (k_m, l_n) are commensurable: $k_m = 2\pi m/L_x$, $l_n = 2\pi n/L_y$, where L_x and L_y are the lengths of the sides of the periodic rectangular domain of coordinates x and y. Here I have assumed that the x coordinate is divided into N_x intervals and the y coordinate is divided into N_y intervals. Inserting the above two-dimensional FFT (Equation (32.7)) into the wave equation (32.6) determines the time evolution of the Fourier coefficients by the set of ODEs:

$$\dot{\eta}_{mn} + i\omega_{mn}\eta_{mn} + i\alpha \sum_{i=-\infty}^{\infty} \sum_{j=-\infty}^{\infty} k_i\eta_{ij}\eta_{m-i,\,n-j} = 0 \tag{32.8}$$

where $\omega_{mn} = c_o k_m - \beta k_m^3 + \gamma l_n^2/k_m$ is the linear dispersion relation. Of course if one eliminates the nonlinear term in Equation (32.8) (by setting $\alpha = 0$), then the Fourier transform is simplified because the time varying coefficients are now simple circular functions $\eta_{mn}(t) = \eta(0)\,e^{-i\omega_{mn}t}$. The analytical temporal evolution of the coefficients $\eta_{mn}(t)$ as the solution to Equation (32.8) in terms of IST parameters is discussed in Section 32.8.

In this chapter, due to the integrability of the KP equation, all smooth, non-singular solutions can be written in terms of Riemann theta functions. To this end the solution to the periodic KP equation has the form (Krichever, 1981):

$$\eta(x,y,t) = \frac{2}{\lambda}\partial_{xx}\ln\theta(x,y,t) = \frac{2}{\lambda}\left[\frac{\theta(x,y,t)\theta_{xx}(x,y,t) - \theta_x^2(x,y,t)}{\theta^2(x,y,t)}\right] \quad (32.9)$$

where

$$\theta(x,y,t) = \sum_{m_1=-\infty}^{\infty}\sum_{m_2=-\infty}^{\infty}\cdots\sum_{m_N=-\infty}^{\infty}\exp\left(-\frac{1}{2}\sum_{m=1}^{N}\sum_{n=1}^{N}m_m m_n B_{mn}\right)$$

$$\times\exp\left(i\sum_{n=1}^{N}m_n\kappa_n x + i\sum_{n=1}^{N}m_n\lambda_n y - i\sum_{n=1}^{N}m_n\omega_n t + i\sum_{n=1}^{N}m_n\phi_n\right)$$

$$(32.10a)$$

or in vector form:

$$\theta(x,y,t|\mathbf{B},\boldsymbol{\phi}) = \sum_{\mathbf{m}\in\mathbb{Z}}q_{\mathbf{m}}e^{i\mathbf{m}\cdot\boldsymbol{\kappa}x+i\mathbf{m}\cdot\boldsymbol{\lambda}y-i\mathbf{m}\cdot\boldsymbol{\omega}t+i\mathbf{m}\cdot\boldsymbol{\phi}}, \quad q_{\mathbf{m}} = e^{-\frac{1}{2}\mathbf{m}\cdot\mathbf{Bm}} \quad (32.10b)$$

The x and y wavenumbers fall on a subset of grid points of the k_m, l_n plane such that $k_n = 2\pi n/L_x$ and $l_n = 2\pi n/L_y$ for periodic boundary conditions (see comments *Nota Bene* below). As before the number of nested sums in the theta function, N, provides the number of nonlinear modes or cnoidal waves in the IST spectrum. Here each diagonal element of the period matrix, B_{nn}, corresponds to a wavenumber pair (κ_n, λ_n) which specifies the direction of a particular cnoidal wave. Interactions among wavenumber pairs are contained in the off-diagonal terms, B_{mn}, $m \neq n$, of the Riemann matrix.

For applications in physical oceanography, it is often useful to write the periodic solutions of the KP equation as a linear superposition of directionally spread cnoidal waves plus nonlinear interactions among them. An identity for the solution of the KP equation that illustrates this idea is (Osborne, 1995a,b):

$$\eta(x,y,t) = \sum_{n=1}^{N}a_n cn^2\{[K(m)/\pi](k_n x + l_n y - \omega_n t + \phi_n|m_n)\} + \eta_{\text{int}}(x,y,t)$$

$$(32.11)$$

Here $cn(k_n x + l_n y - \omega_n t | m_n)$ is the classical Jabcobian elliptic function (Abramowitz and Stegun, 1964) and m_n is its modulus ($0 \le m_n < 1$) which, by abuse of notation, is not to be confused with the summation indices in Equation (32.10). One should not think of the interaction term $\eta_{int}(x, y, t)$ as a perturbation; it is small only if the wave amplitudes are small. For large amplitudes the spectral components are solitons and the interaction terms constitute the usual solitonic phase shifts, albeit with periodic boundary conditions. The cnoidal wave amplitudes, a_n, are normally graphed as a function of κ_n, λ_n or in terms of frequency ω_n and direction $\theta_n = \tan^{-1}(\lambda_n / \kappa_n)$; this latter graph is known as the *nonlinear directional spectrum*. The a_n are computed from the respective nomes, $q_n = \exp(-B_{nn}/2)$, using the diagonal elements of the Riemann matrix, B_{nn}:

$$a_n = \eta_{max} = \eta_n(\pi/k_n, 0) = \frac{4k_n^2}{\lambda} \sum_{m=1}^{\infty} \frac{m q_n^m}{1 - q_n^{2m}} \tag{32.12}$$

The Ursell number of a cnoidal wave, $U_n = 3a_n / 4k_n^2 b^3$, $k_n = \sqrt{\kappa_n^2 + \lambda_n^2}$, is given in terms of the modulus, m_n, by

$$m_n K^2(m_n) = \frac{3\pi^2 a_n}{2k_n^2 b^2} = 2\pi^2 U_n \tag{32.13}$$

where $K(m)$ is the real quarter-period of the elliptic function. Therefore, the modulus, m_n, and the Ursell number, U_n, are two equally good parameters for describing the (Stokes-type) nonlinearity in a cnoidal wave (see Chapter 30 for a discussion).

Nota Bene: The wavenumbers κ_n, λ_n at this juncture appear to be equal to the k_m, l_n of ordinary linear Fourier analysis, but this is not necessarily true. Indeed, κ_n, λ_n are a *subset* of the wavenumbers k_m, l_n. To see how this happens, consider the figure below which is in the k_m, l_n plane of linear Fourier analysis; there are 25 points in the grid, corresponding to 25 pairs of wavenumbers k_m, l_n. Shown is an ellipse in this plane that corresponds to the outer contour of the directional spectrum of interest. Selection of this ellipse corresponds to some small value of the spectrum far from the peak that is, say, less than a small number such as 10^{-8}. Only the spectral amplitudes, a_n, inside the ellipse are considered for further analysis; these are labeled 1, 2, ..., 7.

Continued

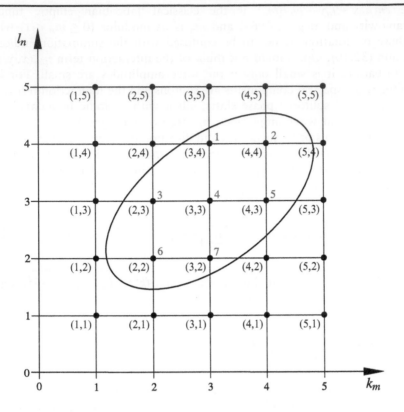

At this point we build Table 32.1 for all the parameters we need to evaluate the theta function.

Table 32.1 Table of Theta Function Parameters

n	κ_n	λ_n	a_n	B_{nn}	φ_n	ω_n
1	3	4	a_1	B_{11}	ϕ_1	ω_1
2	4	4	a_2	B_{22}	ϕ_2	ω_2
3	2	3	a_3	B_{33}	ϕ_3	ω_3
4	3	3	a_4	B_{44}	ϕ_4	ω_4
5	4	3	a_5	B_{55}	ϕ_5	ω_5
6	2	2	a_6	B_{66}	ϕ_6	ω_6
7	3	2	a_7	B_{77}	ϕ_7	ω_7

Shown are the values of the integer wavenumber pairs κ_n, λ_n inside the ellipse, which are labeled 1-7. For example, the amplitudes a_n at each of the seven sites are listed in the table. Using the formulas above for the cnoidal wave one constructs the *diagonal elements of the Riemann matrix*, in the present case 7×7, from the a_n by inverting Equation (32.12). Also shown are the choices for the phases and for the frequencies. To simulate a solution of the KP equation one can, for example, chose the phases to be random. The off-diagonal elements of the Riemann matrix and the frequencies are computed by Schottky uniformization (Chapter 15), loop integrals (Chapter 14), or the Nakamura-Boyd approach (Chapter 16).

32.4 Properties of Riemann Theta Functions and Partial Theta Summations

The Riemann theta functions, for periodic boundary conditions, can be written as ordinary Fourier series with time-dependent coefficients, a result shown in this section. Table 32.2 helps compare the notation for the spectral amplitudes, wavenumbers, frequencies, and phases for the different expressions for the theta function and partial theta summation used herein.

32.4.1 The KdV Equation

Let us first briefly look at the KdV equation. Here the theta function has $1 + 1$ dimensions, see Equation (32.3). In anticipation of the numerical calculations to follow, we convert Equation (32.3) to a *partial summation*:

$$
\theta(x, t | \mathbf{B}, \boldsymbol{\phi}) = \sum_{m_1=-M}^{M} \sum_{m_2=-M}^{M} \cdots \sum_{m_N=-M}^{M} \exp\left(-\frac{1}{2} \sum_{m=1}^{N} \sum_{n=1}^{N} m_m m_n B_{mn} \right)
$$
$$
\times \exp\left(i \sum_{n=1}^{N} m_n \kappa_n x - i \sum_{n=1}^{N} m_n \omega_n t + i \sum_{n=1}^{N} m_n \phi_n \right)
\tag{32.14}
$$

Table 32.2 Notation for Spectral Parameters for the Riemann Theta Function as Written in Three Different Forms: The Classical $1 + 1$ or $2 + 1$ Dimensional Theta Function (Equation (32.3) or (32.10)), the Theta Function with Single Summation (Equations (32.15), (32.27)), and the Ordinary Linear Fourier Transform for the Theta Function (Equations ((32.21), (32.38)) with Time Varying Coefficients

Theta Function Spectral Parameters	Fourier Amplitude	x Wave-Number	y Wave-Number	Frequency	Phase
Theta nested (32.3), (32.10)	q_m	κ_n	λ_n	ω_n	ϕ_n
Theta single (32.15), (32.27)	q_j	K_j	Λ_j	Ω_j	Φ_j
Fourier series (32.21), (32.38)	$\theta_{mn}(t)$	k_m	l_n	ω_{mn}	φ_m

Here M is an arbitrary integer limit in the summation, chosen to replace the theoretical limit of infinity for numerical computations. One could of course use different limits for each summation, M_n, depending on the modulus of the cnoidal wave for each of the degrees of freedom in the theta function, that is, the larger the value of the modulus m_n for a particular degree of freedom, the larger should be the value of M_n in order to insure convergence. Note that together with the definition of the nome $(q_n = \exp(-B_{nn}/2))$ Equations (32.12) and (32.13) link the following variables $a_n \leftrightarrow B_{nn} \leftrightarrow q_n \leftrightarrow m_n \leftrightarrow U_n$ in a natural way so that given a particular period matrix element B_{nn} one can estimate the number of terms necessary for convergence of a single degree of freedom theta function; this turns out to be a practical and conservative way to estimate the limits (see Chapter 22, Introduction for a discussion). The next step is to write the *partial theta sum* (Equation (32.14)) as a single summation that can then replace the nested sum:

$$\theta(x, t|\mathbf{B}, \boldsymbol{\phi}) = \sum_{j=-J}^{J} q_j e^{iK_j x - i\Omega_j t + i\Phi_j} \tag{32.15}$$

where $J = [(2M + 1)^N - 1]/2$ is an integer limit and

$$q_j = \exp\left(-\frac{1}{2}\sum_{m=1}^{N}\sum_{n=1}^{N} m_m^j m_n^j B_{mn}\right) \tag{32.16}$$

$$K_j = \sum_{n=1}^{N} m_n^j \kappa_n \tag{32.17}$$

$$\Omega_j = \sum_{n=1}^{N} m_n^j \omega_n \tag{32.18}$$

$$\Phi_j = \sum_{n=1}^{N} m_n^j \phi_n \tag{32.19}$$

The q_j are referred to as *nomes*. The superscript j on m_n^j emphasizes the one-to-one correspondence between terms in the nested sum (Equation (32.14)) and in the single sum (Equation (32.15)). The number of terms in the partial sums (Equations (32.14) and (32.15)) is given by $(2M + 1)^N$.

Now let us examine the single theta summation (Equation (32.15)). In this formulation, in order to meet the periodic boundary conditions for the spatial variable x $(\theta(x, t|\mathbf{B}, \boldsymbol{\phi}) = \theta(x + L, t|\mathbf{B}, \boldsymbol{\phi}))$, the wavenumbers must be commensurable: $\kappa_n = 2\pi n/L$. This implies that the wavenumber K_j (Equation (32.17)) has the form:

$$K_j = \frac{2\pi}{L}\sum_{n=1}^{N} n m_n^j = \frac{2\pi}{L} I_j, \quad I_j = \sum_{n=1}^{N} n m_n^j \tag{32.20}$$

Here I_j is an integer whose value changes with each value of the index j as it varies in the interval $-J \leq j \leq J$; I_j is *not* ordered with the integers and is often repeated. Osborne (1995a,b) gives a discussion of the behavior of K_j, which is clearly commensurable, as a function of j (see Section 8.2.7). The behavior for K_j (Equation (32.20)) contrasts with that of Ω_j which is not commensurable, because the Ω_j are not commensurable. On this basis the theta function summation (Equation (32.10)) can be written as an ordinary Fourier series:

$$\theta(x, t | \mathbf{B}, \boldsymbol{\phi}) = \sum_{n=-N_x/2}^{N_x/2} \theta_n(t) e^{ik_n x} \tag{32.21}$$

$$\theta_n(t) = \sum_{\{j \in \mathbb{Z} : I_j = n\}} q_j e^{-i\Omega_j t + i\Phi_j} \tag{32.22}$$

The partial theta function summation (Equation (32.14)) occurs for the additional restriction $-J \leq j \leq J$. The vector form for Equation (32.22) is given by

$$\theta_n(t) = \sum_{\{\mathbf{m} \in \mathbb{Z} : n = I(\mathbf{m} \cdot \boldsymbol{\kappa})\}} q_{\mathbf{m}} e^{-i\mathbf{m} \cdot \boldsymbol{\omega} t + i\mathbf{m} \cdot \boldsymbol{\phi}}, \quad q_{\mathbf{m}} = e^{-\frac{1}{2}\mathbf{m} \cdot \mathbf{Bm}}$$

For the partial theta function (Equation (32.14)), one sums over the lattice vector $\mathbf{m}$ inside the N-dimensional cube $(2M + 1)^N$. Note further that

$$I(\mathbf{m} \cdot \boldsymbol{\kappa}) = \frac{L_x}{2\pi} \sum_{n=1}^{N} m_n \kappa_n$$

Equations (32.21) and (32.22) taken together are an identity for the theta function (Equation (32.3)) (provided $\infty < j < \infty$) or the single partial summation (Equation (32.15)) (provided $-J \leq j \leq J$). The derivation of Equations (32.21) and (32.22) is quite simple because the equations follow from the fact that the IST wavenumber K_j is for every j equal to a commensurable wavenumber $k_n = 2\pi n/L_x$ for some particular integer n. Equation (32.22) for $\theta_n(t)$ is therefore a summation of all the Q_j which fall onto a particular wavenumber k_n in the wavenumber plane. The fact that there are *many spectral contributions* to a particular wavenumber k_n arises physically because of nonlinear interactions among all possible combinations of wavenumbers in the solution of the KdV equation.

Note that the time-dependent coefficients $\theta_n(t)$ (Equation (32.22)) are Fourier series over the time variable t, but with *incommensurable frequencies* Ω_j. Note further that for this reason the Fourier coefficients $\theta_n(t)$ are *quasiperiodic* in time. Group theoretic notation is used in Equation (32.22); this is the most natural way to characterize the summation. Note, in the limit $M \to \infty$, that the number of terms in Equations (32.14) and (32.22) approach infinity and the partial sum (Equation (32.14)) becomes an ordinary theta function summation (Equation (32.3)). The results (Equations (32.21) and (32.22)) are an identity for the partial theta summation (Equation (32.14)) and, as will be discussed in detail below, result in considerable savings in computer time.

32.4.2   The KP Equation

Let us now consider the KP equation where the theta function partial sum is in two spatial dimensions:

$$\theta(x,y,t|\mathbf{B},\boldsymbol{\phi}) = \sum_{m_1=-M}^{M} \sum_{m_2=-M}^{M} \cdots \sum_{m_N=-M}^{M} \exp\left(-\frac{1}{2}\sum_{m=1}^{N}\sum_{n=1}^{N}m_m m_n B_{mn}\right)$$

$$\times \exp\left(i\sum_{n=1}^{N}m_n\kappa_n x + i\sum_{n=1}^{N}m_n\lambda_n y - i\sum_{n=1}^{N}m_n\omega_n t + i\sum_{n=1}^{N}m_n\phi_n\right)$$

$$(32.23)$$

Again M is an integer limit taken to be finite since ∞ is not possible in numerical computations. It is theoretically instructive to keep M finite in subsequent calculations. However, at any stage of the calculations one can let $M\to\infty$.

It is also instructive to write the theta function in an alternative vector notation:

$$\theta(x,y,t|\mathbf{B},\boldsymbol{\phi}) = \sum_{\mathbf{m}\in\mathbb{Z}} Q_{\mathbf{m}}(t)e^{i\mathbf{m}\cdot\boldsymbol{\kappa}x + i\mathbf{m}\cdot\boldsymbol{\lambda}y} \qquad (32.24)$$

where

$$Q_{\mathbf{m}}(t) = e^{-\frac{1}{2}\mathbf{m}\cdot\mathbf{B}\mathbf{m} - i\mathbf{m}\cdot\boldsymbol{\omega}t + i\mathbf{m}\cdot\boldsymbol{\phi}} = q_{\mathbf{m}}e^{-i\mathbf{m}\cdot\boldsymbol{\omega}t + i\mathbf{m}\cdot\boldsymbol{\phi}}, \quad q_{\mathbf{m}} = e^{-\frac{1}{2}\mathbf{m}\cdot\mathbf{B}\mathbf{m}} \qquad (32.25)$$

The $Q_{\mathbf{m}}$ solve the linear oscillator equations

$$\dot{Q}_{\mathbf{m}} + i\Omega_{\mathbf{m}}Q_{\mathbf{m}} = 0 \qquad (32.26)$$

where $\Omega_{\mathbf{m}} = \mathbf{m}\cdot\boldsymbol{\omega}$. The nonlinear dynamics of the KP equation has, in this way, been reduced to an infinite number of uncoupled, linear oscillators using the IST (of course one keeps in mind the limit $J = [(2M+1)^N - 1]/2 \to \infty$ as $M\to\infty$).

Is clear from Equation (32.23) that the partial theta summation can be written over a single sum:

$$\theta(x,y,t|\mathbf{B},\boldsymbol{\phi}) = \sum_{j=-J}^{J} q_j e^{iK_j x + i\Lambda_j y - i\Omega_j t + i\Phi_j} \qquad (32.27)$$

where as before $J = [(2M + 1)^N - 1]/2$ and

$$K_j = \sum_{n=1}^{N} m_n^j \kappa_n \qquad (32.28)$$

$$\Lambda_j = \sum_{n=1}^{N} m_n^j \lambda_n \qquad (32.29)$$

$$\Omega_j = \sum_{n=1}^{N} m_n^j \omega_n \qquad (32.30)$$

$$\Phi_j = \sum_{n=1}^{N} m_n^j \phi_n \qquad (32.31)$$

Here I have again used the correspondence $\mathbf{m} \Leftrightarrow j$ (for every vector $\mathbf{m}$ there is a unique integer j) and I have set $\mathbf{m} \rightarrow \mathbf{m}^j$ in order to make the correspondence concrete.

To meet the periodic boundary conditions ($\theta(x, y|\mathbf{B}, \boldsymbol{\phi}) = \theta(x + L_x, y|\mathbf{B}, \boldsymbol{\phi}) = \theta(x, y + L_y|\mathbf{B}, \boldsymbol{\phi})$), the spatial wavenumbers κ_n, λ_n must be integer multiples of $\Delta\kappa = 2\pi/L_x$, $\Delta\lambda = 2\pi/L_y$. This means that

$$K_j = \frac{2\pi}{L_x} I_j, \quad I_j = \frac{L_x}{2\pi} \sum_{n=1}^{N} m_n^j \kappa_n \qquad (32.32)$$

and

$$\Lambda_j = \frac{2\pi}{L_y} J_j, \quad J_j = \frac{L_y}{2\pi} \sum_{n=1}^{N} m_n^j \lambda_n \qquad (32.32)$$

Here of course $I_j, J_j \in \mathbb{Z}$, that is, the positive, negative, and zero integers. This implies that the K_j, Λ_j are commensurable, proportional to the integers, often repeated, but are not ordered with the integers. The single partial theta sum (Equation (32.27)) can be written in the following form:

$$\theta(x, y, t|\mathbf{B}, \boldsymbol{\phi}) = \sum_{j=-J}^{J} Q_j(t) e^{iK_j x + i\Lambda_j y} \qquad (32.34)$$

where

$$Q_j(t) = q_j e^{-i\Omega_j t + i\Phi_j} \qquad (32.35)$$

It is therefore clear once again that the $Q_j(t)$ solve the linear oscillator equations:

$$\dot{Q}_j + i\Omega_j Q_j = 0, \quad -J \leq j \leq J \tag{32.36}$$

The $Q_j(t)$ are thus the *normal coordinates* for the KP equation. The Cauchy initial conditions may be computed from

$$Q_j(0) = q_j e^{i\Phi_j} \tag{32.37}$$

Therefore, the period matrix B_{ij} and phases ϕ_j determine the spectrum for the KP equation.

One is finally and naturally led to the following *ordinary Fourier series for the two-dimensional theta function*:

$$\theta(x, y, t|\mathbf{B}, \boldsymbol{\phi}) = \sum_{m=-N_x/2}^{N_x/2} \sum_{n=-N_y/2}^{N_y/2} \theta_{mn}(t) e^{ik_m x + il_n y} \tag{32.38}$$

where the *time-dependent linear Fourier coefficients* have the form:

$$\theta_{mn}(t) = \sum_{\{j \in \mathbb{Z}: I_j = m, J_j = n\}} Q_j(t) = \sum_{\{j \in \mathbb{Z}: I_j = m, J_j = n\}} q_j e^{-i\Omega_j t + i\Phi_j} \tag{32.39}$$

Equations (32.38) and (32.39) are equivalent forms for the theta function (Equation (32.10)). For the partial theta function sum (Equation (32.23)), one takes the values of the integer j to lie in the range $-J \leq j \leq J$. An alternative form for Equation (32.39) in vector notation is:

$$\theta_{mn}(t) = \sum_{\{\mathbf{m} \in \mathbb{Z}: m = I(\mathbf{m} \cdot \boldsymbol{\kappa}), n = I(\mathbf{m} \cdot \boldsymbol{\lambda})\}} q_{\mathbf{m}} e^{-i\mathbf{m} \cdot \boldsymbol{\omega} t + \mathbf{m} \cdot \boldsymbol{\phi}} \tag{32.40}$$

where the partial summation is over the lattice vector $\mathbf{m}$ inside the N-dimensional cube $(2M + 1)^N$. Note that

$$I(\mathbf{m} \cdot \boldsymbol{\kappa}) = \frac{L_x}{2\pi} \sum_{n=1}^{N} m_n \kappa_n, \quad I(\mathbf{m} \cdot \boldsymbol{\lambda}) = \frac{L_y}{2\pi} \sum_{n=1}^{N} m_n \lambda_n$$

Also $-N_x/2 \leq m < N_x/2$, $-N_y/2 \leq n < N_y/2$. Do not confuse the integers m, n with the vector $\mathbf{m}$ or any of its components; only by abuse of notation do these variables resemble one another. The integers m, n are indices for the x and y wavenumbers k_m, l_n; the $\mathbf{m}$ are the lattice vectors of the theta function (Equation (32.10)): $\mathbf{m} = [m_1, m_2, \ldots, m_N]$. Equations (32.38) and (32.39) taken together are an identity for the partial theta sum (Equation (32.23)) or the single partial summation (Equation (32.27)) provided that the limit on j is taken to be: $-J \leq j \leq J$ for $J = [(2M + 1)^N - 1]/2$.

The derivation of Equations (32.38) and (32.39) is quite simple because the equations follow from the fact that the IST wavenumber pair (K_j, Λ_j) is for every j equal to a commensurable wavenumber pair $(k_m, l_n) = (2\pi m/L_x, 2\pi n/L_y)$ for some particular integer pair (m, n). Equation (32.39) for $\theta_{mn}(t)$ is therefore a summation of all the Q_j that fall onto a particular wavenumber pair (k_m, l_n) in the wavenumber plane. The fact that there are many spectral contributions to a particular wavenumber (k_m, l_n) arises physically because of nonlinear interactions among all possible combinations of the wavenumbers in the solution of the KP equation.

Note that the ordinary Fourier coefficients (Equation (32.39)) of the theta function (Equation (32.38)) are themselves a linear Fourier series with incommensurable frequencies and indeed one has a linear superposition of all the oscillator mode solutions (Equation (32.35)). Equation (32.39) is just a temporal Fourier series over the nome amplitudes q_j for the incommensurable frequencies Ω_j. The linear Fourier coefficients $\theta_{mn}(t)$ in Equation (32.38) are therefore quasiperiodic in time. Equations (32.23)–(32.25), (32.27–32.31), (32.38), and (32.39) are all equivalent forms for the partial theta function summation; see summary in Table 32.2.

Equations (32.38) and (32.39) are among the main results of this Chapter and allow substantial improvement in computer time for the computation of theta functions and hence for the computation of solutions of the KP equation. Numerically, one first computes the $\theta_{mn}(t)$ from Equation (32.39) using standard group theoretic techniques (see below). Then one uses Equation (32.38) to compute the theta function for all space x, y, and all values of time: This latter step is done with a standard two-dimensional FFT. For N_t values of time t one applies the 2D FFT N_t times, together with the transformation (Equation (32.9)), to obtain the space/time evolution of the waves $\eta(x, y, t)$.

It is not difficult to show that when the *amplitudes of the cnoidal wave modes are small* (the diagonal elements of the period matrix are large with respect to the off-diagonal elements) so that they can be described by a simple sine wave, then the time evolution of the linear Fourier modes is given by the following elementary expression:

$$\theta_{mn}(t) = a_{mn} e^{-i\omega_{mn}t + i\varphi_{mn}}$$

The results of this section will be employed for numerical computation of theta functions in Section 32.9.

32.5 Computation of the Spectral Parameters in Terms of Schottky Uniformization

While the approach of Section 32.4 improves our ability to compute theta functions, we are still left with the fundamental problem of selecting the Riemann spectrum such that one has a solution of the KP equation. Algebraic geometric loop integrals (Krichever, 1981), Schottky uniformization (Bobenko and Bordag (1989), and the approach of Nakamura (1980) and Boyd (1984a,b,c) are the methods most commonly used for this purpose. Naturally, one expects that all of these methods satisfy the Novikov conjecture, which states that to ensure physical solutions of the KP equation one must have a Riemann spectrum, which corresponds to an underlying Riemann surface. However, I know of no mathematical proofs that the Nakamura-Boyd approach is known to lie on a Riemann surface. I have found all three methods to be numerically equal. Both the loop integrals and Schottky uniformization begin with holomorphic differentials, so it seems that the Novikov conjecture must be true. The case for the loop integrals was proven by Shiota (1986).

Here we use the *Schottky uniformization of Riemann surfaces* to get access to all smooth, nonsingular, periodic solutions of the KP II equation (Schottky, 1887; Baker, 1897; Krichever, 1981; Shiota, 1986; Bobenko and Bordag, 1989; Burnside, 1892a,b; Belokolos et al., 1994). The Schottky method is now reviewed and results are discussed which are important to the numerical computations in this Chapter.

For Equations (32.9) and (32.10) to constitute a solution to the KP equation one must first compute the appropriate wavenumbers, frequencies, and Riemann matrix. The Schottky procedure provides *Poincaré series* for these parameters as a function of the so-called *uniformization parameters* A_n, ρ_n, $n = 1, 2, \ldots, N$ where N is the number of degrees of freedom (cnoidal waves) in the spectrum. To give a physical interpretation of these parameters (see discussion below), it is enough to remember that at leading order the ρ_n (real numbers) are related to the diagonal elements of the Riemann matrix (or the amplitudes of the cnoidal waves) and the A_n (complex numbers) are related to the wavenumbers κ_n, λ_n, $n = 1, 2, \ldots, N$ (both are real numbers); the exact Schottky relationships specify the wavenumbers, frequencies, and Riemann matrix in terms of the Schottky parameters. In the formulas below the σ's are *linear fractional transformations* written in terms of the *uniformization parameters*, $\sigma_n = \sigma_n(A_n, \rho_n)$. The summations given below are *group theoretic*, using standard group theory notation, and include the identity, $\sigma_0 = I$.

Nota Bene: For the approach given herein the numerical Schottky uniformization procedure may be summarized as follows: Given the diagonal elements of the Riemann matrix B_{nn} and the *commensurable* wavenumbers κ_n, λ_n compute the off-diagonal elements B_{mn} ($m \le n$) and the frequencies ω_n. We now see how this procedure is formulated.

32.5.1   Linear Fractional Transformation

Details and derivations for the Schottky uniformization procedure are given elsewhere (Bobenko and Bordag, 1987, 1989; Belokolos et al., 1994). In what follows we require the linear fractional (Möbius) transformation σ, which has the form:

$$\sigma z = \frac{\alpha z + \beta}{\gamma z + \delta} \tag{32.41}$$

whose constants are given in terms of the uniformization parameters:

$$\alpha = \frac{1}{2}\left(\frac{1+\rho}{\sqrt{\rho}}\right), \quad \beta = -\frac{A}{2}\left(\frac{1-\rho}{\sqrt{\rho}}\right) \tag{32.42}$$

$$\gamma = -\frac{1}{2A}\left(\frac{1-\rho}{\sqrt{\rho}}\right), \quad \delta = \frac{1}{2}\left(\frac{1+\rho}{\sqrt{\rho}}\right) \tag{32.43}$$

The corresponding matrix operator for σ is given by

$$\tilde{\sigma} = \begin{bmatrix} \dfrac{1}{2}\left(\dfrac{1+\rho}{\sqrt{\rho}}\right) & -\dfrac{A}{2}\left(\dfrac{1-\rho}{\sqrt{\rho}}\right) \\[16pt] -\dfrac{1}{2A}\left(\dfrac{1-\rho}{\sqrt{\rho}}\right) & \dfrac{1}{2}\left(\dfrac{1+\rho}{\sqrt{\rho}}\right) \end{bmatrix} \tag{32.44}$$

with inverse

$$\sigma^{-1}z = \frac{\delta z - \beta}{-\gamma z + \alpha} \tag{32.45}$$

$$\tilde{\sigma}^{-1} = \begin{bmatrix} \dfrac{1}{2}\left(\dfrac{1+\rho}{\sqrt{\rho}}\right) & \dfrac{A}{2}\left(\dfrac{1-\rho}{\sqrt{\rho}}\right) \\[16pt] \dfrac{1}{2A}\left(\dfrac{1-\rho}{\sqrt{\rho}}\right) & \dfrac{1}{2}\left(\dfrac{1+\rho}{\sqrt{\rho}}\right) \end{bmatrix} \tag{32.46}$$

The group elements $\sigma_0 = I$, σ_n, and σ_n^{-1} constitute the (Schottky) group over which the Poincaré series for the Schottky spectrum (see below) are summed.

32.5.2 Theta Function Spectrum as Poincaré Series of Schottky Parameters

For a general introduction to Poincaré series see Baker (1897) and for specific applications to KP see Bobenko and Bordag (1989) and Belokolos et al. (1994). You will be quite amazed and pleased to find many of the important formulas below in Baker's classic book from over a century ago; Bobenko has provided us with a great service in linking much of this work, via Schottky uniformization (Schottky, 1887), to modern theories of nonlinear integrable wave equations. Of course it was Krichever (1989) who integrated KP II and his introduction to the latest version of Baker's book is absolutely marvelous.

Here the Schottky problem is revisited in the small-amplitude, oscillatory limit, as opposed to the soliton limit. This is done for convenience in isolating the periodic solutions of the KP equation and for the added advantage that we have the ordinary Fourier transform of the theta function for accelerating numerical codes as discussed in Chapter 9 and in Section 32.4.2 above. The diagonal elements of the Riemann matrix have the following Poincaré series:

$$B_{nn} = \ln \rho_n + \sum_{\sigma \in G_n \backslash G / G_n, \sigma \neq I} \ln \left[\frac{(A_n^* - \sigma A_n^*)(A_n - \sigma A_n)}{(A_n^* - \sigma A_n)(A_n - \sigma A_n^*)} \right] \tag{32.47}$$

The off-diagonal elements have the form:

$$B_{mn} = \sum_{\sigma \in G_m \backslash G / G_n} \ln \left[\frac{(A_m^* - \sigma A_n^*)(A_m - \sigma A_n)}{(A_m^* - \sigma A_n)(A_m - \sigma A_n^*)} \right], \quad m \neq n \tag{32.48}$$

or alternatively

$$B_{mn} = \ln \left[\frac{(A_m^* - A_n^*)(A_m - A_n)}{(A_m^* - A_n)(A_m - A_n^*)} \right]$$
$$+ \sum_{\sigma \in G_n \backslash G / G_n, \sigma \neq I} \ln \left[\frac{(A_m^* - \sigma A_n^*)(A_m - \sigma A_n)}{(A_m^* - \sigma A_n)(A_m - \sigma A_n^*)} \right], \quad m \neq n \tag{32.49}$$

In the last equation I have brought out the term for $\sigma_o = I$. Thus, the identity has been removed from the group theoretic summation; note that this latter summation in Equation (32.49) now excludes the identity term.

The wavenumbers have the form:

$$\kappa_n = -i \sum_{\sigma \in G / G_n} \left(\sigma A_n - \sigma A_n^* \right) \tag{32.50}$$

$$\lambda_n = -ih \sum_{\sigma \in G / G_n} \left[(\sigma A_n)^2 - (\sigma A_n^*)^2 \right] \tag{32.51}$$

The frequency is

$$\omega_n = c_o \kappa_n - 4i\beta \sum_{\sigma \in G / G_n} \left[(\sigma A_n)^3 - (\sigma A_n^*)^3 \right] \tag{32.52}$$

The Equations (32.47)–(32.52) have been written in dimensional form. Therefore, given a set of Schottky parameters (ρ_n, A_n) one can compute all of the parameters of the theta function, namely, B_{mn}, κ_n, λ_n, ω_n for an arbitrary set of phases ϕ_n. All of the Poincaré series given above are summed over the group whose elements are $\sigma_0 = I$, σ_n, and σ_n^{-1}. The notation is a standard one, although one may want to look at Chapter 15 for a brief summary of the summation procedures.

32.6   Leading Order Computation of KP Spectra Using Schottky Variables

Here I discuss physical understanding and numerical computation of the spectral parameters using the method.

Is straightforward to define a simple procedure: One assumes that the diagonal elements of the period matrix B_{ii} and phases ϕ_i are given, together with assumed commensurable wavenumbers $\kappa_n = 2\pi n/L_x$, $\lambda_n = 2\pi n/L_y$ which arise from the periodicity condition for x and y, and then one seeks the off-diagonal elements of the period matrix, B_{ij} $(i \neq j)$, and frequencies, ω_j. To develop this procedure note that since the solution to KP is periodic then so too is the theta function periodic. This implies that the wavenumbers are commensurable, and hence κ_n, λ_n must be integer multiples of $\Delta\kappa = 2\pi/L_x$, $\Delta\lambda = 2\pi/L_y$ where L_x, L_y are the dimensions of the space domain. Since κ_n, λ_n are known one can use Equations (32.50) and (32.51) to give a leading order estimate of the Schottky parameters $A_n = a_n + ib_n$, where a_n, b_n are the real and imaginary parts of A_n. Solving Equations (32.50) and (32.51) for (a_n, b_n) in terms of (κ_n, λ_n) therefore fixes A_n to leading order. To determine the other Schottky parameter ρ_n one uses Equation (32.47); since the B_{nn} are assumed to be known, and A_n (a complex number) is the solution of Equations (32.50) and (32.51), then Equation (32.47) can be solved for the ρ_n (a real number). Generally speaking, solving Equations (32.50) and (32.51) for the A_n and Equation (32.47) for the ρ_n must of course be done numerically as the inversion of the Poincaré series is not easily obtained analytically.

Separating the identity term from the other terms leads to alternative forms for Equations (32.50)–(32.52):

$$\kappa_n = -i\left(A_n - A_n^*\right) - i \sum_{\sigma \in G/G_n, \sigma \neq I} \left(\sigma A_n - \sigma A_n^*\right) \tag{32.53}$$

$$\lambda_n = -ih\left[A_n^2 - A_n^{*2}\right] - ih \sum_{\sigma \in G/G_n, \sigma \neq I} \left[\left(\sigma A_n\right)^2 - \left(\sigma A_n^*\right)^2\right] \tag{32.54}$$

$$\omega_n = c_0\kappa_n - 4i\beta\left[A_n^3 - A_n^{*3}\right] - 4i\beta \sum_{\sigma \in G/G_n, \sigma \neq I} \left[\left(\sigma A_n\right)^3 - \left(\sigma A_n^*\right)^3\right] \tag{32.55}$$

It is not difficult to show that the leading order identity terms dominate the Poincaré series when the waves have small amplitude. Thus, Equations (32.47), (32.49), and (32.53)–(32.55) have a small-amplitude (near linear) limit:

$$B_{nn} \cong \ln \rho_n \tag{32.56}$$

$$B_{mn} \cong \ln\left[\frac{(A_m^* - A_n^*)(A_m - A_n)}{(A_m^* - A_n)(A_m - A_n^*)}\right] \tag{32.57}$$

$$\kappa_n \cong -i(A_n - A_n^*) \tag{32.58}$$

$$\lambda_n \cong -ih(A_n^2 - A_n^{*2}) \tag{32.59}$$

$$\omega_n \cong c_0\kappa_n - 4i\beta(A_n^3 - A_n^3) \tag{32.60}$$

Note that, at this approximation, the nome is given by $q_n = \exp(-B_{nn}/2)$ $\sim 1/\sqrt{\rho_n}$ so that $0 \leq q_n < 1$ and $\infty > \rho_n \geq 0$. Assuming that κ_n, λ_n are given on a rectangular grid in the wavenumber plane (assuming periodic boundary conditions) one can easily solve Equations (32.58) and (32.59), for the A_n (these will provide approximate starter values of the Schottky parameters in the iterative numerical procedure given below). Thus, to leading order the Schottky uniformization variables have the form

$$A_n = a_n + ib_n \simeq \frac{\lambda_n}{2\kappa_n h} + i\frac{\kappa_n}{2} = \frac{p_n}{2} + i\frac{\kappa_n}{2} \tag{32.61}$$

where a_n is the real part of A_n and b_n is the imaginary part:

$$a_n = \frac{\lambda_n}{2\kappa_n h} = \frac{p_n}{2}, \quad b_n = \frac{\kappa_n}{2} \tag{32.62}$$

where $p_n = \lambda_n/\kappa_n h$. These are useful results because they point out the one-to-one relationship between the Schottky uniformization parameters ($A_n = a_n + ib_n$) and the wavenumber pair (κ_n, λ_n) at leading order.

Let us now evaluate Equation (32.60) for the linear dispersion relation. Note that, with Equation (32.61), we have

$$A_n^3 - A_n^{*3} \simeq -\frac{i}{4}\left(\kappa_n^3 - \frac{3\lambda_n^2}{4h^2\kappa_n}\right) \tag{32.63}$$

Then Equation (32.55) has the approximate form

$$\omega_n \simeq c_0\kappa_n - \beta\kappa_n^3 + \frac{c_0}{2}\frac{\lambda_n^2}{\kappa_n} \tag{32.64}$$

The nonlinear dispersion relation (32.55) then becomes

$$\omega_n = c_0 \kappa_n - \beta \kappa_n^3 + \frac{c_0}{2} \frac{\lambda_n^2}{\kappa_n} - 4i\beta \sum_{\sigma \in G/G_n, \sigma \neq I} \left[(\sigma A_n)^3 - (\sigma A_n^*)^3 \right] \tag{32.65}$$

Therefore, the physical interpretation of Equation (32.65) for the frequencies of the nonlinear *dispersion relation* is that one finds linear dispersion at leading order, and then the nonlinear terms, consisting of the nonlinear pairwise interactions of all the degrees of freedom with each other (amplitude dependent corrections to the dispersion), are added by the Poincaré series.

Let us also evaluate the leading-order, off-diagonal elements of the period matrix for KP. This is done by inserting A_n of Equation (32.61) into Equation (32.57):

$$B_{mn} \cong \ln \left[\frac{(\kappa_m - \kappa_n)^2 + \left(\frac{\kappa_n \lambda_m - \kappa_m \lambda_n}{\kappa_m \kappa_n h} \right)^2}{(\kappa_m + \kappa_n)^2 + \left(\frac{\kappa_n \lambda_m - \kappa_m \lambda_n}{\kappa_m \kappa_n h} \right)^2} \right], \quad m \neq n \tag{32.66}$$

An alternative form is

$$B_{mn} \cong \ln \left[\frac{(\kappa_m - \kappa_n)^2 + (p_m - p_n)^2}{(\kappa_m + \kappa_n)^2 + (p_m - p_n)^2} \right], \quad m \neq n \tag{32.67}$$

Note that when $\lambda_n = 0$ for all n (i.e., when there is no directional spreading), the above expressions reduce to the result for the KdV equation:

$$B_{mn} \cong \ln \left[\frac{\kappa_m - \kappa_n}{\kappa_m + \kappa_n} \right]^2, \quad m \neq n \tag{32.68}$$

A brief discussion on the group theoretic summation of Poincaré series is given in Chapter 15.

32.7 The Method of Nakamura and Boyd

An alternative method for determining the Schottky spectrum is to use the method developed by Nakamura (1980) and Boyd (1984a,b,c). Their procedure is here outlined for the KP equation. One first substitutes the transformation (Equation (32.9)) into the KP equation (Equation (32.6)) in order to obtain what might be called the Hirota-KP equation:

$$\theta\theta_{xt} - \theta_x\theta_t + c_0(\theta\theta_{xx} - \theta_x^2) + \beta(3\theta_{xx}^2 - 4\theta_x\theta_{xxx} + \theta\theta_{xxxx}) + \gamma(\theta\theta_{yy} - \theta_y^2) + \frac{\alpha c}{12\beta}\theta^2 = 0 \tag{32.69}$$

This expression can be put into Hirota operator notation to give

$$\left(D_x D_t + c_o D_x^2 + \beta D_{xxxx} + \gamma D_y^2 + \frac{\alpha c}{12\beta}\right)\theta \cdot \theta = 0 \tag{32.70}$$

We use the usual definition of the Hirota operator:

$$D_x^n a \cdot b \equiv \left(\frac{\partial}{\partial x} - \frac{\partial}{\partial y}\right)^2 a(x)b(y)\bigg|_{y=x} = \frac{\partial^n}{\partial y^n} a(x+y)b(x-y)\bigg|_{y=0} \tag{32.71}$$

so that

$$D_x D_t \theta \cdot \theta = 2(\theta\theta_{xt} - \theta_x\theta_t)$$

$$D_x^2 \theta \cdot \theta = 2(\theta\theta_{xx} - \theta_x^2)$$

$$D_x^4 \theta \cdot \theta = 2(3\theta_{xx}^2 - 4\theta_x\theta_{xxx} + \theta\theta_{xxxx}) \tag{32.72}$$

$$D_y^2 \theta \cdot \theta = 2(\theta\theta_{yy} - \theta_y^2)$$

By introducing theta functions with characteristics (Nakamura, 1980; Boyd, 1984a,b,c) (see Chapter 16), one arrives at a set of nonlinear equations written in terms of the period matrix, wavenumbers, and frequencies:

$$\sum_{m_1=-\infty}^{\infty} \sum_{m_2=-\infty}^{\infty} \cdots \sum_{m_N=-\infty}^{\infty} \left\{ \left[2\sum_{j=1}^{N}(m_j - \mu_j/2)\kappa_j\right]\left[2\sum_{j=1}^{N}(m_j - \mu_j/2)\omega_j\right] \right.$$

$$-c_o\left[2\sum_{j=1}^{N}(m_j - \mu_j/2)\kappa_j\right]^2 + \beta\left[2\sum_{j=1}^{N}(m_j - \mu_j/2)\kappa_j\right]^4$$

$$\left. +\gamma\left[2\sum_{j=1}^{N}(m_j - \mu_j/2)\lambda_j\right]^2 - \frac{\alpha c}{12\beta}\right\}\exp\left[-\sum_{j=1}^{N}\sum_{k=1}^{N}(m_j - \mu_j/2)(m_k - \mu_k/2)B_{jk}\right] = 0$$

$$\tag{32.73}$$

In the above equation the characteristics μ_j take on the values 0, 1; this leads to a set of 2^N equations for determining the parameters of the Riemann spectrum. In the present procedure one assumes particular values for the diagonal elements of the period matrix and the wavenumbers, here taken to be commensurable so that $\kappa_n = 2\pi n/L_x$, $\lambda_n = 2\pi n/L_y$. One then seeks to determine the off-diagonal elements of the period matrix and the frequencies so that Equations (32.9) or (32.38) and (32.39) give a solution of the KP equation. It is not difficult to show from Equation (32.73) that to leading order

$$B_{mn} \simeq \ln \left[\frac{(\kappa_m - \kappa_n)^2 + \left(\frac{\kappa_n \lambda_m - \kappa_m \lambda_n}{\kappa_m \kappa_n h} \right)^2}{(\kappa_m + \kappa_n)^2 + \left(\frac{\kappa_n \lambda_m - \kappa_m \lambda_n}{\kappa_m \kappa_n h} \right)^2} \right], \quad m \neq n$$

$$\omega_n \simeq c_0 \kappa_n - \beta \kappa_n^3 + \frac{c_0}{2} \frac{\lambda_n^2}{\kappa_n}$$

coinciding with the leading order terms in the Schottky procedure discussed above. These are of course the starting values in an iterative procedure given in Chapter 16 to determine the off-diagonal elements of the period matrix B_{mn} and frequencies ω_n given the diagonal elements of the period matrix B_{nn} and the commensurable wavenumbers κ_n, λ_n.

32.8 The Exact Solution of the Time Evolution of the Fourier Components for the KP Equation

I now evaluate the KP Fourier series coefficients $\eta_{mn}(t)$ in Equation (32.7) analytically in terms of IST variables. The $\eta_{mn}(t)$ are of course solutions of the ODEs (32.8). The focus of this section is the determination of the solution of the ODEs (32.8) in terms of the inverse scattering parameters of the Riemann spectrum. Note that the solution to the KP equation ((32.9), (32.10)) is expressed as the second x derivative of the logarithm of a theta function. Since the theta function can be written as an ordinary Fourier series by Equations (32.38) and (32.39), it suffices to express the logarithm of a Fourier series as a Fourier series. Then the second x derivative of the latter Fourier series is also a Fourier series and the solution to KP is then a Fourier series itself (Equation (32.7)).

When can we express the logarithm of a Fourier as a Fourier series? This is generally an irresolvable problem, but the results are easily done for Riemann theta functions because their Fourier series are very special, that is, the theta function can never be negative due to the constraints on the properties of the period matrix (always real and positive definite for the KP II equation). It is convenient to write an alternative form for the theta function:

$$\theta(x, y, t) = 1 + 2 \sum_{\mathbf{m}}' q_{\mathbf{m}} \cos(\mathbf{m} \cdot \boldsymbol{\kappa} x + \mathbf{m} \cdot \boldsymbol{\lambda} y - \mathbf{m} \cdot \boldsymbol{\omega} t + \mathbf{m} \cdot \boldsymbol{\phi}) = 1 + X(x, y, t)$$

where the prime indicates that the "1" has been excluded from the summation. The logarithm is then

$$\ln(1 + X) = \sum_{n=1}^{\infty} \frac{(-1)^{n-1}}{n} X^n = X - \frac{X^2}{2} + \frac{X^3}{3} - \frac{X^4}{4} + \cdots \qquad (32.74)$$

Here $X(x, y, t) < 1$ due to this special nature of theta functions and their Fourier series. We see that the logarithm of a theta function Fourier series is easily written:

$$\ln \theta = \ln \left[\sum_{m=-N_x/2}^{N_x/2} \sum_{n=-N_y/2}^{N_y/2} \theta_{mn}(t)e^{ik_m x + il_n y} \right]$$

$$= \ln \left[a_{oo} + \sum_{m=1}^{N_x} \sum_{n=1}^{N_y} a_{mn} \cos(k_m x + l_n y) + b_{mn} \sin(k_m x + l_n y) \right] \quad (32.75)$$

Note that $a_{mn}(t)$ and $b_{mn}(t)$ are then easily computed from the $\theta_{mn}(t)$, Equation (32.39), both of which are related to the parameters of the Riemann spectrum. We then write the logarithm of the latter theta function Fourier series as a Fourier series using recursion:

$$\ln \theta(x, y, t) = \frac{A_o(t)}{2} + \sum_{m=1}^{\infty} \sum_{n=1}^{\infty} [A_{mn}(t) \cos(k_m x + l_n y) + B_{mn}(t) \sin(k_m x + l_n y)]$$

$$(32.76)$$

where

$$A_{oo} = 2 + \sum_{j=1}^{\infty} \frac{(-1)^{j-1}}{j} \alpha_{oo}^{(j)}$$

$$A_{mn} = \sum_{j=1}^{\infty} \frac{(-1)^{j-1}}{j} \alpha_{mn}^{(j)} \quad (32.77)$$

$$B_{mn} = \sum_{j=1}^{\infty} \frac{(-1)^{j-1}}{j} \beta_{mn}^{(j)}$$

The following recursion relations are found to hold:

$$\alpha_{oo}^{(k+1)}(t) = \sum_{i=1}^{\infty} \sum_{j=1}^{\infty} \left(a_{ij}(t)\alpha_{ij}^{(k)}(t) + b_{ij}(t)\beta_{ij}^{(k)}(t) \right) \quad (32.78)$$

$$\alpha_{mn}^{(k+1)}(t) = \frac{1}{2} \sum_{i=1}^{\infty} \sum_{j=1}^{\infty} \left[a_{ij}(t)\left(\alpha_{i+m,j+n}^{(k)}(t) + \alpha_{i-m,j-n}^{(k)}(t) \right) + b_{ij}\left(\beta_{i+m,j+n}^{(k)}(t) + \beta_{j-m,j-n}^{(k)}(t) \right) \right]$$

$$(32.79)$$

$$\beta_{mn}^{(k+1)}(t) = \frac{1}{2} \sum_{i=1}^{\infty} \sum_{j=1}^{\infty} \left[a_{ij}(t)\left(\beta_{i+m,j+n}^{(k)}(t) - \beta_{i-m,j-n}^{(k)}(t) \right) - b_m(t)\left(\alpha_{i+m,j+n}^{(k)}(t) - \alpha_{j-m,j-n}^{(k)}(t) \right) \right]$$

$$(32.80)$$

for the following initial values

$$\alpha_{oo}^{(1)} = 0, \quad \alpha_{mn}^{(1)} = a_{mn}, \quad \beta_{mn}^{(1)} = b_{mn} \tag{32.81}$$

Now use Equation (32.76) in (32.9) to get the Fourier series solution to the KP equation:

$$\eta(x, y, t) = 2\partial_{xx} \ln \theta(x, y, t)$$

$$= -\sum_{m=1}^{\infty} \sum_{n=1}^{\infty} [k_m^2 A_{mn}(t) \cos(k_m x + l_n y) + k_m^2 B_{mn}(t) \sin(k_m x + l_n y)] \tag{32.82}$$

Comparison of this expression with Equation (32.7) means that the nonlinear ODEs (32.8) for the Fourier coefficients have the following exact solution:

$$\eta_{mn}(t) = k_m^2 \sqrt{A_{mn}^2(t) + B_{mn}^2(t)} e^{i \tan^{-1}[B_{mn}(t)/A_{mn}(t)]} \tag{32.83}$$

This completes the exact integration of the ODEs (32.8) using the periodic IST.

32.9 Numerical Procedures for Computing the Riemann Spectrum from Poincaré Series

The numerical procedure for determining an N degree-of-freedom solution to the KP equation requires determination of the theta function parameters B_{mn} and ω_n, referred to here as the *Riemann spectrum*. The approach is necessary for guaranteeing that one has a solution to the KP equation and requires the following steps:

(1) Pick the Riemann matrix diagonal elements, B_{nn}, and then compute the cnoidal wave amplitudes, a_n, by Equation (32.12) ($1 \leq n \leq N$). Alternatively, pick the amplitudes a_n and estimate the diagonal elements B_{nn} by inverting Equation (32.12). Also choose a set of arbitrary phases, ϕ_n, one for each cnoidal wave. For oceanographical applications the selection of uniformly distributed random numbers is often appealing; this parallels a similar procedure using the linear Fourier transform.

(2) Choose the set of wavenumbers κ_n, λ_n corresponding to each of the cnoidal waves selected in step (1). Due to the periodicity condition, these will lie on an evenly spaced grid in the wavenumber domain (κ_m, λ_n).

(3) Given the B_{nn} and the (κ_n, λ_n), use formulas (32.56) and (32.62) to provide an initial estimate of the uniformization parameters $A_n = a_n + ib_n$, ρ_n (32.61).

(4) Then treat Equations (32.47), (32.50), and (32.51) as nonlinear equations which, given the fixed and commensurable κ_n, λ_n, and the B_{nn}, one solves for the $A_n = a_n + ib_n$ and ρ_n iteratively by standard techniques (one has 3N equations in

κ_n, λ_n, and B_{nn} for $3N$ unknowns a_n, b_n, and ρ_n where $n = 1, 2, \ldots, N$). Starting values for a_n, b_n are given by Equation (32.61); the starting value for ρ_n is given by Equation (32.56). Then ρ_n is computed from knowledge of B_{nn} and A_n via Equation (32.47). When the iterations give precise enough values of A_n, ρ_n (i.e., when their values no longer change significantly between iterations) go to the next step.

(5) Compute the off-diagonal elements of the Riemann matrix by Equation (32.49) and the frequencies by Equation (32.52). One now has all the parameters of the Riemann spectrum: period matrix, the wavenumbers, the frequencies, and the phases. This is sufficient information to use formulas (32.9) or (32.38), (32.39) to compute the solution to the KP equation, which is of course associated with a Riemann surface by construction via Schottky uniformization.

32.10   Numerical Procedures for Computing the Riemann Theta Function

The computation of Riemann theta functions is well known, in the multidimensional case, to require vast amounts of computer time. As discussed in Chapter 9 the computer time is proportional to $(2M + 1)^N$, that is, the time required to execute the partial theta sum for a single space-time grid point (i.e., for specific values of x, y, t) increases exponentially with the number of degrees of freedom, N. The algorithm for the present study uses instead the time evolution of the normal coordinates $Q_j(t)$ as given by Equations (32.35) and (32.36). One implements the two-dimensional FFT (Equation (32.38)) together with determination of the linear Fourier modes by Equation (32.39). First one iterates over the index j and accumulates the $\theta_{mn}(t)$ in both x and y wavenumbers (Equation (32.39)); the $\theta_{mn}(t)$, for each value of time, are nothing more than Fourier sums for each (m, n) pair. Once this process is completed the coefficients $\theta_{mn}(t)$ are then used in Equation (32.38) to compute the theta function. The computation of the coefficients $\theta_{mn}(t)$ for all t by Equation (32.39), I call a "preprocessor." Of course Equation (32.38) is computed numerically by a standard two-dimensional FFT algorithm. Why does the process save computer time? Because one only needs to compute the Fourier coefficients $\theta_{mn}(t)$ once as a function of time using the method just described. A theta function subroutine does not need to be called for every x, y, t grid point; this is done only in the FFT operation in Equation (32.38). Thus, effectively, one is computing the theta "constants" (Equation (32.39)) simultaneously with the time evolution of the normal modes for only one grid point (essentially one computes Equation (32.39) by executing the operation $\theta(x = 0, y = 0, t)$). This information is used to compute the Fourier coefficients and, thus, it is only in the FFT step that one computes the values of the theta function at each coordinate grid point, x, y; the two-dimensional FFT is executed for each value of time t desired. How much computer time is saved? Consider a numerical example with a 64×64 grid. Since one computes the *theta constants* only once, one thereby saves the $64^2 = 4096$ calls to a theta function subroutine corresponding to all x, y grid points (for each time step). Of course one must then compute N_t FFTs, but this is a trivial amount of additional computer time as compared to 4096 calls to a theta function subroutine for each time point.

The approach just discussed in Section 32.10 drastically improves computer time for the computation of theta functions. However, the execution times can still be quite large. It is for this reason that I have installed the approach of Chapter 22 in which the summation is limited to points inside an n-ellipsoid. This action is taken only on the preprocessor step in the computation of Equation (32.39). I have also found that the modular transformation discussed in Chapter 8 also improves computer time.

32.11   Numerical Procedures for Computing Hyperfast Solutions of the KP Equation

In Figure 32.2A is a flowchart of a standard program to compute the solution of a *linear* PDE. This provides perspective for the nonlinear KP problem below. Given the directional spectral amplitudes A_{mn}, phases ϕ_{mn}, wavenumbers k_m, l_n, and frequencies ω_{mn}, one computes the "preprocessor" step (see first large box), that is, the *simple* time evolution of the Fourier transform for the PDE: $\eta_{mn}(t) = A_{mn}\, e^{-i\omega_{mn}t + i\phi_{mn}}$. Then the solution to the PDE (see second large box) is given as a two-dimensional FFT for each value of time. Note that the time iterations are shown explicitly in the flowchart of Figure 32.2A; the time iterations are *not* shown in the KP flowchart given below in order to simplify the exposition.

A flow chart of the hyperfast algorithm for the KP equation is given in Figure 32.2B. The preprocessor step, Equation (32.39), is shown inside the first large box. This part of the algorithm computes the linear Fourier coefficients $\theta_{mn}(t)$ for all values of time. Then the two-dimensional FFT is used to compute the theta functions $\theta(x, y, t)$ (see second large box) over all spatial values of x, y for all values of time t. Then the solution of KP is just the Hirota transformation of the theta function. Inside the preprocessor the number of iterations over the integer j is quite large; also the computation of the linear coefficients of the theta function are summed to previous values on each iteration and *indexing* is used to simplify the code. Of course the preprocessor step as actually programed is computed over the n-ellipsoid (Chapter 22) and benefits from a modular transformation of the period matrix (Chapter 8).

A numerical example for a two-soliton solution of the KP equation is given in Figure 32.3. The Mach stem or "dromion" (compare to the beach photograph in Figure 32.1) is seen to reside at the point where the two solitons cross. Mathematically, the Riemann matrix for this case is 2×2. The diagonal elements correspond to the two solitons. The manifestation of the Mach stem/dromion arises from the off-diagonal term in the Riemann matrix. Formally, this off-diagonal term is the source of the nonlinear interactions between the two solitons, which is just the analog of the $1 + 1$ phase shift seen in the KdV equation.

32.12   Numerical Example for KP Evolution

I now show a numerical example of a realization of a random, directionally spread wave train for the KP equation. The results are for a numerical simulation for which the phases are taken to be uniformly distributed random numbers. To compute the *Riemann spectrum* (before accessing the program of the flowchart in Figure 32.2B), I begin with the JONSWAP spectrum of Figure 32.4 with a $\cos^4 \phi$ spreading function in the wavenumber domain. This allows computation of the linear Fourier amplitudes $\eta_{mn}(0)$. I then compute the sea surface function $\eta(x, y, 0)$ from the JONSWAP spectrum using the ordinary linear

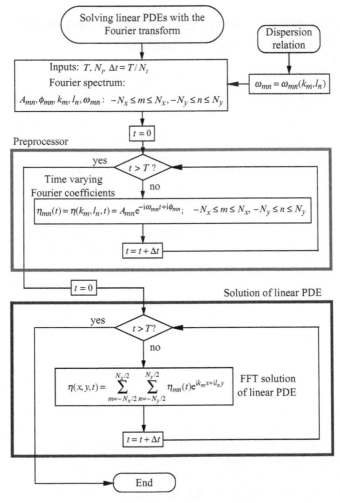

Figure 32.2   (A). Flow chart of solution of a linear PDE equation. (B) Flow chart of hyperfast solution of the KP equation.

Continued

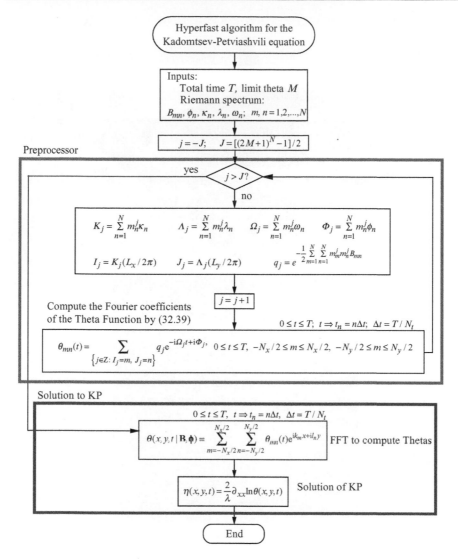

Figure 32.2 Cont'd

Fourier transform (Equation (32.7)). Finally, I use Equation (32.9) in its inverted form

$$\Theta(x, y, 0) = \exp\left[\frac{\lambda}{2}\int\int_x \eta(x', y, 0)\mathrm{d}x'\mathrm{d}x''\right] \tag{32.84}$$

to estimate the theta function. The iterative method of *nonlinear adiabatic annealing on a Riemann surface* (Chapter 23) is used to compute the Riemann matrix and phases, all the while using Schottky uniformization to assure that

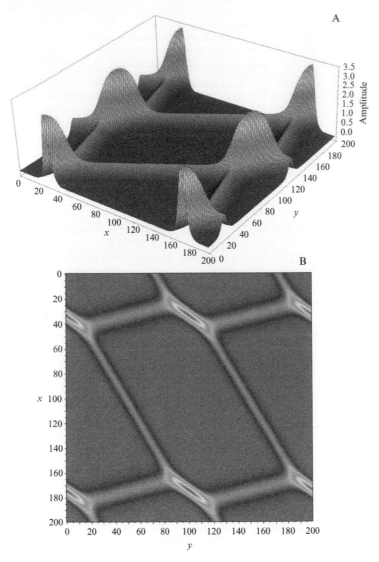

Figure 32.3   Interaction of two soliton trains in the solution of the KP equation. Note the substantial Mach stems where the soliton trains cross. (See color plate).

the Riemann spectrum is associated with the appropriate Riemann surface. In this process the algorithm converges to numerical values for the diagonal elements of the Riemann matrix and phases; the off-diagonal elements have been computed by Schottky uniformization. This provides the full Riemann spectrum B_{mn} and ϕ_n needed for the numerical simulations. Finally, the frequencies, ω_n, are also found by Schottky uniformization. This inversion process leads to the results given in Figure 32.5 after taking the inverse two-dimensional FFT. This figure is a graph of $\theta_{mn}(0)$, the Fourier transform for the theta function,

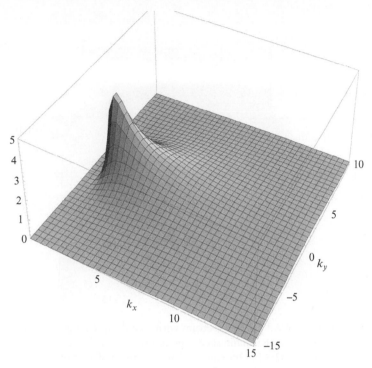

Figure 32.4 A JONSWAP power spectrum with a $\cos^4\theta$ spreading function in the wavenumber domain. One can see that several hundred components are necessary for defining all the spectral amplitudes. (See color plate).

seen here to be quite narrow with respect to the Fourier transform of the sea surface elevation $\eta_{mn}(0)$ in Figure 32.4. This is an important observation, for while I used 400 Fourier components to compute $\eta_{mn}(0)$, only 30 components were necessary to compute $\theta_{mn}(0)$. This means that the "genus" of the Riemann theta function can be made relatively small (~30).

Figure 32.6 shows the numerically computed sea surface for the Cauchy initial condition for the simulation. The water depth is $h = 8$ m. There are 30 components in the Riemann spectrum that has a significant wave height of $H_s = 0.4$ m. Finally, after a considerable amount of time the surface elevation evolved into that in Figure 32.7. Note the rather large wave colored in red (see color plate). This wave has occurred because of the superposition of two Mach stems within the Riemann spectrum of the simulation. This new mechanism for extreme wave generation, via the superposition of several Mach stems in the Riemann spectrum, means that shallow-water rogue waves can occur in a moderately nonlinear sea state for appropriate phasing of the cnoidal waves and their Mach stems. This shallow-water mechanism for rogue wave generation contrasts with the "unstable mode" rogue wave solutions of the NLS equation described in Chapters 12, 18, 24 and 29.

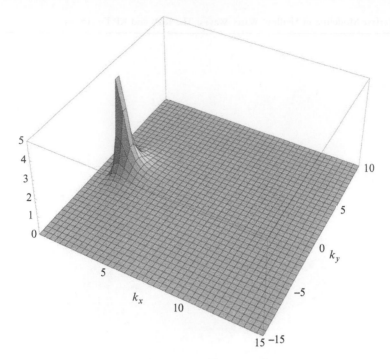

Figure 32.5 The JONSWAP power spectrum with a $\cos^4\theta$ spreading function has been mapped to the domain of the Riemann spectrum. Shown are the linear Fourier modes after the mapping of Equation (32.84). One can see that only about 30 components are necessary for defining all the spectral amplitudes. Thus, the Riemann spectrum contains many fewer spectral components than does the original JONSWAP spectrum. (See color plate).

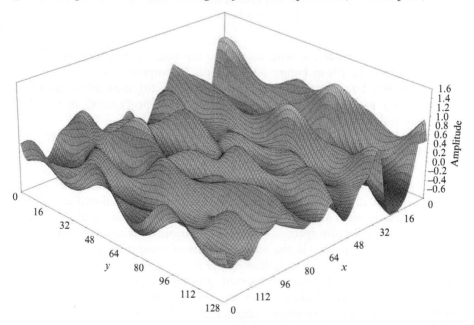

Figure 32.6 Surface elevation from directionally spread sea with 30 degrees of freedom using Riemann theta functions. (See color plate).

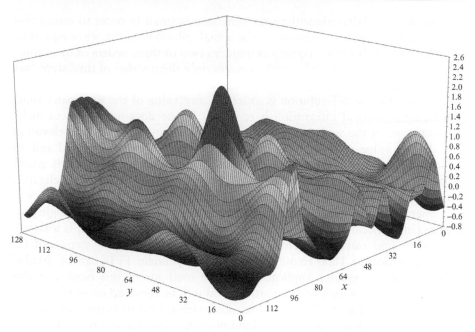

Figure 32.7 The later evolution of the initial condition shown in Fig. 32.1. Note the appearance of an extremely large wave, denoted in red. (See color plate).

Use of the new theta function algorithm described in this chapter serves to save considerable computer time. I used a 512×512 grid to solve the present problem and hence the new algorithm saves approximately a factor of $512^2 = 262,144$ in computer time as compared to a simple brute force approach in which the theta function subroutine is successively called for each of the space/time coordinates (x, y, t). On a single core of a Macintosh Intel computer, for 1000 values of time, a run of this type takes about 32 years of computer time. A Fornberg and Whitham (1978) run takes about 8 h. Using the algorithm introduced herein (Equations (32.38) and (32.39) with summation over an n-ellipsoid and an appropriate modular transform) requires about 20 s of computer time. This speedup occurs because with the brute force algorithm of Chapter 20 this run would require 262,144,000 calls to a theta function subroutine. In the present case the preprocessor equations (Equation (32.39)) are computed (requiring 4 s of computer time) and then one executes (Equation (32.38)) with 1000 calls of a two-dimensional FFT subroutine (requiring 16 s of computer time). It goes without saying that the savings in computer time are considerable although the recipe requires a steep learning curve and considerable programming effort.

It is fruitful to compare further the algorithm introduced here to the more traditional use of the FFT for numerical integration of the nonlinear wave motion. In these latter algorithms the time step is small in order to insure that the numerical integration resembles the actual solutions of the wave equation. An algorithm of this type requires as many as two or three orders of magnitude more time steps than the IST, which requires only the number of time steps (say ~ 1000) required to, perhaps, generate the individual frames of an animation. This is because the IST solution is exact for any value of the parameter time, that is, no numerical integration is required. If one does not desire to make an animation of the wave motion, but only to compute the surface elevation at a desired value of time, then only one FFT evaluation is required and an additional factor of 1000 in computer time (say) is saved. Finally it is worth pointing out that the IST algorithm developed herein is perfectly parallelizable. On a computer with N processors the algorithm can be divided into N independent operations and hence is N times faster than computations on a single processor. This is because (1) the preprocessor step, Equation (32.39), is a simple summation and can be therefore divided into N independent steps and (2) the computation of 1000 two-dimensional, independent FFTs can be divided among 1000 processors. Applications of the methods of this book are also being applied to nonintegrable equations and will be reported on in the sequel. The work of Bona and colleagues is seminal with regard to extensions of shallow water wave dynamics to higher (and often nonintegrable) order (Benjamin et al., 1972; Bona et al., 1981; Bona and Chen, 1999; Bona et al., 2002; Bona et al., 2004).

33 Modeling the 2 + 1 Gardner Equation

33.1 Introduction

This chapter deals with an integrable equation at order higher than the Kadomtsev-Petviashvili (KP) equation that is referred to as the *extended KP equation* (exKP) or as the *2 + 1 Gardner equation*. To my knowledge this equation has not been used by the oceanographic community in the past. It has the amazing properties that in appropriate limits it reduces to the *Korteweg-deVries (KdV) equation*, the *modified KdV (mKdV)* equation, the *Gardner equation*, the *modified KP (mKP)* equation, and the *KP equation*. The 2 + 1 Gardner equation has two main advantages over the KP equation: (1) it describes directional wave trains with much larger spreading angles and (2) it describes higher and more nonlinear waves than KP due to the cubic (Gardner) term. The 2 + 1 Gardner equation is integrable by the inverse scattering transform (Konopelchenko and Dubrovsky, 1984; Konopelchenko, 1991) and it has many types of "coherent structure" solutions: (1) positive solitons similar to those in the KdV equation, (2) positive and negative solitons which appear in the Gardner equation, (3) kink soliton solutions, (4) "table-top" solitons which appear in the Gardner equation (as typically applied to the study of internal waves), and (5) "unstable mode" solutions similar to those in the nonlinear Schrödinger equation (NLS) equation. For applications to internal solitary waves the papers by Grimshaw and colleagues are crucially important (Grimshaw, 2007). The 2 + 1 Gardner equation is one of the richest integrable equations known and characterizing its scattering transform solution with periodic boundary conditions is challenging.

33.2 The 2 + 1 Gardner Equation and Its Properties

The 2 + 1 Gardner equation has the following form:

$$u_t + 6\delta u u_x + u_{xxx} + 3\sigma^2 \partial_x^{-1} u_{yy} - \frac{3}{2}\alpha^2 u^2 u_x - 3\alpha\sigma u_x \partial_x^{-1} u_y = 0 \qquad (33.1)$$

so that, like the KP equation, the 2 + 1 Gardner equation is a nonlocal partial integrodifferential equation. The equation depends on the parameters α, δ, and σ and can also be put into the following form:

$$(u_t + 6\delta u u_x + u_{xxx} - \frac{3}{2}\alpha^2 u^2 u_x)_x + 3\sigma^2 u_{yy} - 3\alpha\sigma(u_x\partial_x^{-1}u_y)_x = 0 \qquad (33.2)$$

The *real solutions* of 2 + 1 Gardner for shallow water waves occur for $\alpha = \sigma = 1$. A normalized form of the 2 + 1 Gardner equation that preserves the standard form of the KdV equation occurs for $\alpha = \sigma = \delta = 1$:

$$u_t + 6u u_x + u_{xxx} + 3\partial_x^{-1}u_{yy} - \frac{3}{2}u^2 u_x - 3u_x\partial_x^{-1}u_y = 0 \qquad (33.3)$$

The Gardner transformation

$$u = \delta v - \frac{1}{2}\alpha v_x - \frac{1}{4}\alpha^2 v^2 - \frac{1}{2}\sigma\alpha\partial_x^{-1}v_y \qquad (33.4)$$

maps the 2 + 1 Gardner equation to the *KP equation*:

$$u_t + 6\delta u u_x + u_{xxx} + 3\sigma^2\partial_x^{-1}u_{yy} = 0 \qquad (33.5)$$

Note that if we also set $\delta = \sigma = 1$ in (33.5) then we get KP in the form studied in Ablowitz and Clarkson (1991):

$$u_t + 6u u_x + u_{xxx} + 3\partial_x^{-1}u_{yy} = 0 \quad \text{(KP II)} \qquad (33.6)$$

which is the so-called KP II equation valid for shallow water waves in the coastal zone (Chapter 32). The KP I equation, $u_t + 6u u_x + u_{xxx} - 3\partial_x^{-1}u_{yy} = 0$, corresponds to surface-tension dominated waves which are thin sheets of water and consequently KP I is not of interest for most oceanographic purposes.

The *transformation* for $\sigma = 0$

$$u = \delta v - \frac{1}{2}\alpha v_x - \frac{1}{4}\alpha^2 v^2 \qquad (33.7)$$

maps the *1 + 1 Gardner equation*

$$u_t + 6\delta u u_x + u_{xxx} = \frac{3}{2}\alpha^2 u^2 u_x \qquad (33.8)$$

to the *KdV equation*:

$$u_t + 6\delta u u_x + u_{xxx} = 0 \qquad (33.9)$$

For $\delta = 0$, $\sigma = 0$ the transformation

$$u = -\frac{1}{2}\alpha v_x - \frac{1}{4}\alpha^2 v^2 \tag{33.10}$$

maps the *mKdV equation*

$$u_t - \frac{3}{2}\alpha^2 u^2 u_x + u_{xxx} = 0$$

to the *KdV equation*:

$$u_t + 6uu_x + u_{xxx} = 0 \tag{33.11}$$

For $\alpha = 0$ the 2 + 1 Gardner equation (33.2) becomes the *KP equation*:

$$u_t + 6\delta uu_x + u_{xxx} + 3\sigma^2 \partial_x^{-1} u_{yy} = 0 \tag{33.12}$$

and for $\delta = \sigma = 1$ we get KP II as

$$u_t + 6uu_x + u_{xxx} + 3\partial_x^{-1} u_{yy} = 0 \tag{33.13}$$

For $\delta = 0$ the 2 + 1 Gardner equation becomes the *mKP equation*:

$$u_t + u_{xxx} - \frac{3}{2}\alpha^2 u^2 u_x + 3\sigma^2 \partial_x^{-1} u_{yy} - 3\alpha\sigma u_x \partial_x^{-1} u_y = 0 \tag{33.14}$$

For $\sigma = 0$ the 2 + 1 Gardner equation becomes the *1 + 1 Gardner equation*:

$$u_t + 6\delta uu_x + u_{xxx} - \frac{3}{2}\alpha^2 u^2 u_x = 0 \tag{33.15}$$

For $\alpha = 0$, $\sigma = 0$ the 2 + 1 Gardner equation becomes the *KdV equation*:

$$u_t + 6\delta uu_x + u_{xxx} = 0 \tag{33.16}$$

For $\delta = 0$, $\sigma = 0$ the 2 + 1 Gardner equation becomes the *mKdV equation*:

$$u_t - \frac{3}{2}\alpha^2 u^2 u_x + u_{xxx} = 0 \tag{33.17}$$

Thus the 2 + 1 Gardner equation contains an amazing number of other physically relevant wave equations. The fact that each of these constituent equations is integrable is a miracle of modern mathematics.

33.3 The Lax Pair and Hirota Bilinear Form

The Lax pair for the 2 + 1 Gardner equation is given by (Konopelchenko, 1991):

$$\sigma \psi_y + \psi_{xx} + \alpha u \psi_x + \delta u \psi = 0 \tag{33.18}$$

$$\psi_t + 4\psi_{xxx} + \alpha u \psi_{xx} + (3\alpha u_x + \frac{3}{2}\alpha^2 u^2 + 6\delta u - 3\alpha\sigma\partial_x^{-1}u_y)\psi_x$$

$$+ (3\delta u_x + \frac{3}{2}\alpha\delta u^2 + 3\delta\sigma\partial_x^{-1}u_y)\psi = 0 \tag{33.19}$$

The compatibility condition between Equations (33.18) and (33.19) gives the 2 + 1 Gardner equation (33.1).

The Hirota dependent variable transformation can often be guessed (Hirota, 2004) and in these cases the Hirota method is quite straightforward. When one encounters a new equation the Hirota transformation may not be so obvious. In these cases the Painlevé analysis suggested by Weiss et al. (1983) leads to the *singular manifold method* for cases which have only a single Painlevé expansion branch and the *two-singular-manifold method* can be useful for the study of integrable equations which have several Painlevé expansion branches (Estévez and Gordoa, 1994; Estévez, 1999). In order to determine the bilinear form it is useful to write the 2 + 1 Gardner equation as a system of two equations:

$$u_t + 6\delta u u_x + u_{xxx} - \frac{3}{2}\alpha^2 u^2 u_x - 3v_y - 3\alpha u_x v = 0 \tag{33.20}$$

$$v_x = u_y$$

To determine the appropriate Hirota dependent variable transformation, expand the solutions of Equation (33.20) in a generalized Laurent series (Weiss et al., 1983; Zhang et al., 2008). The 2 + 1 Gardner equation requires the two singular manifold method (Estévez et al., 1993; Musette and Conte, 1994; Conte et al., 1995):

$$u = \sum_{j=0}^{\infty} u_j\chi^{-a+j}, \quad v = \sum_{j=0}^{\infty} v_j\chi^{-b+j} \tag{33.21}$$

where $\chi = \chi(x,y,t)$ and $u_j = u_j(x,y,t)$, $v_j = v_j(x,y,t)$ are analytical functions in the neighborhood of a noncharacteristic movable singularity manifold $\chi(x,y,t) = 0$. The constants a and b are integers to be determined. An analysis of the leading terms in the series reveals

$$a = 1, \quad b = 1, \quad u_o = 2\varepsilon\frac{\chi_x}{\alpha}, \quad v_o = 2\varepsilon\frac{\chi_y}{\alpha} \tag{33.22}$$

where $\varepsilon = \pm 1$. Thus, u_o, v_o can take on two values so that the system (33.20) has two different Painlevé expansion branches corresponding to the two values of $\varepsilon = \pm 1$. One considers two different singular manifolds ϕ (for $\varepsilon = +1$) and

φ (for $\varepsilon = -1$) (Estévez et al., 1993; Cerveró and Estévez, 1998; Estévez, 1999) and truncates the Painlevé expansion at the constant level term:

$$u' = u + \frac{2}{\alpha}\left(\frac{\phi_x}{\phi} - \frac{\varphi_x}{\varphi}\right)$$

$$v' = v + \frac{2}{\alpha}\left(\frac{\phi_y}{\phi} - \frac{\varphi_y}{\varphi}\right)$$

(33.23)

The last expressions suggest that the following dependent variable transformations

$$u = \frac{2}{\alpha}\left(\frac{g_x}{g} - \frac{f_x}{f}\right) = \frac{2}{a}\left(\ln\frac{g}{f}\right)_x$$

$$v = \frac{2}{\alpha}\left(\frac{g_y}{g} - \frac{f_y}{f}\right) = \frac{2}{a}\left(\ln\frac{g}{f}\right)_y$$

(33.24)

might be used to transform Equation (33.20) into the associated Hirota bilinear forms. Applying Equations (33.24) to Equation (33.20) we obtain the following nonlinear separation of variables:

$$\left(D_t + D_x^3 - 3D_xD_y - 6\frac{\beta}{\alpha}D_y\right)g\cdot f = 0$$

$$\left(D_y + D_x^2 - 2\frac{\beta}{\alpha}D_y\right)g\cdot f = 0$$

(33.25)

which are the bilinear forms for the 2 + 1 Gardner equation in Hirota operator notation.

The *theta function solution* for periodic boundary conditions is then given by

$$u = \frac{2}{\alpha}\partial_x \ln\left(\frac{\theta(x, y, t | \mathbf{B}, \boldsymbol{\phi})}{\theta(x, y, t | \mathbf{B}, \boldsymbol{\delta})}\right)$$

(33.26)

where the phases are different in the numerator and denominator. The period matrix, frequencies, and phases are governed by nonlinear equations obtained after theta functions with characteristics are used in Equation (33.25) (see Chapter 16 for details). These results provide a leading order hyperfast model for describing the nonlinear dynamics of the 2 + 1 Gardner equation.

33.4   The Extended KP Equation in Physical Units

The physical form of the 2 + 1 Gardner equation arises from Equation (33.3):

$$u_t + 6uu_x + u_{xxx} + 3\partial_x^{-1}u_{yy} = \frac{3}{2}u^2 u_x + 3u_x \partial_x^{-1}u_y \qquad (33.27)$$

If we carry out the transformation:

$$u(x,y,t) \rightarrow \lambda'\eta(x,y,t), \quad t \rightarrow \beta't, \quad x \rightarrow a(x+c_o t), \quad y \rightarrow by$$

Then Equation (33.27) becomes:

$$\eta_t + c_o\eta_x + \left(\frac{6\lambda'\beta'}{a}\right)\eta\eta_x + \left(\frac{\beta'}{a^3}\right)\eta_{xxx} + \left(\frac{3\beta'a}{b^2}\right)\partial_x^{-1}\eta_{yy}$$

$$= \left(\frac{3\lambda'^2\beta'}{2a}\right)\eta^2\eta_x + \left(\frac{3\lambda'\beta'}{b}\right)\eta_x\partial_x^{-1}\eta_y$$

The first four coefficients in parenthesis must be equal to their natural physical values (see the discussion on the second equation in the Whitham hierarchy, Chapter 2):

$$\frac{6\lambda'\beta'}{a} = \frac{3c_o}{2h}, \quad \frac{\beta'}{a^3} = \frac{c_o h^2}{6}, \quad \frac{3\beta'a}{b^2} = \frac{c_o}{2}, \quad \frac{3\lambda'^2\beta'}{2a} = \frac{15c_o}{8h^2}$$

for which we find the values:

$$\lambda' = \frac{5}{h}, \quad \beta' = \sqrt{\frac{3}{10}}\frac{c_o}{20h}, \quad a = \sqrt{\frac{3}{10}}\frac{1}{h}, \quad b = \frac{3}{10h}$$

This then gives us the 2 + 1 Gardner equation with physical coefficients:

$$\eta_t + \sqrt{gh}\,\eta_x + \left(\frac{3c_o}{2h}\right)\eta\eta_x + \left(\frac{c_o h^2}{6}\right)\eta_{xxx} + \frac{c_o}{2}\partial_x^{-1}\eta_{yy}$$

$$= \left(\frac{15c_o}{8h^2}\right)\eta^2\eta_x + \sqrt{\frac{15}{8}}\left(\frac{c_o}{h}\right)\eta_x\partial_x^{-1}\eta_y \qquad (33.28)$$

Alternatively in notation I have used before:

$$\eta_t + c_o\eta_x + \alpha\eta\eta_x + \beta\eta_{xxx} + \gamma\partial_x^{-1}\eta_{yy} = \delta\eta^2\eta_x + \rho\eta_x\partial_x^{-1}\eta_y \qquad (33.29)$$

where

$$c_o = \sqrt{gh}, \quad \alpha = \frac{3c_o}{2h}, \quad \beta = \frac{c_o h^2}{6}, \quad \gamma = \frac{c_o}{2}, \quad \delta = \frac{15c_o}{8h^2}, \quad \rho = \sqrt{\frac{15}{8}}\left(\frac{c_o}{h}\right)$$

$$(33.30)$$

and where the coefficients, by abuse of notation, have changed meaning from those in Equation (33.1) to be compatible with the notation used elsewhere in this monograph.

The above result was obtained by ensuring that the constants λ', β', a, and b result in reproducing the known physical coefficients of the four terms:

$$\eta\eta_x, \eta_{xxx}, \partial_x^{-1}\eta_{yy}, \eta^2\eta_x$$

Then the last coefficient for the nonlinear spreading term $\eta_x\partial_x^{-1}\eta_y$ was evaluated as given in Equation (33.29, 30). Thus, the 2 + 1 Gardner equation is not "asymptotic" in the sense that the second equation in the Whitham hierarchy is asymptotic. Indeed 2 + 1 Gardner may be viewed as having an *ad hoc* construction, that is, one adds the Gardner term $\eta^2\eta_x$ to the KP equation and then adds the appropriate spreading term $\eta_x\partial_x^{-1}\eta_y$ to give integrability. Any treatment of this equation to second order as an asymptotic equation derived from the Euler equations by the Whitham procedure would require additional terms from that expansion. One way that one might think of treating the resultant equation would be to use Equation (33.29, 30) "perturbed" up to the second order asymptotic expansion. Then Lie-Kodama transforms would allow the appropriate perturbation analysis. The perturbation of this equation is not treated here, but can easily be done. Here I address the hyperfast numerical integration of the above equation for simulating ocean surface or internal waves.

33.5 Physical Behavior of the Extended KP Equation

It is worth noting that the following equations extracted from Equation (33.29) are all integrable by the inverse scattering transform.

The *KdV equation*:

$$\eta_t + c_o\eta_x + \alpha\eta\eta_x + \beta\eta_{xxx} = 0 \tag{33.31}$$

The *mKdV equation*:

$$\eta_t + c_o\eta_x + \beta\eta_{xxx} = \delta\eta^2\eta_x \tag{33.32}$$

The *Gardner equation*:

$$\eta_t + c_o\eta_x + \alpha\eta\eta_x + \beta\eta_{xxx} = \delta\eta^2\eta_x \tag{33.33}$$

The *KP equation*:

$$\eta_t + c_o\eta_x + \alpha\eta\eta_x + \beta\eta_{xxx} + \gamma\partial_x^{-1}\eta_{yy} = 0 \tag{33.34}$$

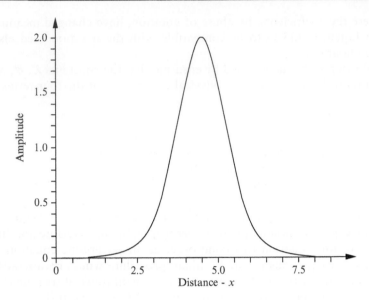

Figure 33.1   KdV pulse-type soliton.

The *KP-Gardner equation*:

$$\eta_t + c_o\eta_x + \alpha\eta\eta_x + \beta\eta_{xxx} + \gamma\partial_x^{-1}\eta_{yy} = \delta\eta^2\eta_x \qquad (33.35)$$

The miracle is that each of these equations is integrable. Many kinds of solitons occur for these equations, including pulse-type solitons, positive and negative solitons, kinks, and table-top solitons. Some of these solutions are shown in Figures. 33.1–33.3.

A hyperfast numerical simulation of the 2 + 1 Gardner equation extends our understanding of shallow water surface waves. All of the nonlinear

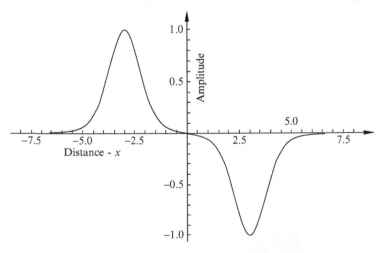

Figure 33.2   Positive and negative solitons.

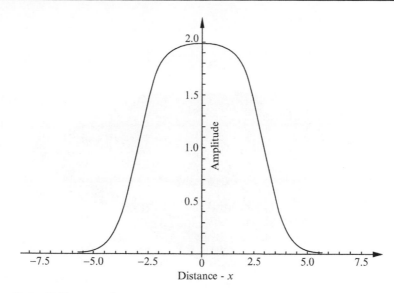

Figure 33.3   Table-top soliton.

components can of course have different directions in the *x-y* plane and are nonlinearly superposed via Equation (33.26) to give the solutions of the 2 + 1 Gardner equation. Figure 33.4 shows a single frame in the time evolution of the equation for a 30 × 30 Riemann matrix. A large wave is preceded by a deep trough. In a separate simulation shown in Figure 33.5, a single large

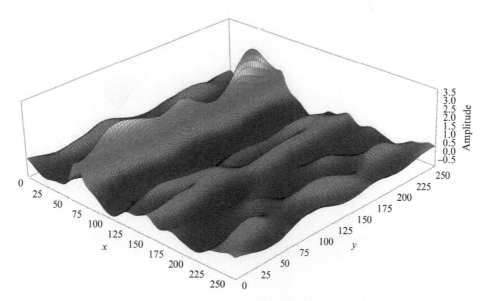

Figure 33.4   A large wave accompanied by a deep trough in a numerical simulation of the 2 + 1 Gardner equation. (See color plate).

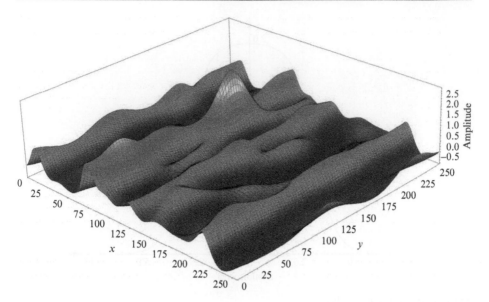

Figure 33.5   The appearance of a Mach stem during the evolution of the 2 + 1 Gardner equation. (See color plate).

Mach stem arises from its two crossing components in the evolution of the wave train. This evolution resembles that of the KP equation (Chapter 32) even though the mathematical structure of the 2 + 1 Gardner equation is different. The hyperfast simulations were about 4000 times faster than the equivalent FFT simulation for the same period of evolution time.

34 Modeling the Davey-Stewartson (DS) Equations

34.1 Introduction

Ocean waves have an enormous, often destructive power over man and his ability to exploit ocean resources. Every type of fixed, floating, or compliant structure used to develop mineral resources from the ocean is subject to ocean waves, currents, and winds (see Figure 34.1). Many kinds of deterministic and stochastic wave models have been used to make scientific and engineering computations of ocean waves and their influence on fixed and floating bodies (Sarpkaya and Isaacson, 1981; Komen et al., 1994; Young, 1999; Janssen, 2004). This chapter discusses a model that to my knowledge has not yet been used for engineering purposes, the Davey-Stewartson (DS) equations. The equations are simpler than many of the more complex models such as the Euler and Boussinesq-type equations, but the DS equations have several advantages, not least of which occurs because the equations are written in terms of two fields, the surface wave elevation $\eta(x, y, t)$ and the "long wave" part of the velocity potential $\Phi(x, y, t)$. The DS model is exactly integrable in shallow water and "almost" integrable in deep water. Furthermore, the model has easily identifiable coherent structures and waves, including solitons, unstable rogue-wave type modes, Stokes waves and the velocity field contains vortices. These waveforms are packets and are the "nonlinear Fourier components" in the theory of the inverse scattering transform and its presumed "slowly varying" or "adiabatic" extensions. Furthermore, the model can easily be modified to include current, winds, and bathymetry. Since the model is based upon the inverse scattering transform we can describe explicitly the physics, develop nonlinear Fourier analysis procedures for analyzing data, and develop a hyperfast numerical model which is two or three orders of magnitude faster than typical FFT-type numerical integrations of the equation. The identifiable coherent structures are nonlinear wave packets whose maximum waves are often referred to as "rogue waves."

34.2 The Physical Form of the Davey-Stewartson Equations

In $2 + 1$ dimensions the Euler equations can be reduced to the DS equations in the fields $\Phi(x, y, t)$ and $\Psi(x, y, t)$, where $\Phi(x, y, t)$ is the (normalized, long wave part of the) velocity potential and $\Psi(x, y, t)$ is the complex envelope of a

Doi: 10.1016/S0074-6142(10)97034-4

Figure 34.1 The Thunder Horse Oil Platform sinking after Hurricane Ivan, July 2005
Source: (courtesy of United States Coast Guard). (See color plate).

narrow-banded wave train which may be modulated in both the x and y directions. As a result a directional, narrow-banded sea state can be accounted for by the DS equations:

$$i\Psi_\tau + \lambda\Psi_{XX} + \mu\Psi_{YY} + \chi|\Psi|^2\Psi = \chi_o\Psi\Phi_X$$
$$\alpha\Phi_{XX} + \Phi_{YY} = -\beta(|\Psi|^2)_X \tag{34.1}$$

where

$$\sigma = \tanh(\kappa h), \quad \kappa = \sqrt{k_o^2 + l^2} \tag{34.2}$$

$$\omega^2 = gh\sigma \geq 0 \tag{34.3}$$

$$\omega_o^2 = g\kappa \tag{34.4}$$

$$\lambda = \frac{\kappa^2(\partial^2\omega/\partial\kappa^2)}{2\omega_o} \tag{34.5}$$

$$\mu = \frac{\kappa^2(\partial^2\omega/\partial l^2)}{2\omega_o} = \frac{\kappa C_g}{2\omega_o} \geq 0 \tag{34.6}$$

$$\chi = -\left(\frac{\omega_o}{4\omega}\right)\left[\frac{(1-\sigma^2)(9-\sigma^2)}{\sigma^2} + 8\sigma^2 - 2(1-\sigma^2)^2\right] \tag{34.7}$$

$$\chi_o = 1 + \frac{\kappa C_g}{2\omega}(1 - \sigma^2) \geq 0 \tag{34.8}$$

$$\alpha = \frac{gh - C_g^2}{gh} \tag{34.9}$$

$$\beta = \left(\frac{\omega}{\omega_o k_o h}\right)\left(\frac{\kappa C_g}{\omega}(1 - \sigma^2) + 2\right) \geq 0 \tag{34.10}$$

$$v = \chi - \frac{\chi_o \beta}{\alpha} \tag{34.11}$$

The following scaled variables have been used:

$$X = \varepsilon k_o(x - C_g t), \quad Y = \varepsilon k_o y, \quad \tau = \varepsilon^2 (gk_o)^{1/2} t \tag{34.12}$$

where $\Phi(x, y, z, t)$ is the *dimensional velocity potential* and $\Psi(x, y, t)$ is the *dimensional envelope function*. The *surface elevation* is computed by

$$\eta(x, y, t) \approx \frac{i\omega\sqrt{gk_o}}{gk_o^2}\Psi(x, y, t)e^{ik_o x - i\omega_o t} + \text{c.c.} + \cdots \tag{34.13}$$

The physical form for the *velocity potential* is

$$\phi(x, y, t) \approx \sqrt{\frac{1}{gk_o^3}}\left(\Phi(x, y, t) + \frac{\cosh k_o(z + h)}{\cosh k_o h}\Psi(x, y, t)e^{ik_o x - i\omega_o t} + \text{c.c.}\right) \tag{34.14}$$

One is thus able to compute the particle velocities by taking the gradient of the potential Equation (34.14). Note that the surface elevation is the solution of a Schrödinger-type equation, the first of Equation (34.1). The velocity potential consists of two parts: (1) a long-wave contribution "averaged" over fast space and time scales due to a "radiation stress" contribution, plus (2) the fast scale oscillations due to the particle velocity oscillations under a single wave. This latter contribution of course decays with depth from the free surface, while the mean flow penetrates the depth without decay.

To include the influence of external forces such as the wind or current and dissipation, one adds an additional "potential" to the first of Equation (34.1):

$$i\Psi_\tau + \lambda\Psi_{XX} + \mu\Psi_{YY} + \chi|\Psi|^2\Psi = \chi_o\Psi\Phi_X + U(x, y, t)\Psi$$

where $U(x, y, t) = U_{\text{wind}}(x, y, t) + U_{\text{current}}(x, y, t) + U_{\text{dissipation}}(x, y, t)$.

> **Nota Bene:** For the cases considered in this book the surface waves are large enough that surface tension can be considered to be negligible. In this case the coefficients have the following signs
>
> $$\lambda < 0, \quad \mu > 0, \quad \chi > 0, \quad \chi_1 > 0, \quad \alpha > 0$$

The DS equations in the *infinite depth limit* become the *nonlinear Schrödinger equation* in $2 + 1$ dimensions (in this case the mean flow vanishes at the order of the NLS equation):

$$i(\psi_t + C_g\psi_x) + \mu\psi_{xx} + \rho\psi_{yy} + v|\psi|^2\psi = 0 \tag{34.15}$$

The associated linear, deep-water dispersion relation is given by $\omega_o^2 = gk_o$ and $C_g = \omega_o/2k_o$, $\mu = -\omega_o/8k_o^2$, $\rho = \omega_o/4k_o^2$, and $v = -\omega_o k_o^2/2$. This equation is *not* integrable by the inverse scattering transform. In the limit that the motion becomes unidirectional one obtains the usual one-dimensional nonlinear Schrödinger equation (Chapter 2, Equation (2.31)), which is integrable.

It is worth pointing out that in shallow-water applications of the DS equations, the second of Equation (34.1) for the mean velocity potential is just the *Poisson equation* with a time varying source. This implies the occurrence of *vortex dynamics* in the current field in the coastal zone, a result that could be important, for example, in the dynamics and erosion of beaches. The *stationary form* of the solutions of this kind of Poisson equation, which clearly must have vortex solutions, is amenable to the analysis of Chapter 27.

Note further that that the *stationary form* of the $2 + 1$ nonlinear Schrödinger equation (Equation (3.24)) is given by

$$\mu\psi_{xx} + \rho\psi_{yy} + v|\psi|^2\psi = 0 \tag{34.16}$$

The first two terms are an operator of hyperbolic form, that is, it is no longer a Poisson equation (the coefficients of the second derivative terms differ in sign). Equation (34.16) is also amenable to the analysis of Chapter 27, in particular Section 27.6. This approach provides the exact form of the *stationary rogue wave solutions* of the above equation. The time evolution of one of these stationary solutions is discussed briefly in a numerical example in Section 34.5, in particular see Figure 34.2.

34.3 The Normalized Form of the Davey-Stewartson Equations

The DS equations (which are the long-wave limit of the Benney-Roskes equation; Benny and Roskes, 1969) are given in normalized form by:

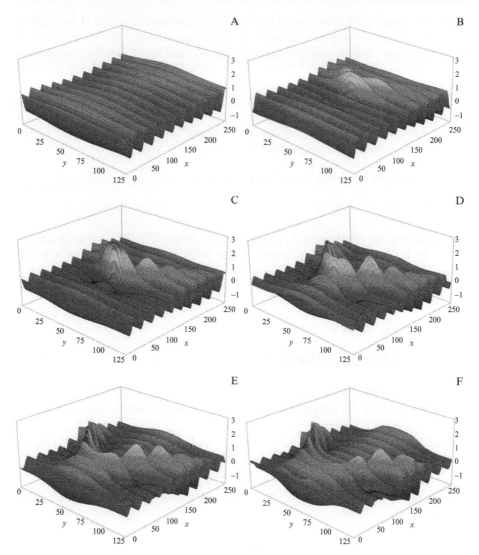

Figure 34.2 Time evolution of a rogue wave packet of the Davey-Stewartson equations. (see color plate).

$$i\psi_t - \sigma\psi_{xx} + \psi_{yy} - \delta|\psi|^2\psi - 2\sigma\delta u\psi = 0, \quad u = \Phi_x$$

$$\sigma u_{xx} + u_{yy} = -(|\psi|^2)_{xx}$$

(34.17)

where $\sigma = \pm 1$, $\delta = \pm 1$. DS I corresponds to $\sigma = -1$, which occurs for water depths where surface tension dominates, that is, for depths less than about 0.5 cm. DS II corresponds to $\sigma = 1$, which occurs for water depths where surface tension is insignificant, that is, for depths much greater than about 0.5 cm. This form of the DS equations is integrable and appropriate for the study of coastal zone dynamics. Of course the application of DS II to shallow water is somewhat limited because is does not include the generation of solitons in very shallow water. Equations like KP and 2 + 1 Gardner do include soliton dynamics.

Here $\psi(x, y, t)$ is the complex envelope of the surface elevation and $u = \Phi_x$ is the horizontal component of the particle velocity along the dominant wave direction, x, and $\Phi(x, y, t)$ is the long-wave averaged velocity potential. Of course Equation (34.17) can be written in terms of the long wave part of the velocity potential at the free surface, $\Phi(x, y, t)$:

$$i\psi_t - \sigma\psi_{xx} + \psi_{yy} - \delta|\psi|^2\psi + 2\sigma\delta\Phi_x\psi = 0$$
$$\sigma\Phi_{xx} + \Phi_{yy} = -(|\psi|^2)_x$$

(34.18)

This leads to the form given in Ablowitz and Segur (1981) corresponding to $\sigma = \delta$ in Equation (34.13) in terms of the velocity potential (where the 2 is absorbed into the velocity potential):

$$i\psi_t - \sigma\psi_{xx} + \psi_{yy} - \delta|\psi|^2\psi + \Phi_x\psi = 0$$
$$\sigma\Phi_{xx} + \Phi_{yy} = -2(|\psi|^2)_x$$

(34.19)

34.4 The Hirota Bilinear Forms

Hietarinta (2002) assumes the form Equation (34.13) for getting bilinear forms. One makes the substitutions (and integrates the second equation):

$$u = \frac{1}{2}\Phi_x, \quad \sigma = \delta$$

(34.20)

to go from Equation (34.13) to Equation (34.19).

Now convert Equation (34.17) to Hirota's bilinear form by using the substitutions:

$$\psi = \frac{G}{F}, \quad u = 2\delta\partial_{xx}\ln F \quad \text{or} \quad \Phi = 2\delta\partial_x \ln F$$

(34.21)

This gives (we normally take F to be real and G to be complex for reasons we note below):

$$(iD_t - \sigma D_x^2 + D_y^2)G \cdot F = 0$$
$$(\sigma D_x^2 + D_y^2)F \cdot F = -|G|^2$$

(34.22)

where we have used the Hirota operator:

$$D_x^k F \cdot G = (\partial_x - \partial_{x'})^k F(x)G(x')|_{x'=x}$$

(34.23)

34.4.1 Davey-Stewartson I—Surface Tension Dominates

DS I corresponds to $\sigma = -1$, which occurs for water depths where surface tension dominates, that is, for depths <0.5 cm. Surface tension dominates in then sheets of water. The equations are

$$(iD_t + D_x^2 + D_y^2)G \cdot F = 0$$
$$(-D_x^2 + D_y^2)F \cdot F = -|G|^2$$

$$(34.24)$$

34.4.2 Davey-Stewartson II—Oceanic Water Waves in Shallow Water with Negligible Surface Tension

DS II is the problem most interesting to oceanographers who study surface waves ($\sigma = \delta = 1$):

$$i\psi_t - \psi_{xx} + \psi_{yy} - |\psi|^2\psi - \Phi_x\psi = 0$$
$$\Phi_{xx} + \Phi_{yy} = -2(|\psi|^2)_x$$

$$(34.25)$$

DS II corresponds to $\sigma = 1$, which occurs for water depths where surface tension can be neglected, that is, for depths >0.5 cm. This is the case for normal oceanic water waves, for which the effects of surface tension are neglected. The bilinear form is

$$(iD_t - D_x^2 + D_y^2)G \cdot F = 0$$
$$(D_x^2 + D_y^2)F \cdot F = -|G|^2$$

$$(34.26)$$

In shallow water the DS II equations are integrable, in deep water they are not. Therefore, the shallow-water problem is amenable to the methods in this book and will not be elaborated on more here in order to help truncate this monograph to a reasonable size.

The deep-water problem, being nonintegrable, must be treated somewhat differently. In this case one is tempted to take the Riemann spectrum as a slowly varying function of time. Therefore, one must first take the holomorphic differentials to be slowly varying in time, leading to a type of "adiabatic" extension for the loop integrals (see Chapter 14), which must also vary as a function of time. Likewise the particular form of the holomorphic differentials (Belokolos et al., 1994) must be modified to allow the Schottky parameters to vary in time. Finally, the method of Nakamura and Boyd (Chapter 16) must allow for time variation in the period matrix, frequencies, and phases in order to describe the time evolution of the system. When we insert the theta functions in the deep-water bilinear form (Equation (34.26)), we *a priori* know that the equations are nonintegrable and therefore we must allow the Riemann spectrum the freedom to vary in time. These considerations are an important part of the second volume in this series of monographs and the reader is referred to this work for further details as they become available. This approach allows for the hyperfast modeling technique to be extended to nonintegrable equations.

34.5 Numerical Examples

A numerical example is shown in Figure 34.2 for the hyperfast numerical integration of the DS equations in deep water. A small-amplitude, two-dimensional sinusoidal modulation of the wave train is shown in Figure 34.2A. The modulation is seen to grow into a nonlinear wave packet and to modify its form as it evolves as seen in Figure 34.2B. In Figure 34.2C the wave packet has reached its maximum height. Note that during the time evolution of the packet there is a considerable transfer of energy and complexity into the direction of propagation (x coordinate) as well as into the lateral direction (y coordinate). In Figure 34.2D the energy of the packet has continued to spread into the lateral directions while decreasing its height somewhat. In Figures 34.2E and F the dynamics continue to spread energy into the lateral direction with a resultant decreasing maximum wave height and increasing energy spread toward the rear of the packet train.

We note that the initial modulation consists of two sine waves of small-amplitude, one in the direction of propagation (x-axis) and the other in the transversal direction (y-axis). The subsequent time evolution brings on "nonlinear focusing" which concentrates the energy locally to give a large-amplitude "extreme" or "rogue" wave while at the same time spreading out the waves in the lateral direction. Clearly, ocean waves, initially constructed in the Cauchy sense to have little directional spreading, will dynamically generate directionally spread waves through nonlinear interactions. Because of the nonlinear Schrödinger structure of the DS II equations, the spread of energy to the highest wavenumbers is nonphysical and exaggerated for large times (Yuen, 1991). A more physical form of the equations can be obtained by including higher order nonlinear terms and by adding linear dispersion to all orders (Trulsen et al., 2000).

We see in Figure 34.3 the nonlinear packet train at its highest amplitude. The largest wave in simulations of this type is often referred to as a rogue wave. The observation point of the wave in Figure 34.3 is taken to lie nearly in the plane of the undisturbed ocean surface. While the maximum crest amplitude is substantial, the "hole" beneath the maximum crest is also quite deep, so that the overall wave height is quite large. Indeed, the maximum wave height is 42 m in the simulation. The background "sine waves" are a memory of the initial conditions; part of these (two) sine waves has been energetically diverted by the modulational instability to "pile up" the waves to their maximum dynamical height, giving the extreme waves in Figures 34.2C and 34.3.

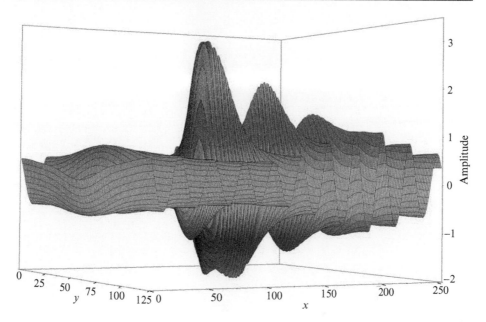

Figure 34.3 A large rogue wave as it evolves in the deep-water Davey-Stewartson equations. (See color plate).

In the sequel to this book I will provide more details about the numerical model for the nonintegrable case and will discuss in detail an analytical expression for the waveform in Figures 34.2 and 34.3 in terms of theta functions that have a particular *time-evolving Riemann spectrum* (with a 2×2 Riemann matrix). The intention is of course to provide numerical results and analytic expressions for extreme packets and waves for engineering design purposes.

References

Ablowitz, M. J., and Clarkson, P. A., *Solitons, Nonlinear Evolution Equations and Inverse Scattering* (Cambridge University Press, Cambridge, 1991).

Ablowitz, M. J., and Fokas, A. S., *Complex Variables: Introduction and Applications* (Cambridge University Press, Cambridge, 1997).

Ablowitz, M. J., and Herbst, B. M., Numerically induced chaos in the nonlinear Schroedinger equation, *SIAM J. Appl. Math.* 59(2), 339, 1990.

Ablowitz, M. J., and Ladik, J., Nonlinear Differential-Difference Equations, *J. Math. Phys.* 16, 598, 1975.

Ablowitz, M. J., and Ladik, J., Nonlinear differential difference equations and Fourier analysis, *J. Math. Phys.* 17, 1011, 1976a.

Ablowitz, M. J., and Ladik, J., A nonlinear difference scheme and inverse scattering, *Stud. Appl. Math.* 55, 213, 1976b.

Ablowitz, M. J., and Ladik, J., On the Solution of a Class of Nonlinear Partial Differential Equations, *Stud. Appl. Math.* 57, 1, 1977.

Ablowitz, M. J., and Segur, H., *Solitons and the Inverse Scattering Transform* (SIAM, Philadelphia, 1981).

Ablowitz, M. J., Kaup, D. J., Newell, A. C., and Segur, H., The Inverse Scattering Transform–Fourier Analysis for Nonlinear Problems, *Stud. Appl. Math.* 249, 1974.

Ablowitz, M. J., Prinari, B., and Trubatch, A. D., *Discrete and Continuous Nonlinear Schroedinger Systems* (Cambridge University Press, Cambridge, 2004).

Abramowitz, M., and Stegun, I. A., *Handbook of Mathematical Functions* (National Bureau of Standards, Applied Mathematics Series 55, 1964).

Akhmediev, N., Nonlinear physics: De'jàvu in optics, *Nature* 413(6853), 267, 2001.

Akhmediev, N., and Korneev, V. I., Modulation instability and periodic solutions of the nonlinear Schroedinger equation, *Teoreticheskaya I Matematicheskaya Fizika* 69(2), 189–194, 1986.

Akhmediev, N., and Ankiewicz, A., *Solitons, Nonlinear Pulses and Beams* (Chapman and Hall, London, 1997).

Akhmediev, N., and Mitskevich, N. V., Extremely high degree of N-soliton pulse compression in an optical fiber, *IEEE J. Quantum Electron.* 27, 849, 1991.

Akhmediev, N., Eleonskii, V. M., and Kulagin, N. E., Generation of periodic trains of picosecond pulses in an optical fiber: exact solutions, *Sov. Phys. JETP* 62, 894, 1985.

Akhmediev, N., Elconskii, V. M., and Kulagin, N. E., Exact first order solutions of the nonlinear Schodinger equation, *Theor. Math. Phys.* 72, 809, 1987.

Akhmediev, N., Heatley, D. R., Stegeman, G. I., and Wright, E. M., Pseudorecurrence in two-dimensional modulation instability with a saturable self-focusing nonlinearity, *Phys. Rev. Lett.* 65, 1423, 1990.

Akhmediev, N., Soto-Crespo, J. M., and Grelu, Ph., Roadmap to ultra-short record high-energy pulses out of laser oscillators, *Phys. Lett. A* 372, 3124, 2008.

Akhmediev, N., Ankiewicz, A., and Taki, M., Waves that appear from nowhere and disappear without a trace, *Phys. Lett. A* 373, 675–678, 2009.

Apel, J. R., A new analytical model for internal solitons in the ocean, *J. Phys. Oceanogr.* 33(11), 2247–2269, 2003.

Apel, J. R., *Principles of Ocean Physics* (Academic Press, San Diego, 1988).

Apel, J. R., et al., Internal solitons in the ocean and their effect on underwater sound, *J. Acoust. Soc. Am.* 121, 695–722, 2007.

Arnold, V. I., *Mathematical Methods of Classical Mechanics* (Springer-Verlag, 1989).

Arnold, V. I., *Geometrical Methods In The Theory Of Ordinary Differential Equations* (Springer-Verlag, 1988).

Arnold, V. I., *Ordinary Differential Equations* (The MIT Press, 1978).

Arnold, V. I., and Avez, A., *Ergodic Problems of Classical Mechanics* (Addison-Wesley, 1989).

Baker, H. F., *Abelian Functions: Abel's Theorem and the Allied Theory of Theta Functions* (Cambridge University Press, Cambridge, 1897).

Baker, H. F., *An Introduction to the Theory of Multiply Periodic Functions* (Cambridge University Press, Cambridge, 1907).

Baldwin, D., Goktas, U., and Hereman, W., Symbolic computation of exact solutions expressible in hyperbolic and elliptic functions for nonlinear PDEs, *J. Symb. Comp.* 37, 669–705, 2004.

Balmforth, N. J., Solitary waves and homoclinic orbits, *Annu. Rev. Fluid Mech.* 27, 335–373, 1995.

Batchelor, G. K., *An Introduction to Fluid Dynamics* (Cambridge University Press, Cambridge, 1967).

Beckmann, P., *Probability in Communication Engineering* (Harcourt, Brace & World, Inc., New York, 1967).

Bellman, R., *A Brief Introduction to Theta Functions* (Holt, Rinehart and Winston, New York, 1961).

Belokolos, E. D., Bobenko, A. I., Enol'skii, V. Z., Its, A. R., and Matveev, V. B., *Algebro-Geometric Approach to Nonlinear Integrable Equations* (Springer-Verlag, Berlin, 1994).

Bendat, J. S., and Piersol, A. G., *Random Data: Analysis and Measurement Procedures* (Wiley-Interscience, New York, 1986).

Benjamin, T. B., and Feir, J. F., The disintegration of wave trains on deep water, *J. Fluid Mech.* 27, 417, 1967.

Benjamin, T. B., Bona, J. L., and Mahoney, J. J., Model equations for long waves in nonlinear dispersive systems, *Philos. Trans. R. Soc. Lond. A* 272, 47–78, 1972.

Benny, D. J., Long nonlinear waves in fluid flows, *Math. J. Phys. Stud. Appl. Math.* 45, 52–63, 1966.

Benny, D. J., and Roskes, G. J., Wave instabilities, *Stud. Appl. Math.* 48, 377–385, 1969.

Bishop, A. R., and Lomdahl, P. S., Nonlinear dynamics in driven, damped sine-Gordon systems, *Physica D* 18, 54, 1986.

Bishop, A. R., Forest, M. G., McLaughlin, D. W., and Overman, E. A. II, A quasi-periodic route to chaos in a near integrable PDE, *Physica D* 23, 293–328, 1986.

Bobenko, A. I., Schottky uniformization and finite-gap integration, *Dokl. Akad. Nauk SSSR* 295, 268–272, 1987.

Bobenko, A. I., and Bordag, L. A., Periodic multiphase solutions of the Kadomsev-Petviashvili equation, *Zap. LOMI* 165, 31–41, 1987.

Bobenko, A. I., and Bordag, L. A., Periodic multiphase solutions of the Kadomtsev-Petviashvili equation, *J. Phys. A Math. Gen.* 22, 1259, 1989.

Bobenko, A. I., and Kubensky, D. A., Qualitative analysis and calculations of the finite-gap solutions of the KdV equation. An automorphic approach, *Teor. Mat. Fiz.* 72, 352–360, 1987.

Boffetta, G., and Osborne, A. R., Computation of the direct scattering transform for the nonlinear Schroedinger equation, *J. Comput. Phys.* 102, 252–264, 1992.

Boiti, M., Leon, J., Martina, L., and Pempinelli, F., Scattering of localized solitons in the plane, *Phys. Lett. A* 132, 432, 1988.

Boiti, M., Leon, J., and Pempinelli, F., A new spectral transform for the Davey-Stewartson I equation, *Phys. Lett. A* 141, 101–107, 1989.

Boiti, M., Leon, J., and Pempinelli, F., On the spectral theory for the Davey-Stewartson equation, In *Inverse Problems in Action*, P. C. Sabatier, ed., pp. 544–551 (Springer-Verlag, Berlin-Heidelberg-New York, 1990).

Boiti, M., Leon, J., and Pempinelli, F., Waves in the Davey-Stewartson equation, *Inverse Probl.* 6, 175–185, 1991.

Bona, J. L., and Chen, H., Comparison of model equations for small-amplitude long waves, *Nonlinear Anal.* 38, 625–647, 1999.

Bona, J. L., Pritchard, W. G., and Scott, L. R., An evaluation of a model equation for water waves, *Philos. Trans. R. Soc. Lond. A* 302, 457–510, 1981.

Bona, J. L., Chen, M., and Saut, J. C., Boussinesq equations and other systems for small-amplitude long waves in nonlinear dispersive media. I: Derivation and linear theory, *J. Nonlinear Sci.* 12, 283–318, 2002.

Bona, J. L., Saut, J. C., and Chen, M., Boussinesq equations and other systems for small-amplitude long waves in nonlinear dispersive media. II: The nonlinear theory, *Nonlinearity* 17, 925–952, 2004.

Borwein, J. M., and Borwein, P. B., *Pi and the AGM: A Study in Analytic Number Theory and Computational Complexity* (Wiley-Interscience Publication, New York, 1987).

Boussinesq, J., Théorie de l'intumescence liquide appelée onde solitaire ou de translation, se propageant dans un canal rectangulaire, *C. R. Acad. Sci. Paris* 72, 755–759, 1871.

Boussinesq, J., Théorie des ondes et des remous qui se propagent le long d'un canal rectangulaire horizontal, en communiquant au liquide contenu dans ce canal des vitesses sensiblement pareilles de la surface au fond, *J. Math. Pures Appl.* 17, 55–108, 1872.

Boussinesq, J., Essai su la théorie des eaux courantes. Académie des Sciences de l'Institut de France, *Mémoires présentés par divers savants* 23(2), 1–680, 1877.

Boyd, J. P., Equatorial Solitary Waves, Part I: Rossby Solitons, *J. Phys. Oceangr.* 10, 1699–1718, 1980.

Boyd, J. P., The double cnoidal wave of the Korteweg-deVries equation: An overview, *J. Math. Phys.* 25(12), 3390, 1984a.

Boyd, J. P., Perturbation series for the double cnoidal wave of the Korteweg-deVries equation, *J. Math. Phys.* 25(12), 3402, 1984b.

Boyd, J. P., The special modular transformation for polycnoidal waves of the Korteweg-deVries equation, *J. Math. Phys.* 25(12), 3415, 1984c.

Boyd, J. P., New directions in solitons and nonlinear periodic waves: Polycnoidal waves, imbricated solitons, weakly nonlocal solitary waves, and numerical boundary value algorithms, *Adv. Appl. Mech.* 27, 1, 1990.

Boyd, J. P., *Weakly Nonlinear Solitary Waves and Beyond-All-Order Asymptotics* (Klewer, Dortrecht, 1998).

Boyd, J. P., The cnoidal wave/corner wave/breaking wave scenario: A one-sided infinite-dimension bifurcation, *Math. Comput. Sim.* 69(3-4), 235–242, 2005.

Boyd, J. P., and Haupt, S. E., In *Nonlinear Topics in Ocean Physics*, A. R. Osborne, ed., (Elsevier, Amsterdam, 1990).

Bryant, P. J., Periodic waves in shallow water, *J. Fluid. Mech.* 59, 625–644, 1973.

Bullough, R. K., "The wave" "par excellence", the solitary, progressive great wave of equilibrium of the fluid—An early history of the solitary wave, In *Solitons: Introduction and Applications*, M. Lakshmanan, ed., pp 7–42 (Springer-Verlag, Berlin, 1988).

Burnside, W., On a class of automorphic functions, *Proc. Lond. Math. Soc.* 23, 49–88, 1892.

Burnside, W., Further note on automorphic funtions, *Proc. Lond. Math. Soc.* 23, 281–295, 1892b.

Byrd, P. F., and Friedman, M. D., *Handbook of Elliptic Integrals for Engineers and Scientists* (Springer-Verlag, Heidelberg, 1971).

Calini, A., and Schober, C. M., Homoclinic chaos increases the likelihood of rogue waves, *Phys. Lett. A* 298, 335–349, 2002.

Calini, A., Ercolani, N. M., McLaughlin, D. W., and Schober, C. M., Mel'nikov analysis of numerically induced chaos in the nonlinear Schroedinger equation, *Physica D* 89, 227–260, 1996.

Calogero, F., and Degasperis, A., *Spectral Transform and Solitons: Tools to Solve and Investigate Nonlinear Evolution Equations. Volume One* (North-Holland, Amsterdam, 1982).

Camassa, R., and Holm, D., An integrable shallow water equation with peakon solitons, *Phys. Rev. Lett.* 71, 1661–1664, 1993.

Camassa, R., and Holm, D., A new integrable shallow water equation, *Adv. Appl. Mech.* 31, 1–33, 1994.

Cavaleri, L., and Zecchetto, S., Reynolds stresses under wind waves, *J. Geophys. Res.* 92, 3894, 1987.

Cerveró, J. M., and Estévez, P. G., Miura transformation between two non-linear equations in 2+1 dimensions, *J. Math. Phys.* 39, 2800, 1998.

Champeney, D. C., *Fourier Transforms and Their Physical Applications* (Academic Press, London, 1973).

Chen, M., Exact solutions of various Boussinesq systems, *Appl. Math. Lett.* 11, 45–49, 1998.

Choi, W., Nonlinear evolution equations for two-dimensional surface waves in a fluid of finite depth, *J. Fluid Mech.* 295, 381–394, 1995.

Choi, W., and Camassa, R., Fully nonlinear internal waves in a two-fluid system, *J. Fluid Mech.* 396, 1–36, 1999.

Chow, K. W., Ko, N. W. M., and Tang, S. K., Solitons in (2+0) dimensions and their applications in vortex dynamics, *Fluid Dyn. Res.* 21, 101, 1997.

Chow, K. W., Ko, N. W. M., and Leung, R. C. K., Inviscid two dimensional vortex dynamics and a soliton expansion of the sinh-Poisson equation, *Phys. Fluids* 10(5), 1111–1119, 1998.

Christov, I., Internal solitary waves in the ocean: Analysis using the periodic, inverse scattering transform, *Math. Comp. Sim.* 80(1), 192–201, 2009.

Christov, I., Internal solitary waves in the ocean: Analysis using the periodic, inverse scattering transform, *Math. Comput. Sim.* (doi: 10.1016/j.matcom.2009.06.005), 2009.

Clarkson, P. A., and Mansfield, E. L., On a shallow water wave equation, *Nonlinearity* 7, 975–1000, 1994.

Coble, A. B., *Algebraic Geometry and Theta Functions* (American Mathematical Society, New York, 1929).

Collins, M. D., Applications and time-domain solution of higher-order parabolic equations in underwater acoustics, *J. Acoust. Soc. Am.* 86, 1097–1102, 1989.

Constantin, A., On the scattering problem for the Camassa-Holm equation, *Proc. R. Soc. Lond. A Math. Phys. Sci.* 457, 953–970, 2001.

Constantin, A., and McKean, H., A shallow water equation on the circle, *Commun. Pure Appl. Math.* 52(8), 949–982, 1999.

Conte, R., Musette, M., and Pickering, A., The two-singular manifold method: II. Classical Boussinesq system, *J. Phys. A* 28, 179, 1995.

Cooley, J. W., An improved eigenvalue corrector formula for solving Schrödinger's equation for central fields, *Math. Comput.* 15, 363, 1961.

Cooley, J. W., and Tukey, J. W., An algorithm for machine calculation of complex Fourier series, *Math. Comput.* 19, 297–301, 1965.

Cooley, J. W., Lewis, P. A. W., and Welch, P. D., The Finite Fast Fourier Transform, *IEEE Trans. Audio Electroacoustics* AU-17, 77, 1969.

Craik, A. D. D., George Gabriel Stokes and water wave theory, *Annu. Rev. Fluid Mech.* 37, 23–42, 2005.

Crawford, D. R., Lake, B. M., Saffman, P. G., and Yuen, H. C., Stability of weakly nonlinear deep-water waves in two and three dimensions, *J. Fluid Mech.* 105, 177, 1981.

Cushman-Roisin, B., *Introduction to Geophysical Fluid Dynamics* (Prentice Hall, Englewood Cliffs, NJ, 1994).

Darrigol, O., The spirited horse, the engineer and the mathematician: Water wave in nineteenth-century hydrodynamics, *Arch. Hist. Exact Sci.* 58, 21–95, 2003.

Darrigol, O., *Worlds of Flow: A History of Hydrodynamics from the Bernoullis to Prandtl* (Oxford University Press, Oxford, 2005).

Date, E., Multi-soliton solutions and quasi-periodic solutions of non-linear equations of sine-Gordon type, *Osaka J. Math.* 19, 125–158, 1982.

Date, E., and Tanaka, S., Periodic multi-soliton solutions of Korteweg-deVries equation and Toda lattice, *Suppl. Prog. Theor. Phys.* 59, 107–126, 1976.

Dauxois, T., and Peyrard, M., *Physics of Solitons* (Cambridge University Press, Cambridge, 2004).

Dauxois, T., Fauve, S., and Tuckerman, L., Stability of periodic arrays of vortices, *Phys. Fluids* 8(2), 487–495, 1995.

Davey, A., and Stewartson, K., On Three-Dimensional Packets of Surface Waves, *Proc. R. Soc. Lond. A* 338, 101, 1974.

Dean, R. G., and Dalrymple, R. A., *Water Wave Mechanics for Engineers and Scientists* (World Scientific, New Jersey, 1991).

Debye, P., *Vorträge öber die Kinetische Theorie der Materie und der Electrizisär* (Leipzig, Germany, 1916).

Deconinck, B., The initial-value problem for multiphase solutions of the Kadomtsev-Petviashvili equation. Ph. D. Thesis, University of Colorado, Department of Applied Mathematics, 1998).

Deconinck, B., and Segur, H., The KP equation with periodic initial data, *Physica D* 123, 123–152, 1998.

Deconinck, B., and van Hoeij, M., Computing Riemann matrices of algebraic curves, *Physica D* 152–153, 28–46, 2001.

Deconinck, B., Heil, M., Bobenko, A., van Hoeij, M., and Schmies, M., Computing Riemann Theta Functions, *Math. Comput.* 73, 1417–1442, 2004.

Deem, G. S., and Zabusky, N. J., Stationary "V-States," interactions, recurrence and breaking, In *Solitons in Action*, K. Lonngren, and A. Scott, eds., (Academic Press, New York, 1978).

Degasperis, A., Nonlinear wave equations solvable by the spectral transform, In *Nonlinear Topics in Ocean Physics*, A. R. Osborne, ed., (Elsevier, Amsterdam, 1991).

Dhanak, M. R., and Marshall, M. P., Motion of an elliptical vortex under applied periodic strain, *Phys. Fluids A* 5, 1224–1230, 1993.

Dickey, L. A., *Soliton Equations and Hamiltonian Systems* (World Scientific, Singapore, 1991).

Dodd, R. K., Eilbeck, J. E., Gibbon, J. D., and Morris, H. C., *Solitons and Nonlinear Wave Equations* (Academic Press, London, 1982).

Dommermuth, D. G., and Yue, D. K. P., A higher-order spectral method for the study of nonlinear gravity waves, *J. Fluid Mech.* 184, 267–288, 1987.

Drazin, P. G., and Johnson, R. S., *Solitons: An Introduction* (Cambridge University Press, Cambridge, 1989).

Dritschel, D. G., A fast contour dynamics method for many-vortex calculations in two-dimensional flows, *Phys. Fluids* A5(1), 173–186, 1993.

Dubrovin, B. A., Theta functions and non-linear equations, *Russ. Math. Surv.* 36(2), 11–92, 1981.

Dubrovin, B. A., Flickinger, R., and Segur, H., Three phase solutions of the Kadomtsev Petviashvili equation, *Stud. Appl. Math.* 92(2), 137, 1997.

Dubrovin, B. A., and Novikov, S. P., Periodic problems for the Korteweg-deVries equation in the class of finite band potentials, *Funct. Anal. Appl.* 9, 215–223, 1975a.

Dubrovin, B. A., and Novikov, S. P., Periodic and conditionally periodic analogues of the many-soliton solutions of the Kortweg-deVries equation, *Sov. Phys. JETP* 40, 1058, 1975b.

Dubrovin, B. A., Matveev, V. B., and Novikov, S. P., Nonlinear equations of Korteweg-de Vries type, finite-zone linear operators and Abelian varieties, *Russ. Math. Surv.* 31, 59–146, 1976.

Duda, T. F., Lynch, J. F., Irish, J. D., Beardsley, R. C., and Ramp, S. R., et al., Internal tide and nonlinear wave behavior in the continental slope in the northern South China Sea, *IEEE J. Ocean. Eng.* 29, 1105–1131, 2004.

Eilenberger, G., *Solitons: Mathematical Methods for Physicists* (Springer-Verlag, Berlin, 1981).

El, G. A., Korteweg-deVries equation: Solitons and undular bores, In *Solitary Waves in Fluids*, R. H. J. Grimshw, ed., (WIT Press, Boston, 2007).

Elgar, S., and Guza, R. T., Nonlinear model predictions of bispectra of shoaling surface gravity waves, *J. Fluid Mech.* 167, 1–26, 1986.

Emmerson, G. S., *John Scott Russell, A Great Victorian Engineer and Naval Architect* (John Murray, London, 1977).

Ercolani, N. M., and Forest, M. G., The geometry of real sine-Gordon wavetrains, *Comm. Math. Phys.* 99, 1–49, 1985.

Estévez, P. G., Darboux transformation and solutions for an equation in 2+1 dimensions, *J. Math. Phys.* 40, 1406, 1999.

Estévez, P. G., and Gordoa, P. R., Double singular manifold method for the mKdV equation, *Teor. Matem. Fizika* 99, 370, 1994.

Faddeev, L. D., and Takhtajan, L. A., *Hamiltonian Methods in the Theory of Solitons* (Springer-Verlag, Berlin, 1987).

Farmer, D. M., and Armi, L., The flow of Mediterranean water through the Strait of Gibraltar, *Prog. Oceanogr.* 21, 1–105, 1988.

Farmer, D., and Armi, L., The generation and trapping of solitary waves over topography, *Science* 283, 188–190, 1999.

Fay, J. D., *Theta Functions on Riemann Surfaces, Lect. Notes Math., Vol. 352* (Springer, Berlin, Heidelberg, 1973).

Ferguson, W. E., Flaschka, H., and McLaughlin, D. W., Nonlinear normal modes for the Toda chain, *J. Comp. Physics* 45, 157, 1982.

Fermi, E., Pasta, J., and Ulam, S., Studies of nonlinear problems, I, Los Alamos Rep. LA1940, 1955; reprod, In *Nonlinear Wave Motion*, A. C. Newell, ed., (American Mathematical Society, Providence, 1974).

Flaschka, H., The Toda lattice. I, *Phys. Rev. B* 9, 1924, 1974a.

Flaschka, H., The Toda lattice. II, *Prog. Theoret. Phys.* 51, 703, 1974b.

Flaschka, H., and McLaughlin, D. W., Canonically conjugate variables for KdV and Toda lattice under periodic boundary conditions, *Prog. Theoret. Phys.* 55, 438–456, 1976.

Flesch, R., Forest, M. G., and Sinha, A., Numerical inverse spectral transform for the periodic sine-Gordon equation: theta function solutions and their linearized stability, *Physica D* 48, 169–208, 1991.

Flierl, G. R., Isolated eddy models in geophysics, *Annu. Rev. Fluid Mech.* 19, 493, 1987.

Fokas, A. S., and Liu, Q. M., Asymptotic integrability of water waves, *Phys. Rev. Lett* 77(12), 2347–2351, 1996.

Fokas, A. S., and Santini, P. M., Coherent structures in multidimensions, *Phys. Rev. Lett.* 63(13), 1329, 1989.

Fokas, A. S., and Santini, P. M., Dromions and a Boundary-value problem for the Davey-Stewartson equation, *Physica D* 44, 99, 1990.

Ford, L., *Automorphic Functions* (McGraw-Hill, New York, 1929).

Fordy, A. P., *Soliton Theory: A Survey of Results* (Manchester University Press, Manchester, 1990).

Forest, M. G., and McLaughlin, D. W., Spectral theory for the periodic sine-Gordon equation: A concrete viewpoint, *J. Math. Phys.* 23, 1248, 1982.

Forest, M. G., and McLaughlin, D. W., Modulations of sinh-Gordon and sine-Gordon wavetrains, *Stud. Appl. Math.* 68, 11–59, 1983.

Forester, C. S., *Hornblower and the Atropos* (Little, Brown and Company, Boston, 1953).

Fornberg, B., and Whitham, G. B., A Numerical and Theoretical Study of Certain Nonlinear Wave Phenomena, *Philos. Trans. R. Soc. Lond. A* 289, 373, 1978.

Fourier, J., *The Analytical Theory of Heat*, English edition of original work from 1822 (Dover Phoenix Editions, New York, 1955).

Friedlander, S. J., *An Introduction to the Mathematical Theory of Geophysical Fluid Dynamics* (Elsevier North-Holland, New York, 1980).

Gaponov-Grekhov, A. V., and Rabinovich, M. I., *Nonlinearities in Action* (Springer-Verlag, Berlin, 1992).

Gardner, C. S., Greene, J. M., Kruskal, M. D., and Miura, R. M., Method for solving the Korteweg-deVries equation, *Phys. Rev. Lett.* 19, 1095, 1967.

Garrett, C., and Munk, W., Internal waves in the ocean, *Annu. Rev. Fluid Mech.* 11, 339–369, 1979.

Gesztesy, F., and Holden, H., *Soliton Equations and Their Algebro-Geometric Solutions* (Cambridge University Press, Cambridge, 2003).

Ginzburg, V. L., and Landau, L. D., *Sov Phys JETP* 20, 1064, 1950.

Ginzburg, V. L., *Sov Phys JETP* 2, 589, 1956.

Ginzburg, V. L., and Pitaevskii, L. P., *Sov Phys JETP* 7, 858, 1958.

Griffiths, P. A., *Introduction to Algebraic Curves* (American Mathematical Society, Providence, 1989).

Griffiths, P. A., and Harris, J., *Principles of Algebraic Geometry* (John Wiley, New York, 1994).

Grimshaw, R., and Melville, W. K., On the derivation of the modified Kadomtsev-Petviashvili equation, *Stud. Appl. Math.* 80, 183–202, 1989.

Grimshaw, R., Pelinovsky, E., and Poloukhina, O., Higher-order Korteweg-deVries models for internal solitary waves in a stratified shear flow with a free surface, *Nonlinear Proc. Geophys.* 9, 221–235, 2002.

Grimshaw, R., Pelinovsky, E., Talipova, T., and Kurkin, A., Simulation of the transformation of internal solitary waves on oceanic shelves, *J. Phys. Oceanogr.* 34, 2774–2791, 2004.

Grimshaw, R., Korteweg-deVries equation, In *Nonlinear Waves in Fluids: Recent Advances and Modern Applications*, R. Grimshaw, ed., (Springer-Verlag, Berlin, 2005).

Grimshaw, R. H. J., Evolution equations for weakly nonlinear long internal waves in a rotating fluid, *Stud. Appl. Math.* 73, 1–33, 1985.

Grimshaw, R. H. J., Internal solitary waves, In *Advances in Coastal and Ocean Engineering*, P.L.F. Liu, ed., Vol. III. pp. 1–30 (World Scientific, Singapore, 1997).

Grimshaw, R. H. J., Internal solitary waves, In *Environmental Stratified Flows*, R. Grimshaw, ed., pp. 1–28 (Kluwer, Boston, 2001).

Grimshaw, R. H. J. ed., In *Solitary Waves in Fluids* (WIT Press, Boston, 2007).

Grimshaw, R. H. J., He, J.-M., and Ostrovsky, L. A., Terminal damping of a solitary wave due to radiation in rotational systems, *Stud. Appl. Math.* 101, 197–210, 1997a.

Grimshaw, R. H. J., Pelinovsky, E., and Talipova, T., The modified Korteweg-deVries equation in the theory of large amplitude internal waves, *Nonlinear Proc. Geophys.* 4, 237–250, 1997b.

Grimshaw, R. H. J., Ostrovsky, L. A., Shrira, V. I., and Stepanyants, Y. A., Long nonlinear surface and internal gravity waves in a rotating ocean, *Surv. Geophys.* 19, 289–338, 1998.

Grue, J., Friis, H. A., Palm, E., and Rusas, P.-O., A method for computing unsteady fully nonlinear interfacial waves, *J. Fluid Mech.* 351, 223–252, 1997.

Grue, J., Jensen, A., Rusas, P.-O., and Sveen, J. K., Properties of large-amplitude internal waves, *J. Fluid Mech.* 380, 257–278, 1999.

Grue, J., Jensen, A., Rusas, P.-O., and Sveen, J. K., Breaking and broadening of internal solitary waves, *J. Fluid Mech.* 413, 181–217, 2000.

Hald, O. H., Numerical solution of the Gel'fand-Levitan equation, *Linear Algebra Appl.* 28, 99, 1979.

Hammack, J. L., A note on tsunamis: their generation and propagation in an ocean of uniform depth, *J. Fluid Mech.* 60, 769–800, 1973.

Hammack, J. L., and Segur, H., The Korteweg-deVries equation and water waves, part 2 Comparison with experiments, *J. Fluid Mech.* 65, 289–314, 1974.

Hammack, J. L., and Segur, H., The Korteweg-deVries equation and water waves, part 3: Oscillatory waves, *J. Fluid Mech.* 84, 337–358, 1978a.

Hammack, J. L., and Segur, H., Modelling criteria for long water waves, *J. Fluid Mech.* 84, 359–373, 1978b.

Hardin, R. H., and Tappert, F. D., Applications of the split-step Fourier method to the numerical solution of nonlinear and variable coefficient wave equations, *SIAM Rev.* 15, 423, 1973.

Hardy, G., On Hilbert transforms, *Messenger Math.* 54, 20–27, 81–88, 1924.

Harris, J., *Algebraic Geometry* (Springer, New York, 1992).

Hasimoto, H., and Ono, H., *J. Phys. Soc. Jpn.* 33, 805–811, 1972.

Hawkins, J. A., Warn-Varnas, A., and Christov, I., Fourier, Scattering and Wavelet Transforms: Applications to Internal Gravity Waves with Comparisons to Linear Tidal Data, In *Nonlinear Time Series Analysis in the Geosciences*, R. V. Donner, and S. M. Barbosa, eds., (Springer, Berlin, 2008).

Helfrich, K. R., Internal solitary wave breaking and run-up on a uniform slope, *J. Fluid Mech.* 243, 133–154, 1992.

Helfrich, K. R., Kuo, A. C., and Pratt, L. J., Nonlinear Rossby adjustment in a channel, *J. Fluid Mech.* 390, 187–222, 1999.

Helfrich, K. R., and Melville, W. K., On long nonlinear internal waves over slope-shelf topography, *J. Fluid Mech.* 167, 285–308, 1986.

Helfrich, K. R., Melville, W. K., and Miles, J. W., On interfacial solitary waves over slowly varying topography, *J. Fluid Mech.* 149, 305–317, 1984.

Helfrich, K. R., and Melville, W. K., Long nonlinear internal waves, *Annu. Rev. Fluid Mech.* 38, 395–425, 2007.

Henderson, K. L., Peregrine, D. H., and Dold, J. W., *Wave Motion* 29, 341, 1999.

Hietarinta, J., Scattering of solitons and dromions, In *Scattering*, R. Pike, P. Sabatier, eds., (Academic Press, New York, 2002).

Hirota, R., *The Direct Method in Soliton Theory* (Cambridge University Press, Cambridge, 2004).

Hirota, R., and Ito, M., *J. Phys. Soc. Jpn.* 52, 744, 1983.

Hirota, R., and Satsuma, J., N-soliton solutions of model equations for shallow water waves, *J. Phys. Soc. Jpn.* 40, 611–612, 2004.

Holloway, P. E., Internal hydraulic jumps and solitons at a shelf break region on the Australian North West Shelf, *J. Geophys. Res.* C95, 5405–5416, 1987.

Holloway, P., Pelinovsky, E., Talipova, T., and Barnes, B., A nonlinear model of the internal tide transformation on the Australian North West Shelf, *J. Geophys. Oceanogr.* 27(6), 871–896, 1997.

Holloway, P., Pelinovsky, E., and Talipova, T., A generalized Korteweg-deVries model of internal tide transformation in the coastal zone, *J. Geophys. Res.* 104(C8), 18, 333–18,350, 1999.

Holloway, P., Pelinovsky, E., and Talipova, T., Internal tide transformation and oceanic internal solitary waves, In *Environmental Stratified Flows*, R. Grimshaw, ed., Chapter 2, pp. 29–60 (Kluwer, Dordrecht, 2001).

Hopfinger, E. J., and Van Heijst, G. J. G., Vortices in rotating fluids, *Annu. Rev. Fluid Mech.* 25, 241, 1993.

Hunkins, K., and Fliegel, M., Internal undular surges in Seneca Lake: A natural occurrence of solitons, *J. Geophys. Res.* 78, 539, 1973.

Igusa, J., *Theta Functions* (Springer-Verlag, Berlin, 1972).

Infeld, E., and Rowlands, G., *Nonlinear Waves, Solitons and Chaos* (Cambridge University Press, Cambridge, 1990).

Iorio, R., and Iorio, V. de. M., *Fourier Analysis and Partial Differential Equations* (Cambridge University Press, Cambridge, 2001).

Islas, A., and Schober, C. M., Predicting rogue waves in random oceanic sea states, *Phys. Fluids* 17, 1–4, 2005.

Its, A. R., and Matveev, V. B., The periodic Korteweg-deVries equation, *Funct. Anal. Appl.* 9(1), 67, 1975.

Jackson, C. R., *An Atlas of Internal Solitary-like Waves and Their Properties* (Global Ocean Associates, Alexandria, Virginia), www.internalwaveatlas.com, 2009.

Janssen, P., *The Interaction of Ocean Waves and Wind* (Cambridge University Press, Cambridge, 2004).

Jensen, F. B., Kuperman, W. A., Porter, M. B., and Schmidt, H., *Computational Ocean Acoustics* (American Institute of Physics Press and Springer Verlag, New York, 2000).

Johnson, R. S., *A Modern Introduction to the Mathematical Theory of Water Waves* (Cambridge University Press, Cambridge, 1997).

Johnson, R. S., On solutions of the Camassa-Holm equation, *Proc. R. Soc. Lond. A* 459, 1687–1708, 2002.

Joseph, R. I., and Egri, R., Another possible model equation for long waves in nonlinear dispersive systems, *Phys. Lett. A.* 61, 429–432, 1977.

Kadomtsev, B. B., and Petviashvili, V. I., On the stability of solitary waves in weakly dispersing media, *Sov. Phys. Dokl.* 15, 539–541, 1970.

Karpman, V. I., *Non-Linear Waves in Dispersive Media* (Pergamon, Oxford, 1975).

Kaup, D. J., A higher-order water-wave equation and the method for solving it, *Prog. Theor. Phys.* 54, 396–408, 1975.

Kay, I., and Moses, H. E., *Nuovo Cimento* 2, 917, 1955.

Kinsman, B., *Wind Waves* (Prentice-Hall, Englewood Cliffs, New Jersey, 1965).

Klymak, J. M., Pinkel, R., Liu, C. T., Liu, A. K., and David, L., Prototypical solitons in the South China Sea, *Geophys. Res. Letters* 33, L11607, 2006.

Kodama, Y., Normal forms for weakly dispersive wave equations, *Phys. Lett.* 112A(5), 193–196, 1985a.

Kodama, Y., On integrable systems with higher order corrections, *Phys. Lett.* 112A(6), 245–249, 1985b.

Komen, G. J., Cavaleri, L., Donelan, M., Hasselmann, K., Hasselmann, S., and Janssen, P. A. E. M., *Dynamics and Modelling of Ocean Waves* (Cambridge University Press, Cambridge, 1994).

Konopelchenko, B. G., Inverse spectral transform for the $(2 + 1)$-dimensional Gardner equation, *Inverse Probl.* 7, 739–753, 1991.

Konopelchenko, B. G., and Dubrovsky, V. G., Some new integrable nonlinear evolution equations in $2 + 1$ dimensions, *Phys. Lett. A* 102, 15, 1984.

Korteweg, D. J., and deVries, G., On the change of form of long waves advancing in a rectangular canal, and on a new type of long stationary waves, *Philos. Mag. Ser. 5*, 39, 422–443, 1895.

Kotljarov, V. P., and Its, A. R., *Dopov. Akad. Nauk. Ukr RSR. A* 11, 965–968 (in Ukranian), 1976.

Kozel, V. O., and Kotljarov, V. P., *Dokl. Akad. Nauk SSSR* 10, 878, 1976.

Krichever, I. M., An algebraic-geometric construction of the Zakharov-Shabat equations and their periodic solutions, *Dokl. Akad. Nauk SSSR* 227(2), 291–294, 1976.

Krichever, I. M., Integration of nonlinear equations by the methods of algebraic geometry, *Funktsional Anal. I Prilozhen.* 2(1), 180–208, 1977a.

Krichever, I. M., The methods of algebraic geometry in the theory of nonlinear equations, *Usp. Mat. Nauk* 32(6), 2(1), 15–31, 1977b.

Krichever, I. M., The periodic problem for the KP-2 equation, *Dokl. Akad. Nauk SSSR* 298(4), 802–806, 1988.

Krichever, I. M., The spectral theory of two-dimensional periodic operators and applications, *Russ. Math. Surv.* 44(2), 121–184, 1989.

Krichever, I. M., Perturbation theory in periodic problems for two-dimensional integrable systems, *Sov. Sci. Rev. C Math. Phys.* 9, 1–103, 1992.

Kuperman, W. A., Hodgkiss, W. S., Chun Song, H., Akal, T., Feria, C., and Jackson, D. R., Phase conjugation in the ocean: Experimental demonstration of an acoustic time-reversal mirror, *J. Acoust. Soc. Am.* 103(1), 25, 1998.

Lamb, G. L., *Elements of Soliton Theory* (John Wiley, New York, 1980).

Lamb, H., *Hydrodynamics* (Dover, New York, 1932).

Landau, L. D., and Lifshitz, E. M., *Quantum Mechanics: Non Relativistic Theory* (Pergamon, Oxford, 1958).

Lamb, K. G., A numerical investigation of solitary internal waves with trapped cores formed via shoaling, *J. Fluid Mech.* 451, 109–144, 2002.

Lamb, K. G., Shoaling solitary internal waves: on a criterion for the formation of waves with trapped cores, *J. Fluid Mech.* 478, 81–100, 2003.

Lamb, K. G., On boundary-layer separation and internal wave generation at the Knight Inlet sill, *Proc. R. Soc. London Ser. A* 460, 2305–2337, 2004.

Lamb, K. G., and Wilkie, K. P., Conjugate flows for waves with trapped cores, *Phys. Fluids* 16, 4685–4695, 2004.

Lamb, K. G., and Yan, L., The evolution of internal wave undular bores: comparison of a fully-nonlinear numerical model with weakly nonlinear theories, *J. Phys. Ocean.* 26, 2712–2734, 1996.

Lax, P. D., Integrals of nonlinear equations of evolution and solitary waves, *Commun. Pure Appl. Math.* 21, 467–490, 1968.

Lax, P. D., Periodic solutions of the Korteweg-deVries equation, *Commun. Pure Appl. Math.* 28, 141–188, 1975.

Lax, P. D., and Levermore, C. D., *Commun. Pure Appl. Math.* XXXVI, The small dispersion limit of the Korteweg-deVries Equation I 253, 1983; The small dispersion limit of the Korteweg-deVries Equation II, 571, 1983.

LeBlond, P. H., and Mysak, L. A., *Waves in the Ocean* (Elsevier, Amsterdam, 1978).

Lee, C. Y., and Beardsley, R. C., The generation of long nonlinear internal waves in a weakly stratified shear flow, *J. Geophys. Res.* 79(3a), 453–462, 1974.

Leibovich S., and Seebass A. R., eds., *Nonlinear Waves* (Cornell University Press, Cornell, 1974).

Leontovich, M., and Fock, V., Solution of the problem of propagation of electromagnetic waves along the earth's surface by the method of parabolic equation, *Zh. Eksp. Teor. Fiz.* 16, 13–24, 1946.

Lesieur, M., *Turbulence in Fluids* (Kluwer, Dordrecht, 1990).

Lighthill, J. M., *Fourier Analysis and Generalised Functions* (Cambridge University Press, Cambridge, 1959).

Lighthill, J. M., Contributions to the theory of waves in nonlinear dispersive systems, *J. Inst. Math. Appl.* 1, 269, 1965.

Lighthill, J. M., Some special cases treated by the Whitham theory, *Proc. R. Soc. Lond.* A 299, 28, 1967.

Lighthill, J. M., *Waves in Fluids* (Cambridge University Press, Cambridge, 1978).

Lighthill, J. M., *An Informal Introduction to Theoretical Fluid Mechanics* (Cambridge University Press, Cambridge, 1986).

Liu, A. K., Chang, Y. S., Hsu, M.-K., and Liang, N. K., Evolution of nonlinear internal waves in the East and South China Seas, *J. Geophys. Res.* 103(C4), 7995–8008, 1998.

Lo, E., and Mei, C. C., A numerical study of water-wave modulation based on a higher-order nonlinear Schroedinger equation, *J. Fluid Mech.* 150, 395–416, 1985.

Long, R. R., Solitary waves in the westerlies, *J. Atmos. Sci.* 21, 156–179, 1964.

Long, C. E., and Resio, D. T., Directional wave observations in Currituck sound, North Carolina, In *8th International Workshop on Wave Hindcasting and Forecasting* (North Shore, Oahu, Hawaii, 14-19 November, 2004).

Long, C. E., and Resio, D. T., Wind wave spectral observations in Currituck sound, North Carolina, *J. Geophys. Res.* 112, C05001, 2007.

Longuet-Higgins, M. S., The changes in amplitude of short gravity waves on steady non-uniform currents, *J. Fluid Mech.* 10, 529–549, 1961.

Longuet-Higgins, M. S., Radiation stress and mass transport in gravity waves, with application to "surf beats", *J. Fluid Mech.* 13, 481–504, 1962.

Longuet-Higgins, M. S., Radiation stresses in water waves; a physical discussion, with applications, *Deep Sea Res.* 11, 529–562, 1964.

Longuet-Higgins, M. S., On the mass, momentum, energy and circulation of a solitary wave, *Proc. R. Soc. Lond.* A 337, 1–13, 1974.

Longuet-Higgins, M. S., and Stewart, R. W., Changes in the form of short gravity waves on long waves and tidal currents, *J. Fluid Mech.* 8, 565–583, 1960.

Lonngren, K., and Scott, A., *Solitons in Action* (Academic Press, New York, 1978).

Magnus, W., Oberhettinger, F., and Soni, R. P., *Formulas and Theorems for the Special Functions of Mathematical Physics* (Springer-Verlag, New York, 1966).

Makhankov, V. G., *Soliton Phenomenology* (Kluwer Academic, Dortrecht, 1990).

Malanotte Rizzoli, P., Planetary solitary waves in geophysical flows, *Advances in Geophysics* 24, 147–224, 1982.

Marteau, D., Cardoso, O., and Tabeling, P., Equilibrium states of two-dimensional turbulence: An experimental study, *Phys. Rev. E* 51, 512, 1995.

Matsuno, Y., *Bilinear Transformation Method* (Academic, New York, 1984).

Maxworthy, T., and Redekopp, L. G., A solitary wave theory of the great red spot and other observed features in the Jovian atmosphere, *Icarus* 29, 261, 1976.

Ma, Y. C., The perturbed plane-wave solutions of the cubic Schrödinger equation, *Stud. Appl. Math.* 60, 43, 1979.

Ma, Y. C., and Ablowitz, M. J., The periodic cubic Schrödinger equation, *Stud Appl. Math.* 65, 113, 1981.

McKean, H. P., The Sine-Gordon and Sinh-Gordon equation on the circle, *Commun. Pure Appl. Math.* 34, 197–257, 1980.

McKean, H. P., Boussinesq's equation on the circle, *Commun. Pure Appl. Math.* 34, 599–691, 1981.

McKean, H. P., and Trubowitz, E., Hill's operator and hyperelliptic function theory in the presence of infinitely many branch points, *Commun. Pure Appl. Math.* 29, 143–226, 1976.

McLaughlin, D. W., and Schober, C. M., Chaotic and homoclinic behavior for numerical discretizations of the nonlinear Schroedinger equation, *Physica D* 57, 447–465, 1992.

McWilliams, J. C., *Fundamentals of Geophysical Fluid Dynamics* (Cambridge University Press, Cambridge, 2006).

Mei, C. C., *The Applied Dynamics of Ocean Surface Waves* (John Wiley and Sons, New York, 1983).

Mei, C. C., and Ünlüata, U., Harmonic generation in shallow water waves, In *Waves on Beaches*, R. E. Meyer, ed., pp. 181–202 (Academic, New York, 1972).

Miles, J. W., Resonantly interacting solitary waves, *J. Fluid Mech.* 79, 171–179, 1977.

Miles, J. W., On the Korteweg-deVries equation for a gradually varying channel, *J. Fluid Mech.* 91, 181, 1979.

Miles, J. W., Solitary waves, *Annu. Rev. Fluid Mech.* 12, 11, 1980.

Miles, J. W., The Korteweg-deVries equation: a historical essay, *J. Fluid Mech.* 106, 131–147, 1981.

Miles, J. W., Solitary Wave Evolution Over a Gradual Slope with Turbulent Friction, *J. Phys. Oceanogr.* 13, 551, 1983.

Mitrool'sky, Y. Z., *Dynamics of Internal Gravity Waves in the Ocean* (Kluwer Academic, Dortrecht, 2001).

Miura, R. M., Korteweg-deVries equation: a survey of results, *SIAM Rev.* 18, 412–459, 1976.

Mori, N., Onorato, M., Janssen, P. A. E. M., Osborne, A. R., Serio, M., On the extreme statistics of long-crested deep water waves: Theory and experiments, *J. Geophys. Res* 112(C9), C09011, 2007.

Morse, P. M., and Feshbach, H., *Methods of Theoretical Physics* (McGraw-Hill, New York, 1953).

Moum, J. N., Farmer, D. M., Smyth, W. D., Armi, L., and Vagle, S., Structure and generation of turbulence at interfaces strained by internal solitary waves propagating shoreward over the continental shelf, *J. Phys. Oceanogr.* 33, 2093–2112, 2003.

Mumford, D., *Tata Lectures on Theta I* (Birkhaeuser, Boston, 1983).

Mumford, D., *Tata Lectures on Theta II* (Birkhaeuser, Boston, 1984).

Mumford, D., *Tata Lectures on Theta III* (Birkhaeuser, Boston, 1991).

Mumford, D., *Selected Papers: On the Classification of Varieties and Moduli Spaces* (Springer, New York, 2004).

Mumford, D., Series, C., and Wright, D., *Indra's Pearls: The Vision of Felix Klein* (Cambridge University Press, Cambridge, 2002).

Munk, W. H., The solitary wave theory and its applications to surf problems, *Ann. N. Y. Acad. Sci.* 51, 376–423, 1949.

Musette, M., and Conte, R., The two-singular manifold method: I. modified Korteweg-de Vries and sine-Gordon equations, *J. Phys. A* 27, 3895, 1994.

Nakamura, A., and Matsuno, Y., Exact One-and Two-Periodic Wave Solutions of Fluids of Finite Depth, *J. Phys. Soc. Jpn.* 48(4), 653–657, 1980.

Nakamura, A., A Direct Method of Calculating Periodic Wave Solutions to Nonlinear Evolution Equations. II. Exact One- and Two-Periodic Wave Solution of the Coupled Bilinear Equations, *J. Phys. Soc. Jpn.* 48(2), 1365–1370, 1980.

Needham, T., *Visual Complex Analysis* (Clarendon Press, Oxford, 1997).

Newell, A. C. Ed., Nonlinear Wave Motion (American Mathematical Society, Providence, 1974).

Newell, A. C., The history of the soliton, *J. Appl. Mech.* 50, 1127–1137, 1983.

Newell, A. C., *Solitons in Mathematics and Physics* (SIAM, Philadelphia, 1985).

Newell, A. C., and Moloney, J. V., *Nonlinear Optics* (Addison-Wesely, 1992).

Ng, K.-C., *J. Comput. Phys.* 16, 396, 1974.

Novikov, S. P., Manakov, S. V., Pitaevskii, L. P., and Zakharov, V. E., *Theory of solitons: The Inverse Scattering Method* (Consultants Bureau, New York, 1984).

Onorato, M., Osborne, A. R., Serio, M., and Bertone, S., Freak waves in random oceanic sea states, *Phy. Rev. Lett.* 86(25), 2001.

Onorato, M., Osborne, A. R., Serio, M., Extreme wave events in directional, random oceanic sea states, *Phys. Fluids* 14(4), 2002.

Onorato, M., Osborne, A. R., Serio, M., Cavaleri, L., and Brandini, C., Observation of strongly non-Gaussian statistics for random sea surface gravity waves in wave flume experiments, *Phys. Rev. E* 70(6), 067302, 2004.

Onorato, M., Osborne, A. R., Serio, M., Cavaleri, L., Modulational instability and non-Gaussian statistics in experimental random water-wave trains, *Phys. Fluids* 17(7), 078101, 2005.

Onorato, M., Osborne, A. R., Serio, M., Cavaleri, L., Brandini, C., Stansberg, C. T., Extreme waves, modulational instability and second order theory: wave flume experiments on irregular waves, *Eur. J. Mech. B Fluids* 25(5), 586–601, 2006.

Osborne, A. R., The simulation and measurement of random ocean wave statistics, In *Topics in Ocean Physics*, A. R. Osborne, and P. Malanotte-Rizzoli, eds. (North-Holland, Amsterdam, 1982).

Osborne, A. R., The spectral transform: Methods for the Fourier analysis of nonlinear wave data, In *Statics and Dynamics of Nonlinear Systems*, G. Benedek, H. Bilz, R. Zeyher, eds., (Springer-Verlag, Heidelberg, 1983).

Osborne, A. R., *Nonlinear Topics in Ocean Physics*, A. R. Osborne, ed. (Elsevier, Amsterdam, 1989).

Osborne, A. R., The generation and propagation of internal solitons in the Andaman Sea, In *Soliton Theory: A Survey of Results*, A. P. Fordy, ed., pp. 152–173 (Manchester University Press, Manchester, 1990).

Osborne, A. R., Nonlinear Fourier analysis, In *Nonlinear Topics in Ocean Physics*, A. R. Osborne, ed. (North-Holland, Amsterdam, 1991a).

Osborne, A. R., Nonlinear Fourier analysis for the infinite-interval Korteweg-deVries equation I: An algorithm for the direct scattering transform, *J. Comput. Phys.* 94(2), 284–313, 1991b.

Osborne, A. R., Construction of nonlinear wave train solutions of the periodic, defocusing nonlinear Schroedinger equation, *J. Comput. Phys.* 109(1), 93–107, 1993a.

Osborne, A. R., Numerical construction of complex, nonlinear wave train solutions of the Periodic Korteweg-deVries equation, *Phy. Rev. E.* 48(1), 296, 1993b.

Osborne, A. R., The behavior of solitons in random-function solutions of the periodic Korteweg-deVries equation, *Phys. Rev. Lett.* 71(19), 3115–3118, 1993c.

Osborne, A. R., The numerical inverse scattering transform for the periodic, defocusing nonlinear Schroedinger equation, *Phys. Lett. A* 176, 75, 1993d.

Osborne, A. R., The Numerical Inverse Scattering Transform: Nonlinear Fourier Analysis and Nonlinear Filtering of Oceanic Surface Waves, In *Proceedings of the Aha Huliko'a Hawaiian Winter Workshop*, P. Müller, D. Henderson, eds., 1993e.

Osborne, A. R., Automatic algorithm for the numerical inverse scattering transform of the Korteweg-deVries equation, *Math. Comput. Simul.* 37, 431–450, 1994.

Osborne, A. R., Soliton Physics and the Periodic Inverse Scattering Transform, *Physica D* 86, 81, 1995a.

Osborne, A. R., Solitons in the periodic Korteweg-deVries equation, the θ-function representation and the analysis of nonlinear, stochastic wave trains, *Phys. Rev. E* 52(1), 1105–1122, 1995b.

Osborne, A. R., Approximate asymptotic integration of a higher order water-wave equation using the inverse scattering transform, *Nonlinear Proc. Geophys.* 4(1), 29–53, 1997.

Osborne, A. R., The random and deterministic dynamics of "rogue waves" in unidirectional, deep-water wave trains, *Mar. Struct.* 14(3), 275–293, 2001.

Osborne, A. R., Nonlinear ocean waves and the inverse scattering transform, In *Scattering*, R. Pike, and P. Sabatier, eds. (Academic Press, New York, 2002).

Osborne, A. R., and Bergamasco, L., The small-amplitude limit of the spectral transform for the periodic Korteweg-deVries equation, *Nuovo Cimento B* 85, 229–243, 1985.

Osborne, A. R., and Bergamasco, L., The solitons of Zabusky and Kruskal revisited: Perspective in terms of the periodic spectral transform, *Physica D* 18, 26–46, 1986.

Osborne, A. R., and Boffetta, G., In *Nonlinear Evolution Equations: Integrability and Spectral Methods*, A. Degasperis, and A. P. Fordy, eds. (Manchester University Press, Manchester, 1989a).

Osborne, A. R., and Boffetta, G., The shallow-water nonlinear Schroedinger equation in Lagrangian coordinates, *Phys. Fluids* A1, 1200, 1989.

Osborne, A. R., and Burch, T. L., Internal solitons in the Andaman Sea, *Science* 208, 451–460, 1980.

Osborne, A. R., and Petti, M., The numerical inverse scattering transform analysis for laboratory-generated surface wave trains, *Phy. Rev. E* 47(2), 1035, 1993.

Osborne, A. R., and Petti, M., Laboratory-generated, shallow-water surface waves: Analysis using the periodic, inverse scattering transform, *Phys. Fluids* 6(5), 1727–1744, 1994.

Osborne, A. R., and Segre, E., Numerical solutions of the Korteweg-deVries equation using the periodic scattering transform μ-representation, *Physica D* 44, 575–604, 1990.

Osborne, A. R., and Segre, E., The numerical inverse scattering transform for the periodic Korteweg-deVries equation, *Phys. Lett. A* 173, 131, 1993.

Osborne, A. R., Burch, T. R., and Scarlet, The influence of internal waves on deep water drilling, *J. Pet. Technol.* 30, 1497, 1978.

Osborne, A. R., Provenzale, A., and Bergamasco, L., On the Stokes wave in shallow water: Perspective in the context of spectral-transform theory, *Nuovo Cimento C* 5, 597, 1982a.

Osborne, A. R., Provenzale, A., and Bergamasco, L., Nonlinear Fourier analysis of localized wave fields described by the Korteweg-deVries equation, *Nuovo Cimento C* 5, 612, 1982b.

Osborne, A. R., Provenzale, A., and Bergamasco, L., Theoretical and numerical methods for the nonlinear Fourier analysis of shallow-water wave data, *Nuovo Cimento C* 5, 633, 1982c.

Osborne, A. R., Provenzale, A., and Bergamasco, L., The nonlinear Fourier analysis of internal solitons in the Andaman Sea, *Lett. Nuovo Cimento* 36, 593, 1983.

Osborne, A. R., Petti, M., Liberatore, G., and Cavaleri, L., In *Computer Modeling in Ocean Engineering*, B. A. Schrefler, and O. C. Zienkiewicz eds., pp. 99 (Balkema, Rotterdam, 1988).

Osborne, A. R., Segre, E., Boffetta, G., and Cavaleri, L., Soliton basis states in shallow water ocean surface waves, *Phys. Rev. Lett.* 67, 592–595, 1991.

Osborne, A. R., Bergamasco, L., Serio, M., Bianco, L., Cavaleri, L., Drago, M., Iovenitti, L., and Viezzoli, D., Nonlinear shoaling of shallow water waves: perspective in terms of the inverse scattering transform, In *Wind and Waves in the Northern Adriatic Sea*, L. Cavaleri, ed. (Società Italiana di Fisica, Bologna, 1996).

Osborne, A. R., Onorato, M., Serio, M., and Bergamasco, L., Soliton creation and destruction, resonant interactions and inelastic collisions in shallow water waves, *Phys. Rev. Lett* 81(17), 3559–3562, 1998a.

Osborne, A. R., Serio, M., Bergamasco, L., and Cavaleri, L., Solitons, cnoidal waves and nonlinear interactions in shallow-water ocean surface waves, *Physica D* 123, 64–81, 1998b.

Osborne, A. R., Onorato, M., and Serio, M., The nonlinear dynamics of rogue waves and holes in deep-water gravity wave trains, *Phys. Lett. A* 275, 386–393, 2000.

Ostrovsky, L., Nonlinear internal waves in a rotating ocean, *Oceanogology* 18(2), 119–125, 1978.

Ostrovsky, L. A., Nonlinear internal waves in a rotating ocean, *Oceanology* 18, 181–191, 1978.

Ostrovsky, L. A., and Grue, J., Evolution equations for strongly nonlinear internal waves, *Phys. Fluids* 15, 2934–2948, 2003.

Ostrovsky, L. A., and Stepanyants, Y. A., Internal solitons in laboratory experiments: Comparison with theoretical models, *Chaos* 15, 037111, 2005.

Pedlosky, J., *Geophysical Fluid Dynamics* (Springer-Verlag, New York, 1982).

Pedlosky, J., *Geophysical Fluid Dynamics* (Springer-Verlag, Berlin, 1998).

Pedlosky, J., *Waves in the Ocean and Atmosphere: Introduction to Wave Dynamics* (Springer-Verlag, Berlin, 2003).

Peregrine, D. H., Calculations of the development of an undular bore, *J. Fluid Mech.* 25, 321, 1966.

Perry, R. B., and Schimke, G. R., Large-amplitude internal waves observed off the northwest coast of Sumatra, *J. Geophys. Res.* 70, 2319, 1965.

Phillips, O. M., On the dynamics of unsteady gravity waves of finite amplitude, *J. Fluid Mech.* 9, 193, 1960.

Phillips, O. M., Nonlinear Dispersive Waves, *Annu. Rev. Fluid Mech.* 6, 93, 1974.

Polishchuk, A., *Abelian Varieties, Theta Functions and the Fourier Transform* (Cambridge University Press, Cambridge, 2003).

Press, W. H., Teukolsky, S. A., Vetterling, W. T., and Flannery, B. P., *Numerical Recipes* (Cambridge University Press, Cambridge, 1992).

Provenzale, A., and Osborne, A. R., Nonlinear Fourier analysis for the infinite-interval Kortweg-deVries equation II, *J. Comput. Phys.* 94, 314–351, 1991.

Rayleigh, L., and Strutt, J. W., On waves, *Phils. Mag. Ser.* 5, 1, 257–279, 1876.

Redekopp, L. G., and Weidmann, P. D., Solitary waves in zonal shear flows, *J. Atmos. Sci.* 35, 790–804, 1978.

Remoissenet, M., *Waves Called Solitons* (Springer, Berlin, 1999).

Ruelle, D., and Takens, F., *Commun. Math. Phys.* 20, 167, 1971.

Russell, J. S., *Report of the committee on waves, Report of the 7th Meeting of British Association for the Advancement of Science, Liverpool*, 417–496, 1838.

Russell, J. Scott, *Report on Waves*, (Murray, London, 311–390, 1844).

Russell, J. S., *The Wave of Translation in the Oceans of Water, Air and Eather* (Truebner, London, 1885).

Saffman, P. G., *Vortex Dynamics* (Cambridge University Press, Cambridge, 1992).

Sarpkaya, T., and Isaacson, M., *Mechanics of Wave Forces on Offshore Structures* (Van Nostrand Reinhold, New York, 1981).

Schober, C. M., Mel'nikov analysis and inverse spectral analysis of rogue waves in deep water, *Eur. J. Mech. B Fluids* 25, 602–620, 2006.

Schottky, F., Uber eine spezielle Funktion, welche bei einer bestimmten linearen Transformation ihres Arguments unverandert bleibt, *J. Reine Angew. Math.* 101, 227–272, 1887.

Scott, A. C., Nonlinear Science: Emergence and Dynamics of Coherent Structures (Oxford University Press, Oxford, 2003).

Scott, A. C., Encyclopedia of Nonlinear Science (Routledge, New York, 2005).

Scott, A. C., Chu, F. Y. F., McLaughlin, D. W., The soliton: A new concept in applied science, *Proc. IEEE* 61, 1443, 1973.

Scotti, A., Beardsley, R. C., and Butman, B., Generation and Propagation of nonlinear internal waves in Massachusetts Bay, *J. Geophys. Res.* 112, C10001, 1–19, 2007.

Scotti, A., Beardsley, R. C., Butman, B., and Pineda, J., Shoaling of nonlinear internal waves in Massachusetts Bay, *J. Geophys. Res.* 113, C08031, 1–18, 2008.

Segur, H., The Korteweg-deVries Equation and Water Waves, Solutions of the Equation Part 1, *J. Fluid Mech.* 59(4), 721–736, 1973.

Segur, H., and Finkel, A., An analytical model of periodic waves in shallow water, *Stud. Appl. Math.* 73, 183, 1985.

Segur, H., In *Topics in Ocean Physics*, A. R. Osborne, P. Malanotte Rizzoli, eds., (North-Holland, Amsterdam, 1982).

Segur, H., Integrable models of waves in shallow water, In *Probability, Geometry and Integrable Systems*, M. Pinski, B. Birnir, eds., (Cambridge University Press, Cambridge, 2007).

Segur, H., and Hammack, J. L., Soliton models of long internal waves, *J. Fluid Mech.* 118, 285–304, 1982.

Segur, H., Henderson, D., Carter, J., Hammack, J., Li, C. M., Pheiff, D., and Socha, K., Stabilizing the Benjamin-Feir instability, *J. Fluid Mech.* 539, 229–271, 2005.

Shermer, L., and Stiassnie, M., In *Nonlinear Topics in Ocean Physics*, A. R. Osborne, ed, North-Holland, Amsterdam, 1991.

Shiota, T., Characterization of Jacobian varieties in terms of soliton equations, *Invent. Math.* 83(2), 333–382, 1986.

Shrira, V. I., Badulin, S. I., and Kharif, C., A model of water wave 'horse-shoe' patterns, *J. Fluid Mech.* 318, 375–405, 1996.

Siegel, C. L., *Topics in Complex Function Theory, Vol. I, Elliptic Functions and Uniformization Theory* (Wiley-Interscience, New York, 1969a).

Siegel, C. L., *Topics in Complex Function Theory, Vol. II, Automorphic Functions and Abelian Integrals* (Wiley-Interscience, New York, 1969b).

Siegel, C. L., *Topics in Complex Function Theory, Vol. III, Abelian Functions and Modular Functions of Several Variables* (Wiley-Interscience, New York, 1969c).

Singleton, R. C., An algorithm for computing the mixed radix fast Fourier transform, *IEEE Trans. Audio Electroacoustics* AU-17, 93, 1969.

Smith, K. B., and Tappert, F. D., *UMPE: The University of Miami Parabolic Equation Model*, MPL Technical Memorandum 432, September, , 1994.

Sneddon, I. N., *Fourier Transforms* (Dover, New York, 1995).

Solomon, R., *Lectures on Geophysical Dynamics* (Oxford University Press, New York, 1998).

Stanton, T. P., and Ostrovsky, L. A., Observations of highly nonlinear solitons over the continental shelf, *Geophys. Res. Lett.* 25, 2695–2698, 1998.

Staquet, C., and Sommeria, J., Internal gravity waves: From instabilities to turbulence, *Annu. Rev. Fluid. Mech.* 34, 559–593, 2002.

Stokes, G. G., On the theory of oscillatory waves, *Trans. Camb. Philos. Soc.* 8, 441–455, 1847.

Stoker, J. J., *Water Waves* (Wiley Interscience, New York, 1957).

Stoker, J. R., and Peregrine, D. H., The current-modified nonlinear Schroedinger equation, *J. Fluid Mech.* 399, 335–353, 1999.

Stuart, J. T., On finite amplitude oscillations in laminar mixing layers, *J. Fluid Mech.* 29 (3), 417–440, 1967.

Sulem, C., and Sulem, P. L., *The Nonlinear Schroedinger Equation* (Springer, Berlin, 1999).

Taha, T. R., and Ablowitz, M. J., Analytical and numerical aspects of certain nonlinear evolution equations I. Analytic, *J. Comput. Phys.* 55, 192–202, 1984a.

Taha, T. R., and Ablowitz, M. J., Analytical and Numerical Aspects of Certain Nonlinear Evolution Equations Part II: Numerical Nonlinear Schrödinger Equation, *J. Comp. Phys.* 55, 203–230, 1984b.

Taha, T. R., and Ablowitz, M. J., Analytical and Numerical Aspects of Certain Nonlinear Evolution Equations Part III: Numerical Korteweg-deVries Equation, *J. Comput. Phys.* 55, 231–253, 1984c.

Tappert, F., The parabolic approximation method, In *Wave Propagation and Underwater Acoustics*, J. B. Keller, and J. S. Papadakis, eds., pp. 224–287 (Springer-Verlag, Berlin, 1977).

Terrones, G., McLaughlin, D. W., Overman, E. A., II, and Pearlstein, A., Stability and bifurcation of spatially coherent solutions of the damped driven nonlinear Schroedinger equation, *SIAM J. Appl. Math.* 50, 791–818, 1990.

Ting, A. C., Tracy, E. R., Chen, H. H., and Lee, Y. C., On the Reality Constraints for the Periodic Sine-Gordon Equation, *Phys. Rev. A, Rapid Commun.* 30, 3355, 1984.

Ting, A. C., Chen, H. H., and Lee, Y. C., Exact solutions of nonlinear boundary value problem: the vortices of the two-dimensional sinh-Poisson equation, *Physica D* 26, 37, 1987.

Titchmarsh, E., Conjugate trigonometrical integrals, *Proc. Lond. Math. Soc.* 24, 109–130, 1925.

Titchmarsh, E., On conjugate functions, *Proc. Lond. Math. Soc.* 29, 49–80, 1928.

Titchmarsh, E., Additional note on conjugate functions, *J. Lond. Math. Soc.* 4, 204–206, 1930.

Titchmarsh, E. C., *Introduction to the Theory of Fourier Integrals* (Oxford, London, 1937).

Tolstov, G. P., *Fourier Series* (Dover, New York, 1962).

Tracy, E. R., *Topics in nonlinear wave theory with applications*, Ph.D. Thesis, University of Maryland, Plasma Preprint UMLPF #85-006, 1984.

Tracy, E. R., and Chen, H. H., Nonlinear Self-modulation: An Exactly Solvable Model, *Phys. Rev. A* 37, 815–839, 1988.

Tracy, E. R., Chen, H. H., and Lee, Y. C., A Study of Quasiperiodic Solutions of the Nonlinear Schrödinger Equation, *Phys. Rev. Lett.* 53, 218, 1984.

Tracy, E. R., Larson, J. W., Osborne, A. R., and Bergamasco, L., On the Nonlinear Schrödinger Equation as an Averaging Theory, In *Nonlinear Evolution Equations and Dynamical Systems*, J. Leon, ed. (World Scientific, Singapore, 1987).

Tracy, E. R., Larson, J. W., Osborne, A. R., and Bergamasco, L., On the Nonlinear Schrödinger Limit of the Korteweg-deVries Equation, *Physica D* 32, 83, 1988.

Tracy, E. R., Larson, J. W., Osborne, A. R., and Bergamasco, L., On the Relationship Between the Spectral Theories for the Periodic Korteweg-deVries and Nonlinear Schrödinger Equations, In *Nonlinear Topics in Ocean Physics*, A. R. Osborne, ed. (Elsevier, Amsterdam, 1991).

Trulsen, K., and Dysthe, K., A modified nonlinear Schroedinger equation for broader bandwidth gravity waves on deep water, *Wave Motion* 24, 281, 1996.

Trulsen, K., and Dysthe, K., Frequency downshift in three-dimensional wave trains in a deep basin, *J. Fluid Mech.* 352, 359–373, 1997a.

Trulsen, K., and Dysthe, K., Freak waves—A three dimensional wave simulation, In Naval Hydrodynamics, Proceedings of the 21st Symposium on Nature, E. P. Rood, ed., (Academic Press, USA, 1997b).

Trulsen, K., Kliakhandler, I., Dysthe, K. B., and Velarde, M. G., On weakly nonlinear modulation of waves on deep water, *Phys. Fluids* 12(10), 2432–2436, 2000.

Tulin, M. P., and Waseda, T., Laboratory observations of wave group evolution, including breaking effects, *J. Fluid Mech.* 378, 197–232, 1999.

Ursell, F., *Proc. Cambridge Phil. Soc.* 49, 685, 1953.

Vallis, G. K., *Atmospheric and Oceanic Fluid Dynamics: Fundamentals and large Scale Circulation* (Cambridge University Press, Cambridge, 2006).

Wang, S., Tang, X. y., Lou, S. Y., Soliton fission and fusion: Burgers equation and Sharma-Tasso-Olber equation, *Chaos Solitons Fractals* 21, 231–239, 2004.

Weigel, R. L., *Oceanographical Engineering* (Englewood Cliffs, N. J., Prentice Hall, 1964).

Weiss, J., Tabor, M., and Carnevale, G., The Painlevé Property for Partial Differential Equations, *J. Math. Phys.* 24, 522, 1983.

West, B. J., Brueckner, K. A., Janda, R. S., Milder, D. M., and Milton, R. L., A New Numerical method for Surface Hydrodynamics, *J. Geophys. Res.* 92, 11,803–11,824, 1987.

Whitham, G. B., *Linear and Nonlinear waves* (John Wiley, New York, 1974).

Whittaker, E. T., and Watson, G. N., *A Course of Modern Analysis* (Cambridge University Press, Cambridge, 1902).

Young, I. R., *Wind Generated Ocean Waves* (Elsevier, Oxford, 1999).

Yuen, H. C., In *Nonlinear Topics in Ocean Physics*, A. R. Osborne, ed. (Elsevier, Amsterdam, 1991).

Yuen, H. C., and Lake, B. M., Nonlinear dynamics of deep-water gravity waves, *Adv. Appl. Mech.* 22, 67–229, 1982.

Zabusky, N. J., Computational Synergetics and Mathematical Innovation, *J. Comput. Phys.* 43, 195, 1981.

Zabusky, N. J., Fermi-Pasta-Ulam, solitons and the fabric of nonlinear and computational science: History, synergetics and visiometrics, *Chaos* 15, 015102, 2005.

Zabusky, N. J., and Galvin, C. J., Shallow water waves, the Korteweg-deVries equation and solitons, *J. Fluid Mech.* 47, 811, 1971.

Zabusky, N. J., and Kruskal, M. D., Interaction of "solitons" in a collisionless plasma and the recurrence of initial states, *Phys. Rev. Lett.* 15, 240, 1965.

Zagrodzinski, J. A., Direct approach to the periodic solutions of the multidimensional sine–Gordon equation, *J. Math. Physics* 24(1), 46–52, 1983.

Zakharov, V. E., Stability of periodic waves of finite amplitude on the surface of a deep fluid, *J. Appl. Mech. Tech. Phys.* USSR, 2, 190, 1968.

Zakharov, V. E., and Kuznetsov, E. A., Multi-scale expansions in the theory of systems integrable by the inverse scattering transform, In *Solitons and Coherent Structures*, D. K. Campbell, A. C. Newell, R. J. Schrieffer, and H. Segur, eds., pp. 455–463 (North-Holland, Amsterdam, 1986).

Zakharov, V. E., and Rubenchik, A. M., Nonlinear interaction of high-frequency and low-frequency waves, *Prikl. Mat. Techn. Phys.* 5, 84–98, 1972.

Zakharov, V. E., and Shabat, A. B., Exact theory of two-dimensional self-focusing and one-dimensional waves in nonlinear media, *Sov. Phys. JETP* 34, 62, 1972.

Zaslavsky, G. M., Sagdeev, R. Z., Usikov, D. A., and Chernikov, A. A., *Weak Chaos and Quasi-Regular Patterns, Cambridge Nonlinear Science Series* (Cambridge University Press, Cambridge, 1991).

Zhang, H. Q., Tian, B., Li, J., Xu, T., Zhang, Y. X., Symbolic-computation study of integrable properties for the $(2 + 1)$-dimensional Gardner equation with the two-singular manifold method, *IMA J. Appl. Math.* 0:hxn024v1-hxn024, 2008.

Zimmerman, W. B., and Haarlemmer, G. W., Internal gravity waves: Analysis using the periodic, inverse scattering transform, *Nonlinear Proc. Geophys.* 6, 11–26, 1999.

Zygmund, A., *Trigonometric Series* (Cambridge University Press, Cambridge, 1959).

International Geophysics Series

EDITED BY

RENATA DMOWSKA

School of Engineering and Applied Sciences,
Harvard University, Cambridge, Massachusetts

DENNIS HARTMANN

Department of Atmospheric Sciences,
University of Washington, Seattle, Washington

H. THOMAS ROSSBY

Graduate School of Oceanography,
University of Rhode Island, Narragansett, Rhode Island

* Out of Print

Index

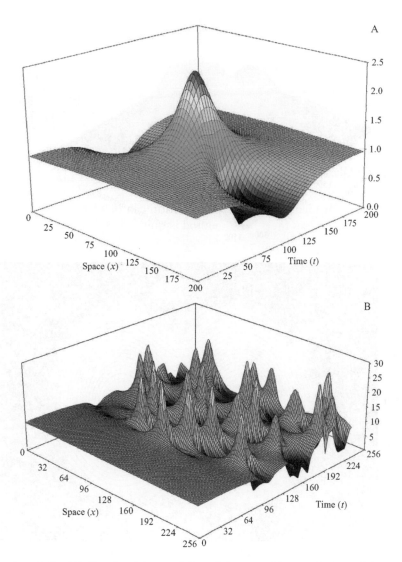

Color plate 1.8 (A) Graph of the modulus of the space/time evolution of the simplest "rogue wave" solution to the sNLS equation given by Equation (1.19). (B) Graph of the modulus of the space/time evolution of a multimodal initial modulation that leads to the generation of many "rogue waves" in a solution of the sNLS equation given by Equation (1.16).

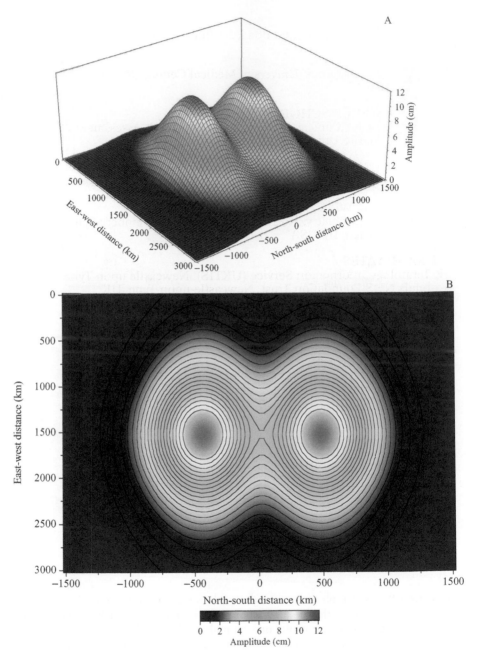

Color plate 1.9 (A) Surface elevation of an equatorial Rossby soliton and (B) contours of the Rossby soliton. Note that the single soliton dynamics are equivalent to a double vortex that sweeps (transports) passive tracers from the East to the West along the equator.

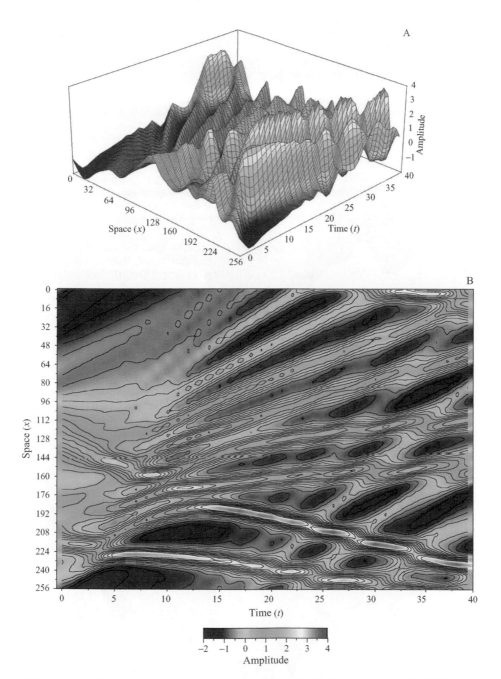

Color plate 1.10 Space-time evolution of a random initial condition for the KdV equation.

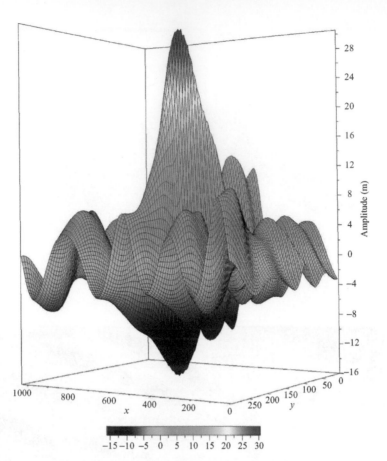

Color plate 1.13 Rogue wave simulation using the Davey-Stewartson equations (Chapter 34).

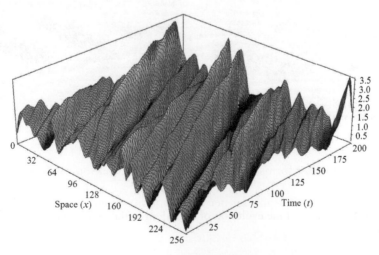

Color plate 1.14 Simulation of the NLS equation for a JONSWAP power spectrum with $H_s = 3$ m and $\gamma = 3$. Extreme waves are shown as they emerge in red.

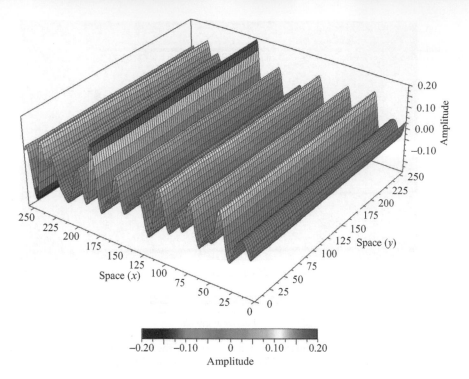

Color plate 9.3 Initial condition for the KdV equation used in the simulations for this chapter. Note that the wave train is unidirectional as required for the KdV equation (this result is an output from a numerical simulation using the KP equation, Chapter 32).

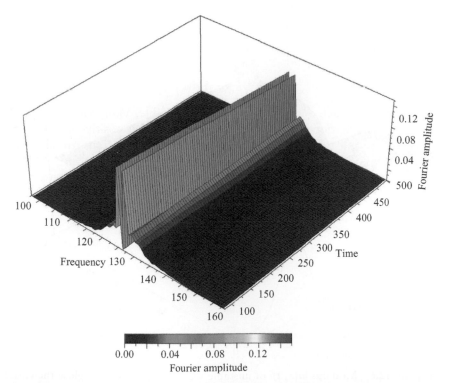

Color plate 9.4 Linear Fourier spectrum of the Riemann theta function as a function of frequency and time for the simulation run for the KdV equation.

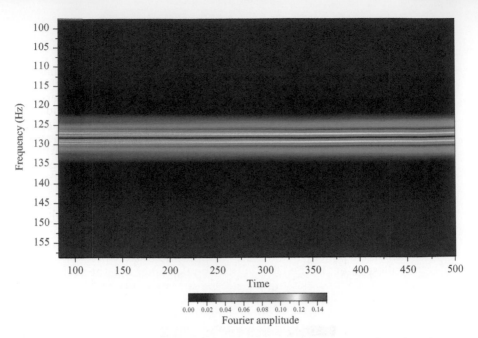

Color plate 9.5 Contours of linear Fourier spectrum of the Riemann theta function as a function of frequency and time for the simulation run for the KdV equation.

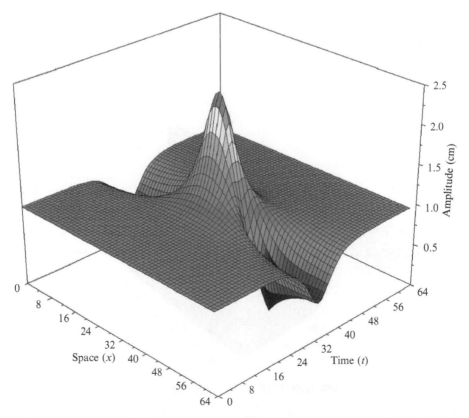

Color plate 12.6 Modulus $|u(x, t)|$ of unstable wave packet that lies below the carrier in the complex lambda plane with spectrum: $\{A, 0, 0, 0, A/\sqrt{2}\}$. The initial condition at time $t = 0$ is seen to be a small-amplitude modulation.

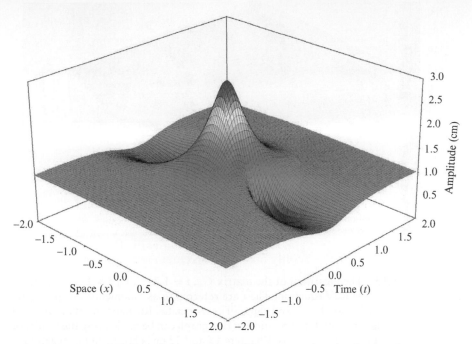

Color plate 12.7 Modulus $|u(x, t)|$ of unstable wave packet that lies on the carrier in the complex lambda plane with spectrum: $\{A, 0, 0, 0, A\}$. The initial condition at time $t = -2$ is seen to be a small-amplitude modulation.

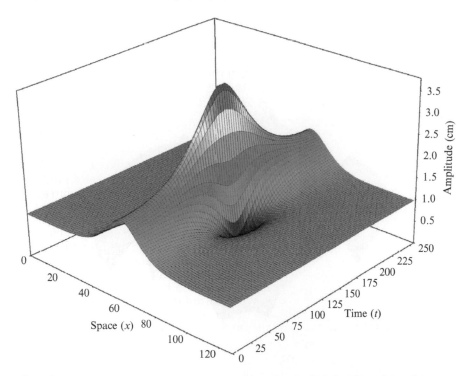

Color plate 12.8 Modulus $|u(x, t)|$ of unstable wave packet that lies above the carrier in the complex lambda plane with spectrum: $\{A, 0, 0, 0, \sqrt{2}A\}$. The initial condition at time $t = 0$ is seen to be a large-amplitude modulation.

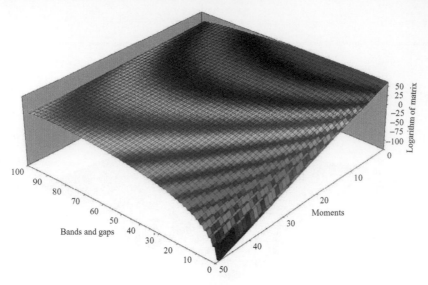

Color plate 19.16 The logarithm of the matrix G_{ij}, $i = 1,2,\ldots, N$, $j = 1,2,\ldots, N$ for a particular case for the KdV equation. The i are referred to as "moments" in the figure and the j are referred to as "bands and gaps." The associated Riemann matrix is 50×50 for this case that is genus 50. Interpretation of this graph can be made using the results of Chapters 14 and 19 on loop integrals, Chapters 15 and 32 on Schottky uniformization, and Chapters 16 and 32 on the Nakamura-Boyd approach. In particular, the practiced eye will notice the power-law behavior of the off-diagonal elements of the Riemann matrix, an approximate, leading order result of all three approaches: $B_{ij} \simeq \ln[(\kappa_i - \kappa_j)/(\kappa_i + \kappa_j)]$.

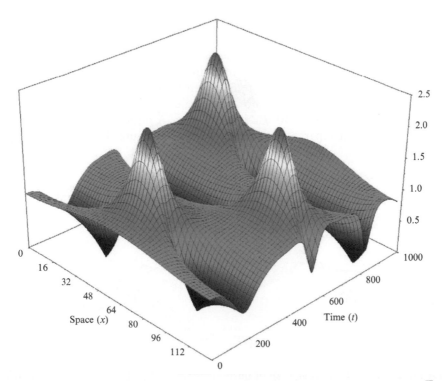

Color plate 24.5 Space/time evolution of a slot-state rogue wave. Here $\lambda = iA/\sqrt{2}$ and $|\varepsilon| = 0.05$. Note that the wave is periodic in space, but alternates its phase along the time axis during the evolution.

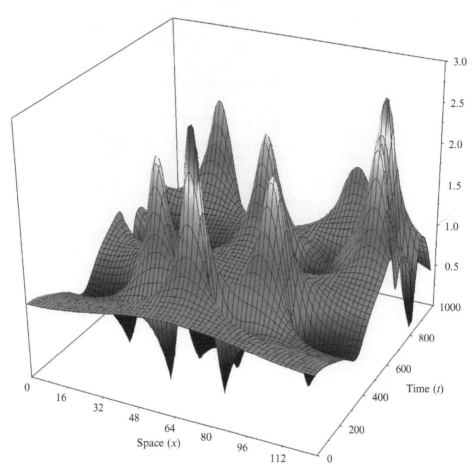

Color plate 24.6 Space/time evolution of two slot-state rogue waves. Note that the maximum amplitude is 3.06. The space/time evolution is quite complex in this case.

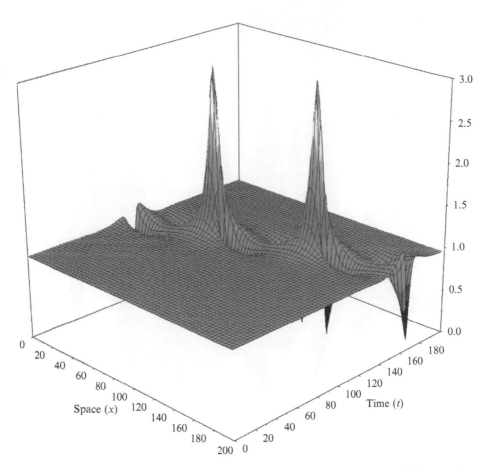

Color plate 24.7 Space/time evolution of a rogue wave solution of the NLS equation for the spectral eigenvalue $\lambda = 0.1 + i$.

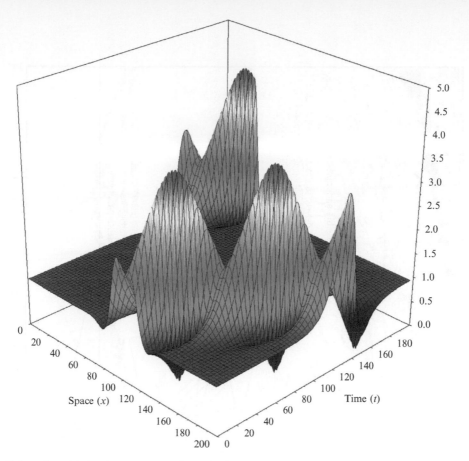

Color plate 24.8 Space/time evolution of a rogue wave solution of the NLS equation for the spectral eigenvalue $\lambda = 1.6 + 1.6i$.

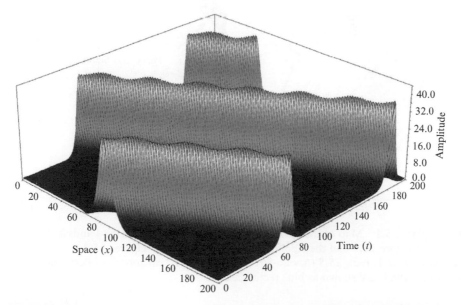

Color plate 24.9 Space/time evolution of a solution of the NLS equation that is near the soliton limit.

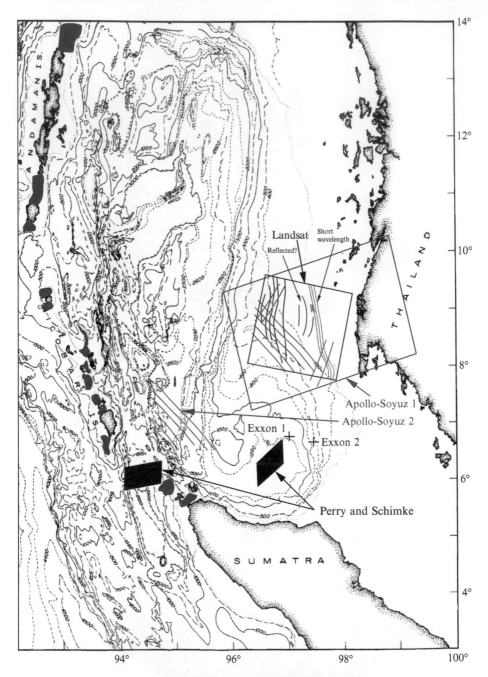

Color plate 25.1 Map of the Andaman Sea and the location of the Landsat (blue) image of Figure 25.3 and the Apollo-Soyuz photographs of Figures 25.4 (Apollo-Soyuz 1, red), 25.5 (Apollo-Soyuz 2, red). The "rip zones" of Perry and Schimke (1965) are shown as blue rectangles.

Color plate 25.2 Russian submarine Victor on the surface in the Straights of Gibraltar. One possible scenario for the damage is that the submarine encountered an upward moving internal wave forcing it to make brisk contact with a surface ship (Office of Naval Research).

Color plate 25.4 Apollo-Soyuz photograph of the surface of the Andaman Sea showing surface striations associated with internal wave activity. North is to the left (Courtesy of NASA, Johnson Spacecraft Center).

Color plate 25.5 Apollo-Soyuz photograph of the surface of the Andaman Sea showing surface striations associated with internal wave activity. This photograph was taken a few minutes before that in Figure 25.4. North is to the left (Courtesy of NASA, Johnson Spacecraft Center).

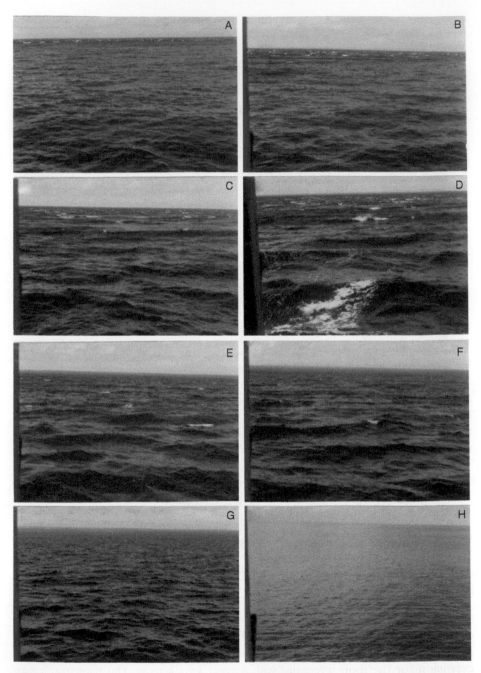

Color plate 25.6 Surface rips associated with long internal wave activity beneath the Andaman Sea surface. This sequence of photographs was taken aboard the survey vessel Oil Creek by the author. The time between each photograph is about 1 min.

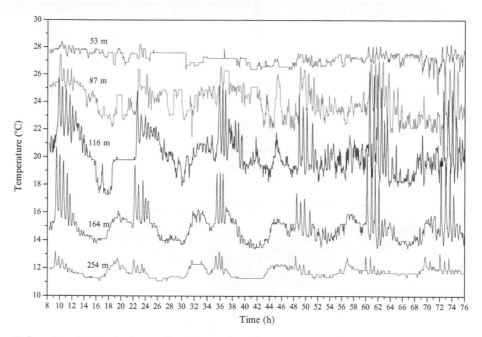

Color plate 25.7 Andaman Sea time series of temperature taken at various depths in the period October 24-28, 1976.

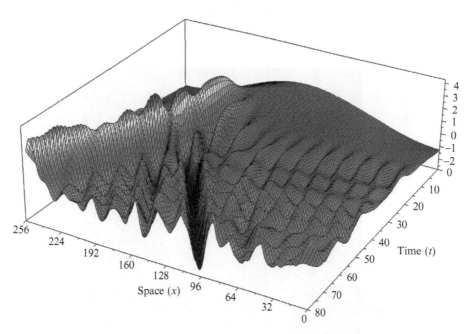

Color plate 25.18 Evolution of a hole state from the W2 equation. The hole is seen as a channel beginning near space coordinate 96 and time 80. A hole state in the internal wave field, when the upper layer is thinner than the lower layer, is a positively buoyant, positive soliton pulse.

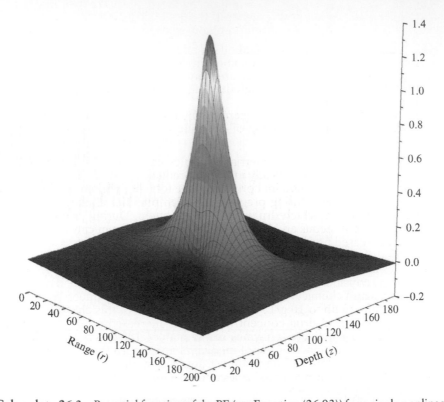

Color plate 26.3 Potential function of the PE (see Equation (26.93)) for a single nonlinear, range-dependent Fourier mode. This case corresponds to an algebraic soliton for the sound wave field (not shown, see discussion with regard Equations (26.89)–(26.95)).

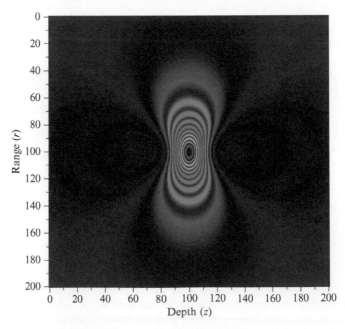

Color plate 26.4 Contours of the potential function of the PE for a single nonlinear Fourier mode as shown in the figure.

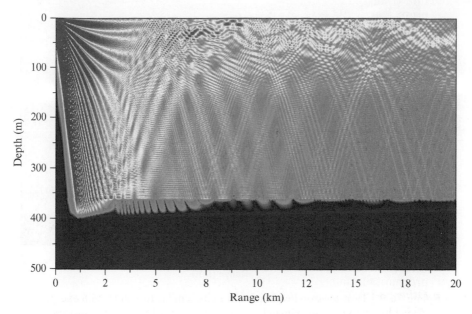

Color plate 26.7 Transmission loss in an environment described by the sound speed profile given in the figure. The central sound frequency is 800 Hz. A simple constant profile bottom and bottom sponge are used.

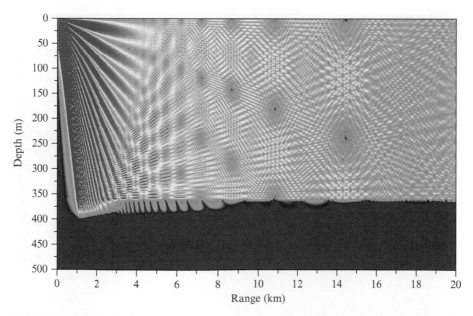

Color plate 26.8 Transmission loss in an isovelocity environment. The central sound frequency is 800 Hz. A simple constant profile bottom and bottom sponge are used.

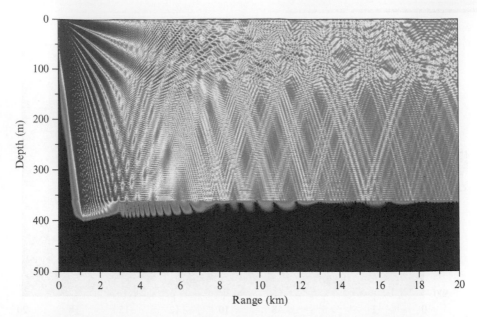

Color plate 26.9 Transmission loss in environment similar to Figure 26.8 except that the profile is filtered to be approximately isovelocity within 150 m of the surface. The central sound frequency is 800 Hz. A simple constant profile bottom and bottom sponge are used.

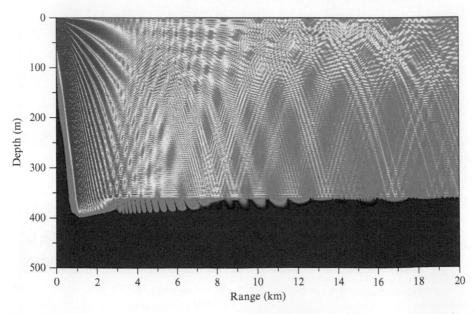

Color plate 26.10 Transmission loss in environment similar to Figure 26.8 except that the sound speed profile is filtered to give a linear sound intensity from 150 m up to the surface. The central sound frequency is 800 Hz. A simple constant profile bottom and bottom sponge are used.

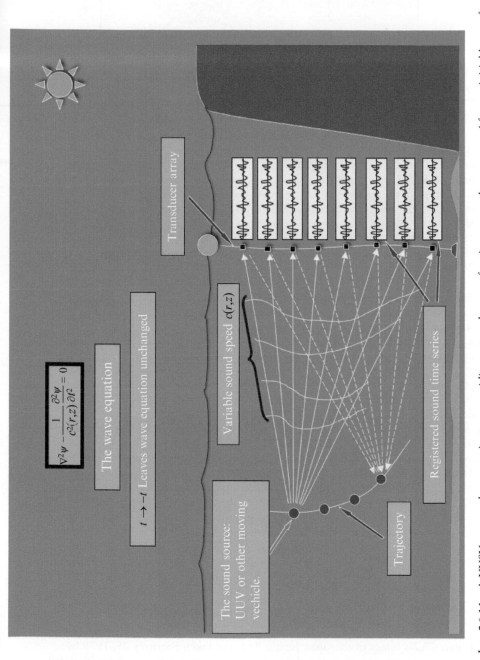

Color plate 26.11 A UUSV moves along a trajectory providing a sound source for time reversal symmetry. After an initial burst of sound the UUV moves silently, but its trajectory is followed as a reflected sound source. Distant communications are maintained by sonar updating of the UUSV trajectory and modification of the IST phases of the exact solution of the PE.

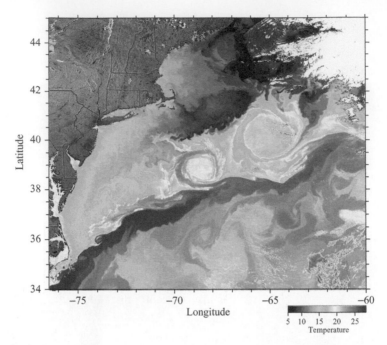

Color plate 27.1 AVHRR (advanced very high resolution radiometer) image taken aboard an NOAA satellite on June 11, 1997. The color-coding indicates temperature (see color bar) where the yellow and orange colors indicate warmer water and blue colors indicate cooler water. Note the two large warm water eddies to the north of the Gulf Stream, seen as a warm front (dark red) on the image. Below and within the front are shown the complex geophysical fluid dynamical turbulence typical of oceanic processes.

Color plate 27.2 Image of the Great Jupiter Red spot taken February 25, 1979 by Voyager 1. The image was taken at a distance of 9.2 million km from Jupiter. Note the complex wave dynamics to the left of the red spot, which is more that 300 years old and is larger than the Earth. Geophysical fluid dynamics on Jupiter is more energetic than on the Earth due to the large size of the Jovian planet and to its fast rotation.

Color plate 27.3   Hurricane Bill off the coast of Florida in August 2009. The hurricane was category 4 at the time of this image taken from a satellite of the National Oceanic and Atmospheric Administration (NOAA).

Color plate 29.1   The wave flume at Marintek, which is 10.5 m × 10 m × 260 m (the last 175 m are 5.6 m deep), as seen from the end of the tank near the wave maker. Many instruments and computers are located on the green carriage suspended above the tank in which a sinusoidal wave can be seen which reflects the ceiling lights. The carriage moves on rails for the full 260 m of the tank.

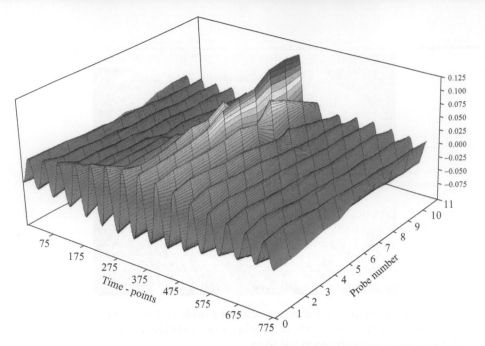

Color plate 29.7 Surface wave evolution of deterministic wave train out to 85 m from the wave maker. Surface constructed from individual time series in Figure 29.4.

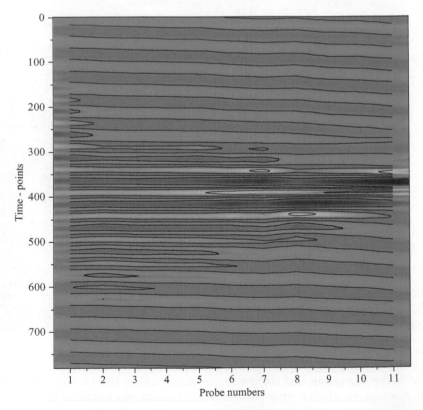

Color plate 29.8 Contours of wave evolution of Figure 29.7 of a deterministic wave train out to 85 m from the wave maker. Surface and contours constructed from individual time series in Figure 29.4.

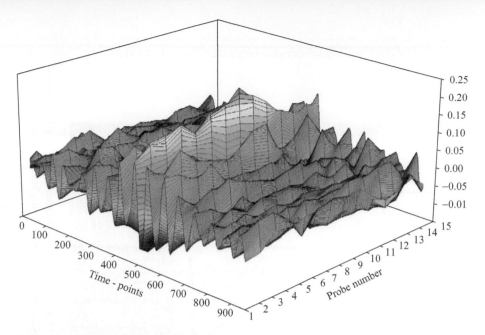

Color plate 29.11 Surface of the random wave packet of Figure 29.10 shown as a function of probe number and time. Note that the redundant probes have been removed.

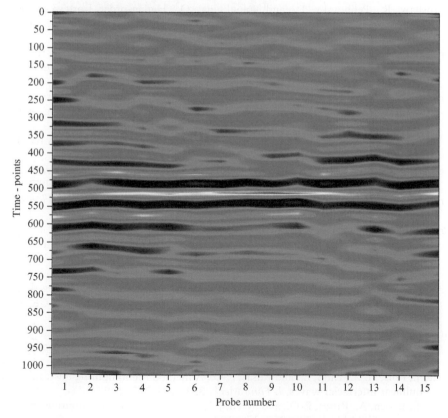

Color plate 29.12 Contours of the random wave packet of Figure 29.10 shown as a function of probe number and time. Note that the redundant probes have been removed.

11.9.2001 08:42

Color plate 32.1 Photograph of a simple shallow water case (San Nicholas, California) that appears to be two crossed cnoidal waves with a Mach stem or dromion at the intersection point. Photograph courtesy of Dr Paul Palo, Naval Facilities Engineering Service Center, Port Hueneme, California.

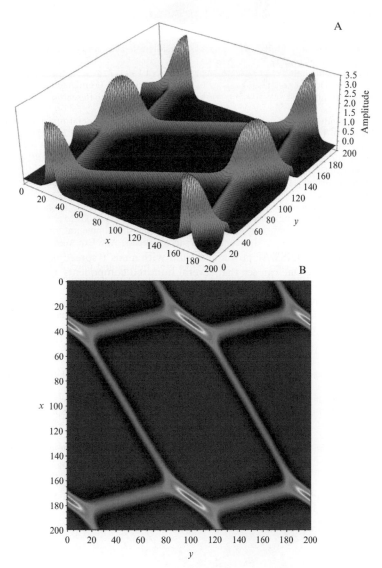

Color plate 32.3   Interaction of two soliton trains in the solution of the KP equation. Note the substantial Mach stems where the soliton trains cross.

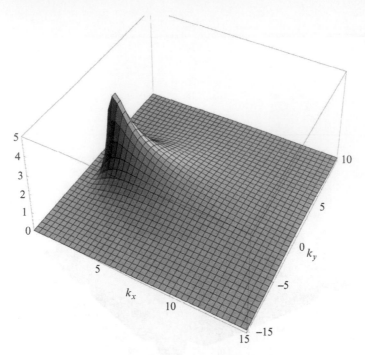

Color plate 32.4 A JONSWAP power spectrum with a $\cos^4\theta$ spreading function in the wavenumber domain. One can see that several hundred components are necessary for defining all the spectral amplitudes.

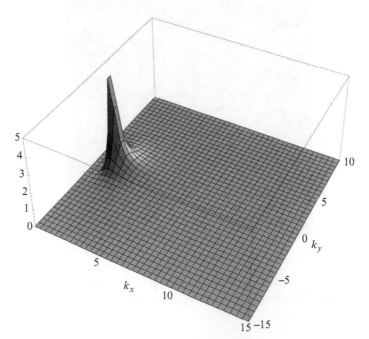

Color plate 32.5 The JONSWAP power spectrum with a $\cos^4\theta$ spreading function has been mapped to the domain of the Riemann spectrum. Shown are the linear Fourier modes after the mapping of eq. One can see that only about 30 components are necessary for defining all the spectral amplitudes. Thus, the Riemann spectrum contains many fewer spectral components than does the original JONSWAP spectrum.

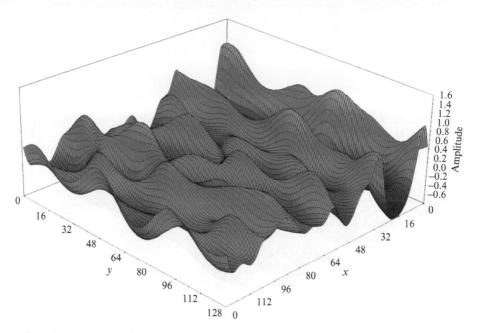

Color plate 32.6 Surface elevation from directionally spread sea with 30 degrees of freedom using Riemann theta functions.

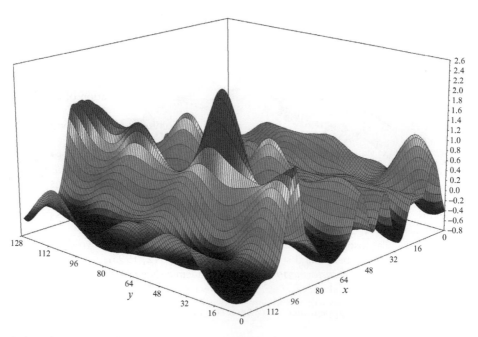

Color plate 32.7 The later evolution of the initial condition shown in Fig. 32.1. Note the appearance of an extremely large wave, denoted in red.

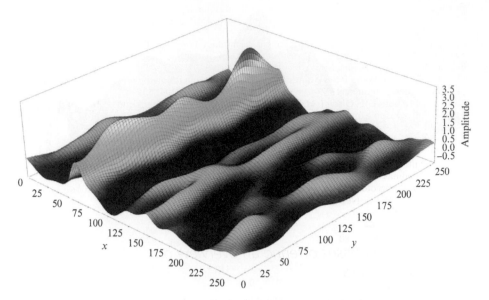

Color plate 33.4 A large wave accompanied by a deep trough in a numerical simulation of the 2 + 1 Gardner equation.

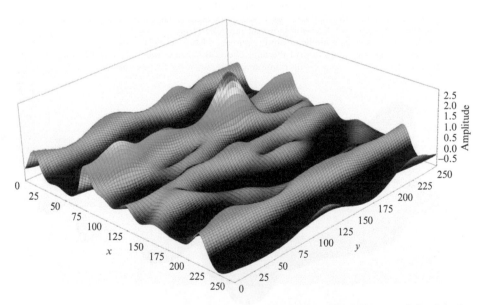

Color plate 33.5 The appearance of a Mach stem during the evolution of the 2 + 1 Gardner equation.

Color plate 34.1 The Thunder Horse Oil Platform sinking after Hurricane Ivan, July 2005 *Source*: (courtesy of United States Coast Guard).

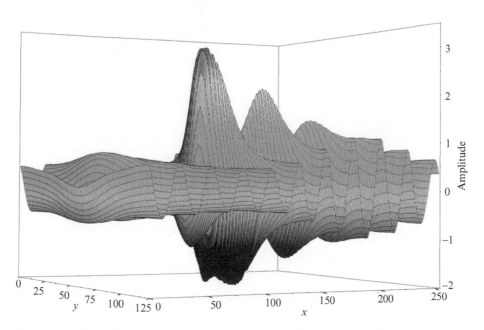

Color plate 34.3 A large rogue wave as it evolves in the deep-water Davey-Stewartson equations.

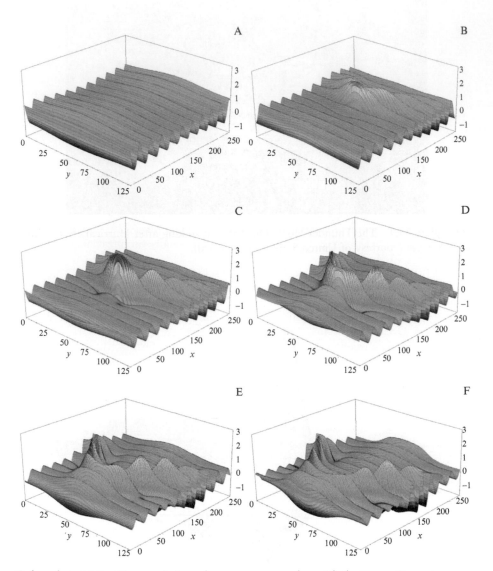

Color plate 34.2 Time evolution of a rogue wave packet with the Davey-Stewartson equations.

Printed and bound by CPI Group (UK) Ltd, Croydon, CR0 4YY

03/10/2024

01040419-0008